W0269313

NUCLEAR RADIATION IN GEOPHYSICS

EDITED BY

H. ISRAËL
TECHNISCHE HOCHSCHULE
AACHEN

A. KREBS
UNIVERSITY OF LOUISVILLE
LOUISVILLE

WITH AN INTRODUCTION BY

R. D. EVANS
MASSACHUSETTS INSTITUTE OF TECHNOLOGY
CAMBRIDGE, MASS.

WITH 112 FIGURES

SPRINGER-VERLAG
BERLIN HEIDELBERG GMBH
1962

KERNSTRAHLUNG IN DER GEOPHYSIK

HERAUSGEGEBEN VON

H. ISRAËL
TECHNISCHE HOCHSCHULE
AACHEN

A. KREBS
UNIVERSITY OF LOUISVILLE
LOUISVILLE

MIT EINER EINFÜHRUNG VON

R. D. EVANS
MASSACHUSETTS INSTITUTE OF TECHNOLOGY
CAMBRIDGE, MASS.

MIT 112 FIGUREN

SPRINGER-VERLAG
BERLIN HEIDELBERG GMBH
1962

ISBN 978-3-642-49046-0   ISBN 978-3-642-92837-6 (eBook)
DOI 10.1007/ 978-3-642-92837-6

Alle Rechte, insbesondere das der Übersetzung in fremde Sprachen, vorbehalten

Ohne ausdrückliche Genehmigung des Verlages ist es auch nicht gestattet, dieses
Buch oder Teile daraus auf photomechanischem Wege (Photokopie, Mikrokopie)
oder auf andere Art zu vervielfältigen

© Springer-Verlag Berlin Heidelberg 1962
Ursprünglich erschienen bei Springer-Verlag OHG / Berlin • Gottingen • Heidelberg 1962
Softcover reprint of the hardcover 1st edition 1962

Library of Congress Catalog Card Number 62-15066

Die Wiedergabe von Gebrauchsnamen, Handelsnamen, Warenbezeichnungen usw.
in diesem Buche berechtigt auch ohne besondere Kennzeichnung nicht zu der
Annahme, daß solche Namen im Sinne der Warenzeichen- und Markenschutz-
Gesetzgebung als frei zu betrachten wären und daher von jedermann benutzt
werden dürften

Vorwort

Die Radioaktivität von Boden, Wasser und Luft ist ein klassisches Forschungsgebiet der Geophysik, aus dessen Ergebnissen diese von jeher reichen Nutzen zieht: Fragen nach der Wärmebilanz des Erdinnern, nach dem Alter der Erde und dem der Gesteine haben erst von hier aus eine befriedigende Lösung gefunden; Hydrologie und Balneologie verdanken der Radioaktivität entscheidende Bereicherung; im Rahmen der Prospektion und Bodenforschung hat sie ihren Platz; in der Physik der Atmosphäre bietet sie die wesentliche Grundlage zum Verständnis der atmosphärisch-elektrischen Erscheinungen; dem Meteorologen gibt sie neue Möglichkeiten zur Bearbeitung atmosphärischer Austausch- und Transportprobleme.

Die Möglichkeit der Injektion gewaltiger Mengen radioaktiven Materials in die Atmosphäre und das Auftreten künstlich-radioaktiver Elemente im geophysikalischen Bereich als Folge von Spaltprozessen oder Wirkungen der kosmischen Strahlung haben dieser engen Verbindung zwischen Radioaktivität und Geophysik neue Impulse verliehen. Die im letzten Jahrzehnt gewonnenen neuen Erkenntnisse und Fortschritte übertreffen bei weitem die in den rund 50 Jahren „klassischer" Periode erworbenen Einsichten und haben dazu neue Probleme, Aufgaben und Möglichkeiten aufgezeigt.

Angesichts dieser raschen Entwicklung schien es an der Zeit, eine Übersicht über die Rolle der natürlichen und künstlichen Radioaktivität — kurz der Kernstrahlungen im weitesten Sinne des Wortes — im geophysikalischen Rahmen zu geben mit dem Ziel, zu informieren und zu neuen Untersuchungen, Fortschritten und Anwendungen anzuregen. Dieser Plan einer umfassenden Darstellung der „Kernstrahlung in der Geophysik" wurde von den zur Mitarbeit angesprochenen Kollegen lebhaft begrüßt. So war es möglich, die einzelnen Teilgebiete jeweils aus der Sicht und Feder eines Spezialisten zur Bearbeitung kommen zu lassen.

Obwohl hierbei mit gelegentlichen Überschneidungen und Verschiedenartigkeiten in der Auffassung und Darstellung zu rechnen war, haben wir dies bewußt in Kauf genommen, um neben der Individualität der Beiträge die Lebendigkeit und Dynamik des gesamten Fragenkomplexes fühlbar werden zu lassen. Um der Originalität der Beiträge willen wurde auch Zweisprachigkeit angestrebt (Deutsch und Englisch), wobei eine Zusammenfassung zu Anfang eines jeden Kapitels in der jeweils anderen Sprache als Erleichterung dienen möge. Ein zweisprachig angelegter Index soll gleichzeitig Wörterbuch für die Fachausdrücke sein.

Wir möchten auch an dieser Stelle allen Mitarbeitern unseren verbindlichsten Dank für die Mühe und Sorgfalt bei der Abfassung ihrer Beiträge aussprechen. Einigen von ihnen danken wir besonders für die Geduld, die sie angesichts der bei einer solchen Gemeinschaftsarbeit leider nicht zu vermeidenden Verzögerungen in der Fertigstellung des Manuskriptes und damit der Drucklegung aufbringen mußten. Unser Dank gilt in gleicher Weise dem Verlag für sein großes Verständnis und sein stets entgegenkommendes Verhalten allen unseren Wünschen gegenüber sowie für die Ausgestaltung des Buches.

Weihnachten 1961

H. Israël, Aachen
A. Krebs, Louisville

Preface

The "radioactivity of soil, water, and air" is a classical research field of geophysics. Many important discoveries and informations are based on its extended use. Questions concerning the heat balance of the earth's interior or the age of the earth and of different rocks have found satisfactory solutions on this basis; hydrologic, oceanographic and balneologic research, prospecting geology, atmospheric-electric phenomena connected with problems of the physics of the atmosphere as well as meteorologic transport- and exchange-problems have profited from and prospered under its protectorate.

The technical possibilities to inject tremendous amounts of radioactive materials into the atmosphere and the occurence of artificial-radioactive elements in the geophysical sphere as a consequence of world-wide application of fission processes and effects of cosmic radiations have given new impulses to this close connection between radioactivity and geophysics. The progress and tremendous advances in this field and its impact on other areas of geophysics in the last decade surpass by far the experiences and information collected during the roughly 40 to 50 years of the "classical" period and have opened new avenues for geophysical research.

It has seemed timely, therefore, to attempt a summary statement on the role of natural, artificial, and man-made radioactivity—of nuclear radiations in the broadest sense of the word—in the geophysical area with the goal to sketch the essential features of our science, the principal directions of current inquiries and of future research.

The response to this plan of a review on "Nuclear Radiation in Geophysics" by the colleagues asked for their advice was so enthusiastic that each individual chapter of the survey could be presented by an expert in the proper field in his own style and his own words. Such an approach could be expected to lead to occasional overlapping, differences in presentation and in interpretation. However, in order to exhibit the dynamic of the whole project and to preserve the originality of the contributions this challenge was accepted, being sure that science is often "most stimulating and convincing, when it is least dogmatic". For the same reason also the two languages—German and English—were accepted; a short summary at the beginning of each chapter in the complementary language and a subject index in the two languages will facilitate the study of the book.

Also here we wish to acknowledge our deep appreciation to all coworkers for their interest, their efforts, their advice and criticism as well as for their patience in connection with some delay in the final formulation of the manuscript. Our sincere thanks are also due to the publisher for helpful understanding and generous support of our many wishes.

Christmas 1961

H. Israël, Aachen
A. Krebs, Louisville

Inhaltsverzeichnis

Biological Aspects. By Professor Dr. A. KREBS, Biology Department, University of
Louisville, Kentucky (USA) and Dr. N. G. STEWART, Head of Health Physics and
Medical Division, Atomic Energy Research Establishment, Harwell, Berks. (Great

Introduction

The tremendous advances of nuclear physics in the last three decades have provided new knowledge and new techniques which have found practical applications in substantially all fields of science and engineering.

The application of radioactive tracers and other nuclear techniques to problems in medicine and the life sciences has produced dramatic advances which affect the lives of nearly all of us. In a parallel manner the application of newer and improved nuclear methods to problems in the earth sciences is rapidly accelerating our acquisition of knowledge and understanding concerning the lithosphere, hydrosphere, atmosphere, and even space. This volume provides a welcome review of many of the newer findings, especially those of the last decade. It also focusses attention on some important problems in the earth sciences whose solutions are tasks for the future.

Reciprocally, we need to be on the alert for observations in the earth sciences which can contribute to the growth of our understanding of nuclear physics. The collaboration between the earth sciences and the nuclear sciences is a two-way street. Recall that the presently accepted value of the half-period of K^{40} (and hence part of our physical theory of highly forbidden β-ray transitions) is the result of reinvestigations which were directly stimulated by FRANCIS BIRCH's comments in 1947 on the geothermal implications of the heat which would have been generated by potassium in Pre-Cambrian times. The more classical examples are of course the discovery of the effects of radioactivity in rocks and minerals even before the discovery in 1896 of radioactivity, and hence before the birth of nuclear physics. These include the discovery by ROSENBUSCH in 1873 of pleochroic halos in mica (explained by JOHN JOLY in 1907 as α-ray damage from minute uranium inclusions) and the discovery through the work of HILLEBRAND in 1891, of RAMSEY and of LOCKYER in 1895 of the first known terrestrial helium in the uranium mineral cleveite (explained by RUTHERFORD and ROYDS' unequivocal identification of α-rays as a helium nuclei, about ten years later).

The extensive bibliographies which are in this book will prove especially valuable, even to veteran workers. This is because the literature in the earth sciences is widely scattered in many periodicals, in technical reports from various atomic energy agencies, in Congressional hearings, in Geneva Conference reports, and in reviews and symposia which have small circulations. The reader will find that much has been accomplished since the appearance in 1954 of the excellent text on *Nuclear Geology* edited by HENRY FAUL. To new workers in the earth sciences a word of warning is in order. Several of the chapters in the present book deal with advances in a given topic during only the last 10 to 15 years, due to page limitations. The earlier literature, some of which is important, will have to be searched out by consulting bibliographies given in the earliest papers which are discussed herein.

There was a marked quickening of the application of nuclear physics to the earth sciences in the 1930's. This was the outgrowth of the discovery in that decade of the neutron, of the positron, of deuterium, of induced radioactivity, and of nuclear fission, together with substantial improvements in instrumentation. It became clearly recognized in the 1930's that, at least in the oceans, the members of a decay series such as uranium, ionium, and radium are not in radioactive equilibrium. Subsequently the quantitative study of cases of "disequilibrium" has illuminated many problems in geochronology and in the study of such dynamic processes as sedimentation, biochemical effects, and oceanic circulation.

The 1940's saw the development of nuclear reactors and of nuclear bombs, with their greatly increased productivity of radioactive nuclides. These nuclides have provided the material for radioactive tracer observations of dynamic processes in the hydrosphere and atmosphere on a global scale. Public concern for an evaluation of the alleged health hazards of some of these radioactive nuclides has made substantial financing available for the study of their distribution, from the stratosphere to the oceanic depths.

The pace of research in the application of nuclear physics to the earth sciences has accelerated tremendously in the 1950's. Among the contributing factors which can be recognized are: (1) an increased scientific interest in geophysics, geochemistry, oceanography, meteorology, and space, (2) the practical problems of exploration for mineral deposits, especially for uranium and for petroleum, (3) the availability of radioactive tracers from tests of nuclear weapons and nuclear devices in the air, in the oceans, and underground, (4) the identification of a number of radioactive nuclides produced in meteorites, in the atmosphere, and in the lithosphere by cosmic rays, including H^3, Be^7, Be^{10}, C^{14}, Na^{22}, P^{32}, P^{33}, and S^{35}, (5) the substantial improvements in mass spectrometers and radiation detection instruments capable of quantifying these nuclides, such as NaI-γ-ray spectrometers and solid-state detector systems, and (6) money.

The earth scientist must beware of taking all quantitative physical and chemical results at face value. Subtle errors of sample selection, sample preparation, analysis, calibration, or interpretation can arise easily. The importance of interchecking results by measuring the same or similar samples in two or more laboratories, and if possible by two or more methods, cannot be overemphasized. It was, for example, a sobering experience when the first such interlaboratory comparisons were made on helium-age measurements and resulted in 1939 in lowering the helium-age scale to about one-half of its former value. The physical sciences and the earth sciences are complex and involve many subtleties. Scientific teams which include profound specialists in both areas usually produce results which involve minimum opportunity for errors in conception, execution, and interpretation.

As knowledge accumulates, the storage and retrieval of scientific information becomes a serious problem, especially in a widely diffused field. Like the nuclides which we study, there is a world-wide distribution of nuclear earth scientists, and at a small but increasing concentration. This book will contribute greatly to the retrieval of information which is currently in storage but which needed to be gathered for reexamination and for the guidance of future research.

October 1961   ROBLEY D. EVANS
                                            Massachusetts Institute of Technology
                                            Cambridge, Massachusetts

Radioactivity of the Lithosphere

by

John A. S. Adams

With 1 Figure

Zusammenfassung

Zahlreiche Bestimmungen des Radioaktivitätsgehaltes der Lithosphäre mittels physikalischer und chemischer Meßverfahren, bei denen etwa eine Messung auf 10^{17} g der zugänglichen Gesteinsphäre kommen, liefern im Mittel folgende Werte: Hauptträger der Gesteinsaktivität sind K^{40} (Häufigkeit $3 \cdot 10^{-6}$ g/g*), U^{238} (etwa $3 \cdot 10^{-6}$ g/g) und Th^{232} ($11-13 \cdot 10^{-6}$ g/g). Rb^{87} und andere vorhandene oder durch natürliche Kernumwandlungen entstehende radioaktive Nuklide kommen in der Lithosphäre nur in sehr viel geringerer Menge vor.

Aus theoretischen Überlegungen und Überschlagsrechnungen läßt sich nach den vorhandenen Erfahrungen eine Beziehung aufstellen, die die mittlere Häufigkeitsverteilung bestimmter Aktivitätskonzentrationen in den Gesteinen beschreibt.

Global betrachtet ist die irdische Radioaktivität im wesentlichen auf die obersten Gesteinsschichten von einigen Zehner-Kilometern Dicke beschränkt. Im einzelnen ist sie, wie die Strahlungsspuren in Mineralien zeigen, in der Regel an relativ widerstandsfähige Mineralien gebunden, aus denen sie nur zum kleinen Teil in den Wasserkreislauf und in die Atmosphäre gelangt.

Zum Schluß werden verschiedene Folgerungen besprochen, so paläontologische Ergebnisse über die verflossenen 600 Millionen Jahre im Zusammenhang mit möglichen Strahlungsvariationen terrestrischen und extraterrestrischen Ursprungs, Änderungen der terrestrischen Isotopenhäufigkeit stabiler Nuklide — vor allem bei Pb und Ar — und die Anwendung dessen bei der Diskussion geologischer und geochemischer Cyclen.

A. General considerations

I. Sampling

The flux of nuclear radiations emitted from the solid rocks of the lithosphere ranges from rare ore deposits where more than half of a few cubic meters may be uranium or thorium to somewhat more common halite (NaCl) deposits where several cubic kilometers may exhibit no radioactivity detectable above instrumental background [Adams et al. (1959)]. Not only does the radioactivity of the lithosphere range widely, the sampling of the accessible lithosphere is sparse. The accessible lithosphere presently consists of the land surfaces, together with samples obtained from borings that may exceed six kilometers on the continents and penetrate the uppermost few meters of sediment at the bottom of the seas. Thus, in principle, present techniques permit the direct sampling of a system that has a mass of approximately 10^{23} grams; no more than 10^5 detailed and accurate determinations of the general level of radioactivity have been made,

---

* Entsprechend 3 „ppm (parts per million)".

yielding no more than one such determination for every 10^{17} grams at best. It should also be noted that some of the material now within six kilometers of the surface, e.g. certain volcanic lavas, must have come originally from some greater depth.

In considering the degree to which the sampling to date has been representative, one must also note that most determinations of radioactivity in the lithosphere have been made either in connection with exploration for unusually high concentrations of uranium or in connection with the use of positive or negative radioactivity anomalies in petroleum exploration. Although both of these major types of information were obtained in a search for *non*-representative concentrations of radioactivity, they both serve to demonstrate indirectly that large volumes of the lithosphere do not vary by more than a factor of two or three in radioactivity. In particular, the thousands of kilometers of low level gamma-ray aerial surveys and the hundreds of kilometers of oil well gamma-ray logging have shown this small variance in both a horizontal and a vertical sense. This small variance in many common rocks has also been indicated by: (1) taking equal weights from many localities and determining the radioactivity of the aggregate [BARANOV et al. (1956)]; (2) making radioactivity determinations on rocks that geologic processes should have homogenized [ADAMS and WEAVER (1958)]; and (3) studying the variance in radioactivity in a common shale on a meter by meter and ten kilometer by ten kilometer basis [PLILER and ADAMS (1959)]. The concept of a wide range of radioactivity combined with a low dispersion is represented qualitatively by Fig. 1.

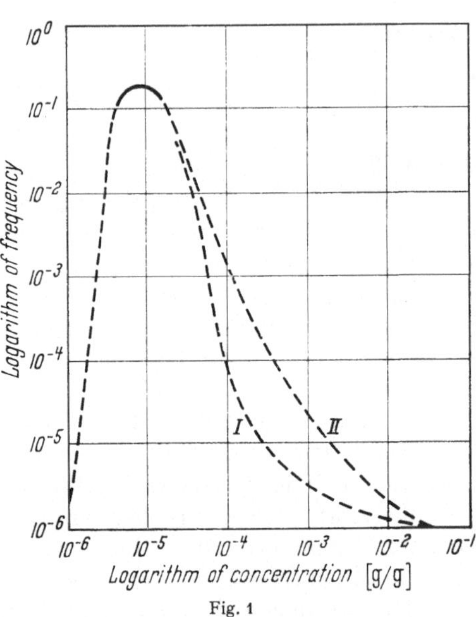

Fig. 1

In Fig. 1 the logarithm of the frequency of a unit mass of rock is plotted against the logarithm of the concentration of a natural source of radioactivity. The mode or most frequent concentration in Fig. 1 is also the best known. Thus, the modes for the most common rocks in the accessible lithosphere—granites and shales—are both 11 to 15 parts per million thorium, 3 to 4 parts per million uranium, and about 3 parts per million potassium-40 [ADAMS et al. (1959)]. These figures also represent the average concentrations for the accessible lithosphere because the statistical weight of other rock types and concentrations are so low. For each major source of natural radioactivity it is possible to select a unit mass so that the upper limits of concentration would be, respectively, 100% ThO_2, 100% UO_2, and 100% KCl, neglecting very minor impurities. The frequency of such high concentrations, although known to be low, remains quite uncertain because the accessible lithosphere is so large and the sampling is so sparse. Thus, the dashed curves I and II in Fig. 1 represent only two of the many possibilities for the true frequency of the higher concentrations. Attempts to show that the frequency distribution of the concentration of a minor element in granite or shale should follow a simple function, e.g. log-normal, have not been convincing to date [MILLER and GOLDBERG (1955)].

The difficulties in sampling are not equally serious for each problem involving the radioactivity of the lithosphere. Thus, for example, in calculating the amount of heat generated by nuclear radiations in the lithosphere, the average concentrations given above are adequate to estimate the heat production to within 20%. For this heat problem it is not necessary to know whether 0.01 or 0.00001% or less of the total inventory occurs in the high economic concentrations; other errors in decay constants and energies of radiation are more important in limiting the accuracy with which the production of heat from nuclear sources can be calculated. On the other hand, the sampling problems are formidable in attempting to estimate the amount of thorium and uranium that might be recoverable for nuclear power generation. The exact shape of the right side of a curve like that in Fig. 1 remains unknown and could only be estimated from the rate of discovery and the acceleration in the rate of discovery, keeping exploration effort constant and equally distributed [ADAMS (1959)].

II. Analytical methods

Quantitative determinations of the radioactivity of rock samples have been made by many physical and chemical techniques, including autoradiography, total and spectral radiometry of untreated rock or of emanations or other extracts, isotope dilution, and various wet chemical methods. It is not appropriate to discuss these various methods fully. However, it should be noted that in the past analytical difficulties have made the distribution of thorium much less well understood then the distributions of uranium and potassium, the other major natural sources of radioactivity in the lithosphere. Modern instrumentation, particularly gamma-ray spectrometers and alpha pulse height analyzers, make it possible to obtain thorium and other determinations with much greater speed and accuracy. The various methods have provided a number of independent cross-calibrations and, in general, the better analytical techniques are more accurate than many of the assumptions regarding the history of the rock sample and the degree to which it is representative.

B. Abundances of natural radioactive nuclei

I. Types of natural radioactive nuclei

At the surface of the lithosphere there are two sources of nuclear radiations —cosmic rays, including cosmic ray induced radioactivities, and manufactured radioactivities—whose importance decreases rapidly with depth and which will be discussed in subsequent chapters. With only the most minor, but interesting, exceptions, the nuclear radiations in the lithosphere are believed to come from radioactive nuclei that have existed in that form from before the formation of the earth. The nuclei listed at the page 15 may be divided into four groups on the basis of their abundance and specific activity: Group I consists of uranium-238, uranium-235, and thorium-232 in secular radioactive equilibrium with their many radioactive daughters, and of potassium-40; the specific activity and abundance of these four nuclei and their radioactive daughters combine to make them the major sources of radioactivity in rocks; Group II contains only rubidium-87, which is relatively very abundant and widely used in absolute age determinations (see following chapter), but whose mode of beta decay and low specific activity permit only a very minor contribution to the nuclear radiation flux in rocks; Group III consists of radioactive nuclei that are known to exist in the lithosphere, but whose low abundances and specific activities combine to make their contributions to the total radiation flux and absolute age determination research very minor. Group IV consists of nuclei whose radioactivity or presence in nature is in doubt.

It should be noted that the abundances and groupings in the table represent the present situation and that several billion years ago uranium-235 and potassium-40 would have been relatively more abundant as a direct consequence of their shorter half-lives. It is also possible that in the distant past there existed short-lived radioactive nuclei that have long since decayed to such an extent that they have not yet been detectable in nature. Kohman and Saito (1954) discussed the possibilities of such extinct radioactivities in general and Rankama (1954) has summarized the various searches for several such nuclei, e.g. caesium-132, which might be analogous to potassium-40 and rubidium-87. Reynolds (1960) recently reported determinations of isotopic xenon ratios in a meteorite that indicate the former presence in nature of iodine-129, which is known to decay to xenon-129 with a half-life of 1.72×10^7 years.

In addition to the radioactive nuclei that are older than the earth and their direct decay products or daughters, radioactive nuclei are being created in the lithosphere by at least two processes—natural fission and the capture of naturally produced neutrons. The natural fission of uranium has been established by the mass spectrometric identification of fission products, particularly isotopes of xenon and krypton [see Fleming and Thode (1953)]. Natural fission is assigned to both spontaneous fission and neutron induced fission. Kurode et al. (1957), for example, estimate that spontaneous fission is three to five times more frequent than neutron induced fission in the pitchblende sample investigated by them. This estimate is in general agreement with the other estimates cited by them. Although each uranium fission event releases a great amount of energy, the frequency of such events in the lithosphere is very low and, thus, natural fission is a very minor source of energy in the lithosphere relative to the radiation flux from thorium, uranium, and potassium [Rankama (1954)]. This is also the case with natural neutron capture processes, whose operation has been established by the detection of plutonium-239 in natural pitchblende [Levine and Seaborg (1951) and Peppard et al. (1951)]. Even though natural fission and neutron capture processes are very rare and unimportant energetically in the lithosphere, they do provide some opportunities to confirm theoretical and experimental predictions about nuclear reactions, despite the complications that arise from variable mineral compositions [see also S. Flügge (1939)].

II. Direct abundance data

Most of the abundance data regarding potassium-40 are not based upon direct determinations of potassium-40, but on determinations of total potassium, assuming that potassium-40 is always 0.0119% [after Nier (1950)] of any natural potassium sample. Rankama (1954) has reviewed the conflicting data on the constancy of the isotopic composition of potassium and in the absence of any subsequent and intensive test of this constancy, one must still conclude that the isotopic composition of potassium may vary within a very restricted range. However, it appears unlikely that any possible variation in isotopic composition will necessitate a revision of the error estimates on the abundance data or on the radiometric determination of potassium in rocks [Adams et al. (1958)]. It should also be noted that important information on the abundance and distribution of potassium has been obtained by the petrographic determination of feldspars and micas that carry an essentially constant concentration of potassium. A comparison of this petrographic method and various chemical and physical methods of potassium determination, as well as a discussion of analytical errors can be found in FAIRBAIRN et al. (1951). All modern estimates from CLARK and WASHINGTON (1924) to RANKAMA and SAHAMA (1950) and GREEN (1959), among others, place the abundance of potassium in the accessible lithosphere between 2.5 and 2.6%, as the metal. Thus, assuming the isotopic abundance given above, potassium-40 has an abundance very close to 3.0 parts per million in the accessible lithosphere.

The quality of the abundance data for uranium isotopes approaches that of the data for potassium-40. RANKAMA (1954) gives 4.0 parts per million, GREEN (1959) 2.6 parts per million, and ADAMS et al. (1959) 3.0 ± 0.6 parts per million. The difference between the last two figures is largely a matter of what estimate is made for the more uncertain uranium concentration in basic rocks (basalts) and what estimate is made for the abundance of such rocks in the accessible lithosphere. The isotopic composition of natural uranium has been shown to be constant to within a small analytical error [DUNNING (1949)], except in certain local situations where recent leaching has partially separated uranium-234 from its parent uranium-238 [BARANOV et al. (1958)]. Neutron capture reactions in the lithosphere give rise to a minute amount (one atom in over 10^{11} atoms of uranium) of uranium-233 and probably uranium-239 [RANKAMA (1954)]. Although 99.28% of natural uranium is uranium-238, it must be noted that its daughter product, uranium-234, with an abundance of only 0.0058%, emits precisely as many alpha particles per unit time per unit rock as its parent uranium-238. This, of course, is also true of the radiation flux from any of the many radioactive daughters in the uranium and thorium disintegration series. Thus, assuming radioactive equilibrium and only one mode of origin, it is possible to relate quantitatively the abundance of any radioactive daughter to the abundance of the parent, the ratio of the half-lives of parent and daughter, and any branching ratio that may be involved. For example, in the entire terrestrial system, the abundance of radium-226 should be 3.5×10^{-6} that of the parent uranium-238, taking 1600 and 4.5×10^9 years for the respective half-lives. Finally, in considering the relative contributions of uranium-238 and uranium-235 to the nuclear radiation flux in the accessible lithosphere, one must note that the lower abundance of uranium-235 (0.72%) is partially offset by its shorter half-life.

Until recently, lack of both interest and suitable analytical techniques have combined to limit the number of thorium determinations made on rock samples. ADAMS et al. (1959) have recently summarized the available data and estimate the thorium abundance in the accessible lithosphere to be 11.4 ± 2.0 parts per million. This estimate is not markedly different from the 11.5 parts per million estimate of RANKAMA (1954) or the 13 parts per million estimate of GREEN (1959). However, some of the agreement may be fortuitous, considering the limited amount of data. All of this thorium is considered to be thorium-232, although its short-lived daughter, thorium-228, and the thorium-234 and thorium-230 daughters of uranium-238, and the thorium-227 daughter of uranium-235 are present in the lithosphere in the low abundances required to maintain secular radioactive equilibrium with their respective parents.

Rubidium-87 is the most abundant of the natural radioactive nuclei in the lithosphere (approximately 75 parts per million), but its long half-life and the fact that its radiation is limited to a single negative beta emission, make its contribution to the nuclear radiation flux in the lithosphere very minor. It also follows from these nuclear properties that its radiometric determination is difficult; the abundance data are largely based upon emission spectrometry methods. RANKAMA (1954) estimates 350 parts per million for natural rubidium's abundance;

Green (1959), taking into consideration the several recent studies which he cites, estimates 280 parts per million. Ahrens (1957) and Horstman (1957) discuss the evidence supporting lower estimates for the abundance of rubidium. No natural isotopic fractionation of rubidium is known and 27.85% of natural rubidium is considered rubidium-87, which would indicate about 75 parts per million of rubidium-87 in the accessible lithosphere.

In comparison to potassium-40, uranium-238, uranium-235, thorium-232, rubidium-87, the other radioactive nuclei listed in the table are of less interest for one or more of the following reasons: (1) their abundances are both lower and less well known; (2) their half-lives are very long or even in doubt; (3) their radiations are generally of low energy; (4) as a consequence of the foregoing, they are relatively rarely used in absolute dating studies (see following chapter) and are unimportant as sources of heat. These minor radioactive nuclei do, however, provide some opportunities for testing predictions about nuclear stabilities [Rankama (1954)], cosmology, and the origin of the elements [Cameron (1959)].

III. Theoretical considerations

The sampling and analytical difficulties outlined above make it imperative to use every possible method to obtain representative samples and to determine any systematics in the abundance data that can be used as criteria for internal consistency. Suess and Urey (1956), Rankama (1954), Cameron (1959), among others, have discussed the relationships among the observed abundances of elements on one hand and what is known about nuclear stabilities and possible origins of the elements on the other hand. These considerations are important for the present discussion because they provide approximate limits for abundance estimates. Thus, for example, samarium with an even proton number of 60 should, according to the rule of Oddo-Harkins [see Cameron (1959)] be more abundant than either element 59 or 61, which is what the abundance data indicate. Furthermore, as a member of the lanthanides (elements 57 to 71), which are very similar chemically, samarium would be expected to have an abundance rather similar to those of the other lanthanides with even proton numbers, which is the case to a good first approximation.

The four major sources of radioactivity in the lithosphere can be divided into two groups on the basis of close chemical similarity, namely, potassium-rubidium and thorium-uranium. Ahrens et al. (1952) discussed the close chemical similarity between rubidium and potassium, especially as regards valence and ionic size, and they suggested that the potassium/rubidium ratio might be quite constant in many samples. Although it is now known that the potassium/rubidium ratio varies widely in nature [Horstman (1957)] and there is some question about the exact potassium/rubidium ratio in meteorites [see Green (1959)], one would still expect that the potassium/rubidium ratio in meteorites might be rather constant and would represent the cosmic ratio as well as the overall ratio in the earth; hence, it would also be expected that the potassium/rubidium ratios observed in the lithosphere should range above and below the cosmic ratio found in meteorites.

The actinides thorium and uranium also form a pair of elements with many chemical similarities, e.g. ionic size. However, because it has an hexavalent state not exhibited by thorium, uranium is strongly fractionated from thorium in the upper lithosphere [see summary in Adams et al. (1959)]. Nevertheless, the following independent lines of evidence all indicate that the thorium/uranium ratio in the cosmos, meteorites, and accessible lithosphere averages very close to 4.0: (1) representative sampling of the lithosphere [ADAMS et al. (1959); MURRAY and ADAMS (1958)]; (2) the most accurate determinations on meteorites [see summary in GREEN (1959)]; (3) calculations of the thorium/uranium ratios required to explain the known lead isotopic data [see RANKAMA (1954), MARSHALL (1957), CANNON et al. (1958), among others]. An overall estimate of the radioactivity of the lithosphere can also be obtained from heat flow data as described in the next section.

C. Heat effects

The generation of heat by radioactive disintegrations in the lithosphere has been estimated by either calculation, using the nuclear abundance figures discussed above and the known energies of the nuclear radiations, or by measuring the heat flow through the accessible lithosphere. It was recognized over 50 years ago that the radioactivity of rocks was high enough to be a major source of heat in the lithosphere. JOLY's early (1909) and quite acceptable abundance determinations were made with the earth's heat balance in mind. RANKAMA (1954), BIRCH [in FAUL (1954)] refined the earlier calculations on the magnitude of heat production by radioactive disintegrations in the lithosphere. Their and most previous estimates of the annual rate of heat production in granite fall in the range of $6 \pm 5 \times 10^{-6}$ calories/gram of rock. UREY (1955) developed models for the radioactivity contribution to the heat balances of the earth, the moon, and mars. JACOBS (1956), VERHOOGEN (1956), and MACDONALD (1959) have also discussed the role of radioactivity in various models of the earth's heat balance. Any of these models for the earth must reconcile the annual rate of heat production given above for granite with the observed rate of heat flow through the earth's surface (about 1×10^{-6} calories/cm²/sec) and with the distribution of granite, the most common of the most radioactive rocks. At first it was thought that because granites appear to be more common under the continental areas, the heat flow from continental areas should exceed that from oceanic areas. However, the measurements of REVELLE and MAXWELL (1952), BULLARD (1954), and BIRCH (1956) refute this simple model and make it necessary to explain why the heat flow from under the oceanic areas equals that from under the continental areas to within experimental error. VERHOOGEN (1956) discussed this problem and how it might be adjusted by assuming a different coefficient of heat conduction, or by assuming heat transfer by some mechanism other than conduction, or by assuming a distribution and abundance of radioactive elements different from those discussed above. BULLARD et al. (1956) have also discussed the problem of having equal heat flow rates for oceanic and continental areas. Although the heat flow data are difficult to interpret, they are of fundamental importance in understanding major terrestrial processes. For example, the observed flow of heat from the earth's interior, small and difficult as it is to measure, represents an amount of energy that is one or two orders of magnitude higher than the more obvious release of energy in earthquakes and volcanoes [VERHOOGEN (1956)]. KENNEDY (1959) and others have suggested that differences in heat flows and geothermal gradients are responsible for many major geologic processes, e.g., mountain building. It has been evident since RAYLEIGH's (1906) pioneer discussion that the radioactive elements must be strongly concentrated in the uppermost levels of the lithosphere, else they would generate much more heat than is observed to flow through the accessible lithosphere. Thus heat flow data provide some limits on the distribution of radioactivity within the earth.

D. Distribution of radioactive elements

I. Distribution on a global scale

VERHOOGEN (1956) concluded that it was difficult to reconcile the essentially equal heat flows in oceanic and continental regions with the conclusion that more radioactive granitic material is dominant in the immediate subsurface of the continental regions and that less radioactive basaltic material is dominant immediately under the oceans. He suggested that either conduction is not the major mechanism of heat transfer or that the continental areas are not really as anomalously high in radioactivity as previously thought. LAWSON and JAMIESON (1958) have discussed various mechanisms of heat transfer and KENNEDY (1959) has suggested some mechanisms that might alter the geothermal gradient. Although it is not clear at present whether or not a square meter of continental surface is underlain by as much or more radioactive material as a square meter of oceanic surface, it is clear that most of the radioactive material must be near the surface [see MACDONALD (1959)]. At present the solution appears to lie with more elaborate models, perhaps compensating the higher radioactivity at the surface of the continents with a depletion at shallow depths so that a square meter on either the continental or oceanic surfaces would be underlain by an equal amount of radioactivity.

The depletion of the mantle and core of the earth in radioactive material is also indicated if one assumes that the entire earth was originally similar in composition to chondritic meteorites. Using this assumption, PATTERSON et al. (1955) concluded that perhaps 90% of the earth's uranium must be concentrated close to the surface in a 5 kilometer thick layer of granitic material underlain by a 25 kilometer thick layer of basaltic material. This model would be adequate to explain the observed lead isotopic data. MACDONALD (1959) has recently discussed this hypothesis that the earth is essentially chondritic in composition and notes that, using the more recent and lower abundances for potassium, uranium, and thorium in chondritic meteorites, the total present rate of heat loss from the earth is quite consistent with this hypothesis. Furthermore, the general chemistry and mineralogy of thorium, uranium, and potassium, as well as the amount and isotopic composition of lead in the accessible lithosphere, are all consistent with a strong enrichment of the major radioactive elements near the surface.

At the land surfaces the most common exposed rocks are shales. Although the thickness of shales and other sediments varies widely, it averages only about 700 meters on the continents [see MASON (1958)]. Immediately below this thin veneer, which is absent over broad areas of the earth's surface, granitic rocks predominate. Shales and granitic rocks together comprise at least $\frac{2}{3}$ and perhaps $\frac{9}{10}$ of the accessible lithosphere and because they have almost the same amounts of radioactive elements [ADAMS et al. (1959)], one might conclude that the radioactivity levels in the accessible lithosphere do not vary widely if one takes samples on a scale of 20 kilometer cubes. It is not clear whether lead isotopic anomalies and greater frequency of commercial concentrations in certain areas, e.g., the Colorado Plateau of USA [see CANNON et al. (1958)], represent parts of the lithosphere that have unusually high average concentrations of uranium or whether they are regions where geologic processes have been unusually effective in redistributing a normal uranium inventory so that more of it occurs in the higher concentrations.

It should be noted that even if there are not broad regional differences where one considers the upper 20 kilometers of the lithosphere, the flux of natural nuclear radiations from the land surfaces varies by at least several orders of magnitude. Thus igneous rocks, e.g. granites exposed at the surface in many regions, and sedimentary rocks, e.g. uraniferous black shales of the Dictyonema type in Scandinavia or the Chattanooga type in the USA, may be a hundred times more radioactive than other igneous rocks, e.g., the basalt flows of the Columbia Plateau in northwestern USA or of the Deccan traps in India, or sedimentary rocks, e.g., the purer limestones or sandstones [see abundance data for different rock types in ADAMS et al. (1959)]. Attempts have been made to relate such differences in surficial radioactivity with the incidence of abnormal births in the human population [GENTRY (1959)]. The wide differences in the radioactivity of foods reported by TURNER et al. (1958) and others may also be relatable to regional rock radioactivity variations in part. It is also possible that certain of the more striking variations in the radioactivity of food may not be primarily a function of rock or soil type. Thus, the observation that Brazil nuts are nearly 20000 times more alpha active than some other foods may be more a function of plant physiology (length of growing time) or of the fact that intensive weathering of all rock types in the tropics gives rise to bauxites and related highly radioactive soils [see ADAMS and RICHARDSON (1957); PLILER and ADAMS (1959)]. Again the differences in the radioactivity of wheat may be a function of rock and soil type and variance, but it may also be a function of whether or not the crops are fertilized with phosphate from one of the highly uraniferous phosphate rocks (see also the Chap. "Biological Aspects").

Aside from the known variations in the concentrations of potassium, thorium, and uranium, the radiation flux observed at or above the land surface is also affected greatly by the sites of these radioactive elements in the rocks. Thus, the radioactive material in foods could also be a function of the availability and not the concentration of radioactive elements in the rocks and soils. KOCZY et al. (1957) concluded from uranium analyses of rivers flowing into the Baltic Sea that uranium is leached more readily from sedimentary rocks than from igneous rocks. In this and other connections it is important to consider the sites or micro-distribution of radioactivity in rocks.

II. Microdistribution of radioactivity

The micro-distribution or exact sites of radioactive elements in rocks affects the rate at which emanations from these elements are released to the atmosphere and at which these elements are dissolved in natural waters and hence enter the natural substrate of organisms. The precise sites, including the chemical state and bonding, are also of interest in connection with the recovery of thorium and uranium from common rocks [BROWN and SILVER (1956)], and in the study of minerals for absolute age determinations [TILTON and NICOLAYSEN (1957) and ADAMS and ROGERS (1961)].

Potassium-40 is greatly diluted by and carried along with the much more abundant and stable isotopes potassium-39 and potassium-41. Hence it occurs as a small, but constant constituent of the very common potassium felspars and micas. Potassium also occurs in important amounts in clay minerals where its exact site and ability to be exchanged has been much investigated [WEAVER (1958)]. Clay minerals are so selectively efficient in fixing potassium that relative to sodium only $\frac{1}{30}$ of the theoretically available potassium supply ever reaches the oceans [MASON (1958)]. Another consequence of this fixation of potassium on clays is the necessity to supply large amounts of potassium in chemical fertilizers for agriculture.

The sites of the small amounts of thorium and uranium in rocks can be determined directly to within 5 microns by autoradiography [see the general summary of Bowie in Faul (1954)] and indirectly by leaching studies. Both of these techniques are necessarily used on small laboratory samples and the results must be considered together with regional studies on river waters, e.g. Koczy et al. (1957) cited above, and with other more representative observations, e.g., the fact that most of the uranium originally in igneous rocks is not oxidized and leached out while igneous rocks are converted to shales. Adams et al. (1959) concluded that at least $\frac{2}{3}$ of the uranium and thorium in the accessible lithosphere is contained in fine grained (mean size measured in tens of microns) minerals like zircon ($ZrSiO_4$) and monazite (mainly $CePO_4$). These and related minerals are both chemically and mechanically stable at the surface of the earth and they constitute but a small fraction of one percent of common rocks. The main reasons for concluding that thorium and uranium occur for the most part in these minor or accessory minerals are: (1) the observation that a considerable amount of the uranium in granite is not leached out during weathering and hence must not be available to natural leaching; (2) autoradiographs of many rocks show the alpha emitters to be concentrated in such rare, accessory minerals; (3) thorium with a valence of $+4$ and uranium with both a valence of $+4$ or $+6$ possible in nature have both charges and sizes that do not permit them to form stable isomorphous series with the common rock forming minerals; thorium and uranium tend to form the most stable phases with the most abundant ions of similar charge and size, namely zirconium and cerium [see Frondel (1956)]. Thus to the extent that chemical equilibrium is approached in rocks, one would expect thorium and uranium to be concentrated in these rarer phases or minerals. It is this lack of isomorphous series with common minerals that gives rise to rare, accessory minerals with high enough concentrations of thorium and uranium to make radiation damage effects readily observable in these minerals.

E. Radiation damage

Radiation damage effects have long been known in two common mineral carriers of thorium and uranium, namely biotite and zircon. In biotite and related micas minute inclusions (commonly less than 1 micron in diameter) of radioactive minerals, e.g., zircon, xenotime, and monazite, produce microscopically visible pleochroic haloes. The concentric rings of these haloes can be related to the range of alpha particles in biotite. Deutsch, Picciotto, and co-workers have made a number of recent studies of this type of radiation damage (see 1954 and 1958 references). In particular they have studied the relationships of the geologic age of the halo to the present flux and the total dosage. Rankama (1954) has enumerated other possible and attempted uses of pleochroic haloes. Pleochroic haloes are known in many minerals, but those in biotite have been most extensively studied.

In zircon ($ZrSiO_4$), thorium and uranium may occupy zirconium positions in the crystal lattice or they may form inclusions of isomorphous phases, e.g. $ThSiO_4$ and $USiO_4$ [see Frondel (1956)]. Concentrations of thorium and uranium up to a few tenths of a percent are not uncommon and these concentrations are high enough for the emitted alpha particles and their recoil nuclei to produce extensive short range damage to the crystal structure. This radiation damage causes the zircon crystal to become less dense, essentially isotropic, and a virtual glass or amorphous material as shown by x-ray diffraction techniques. Holland and Gottfried (1955) made a detailed study of radiation damage in zircons.

Pabst (1952) reviewed the previous mineralogical work on radiation damage of the extreme or metamict type. It is clear that although the radiation flux produced by a few tenths of a percent of thorium and uranium is small, in geologic time a dose can accumulate that is quite comparable to that which produces radiation damage effects in the components of nuclear reactors. Attempts have been made to use the radiation damage to zircons as the basis for absolute age determinations and the fact that much radiation damage can be erased by annealing at moderately elevated temperatures provides the possibility for learning something about the thermal history of a damaged zircon [see Holland and Gottfried (1955)]. Approximately 20 calories/gram of stored energy has been reported in rare metamict minerals containing thorium and uranium (polycrase), but this represents only a minute fraction of the energy of the total dose and there is no direct evidence to indicate that the accumulation of energy in radiation damaged crystals is an important process in considering the earth's heat budget [Rankama (1954)].

In recent years it has been found that minor radiation damage is present in nearly every rock. Aside from the rather rare pleochroic haloes and metamict crystals, discolorations due to irradiation are common in halite, fluorite, and quartz. Upon heating many such naturally irradiated crystals emit visible light far below the incandescence temperature. This light or thermoluminescence has been related to the thermally induced return of trapped electrons to a lower energy state. The effect has also been produced in many minerals by artificial irradiation. Modern instrumentation has made it possible to detect natural thermoluminescence in many minerals and studies have been made to use this property for absolute age determinations, geologic correlation, mineral identification, and dosimetry [see summary and apparatus describtion by Lewis et al. (1959)]. One must also consider that these direct evidences of radiation damage to crystals offer one possibility for experimentally testing some of the speculations regarding the effects of radiation on ancient life (see next section).

F. Fossil record

Evidence of ancient life—particularly in the form of hard skeletal parts—is sufficiently abundant in sedimentary rocks to permit the tracing of the evolutionary lines of descent of many animal groups. These and other results of paleontological investigations offer the possibility of studying what effects—if any—the nuclear radiation flux at the surface of the earth has had on life in the past. The fossil record has detail, abundance, and variance for only about the last 600 million years. Thus the fossil record is only available for the last 15% or so of the earth's existence. There is no evidence that the thorium, uranium, and potassium contents of granites and shales formed more than 600 million years ago are detectably different from those found in granites and shales formed more recently [Adams et al. (1959)]. Thus, radioactive decay is considered to be the main process affecting the magnitude of the nuclear radiation flux from the lithosphere during the last 600 million years. It follows from the respective abundances, half-lives, and energies of radiation that the radiation flux from thorium, uranium, and potassium, and their radioactive daughters would have been 6 or 7% higher 600 million years ago. Earlier in the development of the earth the radiation flux, and hence the heat production would have been higher by a more significant amount. Rankama (1954) calculates 2.2 times higher than present at 3000 million years ago. It is difficult at present to evaluate any effect this slowly declining radiation flux might have had on the origin or evolution of life, because in addition to the contributions from the lithosphere itself, extra-terrestrial sources of radiation must be considered.

Urey (1957), Mason (1958), and many modern studies of the subject conclude that at first the earth did not have an atmosphere comparable to the present atmosphere in mass and composition. Thus, the early atmosphere may not have been so effectual in shielding the earth's surface from direct exposure to cosmic and solar radiations, including an important flux of ultraviolet radiation from the latter. As the atmosphere increased in mass due to the degassing of the earth and changed in composition, particularly by the loss to outer space of hydrogen originally combined with oxygen and nitrogen, the adsorption characteristics of the present atmosphere would be approached. Concurrent with the degassing of the earth, most modern theories have the originally homogeneous earth fractionating radially on the basis of atomic densities, atomic compressibilities, and chemical bonding characteristics. This radial fractionation would result in thorium, uranium, and potassium being strongly concentrated in the uppermost few tens of kilometers. This radial fractionation or differentiation would also give rise to the core of the earth and, hence, to a terrestrial magnetic field, which would in turn affect extra-terrestrial particle radiations. The relative rates of these processes are unknown and it is not possible to say that the total nuclear flux at the surface has declined continuously from some unknown maximum early in the earth's development. One cannot exclude the possibility that the processes mentioned operated to produce the maximum continuous flux at some later period in the earth's development.

A fundamental assumption of the above—and unfortunately largely hypothetical—discussion is that the extra-terrestrial radiation flux has changed as slowly and smoothly as the flux from the lithosphere itself. This may not be the case and at least two mechanisms might greatly increase the extra-terrestrial flux for geologically short periods. Peters (1959) in a review of cosmic ray research discussed the possibility that a super nova explosion might occur close enough to the earth to raise greatly the cosmic ray flux. He estimated that such an event might occur every 200 to 300 million years on the average and suggested that evidence for such rare and cataclysmic events might be sought in the fossil record. Another independent and suggestive line of evidence lies in the interpretations of paleomagnetic data [Runcorn (1955)], which indicate that the earth's magnetic field has reversed direction in the past, perhaps every few tens of millions of years or less. These reversals and perhaps temporary reductions in field strength might affect cosmic and solar particle fluxes. Sudden increases in the extra-terrestrial radiation flux cannot be excluded and if they have occurred, those that caused the highest doses might be indicated by a marked change in the evolutionary processes recorded in the fossil sequences.

A marked change in the fossil record might be somatic in nature with a high radiation flux reducing the population of the most radiation sensitive animals to a point where they became extinct. Radiation sensitivity might be a function largely of reproduction habits and processes, of habitat, or even of size. The geologically sudden disappearance of the main groups of large dinosaurs from the land, sea, and air about 60 million years ago would be a possible example of a somatic type of sudden change in the fossil record. Another type of marked change might be largely genetic, resulting in changes in the rates and directions of evolution. Although there are these marked and not completely explicable changes in the fossil record, there are many alternative hypotheses to the increased radiation hypothesis. The increased radiation hypothesis has the difficulty that no increase in radiation has been high enough in the last 600 million years to cause more than a few groups of animals to become extinct at one time. Not only does the hypothesis require a high degree of selectivity, it also may have occurred only once or twice in the last 600 million years. The increased radiation hypothesis has the possibility of being tested independently by studying radiation damage in crystals, changes in stable isotope ratios, and changes in the earth's magnetic field direction and strength. These speculations are only offerred to illustrate briefly the possibilities of reading another fascinating chapter out of the rock record.

G. Effects on stable isotope abundances

The radioactive decay of thorium and uranium during geologic time has produced a major fraction of the lead isotopes 206, 207, and 208 now found in the accessible lithosphere. The presence of argon-40 in such greater relative abundance than the other stable isotopes of argon and the other inert gases, together with the decay constant and abundance of potassium-40, make it necessary to conclude that nearly all of the argon-40 in the atmosphere (nearly 1% of the atmosphere) has been formed by the decay of potassium-40. Potassium-40 can also decay to calcium-40, but this results in only a small increment in the calcium-40 inventory. Similarly, the production of strontium-87 by the decay of rubidium-87 does not appear to greatly affect the strontium-87/strontium-86 ratio in the older of the available rock samples [GAST (1955)]. In addition to these nuclear processes that are due to internal instabilities, many suggestions have been made regarding the possible ways in which bombardment by nuclear radiations might affect stable isotope ratios [RANKAMA (1954)]. In particular, the markedly lower abundances of lithium, beryllium, and boron, relative to other light elements, has been related to their unusually high capture cross-sections and formation under stellar conditions [CAMERON (1959)]. It is also known that even in geologic time, the present production of carbon-14, tritium-3, and its stable daughter helium-3, and boron-11 by cosmic radiation would consume only a minute amount of the target nitrogen-14. These shorter lived, induced radioactivities have been of great interest in studying the circulation of the atmosphere and hydrosphere; thorium, uranium, and potassium-40, together with their stable and radioactive daughters have provided much information about the slower and less obvious circulation of material in the rock or geochemical cycle.

H. The rock or geochemical cycle

The decay of the natural radioactive elements at known and unvarying secular rates provides many opportunities for studying geologic and geochemical processes. The disruption of secular radioactive equilibrium between uranium and its radioactive daughter products has been used to trace the movement of uranium [ROSHOLT (1958)], and the same techniques offer the possibility of studying the rate and mechanisms of rock weathering with natural radioactive tracers in a manner analogous to corrosion and diffusion studies with artificial radioactive tracers. During weathering uranium is partially separated from thorium [PLILER and ADAMS (1959)] and rubidium from potassium [HORSTMAN (1956) and KEITH and DEGENS (1959)]. If the weathered material was subsequently deposited in the sea and buried under still younger sediments, it may have recrystallized into a metamorphic or igneous rock. Though recrystallized, the rock may still have a thorium/uranium ratio, a potassium/rubidium ratio, and a lead/uranium ratio suggestive of material that had at one time been approaching chemical equilibrium under surficial conditions. There are other geochemical criteria, particularly sulfur isotopic ratios [Ault (1959)] that may also indicate whether or not coarsely crystallized rocks are composed of material that had at one time been fractionated biologically or inorganically at or near the surface of the earth. These radioactive and stable isotope tracers are all consistent with a rock or geochemical cycle where igneous rocks—largely granite—are altered at the surface to sedimentary rocks—largely shales—which in turn are altered at depth back to igneous rock. This continuous, but slow (perhaps once every 200 million years on a crude average) cycle has long been understood from classical geological considerations. The radioactive and stable isotope tracers are beginning to provide a much more quantitative understanding of the development of the earth and the rock cycle.

Each time granite has been transformed into shale and shale back to granite in the rock cycle, both argon-40 from the decay of potassium-40 and helium-4 from the decay of the thorium and uranium series have, in part, been irreversibly lost to atmosphere, and in the case of the helium-4 to outer space. Rankama (1954), Damon and Kulp (1958), Turekian (1959), among other recent studies, have proposed various models using the known rates of argon-40 and helium-4 production to calculate: (1) an upper limit for the amount of completely weathered igneous rock (6462 kg/cm^2 of the earth's surface); (2) the present rate of argon-40 and helium-4 degassing from the earth; (3) an upper limit for the age of the earth; (4) that the hydrosphere and atmosphere resulted from the degassing of the earth during its first billion years of development. These and other conclusions are not equally valid. There is not unanimous agreement on all the details of the various models, particularly in regard to the relative argon-40 contributions from the mantle of the earth and from the uppermost few tens of kilometers. However, many of these calculations lead to more certain and quantitative boundary conditions for the development of the earth, especially where the conclusions are supported independently by other considerations, e.g., lead isotopic data.

The lead isotopes 206, 207, and 208 are largely the stable products of the decay of uranium and thorium and unlike argon-40 and helium-4, all the lead remains in the lithosphere system. The isotopic compositions of lead samples vary widely and are functions of the following properties of the source of the lead: (1) the thorium/uranium ratio; (2) the uranium/lead ratio; (3) the time at which and during which the thorium and uranium decayed to lead. Cannon et al. (1958), Stanton and Russell (1959), Ault and Kulp (1960), among others, have recently discussed the many important conclusions that can be drawn from the lead and sulfur isotopic data, especially in regard to sulfide ore deposits. Again there is not complete agreement upon all the details of each model used for calculation, but the many hundreds of isotopic lead and sulfur analyses now available suggest strongly that much of the material now found in sulfide ores of lead and other metals has been previously processed at or near the surface of the earth. Furthermore, Ault and Kulp (1960) found no inconsistencies between the isotopic lead and isotopic sulfur data. Vinogradov et al. (1958) have shown that lead isotopic compositions are sufficiently distinct to indicate similarities in the genesis and contemporaneity of sulfide mineralizations. A fundamental point in all these studies is that biological and surface or near surface processes, especially those involving oxidation-reduction reactions, fractionate sulfur isotopes and separate thorium and uranium from each other and from their daughter leads. If a large enough volume of the lithosphere were rehomogenized, these fractionations would be erased and a homogeneous source would be created with "normal" isotopic sulfur and lead compositions and normal or cosmic thorium/uranium and potassium/rubidium ratios. Other homogeneous sources are meteorites and the deeper interior of the earth, both representing material that has never been processed at the surface of the earth. The fact that so many isotopic analyses of the accessible lithosphere are not "normal", but „anomalous" supports the conclusion that much of the material in the lithosphere—even if it is 2000 million years old— has been through the rock cycle at least once. The simpler models developed to explain "anomalous" lead and sulfur isotopic compositions may fail because it is necessary not only to consider the number of recyclings, but the degree to which each of the several processes in the rock cycle has been carried to completion and the degree to which the system under consideration is large enough to represent a homogeneous sample of the lithosphere. These present complications offer many possibilities for learning still more about the circulation of material in the lithosphere.

Table 1. *Long-lived naturally radioactive isotopes*

Isotope	Abundance in lithosphere	Half-life in years	Stable decay products
Uranium-238	99.28% of 3.0 ± 0.6 parts per million	4.5×10^9	lead-206 and 8 helium-4
Uranium-235	0.71% of 3.0 ± 0.6 parts per million	7.5×10^8	lead-207 and 7 helium-4
Thorium-232	11.4 ± 2.0 parts per million	1.4×10^{10}	lead-208 and 6 helium-4
Potassium-40	3 parts per million	1.3×10^9	argon-40 and calcium-40
Rubidium-87	75 parts per million	5×10^{10}	strontium-87
Indium-115	0.1 parts per million	6×10^{14}	tin-115
Tellurium-130	0.007 parts per million	approx. 10^{21}	xenon-130
Iodine-129	below present limits of detection	1.7×10^7	xenon-129 [see REYNOLDS (1960)]
Lanthanum-138	0.01 parts per million	7×10^{10}	cerium-138 and barium-138
Samarium-147	1 part per million	approx. 10^{11}	neodymium-143 and helium-4
Lutetium-176	0.01 parts per million	approx. 10^{10}	hafnium-176 and ytterbium-176
Rhenium-187	0.001 parts per million	4×10^{12}	osmium-187
Bismuth-209	0.2 parts per million	2×10^{17}	thallium-205 and helium-4
Vanadium-50	0.2 parts per million	approx. 10^{14}	chromium-50 or titanium-50
Antimony-123	0.1 parts per million	very long	tellurium-123
Neodynium-150	1 part per million	10^{15} (?)	promethium-150 (not positively observed in nature)
Wolfram-178 or wolfram-180	less than 0.001 parts per million	very long	hafnium

Selected bibliography

ADAMS, J. A. S.: Mineral exploration. In: Science and resources. Baltimore: John Hopkins Press 1959.

— J. K. OSMOND and J. J. W. ROGERS: The geochemistry of thorium and uranium. In: Physics and chemistry of the earth. III. London: Pergamon Press 1959.

—, and K. RICHARDSON: Thorium, uranium and potassium contents of bauxites. Bull. Geol. Soc. Amer. **68**, 1693 (1957).

— J. E. RICHARDSON and C. C. TEMPLETON: Determination of thorium and uranium in sedimentary rocks by two independent methods. Geochim. et Cosmochim. Acta **13**, 270 (1958).

—, and J. J. W. ROGERS: Bentonites as absolute time-stratigraphic calibration points. Conf. on geochronology of rock systems, N.Y. Acad. Sci. 390 (1961).

—, and C. E. WEAVER: Thorium to uranium ratios as indicators of sedimentary processes: an example of geochemical facies. Bull. Amer. Assoc. Petrol. Geol. **42**, 387 (1958).

AHRENS, L. H.: A survey of the quality of some of the principle abundance data of geochemistry. In: Physicas and chemistry of the earth. II. London: Pergamon Press 1957.

— W. H. PINSON and M. M. KEARNS: Association of rubidium and potassium and their abundance in common igneous rocks and meteorites. Geochim. et Cosmochim. Acta **2**, 229 (1952).

Ault, W. U.: Isotopic fractionation of sulfur in geochemical processes. In: Researches in geochemistry. New York: John Wiley & Sons, Inc. 1959.
—, and J. L. Kulp: Sulfur isotopes and ore deposits. Econ. Geol. **55**, 73 (1960).
Baranov, V. I., A. B. Ronov and K. G. Kunashova: On the geochemistry of thorium and uranium in clays and carbonate rocks of the Russian platform. Izd. Akad. Nauk SSSR., Moscow No. 3, 3 (1956).
— Yu. A. Surkov and V. D. Vilenskii: Isotopic shifts in natural uranium compounds. Geokhimiya No. 5, 591 (1958).
Birch, F.: Heat from radioactivity. In: Nuclear geology. New York: John Wiley & Sons, Inc. 1954.
— Heat flow at Entiwetok atoll. Geol. Soc. Amer. Bull. **67**, 941 (1956).
Brown, H., and L. T. Silver: The possibilities of obtaining long range supplies of uranium, thorium, and other substances from igneous rocks. U.S. Geol. Survey Profess. Paper **300**, 91 (1956).
Bullard, E. C.: Heat flow through the floor of the ocean. Deep Sea Research **1**, 65 (1954).
— A. E. Maxwell and R. Revelle: Heat flow through the deep ocean floor. In: Advances in geophysics. III. New York: Academic Press 1956.
Cameron, A. G. W.: The origin of the elements. In: Physics and chemistry of the earth. III. London: Pergamon Press 1959.
Cannon Jr., R. S., L. R. Stieff and T. W. Stern: Radiogenic lead in nonradioactive minerals. A clue in the search for uranium and thorium. In: Proc. Second Internat. Conf. on Peaceful Uses of Atomic Energy, Geneva 1958.
Clark, F. W., and H. S. Washington: The composition of the earth's crust. U.S. Geol. Survey Profess. Paper **127** (1924).
Damon, P. E., and J. L. Kulp: Inert gases and the evolution of the atmosphere. Geochim. et Cosmochim. Acta **13**, 280 (1958).
Deutsch, S., D. Hirschberg et E. Picciotto: Mesure des âges géologiques par les halos pléochroïques. Experientia, Basel **11**, 172 (1955).
—, et E. Picciotto: Etude des halos pléochroïques dans le granite de Medel (Massif du St-Gothard). Experientia, Basel **14**, 128 (1958).
Dunning, J. R.: Atomic structure and energy. Amer. Sci. **37**, 505 (1949).
Fairbairn, H. W., et al.: A cooperative investigation of precision and accuracy in chemical, spectrochemical and modal analysis of silicate rocks. U.S. Geol. Survey Bull. No. 980 (1951).
Faul, H.: Nuclear geology. New York: John Wiley & Sons, Inc. 1954.
Fleming, W. H., and H. G. Thode: Neutron and spontaneous fission in uranium ores. Phys. Rev. **92**, 378 (1953).
Flügge, S.: Kann der Energieinhalt der Atomkerne technisch nutzbar gemacht werden? Naturwissenschaften **25,** 402 (1939) (6. Die geologische Frage, p. 409).
Frondel, C.: Mineralogy of thorium. U.S. Geol. Survey Profess. Paper **300**, 567 (1956).
Gast, P. W.: Abundance of strontium-87 during geologic time. Bull. Geol. Soc. Amer. **66**, 1449 (1955).
Gentry, J. T.: Study of congenital malformations in New York State. Amer. J. Publ. Health **49**, No. 4 (1959).
Green, J.: Geochemical table of the elements for 1959. Bull. Geol. Soc. Amer. **70**, 1127 (1959).
Holland, H. D., and D. Gottfried: The effect of nuclear radiation on the structure of zircon. Acta crystallogr. **8**, 291 (1955).
Horstman, E. L.: The distribution of lithium, rubidium, and caesium in igneous and sedimentary rocks. Geochim. et Cosmochim. Acta **12**, 1 (1957).
Jacobs, J. A.: The earth's interior. In Handbuch der Physik, Bd. 47, Geophysics. I. Berlin-Göttingen-Heidelberg: Springer 1956.
Joly, J. J.: Radioactivity and geology. London: Archibald Constable 1909.
Keith, M. L., and E. T. Degens: Geochemical indicators of marine and fresh-water sediments. In: Researches in geochemistry. New York: John Wiley & Sons, Inc. 1959.
Kennedy, G. C.: The origin of continents, mountain ranges, and oceans. Amer. Sci. **47**, 491 (1959).
Koczy, F. F., E. Tomic u. F. Hecht: Zur Geochemie des Urans im Ostseebecken. Geochim. et Cosmochim. Acta **11**, 103 (1957).
Kohman, T. P., and N. Saito: Radioactivity in geology and cosmology. Annual Rev. Nucl. Sci. **4**, 401 (1954).
Kuroda, P. K., R. R. Edwards, B. L. Robinson, J. H. Jonte and C. Goolsby: Chlorine 36 in pitchblende. Geochim. et Cosmochim. Acta **11**, 194 (1957).
Lawson, A. W., and J. C. Jamieson: Energy transfer in the earth's mantle. J. Geology **66,** 540 (1958).

LEVINE, C.A., and G.T. SEABORG: The occurrence of plutonium in nature. J. Amer. Chem. Soc. **73**, 3278 (1951).

LEWIS, D.R., T.M. WHITAKER and C.W. CHAPMAN: Thermoluminescence of rocks and minerals. Part I. An apparatus for quantitative measurement. Amer. Mineral. **44**, 1121 (1959).

MACDONALD, G. J.F.: Chondrites and the chemical composition of the earth. In: Researches in geochemistry. New York: John Wiley & Sons, Inc. 1959.

MARSHALL, R.R.: Isotopic composition of common leads and continuous differentiation of the crust of the earth from the mantle. Geochim. et Cosmochim. Acta **12**, 225 (1957).

MASON, BRIAN: Principles of geochemistry, 2nd ed. New York: John Wiley & Sons, Inc. 1958.

MILLER, R.L., and E.D. GOLDBERG: The normal distribution in geochemistry. Geochim. et Cosmochim. Acta **8**, 53 (1955).

MURRAY, E.G., and J.A.S. ADAMS: Thorium, uranium and potassium in some sandstones. Geochim. et Cosmochim. Acta **13**, 200 (1958).

NIER, A.O.: A redetermination of the relative abundances of the isotopes of carbon, oxygen, nitrogen, argon, and potassium. Phys. Rev. **77**, 789 (1950).

PABST, A.: The metamict state. Amer. Mineral. **37**, 137 (1952).

PATTERSON, C., G.R. TILTON and M. INGHRAM: Concentration of uranium and lead and the isotopic composition of lead in meteoritic material. Phys. Rev. **92**, 1234 (1953).

PEPPARD, D.F., H.M. STUDIER, M.V. GERGEL, G.W. MASON, J.C. SULLIVAN and J.F. MECH: Isolation of microgram quantities of naturally occurring plutonium and examination of its isotopic composition. J. Amer. Chem. Soc. **73**, 2529 (1951).

PETERS, B.: Progress in cosmic ray research since 1947. J. Geophys. Res. **64**, 155 (1959).

PLILER, R., and J.A.S. ADAMS: Distribution of thorium and uranium in the Mancos shale (Cretaceous). Bull. Geol. Soc. Amer. **70**, 1656 (1959).

— — Distribution of thorium and uranium in a Pennsylvanian weathering profile. Bull. Geol. Soc. Amer. **70**, 1657 (1959).

RANKAMA, K.: Isotope geology. London: Pergamon Press 1954.

—, and TH.G. SAHAMA: Geochemistry. Chicago: University of Chicago Press 1950.

RAYLEIGH, Lord (STRUTT, R. J.): On the determination of radium in the earth's crust and on the earth's internal heat. Proc. Roy. Soc. Lond. A **77**, 472 (1906).

REVELLE, R., and A.E. MAXWELL: Heat flow through the floor of the eastern north Pacific Ocean. Nature, Lond. **170**, 199 (1952).

REYNOLDS, J.H.: Determination of the age of the elements. Phys. Rev. Letters **4**, 8 (1960).

ROSHOLT, J.N.: Radioactive disequilibrium studies as an aid in understanding the natural migration of uranium and its decay products. Proc. Second Conf. on Peaceful Uses of Atomic Energy, Geneva 1958.

RUNCORN, S.K.: The permanent magnetization of rocks. Endeavor **14**, 152 (1955).

STANTON, R.L., and R.D. RUSSELL: Anomalous leads and the emplacement of lead sulfide ores. Econ. Geol. **54**, 588 (1959).

SUESS, H.E., and H.C. UREY: Abundances of the elements. Rev. Mod. Phys. **28**, 53 (1956).

TILTON, G.R., and L.O. NICOLAYSEN: The use of monazites in age determination. Geochim. et Cosmochim. Acta **11**, 28 (1957).

TUREKIAN, K.K.: The terrestrial economy of helium and argon. Geochim. et Cosmochim. Acta **17**, 37 (1959).

TURNER, R.C., J.M. RADLEY and W.V. MAYNEORD: The naturally occurring alpha-ray activity of foods. Health Physics **1**, 268 (1958).

UREY, H.C.: The cosmic abundances of K, Th, U, and heat balances of the earth, moon, and mars. Proc. Nat. Acad. Sci., Wash. **41**, 127 (1955).

— Boundary conditions for theories of the origin of the solar system. In: Physics and chemistry of the earth. II. London: Pergamon Press 1957.

VERHOOGEN, J.: Temperatures within the earth. In: Physics and chemistry of the earth. I. London: Pergamon Press 1956.

VINOGRADOV, A.P., S.I. ZYKOV and L.S. TARASOV: The isotopic composition of admixtures of lead in ores and minerals as indicators of their origin and of time of their formation. Geokhimiya No. 6, 653 (1958).

WEAVER, C.E.: The effects and geologic significance of potassium "fixation" by expandable clay minerals derived from muscovite, biotite, chlorite and volcanic material. Amer. Mineral. **43**, 839 (1958).

Radioactivity in Oceanography

by

F. F. Koczy and J. N. Rosholt

With 1 Figure

Zusammenfassung

Die Verwendung radioaktiver Isotope bei ozeanographischen Untersuchungen hat in den letzten 20 Jahren dank der Entwicklung der Radiumchemie und der Verbesserung der Nachweismethoden rasch zugenommen. Im Wasser der Ozeane und in den Tiefsee-Sedimenten sind eine große Zahl von Radioisotopen bekannt, die sich in der Mehrzahl von den Zerfallsreihen des Urans und des Thoriums herleiten. Im Ozeanwasser ist die Radioaktivität im wesentlichen getragen von U^{238}, U^{235}, U^{234}, Ra^{226} und ihren Folgeprodukten, ferner von K^{40}, C^{14} und zu einem sehr geringen Teil vom H^3. Künstlich erzeugte Isotope wie Sr^{90}, Cs^{137}, Ce^{144} und Pr^{147} sind nur in geringem Maße vorhanden, nehmen aber zu. Die Radioaktivität junger Tiefsee-Sedimente ist in der Hauptsache von folgenden Isotopen getragen: Th^{232} und Th^{230} mit ihren Folgeprodukten, Pa^{231} mit seinen Tochterprodukten, K^{40}, C^{14} und Uran-Isotope. Die einzelnen Isotope sind absteigend nach dem Anteil ihres Beitrages zur Gesamtaktivität geordnet.

Wichtigste Anwendung radioaktiver Untersuchungen im Rahmen der Ozeanographie ist das Studium der Vermischung verschiedener Wassermassen, das durch Messungen des C^{14} und des Ra^{226} ermöglicht wird. Zum Studium von Diffusionsprozessen im Ozean und in den Sedimenten dient Ra^{226}. Altersbestimmungen der Sedimente können für einen Altersbereich zwischen etwa 10000 bis 150000 Jahren aus dem Verhältnis Pa^{231}/Th^{230} abgeleitet werden. C^{14}-Datierungen in Carbonat-Sedimenten können bis zu maximalen Alterswerten von 40000 Jahren verwandt werden. Weitere Methoden beruhen auf der Bestimmung des Th^{230}/Th^{232}-Verhältnisses, des Th^{230}-Zerfalls oder der Ra^{226}-Verteilung in den Sedimenten.

Die Arbeit berichtet außerdem über wichtige neuere Erkenntnisse folgenden Inhalts: Die mittlere Aufenthaltszeit eines CO_2-Moleküls in der Atmosphäre bis zu seinem Austausch mit Oberflächenwasser des Ozeans ist zu 7 Jahren zu schätzen. Das Verhältnis C^{14}/C^{12} in mittleren Ozeantiefen läßt auf Verweilzeiten des Wassers in diesen Tiefen von 300 bis 1000 Jahren schließen, während sich aus dem gleichen Verhältnis die Verweilzeit von Tiefenwasser zwischen etwa 600 und 1500 Jahren ergibt. Die genannten Werte sind von Ozean zu Ozean verschieden. Aus C^{14}-Datierungen läßt sich schließen, daß der letzte große Klimawechsel vor etwa 11000 Jahren stattgefunden hat, während Pa^{231}/Th^{230}-Datierungen in den Ozean-Sedimenten erkennen lassen, daß die letzte Zwischeneiszeit vor etwa 100000 Jahren begann und etwa vor 65000 Jahren endigte.

A. Introduction

The importance of radioactive elements for oceanography has steadily increased during the last twenty years as a consequence of the development of radiochemistry and improvements in detection techniques. In the late thirties only a few of the natural radioactive elements were studied (PIGGOT 1933, PETTERSSON 1937, PIGGOT and URRY 1939); whereas now, nearly the whole spectra of radioactive elements existing in nature have been measured in the ocean and several of them have been used to study oceanographic problems.

There are several reasons for this rapid development. The radioactive elements lend themselves extremely well to the study of diffusion processes in ocean water and sediments as they can be detected in minute concentrations by their radioactive disintegration. For the same reason, they are used for tagging water masses, a technique by which the path of the water masses can be followed over long distances. Similar techniques allow the description of the circulation and mixing of ocean water and the determination of the time element involved in the overturn of ocean water. Geochemical processes affect the members of natural radioactive series to different extents. In fact, the ocean is certainly the largest example of disequilibrium of radioactive series in nature, not only in extension but also in magnitude. The disequilibrium of these series is used to study some of the geochemical processes involved. An important use of radioactive elements is found in their application to determine the age of m nerals and sediment layers. Another impetus to study radioactive elements comes from the increasing use of nuclear power and radioactive tracers. Highly radioactive waste products must be disposed of and it has been proposed that an economic way to do this would be to use the ocean. Many studies must be carried out to arrive at a safe solution in order to avoid health hazards and unwarranted oceanic contamination.

The sources and occurences of radioactivity in the ocean are many and varied. To simplify the description of these sources and their uses, a geochemical classification can be used. The oceanic envelope consists of three interfaces: that with the atmosphere, that with the basin undergoing deposition of sediments, and that with the continents. From the atmosphere, cosmic-ray produced radioactive nuclides enter the ocean phase and some have sufficient half life to enter the sediment phase. Some countercurrent exchange takes place at the atmosphere-water interface with the evaporation of water and exsolution of carbon dioxide. Fission-produced radioactive isotopes also enter the ocean through fall-out from the atmosphere. The largest amount of radioactivity, that from the three natural radioactive series, is contributed to the oceans from the continents, by river influx, ground water supply, and airborne dust. Radioactive tracers and fission products contributed by human emplacement and waste disposal are increasing in amount year by year. At the bottom sediment interface, a countercurrent exchange of radioactive isotopes takes place with their precipitation from solution in the ocean, from particle absorption and subsequent sedimentation, and from solution of sediments.

Contributions of radioactivity from the atmosphere and that redissolved from sediments at the base of the ocean have been found to be the most useful for estimating mixing rates and tracing water masses. The precipitation and sedimentation of radioactive elements associated with particulate matter have been used primarily for estimating sedimentation rates and age determinations.

B. Occurrence of radioactive elements in the oceanic envelope

Radioactive elements can be classified by their origin into those that were originally present at the formation of the earth and those that are being continually produced by nuclear reactions. The primordial isotopes can be divided further into those belonging to the three radioactive series, headed by U^{238}, U^{235}, and Th^{232}, and those which decay directly to a stable isotope. The isotopes belonging to the three radioactive series are listed in Table 1 and the nonseries isotopes are listed in Table 2. The continually produced isotopes can be divided into those that result from nuclear reactions occurring in nature, caused by cosmic-ray collision with elements in the atmosphere, and the artificial isotopes produced by nuclear bombs and atomic piles. Table 3 lists the more important long-lived artificial isotopes as well as all isotopes produced in the atmosphere whose oc-curence has been detected. The half-life period, type of decay, source, and esti-mated concentrations are included in the three tables.

It may be noted that nearly all of the primordial isotopes are contributed to the oceanic envelope from the continental phase, whereas the cosmic-ray pro-duced isotopes are contributed from the atmospheric phase along with most of the artificial isotopes considered.

I. Primordial radioactive isotopes

Only a relatively few of those listed have been measured directly. In the case of the radioactive series, those that have not been measured are estimated from equilibrium concentrations with their most immediate parent which has been determined. Estimates of the other radioactive elements are based on the abund-ance relationships between the radioactive and stable isotopes that have been determined in minerals in which these elements are relatively abundant.

Uranium is the only element which has been analyzed primarily by chemical methods. It alone has a concentration in sea water sufficient to be detected by chemical methods, whereas nearly all the other elements are present in such trace amounts that the more sensitive detection, using radioactivity, is required.

1. Uranium

Studies of the occurrence of uranium in ocean water and the sediments were started when the fluorescence method for its determination was introduced by HERNEGGER and KARLIK (1934) which allowed the detection of uranium concentra-tion in the range of parts per billion. The fluorimetric techniques have been improved by NAKANISHI (1951), SMITH and GRIMALDI (1954), TOMIC, LADEN-BAUER and POLLACK (1958). Other methods that have been used recently are isotope dilution techniques by RONA, GILPATRICK, and JEFFREY (1956), fission count analysis by STEWART and BENTLEY (1954), and polarography by HECHT, KORKISCH, PATZAK, and THIARD (1956). All of these methods have been used in oceanography.

The concentration of uranium in ocean water is believed to be rather uniform with values of 2.7 to 3.4×10^{-6} gU/L. Deviations from these values are found in enclosed basins with reduced salinity or under chemically reducing conditions (STRØM 1948; KOCZY, TOMIC, and HECHT 1956).

The isotopic ratio of U^{238} to U^{235} has not been measured in the ocean but it is assumed to be the same as that found elsewhere on the continents (NIER 1939a; SENFTLE, STIEFF, CUTTITTA, and KURODA 1957; HAMER and ROBBINS 1960). The most quoted values of the concentration of uranium in sea water are based on the assumption of constant isotopic abundance required for analysis by the method of isotope dilution.

Uranium in sediments is derived partly from precipitation from ocean water and partly included in minerals carried by river influx and airborne dust derived from the continents. A smaller component is contributed from disintegration of original oceanic rock. The concentration varies from 0.4 parts per million to 80 parts per million in recent sediments and may even exceed 300 parts per million in ancient shales and phosphates.

2. Protactinium

One recent attempt to measure the Pa^{231} content in sea water by its inherent alpha particle activity was made by SACKETT (1960). The results did not show the existence of a detectable quantity in 76 liters of Pacific Ocean water off the coast of Southern California. It must be concluded, therefore, that less than 3 % of the Pa^{231} required for radioactive equilibrium with a uranium content of 3 μg/L exists in ocean water. For this limit of concentration, the residence time of protactinium in sea water will be less than 1500 years.

Pa^{231} is contained primarily in the sediments, where concentrations greater than 50 times the amount required for radioactive equilibrium with the uranium parent present have been found. SACKETT also analyzed several surface layers of deep sea calcareous ooze cores from the Pacific and found Pa^{231} concentrations, in uranium equivalent units, of 30 to 76 equivalent parts per million. The uranium-unsupported Pa^{231} content showed the expected decrease with depth in one core. The maximum Pa^{231} content of deep-sea calcareous ooze cores from the Central Caribbean and North Atlantic would not exceed 40 equivalent parts per million based on analyses of samples from several layers in three cores (ROSHOLT, EMILIANI, KOCZY, GEISS, and WANGERSKY 1961). Compared on a carbonate-free basis, the only feasible means of intercomparison between core samples, the Pacific Ocean cores show a Pa^{231} content from 2 to 10 times as high as that found in the Caribbean Sea or Atlantic Ocean cores. The higher concentrations in the Pacific are consistent with a lower rate of sedimentation in that area.

3. Thorium

The determination of the thorium content of ocean water has been tried several times, but all attempts have been unsuccessful and only upper limits on its concentration are known. The first attempts were made by FØYN, KARLIK, PETTERSSON and RONA (1939), and the later by KOCZY, PICCIOTTO, POULAERT and WILGAIN (1957) using photographic emulsion techniques, and by SACKETT, POTRATZ and GOLDBERG (1958) using alpha proportional counter measurements. The modern methods employed were based on the coprecipitation of thorium from the sea-water sample with ferric hydroxide and chemical isolation by ion-exchange techniques. The final measurement carried out by photographic techniques makes it possible to distinguish between Th^{232}, Th^{230}, Th^{228}, and Th^{227} alpha activity. The proportional counter method for total alpha particle activity will not allow separation of activities by decay scheme patterns at the low level found in the deep Pacific Ocean waters analyzed by SACKETT and his coworkers. The upper limit values obtained for three alpha-emitting thorium isotopes are given in Table 1. A specific concentration for Th^{228} was found in the coastal waters in the Skagerak and Gullmerfjord, indicating Ra^{228} in solution in sea water as its source. KOCZY et al. (1957) have shown the residence time of thorium in sea water to be less than 300 years.

Oceanic sediments have relatively high Th^{232} content. It is measured by emulsion technique (Picciotto and Wilgain 1954), by radioactivity using an alpha-ray spectrometer (Goldberg and Koide 1958) or by decay scheme analyses (Rosholt 1957), and colorimetric techniques (Korkisch and Antal 1960; Starik, Kuznetsov, Grashchenko and Frenklikh 1958; Baranov and Kuzmina 1957; Goldberg and Picciotto 1955). The Th^{232} content is similar to that of continental rocks, and the majority of analyses reported on sediments from all three major oceans, both red clay and Globigerina ooze, show values near 5 parts per million. The extremes reported are less than 0.17 parts per million of Th^{232} in coral limestones from the Pacific (Barnes, Lang and Potratz 1954) and a value of 16 parts per million reported by Starik et al. (1958) in a calcareous ooze from the Indian Ocean.

No determination of the equilibrium between Th^{232} and Th^{228} has been made in sediments. The Th^{228} content in sea water in some locations is found to be greater than the equilibrium amount of Th^{232}. As Th^{228} should be in equilibrium with Ra^{228}, the latter must be formed in some ocean water in excess of the Th^{232} equilibrium amount; consequently the Ra^{228} and Th^{228} content in some surface sediments must be deficient. Because of the low concentration of all daughter products in the thorium series, the extent of the nonequilibrium has not yet been established.

The Th^{230} content in sediments is relatively high with contents reported up to 2×10^{-9} g/g for slowly settling red clay from the Pacific Ocean (Isaac and Picciotto 1953). The concentration depends in part upon the rate of sedimentation of the bulk sediment. Approximately 500000 years are required until the Th^{230} is decayed to the amount, approximately 2×10^{-14} g/g, which is required for equilibrium with the uranium in deep-sea clay. More recent results indicate that values in the Pacific Ocean sediments (Sackett 1960) are considerably higher than those in the Indian Ocean (Baranov and Kuzmina 1958), Caribbean Sea, and the Atlantic Ocean (Rosholt et al. 1961). Th^{227} has been measured in Caribbean and Atlantic Ocean cores, assumed to be in equilibrium with Ac^{227} and Pa^{231}, and thus used for the determination of the Pa^{231} content. In the same cores, Th^{228} is assumed to be in equilibrium with Ra^{228} and Th^{232} at subsurface depths, and used in the determination of the Th^{232} content.

The number of measurements of the thorium isotopes directly or the ratios of the more abundant isotopes, Th^{232} and Th^{230}, in marine environments is increasing rapidly and will soon overtake the large number of radium analyses made during the past 25 years. This increase has been due to the favorable geochemistry of thorium which lends itself well for potential dating methods or estimates of sedimentation rates. Since the geochemistry of protactinium and thorium is very similar, age determinations in deep-sea cores using the Pa^{231}/Th^{230} or Th^{227}/Th^{230} ratios have become possible. The use of Th^{230}/Th^{232} ratios are also being attempted for this purpose.

4. Radium

Ra^{226} was the first radioactive isotope determined in sediments or ocean water. Generally, it was determined by the emanation method with Rn^{222} daughter collected in ion chambers and measuring the produced ionization with an electroscope or electronic amplifiers, or using alpha-pulse counting techniques.

Because of its low concentration in sea water, radium is precipitated together with the barium sulfate formed after the addition of barium chloride at a concentration of 0.1 gram/liter of sea water which has been acidified with hydrochloric acid. The precipitate is fused with 5 times the amount of potassium-sodium carbonate to obtain a solution of the radium and barium. This fusion mixture is used to obtain a lower melting point. Potassium hydroxide is added in the amount of approximately 50 to 100 mg per gram of barium sulfate. FØYN, KARLIK, PETTERSSON and RONA (1939) found that 96% of the radium in sea water is precipitated with barium sulfate.

The radium content of sea water varies with location and depth. The vertical distribution shows generally a surface value of 0.5 to 0.9×10^{-13} g/L, decreases to a minimum at about 500 m depth, from where it increases, first rapidly, then more slowly with depth. The values in the bottom water are of the order of 1.0 to 1.8×10^{-13} g/L.

Ra^{226} in sea water is not in equilibrium with U^{238} or Th^{230} because it is always found in excess of Th^{230}; consequently it must be supplied to sea water from other sources. Ra^{228} has not been measured but is expected to be present in excess over that required for equilibrium with Th^{232}, at least near the sediment surface. Ra^{224} and Ra^{223}, on the other hand, because of their short half lives, are assumed to be in equilibrium with Th^{228} and Th^{227}, respectively.

Radium in sediments is usually determined after total solution of the sediment sample is made; however, KRÖLL (1955) found that an extraction with hydrochloric acid is generally satisfactory to bring the radium into solution from ocean sediments. Determinations have been concentrated on the Ra^{226} isotope (PETTERSSON 1954 and earlier papers, KRÖLL 1955 and earlier papers, URRY 1950 and earlier papers, PIGGOT 1944 and earlier papers, VOLCHOK and KULP 1957). The highest values found, excluding manganese nodules, are about 60×10^{-12} gRa/g in carbonate-free deep-sea clay. Carbonate shells contain about 0.2 to 0.1×10^{-12} gRa/g $CaCO_3$ (KOCZY and TITZE 1958).

5. Short-lived daughter products

No determinations of any of the other radioactive daughter products in the three natural series listed in Table 1 have been reported. The concentrations shown are calculated under the assumption that they exist in radioactive equilibrium with their most immediate long-lived parent. For example, Ac^{227} is expected to be in equilibrium with Pa^{231} and Ac^{228} is expected to be in equilibrium with Ra^{228}. All of the members of the U^{238} series from Rn^{222} to Po^{210} are assumed to be in equilibrium with Ra^{226}; all of the members of the Th^{232} series from Th^{228} to Tl^{208} are assumed to be in equilibrium.

Although for purposes of calculation, equilibrium was assumed for the preceding cases, there may be some notable exceptions. Radon may be less absorbed than radium on sediment particles and it could be found enriched in waters close to the ocean floor. Even though there have been no measurements reported for most of these members, this does not preclude that some specific isotopes may prove useful for further research applications. Lead-210, because of its long half life of 22 years, may not be in equilibrium and could be measured by its shorter-lived daughters, Bi^{210} or Po^{210}, which have characteristic and strong beta and alpha emissions. Ac^{227}, like Pb^{210}, has a half life of the order of magnitude useful for water turnover studies. Recently RAMA, KOIDE, and GOLDBERG (1961) have measured Pb^{210} in natural waters.

6. Stable lead isotopes

The lead isotopes, Pb^{206}, Pb^{207}, and Pb^{208}, being the last decay products of the natural radioactive series along with nonradiogenic Pb^{204}, have attracted the interest of many research workers, since they are used for age determination studies (RUSSELL and FARQUHAR 1960). The abundances of the stable isotopes of lead are determined by mass spectrometry (Nier 1939b, Chow and McKinney 1958). Slight variations in these abundances can be very useful for studying the previous geochemical history of the lead (Patterson 1956; Cannon, Stieff, and Stern 1958).

Recently, Chow (1958), by stable isotope dilution methods, has determined the lead concentration in sea-water samples from the San Juan Channel, Washington, and found 0.1 to 0.2 µg Pb/L. The isotopic composition of lead in sea water has not yet been determined because of this very low lead concentration and accompanying contamination difficulties.

The lead concentrations in the sediments were proportional to those in adjacent manganese nodules in one Atlantic and two Pacific cores, and averaged about 30 parts per million when the manganese content was low. The relative abundances of the lead isotopes in these sediments showed a similar distribution to that of terrestrial lead. Chow and Patterson (1959) determined the isotopic distribution in a large number of manganese nodules from Atlantic and Pacific Ocean bottoms. Both investigations demonstrated that the relative abundance of Pb^{206} is lower in the Pacific than in the Atlantic. This is believed to result from the general composition of the rocks in the two areas; the Pacific consisting of more young basaltic rocks and the Atlantic of older granitic rocks. Previously, Patterson, Goldberg, and Inghram (1953) had obtained lead isotopic distribution in marl, a manganese nodule, and clay and proposed that manganese nodules reflect the isotopic composition of the water. Another school of thought contends that the manganese nodules reflect the isotopic composition of the total surrounding sediment and may indicate the origin of the sediments.

7. Rhenium-187

Arrhenius (1959) proposed the possibility of the use of radioactive Re^{187} for age determinations in some marine sediments. Re^{187} decays by beta emission to Os^{187} with an estimated half life of 6.2×10^{10} years (Herr and Merz 1958). Potentially this decay system could be useful for ocean sediments older than 25 million years if the geochemistry of rhenium and radiogenic osmium are favorable and sufficient concentrations for radioactive detection and mass spectroscopy are found.

No measurements of the concentrations of either of these elements in marine environments are known; however, radioactive rhenium constitutes 62.9% of the isotopic composition of the element. Osmium may prove favorable since its isotopic abundance is very low, only 1.6% of the element.

8. Rare earths and other heavy elements

Several other natural radioactive isotopes with measurable half lives have been reported. No appreciable interest has been shown for these isotopes because either their half life is too long or their isotopic abundance is too small. Lutecium-176, samarium-147, and cerium-142 do not exceed 15% in isotopic abundance of their respective elements. Neodymium-144 and indium-115 have appreciable isotopic abundances but their half lives are of the order of 10^{15} years which limits their usefulness in marine environments.

9. Rubidium-87

The Rb^{87}-Sr^{87} beta decay system with a 4.7×10^{10} year half life (Flynn and Glendenin 1959) has been used extensively for age determination on alkali minerals of continental origin (Aldrich, Wetherill, Tilton, and Davis 1956).

No attempts are known of the use of this method for ocean sediments, the age of which would probably have to exceed 25 million years. The relatively high concentration of strontium-87 in sediments, of which only a very small fraction is radiogenic strontium, limits its use considerably, due to the large correction required for nonradiogenic strontium.

Analyses for rubidium in sea water have been made by neutron activation (SMALES and SALMON 1955) and isotope dilution (SMALES and WEBSTER 1957) indicating a concentration of 120 µg/L of which 28% should be radioactive rubidium.

10. Potassium-40

The potassium content of sea water is 0.035% (JENTOFT and ROBINSON 1956) and varies very little with location. GOLDBERG and ARRHENIUS (1958) list several Pacific pelagic sediments where the potassium content varies from 0.7 to 3.5% as determined by emission spectroscopy. Concentrations as high as 8% in glauconite in modern foraminiferal sediments have been reported (EHLMANN 1960), while concentrations less than 1% are normal for limestones. The isotopic abundance of radioactive K^{40} is 0.0119% (NIER 1950).

The K^{40}-Ar^{40} electron-capture decay system with a 1.2×10^9 years half life (WETHERILL 1957), like the rubidium-strontium method, has been used extensively for age determinations in mica, feldspar, glauconite and other minerals (WASSERBURG, HAYDEN and JENSEN 1956). Dates on some deep-sea calcareous clays from the Caribbean Sea and Atlantic Ocean and pelagic red clays from the Pacific and Atlantic Oceans have been made by HURLEY and others (1960). The results and their implications will be discussed later in this chapter.

II. Cosmic-ray produced radioactive isotopes

Apart from the radioactive elements which have been present on the earth since its origin, there is a significant and relatively steady-state production of shorter-lived radioactive isotopes in the upper atmosphere by nuclear reactions. Cosmic radiation in the atmosphere, composed of a variety of low and high energy particles, produce several different types of nuclear transformations upon collision with nitrogen, oxygen, and argon. Several of the radioactive isotopes produced have been detected in rain water, aerosols, fresh water, ocean water and even deep-sea sediments. Those isotopes with sufficient half life to be detected are shown in Table 3.

The great oceanographic interest in these isotopes prevalent in the last few years stems from two reasons. Physicists concerned with the production of the elements turned to oceanography in order to obtain information about the constance of production. Oceanographers, on the other hand, assuming the production to be constant, wanted to use these tracer elements to study the time relationships in oceanographic processes such as circulation, diffusion, formation of minerals, and rates of sedimentation. The production rates of the more important cosmic-ray produced nuclides as determined by several investigators are tabulated in Table 4.

1. Short half-lived isotopes

Chlorine-39, sulfur-35, phosphorus-33, phosphorus-32, and sodium-22 have been detected in rain water in concentrations up to 0.3×10^6 atoms/L for S^{35} (PETERS 1959). The more abundant and more readily detected nuclides, S^{35}, P^{33}, P^{32} together with Be^7, have been described by a number of investigators recently with determinations of concentrations in rain water and their production rates calculated (GOEL, NARASAPPAYA, PRABHAKARA, RAMA and ZUTSHI 1959; LAL, ARNOLD, and HONDA 1960; LAL, MALHOTRA, and PETERS 1958; LAL, RAMA, and ZUTSHI 1960; and RAMA 1960). Cl^{39} probably has too short a half life to be useful for atmosphere circulation studies, and Na^{22} awaits improved detection techniques for its weak beta emission.

2. Silicon-32

Si^{32} is formed in the atmosphere with a half life of approximately 710 years from the nuclear spallation of argon by cosmic rays. It decays by beta emission to P^{32} which also decays by beta emission with a 14.3 day half life to S^{32}. LAL, GOLDBERG, and KOIDE (1960) developed a method of radiochemical analyses which consists of milking and counting the P^{32} daughter from large amounts of silicon. Their analyses of three oceanic siliceous sponges taken from the Gulf of California gave an average specific activity of 19.6 ± 1.3 d/m of Si^{32}/kg Si. This is equivalent to 6.8×10^{-17} atoms Si^{32}/atom of Si which is of the same order as the specific activity of tritium. Concentration in sea water was estimated to be 2.6×10^{-5} disintegrations per minute per liter and the production rate, in Table 4, was estimated by comparisons with the production rates of P^{32}, P^{33}, and S^{35} previously determined.

3. Carbon-14

In the atmosphere, the reactions of cosmic ray produced neutrons on nitrogen yield C^{14} by the reaction:

$$N^{14} + n \rightarrow C^{14} + H^1.$$

Where the C^{14} is produced at the rate of approximately 2 atoms/cm² sec and it decays to N^{14} by weak beta emission with a 5568 ± 30 year half life (accepted value; LIBBY 1955, p. 36); a recent value for the half life is 5760 years (NBS 1961). The C^{14} combines with oxygen, mixes with ordinary carbon dioxide, and is distributed in the lower atmosphere by eddy diffusion.

The presence of natural radiocarbon was predicted by LIBBY (1946). Subsequently the "radiocarbon method" was developed by LIBBY and his coworkers and rapidly became the most versatile and most used technique for dating work in oceanography. A variety of measurement techniques have been developed since this time; however, some steps in the sample preparation have remained the same. The sample is converted to CO_2, by combustion of organic material or by acid treatment on carbonate sediments, shells, or ocean-water bicarbonate. Nearly all present day methods utilize gas counting techniques to determine the C^{14} activity. Either the CO_2, under 1 to 3 atmospheres pressure is counted (DE VRIES and BARENDSEN 1953); or the CO_2 is converted to acetylene (SUESS 1954, BARKER 1953, CRATHORN 1953), or converted to methane (BURKE and MEINSCHEIN 1955) prior to filling and measurement in proportional counters. Scintillation counter techniques have been developed (ARNOLD 1954), but are not used to any degree in oceanographic studies. Background activity, which is the greatest obstacle for the measurement of C^{14}, is reduced by the use of anticoincidence counting systems and enclosing the counter with several tons of iron shielding incorporating a "neutron trap" consisting of a mixture of paraffin and boric acid. The obtainable age limit is about 50000 years for large samples in which the CO_2 is measured without enrichment of C^{14}. An isotopic enrichment technique has been described by HARING, DE VRIES, and DE VRIES (1958) where C^{14} is concentrated, by thermal diffusion from a very large sample, to a factor of 16. The degree of enrichment is monitored by the fractionation of O^{18}. Thus, the range for radiocarbon dating has been extended to about 70000 years. A relatively large quantity of material is required to date samples approaching the age limits cited, and even a minute contamination with recent carbon will give ages much too young. Some of these effects have been described briefly by DE VRIES (1959).

The most detailed description of sample collection and radiocarbon analysis of oceanic CO_2 is given by BROECKER, TUCEK, and OLSON (1959) and FONSELIUS and ÖSTLUND (1959). Natural radiocarbon variations are expressed by differences in the C^{14}/C^{12} ratio from that of 18th century wood. The per mil differences from the oxalic acid counting standard are calculated by an expression of the form:

$$\delta C^{14} = \frac{A_{\text{sample}} - 0.950\, A_{\text{standard}}}{0.950\, A_{\text{standard}}}$$

where A_{sample} is the net C^{14} activity of the sample corrected for radioactive decay from the time of collection; the factor 0.950 is chosen to adjust standard reference activity to the value of the 18th century wood and has been accepted by nearly all of the C^{14} laboratories to be used with all measurements made 1960 and later. For cited limits of error, nearly all laboratories use 1 sigma unit (67% chance of being within the stated limits) based on the statistical deviation of the count rate.

LIBBY (1952) and CRAIG (1953) pointed out that differences in the isotopic abundances of carbon can be produced by natural fractionation of the isotopes (normally, 98.9% C^{12}, 1.1% C^{13}, and about 10^{-10}% C^{14}). Possible fractionation of the isotopes is corrected by "normalization to the same C^{13}/C^{12} ratio" as described by BROECKER and OLSON (1959), and the final results reported as per mil differences.

The corrections for C^{14} activity must be made because variations in its concentration with time have been found. By analyzing tree rings, SUESS (1955) showed that the C^{14} activity in the atmosphere had decreased by about 1.73% between 1850 and 1950. This decrease has been attributed to the dilution of the atmosphere with inactive carbon produced by combustion of fossil fuel, known as the "Suess effect." ARNOLD and ANDERSON (1957) have predicted the dilution to be 2.1% and, on limited data from the southern hemisphere, calculated that the dilution in that area is less than 0.5%. CRAIG (1957a) predicted a 1.75% dilution. DE VRIES (1958) determined a dilution factor of 1.5% between 1845 and 1935 by tree-ring investigations, and MUNNICH (1957) determined 3% dilution for a longer period. BRANNON, DAUGHTRY, PERRY, WHITAKER and WILLIAMS (1957) also made assays of tree rings of known ages and obtained a higher dilution value of 3.5% for the period of 1860 to 1954. Their calculations of the cumulative mass of fossil carbon released to the atmosphere during this period amounted to 3.3×10^{17} g of carbon and equivalent to 14% of the CO_2 in the atmosphere. SUESS and REVELLE (1957) have estimated the cumulative addition to be 12% and the dead CO_2 produced equivalent to a constant production rate of 0.25% of the atmosphere CO_2 for the past 40 years. ARNOLD and ANDERSON (1957) have concluded that if all the CO_2 were to remain in the atmosphere, the world-wide depression of C^{14} activity of CO_2 would be about 12%; if it mixed with the atmosphere, land life, and humus, a 4% depression would result; and if diluted with the entire exchange reservoir, the effect would be a 0.2% depression.

In the last decade, the activity of C^{14} has been increasing as a result of atomic bomb tests (MÜNNICH and VOGEL 1958). RAFTER and FERGUSSON (1957) have reported an increase of 4.1% in the southern hemisphere from February 1955 to March 1957. DE VRIES (1958b) has found a 4.3% increase in the northern hemisphere between November 1953 and June 1957. From the data of BROECKER and Olson (1960), the atmospheric CO_2 in 1959 contained about 27% more C^{14} than the prebomb estimates and the surface ocean water had increased about 1%.

After suitable correction for isotopic fractionation, the results of C^{14} analyses on ocean water vary from -4 per mil in the Barents Sea surface water (Fonselius and Östlund 1959) to -243 per mil in deep Pacific waters (Bien, Rakestraw, and Suess 1960) expressed in the accepted NBS scale. Broecker, Gerard, Ewing, and Heezen (1960) calculated the average per mil difference for the entire ocean reservoir to be -160. Further details are given in the section on ocean phase.

Even though C^{14} is primarily accumulated in the ocean water, the sediments contain a significant amount of radiocarbon. Arnold and Anderson (1957) have calculated, from the cosmic-ray production rate and the decay rate that a maximum of 30% of the C^{14} inventory is contained in the sediments, and that the mixing half-time between the ocean carbonate and the sediments is 15 000 years. Using the average value for the C^{14} content of the entire ocean reservoir, the distribution of the C^{14} inventory, including the bomb-produced radiocarbon which is predicted to exist in 1961 (Broecker and Olson 1961), is calculated as: 66% in ocean water; 29% in sediments; and 5% in the atmosphere and biosphere.

4. Beryllium

Two radioactive isotopes of beryllium are produced in the atmosphere by the bombardment of nitrogen with cosmic-ray produced particles. The half life of Be^{10} is 2.5×10^6 years; it is a beta emitter of 560 kev energy, emits no gamma rays, and decays to B^{10}. On the other hand, Be^7, with a half life of 53 days, is an orbital electron-capture species with a single 579 kev gamma ray and Li^7 is its stable decay product. With these vastly different modes of decay and half lives, there should be little difficulty distinguishing between the two isotopes.

The beryllium content of several Pacific Ocean water samples has been determined by Merrill, Honda, and Arnold (1958) and Merrill, Lyden, Honda, and Arnold (1960) using ion-exchange separation and solvent extraction techniques, with spectrophotometric measurement of beryllium. Concentrations between 2.2 and 8.8×10^{-10} g Be/kg sea water have been found with $\frac{1}{6}$ to $\frac{2}{3}$ of that total occurring as particulate beryllium separated on 0.45 micron pore size millipore filters. The average concentration of Be in solution is given as 3.9×10^{-13} parts per part and particulate Be at 1.8×10^{-13} parts per part. Beryllium concentrations in Pacific red clay cores and one Atlantic core were rather uniform between 2.2 and 3.0 parts per million.

No measurement of the content of either radioactive isotope has been made in sea water. Arnold (1956) measured the Be^{10} activity of several layers in two cores from the eastern Pacific using BeO preparations on small thin-walled counters inside an anticoincidence G-M ring. Very low activities, ranging from 7.6 down to 3.8×10^{-3} disintegrations per minute per cubic centimeter of clay, were found together with a general decrease of activity with depth in the cores. Goel, Kharkar, Lal, Narsappaya, and Peters (1957) also measured Be^{10} in Pacific cores and found a similar range of activity which showed a decrease with depth in the core. An average concentration of 1.6×10^{-10} atoms of Be^{10}/g of core was obtained. Later analyses by Merrill et al. (1960) generally showed the same range of activity in two additional Pacific cores except for one near-surface sample with an activity an order of magnitude higher than the rest. With the exception of the later sample, a decrease with depth is not apparent in these two cores.

Except for detection in aerosols (CRUIKSHANK, COWPER, and GRUMMITT 1956), all of the Be[7] measurements have been made on rain water collected at several different latitudes and elevations (RAMA 1960; LAL, RAMA, and ZUTSHI 1960; GOEL, NARASAPPAYA, PRABHAKARA, RAMA, and ZUTSHI 1959; RAMA and ZUTSHI 1958; GOEL, JHA, LAL, RADHAKRISHNA, and RAMA 1956; ARNOLD and AL-SALIH 1955). Concentrations up to 10^{-16} g Be[7]/L of rain water have been found.

5. Hydrogen-3

In nature the heaviest isotope of hydrogen, H^3 or tritium (T), is continually formed in the atmosphere by cosmic radiation and its principal source is from the nuclear disintegration of nitrogen and oxygen (CURRIE, LIBBY, and WOLFGANG 1956). A large fraction of the tritium combines with hydrogen and oxygen to form the HTO molecule and then enters the water cycle. After 1952 at least as much or more artificial tritium has been contributed to the atmosphere by man-induced nuclear reactions. The half-life of tritium is 12.26 years (JONES 1955) and it decays by weak beta emission to He^3.

A large number of estimations and calculations of the mean global production rate of cosmic-ray produced tritium have been made since its discovery in molecular hydrogen in the atmosphere by V. FALTINGS and HARTECK (1950) and in natural waters by GROSSE, JOHNSTON, WOLFGANG, and LIBBY (1951). The earlier estimates of the production rates shown in Table 4 were calculated from the tritium inventory in water. More recently WILSON and FERGUSSON (1960) have calculated production rates from the tritium content of meteorites, and cosmic-ray flux data and nuclear evaporation theory. Finally, CRAIG and LAL (1961) have calculated lower production rates of 0.35 atom T/cm^2 sec from the geochemical inventory and 0.25 atom T/cm^2 sec from cosmic-ray and nuclear cross-section data.

Tritium activity in most water samples is very low and the ratio of T/H must be enriched for the present measurement techniques, using proportional counting of hydrogen gas (GILETTI, BAZAN, and KULP 1958). Before converting the water hydrogen to hydrogen gas, the water sample is electrolized, liberating light hydrogen preferentially to tritium by the method of KAUFMAN and LIBBY (1954). The degree of enrichment attained is monitored by deuterium analysis. Concentrations are usually reported in tritium units (T.U.) where 1 T.U. $= T/H \times 10^{18}$.

KAUFMAN and LIBBY (1954) and V. BUTTLAR and LIBBY (1955) reported tritium analyses on oceanic mixed layer samples with concentrations ranging from 0.2 to 1.6 T.U. GILETTI, BAZAN, and KULP (1958) reported analyses on several surface samples from the Atlantic ranging from 0.6 to 4 T.U.; samples taken at depth, just below the thermocline, showed a marked decrease to about 0.25 T.U.; and north Atlantic deep-water samples indicated concentrations less than 0.05 T.U. BEGEMANN and LIBBY (1957) also found tritium in Atlantic surface waters in the same range of concentrations. However, their analyses on 2 out of 5 Pacific surface samples collected in 1952 indicated tritium concentrations of 10 and 16 T.U. Excluding the latter two samples, all of the data indicated an average concentration in the oceanic mixed layer of about 1 T.U. After the spring of 1954, tritium fall-out from marine thermonuclear testing during Operation Castle increased the surface ocean water about 2 T.U. according to BEGEMANN and LIBBY data and that of BROWN and GRUMITT (1956).

Calculations by CRAIG and LAL (1961), based on both deuterium and tritium analyses, indicated that fall-out of artificial tritium was already significant in 1953 when most of the pre-Castle samples were obtained. In the fall of 1952, the much smaller marine nuclear testing of Operation Ivy took place, thus the actual prebomb tritium concentrations in the ocean are subject to some uncertainty. The recent calculation of the production rate of 0.35 atom T/cm² sec indicates that the tritium concentration in the surface ocean is too high to be accounted for by the mixing models of Broecker, Gerard, Ewing, and Heezen (1960).

III. Artificial radioactivity

Artificial isotopes of a large number of elements are produced by thermo-nuclear bombs and by atomic power reactors. No attempt will be made to discuss all of the radionuclides introduced into the ocean since most of them decay rather rapidly and others constitute only a small percentage of the total activity. As an indication of the magnitude of such sources, Libby (1956) estimated that the quantity of artificially produced Sr^{90} in the stratosphere in 1956 was approximately 10 millicuries per square mile of the earth's surface. The more important isotopes are listed in Table 3 (Pm^{147}, Ce^{144}, Cs^{137}, Sr^{90}, Cl^{36}, C^{14}, and H^{3}). Artificial H^3 and C^{14} have been discussed in the section on atmospheric radioactivity.

Several investigations of gross radioactivity related to sea water have been made with no determination of separate radioisotopic components. Nakai, Hattori, Honjo, Okutani, and Kidachi (1959) measured total beta activity in marine organisms from waters of the western Pacific adjacent to Japan, and their results showed that mixed plankton from the surface contained about 10 times as much activity as benthos from depths of 500 to 2000 m. Nelepo (1959) reported measurements of the gross radioactivity of a large area of Pacific waters adjacent to the Antarctic continent and found uniform distribution of the radio-activity of the ocean, with highest activity in the upper mixed layer of 50 m depth, with considerable decrease of activity with depth between 50 and 150 m. In the Pacific North Equatorial Current region, Higano (1959) has reported measurable amounts of artificial radioactivity from iodine, antimony, mercury, bismuth, cadmium, copper, arsenic, tin, ruthenium, zirconium-niobium, uranium, strontium, yttrium, barium, cesium, and rare earths.

1. Strontium-90, cerium-144, promethium-147, and cesium-137

In addition to artificial Sr^{90}, a natural source produced from spontaneous fission of U^{238} and thermal neutron induced fission has been found by Heydegger and Kuroda (1959). Contributions from this source appear negligible compared to the artificially produced Sr^{90} which has accumulated in a quantity about a million times as large.

Analyses have been made for the separate radioisotopic components of cesium, cerium, promethium, and strontium in sea water and marine organisms. Descriptions of the collection techniques, counting procedures, and methods of radiochemical identification have been given by Bowen and Sugihara (1957) and Sugihara, James, Troianello, and Bowen (1959). 50 to 200 liters of sea water are usually required.

Sr^{90} was found in fairly uniform distribution in surface waters over shelf areas and deep tropical Atlantic waters collected in 1956, 1957, and 1958 (Bowen and Sugihara 1958, 1960). For the shelf areas, the range of radioactivity was between 0.03 and 0.14 μμ curies/L, with an average of 0.06 μμ curies/L for 11 samples. Surface values for deep waters showed a significantly lower range from 0.01 to 0.06 μμ curies/L, with an average of 0.03 μμ curies/L for 26 samples. In several of the same samples, the range of Ce^{144} activity was from 0.007

to 0.11 $\mu\mu$ curies/L, and Pm^{147} activity from 0.003 to 0.03 $\mu\mu$ curies/L. The Sr^{90} and Ce^{144} activities showed a definite decrease with depth in the water column, however Pm^{147} activity showed no apparent decrease with depth. Very little Sr^{90} activity appeared at depths below 800 m. YAMAGATA and MATSUDA (1959) have calculated that the Cs^{137} activity in the coastal waters of Japan is 0.07 to 0.15 $\mu\mu$ curies/L from measurements of bittern concentrated sea water in 1958.

2. Chlorine-36

It has been suggested by SCHAEFFER, THOMPSON, and LARK (1960) that Cl^{36} (306000 year half life) has been produced by neutron irradiation of sea water from marine nuclear explosions since the observed Cl^{36} activity in rain water is about 3000 times as large as that predicted from cosmic-ray production rates. Their analyses are made by ion-exchange concentration, final separation as ammonium chloride, and detection by anticoincidence counting in a screen-wall counter. Concentrations as high as 10^{-16} g Cl^{36}/L were reported in rain water, with an average of 0.05×10^{-16} g/L. The isotope was also detected in ground water, however detection in sea water is unlikely because of dilution with the large amount of stable chloride.

C. Applications of radioactivity data to oceanographic studies

The measurement of artificial and cosmic-ray produced isotopes in seawater and in the atmosphere has been most popular in the last decade, whereas the measurement in sea water of the lower members of the natural radioactive series has been much more limited. Such a lack of interest, however, should not exist from the standpoint of radioactivity levels. For example, the reported activity of Cs^{137} in sea water is almost indentical to that which should be found for Pb^{210}, if it exists approximately in equilibrium with Ra^{226}, as indicated in Tables 1 and 3. Parallel with the increase in measurements, the large fraction of recent applications in the ocean phase has been with the use of cosmic-ray produced isotopes and artificial radioactivity.

I. Atmosphere-ocean interface

The short-lived, less abundant radioisotopes produced by cosmic radiation have been applied primarily to atmospheric circulation studies and other meteorological problems discussed in a separate chapter. Their use in ocean-water studies is limited because dilution with such a large volume of water renders their detection exceedingly difficult if not impossible.

The primary oceanographic investigation at this interface concerns the measurement of the exchange rate of CO_2 between the ocean and the atmosphere, and the residence time of radiocarbon and tritium in the atmosphere. Knowledge of the time, since production, required for these radioisotopes to invade the ocean is desirable as an aid in establishing the age of surface waters based on the concentrations of the radioisotopes. In the case of CO_2, the residence time, τ, can be defined (SUESS and REVELLE 1957) as the time it takes on the average for a CO_2 molecule, as a member of the atmospheric carbon reservoir, to be absorbed by the sea. Since $\tau = 1/k$, where k is the rate of CO_2 transfer per year, the exchange rates of CO_2 can be expressed in the usual unit, moles CO_2/m^2 year calculated from the residence time with the introduction of the additional quantities of total CO_2 in the atmosphere and the surface area of the sea.

Two different and somewhat complicated approaches have been used to estimate the residence time, τ, assuming a steady state C^{14} exchange between the two reservoirs. One depends upon the average C^{14}/C^{12} ratio of ocean water and the difference in the C^{14}/C^{12} ratio between the atmosphere and the ocean surface water, with corrections for isotopic fraction between the reservoirs. The other method depends upon the total carbon inventory of the atmosphere and the surface ocean used together with an additional atmospheric parameter; either the magnitude of the industrial effect, the addition of radiocarbon by nuclear bombs, or the C^{14} cosmic-ray production rate. Investigators have used combinations of the two approaches (Suess and Revelle 1957, Craig 1957a), while the former method was used by Broecker, Gerard, Ewing, and Heezen (1960), and the latter method by Arnold and Anderson (1957) and Fergusson (1958). Rafter and Fergusson (1958) using the atomic bomb effect estimated a somewhat lower residence time of 3.3 years, whereas all of the other methods yielded residence times of 7 to 10 years. The best estimate from these investigators appears to be about 7 years for the residence time of CO_2 between the atmosphere and absorption in the ocean and an exchange rate of about 20 moles CO_2/m^2 year. Bolin (1960) has found general agreement in the transfer rate of CO_2 in the sea deduced from molecular and turbulent transfer and the rate of exchange estimated from the C^{14}/C^{12} data in the sea and the atmosphere.

Craig and Lal (1961) evaluated the latitudinal variation in both residence time and stratospheric production rate of tritium and estimated that the mean stratospheric tritium residence time is about 1.6 years.

By the use of mathematical models of the turbulent exchange of tritium in the atmosphere, Bolin (1958) has shown that tritium concentrations can be used to estimate the average global evaporation based on the vertical gradient of water vapor above the ocean.

II. Continent-ocean interface

Since most of the primordial radioisotopes are contributed to the ocean water and the ocean floor from sources on the continents, it is to these elements that further radioactive investigations will be directed to the study the interchange of matter specifically across this interface.

One important problem that is little understood is the different effects that glacial and interglacial environments have on the contribution of sediment to the ocean. When we study the total amount of thorium and uranium in deep-sea clays it is remarkable that uranium shows a lower concentration than the terrigenous clays. This would indicate that sea water does not supply uranium to the deep-sea sediment. It is evident that an exchange reaction takes place in the process of transferring terrigenous material to the deep sea, and primarily uranium is dissolved and reprecipitated on the shelf where reducing areas with high rate of sedimentation exist. In the case of thorium it seems evident, from the fact that river and sea water have an extremely low thorium content, that the main part of the thorium is transported to the sea floor with the sediment particles. Uranium is not preferentially precipitated in most nonreducing areas and in such locations on the shelf with a shallow water column above, very low contributions of Th^{230} and Pa^{231} from the disintegration of uranium in solution in the sea water take place and the occurrence of these two isotopes will be essentially of terrigenous origin, never having been divorced from the sediment particles. The terrigenous Th^{230} and Pa^{231} should exist in excess over that required for radioactive equilibrium with uranium but not to the extent that occurs in deep-sea sediments. STARIK et al. (1958) give data on a core taken at 90 m depth from the Bering Sea which substantiate this conclusion. Their results include analyses in closely spaced sediment layers for uranium and thorium together with separate determinations of the isotopes, Th^{230} and Ra^{226}. If an additional determination for Pa^{231} were made, an estimation of the sedimentation rate could be possible using a modification of the method for deep-sea sediments (ROSHOLT et al. 1961) based on Pa^{231}/Th^{230} ratios decreasing from an equilibrium ratio rather than an enriched Pa^{231}/Th^{230} ratio as used in the case of deep-sea sediments. Using the time element deduced from the estimation of the sedimentation rate, the variation of the amount of excess Th^{230} or Pa^{231} over the uranium-equilibrium concentration would provide one parameter related to the interface exchange effects of at least one glacial-interglacial cycle.

III. Ocean phase

The use of radiocarbon to study oceanic mixing has proved to be one of the most valuable tracer techniques that has been applied to oceanography. The results of C^{14} radiochemical analyses that have recently been contributed by several different investigators (BIEN et al. 1960; BROECKER, GERARD et al. 1960; FONSELIUS and ÖSTLUND 1959; FERGUSSON 1958) indicate quite a consistent pattern of radiocarbon age (C^{14}/C^{12} ratio) in the water masses studied.

The Atlantic surface ocean water has a slight increase in the C^{14}/C^{12} ratio from 50° S latitude to 50° N latitude, and a rapidly decreasing ratio at higher latitudes in the South Pacific and Antarctic water. One sample of Arctic surface water also shows a slightly lower ratio.

The central water has a radiocarbon age intermediate between the deep waters and the surface water. The deep water of the Atlantic originating in the higher latitudes of the southern ocean has a lower ratio than that originating in the northern hemisphere, which is consistent with the distribution found in the surface water. A tongue of water in the western North Atlantic between 1200 and 2400 m has a slightly higher ratio, which is attributed to a wedge of younger water penetrating the North Atlantic deep water.

The bottom water on the eastern side of the mid-Atlantic ridge has an apparent age greater than the adjacent water on the opposite side of the ridge, and greater than the overlying North Atlantic deep water. The deep water in the Pacific appears to be the oldest found in the oceans with an apparent age of about 1500 years.

Since the circulation of water in the oceans is quite complex with all the water masses interacting with each other, the radiocarbon distribution cannot be interpreted fully without the aid of additional oceanographic evidence. One of the most useful parameters by which the radiocarbon data can be interpreted is that of mixing rates between water masses whereby the residence time of an H_2O molecule within a given water mass is estimated. "Box models" are often designed to evaluate the various assumptions concerning the distribution of radiocarbon required to estimate the residence times. One such model is shown in Fig. 1. In this figure, the mean residence time, τ, is defined by BROECKER et al. (1960) as the time required for a volume of water equal to that contained in the deep-water masses to be supplied from the surface source region. The source areas for the water masses and their location are given in SVERDRUP, JOHNSON, and FLEMING (1942). The estimates of the mean residence times of water molecules

34   F.F. Koczy and J.N. Rosholt: Radioactivity in Oceanography

on the subsurface water masses were based on the following assumptions: (1) circulation in the oceans is at a steady state; (2) the corrected C^{14}/C^{12} ratio for surface water in the convergence area is the ratio in the new water which is continuously being added to the subsurface water masses; (3) mixing between adjacent subsurface water masses does not measurably alter their respective C^{14}/C^{12} ratios; and (4) CO_2 supplied to a deep-water mass through solution of carbonate sediment and through oxidation of organic material does not measurably alter the C^{14}/C^{12} ratio of the CO_2 initially present in the water mass.

Using the model shown in Fig. 1, Broecker et al. (1960) have calculated the rate of transfer of CO_2 between the major reservoirs by the solution of a set of seven simultaneous equations involving the following functions for each of the seven individual oceanic reservoirs: (1) the rate of exchange of CO_2 between the atmosphere and the surface ocean; (2) the area of a given reservoir exposed to the atmosphere; (3) the quantity of CO_2 contained in a given reservoir; (4) the differences in the C^{14}/C^{12} ratios between reservoirs as indicated by the arrows in Fig. 1; and (5) the decay constant of C^{14}. The annual rates of transfer of moles of CO_2 between reservoirs in the Atlantic cycle of 1.0×10^{15}, exchange between Pacific plus Indian surface water and Antarctic water of 1.6×10^{15}, and exchange between Pacific plus Indian deep water and Antarctic water of 2.5×10^{15} were calculated.

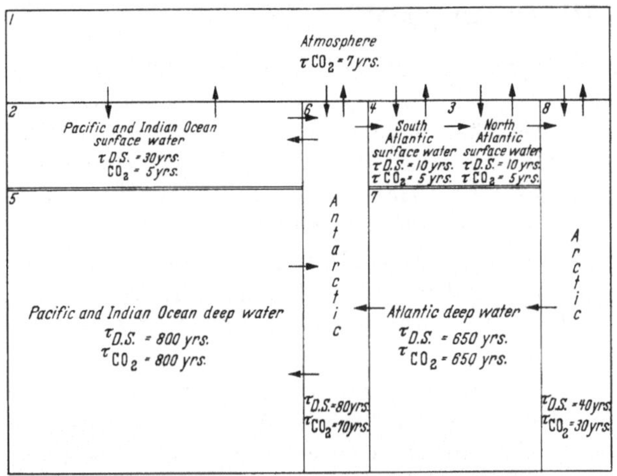

Fig. 1. Model of the steady-state CO_2 cycle in the oceans and atmosphere. (Redrawn from Broecker, Gerard, Ewing, and Heezen 1960.) The CO_2 and DS (dissolved solid) subscripts mean residence times differ because significant quantities of CO_2 are transferred through the atmosphere. The arrows indicate possible modes of transfer between reservoirs

A somewhat similar "box model" has been designed by B. Bolin and H. Strommel (written communication) in which the water-mass fluxes between reservoirs and the residence time in a given reservoir are calculated from a set of four simultaneous equations in which the geothermal heat flux through the ocean sediment interface, the variation in salt content and temperature between reservoirs, and the available C^{14} data are used. The model reservoirs used were the common water in the Pacific, the Antarctic bottom water, the Pacific and Indian intermediate water, and the north Atlantic deep water. Thus, the assumptions of the previous model relating to the mixing of surface waters and the exchange rates CO_2 from the atmosphere are eliminated. Although there are some differences between the two models concerning the contributions of water masses to the Pacific common water, the results do not differ greatly. The latter models indicated a residence time of about 1200 years for the Pacific common water and about 300 years for the Atlantic intermediate water. For the Atlantic intermediate layer, a net upward motion into the Atlantic surface water and a downward motion into the deep water of 1 to 3 m/year was estimated. It was calculated that the Antarctic water flows northward underneath the Atlantic surface water at a rate of 0.03 to 0.1 cm/sec at latitudes 40 to 50° S.

Based on the few measurements of Sr^{90} at various depths in the Atlantic, BROECKER (written communication) has estimated that less than 5 years is indicated for the time required for mixing to depths of 500 m. Since the C^{14}/C^{12} ratios for waters between 200 and 500 m are about median between those in the water above and the water below this depth, a vertical mixing rate which would be an order of magnitude higher than that calculated from the C^{14} models would be required to equate the observed C^{14} and Sr^{90} distribution. This possibility is considered unlikely, and additional Sr^{90} and other artificial isotope measurements must be made before their true usefulness can be evaluated.

The concentration of Ra^{226} in the surface waters and the mixed layer is about one-half of that in the deep water to which radium is added by diffusion from the sediments at a nearly constant rate over the whole deep-sea floor (KOCZY 1958). If this decrease results from radioactive decay only, it would indicate that either the residence time for the combined surface plus an intermediate reservoir would be about one half life of Ra^{226}, about 1600 years, or that the net transport of bottom waters is much slower than calculated from the radio-carbon models. To account for the slow upward transport, CRAIG (1958) has suggested that the thermocline, because of its great stability, represents a major barrier between the deep waters and the intermediate and surface waters. To accomodate both the C^{14} and Ra^{226} data, BROECKER (written communication) has proposed a four-reservoir model, characterized by two nearly independent circulation systems that consist of an intermediate ocean with exchange between a surface ocean and slow exchange with a "polar ocean", and a deep ocean which exchanges with the "polar ocean" only. The results from such a model indicate that the residence times are slightly longer than those shown in Fig. 1, except for the intermediate ocean of about 300 years, all of which are in closer agreement with the model proposed by BOLIN and STROMMEL (written communication). The vertical water velocity of 0.6 to 2 m/year calculated by KOCZY (1958) from the Ra^{226} data agrees well with that estimated for the Atlantic intermediate layers by BOLIN and STROMMEL. Thus it appears that a somewhat impermeable barrier exists between the intermediate and deep waters according to all of the radioactivity data accumulated thus far.

In addition to the estimation of vertical velocity, a similar mathematical treatment of diffusion equations based on the vertical distribution of radium and an estimate of eddy diffusivity at any level will permit the calculation of the variation of eddy diffusivity with depth. From this study it has been concluded that the eddy diffusivity increases toward the bottom of the ocean and the magnitude of its vertical component is related to the stability of the water mass. The vertical transport through the minimum diffusivity layer at about 1500 m depth is primarily a function of the vertical component of the current.

The use of natural tritium is very limited for the study of subsurface circulation because (1) its concentration below the mixed layer is considerably less than the present limits of detection, and (2) the prebomb concentration levels in the ocean are too uncertain. It appears likely that its future use will be restricted to artificial tritium tracer, except in the polar seas.

A most probable application of artificial radioactivity will be that of the study of the distribution of radioactivity between seawater on the one hand and fish, animal, and plant life in the ocean on the other. ATEN (1958) has proposed several mathematical models to evaluate the distribution while on a theoretical basis, KETCHUM and BOWEN (1958) have compared the biological transport of radioactive elements to the physical transport of elements in the sea.

IV. Ocean-sediment interface

Contrary to the migration of Ra^{226} and C^{14} by diffusion across the interfaces into the ocean, the primary direction of movement of radioactive elements is that connected with particulate matter, which is slowly accumulated on the deep-sea floor. Thorium contributes the great majority of the radioactivity accounted for by this process with the inclusion of the Th^{232} and Th^{230} isotopes and the subsequent growth of their daughter products. These together with the small amount of C^{14} incorporated in calcareous shells of marine organisms and the much smaller amount of K^{40} radioactivity in the clay account for 98% of the total radioactivity in the uppermost portion of the deep-sea sediment column. All of these three major radioactivity components have been used in oceanographic studies with primary application to Pleistocene chronology.

C^{14} was used initially in oceanography to provide a time calibration of other parameters related to the sediment which are believed to be time-independent and would, in that case, provide an extension to the time scale. A definite horizon in the upper part of the sediment column was dated by C^{14} assay of a suitable carbonate component; from that point downward in the sediment column, the parameter was employed by assuming its time variation to be constant in order to obtain ages of the deeper strata. Calibration of a titanium parameter was used by Arrhenius (1952) assuming a constant rate of accumulation of titanium in the sediments to obtain a climatic record, based on the variation of biogenous components in Pacific cores. Emiliani (1955, 1958) has used C^{14} dates on Foraminifera in several Atlantic and Caribbean cores, determined by Rubin and Suess (1955) and Rubin (1956), to calibrate the oscillations of temperature which were assumed to be of constant oscillation period, and to construct a generalized paleotemperature curve for the surface of the ocean during the last 300000 years. The dating was tentative but proved to be nearly correct as later measurements have shown. The paleotemperature is deduced from the variation of the O^{18}/O^{16} ratio in pelagic Foraminifera in cores of Globigerina ooze. Ericson, Broecker, Kulp, and Wollin (1956) obtained C^{14} dates, at several intervals, in each of nine Atlantic and Caribbean cores and correlated these dates with paleotemperatures of the surface ocean, based on the relative foraminiferal abundance. Later, using these data coupled with evidence from continental deposits, Broecker, Ewing, and Heezen (1960) have shown that there was an abrupt change in the climate close to 11000 years ago, with the advent of a warm period. In addition, their C^{14} ages indicate that the change occurred in less than about 1000 years with a substantial decrease in both the clay and $CaCO_3$ sedimentation rates in the Atlantic and Caribbean deep sea.

The decay of uranium-unsupported Th^{230} (ionium) contained in deep-sea sediments has been used in the attempt to date sediments up to 400000 years old; Volchok and Kulp (1957) have presented the historical development of this method and a critical review of the assumptions required. Most of the Th^{230} occurring in deep-sea sediments is produced in the water column above by the decay of U^{238} and U^{234}, and is scavenged by the particles settling to the bottom (Pettersson 1937). The concentration of uranium-unsupported Th^{230} in modern sediments depends upon the amount produced in and scavenged from the water column above it, and its dilution in the sediment. If both factors remain constant through time, the concentration of Th^{230} in the sediment also depends upon the concentration of the parent element uranium in sea water, which, in turn, depends upon (a) the rate of supply from land; (b) the rate of removal by trapping in nearshore reducing environments and direct sedimentation on the ocean floor;

and (c) the rate of reworking from nearshore basins to the open ocean, especially during times of low sea levels. The uranium content in ocean water, therefore, may have changed from time to time in a rather complex manner, with resulting changes in the rate of sedimentation of Th^{230}. In addition, there are reasons to believe that the sedimentation rate of the noncarbonate component did not remain constant throughout the time interval to be dated, according to the C^{14} results described in the last paragraph. In the early use of this method, the concentration of Th^{230} was deduced by measuring the concentration of the immediate daughter, Ra^{226}, which requires a simpler analytical technique, however, with the additional assumption that radium does not migrate away from the loci where it is formed. This assumption proved untenable especially in deep-sea cores with rates of sedimentation less than about 1 cm/1000 years.

Almost exactly the same considerations and use for an age method as that stated for Th^{230} could be reiterated for Pa^{231} which is produced in the water column by the decay of U^{235}. The Pa^{231} would be contained in the same sediment particles as the Th^{230}, but would decay independently with a 34300-year half life. Thus, the use of both of these isotopes, in combination, will allow the determination of ages of deep-sea sediments, independent of changes in sedimentation rates and uranium concentration in the ocean because the ratio Pa^{231}/Th^{230}, with a "half life" of 60100 years, will be a function of time only. The initial Pa^{231}/Th^{230} ratio in the sediment will be a fixed value, dependent upon the relative abundances of the uranium isotopes in the sea water which are assumed to be constant. If the residence time of the two isotopes is not greater than the order of about 1000 years, the ratio will be enriched in Pa^{231} by a factor of 2.33 over that required for equilibrium with uranium, due to the faster production rate of Pa^{231} from U^{235} in the seawater. Eventually, the equilibrium ratio would be obtained after uranium-unsupported decay of the isotopes in the sediment for 73500 years. According to the decay equation, the age of the sediment will be:

$$t = 8.66 \ln \left[2.33 \, \frac{Th^{230} - U}{Pa^{231} - U} \right] \times 10^4 \text{ years}$$

where the isotope concentrations are measured in uranium equivalent units, and where U is the uranium content of the sediment and must be subtracted to obtain the uranium-unsupported isotope concentrations.

In the case of negligible contributions of uranium or dating close to the 73500-year horizon, the Th^{230}/Th^{227} ratio could be used, eliminating the requirement of quantitative analyses for thorium isotopes with simply the determination of their ratios in the sediment.

Two Caribbean cores and one Atlantic core have been dated using this method (ROSHOLT et al. 1961) and correlated with paleotemperature curves, based on oxygen isotope analyses from the same calcareous sediment cores. The two deep-sea cores from the Caribbean, about 600 km apart, gave a set of dates which was coincident with the C^{14} chronology over its range and was internally consistent in regard to the paleotemperatures. This set of dates is believed to provide a reliable, absolute time scale extending from the present to about 175000 years ago. The temperature minimum of the Early Würm was dated at about 60000 years; the temperature maximum of the last interglacial was dated at about 95000 years; and the preceding temperature minimum was dated at about 106000 years. The last interglacial appears to have lasted about 35000 years. The sedimentation rates during the last 11000 years were lower than during the previous time intervals. The data from the North Atlantic core give ages which were consistently about 30000 years greater than those obtained from the Caribbean cores and the C^{14} chronology. This is believed to result from the addition of reworked clay, an effect which may exist in most deep-sea cores and indicating that only undisturbed cores will give reliable dates. Sackett (1960) determined the Pa^{231}/Th^{230} ratio at one layer, apparently belonging to the last interglacial in a deep-sea core from the eastern equatorial Pacific, giving an age of 95000 years.

Picciotto and Wilgain (1954) suggested using the ratio of Th^{230}/Th^{232} as an improvement on the old "ionium" method to provide corrections for possible variations of the sedimentation rate. On the same basis, Baranov and Kuzmina (1958) estimated rates of sedimentation in Indian Ocean cores using the ratio of uranium-unsupported Th^{230} to MnO and Fe_2O_3; Goldberg and Koide (1958) estimated rates of sedimentation in three Eastern Pacific cores using the ratio of Th^{230}/Th^{232} in preferentially dissolved nondetrital material; and Almodovar (1960) estimated the ages of various levels, back to 600000 years, in a red-clay core from the central Pacific Ocean, using the ratio of uranium-unsupported Th^{230} to Th^{232}. The validity of these methods is questionable because Th^{230} produced in sea water by the decay of uranium has a geochemical history different from that of Th^{232}, MnO, and Fe_2O_3. Another difficulty in dating deep-sea cores by the decay of Th^{230} alone is that the concentration of uranium-unsupported Th^{230} at time zero (top of the core) is not a fixed parameter. Plotting the ratio of uranium-unsupported Th^{230} to Th^{232}, from the data on the Caribbean cores (Rosholt et al. 1961) calibrated with a C^{14} date at 10500 years, gives an age for the last interglacial which is 50% older than that obtained from the Pa^{231}/Th^{230} ratio. This large difference is primarily due to significant changes in the sedimentation rate at a time earlier than that represented by the horizon from which the C^{14} date was obtained.

Goldberg, Patterson, and Chow (1958) have determined both the Th^{230}/Th^{232} ratio in surface layers of recent sediments from the Pacific Ocean along with the stable lead isotope ratios in manganese nodules from approximately the same locations. Using such data, they have initiated an investigation to study the possibilities of correlating the distribution of thorium and lead isotopes and comparing these distributions with oceanic circulation patterns. An interesting result from this preliminary study showed that the Th^{230}/Th^{232} ratios in the surface sediments of the eastern area of the Pacific were about twice as large as those in the western Pacific; and that greater Pb^{206} relative abundances existed in the eastern Pacific.

Preliminary studies of the age of potassium-phases in deep-sea sediments have been initiated by Hurley and his coworkers (1960). The potassium-argon age data indicate that a large fraction of the potassium in the deep-sea sediments is contained in detrital potassium-bearing phases which, in a large part, may be wind-borne mica. From such studies, it is anticipated that some knowledge can be gained concerning the hemispheric circulations of dust, and possibly its principal source region. Although data were obtained at various depths in an Atlantic equatorial core, no correlations with ice ages and their accompanying changes in atmospheric circulation were obtained. Ar^{40}/K^{40} ratios were also determined at four different glacial and interglacial horizons in the same Caribbean core from which the Pa^{231}/Th^{230} Pleistocene chronology was obtained. These results showed an extremely constant ratio, indicating a single source of uniform composition for the potassium-bearing minerals or very complete mixture of detrital minerals prior to sedimentation. A valuable extension of the time scale in deep-sea sediments would be possible if a sufficient amount of clay or other suitable minerals of known authigenic origin with a significant potassium content could be found in deep-sea sediment components.

In areas where thorium contribution to the sediment is negligible, such as very pure carbonate deposition, the time-related growth of uranium daughter products following the synchronous deposition of uranium can possibly be used for age determinations. BARNES, LANG, and POTRATZ (1956) have suggested such a method using the Th^{230}/U ratio in coral limestone. In this study, they have analyzed 17 samples at various depths to 190 feet in an Eniwetok Island core. Uranium concentrations from 1 to 5 parts per million were found with the Th^{230}/U ratio generally increasing steadily with depth. Age estimations extending to about 300000 years potentially could be obtained where this method is applicable. A refinement of this method could be made if the Pa^{231}/U ratios were also determined, providing the requirement of concordant ages of both ratios in order to evaluate the validity of the result as described by ROSHOLT (1958).

The science of age determinations in a variety of marine sediments is approaching the stage where the entire range of time can be studied. The C^{14} dates can now be overlapped by the Pa^{231}/Th^{230} chronology, which is somewhat extended in time by the potential Th^{230}/U and Th^{230}/Th^{232} chronologies. The determination of Ar^{40}/K^{40} ages down to the lower limit of the order of 100000 years is now a definite possibility, assuming the availability of authigenic potassium-bearing minerals formed in the ocean. The Be^{10} age method described by PETERS (1957), but still in its infancy may be a potentiality for the extension in time studies to a few million years. Studies of greater time extension may be restricted to determining the age of source material, where, overlapping the K-Ar age methods, the geochronology of the lead isotope methods reviewed by TILTON and DAVIS (1959) complete the spectrum.

D. Future application of radioactivity in oceanography

Many of the described methods must be refined and additional measurements obtained in order to solve finally the problems under consideration. In the near future we will see an increased number of better measurements of carbon-14, radium-226, and hydrogen-3 in the ocean. These determinations together with all the other oceanographic parameters should give a clear picture of circulation in the ocean and the exchange of water masses in vertical and horizontal directions. The differences among the oceans and between oceans and mediterraneans with respect to intensity of circulation and exchange of water masses will be obtained.

Increasing numbers of age determinations will be carried out in deep-sea sediments and studies of the different methods applied will demonstrate their limitations and the meaning of the results. These results are necessary to establish the historic sequence of events during the Pleistocene and to arrive at a well defined sedimentation rate.

The application of the potassium-argon dating method in oceanography is now possible. Separation methods for authigenic minerals will allow age determination extending the time scale beyond the protactinium-ionium method. The potassium-argon ratio has been used and its use should be intensified for the determination of the age of source material for deep-sea sediments.

The influence of the organic life in the ocean on the distribution and the redistribution of radioactive elements must be carefully studied. The organisms participate to a considerable degree in the transfer and concentration of trace elements by the formation of skeletons and organic matter. These studies will give further information about the organic processes in the ocean and their influence on the geochemistry. Their effect must be known before final conclusions about the limits of the application of radioactivity in geophysics can be obtained. In many cases, the study of skeletons may result in a new application of radioactivity in oceanography.

A very important application of radioactive elements just started is the use of radioactive tracers for studying properties of deep-sea sediments and of slow processes in sediments investigated in the laboratory. Diffusion rates in the deep sea and the transformation of minerals must be studied by the use of tracers, since the rate of these processes at low temperatures makes ordinary chemical or physical-chemical methods not easily applied during the short time available in laboratory experiments.

Table 1. *Oceanic concentrations of the radioactive isotopes belonging to the three natural radioactive series*

Isotope	Half-life[1]	Mode of decay	Estimated average concentration in seawater (g/l)	Estimated average concentration in surface sediments (g/g)	Range of concentration in surface sediments (g/g)
U^{238}	4.5×10^9 yr	α	3.0×10^{-6}	1×10^{-6}	$(0.4-80) \times 10^{-6}$
U^{235}	7.13×10^8 yr	α	2.1×10^{-8}	7.1×10^{-9}	
U^{234}	2.48×10^5 yr	α	1.6×10^{-10}	8.1×10^{-11}	
Pa^{234}	1.14 min	β	1.4×10^{-19}	4.7×10^{-20}	
Pa^{231}	3.43×10^4 yr	α	$<2 \times 10^{-12}$	1×10^{-11}	$(0.08-9) \times 10^{-11}$
Th^{234}	24.1 dy	β	4.3×10^{-17}	1.4×10^{-17}	
Th^{232}	1.42×10^{10} yr	α	$<2 \times 10^{-8}$	5.0×10^{-6}	$(1-16) \times 10^{-6}$
Th^{231}	25.6 hr	β	8.6×10^{-20}	2.9×10^{-20}	
Th^{230}	8.0×10^4 yr	α	$<3 \times 10^{-13}$	2.0×10^{-10}	$(0.3-20) \times 10^{-10}$
Th^{228}	1.91 yr	α	4.0×10^{-18}	7×10^{-16}	
Th^{227}	18.17 dy	α	$<7.0 \times 10^{-20}$	1.3×10^{-17}	
Ac^{228}	6.13 hr	β	1.5×10^{-21}	2.4×10^{-19}	
Ac^{227}	21.6 yr	β, α	$<1 \times 10^{-15}$	5.9×10^{-15}	
Ra^{228}	6.7 yr	β	1.4×10^{-17}	2.3×10^{-15}	
Ra^{226}	1.622×10^3 yr	α	1.0×10^{-13}	4.0×10^{-12}	$(0.3-40) \times 10^{-12}$
Ra^{224}	3.64 dy	α	2.1×10^{-20}	3.4×10^{-18}	
Ra^{223}	11.68 dy	α	$<4.4 \times 10^{-20}$	8.5×10^{-18}	
Fr^{223}	22 min	β	$<7.0 \times 10^{-24}$	1.4×10^{-21}	
Rn^{222}	3.823 dy	α	6.3×10^{-19}	2.5×10^{-17}	
Rn^{220}	51.5 sec	α	3.3×10^{-24}	5.4×10^{-22}	
Rn^{219}	3.92 sec	α	$<1.7 \times 10^{-25}$	3.1×10^{-23}	
Po^{218}	3.05 min	α	3.4×10^{-22}	1.4×10^{-20}	
Po^{216}	0.158 sec	α	1.0×10^{-26}	1.7×10^{-24}	
Po^{215}	1.83×10^{-3} sec	α	$<8.1 \times 10^{-29}$	1.4×10^{-26}	
Po^{214}	1.64×10^{-4} sec	α	3.0×10^{-28}	1.1×10^{-27}	
Po^{212}	3.04×10^{-7} sec	α	1.2×10^{-32}	2.4×10^{-29}	
Po^{211}	0.52 sec	α	$<6.8 \times 10^{-29}$	1.2×10^{-26}	
Po^{210}	138.4 dy	α	2.2×10^{-17}	8.8×10^{-16}	
Bi^{214}	19.7 min	β	2.1×10^{-21}	8.8×10^{-20}	
Bi^{212}	60.5 min	β, α	2.2×10^{-22}	3.7×10^{-24}	
Bi^{211}	2.16 min	α, β	$<5.6 \times 10^{-24}$	1.0×10^{-21}	
Bi^{210}	5.01 dy	β	7.8×10^{-19}	3.1×10^{-17}	
Pb^{214}	26.8 min	β	2.9×10^{-21}	1.2×10^{-19}	
Pb^{212}	10.6 hr	β	2.4×10^{-21}	3.9×10^{-19}	
Pb^{211}	36.1 min	β	$<9.0 \times 10^{-23}$	1.6×10^{-20}	
Pb^{210}	19.4 yr	β	1.1×10^{-15}	4.5×10^{-14}	
Tl^{208}	3.10 min	β	4.1×10^{-24}	6.7×10^{-22}	
Tl^{207}	4.79 min	β	$<1.2 \times 10^{-23}$	2.1×10^{-21}	

[1] Half-life taken from Strominger, Hollander, and Seaborg (1958).

Table 2. *Oceanic concentration of non-series primordial isotopes*

Isotope	Half-life (years)	Mode of decay	Percent abundance	Estimated average concentration in seawater (g/l)	Estimated concentration in sediments (g/g)
Re^{187}	5.0×10^{10}	β	62.9	—	—
Lu^{176}	2.4×10^{10}	β	2.6	—	—
Sm^{147}	1.3×10^{11}	α	15.0	—	—
Nd^{144}	5×10^{15}	α	23.87	—	—
Ce^{142}	5×10^{15}	α	11.07	—	—
In^{115}	6×10^{14}	β	95.77	—	—
Rb^{87}	4.7×10^{10}	β	27.85	3.3×10^{-5}	—
K^{40}	1.25×10^{9}	β, electron capture	0.0119	4.2×10^{-5}	$(0.8 - 4.5) \times 10^{-6}$

Table 3. *Oceanic concentration of cosmic ray produced and artificial isotopes*

Isotope	Half-life	Estimated concentration in surface seawater (g/l)	Estimated concentration in surface sediments (g/g)
		Cosmic ray produced isotopes	
Cl^{39}	1 hr	—	—
S^{35}	87 dy	$< 1.8 \times 10^{-18}$*	—
P^{32}	14.3 dy	$< 1.5 \times 10^{-18}$*	—
P^{33}	25 dy	$< 3.1 \times 10^{-18}$*	—
Si^{32}	710 yr	5×10^{-19}	$(0 - 2) \times 10^{-16}$
Na^{22}	2.6 yr	—	—
C^{14}	5570 yr	$(2 - 3) \times 10^{-14}$	$(0.1 - 1) \times 10^{-13}$
Be^{10}	2.5×10^{6} yr	$(0.7 - 8) \times 10^{-17}$	$(1 - 3) \times 10^{-13}$
Be^{7}	53 dy	$< 4.9 \times 10^{-17}$*	—
H^{3}	12.26 yr	$(0.7 - 5) \times 10^{-16}$	—
		Artificial isotopes	
Pm^{147}	2.8 yr	$(0.2 - 3) \times 10^{-17}$	—
Ce^{144}	285 dy	$(0.1 - 2.5) \times 10^{-17}$	—
Cs^{137}	28 yr	$(0.5 - 1.2) \times 10^{-15}$	—
Sr^{90}	28 yr	$(0.6 - 7) \times 10^{-15}$	—
Cl^{36}	3.1×10^{5} yr	$< 5 \times 10^{-18}$*	—
C^{14}	5570 yr	$\sim 3 \times 10^{-16}$	—
H^{3}	12.26 yr	$(0.1 - 1) \times 10^{-15}$	—

* The maximum limits of concentration for these isotopes are taken to be the average concentration in rain water reported by PETERS (1959) and SCHAEFFER, THOMPSON, and LARK (1960).

Table 4. *World-wide production rates, Q, of cosmic ray produced radioisotopes*

Nuclide	Q, atoms/cm² · sec	Reference
H^{3}	0.35 ± 0.20	CRAIG and LAL 1961
	$0.6 - 1.3$	WILSON and FERGUSSON 1960
	1.06	BEGEMANN 1958
	0.9	BOLIN 1958
	0.75	GILETTI, BAZAN, and KULP 1958
	1.2	BEGEMANN and LIBBY 1957
	1.2	CRAIG 1957b
	0.14	v. BUTTLAR and LIBBY 1955
	0.12	KAUFMAN and LIBBY 1954
Be^{7}	0.021	LAL, MALHOTRA, and PETERS 1958
	0.035	ARNOLD and AL-SALIH 1955

Table 4 (continued)

Nuclide	Q, atoms/cm$^2 \cdot$ sec	Reference
Be^{10}	0.08	LAL, MALHOTRA, and PETERS 1958
	0.04	ARNOLD 1956
	0.05—0.1	PETERS 1955
C^{14}	2.0 ± 0.5	CRAIG 1957a
	2	ARNOLD and ANDERSON 1957
	2.6	LIBBY 1955
Si^{32}	2×10^{-4}	LAL, GOLDBERG, and KOIDE 1960
P^{32}	1×10^{-4}	LAL, MALHOTRA, and PETERS 1958
P^{33}	1×10^{-4}	Ibid.
S^{35}	2×10^{-4}	Ibid.

References

ALDRICH, L. T., G. W. WETHERILL, G. R. TILTON and G. L. DAVIS: Half-life of Rb^{87}. Phys. Rev. **103**, 1045—1047 (1956).

ALMODOVAR, I.: Thorium isotope method for dating marine sediments. Thesis, Carnegie Institute of Technology, Dept. of Chemistry 1960.

ARNOLD, J. R.: Scintillation counting of natural radiocarbon. Science **119**, 155—158 (1954).

— Beryllium-10 produced by cosmic rays. Science **124**, 584—585 (1956).

—, and H. A. AL-SALIH: Beryllium-7 produced by cosmic rays. Science **121**, 451—453 (1955).

—, and E. C. ANDERSON: The distribution of C^{14} in nature. Tellus **9**, 28—32 (1957).

ARRHENIUS, G.: Sediment cores from the east Pacific. Swedish Deep Sea Exped. 1947—1948. Repts. **5**, fasc. I, 227 p. (1952).

— Sedimentation on the ocean floor. In: Researches in geochemistry, P. H. ABELSON, ed., p. 1—24. New York: John Wiley & Sons 1959.

ATEN jr., A. H. W.: Radioactivity in marine organisms. In: United Nations Internat. Conf. Peaceful Uses of Atomic Energy, 2nd, Geneva, 1958, P/402, proc., vol. 18, p. 414—418 (1958).

BARANOV, V. I., and L. A. KUZMINA: Radiochemical analysis of deep sea sediments in connection with the determination of the rate of sediment accumulation. In: Radioisotopes in scientific research. First UNESCO Internat. Conf., Paris, 1957, Proc. vol. II, Chemistry and Geology, p. 601—618 (1957).

— — The rate of silt deposition in the Indian Ocean. Geochemistry (English translation of Geokhimiya) No. 2, 131—140 (1958).

BARKER, A.: Radiocarbon dating; large scale preparation of acetylene from organic material. Nature, Lond. **172**, 631—632 (1953).

BARNES, J. W., E. J. LANG and H. A. POTRATZ: A radiochemical procedure for thorium and its applications to the determination of ionium. U.S. Atomic Energy Commission Report, LA-1845, 19 p. (1954).

— — — Ratio of ionium to uranium in coral limestone. Science **124**, 175—176 (1956).

BEGEMANN, F.: New measurements on the world-wide distribution of natural and artificially produced tritium. United Nations Internat. Conf. Peaceful Uses of Atomic Energy, 2nd, Geneva, 1958, P/1963, Proc., vol. 18, p. 545—550 (1958).

—, and W. F. LIBBY: Continental water balance, ground water inventory and storage time, surface ocean mixing rates, and world-wide water circulation patterns from cosmic ray and bomb tritium. Geochim. et Cosmochim. Acta **12**, 277—296 (1957).

BIEN, G. S., N. W. RAKESTRAW and H. E. SUESS: Radiocarbon concentration in Pacific Ocean water. Tellus **12**, 436—443 (1960).

BOLIN, B.: On the use of tritium as a tracer for water in nature. United Nations Internat. Conf. Peaceful Uses of Atomic Energy, 2nd, Geneva, 1958, P/176, Proc., vol. 18, p. 336—343 (1958).

— On the exchange of carbon dioxide between the atmosphere and the sea. Tellus **12**, 274—281 (1960).

BOWEN, V. T., and T. T. SUGIHARA: Strontium-90 in North Atlantic surface waters. Proc. Nat. Acad. Sci., Wash. **43**, 576—580 (1957).

— — Marine geochemical studies with fallout radioisotopes. United Nations Internat. Conf. Peaceful Uses of Atomic Energy, 2nd, Geneva, 1958, P/403, Proc., vol. 18, p. 431—438 (1958).

— — Strontium-90 in the "mixed layer" of the Atlantic Ocean. Nature, Lond. **186**, 71—72 (1960).

BRANNON, R.R., A.C. DAUGHTRY, D. PERRY, W.W. WHITAKER and M. WILLIAMS: Radiocarbon evidence on the dilution of atmospheric oceanic carbon by carbon from fossil fuels. Trans. Amer. Geophys. Un. **38**, 643—650 (1957).

BROECKER, W.S., M. EWING and B.C. HEEZEN: Evidence for an abrupt change in climate close to 11000 years ago. Amer. J. Sci. **258**, 429—448 (1960).

— R. GERARD, M. EWING and B.C. HEEZEN: Natural radiocarbon in the Atlantic Ocean. J. Geophys. Res. **65**, 2903—2931 (1960).

—, and E.A. OLSON: Lamont radiocarbon measurements. VI. Amer. J. Sci. Radiocarbon Suppl. **1**, 111—132 (1959).

— Radiocarbon from nuclear tests, II. Science **32**, 712—721 (1960).

— C.S. TUCEK and E.A. OLSON: Radiocarbon analysis of oceanic CO_2. Appl. Radiation and Isotopes **7**, 1—18 (1959).

BROWN, R.M., and W.E. GRUMMITT: The determination of tritium in natural waters. Canad. J. Chem. **34**, 220—226 (1956).

BURKE jr., W.H., and W.G. MEINSCHEIN: C^{14} dating with a methane proportional counter. Rev. Sci. Instrum. **26**, 1137—1140 (1955).

BUTTLAR, H. v., and W.F. LIBBY: Natural distribution of cosmic ray produced tritium. II. J. Inorg. and Nucl. Chem. **1**, 75—91 (1955).

CANNON jr., R.S., L.R. STIEFF and T.W. STERN: Radiogenic lead in nonradioactive minerals. United Nations Internat. Conf. Peaceful Uses of Atomic Energy, 2nd, Geneva, P/773, Proc., vol. 2, p. 215—223 (1958).

CHOW, T.J.: Lead isotopes in sea water and marine sediments. J. Mar. Res. **17**, 120—127 (1958).

—, and C.R. McKINNEY: Mass spectrometric determination of lead in manganese nodules. Analyt. Chem. **30**, 1499—1503 (1958).

—, and C.C. PATTERSON: Lead isotopes in manganese nodules. Geochim. et Cosmochim. Acta **17**, 21—31 (1959).

CRAIG, H.: The geochemistry of the stable carbon isotopes. Geochim. et Cosmochim. Acta **3**, 53—92 (1953).

— The natural distribution of radiocarbon and the exchange time of carbon dioxide between the atmosphere and sea. Tellus **9**, 1—17 (1957a).

— Distribution, production rate, and possible solar origin of natural tritium. Phys. Rev. **105**, 1125—1127 (1957b).

— Study of mixing rates in oceans and air using carbon-14. United Nations Internat. Conf. Peaceful Uses of Atomic Energy, 2nd, Geneva, 1958, P/1979, Proc., vol. 18, p. 358—363 (1958).

—, and D. LAL: The production rate of natural tritium. Tellus **13**, 85—105 (1961).

CRATHORN, A.R.: Use of an acetylene-filled counter for natural radiocarbon. Nature, Lond. **172**, 632—633 (1953).

CRUIKSHANK, A.J., G. COWPER and W.E. GRUMMITT: Production of Be^7 in the atmosphere. Canad. J. Chem. **34**, 214—219 (1956).

CURRIE, L.A., W.F. LIBBY and R.L. WOLFGANG: Tritium production by high energy photons. Phys. Rev. **101**, 1557—1563 (1956).

EHLMANN, A.J.: Stages of glauconite formation in modern foraminiferal sediments. Geol. Soc. Amer., annual meeting, Denver, Colo 1960.

EMILIANI, C.: Pleistocene temperatures. J. Geology **63**, 538—578 (1955).

— Paleotemperature analysis of core 280 and pleistocene correlations. J. Geology **66**, 264—275 (1958).

ERICSON, D.B., W.S. BROECKER, J.L. KULP and G. WOLLIN: Late-pleistocene climates and deep-sea sediments. Science **124**, 385—389 (1956).

FALTINGS, V. v., u. P. HARTECK: Der Tritiumgehalt der Atmosphäre. Z. Naturforsch. **5a**, 438—439 (1950).

FERGUSSON, G.J.: Reduction of atmospheric radiocarbon concentration by fossil fuel carbon dioxide and the mean life of carbon dioxide in the atmosphere. Proc. Roy. Soc. Lond. A **243**, 561—564 (1958).

FLYNN, K.F., and L.E. GLENDENIN: Half-life and beta spectrum of Rb^{87}. Phys. Rev. **116**, 744—748 (1959).

FØYN, E., B. KARLIK, H. PETTERSSON and E. RONA: The radioactivity of sea water. Göteborgs Kgl. Vetenskaps-Vitterhets-Samhäll Handl., Ser. B **6**, No. 12, 44 p., also, Nature, Lond. **143**, 275—276 (1939).

FONSELIUS, S., and G. ÖSTLUND: Natural radiocarbon measurements on surface water from the North Atlantic and Arctic Sea. Tellus **11**, 77—82 (1959).

GILETTI, B.F., F. BAZAN and J.L. KULP: The geochemistry of tritium. Trans. Amer. Geophys. Un. **39**, 807—818 (1958).

GOEL, P.S., S. JHA, D. LAL, P. RADHAKRISHNA and RAMA: Cosmic ray produced beryllium isotopes in rain water. Nuclear Phys. **1**, 196—201 (1956).

Goel, P. S., D. P. Kharkar, D. Lal, V. Narasappaya, B. Peters and V. Yatirajam: The beryllium-10 concentration in deep sea sediments. Deep Sea Research 4, 202—210 (1957).
— N. Narasappaya, C. Prabhakara, Rama and P. K. Zutshi: Study of cosmic ray produced short-lived isotopes P^{32}, P^{33}, Be^7 and S^{35} in tropical latitudes. Tellus 11, 91—100 (1959).
Goldberg, E. D., and G. Arrhenius: Chemistry of Pacific pelagic sediments. Geochim. et Cosmochim. Acta 13, 153—212 (1958).
—, and M. Koide: Ionium-thorium chronology in deep sea sediments of the Pacific. Science 128, 1003 (1958).
— C. C. Patterson and T. J. Chow: Ionium-thorium and lead isotope indicators of oceanic water masses. United Nations Internat. Conf. Peaceful Uses of Atomic Energy, Geneva, 2nd, 1958, P/1980, vol. 18, p. 347—350 (1958).
—, and E. Picciotto: Thorium determinations in manganese nodules. Science 121, 613—614 (1955).
Grosse, A. V., W. H. Johnston, R. L. Wolfgang and W. F. Libby: Tritium in nature. Science 113, 1—2 (1951).
Hamer, A. N., and E. J. Robbins: A search for variations in the natural abundance of uranium-235. Geochim. et Cosmochim. Acta 19, 143—145 (1960).
Haring, A., A. E. de Vries and Hl. de Vries: Radiocarbon dating up to 70000 years by isotopic enrichment. Science 128, 472—473 (1958).
Hecht, R., J. Korkisch, R. Patzak u. A. Thiard: Bestimmung kleinster Uranmengen in Gesteinen und natürlichen Wassern. Mikrochim. Acta (Wien) 7/8, 1283—1309 (1956).
Hernegger, F., u. B. Karlik: Die quantitative Bestimmung sehr kleiner Uranmengen und der Urangehalt des Meerwassers. Sitzungsber. Akad. Wiss. Wien, Math.-naturw. Kl., Abt. II a 144, 21 (1934).
Herr, W., u. Merz: Zur Bestimmung der Halbwertszeit des ^{187}Re. Weitere Datierungen nach de Re/os Methode. Z. Naturforsch. 13a, 231—233 (1958).
Heydegger, H. R., and P. K. Kuroda: Natural occurrence of the short-lived barium and strontium isotopes. J. Inorg. and Nucl. Chem. 12, 12—17 (1959).
Higano, R.: Radiochemical analysis of the equatorial Pacific surface water. Internat. Oceanographic Congr., New York 1959, Preprints, p. 815—816.
Hurley, P. M., and others: Studies of the age of K-phases in deep ocean sediments, in, variations in isotopic abundances of strontium, calcium and argon and related topics. U.S. Atomic Energy Commission Rept., NYO-3941, p. 267—272 (1960).
Isaac, N., and E. Picciotto: Ionium determination in deep sea sediments. Nature, Lond. 171, 742—743 (1953).
Jentoft, R. E., and R. J. Robinson: The potassium-chlorinity ratio of ocean water. J. Mar. Res. 15, 170—180 (1956).
Jones, W. M.: Half-life of tritium. Phys. Rev. 100, 124—125 (1955).
Kaufman, S., and W. F. Libby: The natural distribution of tritium. Phys. Rev. 93, 1337—1344 (1954).
Ketchum, B. H., and V. T. Bowen: Biological factors determining the distribution of radioisotopes in the sea. United Nations Internat. Conf. Peaceful Uses of Atomic Energy, 2d, Geneva, 1958, P/402, Proc., vol. 18, p. 429—433 (1958).
Koczy, F. F.: Natural radium as a tracer in the ocean. United Nations Internat. Conf. Peaceful Uses of Atomic Energy, 2d, Geneva, 1958, P/2370, vol. 18, p. 351—357 (1958).
— E. Picciotto, G. Poulaert et S. Wilgain: Mesure des isotopes du thorium dans l'eau de mer. Geochim. et Cosmochim. Acta 11, 103—129 (1957).
—, and H. Titze: Radium content of carbonate shells. J. Mar. Res. 17, 302—311 (1958).
— E. Tomic u. T. Hecht: Zur Geochemie des Urans im Ostseebecken. Geochim. et Cosmochim. Acta 11, 86—102 (1957).
Korkisch, J., u. P. Antal: Zur Bestimmung von Mikrogrammengen Thorium in Silicatgesteinen, Sedimenten und anderen Materialien nach vorangehender Anreicherung des Thoriums mittels Ionenaustausches. Z. Anal. Chem. 173, 126—138 (1960).
Kröll, V.: The distribution of radium in deep sea cores. Swedish Deep Sea Exped., 1947 to 1948, Repts. 10, fasc. I, 32 p. (1955).
Lal, D., J. R. Arnold and M. Honda: Cosmic ray production rates of Be^7 in oxygen and P^{32}, P^{33}, S^{35} in argon at mountain latitudes. Phys. Rev. 118, 1626—1632 (1960).
— E. D. Goldberg and M. Koide: Cosmic ray produced silicon-32 in nature. Science 132, 332—337 (1960).
— P. K. Malhotra and B. Peters: On the production of radioisotopes in the atmosphere by cosmic radiation and their application to meteorology. J. Atmosph. Terr. Phys. 12, 306—328 (1958).
— Rama and P. K. Zutshi: Radioisotopes P^{32}, Be^7 and S^{35} in the atmosphere. J. Geophys. Res. 65, 669—674 (1960).
Libby, W. F.: Atmospheric helium three and radiocarbon from cosmic radiation. Phys. Rev. 69, 671—672 (1946).

Libby,W. F.: Radiocarbon dating, 2nd ed., 1955, p. 175. Chicago: Chicago University Press 1952.
— Radioactive strontium fallout. Proc. Nat. Acad. Sci., Wash. **42**, 365—390 (1956).
Merrill, J. R., M. Honda and J. R. Arnold: Beryllium geochemistry and beryllium-10 age determinations. United Nations Internat. Conf. Peaceful Uses Atomic Energy, 2d, Geneva, 1958, P/412, Proc., vol. 2, p. 251—254 (1958).
— E. F. X. Lyden, M. Honda and J. R. Arnold: The sedimentary geochemistry of the beryllium isotopes. Geochim. et Cosmochim. Acta **18**, 108—129 (1960).
Münnich, K.O.: Heidelberg natural radiocarbon measurements. I. Science **126**, 194—199 (1957).
—, u. J. C. Vogel: Durch Atomexplosionen erzeugter Radiokohlenstoff in der Atmosphäre. Naturwissenschaften **45**, 327—329 (1958).
Nakai, Z., S. Hattori, K. Honjo, T. Okutani and T. Kidachi: The present radioactive states of marine organisms in the sea adjacent to Japan. Internat. Cosmographic Congr. New York 1959. Preprints, p. 973—974.
Nakanishi, M.: Fluorimetric mecrodetermination of uranium. V. The uranium content of sea water. Bull. Chem. Soc. Japan **24**, 36—39 (1951).
NBS (National Bur. of Standards): Redetermination of the half-life of carbon-14. U.S. Natl. Bur. of Standards Tech. News Bull. **45**, p. 21 (1961).
Nelepo, B.A.: Direct determination of radioactivity in the water of the Pacific Antarctic. Internat. Oceanographic Congr., New York 1959. Preprints, p. 820.
Nier, A.O.: The isotopic constitution of uranium and the half-lives of the uranium isotopes. Phys. Rev. **55**, 150—153 (1939a).
— The isotopic constitution of radiogenic leads and the measurement of geologic time. II. Phys. Rev. **55**, 153—163 (1939b).
— A redetermination of the relative abundances of the isotopes of carbon, nitrogen, oxygen, argon, and potassium. Phys. Rev. **77**, 789—793 (1950).
Patterson, C.C.: Age of meteorites and the earth. Geochim. et Cosmochim. Acta **10**, 230—237 (1956).
— E. D. Goldberg and M. G. Inghram: Isotopic compositions of Quaternary leads from the Pacific Ocean. Bull. Geol. Soc. Amer. **64**, 1387—1388 (1953).
Peters, B.: Radioactive beryllium in the atmosphere and on the earth. Proc. Ind. Acad. Sci. **41**, 67—71 (1955).
— Über die Anwendbarkeit der Be⁻¹⁰-Methode für Messung kosmischer Strahlungsintensität und der Ablagerungsgeschwindigkeit von Tiefseesedimenten vor einigen Millionen Jahren. Z. Physik **148**, 93—111 (1957).
— Cosmic ray produced radioactive isotopes as tracers for studying large-scale atmospheric circulation. J. Atmosph. Terr. Phys. **13**, 351—370 (1959).
Pettersson, H.: Das Verhältnis Thorium zu Uran in den Gesteinen und im Meer. Ann. Akad. Wiss. Wien, math.-naturw. Kl. 127—128 (1937).
— Radioactive elements in ocean waters and sediments. In: Nuclear geology, H. Faul, ed., p. 115—120. New York: John Wiley & Sons; London: Chapman & Hall Ltd. 1954.
Picciotto, E., and S. Wilgain: Thorium determination in deep sea sediments. Nature, Lond. **173**, 623—633 (1954).
Piggot, C. S.: The radium content of ocean bottom sediments. Amer. J. Sci. **25**, 229—238 (1933).
— Radium content of ocean bottom sediments. Carnegie Inst. Wash. Publ. **556**, Oceanography II, pt. 2, 183—193 (1944).
—, and W. D. Urry: The radium content of an ocean bottom core. J. Wash. Acad. Sci. **29**, 405—415 (1939).
Rafter, T.A., and G. J. Fergusson: "Atom bomb effect"—recent increase of carbon-14 content of the atmosphere and biosphere. Science **126**, 557—558 (1957).
— — Atmospheric radiocarbon as a tracer in geophysical circulation problems. United Nations Internat. Conf. Peaceful Uses of Atomic Energy, 2d, Geneva 1958, P/2128, Proc., vol. 18, p. 526—532 (1958).
Rama: Investigations of the radioisotopes Be⁷, P³², and S³⁵ in rain water. J. Geophys. Res. **65**, 3773—3776 (1960).
—, M. Koide, and E. D. Goldberg: Lead-210 in natural waters. Science **134**, 98—99 (1961).
—, and P. K. Zutshi: Annual deposition of cosmic ray produced Be⁷ at equatorial latitudes. Tellus **10**, 99—103 (1958).
Rona, E., L. O. Gilpatrick and L.M. Jeffrey: Uranium determination in sea water. Trans. Amer. Geophys. Un. **37**, 697—701 (1956).
Rosholt, J.N.: Quantitative radiochemical determination for the sources of natural radioactivity. Analyt. Chem. **29**, 1398—1408 (1957).
— Radioactive disequilibrium studies as an aid in understanding the natural migration of uranium and its decay products. In: United Nations Internat. Conf. Peaceful Uses of Atomic Energy, 2nd, Geneva, 1958, P/772, proc., vol. 2, p. 230—236 (1958).

Rosholt, J.N., C. Emiliani, J. Geiss, F. F. Koczy and P. J. Wangersky: Absolute dating of deep-sea sediments by the Pa^{231}/Th^{230} method. J. Geology 69, 162—185 (1961).

Rubin, M.: U.S. Geological Survey radiocarbon dates. III. Science 123, 442—448 (1956).

—, and H.E. Suess: U.S. Geological survey radiocarbon dates. II. Science 121, 481—488 (1955).

Russell, R.D., and R.M. Farquhar: Lead isotopes in geology. 243 pp. New York: Interscience Publishers 1960.

Sackett, W.M.: The protactinium-231 content of ocean water and sediments. Science 132, 1761—1762 (1960).

— H. Potratz and E.D. Goldberg: Thorium content of ocean water. Science 128, 204—205 (1958).

Schaeffer, O.A., S.O. Thompson and N.L. Lark: Chlorine-36 radioactivity in rain. J. Geophys. Res. 65, 4013—4016 (1960).

Senftle, F.E., L.R. Stieff, F. Cuttitta and P.K. Kuroda: Comparison of the isotopic abundance of U^{235} and U^{238} and radium activity ratios in Colorado Plateau uranium ores. Geochim. et Cosmochim. Acta 11, 189—193 (1957).

Smales, A.A., and L. Salmon: Determination by radioactivation of small amounts of rubidium and caesium in sea water and related materials of geochemical interest. Analyst 80, 37—50 (1955).

—, and R.K. Webster: The determination of rubidium in sea water by stable isotope dilution method. Geochim. et Cosmochim. Acta 11, 139 (1957).

Smith, A.P., and F.S. Grimaldi: The fluorimetric determination of uranium in nonsaline and saline waters. In F.S. Grimaldi et al., Compilers, Collected papers on methods for uranium and thorium. U.S. Geol. Survey Bull. 1006, p. 125—131 (1954).

Starik, I.E., Y.V. Kuznetsov, S.M. Grashchenko and M.S. Frenklikh: The ionium method of determination of age of marine sediments. Geochemistry (English translation of Geokhimiya) No. 1, 1—15 (1958).

Stewart, D.C., and W.C. Bentley: Analysis of uranium in sea water. Science 120, 50—52 (1954).

Strøm, K.: A concentration of uranium in black muds. Nature, Lond. 162, 922 (1948).

Strominger, D., J.M. Hollander and G.T. Seaborg: Table of isotopes. Rev. Mod. Phys. 30, 585—904 (1958).

Suess, H.E.: Natural radiocarbon measurements by acetylene counting. Science 120, 5—7 (1954).

— Radiocarbon concentration in modern wood. Science 122, 415—417 (1955).

—, and R. Revelle: Carbon dioxide exchange between atmosphere and ocean and the question of an increase of atmospheric CO_2 during the past decade. Tellus 9, 18—27 (1957).

Sugihara, T.T., H.I. James, E. J. Troianello and V.T. Bowen: Radiochemical separation of fission products from large volumes of sea water, strontium, cesium, cerium, promethium. Analyt. Chem. 31, 44—49 (1959).

Sverdrup, H.U., M.W. Johnson and R.H. Fleming: The oceans, their physics, chemistry, and general biology. New York: Prentice-Hall, Inc. 1942, 1087 p..

Tilton, G.R., and G.L. Davis: Geochronology. In: Researches in Geochemistry, P.H. Abelson, ed., p. 190—213. New York: John Wiley & Sons 1959.

Tomic, E., I.M. Ladenbauer u. M. Pollack: Beitrag zur Trennung des Uran von Thorium mittels Ionenaustausches und zur fluorimetrischen Uranbestimmung. Z. analyt. Chem. 161, 28—38 (1958).

Urry, W.D.: Radioactivity in ocean sediments. VII. Rate of deposition of deep sea sediments. J. Marine Res. 7, 618—634 (1950).

Vries, H. de: Variation in concentration of radiocarbon with time and location on earth. Proc. Kon. Ned. Acad. Wet., Ser. B 61 (1958a).

— Atomic bomb effect, the natural activity of radiocarbon in plants, shells, and snails in the past 4 years. Science 128, 250—251 (1958b).

— Measurement and use of natural radiocarbon. In: Researches in geochemistry, P.H. Abelson, ed., p. 169—189. New York: John Wiley & Sons 1959.

—, and G.W. Barendsen: A new technique for radiocarbon dating by a proportional counter filled with carbon dioxide. Physica, Haag 19, 987—1003 (1953).

Volchok, H.L., and J.E. Kulp: The ionium method of age determination. Geochim. et Cosmochim. Acta 11, 219—246 (1957).

Wasserburg, G.L., R. J. Hayden and K. J. Jensen: $A^{40}-K^{40}$ dating of igneous rocks and sediments. Geochim. et Cosmochim. Acta 10, 153—165 (1956).

Wetherill, G.W.: Radioactivity of potassium and geologic time. Science 126, 545—549 (1957).

Wilson, A.T., and G. J. Fergusson: Origin of terrestrial tritium. Geochim. et Cosmochim. Acta 18, 273—277 (1960).

Yamagata, N., and S. Matsuda: Cerium-137 in the coastal waters of Japan. Internat. Cosmographic Congr., New York 1959, Preprints, p. 825—827.

Radioactivity in Hydrology

by

ERIK ERIKSSON

With 2 Figures

Zusammenfassung

Die natürliche Radioaktivität der Grundwässer wird vorwiegend getragen von den natürlich-radioaktiven Isotopen der Uran- und Thorium-Familie. Unter ihnen überwiegt in der Regel das Radon. Nähere Informationen über die Grundwasser-Aktivität liegen nur in geringem Maße und nur in Gestalt sporadischer Erfahrungen vor. Es ist zu erwarten, daß systematische Studien dessen Informationen über die Grundwasser-Bewegungen geben werden.

Eine wichtige Rolle spielt in der Grundwasser-Forschung die Untersuchung des Tritiums, das durch die Wirkung der kosmischen Strahlung gebildet wird. Der ursprünglich vorhandene Gleichgewichtswert des Tritium-Gehaltes ist zur Zeit ernsthaft gestört durch das bei Kernwaffenexplosionen freigewordene Tritium. Es wäre von besonderer Bedeutung gewesen, wenn man vor diesen Bombenversuchen ein genaues Programm für die Untersuchung von Regen- und Flußwasser ausgearbeitet hätte, um über das durch Wasserstoff-Explosionen erzeugte Tritium Klarheit zu schaffen.

Künstlich-radioaktive Isotope werden in steigendem Maße als Spurenelemente in der Hydrologie verwandt; besonders wichtig sind dabei die Glieder der Halogen-Gruppe.

Es ist möglich, die wichtigsten Eigenschaften des Grundwassers durch entsprechend geführte Experimente unter Verwendung der Isotope als Spurenelemente zu ermitteln. Im folgenden Beitrag wird die Theorie für solche Versuche im einzelnen dargelegt und diskutiert.

Weitere Anwendungsmöglichkeiten radioaktiver Isotope bei hydrologischen Problemen sind ihre Verwendung bei Untersuchungen der Vermischung in Seen und Flüssen, des Austausches zwischen Seen und der Atmosphäre, der Grundwasser-Bewegungen, der Bewegungen des Flußbettes und der Verdunstung vom Boden. Die Methoden befinden sich noch im Entwicklungsstadium, man kann aber schon jetzt eine stark zunehmende Anwendung der Radioaktivität im Rahmen der Grundwasser-Forschung voraussehen.

A. Natural radioactivity of waters

I. Radioactivity from the uranium and thorium series in rocks

The two most important radioactive series in rocks are the uranium and the thorium series. The presence of the gaseous member in these series, radon, has been known for a long time but lack of suitable analytical methods has prevented search for the first members of the series, U-238 and Th-232. In recent years a good deal of information has been gathered on these and is summarized recently by Adams et al. (1959). They are, of course, the most abundant of all the elements in the series so that uranium and thorium analyses represent practically these two isotopic species. The average concentration in rocks is given by Adams et al. and is

$$\textit{Igneous rocks:} \quad \text{Uranium} \quad 11.4 \text{ ppm}$$
$$\text{Thorium} \quad 3.0 \text{ ppm}$$
$$\textit{Sedimentary rocks:} \quad \text{Uranium} \quad 9.9 \text{ ppm}$$
$$\text{Thorium} \quad 3.2 \text{ ppm}$$

where ppm presumable is given on a weight basis. The ratio U/Th is consequently about 3.5 as an average for all rocks being about the same in igneous and sedimentary rocks.

In the ground these isotopes and their decay product will be dissolved by the water to varying degrees. Only recently have river water analyses for such substances, notably uranium and radium been made.

From a chemical point of view U-238 and Th-232 make up the main mass of radioactive substances in rocks and consequently also in river water. From a radiation point of view each member in the series should be of equal importance if radiation equilibrium was strictly attained. As to the different series, considering the ratio U/Th and the half lifes of U-238 and Th-232 the rate of decay of each member of the series at equilibrium should be about 10 times greater for the uranium series. Now, the chemical properties of the different members are, of course, different so that one cannot expect that they are dissolved at the same rate. One can expect uranium to be easier dissolved than thorium, because of its ability to form a stable anion in presence of oxygen, a property lacking in the other elements. It can also be expected that the gaseous members, the radon isotopes, escape much more easily into the water phase than any of the other members. Radiation equilibrium can therefore never be expected either in the soil and rock in contact with water, or in the water, or in the mixture of rock and water as long as the water is moving.

The more prominent elements in the radioactive series are U-238, Th-230 (ionium), Ra-226 and Rn-222 (radon) in the uranium series and Th-232 and Rn-220 (thoron) in the thorium series. The chemical properties of the decay products of radon are such that they will be substantially absorbed by the rock material, and as they are short-lived there are consequently no measurements available on them. Koczy (1956), in discussing the geochemical balance of the more prominent members in connection with ocean water content of radioactive substances, computed the average concentration of U-238, Th-230, and Th-232 in river water using a figure of $0.7 \cdot 10^{-16} \text{ g} \cdot \text{ml}^{-1}$ for the radium concentration in river water.

In Table 1 relevant data are collected on these isotopes as well as their half-lifes and their equilibrium concentration. The figure for Rn-222 is an order of magnitude figure likely to be found in groundwaters. When this comes in contact with the atmosphere a good deal of it will be lost to the atmosphere.

It is seen that again U-238 and Th-232 dominate but that the ratio between U-238 and Th-232 is very much higher than in rocks, indicating the difference in chemical character between these two elements. It is also seen that the actual concentration of Th-230 and Ra-226 is lower than the equilibrium concentration, again showing the greater solubility of uranium compared to the other two. However, radon, Rn-222, behaves differently being enriched by several orders of magnitude relative to the other members as predicted. Because of this enrichment radon will contribute practically alone to the radioactivity of groundwaters.

Thoron may contribute some but because of its short half-life it would disappear very quickly from groundwater which enters streams and other water courses.

The high figure for Rn-222 in groundwater can be anticipated considering the rate of decay of the uranium series in rocks on a equilibrium basis. The U-238 content is approximately $1 \cdot 10^{-5}\,\mathrm{g} \cdot \mathrm{ml}^{-1}$ of rock and at equilibrium the

Table 1. *Average concentration of various radioactive elements in continental waters*

Elements	Conc. in g·ml⁻¹	Half-life	Equilibrium conc. g·ml⁻¹
Uranium series			
U-238	$1.3 \cdot 10^{-9}$	$4.5 \cdot 10^{9}$ years	$1.3 \cdot 10^{-9}$
Th-230	$1.5 \cdot 10^{-14}$	$8.0 \cdot 10^{4}$ years	$2.3 \cdot 10^{-14}$
Ra-226	$0.7 \cdot 10^{-16}$	1622 years	$4.7 \cdot 10^{-16}$
Rn-222	$\sim 10^{-17}$	3.825 days	$3.0 \cdot 10^{-21}$
Thorium series			
Th-232	$2 \cdot 10^{-11}$	$1.389 \cdot 10^{10}$ years	
Rn-220		54.5 sec	

corresponding concentration of Rn-222 would be $2.3 \cdot 10^{-17}\,\mathrm{g} \cdot \mathrm{ml}^{-1}$. If this is concentrated in the volume of water in contact with unit volume of rock very high concentrations of Rn-222 can build up.

As the half-life of radon is fairly short a stationary state in groundwater can be readily established in which the rate of decay of radium atoms per unit volume of the rock equals the rate of decay of radon atoms in the same volume. If most radon is found in the water phase it will thus give an indication on the volume concentration of radium in the rocks in contact with the water. This is practically the only information one can derive by radon analyses of ground-waters. If the concentration of radium in the rock material, which is in contact with water, was known one could, of course, compute the water content of the rock material (e.g. in porous sandstones) but, to what extent this can be done practically is difficult to decide.

It was mentioned earlier that radiational equilibrium of the bulk material can be reached if the water moves slowly. Taking the radium-radon equilibrium a storage time in a uniform material of a couple of weeks would lead to equilibrium. Departure from this equilibrium would thus indicate a shorter storage or turnover time for the water. In the case of Th-230—Ra-226 a turnover time of several thousands of years would give equilibrium but in actual computations in the case of non-equilibrium one must also consider the rates of solution of these two isotopes. Non-equilibrium between two isotopes in a groundwater region can only be established because of differences in rates of dissolution of the isotopes.

There may be more unexplored possibilities using other members of the uranium series for the same purpose for instance RaD—RaE where RaD has a half-life of 25 years and RaE of 5 days. Being chemically different elements their solubility properties may differ and hence non-equilibrium may be established provided the turnover time for the groundwater is of the order of a few decades. In principle it should thus be possible to "date" groundwater using suitable member pairs of the uranium series.

In rivers radon can escape to the atmosphere and would most probably leave it rather rapidly because of the generally turbulent conditions in rivers. Additions, of course take place from the bottom and incoming groundwater so that a stationary state may be reached. Surveys of radon in river waters do not seem to have been made.

In lakes the additions of radon take place through tributaries, submerged springs and from bottom material while subtractions take place through the surface and by outlets of water and decay. It may be possible that the distribution of radon in a lake can give information on the circulation of water in a lake and mixing phenomena but this possibility does not seem to have been explored yet.

II. Tritium

Tritium is the heaviest of the hydrogen isotopes with a half-life of 12.5 years and seems to be produced in the atmosphere by cosmic ray neutrons combarding nitrogen molecules—(LIBBY 1946) the same process which yields carbon-14. The formed tritium is, of course, combined with oxygen into tritiated water, mainly HTO which follows the water in its circulation in nature. For this reason, and because of a half-life that is equivalent to the turnover times of groundwater bodies its usefulness in hydrology is being more and more realized. Its disadvantage is its weak β-radiation which requires expensive counting devices for measuring.

The rate at which the cosmic ray tritium is precipitated over the earth's surface varies geographically and possibly also from year to year being a climatic element like precipitated water. The concentration level in precipitation in continental areas is about $10 \cdot 10^{-18}$ T/H (10^{-18} T/H is called a tritium unit, TU.). In lake waters the concentration is less and depends partly on the turnover time of water in the lake, i.e. the rate at which the water is replaced. Gas exchange with the atmosphere causes a net flux of tritium from the air to lakes and other exposed water bodies compensating partly the loss by decay in the water itself.

Most of the precipitation enters the groundwater region where it spends a certain time before it enters rivers. During this time the gaseous exchange with the atmosphere is negligible and the concentration of tritium is only dependant on the original concentration and the time elapsed since the water arrived from the atmosphere. River waters have consequently lower tritium concentrations than precipitation and the difference may be used to compute its average age and thus the groundwater storage, a subject that will be discussed more in detail later.

Some actual figures on the tritium concentration in various waters prior to 1954 are shown in Table 2 taken from KAUFMAN and LIBBY (1953), von BUTTLAR

Table 2. *The concentration of tritium in various waters*

Source	Conc. in TU ($=10^{-18}$ T/H)	Source	Conc. in TU ($=10^{-18}$ T/H)
Chicago rain	8.1	Chile rain	4.3
Lamont rain	8.4	Norway lake water . . .	2.4
Ottawa rain	26.6	Atlantic rain	3.3
Livingstone river water .	6.5	Puerto Rico river	0.9
Cape Province river water	2.8	New Zealand rain	1.7
Spanish wines	3.1	Phillipines rain	0.9
Irish rivers	3.0	Hawaii rain	0.83
European rivers.	2.17	Atlantic surface water . .	1.13
California rain	4.0	Pacific surface water . .	1.02
Japan rain	6.5		

and LIBBY (1957), BEGEMANN and LIBBY (1957), GILETTI (1957) and BROWN and GRUMMIT (1956). The variation is as seen great and most notable is the higher continental values for rain water. The figures are, however, not entirely representative as some of them are averages of only a few samples, but the order of magnitude is clearly seen.

In 1954 the first extensive hydrogen bomb tests took place increasing the tritium content of nature at least by an order of magnitude and similar additions have also occurred later. Despite this fact groundwater samples taken after 1954 can sometimes show extremely low concentrations indicating that the water has not been renewed since 1954 and may be even older.

Rather few data exist on the tritium concentration in waters prior to 1954. Even if all hydrogen bomb tests stopped presently it would take several half-lifes of tritium before the normal level was established.

III. Carbon-14

The cosmic ray produced carbon-14 appears in the carbon dioxide in the atmosphere and is brought down to the ground by precipitation, gaseous exchange and assimilation of plants following closely the circulation of carbon in nature. The half-life of this isotope is long—about 5600 years. It is to some extent dependent on the water circulation in nature and has been used for dating ocean water. On land there is a large reservoir of carbon in limestone minerals—far greater than the reservoir of carbon lakes, rivers and groundwater—and the usefulness of carbon-14 in hydrological research is therefore severely limited. Exchange of carbon dioxide between carbonates in groundwater and carbonate minerals will take place giving apparent ages of groundwater far above the turnover times of groundwater bodies. This can be interpreted to mean that the carbon-14 in groundwater gives information on the turnover rate of carbon dioxide in the ground in contact with the water, not on the water itself.

IV. Other cosmic ray produced radioactive isotopes

The chemical properties of other radioactive isotopes produced by cosmic radiation are such that they are mostly strongly absorbed by the solid material in the ground. They are thus of but little interest in hydrology. There is one exception, chlorine-39, but its half-life is too short to be of any use.

B. Artificially produced radioactive isotopes

I. Bomb produced

Most of the radioactive isotopes emitted into the atmosphere are produced in the explosion of atomic bombs. The fission bombs produce an enormous number of radioactive isotopes of which, however, only a few have a half-life comparable to the turnover times of water bodies. Best known of these are strontium-90 and cesium-137 which are found in precipitation, mostly in solution. They are, however, absorbed strongly in the ground so that groundwater is practically free from these. Even in lakes an absorption of these isotopes would take place in the bottom mud. They are thus of less interest in hydrology.

In the fusion or thermonuclear bombs large amounts of tritium are produced. The usefulness of this tritium does not lie in its radioactivity—except from an analytical point of view—but as a tracer with properties very similar to those of ordinary water. It will be considered more in a later section.

Also carbon-14 seems to be released during hydrogen bomb explosions but because of its chemical properties is of but little use even as a tracer in hydrology except in the case of very large inland lakes.

II. Commercially available radioactive isotopes
applied to water for tracing purpose

The application of tracers in hydrology is older than radioactive tracers but inactive tracers do not normally have the enormous analytical advantage of radioactive tracers.

It is obvious from preceeding paragraphs that only a few radioactive isotopes are suitable for tracing purposes. All tracers in cationic form are unsuited because of their absorption in the solid material of the ground. Experiments have been made to put such cations into anionic forms by complex formation but not much success has been achieved yet. However, Halevy and Nir (1960) report that the cobalto-cyanide complex works well in calcareous ground and is, of course, easy to detect in situ. Of other anionic tracers only iodine-131 and bromine-82 seem to be suitable as they are much less absorbed by minerals than other tracers except tritium. An extensive review of these questions has been given recently by Hours (1960).

Tritiated water is, of course, the most ideal tracer for water but its weak β-radiation excludes the possibility of in situ measurements.

C. Water bodies as reservoirs and their characteristics
as revealed by radioactive tracer studies

I. General

The circulation of water in its various forms in nature is never strictly stationary as climatic elements vary with time. The variations are of periodic nature of which the seasonal variation is the most conspicous. Longer periods, however, are also present and are normally referred to as secular changes. Superimposed on these are shorter rather irregular year to year variations. Such variations may affect the size of ground water bodies.

These periodicities are normal features and are, of course, met with also in meteorology. They are theoretically difficult to tackle and one way of avoiding them in theoretical considerations is to neglect them defining the state as quasi-stationary hoping that the variations shall not influence materially the validity of the theory. Sometimes the errors introduced by such assumptions can be evaluated qualitatively, in other cases not. In the present discussion the circulation of water in nature will be regarded as quasi-stationary but the effect of periodical variations will be evaluated whenever possible.

Any theoretical approach to hydrological problems has ultimately in mind to find some common properties in different systems expressable as parameters of the model used in the theoretical approach for the actual system. There are two properties in connection with water bodies which are of immediate interest namely size and turnover characteristics. In planning utilization of groundwater it is highly desirable to know both.

The areal extent of groundwater bodies can often be estimated from topographic maps and a knowledge of the geology of the region. Knowing the areal extent the size of a groundwater body is apparently the area times the average depth of groundwater expressed in terms of the height of a water column. For the water economy the amount of groundwater is most suitably expressed in this way as also other items in the water economy are expressed in the same way for instance precipitation and evaporation. Denoting precipitation by P and evaporation by E then the run-off, R_0 in a stationary state is simply $P - E$.

If the groundwater storage, i.e. the average height of the groundwater, is denoted by V_0 then one can define the turnover time T simply as

$$T = \frac{V_0}{P - E} = \frac{V_0}{R_0}.\tag{1}$$

This definition, however, does not tell anything on how rapid the water is replaced in various parts of the reservoir. Deeper water may be replaced very slowly compared to more superficial water. A closer inspection of this problem seems therefore worth while to undertake.

When the water enters the soil as precipitation part of it will be evaporated, mainly through plants, before it sinks deeper into the soil. The upper part of the ground, the soil proper, contains organic matter and has a fairly large capacity for storing water in pores and capillaries and this water is the reserve the plants use between rains. The water that does not evaporate sinks further and reaches the part of the ground where all pore space is filled with water, i.e. the groundwater region. In this region the motion of water is described as laminar, i.e. it takes place without formation of eddies because frictional forces dominate

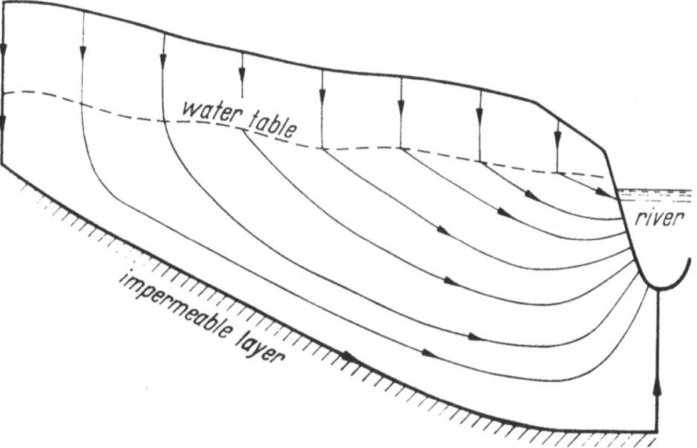

Fig. 1. Generalized flow of groundwater in soil with indicated stream-lines

over inertial forces. With this condition of motion the trajectories of the water molecules are fixed and determined entirely by the continuity equation and the geometry of the groundwater region. The flow can be visualized as taking place along fixed streamlines, stream-tubes or stream-sheets. According to the continuity condition streamlines can never cross nor intertwine. The same is, of course, true also for stream-tubes and stream-sheets. In general, the picture of streamlines will look like that in Fig. 1. The deepest groundwater is, as seen, derived from the water divides. Some longitudinal mixing will take place along streamlines due to the arrangement of pores but perpendicular to the streamlines only molecular diffusion will cause any mixing. This being very moderate in water one can expect a certain stratification of groundwater chemically.

A tracer added with precipitation has thus no obvious means to mix into the groundwater body. Yet, it can be anticipated from Fig. 1 because of the arrangement of stream-sheets that it may appear as well mixed when the tracer arrives in the outlet, i.e. a stream, and mixes with the rest of the water. This is of considerable interest in the case of application of tracers to groundwater studies as pointed out by ERIKSSON (1958a). But on the other hand it may be somewhat misleading to accept this apparent mixing under all circumstances. A better description can be made as follows.

Consider the trajectories of water molecules from the groundwater surface to the outlet in a region. These trajectories—or tiny stream-tubes—can be characterized by the average time τ it takes for a water molecule to travel along them. We can, consequently, divide the whole groundwater body into tiny stream-tubes and label each tube with its travel time. To each of the stream-tube passes a part of the run-off, R_0, from the region. Thus we can also label each stream-tube with a small run-off amount δR.

Next, arrange the run-off elements according to increasing travel times in a cumulative fashion, summing the δR's. In this way we arrive at a function R of τ where R is the rate of run-off from those parts of the region where the travel times of water molecules are shorter than or equal to τ. Such a function may look like curve A in Fig. 2. For very long travel times R approaches R_0, the total run-off rate.

Now, one can define a volume element δV associated with each run-off element δR by

$$\delta V = \tau \, \delta R \tag{2}$$

which follows from simple continuity considerations. Consequently, the area in Fig. 2 between the ordinate axis and the curve in the figure is equal to the volume V_0. Curve A in Fig. 2 thus gives a complete description of the storage properties of the groundwater reservoir.

A special case of the function R is

$$R = R_0(1 - e^{-k\tau}) \tag{3}$$

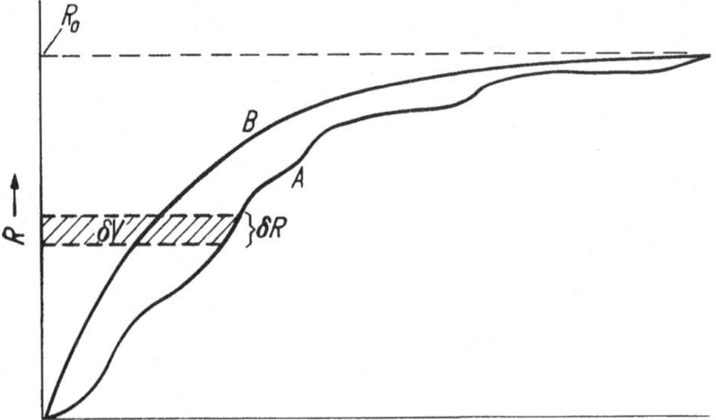

Fig. 2. The function R of τ. Curve A generalized function, curve B the exponential model

which can be called the *exponential model* of the groundwater reservoir. Any tracer added to the surface of such a reservoir will appear in the outlet as if thoroughly mixed into the whole reservoir. Its volume is consequently

$$V_0 = R_0/k \tag{4}$$

and $1/k$ is its turnover time. The function in Eq. (3) is shown by curve B in Fig. 2. Though not unreasonable as a model it may in cases be too simple to apply and be misleading on the actual volume stored. This is especially true where a large amount of deep groundwater is present. One can anticipate that the application of this model to tracer experiments will give a minimum volume of groundwater storage.

II. Stationary addition of a radioactive tracer

Before 1954 when the addition of tritium could be regarded as a stationary process it would have been possible to estimate groundwater storage based on the exponential model because it behaves as a thoroughly mixed groundwater body. If the concentration of tritium in precipitation is c_P and in run-off c_R and the volume of the reservoir is V_0 then continuity consideration give

$$R_0(c_P - c_R) = \lambda c_R V_0 \tag{5}$$

where λ is the decay constant of tritium. Thus V_0 can be computed from available data. Eriksson (1958a) applied this to the data on Chicago rainwater and Mississippi river water data prior to 1954 (Begemann and Libby 1957) and computed an average groundwater height of about 2 m thus close to that computed from the 1954 data when bomb tritium was added. But the volume may be greater than that in this area. At present there are no possibilities to utilize this method for computing groundwater storage.

III. Transient additions

Suppose that at time zero precipitation contains an ideal tracer for a short time enough to give a thin layer of labelled water on top of the groundwater reservoir. The amount added is denoted by m per cm^2. At time τ a certain part of the tracer has left the region corresponding to the functional relationship in Fig. 2. The rate at which it is added to the outlet is now $\dfrac{m}{R_0} \cdot \dfrac{dR}{d\tau}$. If its concentration in the outlet is c, the rate at which it leaves the area is $c R_0$. Thus we have

$$c R_0 = \frac{m}{R_0} \cdot \frac{dR}{d\tau}. \tag{6}$$

From this we obtain

$$R = \frac{R_0^2}{m} \int_0^\tau c \, dt. \tag{7}$$

It is thus possible to reconstruct the function R of τ by following the concentration in the outlet as a function of time τ. By such an experimental arrangement the complete groundwater characteristics is revealed.

Using Eq. (2) we get

$$V_0 = \frac{R_0^2}{m} \int_0^\infty c t \, dt \tag{8}$$

and thus the total volume of the reservoir.

Applying Eq. (6) to the exponential model gives

$$c = \frac{m}{V_0} e^{-k\tau}$$

which shows that it behaves as if thoroughly mixed.

To eliminate as much as possible the effect of seasonal fluctuations c should be determined at normal run-off rates. During high run-off a lot of the water is surface run-off thus diluting the water that comes from the groundwater reservoir. During lower than normal run-off the concentration c may be greater than during a normal state.

If the tracer is radioactive a correction of c has to be made based on the length of time elapsed since the addition.

As the turnover time of most water bodies can be counted in years such a determination of groundwater characteristics would take a fairly long time. This, however, cannot be regarded as any greater disadvantage. And already after the addition a rough idea can be obtained applying the exponential model. This was done by BEGEMANN and LIBBY (1957) after the 1954 bomb tests assuming rapid mixing of the Mississippi groundwater. If a proper deduction of the re-evaporated tritium is made in their computations the groundwater storage becomes about 2 m (cf. ERIKSSON 1958a), i.e. about the same as computed from stationary state addition of cosmic tritium.

More or less continuous but varying additions of a tracer can be utilized for obtaining groundwater characteristics. Suppose that at time τ' the concentration in the water added to the groundwater surface is $c_i(\tau')$. Then during a time dt is added $c_i(\tau')\,dt$. The contribution to the outlet at time τ from this addition is thus

$$R_0\,dc = \frac{c_i(\tau')\,R_0}{R_0}\,\frac{dR(\tau-\tau')}{d\tau}\,d\tau' \tag{9}$$

and by integration

$$c = \frac{1}{R_0}\int\limits_0^\tau c_i(\tau')\,\frac{dR(\tau-\tau')}{d\tau}\,d\tau' \tag{10}$$

just summing up all additions done up to time τ. The equation above can be solved by using finite differences. Denote by $c_1, c_2, \ldots, c_n$ the yearly average concentrations during the first, second, etc. year and by m_1, m_2 etc. the additions made during the same years. Finally, denote by φ_1, φ_2 etc. the average values of $dR/d\tau$ for successive years. Then one can write

$$\left.\begin{aligned}
c_1 &= \frac{1}{R_0^2}\,m_1\,\varphi_1 \\[4pt]
c_2 &= \frac{1}{R_0^2}\,[m_2\,\varphi_1 + m_1\,\varphi_2] \\[4pt]
&\cdot\ \cdot\ \cdot\ \cdot\ \cdot\ \cdot\ \cdot\ \cdot\ \cdot\ \cdot\ \cdot\ \cdot\ \cdot\ \cdot\ \cdot \\[4pt]
c_n &= \frac{1}{R_0^2}\,[m_n\,\varphi_1 + m_{n-1}\,\varphi_2 + \cdots + m_1\,\varphi_n]
\end{aligned}\right\} \tag{11}$$

Thus, φ_1, φ_2 etc. can be solved for successively. In a laboratory experiment it was possible to solve up to φ_{30} which was approaching zero. At that time, however, φ started to oscillate around zero.

In the case of a radioactive tracer m_1, m_2 etc. have to be corrected for radioactive decay.

In the case of an exponential model the treatment is simplified by regarding the reservoir as well mixed. Brown (1959) has done such an analysis based on Ottawa rain and river water analyses on tritium after 1954 and obtained interesting results.

The possibility of utilizing these ways of estimating groundwater storage is at present somewhat limited on a larger scale. Some idea of the groundwater storage can still be arrived at from more recent determinations of tritium in rain and river water. In future, however, provided no further bomb tests are made, there is a possibilitiy of more organized global experiments aiming at assessing groundwater characteristics.

D. Other applications of radioactive tracers to hydrological problems

I. Groundwater studies

A considerable work has been done in recent years for tracing groundwater movements with the help of radioactive tracers. The possibility is, however, somewhat limited by the longitudinal dispersion along the flow, studied theoretically and experimentally by Kaufman and Orlob (1956), Day (1956), de Josselin de Jong (1958) and Eriksson (1958b). The problem is, however, complex when considering porous systems which are anisotropic in the permeability characteristics. The longitudinal dispersion blurs the front and makes it difficult to compute average flow velocities. Tracing the groundwater for longer distances therefore requires rather high initial concentrations.

Some use of the bomb produced tritium has been made in order to study the flow of groundwater by sampling groundwater at different times after a major bomb test. BEGEMANN and LIBBY (1957) and BEGEMANN (1958) sampled well waters in a few places to determine the time lag between the explosion and the appearance in the well water. BEGEMANN obtained a time lag of about one month and a dilution of the rainwater by the groundwater about 50 times. v. BUTTLAR and WENDT (1958) reported similar investigations in New Mexico sampling from drilled wells about two years after the 1954 tests. Some of the water samples did not show any appreciable increase while others were definitely affected by the bomb tritium. In a special series of wells sampling over a year revealed rather interesting variations though somewhat difficult to interpret. This can be understood from Fig. 1 which by and large may give an idea about the layered structure of groundwater. A shallow well will probably get more recent water than a deeper one. Drilled wells with perforated casings are not ideal for such a study.

A more direct method of measuring groundwater flow rates from wells has been developed by HALEVY and coworkers in Israël (see HALEVY and NIR 1960). They developed a method which has been called the single-well pulse method. The idea behind it is not new but its application to wells with perforated casings is new. If a well had a casing which was equally or more permeable than the soil, addition of a tracer to the well water would cause a flow out of the well. If stirred suitably, measurements of the decrease of tracer in the well would permit a computation of the horizontal rate of flow of the groundwater. Mostly, however, casings with rather low permeability are used and in this case the method above cannot be applied. HALEVY et al. overcame this difficulty by forcing the tracer out of the well and then investigating the recovery by pumping water from the well. A loss indicates a horizontal current which can be computed from experimental data.

A similar method can be used for the study of groundwater movements in more humid regions where, sometimes, impeded groundwater flow causes high water tables which are retarding plant growth. In such cases the groundwater is readily reached by a sonde. If a tracer is injected through the sonde it will form a sphere around the tip. By pumping up the same volume as injected some time later and examining the recovery of the tracer the rate of flow can be computed. Such a method would be extremely suitable for studying slow horizontal flows. Coupled with gradient measurements at the measuring site it also allows a computation of the horizontal permeability coefficient of the soil in situ.

II. Mixing problems in lakes and rivers, determinations of stream flow and bed load transports

Radioactive tracers seem to be exceedingly well suited for studies of mixing rates in reservoirs like lakes, of considerable interest in connection with waste disposal. Water in a lake during the warm season is often stratified which causes a vertical resistance towards mixing. But even longitudinal and lateral mixing are of considerable interest and the only generating force in stable stratification is the wind action on the surface. Such problems are very little studied yet but will no doubt receive much more attention in future as the use of radioactive tracers becomes more widespread. For such purposes the most suitable tracers are iodine-131 and bromine-82, γ-emitters which can be detected in situ by suitable probes. Bromine-82 is reported to be detectable above background down to

10^{-8} curie/m³ (Ljunggren et al. 1959). An investigation on the mixing in a lake was made recently by Svantesson and Sundberg-Falkenmark (1959). They found that in the lake studied a dilution of only about 10 times was obtained after a distance of 1.5 km despite an average current velocity of 20 cm/sec.

In rivers a tracer method of considerable interest for measurement of discharge rate has recently been adopted for radioactive tracers by Hull (1958) who has named it the total count method. A stream is labelled by a tracer as uniformly as possible at a point. Downstream after vertical and lateral mixing has been completed, the concentration of the radioactive tracer is measured continuously as long as the tracer is passing. If the discharge rate is Q, the amount of added tracer is q and the measured concentration is c one has apparently

$$q = Q \int\limits_0^\infty c \, dt. \tag{12}$$

For a radioactive tracer the integral represents the number of counts minus the background. A simple calibration of the counting device is therefore the only necessary prerequisite.

Basically, the problem is, however, more complex because one has to use an adequate amount of tracers. But this amount depends on the longitudinal mixing which is probably related to the geometry of the cross section of the river. A useful concept in this case can be a mixing length for the longitudinal mixing. If this is denoted by l and the distance between the labelling and measuring point is L then the dispersion of the originally sharp front, expressed as a standard deviation, will be $\sigma = \sqrt{2lL}$, i.e. proportional to the square root of the distance. As l, the mixing length, must be related to the geometrical dimensions of the river, a study of longitudinal mixing in rivers of various shapes could be rewarding. In the fairly small river studied by Hull and Macomber (1958) the longitudinal mixing length seems to have been about 30 m. In Mohawk river in the N.E. United States Parker (1958) studied the dispersion of radioactive tracers and from his data a mixing length of about 50 m is obtained at 2.8 and 7.5 miles downstream at a discharge rate of 30 m³ · sec⁻¹. It is possible that the mixing length is variable and model experiments by Bryant and Geyer (1958) on the Delaware river estuary indicate increasing mixing length with the distance from the point of injection. This, however, is an estuary where tidal motions are strong and with a typical estuarine type of water circulation, i.e. a saline counter current mixing gradually with the outgoing freshwater.

Another method for discharge measurements called the dilution method is simpler in theory. A tracer is injected at a steady rate at one point of a stream and downstream, after mixing is complete, its concentration is measured. From this and the rate of injection the discharge rate of water is computed.

The use of radioactive tracers for studying bed load transports in rivers has recently been suggested by the author. Bed load movement seems to be very important for silt and sand movements and occurs through rolling and saltation of sand particles along the river bed, often giving rize to moving dunes. There is no quantitative experimental work with the aid of labelled sand done sofar, but the theory is fairly simple and deserves to be mentioned. One prerequisite for such measurements is that it is done in a place along a river where no net accumulation or depletion of sandy bed material takes place. If the bed material is seeded across the bed with radioactively labelled grains made for instance of irradiated glass containing iridium and of appropriate size then the transport rate could be determined in the same way as for water, i.e. by following the activity of the moving sand at some point downstream during the passage of the active sand. One condition has to be satisfied, however, namely that the labelled grains are uniformly incorporated into the whole moving layer. This may require adequate lateral mixing upstream of the measuring point. Again the problem of longitudinal mixing enters for a proper dosage of labelling material. At present it is difficult to say how this method will work in practice as it is still in the experimental stage.

III. Miscellaneous applications

Radioactive tracers can also be used for studying infiltration rates into soil of water. Again a longitudinal dispersion will be encountered but on the other hand it may also give valuable information on the pore size distribution in soils in situ provided a proper theory is worked out, covering also cases of unsaturated soils.

A method for determining evapo-transpiration from ground has been suggested involving the use of for instance iodine to determine the turbulent transport velocity of the air above ground. If the rate of absorption of iodine-131 from air is measured and related to its concentration difference at two levels a vertical turbulent transport velocity can be computed which applies to other gases, thus also to water vapor. Measuring water vapor density simultaneously at the same levels will therefore make it possible to compute the rate of evapo-transpiration if any.

Another application of radioactivity in hydrology is fairly well-known and established, namely determination of the moisture content in soils by measuring neutron scattering. Equipments for such measurements are commercially available.

E. Conclusions

The application of radioactive tracers to hydrological problems seems to be a promising field and may help to solve many problems which are difficult to tackle with classical methods. The work in this field is, however, fairly recent and although many applications can be suggested a considerable amount of laboratory work has to be carried out to test these.

F. Suggestions on future research

It is apparent from Sect. A that the possible use of radioisotopes of the uranium series should be explored further by systematic investigations on various isotopes in groundwater and a study of the absorption equilibrium between these isotopes and the solid material in the solid material of a groundwater body. One can expect the absorption to be due to hydrated mineral of clay or zeolitic nature so that results can be generalized. Such a study will reveal any possibilities of using pairs of elements in the uranium series for age determinations of ground water.

Another important field of research concerns methods suggested in Sect. C. Here, small scale experiments should be carried out preferably using tritium as a tracer.

The application of radioactive tracers to stream flow measurements is fairly well known and may be used as an alternative to other well established methods. In the case of bed load transports, however, there are no satisfactory methods at present and proper tests of the suggested method seems therefore urgent, considering both the fundamental and the practical aspects of bed load transports.

References

Adams, J.A.S., J.K. Osmond and J.J.W. Rogers: The geochemistry of thorium and uranium. Physics and Chemistry of the Earth **3**, 298—348 (1959).

Begemann, F.: New measurements on the world-wide distribution of natural and artificially produced tritium. U.N. Sec. Int. Conf. on the Peaceful Uses of Atomic Energy, P/1963, Geneve 1958.

—, and W.F. Libby: Continental water balance, groundwater inventory and storage times, surface ocean mixing rates and world-wide water circulation patterns from cosmic ray and bomb tritium. Geochimica et Cosmochimica Acta **12**, 277—296 (1957).

Brown, R.M.: Hydrology of tritium in the Ottawa valley. Contr. Defense Research Chem. Laboratories Ottawa, Canada. 1959. (In manuscript.)

—, and W.E. Grummit: The determination of tritium in natural waters. Canad. J. Chem. **54**, 220—226 (1956).

Bryant, G.T., and J.C. Geyer: The travel times of radioactive wastes in natural waters. Trans. Amer. Geophys. Un. **39**, 440—445 (1958).

Buttlar, H. v., and J. Wendt: Groundwater studies in New Mexico using tritium as a tracer. U.N. Sec. Int. Conf. on the Peaceful Uses of Atomic Energy, P/1954, Geneve 1958.

Day, P.R.: Dispersion of a moving salt-water boundary advancing through saturated sand. Trans. Amer. Geophys. Un. **37**, 595—601 (1956).

Eriksson, E.: The possible use of tritium for estimating groundwater storage. Tellus **10**, 472—478 (1958a).

— A note on the dispersion of a salt-water boundary moving through saturated sand. Trans. Amer. Geophys. Un. **39**, 937—938 (1958b).

Giletti, B.J.: The geochemistry of tritium. Thesis, Columbia University, June 1957.

Halevy, E., and A. Nir: Use of radio-isotopes in studies of groundwater flow. Tahal, Tel-Aviv, April 1960.

Hours, R.: Applications de la radioactivite a l'hydraulique souterraine. Commissariat a l'energie atomique — Centre d'etudes nucléares de Saclay, Mai 1960.

Hull, D.E.: The total count technique: a new principle in flow measurements. Int. J. Appl. Rad. and Isotopes **4**, 1—15 (1958).

—, and M. Macomber: Flow measurements by the total count method. U.N. Sec. Int. Conf. on the Peaceful Uses of Atomic Energy, P/817, Geneve 1958.

Josselin de Jong, G. de: Longitudinal and transverse diffusion in granular deposits. Trans. Amer. Geophys. Un. **39**, 67—74 (1958).

Kaufman, W.J., and G.T. Orlob: An evaluation of groundwater tracers. Trans. Amer. Geophys. Un. **37**, 297—306 (1956).

Koczy, F.F.: Geochemical balance in the hydrosphere. In: Nuclear Geology, ed. H. Foul, pp. 120—127. New York: Wiley & Sons; London: Chapman & Hall Ltd. 1956.

Libby, W.F.: Atmospheric helium three and radio-carbon from cosmic radiation. Phys. Rev. **69**, 671—672 (1946).

Ljunggren, K., L.G. Erwall, J. Rennerfelt and T. Westermark: Tracing of water flow by means of radioactive isotopes and scintillation counters. Int. J. Appl. Rad. and Isotopes **5**, 204—212 (1959).

Parker, F.L.: Radioactive tracer studies. Trans. Amer. Geophys. Un. **39**, 434—439 (1958).

Svantesson, N.L., och M. Sundberg-Falkenmark: Spridningsproblem i sjöar och vattendrag. Tekn. Tidskr. **42**, 1169—1174 (1959).

Radioactive Methods of Age Determination

by

WALTER R. ECKELMANN

With 1 Figure

Zusammenfassung

Zur Altersbestimmung auf radioaktiver Grundlage kommen vorwiegend die vier folgenden Methoden zur Anwendung: Kalium-Argon-Methode, Rubidium-Strontium-Methode, Uran-Blei-Methode und Kohlenstoff 14-Methode. — Der Artikel gibt einen Überblick über Grundlagen, Anwendungen und Grenzen der einzelnen Methoden. Die Hauptergebnisse werden kurz besprochen.

A. Introduction

In the past, many attempts have been made by scientists to evaluate the age of rocks in order to better understand earth processes that have taken place over the range of geologic time. Methods have varied from studies of varves and rates of sedimentation to the use of radioactive decay schemes. The discovery of radioactivity in the 1890's which provided the basis for truly quantitative chronometry was immediately followed by the publication of so-called "chemical ages" (BOLTWOOD 1907). In fact, the uranium and lead analyses of HILLEBRAND (1891) and v. FOULLON (1883) on uraninite from the Spruce Pine Pegmatite District of North Carolina have been shown to yield rather acceptable ages in view of more recent isotopic data (ECKELMANN and KULP 1957). However, the early data were, in general, subject to considerable errors since the measurements did not include isotopic analyses of lead nor any correction for the thorium decay scheme which can also account for the production of one of the lead isotopes, Pb^{208}.

Qualitatively, the early measurements established the extensive length of geologic time but it was not until the classic work of ASTON (1933) and NIER (1939) that geochronometry entered into the era of truly quantitative measurements.

The availability of enriched stable isotopes and of radioactive materials as well as advances in instrumentation involved in the measurements of isotopes and their radioactive emanations have considerably aided in the development and understanding of most of the radioactive methods of age determination presently in use.

Numerous decay schemes are currently being investigated by laboratories throughout the world today. However, a number have, as yet, not been established as reliable chronometers and will not be considered in this paper. The only decay schemes that will be discussed in view of recent developments will be the carbon-14 decay to nitrogen-14, the uranium to lead decay scheme, the rubidium[87]-

strontium[87] age method, and the potassium[40]-argon[40] decay scheme. These chronometers, in use at laboratories throughout the world today, have considerable value in the studies of sedimentary, metamorphic and igneous rocks and their relationship to earth history.

B. The carbon-14 chronometer

The discovery of natural radiocarbon by Libby and his coworkers (Anderson et al. 1947) produced one of the most significant advances in geochronometry subsequent to the discovery of radioactivity by Becquerel. Radiocarbon has been particularly useful in dating relatively recent events because of its short $(5568 \pm 30$ yr) half-life and it appears that the maximum range of this technique with the most refined methods will permit dating to approximately 70000 years.

The interaction of cosmic rays with the upper atmosphere causes numerous nuclear reactions to take place. During these events, neutrons are produced some of which in turn interact with available nitrogen-14 to produce carbon-14. The carbon-14 atoms thus formed become rapidly mixed in the atmosphere and the oxidized carbon-14 enters into the photosynthetic cycle of plants, animals and air and equilibrates with the bicarbonate in surface ocean water.

Therefore, when a substance is removed from its source of carbon-14 with which it has equilibrated, the equilibrium concentration decays away at a known half-life. The ratio of C^{14}/C^{12} of a substance at any time relative to its equilibrium ratio is a measure of the time interval that has elapsed.

Libby (1952) has summarized his classic research on this subject in his book entitled "Radiocarbon Dating". Since then numerous analyses have been published in a large variety of journals. The American Journal of Science now publishes a supplement which lists the data of laboratories currently active in this field.

Inherent to the carbon-14 method are a number of assumptions which have been investigated in recent years. First, mixing of the cosmic ray produced C^{14} is rapid compared to the minimum age limit of the method. Data of Libby (1952) support the contention that living materials such as trees, etc., from widely different latitudes and altitudes have similar C^{14}/C^{12} ratios within 5% or so suggesting rapid exchange compared to the half-life of radiocarbon. Recent data published by Hagemann et al. (1959) resolve this question with rather high precision. Nuclear devices produce carbon-14 because of the accessibility of neutrons produced during fission and their subsequent interaction with available nitrogen-14. On the basis of high altitude balloon tests used to sample the stratosphere, Hagemann et al. (1959) conclude that bomb produced C^{14} has a five year atmospheric mean residence time. This value was obtained by comparing the annual addition of bomb produced C^{14} to the troposphere with the stratospheric inventory. Data reported by Feeley (1960) on the strontium-90 content of the stratosphere suggest an 18 month mean residence time. Which ever of these values more closely approximates the true value, there appears little question that the carbon-14 produced in the upper atmosphere is rapidly brought into the troposphere where prevailing winds would cause it to be mixed. Fergusson (1958), using data relating to the distribution of industrial CO_2, has concluded that the mixing rate between the Northern and Southern Hemispheres is less than two years. The comparable increase in the carbon-14 activity of vegetation and atmospheric CO_2 collected in areas away from industrial produced CO_2 as reported by Broecker and Walton (1959) infers rapid exchange between the atmosphere and vegetation.

Another assumption that has been made in carbon-14 dating is that the cosmic ray flux, which is a controlling factor in the production of carbon-14 in the upper atmosphere, has been constant. ARNOLD and LIBBY (1949) demonstrated by measurements of historically and tree ring dated samples that variations in the flux have not exceeded some 5% over the past 5000 years.

However, ELSASSER, NEY and WINCKLER (1956) have estimated that the magnetic field intensity has been sufficient to reduce the carbon-14 production rate thus altering apparent ratios of C^{14}/C^{12} in the past. Inherent in this estimate is the assumption that residual magnetism measurements made by THELLIER are representative of the entire earth. W.S. BROECKER, E.A. OLSON, and J. BIRD (1959) have discussed this problem and have reported a number of carbon-14 ages for well dated samples. Most of the samples were of wood that was dated by ring counting as well as by carbon-14. They concluded that for samples younger than 2000 years, the radiocarbon age will lie within 250 years of the true age. The inability to truly define the extent of the variations due to cosmic ray flux changes is mainly due to the limited number of historically and tree ring dated samples. A parallel method of age dating independent of the cosmic ray flux and with comparable precision would provide a unique check of this assumption. This would be particularly helpful for the period 5000 years B.P. and beyond where there is essentially no data to demonstrate the constancy of the cosmic ray flux.

Another assumption that has been made is that the carbon-12 reservoir with which the carbon-14 is diluted has remained constant. Variations in the reservoir could also alter the so-called contemporary value (specific activity) of modern carbon thus producing the same apparent effect as that resulting from fluctuations in the cosmic ray flux. One known cause of addition of carbon-12 is the effect of industrial fuel. This has been termed the "Suess" effect (SUESS 1953, 1955). SUESS noted that the C^{14} specific activity (dpm/g) had decreased over the last 50 years as shown by measurements of wood. This is known to be due to the introduction of C^{14}-free CO_2 into the atmosphere by industrial fuel combustion. REVELLE and SUESS (1957) have compared C^{14}/C^{12} and C^{13}/C^{12} ratios in wood and in marine material and have concluded that the average lifetime of a CO_2 molecule in the atmosphere before it is dissolved in the sea is approximately 10 years.

Another mechanism that could cause a change in the atmospheric C^{14}/C^{12} ratio would be varying rates of surface ocean uptake. If the rate of exchange of atmospheric CO_2 with the surface waters and terrestrial organisms is rapid compared to the rate of circulation of surface and deep water, then this is a minor problem. On the other hand, if the rate of formation of low C^{14}/C^{12} waters can be accelerated at various times, and these waters brought to the surface, the atmospheric C^{14}/C^{12} ratio would show a corresponding drop. BROECKER et al. (1958) have observed no C^{14}/C^{12} changes which relate to this latter mechanism.

LIBBY (1952) reported C^{14}/C^{12} ratios which were consistent with historical ages ascribed to the samples. This pointed out in a very general way that contamination of various sorts was not a serious problem. However, improved counting techniques reported by numerous investigators plus the considerable extension of the range of this dating method since LIBBY's pioneering work have made an evaluation of this problem a real necessity. HARING et al. (1958) have enriched samples using thermal diffusion columns in order to extend the maximum age of their dating apparatus. They report apparent ages up to 73000 years. The importance of a control on possible contamination in this range is an absolute necessity. For example, a sample with a true age of 40000 years which is contaminated by 20% modern carbon will give an apparent age of approximately 12900 years. This effect becomes more pronounced as the true age increases.

Olson and Broecker (1958) have discussed methods of detecting contamination in bone carbon, carbonates, organic rich soils and wood. Although multiple sampling and C^{14}/C^{12} analyses provide cross checks, this method of approach does not necessarily establish the true age. The authors report data on humic acid ages as compared to the treated sample ages for soils. They also compare multiple samples from locations where the samples in a number of cases were pretreated prior to isotopic analyses. Olson and Broecker (1958) conclude that several per cent contamination of buried samples is the exception and not the rule. However, since the effect of contamination is dependent on the age of a particular sample, it must be demonstrated that contamination is not present or has been removed particularly for samples older than 25000 years.

The early data of Libby (1952) and his co-workers as well as that of other laboratories was obtained by counting solid carbon mounted inside a Geiger tube. This method had an apparent age limitation of 20000 years although the addition of a mercury shield inside an anticoincidence ring as described by Kulp and Tryon (1952) extended the range to 30000 years. Present day techniques as described by Fergusson (1955), de Vries (1958), and Broecker (1958) generally consist of determining the natural carbon-14 by gas proportional counting of CO_2. Methane also is used and generally gives longer counting plateaus. However, methane, with its four hydrogens, gives a higher background than CO_2 due to interactions with neutrons. CO_2 is also a more convenient gas to work with for high precision analyses because it can be run directly into a mass spectrometer for C^{13}/C^{12} measurements. Suess (1955) has used acetylene gas for C^{14} analyses but this has the disadvantage of being explosive at higher pressures due to the triple bond in the acetylene structure.

Because of the relatively short time interval in which the carbon-14 method can be applied (present — 45,000 to 50,000 years), it finds considerable application to studies of glacial geology, oceanography, recent sedimentation, archaeology and anthropology to name a few.

One interesting use of C^{14} data has been in dating bones. Certainly the precise definition of the age of many of the better skeletal finds has proved rather difficult. In 1950 the Piltdown mandible and cranial bones were shown to be much younger than the Lower and Middle Pleistocene ages assigned to them (K. P. Oakley and C. R. Haskins 1950). de Vries and Oakley (1959) have reported carbon-14 analyses of the bone protein (collagen). After correcting for the Suess effect, the Piltdown mandible gave an age of 500 ± 100 years while the Piltdown skull gave a value of 620 ± 100 years. These data support the initial statements that the Piltdown finds were part of a hoax. Even so, some investigators claimed that mandible and skull were mid-Pleistocene age but the C^{14} analyses have apparently confirmed the flourine data of Oakley.

Another application of carbon-14 has been the dating of deep sea sediments. Broecker et al. (1958) have reported C^{14} ages on a mid-Atlantic core for which there was no evidence of erosion, carbonate solutioning or turbidity currents. Erickson and Wallin (1956) have shown temperature variations as a function of depth and broadly consistent with O^{18}/O^{16} paleotemperature measurements reported by Emiliani (1955). The major temperature change over the last 30000 years occurred some 11000 years ago and it was accompanied by a change in both the sedimentation rates of the carbonate and clay fractions. Consequently, any attempts at dating Pleistocene deep ocean sediments on the basis of an average rate of deposition result in eroneous data. Broecker et al. (1958) conclude that although rates of sedimentation do vary among the cores of the Atlantic equatorial region, a constant relationship between the rates of sedimentation during glacial and non-glacial periods exists. They conclude that the beginning of the last cold period took place some 65000 to 81000 B.P. They also estimate the beginning of the warm period preceeding the last cold water stage at 150000.

One of the most interesting areas to which carbon-14 can be applied is that of ocean water circulation. The first values for deep ocean water were reported by KULP et al. (1952). Of the three samples listed, a surface water at $54°35'41°$W gave a modern value by the black carbon counting method. Two other samples ($58°19'$N $32°56.8'$W at 1829 m depth and $53°52.6'$N $21°06'$W at 2743 m) gave "ages" of 1600 ± 130 and 1750 ± 150 years. Although later work showed these values to be subject to some error resulting from CO_2 extraction problems involving ascarite, the data did point up the existence of deep ocean waters decidedly not in equilibrium with the atmosphere.

BROECKER et al. (1959) have discussed methods of analyzing oceanic bicarbonate. By applying the radiocarbon method to various ocean water masses, it will be possible to define the average residence times, rates of mixing, and rates of travel of specific water bodies. The increased C^{14} due to nuclear weapons testing should be very useful in evaluating the problem.

As has been pointed out earlier, the lowering of the specific activity of carbon as the sample becomes progressively older essentially limits the maximum age that can be determined by carbon-14. HARING et al. (1958) converted the combusted CO_2 to CO. The CO was then enriched in a series of five thermal diffusion columns connected in parallel. The enrichment was calculated from the O^{18} abundance in the CO. They then converted the gas back to CO_2 and counted it in the proportional region. They were able to enrich the C^{14} by a factor of 16 over a period of 2 months. Using this technique, they were able to date samples in the 60000 to 70000 year range. Certainly, the enrichment procedure is well justified in some cases but it is also very time consuming and expensive. It is plagued by problems of contamination.

C. Potassium-40, argon-40 dating

Although radioactivity was discovered in the latter 1890's, it was not until 1937 that WEISZACKER (1937) suggested the possibility of the formation of A^{40} from K^{40}. He postulated that the high A^{40} in the atmosphere relative to other inert gases was the result of the K^{40} which was known to occur in igneous rocks.

Since the suggestion of WEISZACKER, there has been extensive research on both the constants involved in the K^{40} decay scheme and the A^{40} and Ca^{40} produced by K^{40}. The isotope K^{40} decays dually producing stable Ca^{40} by negative beta emission and stable A^{40} by K capture to an excited state of A^{40}, followed by gamma ray emission to its ground state (Fig. 1). WETHERILL (1957) reviewed previous physical measurements and also measured the γ/g sec value for K^{40}. He determined the efficiency of the counter for the K^{40} produced gammas by calibrating the counter with gamma rays from Co^{60} and Na^{24} sources of known specific activity. WETHERILL's (1957) reported value of $3.39 \pm 0.12\ \gamma/g$ sec is the generally accepted value at the present time. This gives a value for λ_e, the decay constant for electron capture to A^{40}, of 0.585×10^{-10} y^{-1}. The generally accepted value of the specific beta activity is $27.6\ \beta/g$ sec (WETHERILL 1957) which is equivalent to a partial beta decay constant of 4.72×10^{-10} y^{-1}. The choice of this constant is not very critical to the age. This point has been demonstrated graphically by ALDRICH and WETHERILL (1958).

Potassium has three naturally occurring isotopes, K^{39}, K^{40}, and K^{41}. Based on a mass spectrometric absolute determination of the relative abundances, Nier (1950) reports values of $K^{39} = 93.08\%$, $K^{40} = 0.0119\%$, and $K^{41} = 6.91\%$. K^{40} is the only radioactive isotope of the three.

The value of t in the calculation of the age can be expressed in the following manner:

$$t = \frac{2.303}{\lambda_\beta + \lambda_e} \log\left(1 + \frac{\lambda_\beta + \lambda_e}{\lambda_e}\, \frac{A^{40}}{K^{40}}\right)$$

where λ_e is the decay constant for electron capture to A^{40} and λ_β is the decay constant for the beta decay of K^{40} to Ca^{40}. The K^{40} content is calculated on the basis of the ratio of K^{40}/total K of 0.0119%. Investigators have assumed that this value is a constant. This is substantiated by measurements of other isotopes in the same general mass range which undergo valence changes but show little if any isotopic variations. However, data of Taylor and Urey (1938) on zeolites suggest that slight fractionation may take place. This point has been checked (Kendall 1960).

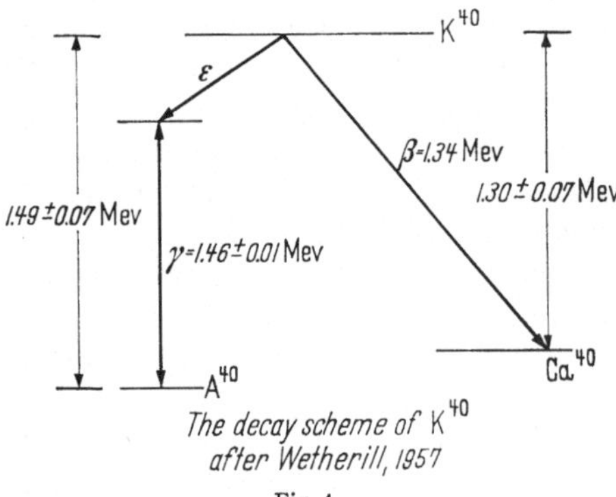

The decay scheme of K^{40}
after Wetherill, 1957

Fig. 1

Since the K^{40} content can be calculated on the basis of the total potassium, conventional methods of analyses can be used. Many laboratories have used flame photometers with internal standards. However, highly accurate analyses are somewhat difficult and this has led numerous laboratories to do their potassium analyses by isotope dilution techniques employing solid source mass spectrometry.

Most laboratories now utilize the isotope dilution technique when analyzing argon. This procedure has been described by Wasserburg (1954). The method involves utilizing a spike gas enriched in argon 38 such that the A^{38} constitutes some 90% or more of the gas. Air contamination is measured by the relative abundance of argon-36. The release of the argon from the mineral is generally carried out with a NaOH fusion system or by R.F. induction furnace heating. Both methods appear to give comparable results.

The decay of K^{40} to A^{40} is particularly attractive as an isotopic chronometer because of the relatively widespread occurrence of potassium minerals in continental rocks and the relatively high concentration of potassium in basalt. Also, the half-life is ideal for dating almost the entire range of geologic time and the weak nature of the disintegrations should not result in accelerated argon leakage due to radiation damage.

In recent years, there has been particular emphasis on the evaluation of various minerals as precise indicators of the age of a particular rock or mineral. Recently Evernden et al. (in press) and Hurley et al. (in press) have published data on the reliability of glauconite and illite as means of dating sediments. Hurley et al. (1959) report that illite and potassium bearing materials in the Rappahannock River sediments have an age greater than 300 m.y. while aggregate materials of the Mississippi Delta show ages in excess of 200 m.y. On the basis of K/A analyses of deep ocean cores, the authors also suggest that the illite observed in these sediments has been air borne from continental areas.

Generally speaking, the bulk of the literature suggests an authigenic origin for glauconite. This is not true in the case of clay minerals in shales (Weaver

1958). WEAVER believes that most of the clay minerals found in sedimentary rocks are of a detrital origin and that the clays are only slightly colored subsequent to deposition.

HURLEY et al. (in press), reporting a total of 51 samples from the Quaternary through the lower Paleozoic, note that the average potassium content of the Lower Paleozoic glauconites is 6.1 % and that the average value decreases regularly to the Quaterary (only 2 reported samples) with an average K content of 3.0%. HURLEY, on the basis of these data, states that the possibility of changes in the structure and composition of glauconites during later diagenesis cannot be overlooked. The addition of potassium to a structure with time would make the age appear too young. The data of EVERNDEN et al. (in press) support this observed trend. Addition of K in the amounts suggested by HURLEY and EVERNDEN's data and added over a long period of time would only alter the ages by some 3 to 4%.

EVERNDEN et al. (1960) have demonstrated by studies of diffusion rates of argon that the addition of water and potassium to glauconite accelerates the diffusion and, therefore, loss of argon. Some authors have attempted to compare the rubidium-strontium and potassium-argon ages of older samples to check this point. However, not very much is understood of the influence of ground waters on rubidium-strontium ages and the corresponding effect on potassium-argon ratios.

EVERNDEN et al. (1960) have studied the influence of depth of burial and the associated increase in temperature. They concluded that it is possible to obtain useful potassium-argon ages from glauconites over the past 500 m.y., but point out the need for fulfilling certain specific conditions of geologic history. Potassium-argon ages of illites analyzed at California do not give exceptionally old ages. In general, they yield ages within 10% of the apparent true age or younger.

LONG et al. (1959) have dated samples in a 10 mile transition zone across a structural trend in Western North Carolina and find that where complete recrystallization has occurred the potassium-argon, uranium-lead and rubidium-strontium ages (ECKELMANN and KULP 1957; ALDRICH, WETHERILL, DAVIS and TILTON 1958) show good agreement at 360 m.y. The average potassium-argon age appears slightly less. However, proceeding N-NW toward Pardee Point and cutting across the structural trend, both potassium-argon and rubidium-strontium ages increase reaching 890 m.y. for the rubidium-strontium values, and 800 m.y. for potassium-argon. It appears that the original metamorphic complex in the Southern Appalachians formed some 1100 m.y. ago at approximately the same time as the Grenville Province of the Canadian Shield (ECKELMANN and KULP 1957, ALDRICH et al. 1958). LONG et al. (1959) also conclude that the major metamorphic event now displayed in the plutonic metamorphic complex of the Southeast Piedmont and Blue Ridge occurred about 350 m.y. ago, slightly later than the major Paleozoic orogeny in the Northern Appalachians which appears to have begun about 300 m.y. ago.

The continued effort in dating potassium bearing rocks and minerals has pointed out the sensitivity of mineral "ages" to their crystal structure and to their geological history. A single potassium-argon date simply sets a minimum age for the rock unit with which it is associated. Multiple samples permit a detailing of the history of an area.

One of the outgrowths of potassium-argon and other radioactive dating methods has been the further evolution of the so-called geological time scale. Because of recent data, the Holmes (1947) time scale has had to be modified. KULP (1959), KULP et al. (1959) and HOLMES (1960) have published time scales which make use of the most recent data. The effect has been to give a scale about 15% longer than the 1947 Holmes scale before the Cretaceous and to considerably increase the estimated lengths of the Jurassic, Permian and Carboniferous.

As pointed out earlier, one of the advantages of the potassium-argon method is its application to Pleistocene geology. As an outgrowth of the work of Reynolds (1954) in developing a high sensitivity mass spectrometer useful for inert gases, investigations into dating of Pleistocene rocks have begun (Evernden 1959).

Over the last 5 years or so, a considerable amount of potassium-argon, rubidium-strontium and uranium-lead data has been published on the age of various types of meteorites. These data seem to point to an age of formation of chondrites of approximately 4.6×10^9 years. Recently, Reynolds (1960) has observed the presence of Xe^{129} in the Richardton chrondritic stone meteorite. Assuming that I^{129} and I^{127} were formed in equivalent abundances, I^{129} decay to Xe^{129} as observed by Reynolds (1960) is sufficient to account for a time interval between nucliogenesis and formation of the meteorite of $(0.35 \pm 0.06 \times 10^9)$ years. This would point to an age of the elements of close to 4.95×10^9 years.

D. Rb-Sr dating

Four main isotopes of strontium occur in nature. They are Sr^{84}, Sr^{86}, Sr^{87}, and Sr^{88}. Bainbridge and Nier (1950) and Herzog et al. (1953) have determined their relative abundances. These data were published following the suggestion of Goldschmidt (1937) that the Rb^{87} to Sr^{87} decay could be used as a geochronometer.

One of the main problems in using the Rb^{87} decay scheme has been an accurate value of the associated half-life. Aldrich and Wetherill (1958) have reviewed the literature showing the rather wide range of values obtained by physical measurements. The value that has been accepted by most laboratories is that of Aldrich et al. (1956). Recently Flynn and Glendenin (1959) have redetermined the half-life using a 5 inch liquid scintillation spectrometer for their differential pulse height analysis of the Rb^{87} beta spectrum. They obtained a Rb^{87} half-life value of $47.0 \pm 0.5 \times 10^9$ years. This corresponds to a Rb^{87} decay constant of 1.47×10^{-11} y^{-1}. It would appear at this time that the 47.0×10^9 year value is the most reliable Rb^{87} half-life value published in the literature. Most laboratories are currently using the 50.0×10^9 year half-life.

Table 1 lists a number of K-A, Rb-Sr and uranium-lead ages from localities where the absolute age of the rocks are well known. In all cases the samples represent areas where concordant or nearly concordant ages have been obtained and the apparent age via lead-uranium data has been listed under the U-Pb age heading. From these and other data, it is quite apparent that the potassium-argon ages are lower than the corresponding rubidium-strontium values. In fact a review of the literature shows that the potassium-argon ages are generally 6 to 9% lower than Rb-Sr ages (Gast et al. 1958, Wetherill et al. 1956, Wasserburg et al. 1956). This difference may be in part due to uncertainties in the decay constants now in use. In areas such as the Bighorn Basin, Gast et al. (1958) have also observed very real differences between the potassium-argon and rubidium-strontium ages suggesting argon loss. In this case it seems likely that at least part of the argon loss may be due to plutonic or orogenic activity in the area during late Mesozoic and early Cenozoic time. Generally, potassium-argon ages can be viewed as minimum ages. To establish absolute ages it is necessary to compare different mineral systems with one another by several age methods.

Table 1. *A comparison of Rb-Sr, K-A, and U-Pb ages (in m. y.)*

Locality	U-Pb	K-A	Rb-Sr
Spruce Pine, N.C. . . .	361 ± 28[1]	musc. 340 ± 15[2]	375[2] musc. 385[2] feldspar
Bob Ingersoll Mine, Keystone, S.D.	1620[1]	musc. 1550[3] Lep. 1380[3]	1650[3] musc. 1730[3] Lep.
Viking Lake pegmatite, Saskatchewan, Canada.	1910[1,2]	biotite 1780[3]	1970 biot.[3]
Huron Claim, Manitoba .	2650[1,4,5]	musc. 2150[5] albite 2030[5]	

[1] Average of eight samples (ECKELMANN and KULP 1957).
[2] LONG et al. (1959).
[3] ALDRICH et al. (1958).
[4] NIER et al. (1941).
[5] GAST et al. (1958).

Another example of comparable rubidium-strontium and potassium-argon ages within the limits of the experimental errors has been reported by KULP et al. (1960) for Hercynian and Caledonian granites. A Rb[87] half-life of $4.7 \pm 0.1 \times 10^{10}$ years was used rather than the 5.0×10^{10} years used by most laboratories.

ALDRICH and WETHERILL (1960) have recently published potassium-argon and rubidium-strontium ages of lepidolites and biotites from rocks in Ontario and Northern Minnesota. Samples of biotite give comparable rubidium-strontium and potassium-argon ages with the exception of a single lepidolite from the Silverleaf Mine pegmatite where the potassium-argon age is considerably younger. The data show that 2600 m.y. rocks extend over the entire width of the Province of Ontario. Similar ages have been reported for parts of Manitoba.

HART (1960) has reported data on the influence of metamorphism on the common strontium correction. Biotites and feldspars recrystallized during a period of metamorphism have higher than "normal" Sr^{87}/Sr^{86} ratios after recrystallization. Using normal procedures for calculating the contribution of Sr^{87} not produced in the mineral yields discrepant dates. HART (1960) has demonstrated that the age of metamorphism can be determined from a plot of the 87/86 ratios of various minerals versus age, and an age of crystallization can be obtained by a gross analysis of the whole rock. This age is valid if the rock has been maintained as a closed system which started with a normal Sr^{87} abundance. He has also pointed out the apparent usefulness of hornblende for age dating. It may also be true that tremolite is an excellent mineral for dating purposes.

GOLDICH et al. (1959) have reported potassium-argon and rubidium-strontium ages on illitic shale from the Siyeh formation of the Belt series, Montana. The sample was completely converted to illite and gave potassium-argon and rubidium-strontium ages of 740 and 780 m.y. respectively. These ages significantly differ from uranium-lead and lead-lead isotopic ages reported on samples of pitchblende which intrude the Belt Series at the Sullivan Mine, Idaho (ECKELMANN and KULP 1957) (Table 2). In both cases, the Pb^{206}/U^{238} and Pb^{207}/U^{235} were discordant. FARQUHAR and CUMMINGS (1954) have reported the isotopic composition of lead from galenas over a vertical range of 3500 feet in the Sullivan Mine. The common lead data suggest an age of 1030 ± 290 m.y. It appears, therefore, that the 1190 m.y. pitchblende Pb^{207}/Pb^{206} age is a minimum age for the Belt Series and that the rubidium-strontium and potassium-argon ages register the effect of some later metamorphic event.

Table 2. *Isotopic analyses of Belt series samples*

Sample	Location	Rb-Sr *	K-A*	Pb^{207}/Pb^{206} **
Pitchblende	Sullivan Mine, Idaho	—	—	1190 ± 30 m.y.
Pitchblende	Sullivan Mine, Idaho	—	—	1035 ± 35 m.y.
Illite	Siyeh Frm., Montana	740 m.y.	780 m.y.	—

* GOLDICH et al. (1959).
** ECKELMANN and KULP (1957).

The more detailed knowledge of the rubidium half-life plus continued research on the distribution of strontium in various minerals will greatly expand the usefulness of rubidium-strontium and potassium-argon ratios as indicators of not only the age but of the history of minerals and rock systems. Bulk rock analyses will also add considerable information.

E. Uranium lead method of age determination

The advances in the understanding of the uranium-lead method of age determination that have occurred during the past ten years have made this method the primary means of calibration for rubidium-strontium and potassium-argon dating. The relative importance of lead, uranium, and thorium loss, radon leakage and the common lead correction can be well defined in many cases, making it possible to interpret discordant ages as well as concordant ages in terms of the time of formation of the mineral and its subsequent history. Since each radioactive mineral appears to show a unique pattern of discordance if altered (KULP and ECKELMANN 1957), the analysis of more than one mineral type from a locality provides much interlocking data which can be used to estimate times of formation and alteration.

Table 3. *Equations and constants used in the uranium-lead method*

$$\frac{N_{207}}{N_{206}} = \frac{N_{235}}{N_{235}} \frac{e^{\lambda_{235} T} - 1}{e^{\lambda_{238} T} - 1}$$

$$= \frac{1}{137.7} \frac{e^{+.9802 \times 10^{-9} T} - 1}{e^{+.1543 \times 10^{-9} T} - 1}$$

$$\frac{N_{206}}{N_{238}^{0}} = 1 - e^{-\lambda_{238} T} = 1 - e^{-.1543 \times 10^{-9} T}$$

$$\frac{N_{207}}{N_{235}^{0}} = 1 - e^{-\lambda_{235} T} = 1 - e^{-.9802 \times 10^{-9} T}$$

$$\frac{N_{208}}{N_{232}^{0}} = 1 - e^{-\lambda_{232} T} = 1 - e^{-.4986 \times 10^{-10} T}$$

$T_{\frac{1}{2}}(U^{238}) = 4.49 \pm 0.01 \times 10^{9}$ years

$T_{\frac{1}{2}}(U^{235}) = 7.07 \pm 0.16 \times 10^{8}$ years

$\dfrac{N_{238}}{N_{235}} = 137.7 \pm .5$

$T_{\frac{1}{2}}(Th^{232}) = 1.39 \pm 0.02 \times 10^{10}$ years

N_{238}^{0}, N_{235}^{0}, — number of atoms at $T = 0$

N_{238}, N_{235}, etc. — numbers of atoms at present

Because of the radioactive decay of U^{238} to Pb^{206}, U^{235} to Pb^{207}, and Th^{232} to Pb^{208}, via intermediate daughters, the ratio of any of the parent nuclides to the final stable lead isotope is a function of the age of a mineral (NIER 1939). The equations and constants used in the calculations of the various isotopic ages are listed in Table 3. Since the lead isotopes produced by the parent nuclide decay form at different rates to one another, the ratio of Pb^{207}/Pb^{206} is also an index of the age.

In practice, Pb^{204}, the only stable lead isotope not produced by the radioactive decay of a uranium or thorium isotope, is observed in the isotopic analysis.

This is, in part, due to slight lead contamination during physical and chemical processing. Some of the Pb^{204} and therefore some Pb^{206}, Pb^{207} and Pb^{208} are incorporated into the lattice of the uranium mineral. Some of the observed "contamination" is due to associated minerals which are not completely separated in some cases. Since there is observable Pb^{204} in the isotopic analyses of lead separated from its associated uranium bearing mineral, it is necessary to correct the Pb^{206}, Pb^{207}, and Pb^{208} abundances for that not produced by parent nuclide decay in the mineral. In numerous cases, associated lead minerals can give the so-called common lead correction. Where this is not possible, an estimate of the age can permit the evaluation of the common lead correction based on the present day knowledge of the variations of common lead throughout geologic time (BATE et al. 1957). Such a correction assumes the common lead constitution

Table 4. *Isotopic ages from some pitchblends and uraninites*

Locality	206/238	207/235	207/206	208/232	Reference
Colorado Front Range[1] (Sunshine Mine)	805	860	1035	—	ECKELMANN and KULP (1956)
Great Bear Lake[1] . . (Eldorado Mine)	1220	1275	1395	—	NIER (1939)
Belgian Congo[1] . . . (Katanga)	616	614	610	—	NIER (1939)
(Katanga)	575	595	630	—	ECKELMANN and KULP (1956)
South Dakota[2] . . . (Black Hills)	1580	1600	1630	1440	TILTON (1956)
S.E. Manitoba[2] . . .	1564	1985	2475	1273	NIER (1939)

[1] Samples of pitchblende.
[2] Samples of uraninite.

for a given period of geologic time to be representative of the common lead associated with the mineral analysis. Where no thorium is present with the uranium mineral, the Pb^{208} abundance can be used for a more accurate assessment of the common lead correction.

Table 4 lists uranium-lead ages that have been reported in the literature for a number of locations. Although these are a limited number of localities and mineral types, they illustrate that concordant ages via the various decay schemes are rarely found and any attempt to use a single U-Pb age measurement as an index of the absolute age can lead to a completely false interpretation of the geologic framework. In general, uraninites give the most consistent agreement between the Pb^{207}/Pb^{206}, Pb^{206}/Pb^{238}, and Pb^{207}/Pb^{235} ages. Thorium-lead ages are anomalous and have never been satisfactorily explained. In some cases, the thorium age is subject to a large error introduced by the common lead correction, but this explanation is unsatisfactory for a number of determinations that have been made.

The concordance of the Pb^{206}/U^{238}, Pb^{207}/U^{235} and Pb^{207}/Pb^{206} ages gives a rather high degree of confidence in such data. However, as pointed out, most uranium bearing minerals do not show a concordant pattern.

An example of widely divergent ages from a given locality has been demonstrated by ECKELMANN and KULP (1956) in a study of the Lake Athabasca pitchblende deposits. A suite of samples of pitchblende, clausthalite and galena from various mines in the Lake Athabasca uranium province were analyzed isotopically and chemically to obtain the various uranium-lead and lead-lead ages. Pb^{210}/Pb^{206} measurements (KULP et al. 1954) were made to check equilibrium in the decay series over the last 100 years. Apparent isotopic ages ranged from 220 to 1860 m.y. The samples showed no outward signs of weathering and essentially all of the comparative ages showed effects of apparent loss of radiogenic lead. Support of this point is evidenced by the appearance of moderate amounts of radiogenic lead in clausthalites associated with the pitchblende.

The diverse ages that were observed can be calculated to have been the result of a single period of pitchblende deposition at 1.90 ± 0.04 b.y. ago followed by two periods of exsolution of lead at about 1.2 and 1.5 b.y. This hypothesis is consistent with the Pb^{207}/Pb^{206} ratio of radiogenic lead in the clausthalites and galenas which apparently formed at the time of recrystallization. The data, however, do not preclude more than two periods of lead exsolution. It is quite certain, however, that the initial uranium deposition occurred about 1.9 b.y. ago, that there have been at least two periods of lead exsolution, that one period of lead exsolution must have occurred subsequent to 0.22 b.y. ago and that another period of lead exsolution probably occurred about 1.2 b.y. ago.

Geologically, the results of this study suggest that the basement complex of the Lake Athabasca region is older than data assigned to the initial pitchblende intrusion (1900 m.y.).

Wetherill (1956) has demonstrated a means for graphically plotting so-called "concordia" and the locus of points representing discordant ages resulting from varying amounts of uranium and lead loss or addition. The locus of points for the plot of the mole ratios of Pb^{207}/U^{235} on the abscissa and Pb^{206}/U^{238} on the ordinate is actually a straight line (Wetherill 1956). It is apparent that for samples with rather young ages, the ages will appear concordant and extreme caution in interpretation of these ages must be practiced.

Tilton (1956) has shown via leaching experiments of thorium and uranium minerals that a certain lead isotope can be leached preferentially and that the leached material is isotopically and/or chemically consistent with the distribution of ages observed in the mineral. His data parallels the concept of parent-daughter fractionation occurring in many minerals previously analyzed for their age.

Aldrich et al. (1958) have reported potassium-argon and rubidium-strontium ages on biotites, muscovites and lepidolites as well as uranium-lead data on numerous samples of zircon, monazite and microlite. In almost all cases the uranium-lead data are discordant but potassium-argon and rubidium-strontium ages of mica from pegmatites are found to agree with concordant uranium-lead ages of associated zircons in granites or uraninites and monazites.

F. Conclusion

The field of absolute age dating has continued to be an active field of research which bears on many geological, geophysical and geochemical problems. The improved decay constants, a greater understanding of discordant age patterns as well as the extended ranges of carbon-14 and potassium-argon dating continue to make this field of research very important to the understanding of the evolution of the earth.

References

Aldrich, L. T., and G. W. Wetherill: Geochronology by radioactive decay. Annual Rev. Nucl. Sci. **8**, 257 (1958).
— — G. L. Davis and G. R. Tilton: Radioactive ages of micas from granitic rocks by Rb-Sr and K-A methods. Trans. Amer. Geophys. Un. **39**, 1124 (1958).
Anderson, E. C., W. F. Libby, S. Weinhouse, A. P. Reid, A. D. Kirshenbaum and A. v. Grosse: Radiocarbon from cosmic radiation. Science **105**, 576 (1947).

ARNOLD, J.R., and W.F. LIBBY: Age determinations by radiocarbon content. Checks with samples of known age. Science **110**, 678 (1949).

ASTON, F.W.: The isotopic constitution and atomic weights of lead from different sources. Proc. Roy. Soc. Lond. **140**, 535 (1939).

BAINBRIDGE, K.T., and A.O. NIER: Relative isotopic abundances of the elements. Preliminary reports No. 9, Nuclear Science Series. Nat. Res. Council, U.S., Washington, D.C., 1950.

BATE, G.L., D.S. MILLER and J.L. KULP: Isotopic analysis of tetramethyl lead. Analyt. Chem. **29**, 84 (1957).

BROECKER, W.S.: Ph.D. Thesis, Columbia University, 1957.

— M. EWING, R. GERARD, B.C. HEEZEN and J.L. KULP: U.S. Nat. Acad. Sci. and National Res. Council, Publ. **572**, 118 (1958).

— E.A. OLSON and J. BIRD: Radiocarbon measurements on samples of known age. Nature, Lond. **183**, 1582 (1959).

— K.K. TUREKIAN and B.C. HEEZEN: The relation of deep sea sedimentation rates to variations in climate. Amer. J. Sci. **256**, 503 (1958).

—, and A. WALTON: Radiocarbon from nuclear tests. Science **130**, 309 (1959).

BOLTWOOD, B.B.: On the ultimate disintegration products of the radioactive elements. Amer. J. Sci. **23**, 77 (1907).

ECKELMANN, W.R., and J.L. KULP: Uranium-lead method of age determination. Part I. Lake Athabasca Problem. Bull. Geol. Soc. Amer. **67**, 35 (1956).

— — Uranium-lead method of age determination. Part II. North American localities. Bull. Geol. Soc. Amer. **68**, 1117 (1957).

ELSASSER, W., E.P. NEY and J.R. WINCKLER: Cosmic ray intensity and geomagnetism. Nature, Lond. **178**, 1226 (1956).

EMILIANI, C.: Pleistocene temperature. J. Geology **63**, 538 (1955).

ERICKSON, D.B., and G. WALLIN: Correlation of six cores from the equatorial Atlantic and Caribbean. Deep Sea Research **3**, 104 (1956).

EVERNDEN, J.F.: Dating of tertiary and pleistocene rocks by the potassium-argon method. Proc. Geol. Soc. Lond. **1565**, 17 (1959).

— G.H. CURTIS, J. ABRADOVICH and R. KICHTER: On the evaluation of glauconite and illite for dating sedimentary rocks. (in Press 1960).

— — R.W. KISTLER and J. ABRADOVICH: Argon diffusion in glauconite, microcline, sanidine, leucite, and phlogopite. Amer. J. Sci. (in Press 1960).

FARQUHAR, R.M., and G.L. CUMMINGS: Isotopic analyses of anomalous lead ores. Trans. Roy. Soc. Canada **48** (1954).

FEELEY, H.W.: Strontium-90 content of the stratosphere. Science **131**, 645 (1960).

FERGUSSON, G.J.: Radiocarbon dating system. Nucleonics **13**, 18 (1955).

— Reduction of atmospheric radiocarbon concentration by fossil fuel CO_2 and the mean life of CO_2 in the atmosphere. Proc. Roy. Soc. Lond. A **243**, 561 (1958).

FOULLON, H. VAN: Über Verwitterungs-Producte des Uranpecherzes. Verh. geol. Reichsanst. **6**, 95 (1883).

FLYNN, K.F., and L.E. GLENDENIN: Half-Life and Beta Spectrum of Rb^{87}. Phys. Rev. **116**, No. 3, 744 (1960).

GAST, P.W., J.L. KULP and L.E. LONG: Absolute age of early precambrian rocks in the Bighorn basin of Wyoming and Montana and southeastern Manitoba. Trans. Amer. Geophys. Un. **39**, 322 (1958).

GOLDICH, S.S., H. BAADSGAARD, G. EDWARDS and C.E. WEAVER: Investigations in radioactivity dating of sediments. A.A.P.G. **43**, 654 (1959).

GOLDSCHMIDT, V.M.: Geochemische Verteilungsgesetze der Elemente. IX. Die Mengenverhältnisse der Elemente und der Atom-Arten. Skr. naturv. Kl. No. 4 (1937).

HAGEMANN, F.T., J. GRAY jr., L. MACHTA and A. TURKEVICH: Stratospheric carbon-14, carbon dioxide and tritium. Science **130**, 542 (1959).

HARING, A., A.E. DE VRIES and H. DE VRIES: Radiocarbon dating up to 70000 years by isotopic enrichment. Science **128**, 472 (1958).

HART, S.R.: Paper presented at the conference on geochronology of rock systems. New York Acad. Sci., March, 1960.

HERZOG, L.F., L.T. ALDRICH, W.K. HOLYK, F.B. WHITING and L.H. AHRENS: Variations in strontium-87 in biotite, feldspar, and celestite. Trans. Amer. Geophys. Un. **34**, 461 (1953).

HILLEBRAND, W.F.: On the occurrence of nitrogen in uraninite and on the composition of uraninite in general. Amer. J. Sci. **40**, 384 (1891).

HOLMES, A.: The construction of a geological time scale. Trans. Geol. Soc. Glasgow **21**, 117 (1947).

— A revised geological time scale. Trans. Edinburgh Geol. Soc. **17**, 183 (1960).

Hurley, P.M., R.F. Cormier, J. Hower, H.W. Fairbairn and W.H. Pinson jr.: Reliability of glauconite for age measurements by the K-Ar and Rb-Sr methods. A.A.P.G. (in Press 1960).
— S.R. Hart, W.H. Pinson and H.W. Fairbairn: Authigenic versus detrital illite in sediments. Bull. Geol. Soc. Amer. **70**, 1622 (1959).
—, and L.E. Tryon: Extension of the carbon-14 age method. Rev. Sci. Instrum. **23**, 296 (1952).
Kulp, J.L.: Geological time scale. Bull. Geol. Soc. Amer. **70**, 1634 (1959).
— W.S. Broecker and W.R. Eckelmann: Age determination of uranium minerals by the Pb^{210} method. Nucleonics **11**, 19 (1954).
—, and W.R. Eckelmann: Discordant U-Pb ages and mineral type. Amer. Mineral. **42**, 154 (1957).
— L.E. Long, C.E. Giffin, A.A. Mills, R.J. Lambert and B.J. Giletti: Potassium-argon and rubidium-strontium ages of some granites from Britain and Eire. Nature, Lond. **185**, 495 (1960).
Libby, W.F.: Radiocarbon dating. Chicago: Chicago University Press 1952.
Long, L.E., J.L. Kulp and F.D. Eckelmann: Chronology of major metamorphic events in southeastern United States. Amer. J. Sci. **257**, 585 (1959).
Nier, A.O.: The isotopic composition of radiogenic leads and the measurement of geological time. II. Phys. Rev. **55**, 153 (1939).
— A re-determination of the relative abundances of the isotopes of carbon, nitrogen, oxygen, argon and potassium. Phys. Rev. **77**, 789 (1950).
— R.W. Thompson and B.F. Murphy: The isotopic composition of lead and the measurement of geologic time. II. Phys. Rev. **60**, 112 (1941).
Oakley, K.P., and C.R. Haskins: New evidence on the antiquity of piltdown man. Nature, Lond. **165**, 379 (1950).
Olson, E.A., and W.S. Broecker: Sample contamination and reliability of radiocarbon dates. Trans. N.Y. Acad. Sci. **20**, 593 (1958).
Patterson, C., H. Brown, G. Tilton and M. Inghram: Abundances of uranium and the isotopes of lead in the earth's crust and meteorites. Bull. Geol. Soc. Amer. **64**, 1461 (1953).
Revelle, R., and H.E. Suess: Carbon dioxide exchange between atmosphere and ocean and the question of an increase of atmospheric CO_2 during the past decades. Tellus 9, 18 (1957).
Reynolds, J.H.: A high sensitivity mass spectrometer. Phys. Rev. **98**, 283 (1954).
— Determination of the age of the elements. Phys. Rev. Letters 4, 8 (1960).
Suess, H.E.: Radiocarbon concentration in modern wood. Science **122**, 415 (1955).
— Nuclear processes in geologic settings, vol. 52. Washington, D.C.: Nat. Res. Council 1958.
Taylor, T.I., and H.C. Urey: Fractionation of the lithium and potassium isotopes by chemical exchange with zeolites. J. Chem. Phys. **6**, 429 (1938).
Tilton, G.R.: The interpretation of lead age descripancies by acid washing experiments. Trans. Amer. Geophys. Un. **37**, 224 (1956).
Vries, H.L. de: The removal of radon from CO_2 for use in C^{14} age measurements. Appl. Sci. Res. **6**, 461 (1958).
—, and K.P. Oakley: Radiocarbon dating of the piltdown skull and jaw. Nature, Lond. **165**, 379 (1950).
Wasserburg, G.J.: Argon-40: Potassium-40 dating. Nuclear Geology, ed. by H. Faul. New York: Wiley 1957.
Weaver, C.E.: Geologic interpretation of argillaceous sediments. Part I. Origin and significance of clay minerals in sedimentary rocks. A.A.P.G. **42** (1958).
Weizsacker, C.F. v.: Über die Möglichkeit eines dualen B-Zerfalls von Kalium. Phys. Z. **38**, 623 (1937).
Wetherill, G.W.: Discordant uranium-lead ages. I. Trans. Amer. Geophys. Un. **37**, 320 (1956).
— Radioactivity of potassium and geologic time. Science **126**, 545 (1957).

Some additional references[1]

Craig, H.: Mass-spectrometer analyses of radiocarbon standards. Amer. J. Sci., Radiocarbon Suppl. B 1 (1961).

---

[1] Added September 1961. Because of a delay in publication, "Some additional references" have been added to update the reader on some of the recent publications in the field of geochronometry. These references illustrate that isotope geochemistry is a rapidly expanding field of fundamental research.

FAIRBAIRN, H.W., P.M. HURLEY and W.H. PINSON: The relation of discordant Rb-Sr mineral and whole rock ages in an igneous rock to its time of crystallization and the time of subsequent Sr^{87}/Sr^{86} meta. Geochim. et Cosmochim. Acta **23**, 135 (1961).

FERGUSSON, G.J., and F.B. KNOX: The possibilities of natural radiocarbon as a ground water trace in thermal areas. Nucl. Sci. and Engng. Abstract **15**, 1872 (1961).

FOLINSBEE, R.E., H. BAADSGAARD and J. LIPSON: Potassium-argon time scale; Rept., Proc. 21st Session Int. Geol. Congr., Copenhagen; Sec. 3, Pre-Quaternary Absolute Age Determinations 1961

GAST, P.W.: The rubidium-strontium method. Ann. New York Acad. Sci. **91**, 181 (1961).

Geochronology of Rock Systems: Ann. New York Acad. Sci. **91** Art. 2 (1961).

GILETTI, B.J., and P.E. DAMON: Rubidium-strontium ages of some basement rocks from Arizona & Northwestern Mexico. Bull. Geol. Soc. Amer. **72**, 639 (1961).

GLENDENIN, L.E.: Present status of the decay constants. Ann. New York Acad. Sci. **91**, 166 (1961).

HURLEY, P.M.: Glauconite as a possible means of measuring the age of sediments. Ann. New York Acad. Sci. **91**, 294 (1961).

— R.F. CORMIER, J. HOWER, H.W. FAIRBAIRN and W.H. PENSIEN jr.: Reliability of glauconite for age meas. by K-Ar and Rb-Sr methods. Bull. A.A.P.G. **44**, 1793 (1960).

KENDALL, R.F.: Isotopic composition of potassium. Nature, Lond. **186**, 225 (1960).

KULP, J.L.: Geologic time scale. Science **133**, No. 3459, 1105 (1961).

RUSSELL, R.D., and R.M. FARQUHAR: Lead isotopes in geology. New York: Interscience 1960.

— T.J. ULRYCH and F. KOLLAR: Anamolous leads from Broken Hill Australia. J. Geophys. Res. **66**, 1495 (1961).

THATCHER, L., M. RUBIN and G.F. BROWN: Dating ground waters, USGS. Science **134**, 105 (1961).

WILLIS, E.H., H. TAUHER and K.O. MUNNICH: Variations in the atmospheric radiocarbon concentration over the past 1300 years. Amer. J. Sci., Radiocarbon Suppl. **2**, 1 (1960).

Date received of this manuscript: the 17th of may 1960.

Die natürliche und künstliche Radioaktivität der Atmosphäre

von

H. ISRAËL

Mit 7 Figuren

Summary

The atmosphere's content of natural-radioactive material may be understood and its amount may be estimated as the consequence of a circulation process:

Each of the three radioactive decay-series contains as one of its decay products an inert gas ("emanation") which partially diffuses into the atmosphere from the upper layers of the soil. The changes of the emanation in the soil is caused by the diffusion.

After the exhalation from the soil's capillaries the emanations follow the pattern of the motions of the atmosphere. Their transport is controlled by the apparent diffusion of the atmospheric "austausch". Under this control they arrive at a certain mean distribution with height. They are returned to the ground by precipitation and other processes, mostly after having changed into their decay products.

As a consequence of this it must be expected that the deficit of emanations in the upper layers of the ground is equal to the atmosphere's total content of emanations—a fact confirmed by comparison of measurements and estimations.

The atmosphere's radioactivity is mainly represented by radon and thoron and their decay products; this fact due to the existence of the mother materials U^{238}, U^{235}, and Th^{232} in the soil. The daughter-products of actinium do not play a role in comparison to the activities coming from radon and thoron products.

Data on the content of emanations in the soil's air, on exhalation, and on the concentration of radon (and thoron) in the atmosphere are completing the survey.

The second part of the paper communicates a short survey on the artificial radioactivity of the atmosphere.

The composition of the artificial radioactivity in the atmosphere is derived by the frequency-distribution of fission products depending on their mass-numbers. The variation of this composition with time may be derived from the different decay-constants. Besides this some experiences on the height-distribution and the variation of fission-product-mixture with time are given. The test-stop in fall 1958 had the consequences, that, beginning in the middle of the year 1959, the artificial radioactivity in air and precipitation decreased and arrived almost at zero at the end of 1960.

A. Natürliche Radioaktivität[1]

I. Übersicht

Die Lebensgeschichte der in der Lithosphäre enthaltenen langlebigen radioaktiven Muttersubstanzen U^{238}, U^{235} und Th^{232} führt jeweils über eine Reihe von Zwischenprodukten zu den inaktiven Bleiisotopen Pb^{206}, Pb^{207} und Pb^{208}.

Im Erdinnern spielen sich diese Umwandlungsprozesse in der Regel im kompakten Gesteinsgefüge am gleichen Ort ab. In der Erdkruste dagegen können durch geologisch-vulkanologische Vorgänge oder durch Wasser- und Gasbewegungen Verlagerungen stattfinden, durch die die radioaktiven Substanzen entweder als ganze Familien mit ihren Trägergesteinen oder als Familienteile verfrachtet werden.

Wir betrachten im folgenden den Gastransport, aus dem sich die natürliche Radioaktivität herleitet und erklärt.

Jede der drei natürlich-radioaktiven Reihen enthält bekanntlich in der betreffenden Emanation ein gasförmiges Zwischenglied von Edelgascharakter. Ohne diese Emanationen würde sich das gesamte Zerfallsgeschehen *im* Erdboden und Erdinnern abspielen und damit eine „Aktivierung" der Lufthülle unmöglich sein. Das bedeutet zugleich, daß im atmosphärischen Bereich *nur* die Emanationen und die sog. radioaktiven Induktionen, d.h. die Produkte von den Emanationen abwärts, vorkommen können[2].

Der Kreislauf beginnt damit, daß die in der Nähe der Erdoberfläche — in den verwitterten Gesteinsoberflächen ebenso wie in lockeren Bodenschichten — freiwerdenden Emanationen durch Diffusion an die Erdoberfläche herangebracht und hier nach außen abgegeben („exhaliert") werden. Einmal in den atmosphärischen Bereich gebracht, nehmen sie dann an den atmosphärischen Luftbewegungen in horizontaler und vertikaler Richtung teil, bis sie — nach Umwandlung in die Folgeprodukte — durch Niederschläge und andere Vorgänge wieder zum Boden zurückgebracht werden.

II. Der atmosphärische radioaktive Kreislauf

Rn-Bestimmungen in der Atmosphäre liegen in großer Zahl vor (vgl. z.B. die Zusammenstellungen bei St. Meyer und E. v. Schweidler 1927 und H. Israël 1951 und 1957). Man entnimmt ihnen, daß in der unteren, bodennahen Atmosphäre über dem Festland ein mittlerer Rn-Gehalt von etwa 100 bis $150 \cdot 10^{-18}$ Curie/cm³ besteht; über den Ozeanen nimmt dieser mit der Entfernung vom Land rasch auf etwa $\frac{1}{100}$ des Festlandwertes ab, ebenso wird er mit zunehmender Höhe rasch kleiner.

---

[1] Ohne den durch die kosmische Strahlung erzeugten Anteil (s. dazu den folgenden Beitrag von O. Haxel und G. Schumann).

[2] Die *vor* diesen Emanationen stehenden Elemente der Zerfallsreihen sowie die übrigen im Boden enthaltenen natürlich-radioaktiven Elemente können nur mit den von der Erdoberfläche in die Atmosphäre gelangenden Staub in diese eindringen. Man kann sich leicht überzeugen, daß dieser Anteil der atmosphärischen Radioaktivität neben dem des Emanations-Induktions-Kreislaufes zu vernachlässigen ist: Nimmt man z.B. zur Abschätzung des Radiumgehaltes der Atmosphäre den besonders günstigen Fall eines Staubgehaltes von $2{,}9 \cdot 10^{-10}$ g/cm³ Luft an, wie er im Winter in Großstadtluft herrschen kann (A. Kratzer 1937) und nimmt weiter für diesen Staub einen Ra-Gehalt von $1{,}7 \cdot 10^{-12}$ g/g an, wie er in sauren Eruptivgesteinen gefunden wird (R. D. Evans and C. Goodman 1934), so ergibt sich selbst in diesem Extremfall ein Ra-Gehalt von nur rund $5 \cdot 10^{-22}$ g (Curie) pro cm³ Luft — ein Wert, der neben den rund 10^{-16} Curie Rn/cm³ vernachlässigt werden kann. — Die wirklichen Mittelwerte des Gehaltes der Luft an Radium oder anderen nicht zum Emanations-Induktions-Kreislauf gehörigen radioaktiven Elemente dürften noch um Zehnerpotenzen niedriger liegen.

Die Einzelmeßwerte unterliegen sowohl von einem Ort zum anderen wie auch in ihrer zeitlichen Aufeinanderfolge erheblichen Variationen, in denen zum Teil geologisch bedingte Exhalationsunterschiede (s. z.B. J. Jaufmann 1907), vor allem aber meteorologische Einflüsse sichtbar werden. Wir wollen diese Variationen, die in späteren Abschnitten noch näher behandelt werden, hier im wesentlichen beiseite lassen und uns auf die Betrachtung der mittleren Verhältnisse beschränken.

Der radioaktive Kreislauf beginnt mit der Abgabe der Emanationen in den oberen Gesteins- und Bodenschichten an die in Gesteinsspalten und Bodencapillaren enthaltene Luft. Zur Abschätzung knüpfen wir an den mittleren Radiumgehalt der Gesteine und Böden an und verwenden im übrigen die Diffusionsgesetze.

Nach einer in einer großen Anzahl von Einzeluntersuchungen immer wieder bestätigten Erfahrung steigt der Rn-Gehalt in Bodenluft mit der Tiefe zunächst rasch und dann immer langsamer an, bis er sich in einigen Meter Tiefe einem konstanten Endwert nähert, der um mehrere Zehnerpotenzen höher ist als die in der Atmosphäre vorhandene Konzentration. [Daß der Rn-Gehalt in den obersten Bodenschichten zeitlichen Schwankungen unterworfen ist, die in geringer Tiefe am größten sind und mit zunehmender Tiefe allmählich verschwinden, sei hier nicht berücksichtigt. — Ausführliche Literaturzusammenstellungen über Rn-Untersuchungen in Bodenluft finden sich z.B. bei St. Meyer und E. v. Schweidler (1927) und bei H. Israël und F. Becker (1936); über die Variabilität in verschiedenen Tiefen s. z.B. H. Bender (1934) und H. Israël und J. Saldukas (1939).]

Die offensichtlich durch Diffusion hervorgerufene Konzentrationsabnahme zur Erdoberfläche hin läßt sich wie folgt mathematisch erfassen:

Wir gehen aus vom zweiten Fickschen Diffusionsgesetz

$$\frac{\partial c}{\partial t} = D\,\Delta c, \qquad c = \text{Konzentration}; \; D = \text{„Diffusionskoeffizient''} \tag{1}$$

das durch zwei Glieder zu erweitern ist, die der Nachlieferung und dem Zerfall der betreffenden Emanation Rechnung tragen:

$$\frac{\partial c}{\partial t} = D\,\Delta c + a - \lambda c, \tag{2}$$

$a =$ Konzentrationszunahme durch Nachlieferung im betrachteten Raumelement des Bodens[1];

$\lambda =$ Zerfallskonstante; also:

$\lambda c =$ Konzentrationsabnahme durch Zerfall im betrachteten Raumelement.

---

[1] Es wird zur Vereinfachung zunächst angenommen, daß die Diffusion in den Bodencapillaren in genau der gleichen Weise vor sich geht, wie in einem freien Luftraum. Dies dürfte zutreffen, solange der Capillarendurchmesser groß bleibt gegenüber der mittleren freien Weglänge und solange die festen Bodenelemente (Capillarenwandungen) keine besonderen Adsorptionskräfte auf die Emanationen ausüben. Insbesondere muß also innerhalb weiter Grenzen die Konzentration der Emanation wesentlich unabhängig sein von der Korngröße des Bodenmaterials. Findet merkliche Adsorption an den Bodenelementen statt, so wird sich dies durch eine Verlangsamung der Diffusion, also eine Verkleinerung des Diffusionskoeffizienten D bemerkbar machen. Über die Veränderung der theoretischen Ansätze bei Berücksichtigung der „Porosität" s. Yu.P. Bulashevich (1958) und Yu.P. Bulashevich und R.K. Khairitdinov (1959). — Einen sehr merklichen Einfluß auf die Diffusion im Boden kann sein Feuchtigkeitsgrad gewinnen: Solange die Bodenfeuchte die Capillare noch genügend offenläßt, wird sie durch stärkere Adsorption an den Capillarenwänden die Diffusion merklich verlangsamen, sofern sie die Capillaren ganz verschließt, diese praktisch ganz unterbinden können, da die Emanation in Wasser rund 7000mal langsamer diffundiert als in Luft.

In einem homogenen Boden gleichmäßigen Aktivitätsgehaltes und Emanierungsvermögens braucht nur die Variation in der Vertikalen betrachtet zu werden. Im Gleichgewichtsfall $(\partial c/\partial t = 0)$ gilt also

$$\frac{d^2 c}{dz^2} = \frac{\lambda}{D}\, c - \frac{a}{D} \qquad z = \text{Tiefe (positive Richtung nach unten).} \qquad (3)$$

Integration und Konstantenbestimmung liefert für die Konzentration der betreffenden Emanation in Abhängigkeit von der Tiefe die Beziehung

$$c(z) = \frac{a}{\lambda}\left[1 - \exp\left(-\left|\sqrt{\lambda/D}\cdot z\right|\right)\right] + c_0 \exp\left(-\sqrt{\lambda/D}\cdot z\right), \qquad (4)$$

wo c_0 die an der Erdoberfläche bzw. in der untersten bodennahen Luftschicht herrschende Konzentration bedeutet.

Die Konzentration der betreffenden Emanation nimmt also in der Bodenluft von einem in der Tiefe herrschenden Wert a/λ zur Erdoberfläche hin auf einen Wert c_0 ab. c_0 ist, wie die Erfahrung zeigt, um einige Zehnerpotenzen kleiner als a/λ und kann deshalb in Gl. (4) vernachlässigt werden. In Fig. 1 ist dieser Verlauf wiedergegeben.

Die „Halbwertstiefe" T_H ergibt sich nach Gl. (4) zu

$$T_H = \frac{\ln 2}{\sqrt{\lambda/D}}. \qquad (5)$$

Die „Exhalation" E, d.h. der pro Flächeneinheit und Sekunde aus dem Boden austretende Emanationsbetrag ergibt sich dann nach dem ersten Fickschen Diffusionsgesetz zu

$$E = \left\{D\,\frac{dc}{dz}\right\}_{z=0} = a\,\sqrt{D/\lambda}. \qquad (6)$$

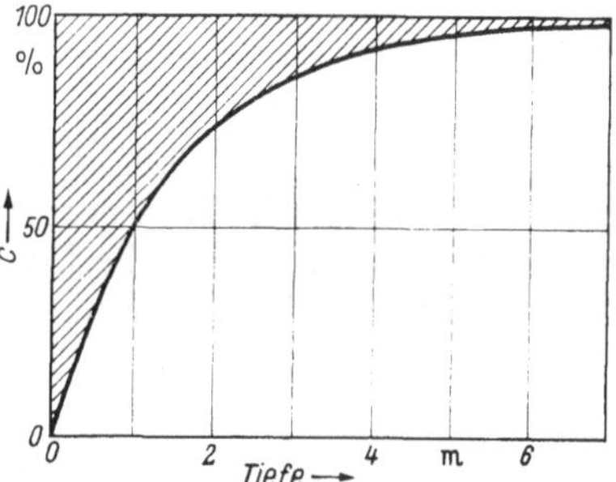

Fig. 1. Konzentration der Radium - Emanation (Rn) in der Bodenluft in homogenem Boden gleichmäßigen Radium-Gehaltes und Emanierungsvermögens in Abhängigkeit von der Tiefe bei Annahme eines Diffusionskoeffizienten in Bodenluft von $D = 0{,}05\ \mathrm{cm^2\,sec^{-1}}$

Da nun angenommen werden kann, daß sich im Mittel der atmosphärische Radioaktivitätsgehalt weder vermehrt noch vermindert, und da weiter bei einer Abschätzung, wie der hier beabsichtigten, vom atmosphärischen Transportgeschehen im einzelnen abgesehen werden kann, muß der mittlere Rn-Gesamtgehalt in einer vertikalen Luftsäule zur Exhalation durch ihre Basisfläche in Beziehung stehen. Durch die Exhalation steigt der Emanationsgehalt in der darüber befindlichen Luftsäule solange an, bis die in der Zeiteinheit zerfallende Menge den Nachlieferungsbetrag erreicht. Dieser Gleichgewichtswert G ist nach den Zerfallsgesetzen definiert durch die Beziehung

$$G\,\lambda = E, \qquad (7)$$

bzw.

$$G = \frac{a}{\lambda}\,\sqrt{D/\lambda}. \qquad (8)$$

Der in Gl. (4) gegebene Konzentrationsverlauf bedeutet eine Verarmung der obersten Bodenschichten an Emanation. Das „Defizit" berechnet sich zu

$$\int_0^\infty [1 - c(z)]\,dz = \frac{a}{\lambda}\,\sqrt{D/\lambda} \qquad (9)$$

und stimmt, wie zu erwarten, mit dem in Gl. (8) gegebenen Gleichgewichtswert überein.

Nach der Exhalation aus den Bodencapillaren gelangen die Emanationen in den Bereich der atmosphärischen Luftbewegungen. Damit tritt an Stelle der gaskinetischen Diffusion die „Schein-Diffusion" des atmosphärischen Massenaustausches.

Zur theoretischen Behandlung kann man auch hier formal von den Grundgleichungen der Diffusion ausgehen, wenn man an Stelle des Diffusions-Koeffizienten D den „Schein-Diffusions-Koeffizienten" K der Turbulenzbewegung einführt und diesem entsprechende räumliche und zeitliche Abhängigkeiten zuschreibt.

Da wir nur die mittleren Verhältnisse betrachten, können wir uns auch hier bei der mathematischen Behandlung auf den eindimensionellen Fall beschränken.

Für die Emanations-Konzentration $c(h)$ in Abhängigkeit von der Höhe h, ergibt sich dann die Differentialgleichung

$$K \frac{d^2 c}{d h^2} + \frac{dK}{dh} \frac{dc}{dh} - c \lambda = 0. \tag{10}$$

Diese Gleichung hat für ein höhenunabhängiges K, wie wir es in vereinfachender Näherung hier annehmen wollen[1], die Lösung

$$c(h) = c_0 \exp\left(- \sqrt{\lambda/K} \cdot h\right). \tag{11}$$

c_0 bedeutet wieder die Konzentration in der Höhe 0 und bestimmt sich nach unserer oben skizzierten Bilanzüberlegung (Gleichgewicht von Emanationsdefizit im Boden und Gesamt-Emanationsgehalt der darüber befindlichen vertikalen Luftsäule) wie folgt:

$$\int\limits_0^\infty c(h) \cdot dh = c_0 \sqrt{K/\lambda} = \frac{a}{\lambda} \sqrt{D/\lambda}, \tag{12}$$

$$c_0 = \frac{a}{\lambda} \sqrt{D/K}. \tag{13}$$

Die Betrachtungen gelten in dieser Form nur für die gasförmigen Emanationen. Eventuelle Veränderungen der durch die Gln. (4) und (11) gegebenen Vertikalprofile im Boden und in der Atmosphäre durch Niederschläge sind dabei nicht berücksichtigt, da die Emanationen nur in geringerem Maße von den Niederschlägen aufgenommen werden und sich außerdem nur zu einem geringen Teil an atmosphärische Suspensionen anzulagern vermögen (s. z.B. H. Israël 1934 und 1936).

Anders liegen die Verhältnisse bei den Folgeprodukten (Induktionen), da sich diese im Gegensatz zu den Emanationen alsbald vollständig an Aerosolteilchen anlagern (H. Israël 1934, O. Haxel 1953).

Zur Behandlung dieser Probleme muß zunächst in Gl. (10) der Nachlieferung des betreffenden Elementes (i) aus dem Vorhergehenden ($i-1$) Rechnung getragen werden. An Stelle von Gl. (10) tritt damit die Gleichung

$$K \cdot \frac{d^2 c_i(h)}{d h^2} + \frac{dK}{dh} \cdot \frac{d c_i(h)}{dh} + c_{i-1}(h) \cdot \lambda_{i-1} - c_i(h) \cdot \lambda_i = 0. \tag{14}$$

Außerdem ist der Anlagerungsgeschwindigkeit und dem Aerosolkreislauf Rechnung getragen. Genaueres dazu s. unten im Abschnitt „Transfer and circulation

---

[1] Lösungen für einfache Ansätze für eine Höhenabhängigkeit von K s. bei W. Schmidt (1926) (Potenzgesetz für K) und H. Lettau (1941) (lineare Höhenzunahme von K) [s. dazu auch W. Jacobi (1960)]. Über eine Lösung bei höhenkonstantem aber zeitabhängigem Scheindiffusions-Koeffizienten s. B. Iswekow (1929). Ansätze für noch allgemeinere Lösungen s. bei H. Lettau (1951) und O. G. Sutton (1951).

of radioactivity in the atmosphere" von B. BOLIN, ferner bei W. JACOBI, A. SCHRAUB, K. AURAND und H. MUTH (1959), W. M. BURTON und N. G. STEWART (1960) und W. JACOBI (1960).

III. Abschätzungen und Bilanz

Zur quantitativen Abschätzung gehen wir vom Gehalt des Untergrundes an langlebiger radioaktiver Muttersubstanz aus.

Der Gehalt der Gesteine an Uran und Thorium schwankt zwar je nach Gesteinsart und Fundort (vgl. z. B. die Zusammenstellungen bei R. D. EVANS und C. GOODMAN 1941, im Landolt-Börnstein 1952, oder bei J. T. WILSON, R. D. RUSSEL und R. M. FARQUHAR 1956), doch kann man für die Gesteine und Böden der Festlandoberfläche in befriedigender Annäherung im Mittel etwa die in Tabelle 1 zusammengestellten Werte annehmen.

Tabelle 1. *Mittlerer Gehalt der Gesteine und Böden der Festland-Oberfläche an langlebiger Muttersubstanz der drei natürlich-radioaktiven Zerfallsreihen*

Substanz	Konzentration	
	in g/g	in Curie/g
U^{238}	$3 \cdot 10^{-6}$	$1,0 \cdot 10^{-12}$
U^{235}	$2 \cdot 10^{-8}$	$4,3 \cdot 10^{-14}$
Th^{232}	$1 \cdot 10^{-5}$	$1,1 \cdot 10^{-12}$

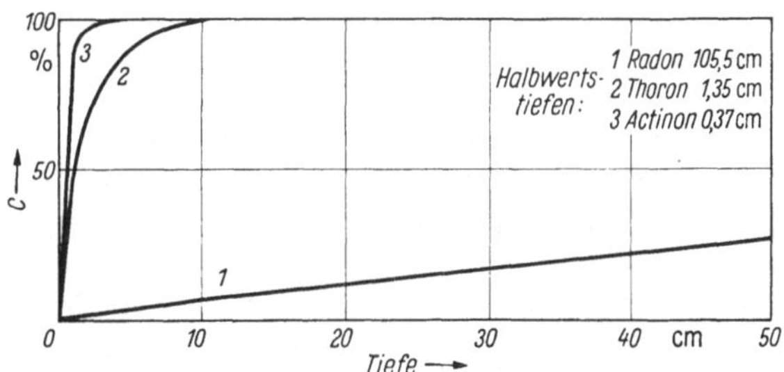

Fig. 2. Änderung der Konzentration der drei Emanationen mit der Tiefe in Böden gleichmäßiger Aktivität und gleichmäßigen Emanierungsvermögens bei Annahme eines Diffusions-Koeffizienten in Bodenluft von $D = 0,05$ cm² sec⁻¹

Hieraus lassen sich folgende Relationen ableiten:

Die Umrechnung der mittleren Gewichtsanteile von Uran und Thorium in Strahlungsmaß (letzte Kolonne der Tabelle) gibt zugleich die Emanationsmengen an, die im radioaktiven Gleichgewicht je g Gestein bzw. Boden zu erwarten sind. Danach verhalten sich die in Curie ausgedrückten a/λ-Werte von Rn, Tn und An im Boden im Mittel etwa wie folgt:

$$\left(\frac{a}{\lambda}\right)_{\mathrm{Rn}} : \left(\frac{a}{\lambda}\right)_{\mathrm{Tn}} : \left(\frac{a}{\lambda}\right)_{\mathrm{An}} = 100 : 110 : 4,3\,. \tag{15}$$

Für die „Defizit"-Beträge leiten sich dann nach Gl. (8) folgende Verhältniszahlen ab:

$$G_{\mathrm{Rn}} : G_{\mathrm{Tn}} : G_{\mathrm{An}} = 100 \cdot \lambda_{\mathrm{Rn}}^{-\frac{1}{2}} : 110 \cdot \lambda_{\mathrm{Tn}}^{-\frac{1}{2}} : 40,3 \cdot \lambda_{\mathrm{An}}^{-\frac{1}{2}} = 100 : 1,38 : 0,015\,, \tag{16}$$

während sich für die in der bodennahen Atmosphäre zu erwartenden Emanations-Konzentrationen nach Gl. (13) wieder die Zahlen

$$(c_0)_{\mathrm{Rn}} : (c_0)_{\mathrm{Tn}} : (c_0)_{\mathrm{An}} = 100 : 110 : 4,3 \tag{17}$$

ergeben.

Zur Veranschaulichung dessen sind in Fig. 2 für einen Wert des Diffusions-Koeffizienten von $D = 0,05$ cm² sec⁻¹ für die drei Emanationen Tiefenprofile nach Gl. (4) dargestellt. Der Bezugswert („100") ist in allen drei Fällen der a/λ-Wert, der sich in größerer Tiefe einstellt. Die „Halbwertstiefen" verhalten sich umgekehrt proportional zur Wurzel aus der Zerfallskonstante; für $D = 0,05$ cm² sec⁻¹ errechnen sie sich zu 105,5 cm für Rn, 1,35 cm für Tn und 0,37 cm für An.

Entsprechend ergeben sich dann für die Höhenabhängigkeit in der Atmosphäre gemäß Gl. (11) für einen Scheindiffusions-Koeffizienten des Betrages $K = 4\,(8) \times 10^4$ cm² sec⁻¹ „Halbwertshöhen" von 960 (1350) m für Rn, 12,3 (17,3) m für Tn und 3,3 (4,6) m für An.

Die Abkömmlinge des Uran[235] (An-Folgeprodukte) treten also in jedem Fall in der Atmosphäre stark hinter denen des Uran[238] und des Thorium[232] zurück und können hier deshalb in der Regel neben diesen vernachlässigt werden.

Die tatsächlich meßbaren Emanations-Konzentrationen in Bodenluft sowie die daraus resultierenden Defizitbeträge sind kleiner als die nach Tabelle 1 errechneten Werte, da sie noch vom Emanierungsvermögen des Bodenmaterials und von den Diffusions-Koeffizienten in der Bodenluft abhängen.

Nach J. Satterly (1911) liegt das Emanierungsvermögen verschiedener Bodenmaterialien je nach Bodenart und Feuchtigkeit zwischen etwa $\frac{1}{6}$ und $\frac{1}{20}$ (vgl. auch B. B. Boltwood 1909). Feuchter Boden emaniert besser als trockener. Wir wollen danach mit einem mittleren Emanierungsvermögen des Bodenmaterials von etwa $\frac{1}{10}$ rechnen und dies für alle drei Emanationen gleich annehmen.

Für die Größe des Diffusions-Koeffizienten D in Bodenluft geben V. J. Baranow und E. G. Gratschewa (1933) nach Laboratoriumsuntersuchungen einen Wert von $D = 0{,}035$ cm² sec⁻¹ an. H. Israël und F. Becker (1936) schließen aus Messungen der Luftdruckabhängigkeit von Radon-Tiefenprofilen über einer stark emanierenden tektonischen Störung auf einen Wert von $D = 0{,}05$ cm² sec⁻¹. Untersuchungen von D. A. De Vries (1950) und C. M. H. van Bavel (vgl. auch W. Jost) ergeben ein $D = 0.072$ cm² sec⁻¹. E. Budde (1958) dagegen leitet aus Laboratoriumsversuchen mit feuchter Bodensubstanz je nach der Feinkörnigkeit des Materials sehr viel kleinere Werte des Diffusions-Koeffizienten ab ($D > 10^{-3}$ bis $D < 10^{-5}$ cm² sec⁻¹). Vgl. dazu auch Yu. P. Bulashevich und R. K. Khairitdinov (1959).

Eine Entscheidung darüber, welcher dieser D-Werte die tatsächlichen radioaktiven Verhältnisse in Bodenluft und Atmosphäre im Mittel am besten wiedergibt, ergibt sich aus dem Vergleich zwischen Berechnung und Erfahrung:

In Tabelle 2 sind dazu die Werte für Defizit, Exhalation und Konzentration in der bodennahen Atmosphäre für Radon zusammengestellt, die sich nach den

Tabelle 2. *Radon-Defizit im Boden, Exhalation und Rn-Konzentration in der bodennahen Atmosphäre für verschiedene Werte des Diffusions-Koeffizienten D in Bodenluft, berechnet für* $a/\lambda = 2 \cdot 10^{-13}$ C/cm³ *

D	Defizit in Curie	Exhalation in C/cm² sec	Konzentration in der bodennahen Atmosphäre in C/cm³	
			(K_1)	(K_2) **
0,05	$3{,}1 \cdot 10^{-11}$	$65 \cdot 10^{-18}$	$224 \cdot 10^{-18}$	$158 \cdot 10^{-18}$
0,035	2,6	54	94	66
10^{-3}	$4{,}4 \cdot 10^{-12}$	9	32	22
10^{-4}	1,4	3	10	7
10^{-5}	$4{,}4 \cdot 10^{-13}$	1	3	2

* Das spezifische Gewicht unverfestigten Bodenmaterials schwankt je nach der Bodenart in gewissen Grenzen um den Wert 2 (s. z. B. Landolt-Börnstein l. c., S. 325). Bei 10% Emanierungsvermögen ergibt sich also als Ausgangswert ein mittleres $a/\lambda = 2 \cdot 10^{-13}$ Curie/cm³.

** Berechnet für die beiden Grenzwerte eines (höhenkonstant angenommenen) Schein-Diffusionskoeffizienten

$$K_1 = 4 \cdot 10^4 \text{ cm}^2 \text{ sec}^{-1}$$

und

$$K_2 = 8 \cdot 10^4 \text{ cm}^2 \text{ sec}^{-1}.$$

Die durch diese Werte bestimmten Höhenverteilungen des Radons in der Atmosphäre schließen die tatsächlich beobachtete Verteilung ein.

Tabelle 3. *Radon-Gehalt in Bodenluft (Messungen meist in 1 m Tiefe)*

Meßort	Bodenart	Rn-Gehalt in 10^{-13} C/cm³	Autor
Freiburg (Schweiz)	über Moränenschotter	0,7 bis 2,8	A. Gockel 1908
Yale University .	verwitterter Rotsandstein	2,4	J. C. Sanderson 1911
Cambridge . . .	verschiedene Böden	0,7 bis 2,3	J. Satterly 1911b
Dublin		1,6	L. B. Smyth 1912
Potsdam	Sand	0,1	K. Kähler 1914
Manila		3,0	J. R. Wright and O. F. Smith 1915
Freiberg i. Sa. . .	Lehm	12,0	J. Olujić 1918
	Lehm (feucht)	2,5	
Schneeberger Gruben. . . .	uranhaltiges Gestein	3,0 bis 182	P. Ludewig und E. Lorenser 1924
Innsbruck . . .	über alluvialem alpinem Schotter	5,4	H. Bender 1934
Bad Nauheim . .	über Taunus-Quarzit und Taunusrandverwerfung	etwa 15 bis über 200	H. Israël und F. Becker 1935
Potsdam	Sand	1,1	H. Israël und J. Saldukas 1938
Harz	a) über Granit	130	H. Israël 1938
	b) über Hornfels	14	
Aachen		1,8	H. Israël (unveröffentlicht)

Tabelle 4. *Radon-Exhalation aus dem Boden*

Ort	Exhalation in 10^{-18} C/cm² sec	Autor
Dublin.	73	L. B. Smyth 1912
Manila.	21	J. R. Wright and O. F. Smith 1915
Innsbruck	23	P. R. Zupancic 1934
Innsbruck	50	P. R. Zeilinger 1935
Hötting bei Innsbruck .	35	P. R. Zupancic 1934
Graz (1933)	54	W. Kosmath 1935
Graz (1934)	25	W. Kosmath 1935
Slutzk	13	E. S. Merkulowa 1937
New York	6	S. A. Jaki and V. F. Hess 1958
Socorro	90	M. H. Wilkening and J. E. Hand 1960

Tabelle 5. *Mittlere Werte des Radon-Gehaltes der unteren Atmosphäre.*
(Nach den Zusammenstellungen von H. Israël 1951, und I. H. Blifford u. a. 1956)

Meßgebiet	Radon-Gehalt in 10^{-18} C/cm³
Festland geringer Meereshöhe (ohne Alpen- und Küstengebiete) .	etwa 140
Alpengebiet	etwa 300
Küstengebiet	etwa 10
Ozeane und landferne kleine Inseln	etwa 1

Gln. (6), (9) und (13) für verschiedene Werte des Diffusions-Koeffizienten D in Bodenluft errechnen. Dem stehen nach den bisherigen Meßerfahrungen etwa die folgenden Ergebnisse gegenüber:

In Tabelle 3 sind für eine Reihe von Orten mittlere Werte des Radon-Gehaltes der Bodenluft zusammengestellt. Tabelle 4 zeigt mittlere Radon-Exhalationswerte. Tabelle 5 schließlich enthält Meßergebnisse des bodennahen atmosphärischen Aktivitätsgehaltes.

Trotz der starken Schwankungen der Einzelwerte kann man — wenn man in Tabelle 3 die hohen Zahlenwerte nicht berücksichtigt, die offensichtlich Besonderheiten darstellen (Schneeberg, Bad Nauheim, Harz) und sich in Tabelle 5 auf Festlandverhältnisse ohne Alpen- und Küstenbereich beschränkt — etwa folgende Angaben für die mittlere Rn-Verteilung machen (Tabelle 6):

Tabelle 6. *Mittlere Radon-Verhältnisse*

Rn-Gehalt in Bodenluft	etwa $3 \cdot 10^{-13}$ C/cm^3
Rn-Exhalation	etwa $40 \cdot 10^{-18}$ C/cm$^3 \cdot$ sec
Rn-Gehalt in der unteren Atmosphäre	etwa $140 \cdot 10^{-18}$ C/cm^3

Nach diesen Zahlen ergibt sich durch Vergleich mit Tabelle 2, daß man durch Annahme eines im Bereich von 0,05 cm^2 sec^{-1} oder nur wenig darunter liegenden mittleren Wertes für den Radon-Diffusions-Koeffizienten D in Bodenluft den tatsächlichen mittleren Verhältnissen am nächsten kommt. — Da sich die drei Emanationen nur wenig im Atomgewicht voneinander unterscheiden, können wir diesen für Radon abgeschätzten mittleren D-Wert auch für die beiden anderen Emanationen (Tn und An) annehmen.

Zur Beteiligung des Thorons ist zu sagen, daß nach den Ausgangswerten der Tabelle 1 sowohl in Bodenluft wie in der unteren Atmosphäre die Tn-Konzentration — ausgedrückt in Strahlungsmaß — größer sein sollte als die Rn-Konzentration. In der Bodenluft ist dies auch der Fall, soweit man dies aus den bisher allerdings nur sehr spärlichen Messungen dessen wissen kann (s. z.B. J.C. Sanderson 1911, J. Satterly 1911b, H. Israël und Mitarbeiter 1935 und 1962 u.a.). Nach Messungen von J. Satterly in verschiedenen Tiefen (Oberflächennähe 1,50 und 4 m), nimmt außerdem das Verhältnis Rn/Tn deutlich mit der Tiefe zu, wie es nach unseren Überlegungen über die Diffusion im Boden erwartet werden muß [vgl. Gln. (3) und (5) und Fig. 2].

In der Atmosphäre dagegen ist das Verhältnis Rn/Tn durchweg größer als 1: In Bodennähe liegt sein Wert bei etwa 1,3 bis 1,7 (V.F. Hess und R.T. Vancour 1950 und 1953; A. F. El-Nadi und H.M. Omar 1960); mit zunehmender Höhe muß es nach Gl. (11) rasch ansteigen.

Die betrachteten Größen durchlaufen unter der Wirkung meteorologischer Einflüsse ständig regelmäßig wiederkehrende und unregelmäßige Variationen. Sie lassen sich kurz wie folgt zusammenfassen:

Bei fast allen Arbeiten über Bodenemanation zeigen sich wie schon gesagt meteorologisch bedingte Schwankungen der Konzentration. Diese erklären sich aus Veränderungen des Diffusions-Geschehens und sind demgemäß am größten in Erdoberflächennähe; mit zunehmender Tiefe nehmen sie rasch ab.

Die Exhalation besitzt offenbar eine einfache Tagesperiode mit einem Höchstwert am frühen Nachmittag, doch wird dieser allgemeine Verlauf durch meteorologische Einflüsse stark modifiziert. — Im Jahresverlauf fällt das Maximum der Exhalation auf den Spätsommer. Niederschläge setzen die Exhalation stark herab, während Sonnenbestrahlung und Temperaturzunahme sie ansteigen lassen. Fallender Luftdruck hat Anstieg steigender Abnahme zur Folge. Frost erniedrigt im allgemeinen die Exhalation und kann sie völlig unterbinden (s. z.B. H. Bender 1934 und W. Kosmath 1935).

Der atmosphärische Rn-Gehalt zeigt im allgemeinen Tagesgänge mit Höchstwerten nachts bzw. gegen Morgen und niedrigsten Werten in den Nachmittagsstunden. In Gebirgslagen sind diese Gänge weniger ausgeprägt und offensichtlich entscheidend durch das Austauschgeschehen beeinflußt insofern, als auf

Bergstationen nur bei kräftigem konvektivem Luftaustausch mit den tieferen Schichten klar ausgeprägte regelmäßige Tagesgänge zustande kommen. — Der Jahresgang des atmosphärischen Rn-Gehaltes ist uneinheitlich, weil sich in ihm mehrere steuernde Faktoren überschneiden.

Sehr ausgeprägt sind meteorologische Einwirkungen: Fallender Luftdruck erhöht, steigender verringert im allgemeinen den Rn-Gehalt, wie es nach dem Luftdruckeinfluß auf die Exhalation zu erwarten ist. Der Windeinfluß ist ein doppelter: Mit zunehmender Windstärke tritt zunächst — infolge erhöhter Exhalation! — Vermehrung, dann — infolge stärkerer Aufwärtsverfrachtung durch Austausch — eine Verminderung ein. Die Windrichtung hat insofern Einfluß, als

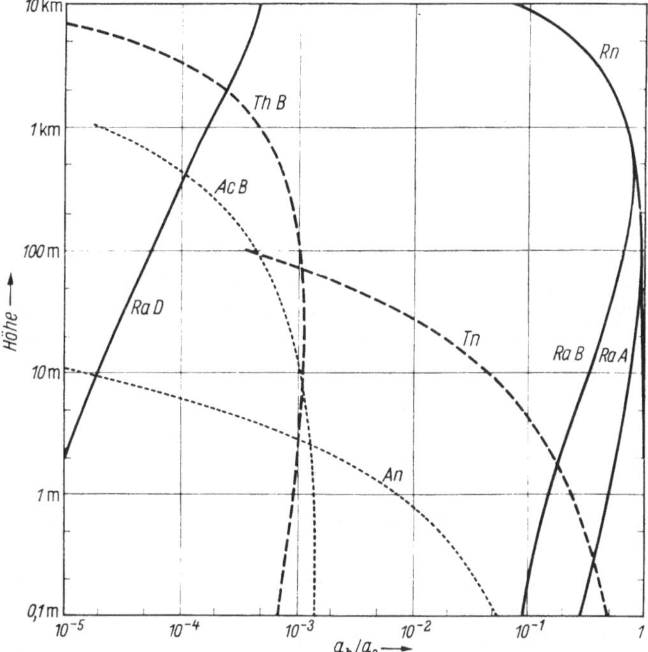

Fig. 3. Erwartungswerte für die mittleren Höhenverteilungen der radioaktiven Emanationen und Induktionen nach J. JACOBI (1960)

Luftkörper maritimen Ursprungs geringeren Rn-Gehalt zeigen als solche kontinentalen Ursprungs. Auch kann in der Abhängigkeit des Rn-Gehaltes von der Windrichtung verschiedene Exhalation sichtbar werden, wie es z.B. die Ergebnisse von J. JAUFMANN (1907) auf der Zugspitze zeigen. — Niederschläge, vor allem solche von längerer Dauer, verringern den Rn-Gehalt, was die Folge eines gewissen „Auswaschens" der Atmosphäre sein dürfte.

Die mittlere Höhenverteilung der Emanationen und Induktionen in der Atmosphäre bestimmt sich gemäß Gl. (14) aus dem Zusammenwirken von Austausch, Nachlieferung und Zerfall. Messungen der Emanationen und Induktionen in verschiedenen Höhen liegen nur in verhältnismäßig geringer Zahl vor. Sie zeigen erwartungsgemäß rasche Abnahme der Radioaktivität mit der Höhe, genügen aber nicht, um diese für die einzelnen Partner getrennt ermitteln zu können. — Die rechnerische Behandlung dessen auf der Basis der Gl. (14) liefert unter der Annahme eines bis zu 1 km Höhe etwa linear auf den Wert $5 \cdot 10^5$ cm² sec⁻¹ ansteigenden und dann konstant bleibenden Scheindiffusions-Koeffizienten K nach J. JACOBI (1960) die in Fig. 3 wiedergegebene mittlere Höhenverteilung der einzelnen Substanzen.

Aus den Jacobischen Rechnungen folgt unter anderem, daß in den unteren Atmosphärenschichten radioaktives Gleichgewicht zwischen Radon und seinen kurzlebigen Induktionen erst in 100 und mehr Meter Höhe zu erwarten ist. Experimentelle Untersuchungen dessen durch K. Schwalb (1941) haben indes seinerzeit ergeben, daß häufig auch schon in Bodennähe radioaktives Gleichgewicht zwischen Rn und RaA bis RaC besteht. Vgl. dazu auch M. Kawano und S. Nakatani (1959).

B. Künstliche Radioaktivität

I. Übersicht

Neben den natürlich-radioaktiven Substanzen und den durch die kosmische Strahlung gebildeten radioaktiven Isotopen tritt seit etwa einem Jahrzehnt in der Atmosphäre eine dritte Art von Radioaktivität in Erscheinung in Gestalt der instabilen Elemente, die bei der Uran- und Plutonium-Spaltung entstehen. Für diesen Radioaktivitätsanteil läßt sich ein ähnlicher zusammenfassender Überblick, wie er im vorstehenden für die natürliche Radioaktivität der Atmosphäre gegeben werden konnte, nicht ohne weiteres aufstellen. Dafür ist vor allem der Aktivierungsvorgang in beiden Fällen zu verschieden: Anstelle der „gleichmäßig fließenden Flächenquellen" für die natürliche Aktivität in Gestalt der Landoberflächen treten jetzt Punktquellen in unregelmäßiger örtlicher und zeitlicher Verteilung, die zudem ihren Sitz in der Regel in höheren Schichten der Atmosphäre haben. Diese hat z.B. zur Folge, daß die künstliche Radioaktivität der Atmosphäre und ihre Wanderung in sehr viel stärkerem Maße von der Wetterlage und -entwicklung abhängen als dies bei der mehr klimatisch gebundenen natürlichen atmosphärischen Radioaktivität der Fall ist.

Damit wird zwangsläufig die Behandlung dieser Dinge bei der künstlichen Radioaktivität in sehr viel stärkerem Maße zur Betrachtung von Einzelfällen und ihren Folgen, was eine Darstellung mittlerer Verhältnisse natürlich erschwert[1].

Wir wollen im folgenden Vorkommen, stoffliche Zusammensetzung und Variabilität dieser künstlichen Radioaktivität in der Atmosphäre durch einige beobachtende Erfahrungen aus dem letzten Jahrzehnt charakterisieren.

II. Die Spaltprodukte

Fig. 4 zeigt die bekannten Häufigkeitskurven für die Verteilung der Spaltprodukte in Abhängigkeit von der Massenzahl A, die bei der Spaltung von U^{235} bzw. Pu^{239} auftreten. Dargestellt ist ihre Ausbeute η_{sp} in Prozent der gespaltenen U^{235}- bzw. Pu^{239}-Atome.

Tabelle 7 gibt in Ergänzung zu Fig. 4 einen Überblick über die durch Kernspaltung entstehenden Isotope in ihrer Ordnung zu radioaktiven Zerfallsreihen. Die einzelnen Reihenmitglieder entstehen dabei unmittelbar primär bei der Spaltung, zum Teil durch Zerfall der vorhergehenden Reihenglieder.

Tabelle 8 schließlich gibt die radioaktiven Daten einiger wichtiger Spaltprodukte an, die entweder in besonders großer Häufigkeit entstehen oder durch ihre lange Halbwertszeit besondere Bedeutung haben.

Bei den Kernwaffenversuchen werden die Spaltprodukte mit dem Explosionspilz bei kleineren Explosionen bis an die Stratosphärengrenzen, bei größeren bis

---

[1] Auf der anderen Seite ist dieser Zwang zur Behandlung von Einzelergebnissen insofern von besonderer Bedeutung, als er für die Meteorologie völlig neue Möglichkeiten zum Studium der atmosphärischen Zirkulation bietet.

Tabelle 7. *Übersicht über die bei der Spaltung von U^{233}, U^{235}, Pu^{239} und U^{238} entstehenden künstlich-radioaktiven Isotope in ihrer Anordnung zu Zerfallsreihen, dargestellt für die Massenzahlen 62 bis 171 mit Angabe der Ausbeute* (nach RIEZLER-WALCHER 1958)

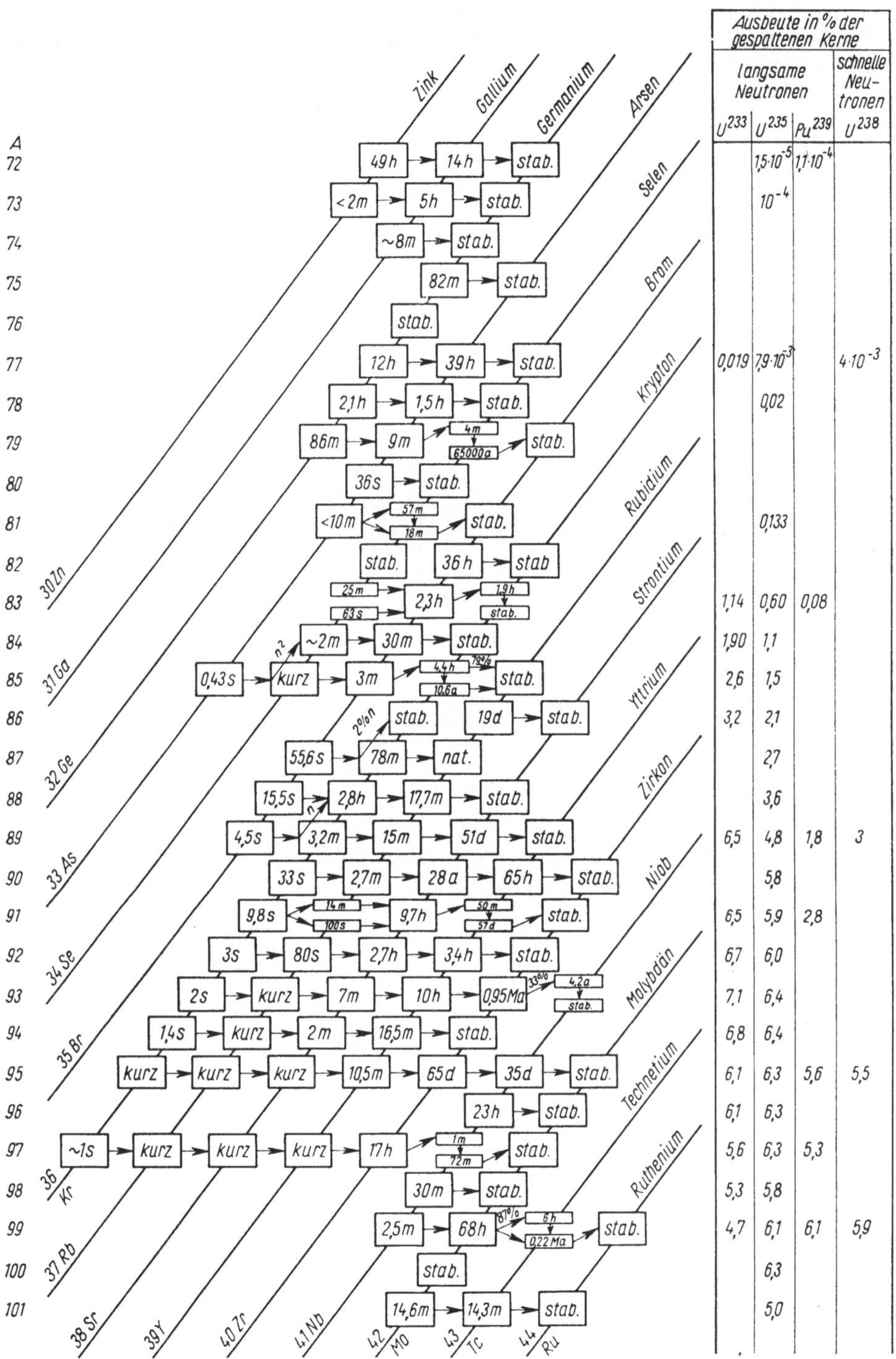

A				Ausbeute in % der gespaltenen Kerne			
				langsame Neutronen			schnelle Neutronen
				U^{233}	U^{235}	Pu^{239}	U^{238}
72	49h → 14h → stab.				$1{,}5\cdot10^{-5}$	$1{,}1\cdot10^{-4}$	
73	<2m → 5h → stab.				10^{-4}		
74	~8m → stab.						
75	82m → stab.						
76	stab.						
77	12h → 39h → stab.			0,019	$7{,}9\cdot10^{-3}$		$4\cdot10^{-3}$
78	2,1h → 1,5h → stab.				0,02		
79	86m → 9m → 4m / 65000a → stab.						
80	36s → stab.						
81	<10m ← 57m / 18m → stab.				0,133		
82	stab. → 36h → stab						
83	25m / 63s → 2,3h → 1,9h / stab.			1,14	0,60	0,08	
84	~2m → 30m → stab.			1,90	1,1		
85	0,43s → kurz → 3m → 4,4h / 10,6a → stab.			2,6	1,5		
86	stab. → 19d → stab.			3,2	2,1		
87	55,6s → 78m → nat.				2,7		
88	15,5s → 2,8h → 17,7m → stab.				3,6		
89	4,5s → 3,2m → 15m → 51d → stab.			6,5	4,8	1,8	3
90	33s → 2,7m → 28a → 65h → stab.				5,8		
91	9,8s → 14m / 100s → 9,7h → 50m / 57d → stab.			6,5	5,9	2,8	
92	3s → 80s → 2,7h → 3,4h → stab.			6,7	6,0		
93	2s → kurz → 7m → 10h → 0,95Ma → 4,2a / stab.			7,1	6,4		
94	1,4s → kurz → 2m → 16,5m → stab.			6,8	6,4		
95	kurz → kurz → kurz → 10,5m → 65d → 35d → stab.			6,1	6,3	5,6	5,5
96	23h → stab.			6,1	6,3		
97	~1s → kurz → kurz → kurz → 17h → 1m / 72m → stab.			5,6	6,3	5,3	
98	30m → stab.			5,3	5,8		
99	2,5m → 68h → 6h / 0,22Ma → stab.			4,7	6,1	6,1	5,9
100	stab.				6,3		
101	14,6m → 14,3m → stab.				5,0		

Tabelle 7 (Fortsetzung)

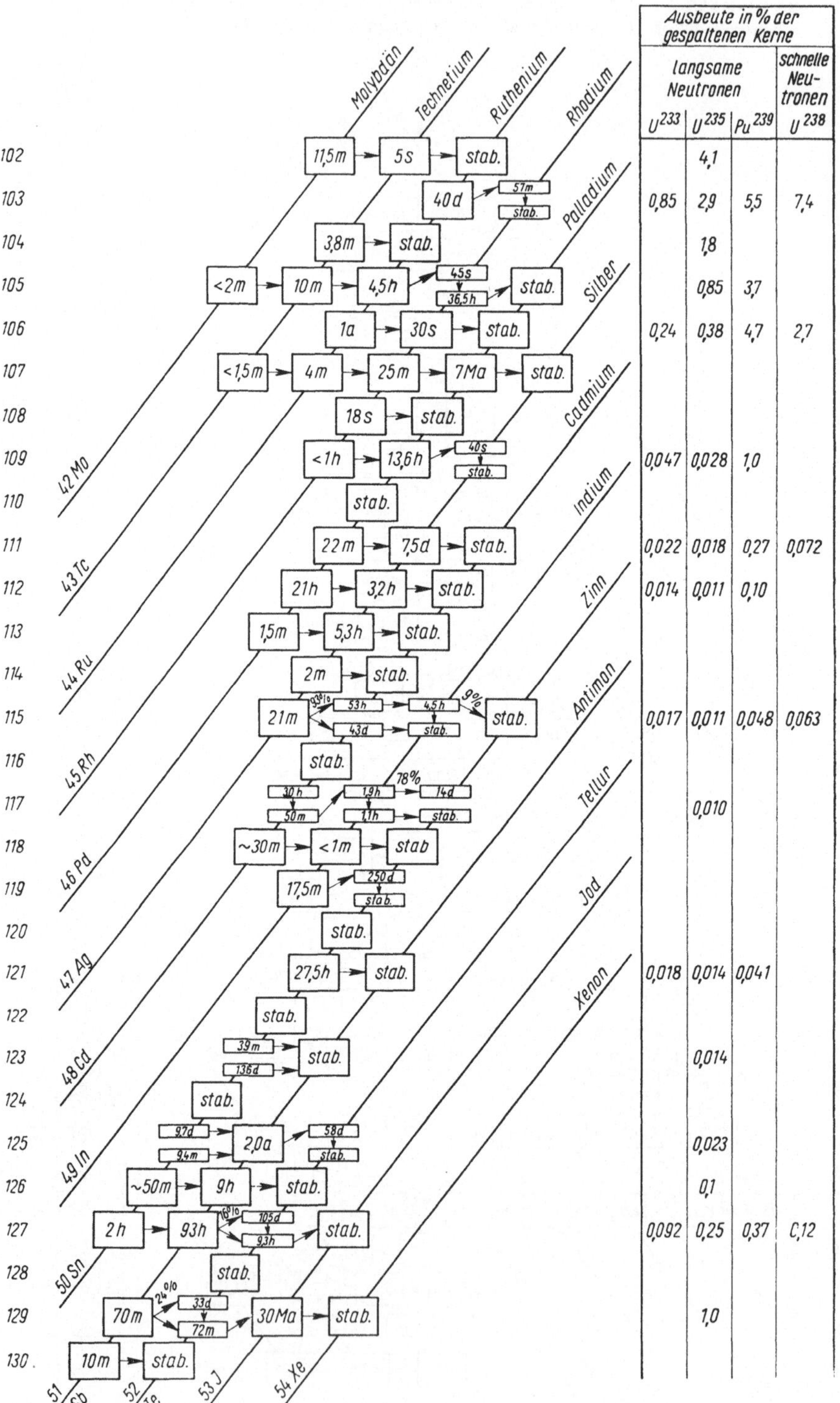

Tabelle 7 (Fortsetzung)

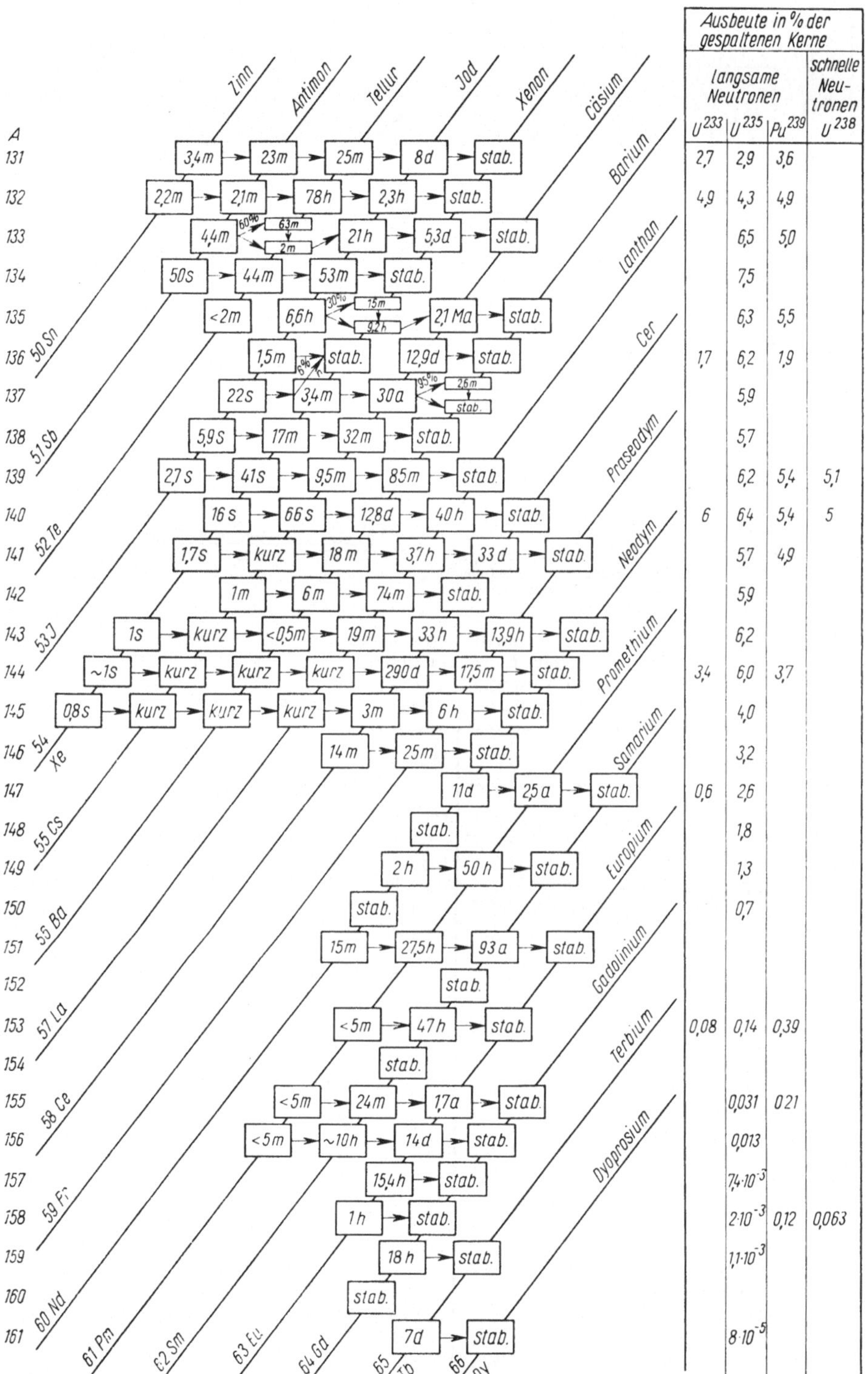

A	Ausbeute in % der gespaltenen Kerne			
	langsame Neutronen			schnelle Neutronen
	U^{233}	U^{235}	Pu^{239}	U^{238}
131	2,7	2,9	3,6	
132	4,9	4,3	4,9	
133		6,5	5,0	
134		7,5		
135		6,3	5,5	
136	1,7	6,2	1,9	
137		5,9		
138		5,7		
139		6,2	5,4	5,1
140	6	6,4	5,4	5
141		5,7	4,9	
142		5,9		
143		6,2		
144	3,4	6,0	3,7	
145		4,0		
146		3,2		
147	0,6	2,6		
148		1,8		
149		1,3		
150		0,7		
151				
152				
153	0,08	0,14	0,39	
154				
155		0,031	0,21	
156		0,013		
157		$7,4 \cdot 10^{-5}$		
158		$2 \cdot 10^{-3}$	0,12	0,063
159		$1,1 \cdot 10^{-3}$		
160				
161		$8 \cdot 10^{-5}$		

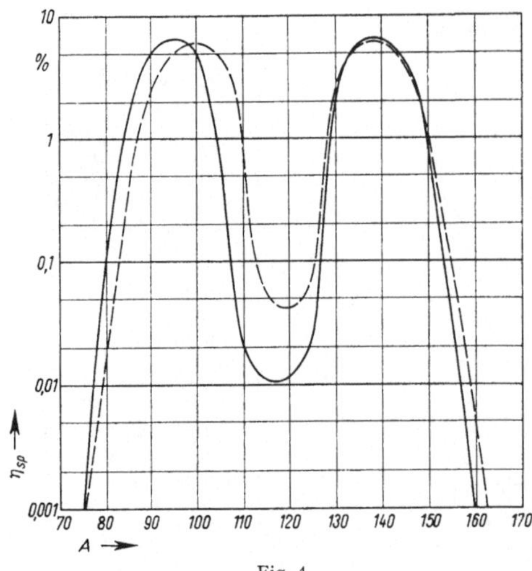

Fig. 4

Fig. 4. Spaltausbeute η_{sp} in Prozent der gespalteten Pu^{235}-(ausgezogene Kurve) bzw. U^{239}-Atome (gestrichelte Kurve) in Abhängigkeit von der Massenzahl A (nach Riezler-Walcher 1958)

Fig. 5. Mittlere Werte des Gehaltes der Luft an Kernspaltungsprodukten in verschiedenen Höhen nach englischen Flugzeugmessungen im Herbst 1954 (nach Stewart 1957)

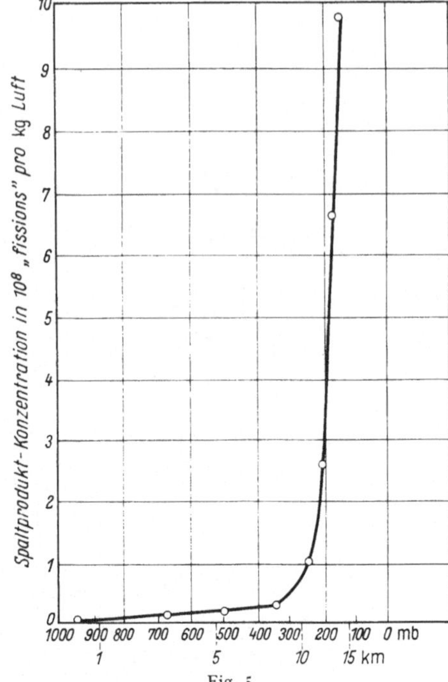

Fig. 5

Tabelle 8. *Radioaktive Eigenschaften einiger wichtiger Spaltprodukte*
A Massenzahl; HWZ Halbwertszeit; E_β Maximalenergie der β-Strahlung in MeV, k bedeutet dabei komplexes β-Spektrum; E_γ Energie der γ-Quanten in MeV; % Häufigkeit in Prozent gespaltener U^{235}-Atome (thermische Neutronen) (nach Riezler-Walcher 1958)

A	Element	HWZ		E_β	E_γ	%	A	Element	HWZ		E_β	E_γ	%
85	Kr	4,4	h	0,85	0,15	1,2	129	Te	1,2	h	1,8	0,3	1,0
85	Kr	10,6	a	0,7	—	0,32		J	30	Ma	0,15	0,038	
87	Kr	1,3	h	3,2	—	2,7	131	J	8	d	0,6 (k)	0,36 u.a.	2,9
88	Kr	2,77	h	2,7 u.a.	2 u.a.	3,6	132	Te	78	h	0,22	0,23	4,3
	Rb	17,7	m	5,3 (k)	1,85 u.a.			J	2,3	h	—	—	
89	Sr	51	d	1,5	—	4,8	133	Xe	5,3	d	0,34	0,08	6,5
90	Sr	28	a	0,6	—	5,8	135	J	6,65	h	1,4 (k)	1,38 u.a.	
	Y	65	h	2,2	—			Xe	9,2	h	0,91	0,25	6,3
91	Sr	9,7	h	3,2 (k)	1,41 u.a.			Cs	2,1	Ma	0,21	—	
	Y	50	m	—	0,551	5,8	137	Cs	30	a	0,52	—	5,9
	Y	57	d	1,5	—			Ba	2,6	m	—	0,66	
92	Sr	2,7	h	—	—	5,0	139	Ba	85	m	2,38(k)	0,165	6,2
	Y	3,4	h	3,6	0,94 u.a.	6,0	140	Ba	12,8	d	1,02(k)	0,16 bis 0,54	6,4
93	Y	10	h	3,1	0,7	6,3		La	40	h	2,2	0,09 bis 2,5	
	Zr	0,95	Ma	0,063	—		141	La	3,7	h	2,43	—	5,7
95	Zr	65	d	0,37	0,73	6,4		Ce	33	d	0,58	0,14	
	Nb	35	d	0,16	0,95		142	La	74	m	2,5	0,63; 0,87	5,9
96	Nb	23	h	0,750	0,77 u.a.	6,3	143	Ce	33	h	1,40(k)	0,057 bis 1,10	6,2
97	Nb	1,2	h	1,27	0,66	6,3		Pr	13,9	h	0,93	—	
99	Mo	68	h	1,23	0,04 bis 0,78		144	Ce	290	d	0,30(k)	0,3 bis 0,13	6,0
	Tc	6	h	—	0,14	6,1		Pr	17,5	m	2,97	2.2 u.a.	
	Tc	0,22	Ma	0,29	—		145	Pr	6	h	1,7	—	4,0
103	Ru	40	d	0,22	0,50	2,9	147	Nd	11	d	0,8 (k)	0,52 u.a.	2,6
	Rh	57	m	—	0,04			Pm	2,5	a	0,22	—	
106	Ru	1	a	0,039	—	0,38	149	Nd	2	h	1,5 (k)	0,030 bis 0,65	1,3
	Rh	30	s	3,5 (k)	0,5 u.a.			Pm	50	h	0,97	0,285	
107	Pd	7	Ma	0,04	—	(0,1)	151	Sm	93	a	0,076	0,019	(0,5)
							155	Eu	1,7	a	0,24(k)	0,018 bis 0,102	0,031

in die untere Stratosphäre emporgetragen. Da ihre Ausscheidung zum Boden eine mit der Höhe wachsende Zeit beansprucht, ist eine Konzentrationszunahme von Spaltprodukten mit der Höhe zu erwarten und durch gelegentliche Flugzeugmessungen auch bestätigt (vgl. z.B. Fig. 5).

III. Zeitliche Variationen

Da ein bei der Kernspaltung zu einem bestimmten Zeitpunkt entstandenes Gemisch von Spaltprodukten aus Isotopen der verschiedensten Zerfallskonstanten besteht, klingt seine Strahlung J im ganzen nach einem eigenen Gesetz ab, das sich innerhalb gewisser zeitlicher Grenzen in der Form

$$J = J_0 \cdot t^{-(1+x)} \tag{18}$$

darstellen läßt. Für x ergibt sich dabei in der Regel ein Wert von etwa 0,2[1].

Tabelle 9. *Beiträge zur Spaltprodukt-Aktivität bei momentaner Spaltung von U^{235} durch thermische Neutronen nach* HUNTER *und* BALLOU *1951*
Die Angaben sind Aktivitätsprozente, nicht Gewichtsprozente. Die jeweils vorherrschenden Anteile sind durch Kursivdruck hervorgehoben. Der zeitlichen Änderung des Verhältnisses $Sr/Y^{90}:Cs/Ba^{137}$ kommt keine Bedeutung zu, da das Verhältnis der Halbwertszeiten von Sr^{90} und Cs^{137} noch immer ungenügend bekannt ist

Element	Alter in Tagen												
	10	20	30	40	50	60	80	100	150	200	250	300	365
Ba^{140}, La^{140}	*22,6*	*25,9*	*23,2*	*18,9*	14,6	11,0	5,4	2,2					
Xe^{133}	11,5	6,0	2,5	1,0									
Te^{132}, I^{132}	10,4	2,6											
Pr^{143}	10,0	12,0	11,2	9,6	7,8	6,2	3,4	1,4					
I^{131}	6,8	5,6	3,6	2,3	1,4								
Mo^{99}	6,8	1,1											
Ce^{141}	6,3	9,7	11,2	11,7	11,3	10,8	9,3	7,8	4,5	1,8			
Ru^{103}, Rh^{103}	5,1	8,6	11,2	13,0	14,0	14,6	14,6	13,8	10,5	7,1	4,5	2,9	
Nd^{147}, Pm^{147}	4,8	5,0	4,1	3,1	2,1	1,4			1,4	2,2	3,2	4,2	5,8
Zr^{95}, Nb^{95}	3,3	8,1	12,2	16,6	*20,5*	*24,5*	*30,3*	*34,5*	*39,2*	*39,4*	*35,4*	29,5	22,5
Y^{91}	3,2	5,6	7,6	9,0	10,2	11,2	12,3	12,8	12,0	10,3	9,5	6,5	3,8
Sr^{89}	2,9	5,0	6,7	8,0	9,0	9,8	10,3	10,5	9,6	7,9	6,1	4,5	2,7
Ce^{144}, Pr^{144}		2,6	4,0	5,4	6,6	8,0	10,6	13,4	19,6	26,0	33,6	*42,0*	*53,0*
Ru^{106}, Rh^{106}											3,0	3,8	4,8
Sr^{90}, Y^{90}											2,0	2,6	3,8
Cs^{137}, Ba^{137}												2,0	3,0

Element	Alter in Jahren										
	1	2	3	4	6	8	10	15	20	50	100
Zr^{95}, Nb^{95}	22,0	1,0									
Y^{91}	3,8										
Sr^{89}	2,7										
Ce^{144}, Pr^{144}	*53,9*	*60,0*	*41,0*	*27,0*	8,6	1,9					
Pm^{147}	5,8	13,5	19,0	21,3	21,2	18,8	15,5	8,0	3,4		
Ru^{106}, Rh^{106}	4,8	6,8	6,0	4,7	2,3						
Sr^{90}, Y^{90}	3,8	10,4	17,4	24,0	*34,0*	*40,0*	*43,8*	*47,6*	*48,5*	44,0	34,0
Cs^{137}, Ba^{137}	3,0	8,0	13,6	18,8	27,4	33,0	36,4	41,8	44,0	*54,0*	*64,0*
Sm^{151}			1,2	1,5	2,0	2,3	2,5	2,6	2,6	2,0	1,1
Kr^{85}			1,2	1,5	1,7	1,7	1,5	1,2			

[1] Zu Erklärung und Gültigkeitsbereich s. z.B. K. WAY und E.P. WIGNER (1948). — Die Anwendung dieses Zerfallsgesetzes gestattet bekanntlich durch Extrapolation auf $t = 0$ die Altersbestimmung eines durch Kernspaltung entstandenen Isotopengemisches (vgl. z.B. O. HAXEL und G. SCHUMANN 1953).

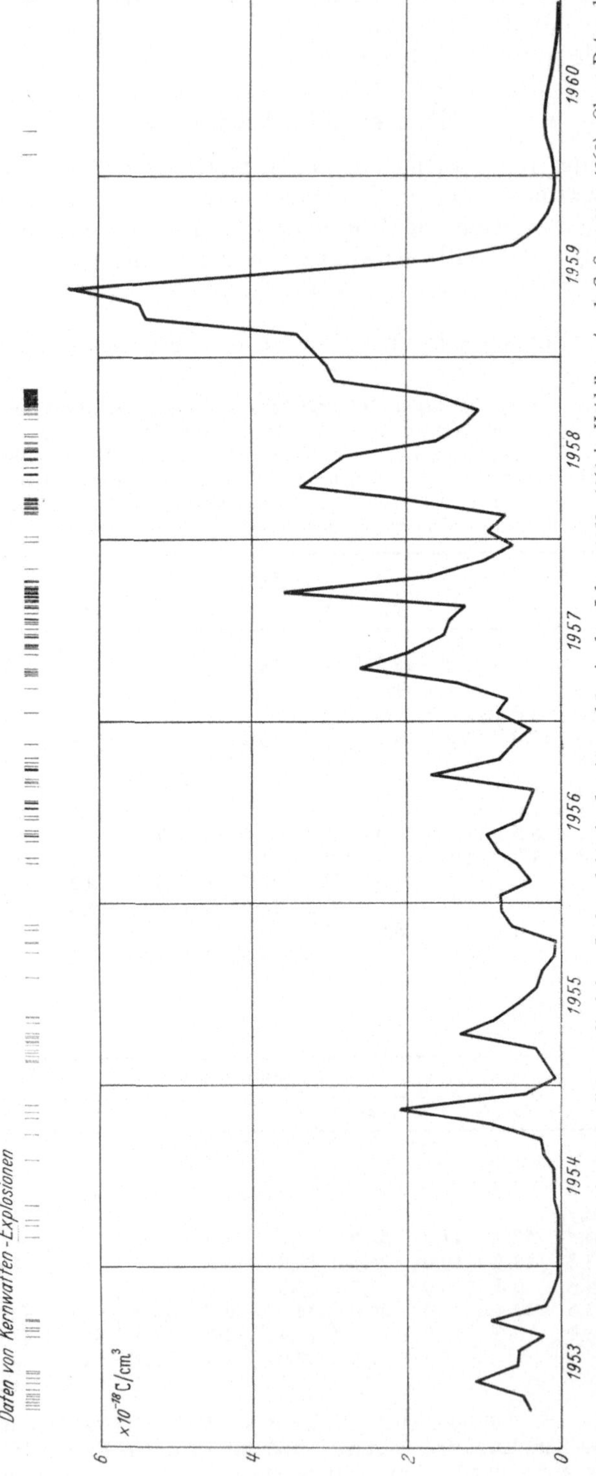

Fig. 6. Die Konzentration langlebiger[1] künstlicher radioaktiver Spaltprodukte in der Atmosphäre in den Jahren 1953—1960 in Heidelberg (nach G. Schumann 1960). Oben: Daten der Kernwaffenexplosionen

[1] „langlebig" im Vergleich zu den kurzlebigen Ra- und Th-Induktionen RaA bis RaC und ThA bis ThC.

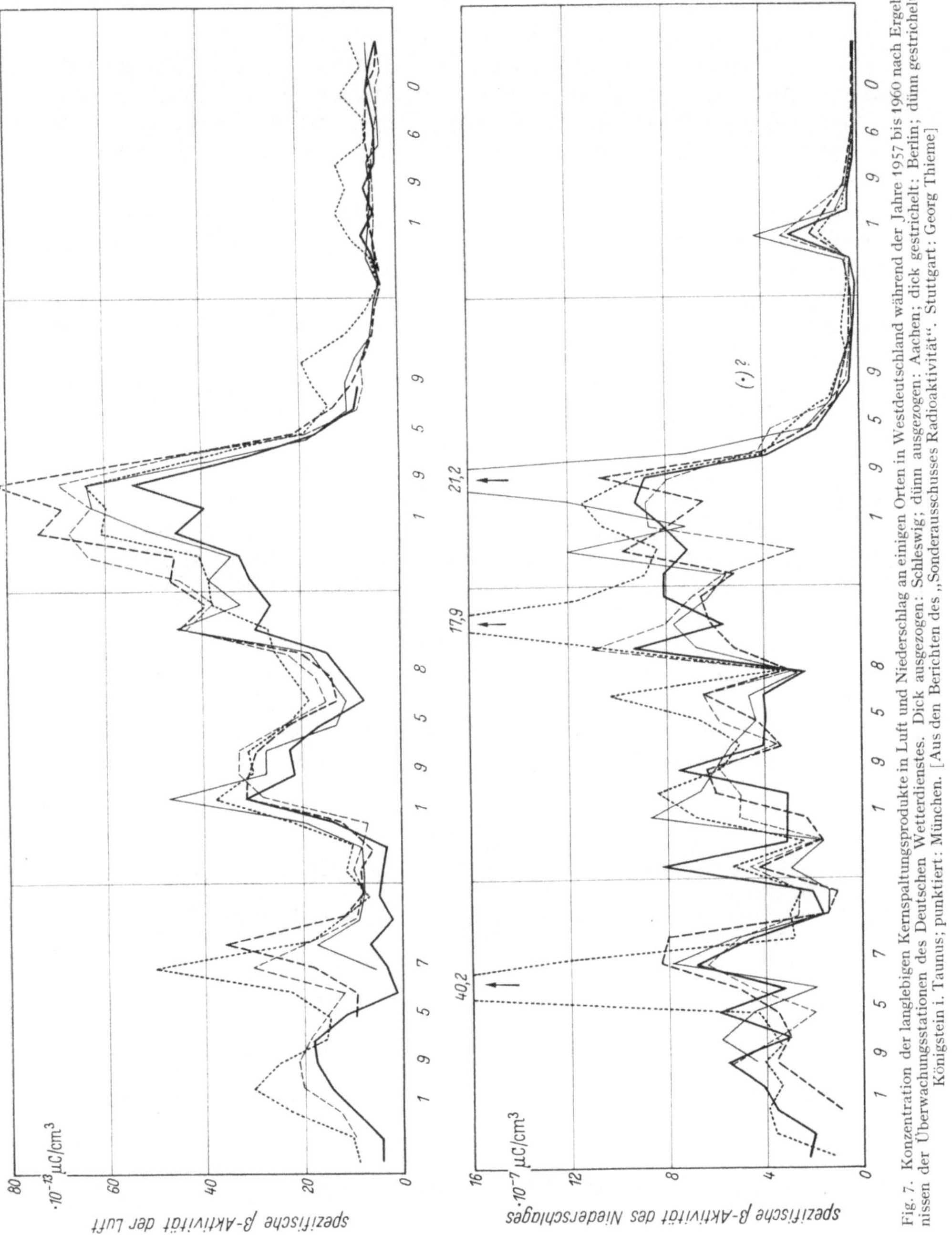

Fig. 7. Konzentration der langlebigen Kernspaltungsprodukte in Luft und Niederschlag an einigen Orten in Westdeutschland während der Jahre 1957 bis 1960 nach Ergebnissen der Überwachungsstationen des Deutschen Wetterdienstes. Dick ausgezogen: Schleswig; dünn ausgezogen: Aachen; dick gestrichelt: Berlin; dünn gestrichelt: Königstein i. Taunus; punktiert: München. [Aus den Berichten des „Sonderausschusses Radioaktivität". Stuttgart: Georg Thieme]

Zur Beurteilung der prozentualen stofflichen Zusammensetzung eines solchen Isotopengemisches nach bestimmter Zerfallszeit kann man nach den Rechnungen von H. F. Hunter und N. E. Ballou (1951) etwa die in Tabelle 9 gegebene Übersicht gewinnen.

Über die Auswirkungen der rund 500 Explosionen, die bis zum Versuchsstop im Oktober 1958 insgesamt ausgelöst worden sind, ist aus den Ergebnissen der Verseuchungsmessungen in allen Teilen der Welt folgendes zu entnehmen:

Die seit 1950 ständig ansteigende Zahl der Versuchsexplosionen hat ein entsprechendes Ansteigen der „Verseuchung" der Atmosphäre zur Folge gehabt. Fig. 6 zeigt dies nach Messungen von G. Schumann (1960) in Heidelberg für die Jahre 1953 bis 1960. Die Figur zeigt außer dem durch einen überlagerten Jahresgang modifizierten stetigen Ansteigen der Werte von 1953 an sowie den raschen Rückgang der atmosphärischen „Verseuchung", der etwa $\frac{1}{2}$ Jahr nach dem Kernwaffenversuchsstop eingesetzt hat.

Fig. 7 zeigt für fünf westdeutsche Stationen (Schleswig, Aachen, Berlin, Königstein i. Taunus und München) monatliche Mittelwerte der von langlebigen Kernspaltungsprodukten in Luft und Niederschlag herrührenden β-Strahlung.

Daß in Fig. 7 die Kurven der Luftaktivität nicht auf den Wert Null absinken, sondern nur bis auf einen Wert von etwa $4\cdot10^{-19}$ Curie/cm³ zurückgehen, erklärt sich dadurch, daß bei diesen Messungen die Wartezeit, die man zwischen Sammlung und Messung des atmosphärischen Staubniederschlages einschalten muß, um die natürliche Radioaktivität abklingen zu lassen, nur 48 Std betrug. Infolgedessen kommt der ThB-Gehalt der Atmosphäre mit etwa diesem Wert von $4\cdot10^{-19}$ Curie/cm³ zur Auswirkung (H. Israël 1961). — Die relativ hohen Werte des ThB in München (punktierte Kurve), die im unteren Teil der Fig. 7 besonders ins Auge fallen, bringt A. Maas (1961) mit der Nähe der Alpen in Verbindung (höhere Exhalation des Urgesteinmassivs?).

Bei den Niederschlagsaktivitäten zeichnet sich in den Märzwerten 1960 die Wirkung des Sahara-Kernwaffenversuchs vom Februar 1960 ab.

Weitere Zahlenangaben zur Verbreitung von Kernspaltungsprodukten in Luft, Wasser und Boden enthält die Veröffentlichung Environmental contamination from weapon tests (1958).

Literatur

Baranow, W. J., u. E. G. Gratschewa: Zur Theorie der Emanationsvermessung. [Russisch.] Tr. Gos. radievogo in-ta **2** (1933).
— — Über den radioaktiven Gasaustausch zwischen Boden und Atmosphäre. [Russisch.] Izv. Acad. Nauk SSSR., Ser. geogr. **1** (1933).
Bavel, C. M. H. van: Gaseous diffusion and porosity in porous media. Soil Sci. **73**, 73—80 (1952).
Bender, H.: Über den Gehalt der Bodenluft an Radium-Emanation. Gerlands Beitr. Geophys. **41**, 401—415 (1934).
Berichte des „Sonderausschuß Radioaktivität", Bundesrepublik Deutschland. Stuttgart: Georg Thieme 1958, 1959 und später.
Blifford, I. H., L. B. Lockhardt and R. A. Baus: Geographical and time distribution of radioactivity in the air. J. Atmosph. Terr. Phys. **9**, 1—17 (1956).
Boltwood, B. B.: On the radioactivity of uranium minerals. Amer. J. Sci. **25**, 269—298 (1908).
Budde, E.: Bestimmung des Beweglichkeits-Koeffizienten der Radium-Emanation in Lockergesteinen. Z. Geophys. **24**, 96—105 (1958). Siehe auch Diss. Münster 1957, sowie Z. Geophys. **26**, 72—76 (1960) [s. dazu H. Israël, Z. Geoph. **25** (1959) und **27** (1961)].
Bulashevich, Yu. P.: Method of determining the emanation coefficient of rocks in their natural occurrence. [Russisch.] Izv. Akad. Nauk SSSR., Geoph. Ser. **1958**, 1383—1388.
—, and R. K. Khairitdinov: On the theory of emanation diffusion in porous media. [Russisch.] Izv. Akad. Nauk SSSR. **1959**, 1787—1792.
Burton, W. H., and N. G. Stewart: The radiochemical analysis of long-lived radon decay products and their use as natural atmospheric tracers. AERE-HP/R 2084, Harwell, Berkshire 1960.
El-Nadi, A. F., and H. G. Omar: Radon and thoron content in atmospheric air at Hiza, Egypt. Geofis. pura e appl. **45**, 261—266 (1960).

Environmental contamination from weapon tests. United States Atomic Energy Commission, Health and Safety Laboratory, New York Operations Office, October 1958.

EVANS, R.D., and C. GOODMAN: Radioactivity of rocks. Bull. Geol. Soc. Amer. **52**, 459ff. (1941).

GOCKEL, A.: Über den Gehalt der Bodenluft an radioaktiver Emanation. Phys. Z. **9**, 304ff. (1908).

HAXEL, O.: Eine einfache Methode zur Messung des Gehaltes der Luft an radioaktiven Substanzen. Z. angew. Phys. **5**, 241—242 (1953).

—, u. G. SCHUMANN: Über die radioaktive Verseuchung der Atmosphäre. Naturwissenschaften **40**, 458—459 (1953).

HESS, V.F.: Radon, thoron, and their decay products in the atmosphere. J. Atmosph. Terr. Phys. **3**, 172—177 (1953).

—, and R.P. VANCOUR: The ionization balance of the atmosphere. J. Atmosph. Terr. Phys. **1**, 13—25 (1950).

HUNTER, H.F., and N.E. BALLOU: Fission-product decay rates. Nucleonics **9**, Nr. 5, C2—C7 (1951).

ISRAËL, H.: Emanation und Aerosol. Gerlands Beitr. Geophys. **42**, 385—408 (1934).

— Zur Frage der Adsorption von Radium-Emanation an Aerosolteilchen (Bemerkungen zu den Arbeiten von G. ALIVERTI und G. ROSA über das gleiche Thema). Gerlands Beitr. Geophys. **46**, 413—417 (1936).

— Die Radioaktivität als Klimafaktor. Forsch. u. Fortschr. **14**, 103—105 (1938).

— Radioactivity of the atmosphere. Compendium of Meteorology (American Meteorological Soc.), pp. 155—161, Boston, Mass. 1951.

— Atmosphärische Elektrizität. Teil I. Grundlagen, Leitfähigkeit, Ionen. IX u. 370 S. (Probleme der kosmischen Physik, Bd. 29.) Leipzig: Akademische Verlags-Gesellschaft 1957.

— Zur Vergleichbarkeit von Radioaktivitätsmessungen. Atomkernenergie **6**, 218—222 (1961).

—, u. F. BECKER: Die Bodenemanationen in der Umgebung der Bad Nauheimer Quellenspalte. Gerlands Beitr. Geophys. **44**, 40—55 (1935).

— — Die Emanationsverhältnisse in der Bodenluft. Gerlands Beitr. Geophys. **48**, 13—58 (1936).

—, S. BJÖRNSSON u. S. STILLER: Emanometrische Messung von Radon und Thoron in Bodenluft. Z.f. Geoph. 1962 (im Druck).

—, u. J. SALDUKAS: Boden-Emanation und Gammastrahlung in diluvialen Sandboden (Potsdam). Meteor. Z. **56**, 39—41 (1939).

ISWEKOW, B.: Zur Frage des täglichen Windganges. Meteor. Z. **46**, 1—7 (1929).

JACOBI, W.: Natural atmospheric radioactivity. Hahn-Meitner-Institut für Kernforschung, Berlin-Wannsee 1960.

— A. SCHRAUB, K. AURAND u. H. MUTH: Über das Verhalten der Zerfallsprodukte des Radons in der Atmosphäre. Beitr. Phys. Atmosph. **31**, 244—257 (1959).

JAKI, S.L., and V.F. HESS: A study of the distribution of radon, thoron, and their decay products above and below the ground. J. Geophys. Res. **63**, 373—390 (1958).

JAUFMANN, J.: Untersuchungen über den radioaktiven und elektrischen Zustand der Atmosphäre. Dtsch. meteor. Jb. Bayern **29** (1907).

JOST, W.: Diffusion. New York: Academic Press 1952.

KÄHLER, K.: Registrierungen des Emanationsgehaltes der Bodenluft in Potsdam mit dem Benndorf-Elektrometer. Phys. Z. **15**, 27—31 (1914). Siehe auch Ergebnisse der meteorologischen Beobachtungen in Potsdam im Jahre 1912. Berlin 1913.

KAWANO, M., and S. NAKATANI: The absolute measurement of the concentration of the radioactive substances in the atmosphere in Tokyo. J. Geomagn. Geoelectr., Japan **10**, 56—63 (1959).

KOSMATH, W.: Die Exhalation der Radiumemanation aus dem Erdboden und ihre Abhängigkeit von den meteorologischen Faktoren. Gerlands Beitr. Geophys. **43**, 258—279 (1935).

KRATZER, A.: Das Stadtklima. 1937.

LANDOLT-BÖRNSTEIN: Zahlenwerte und Funktionen, 6. Aufl., Bd. 3, S. 307ff. 1952.

LETTAU, H.: Anwendung neuerer Ergebnisse der Austauschlehre auf zwei luftelektrische Fragen. Gerlands Beitr. Geophys. **57**, 365—383 (1941).

— Diffusion in the upper atmosphere. Compendium of Meteorology, S. 320—333. Boston, Mass. 1951.

LUDEWIG, P., u. E. LORENSER: Untersuchungen der Grubenluft in den Schneeberger Gruben auf deren Gehalt an Radium-Emanation. Z. Physik **22**, 178—185 (1924).

MAAS, A.: Sur la proportion des produits de filiation du thoron dans l'activité des aérosols dans l'air a proximité du sol et measures comparatives au moyen d'appareils de différente fabrication. Atompraxis **5**, 173—177 (1961).

Merkulowa, E. S.: Exhalation der Radium-Emanation aus dem Boden nach Beobachtungen in Slutzk. [Russisch.] 1937. Zit. nach B. Styra, Questions of nuclear meteorology. Akad. Lithuanien SSR, Monographic Series, I, Vilnius, 1959, 414 S.

Meyer, St., u. E. v. Schweidler: Radioaktivität. VII. Kap. Die Radioaktivität in Geophysik und kosmischer Physik, S. 545—624. Leipzig 1927.

Olujić, J.: Beiträge zur Messung der Radiumemanation in der Atmosphäre. Jb. Radioakt. u. Elektronik **15**, 158—194 (1918).

Riezler, W., u. W. Walcher: Kerntechnik; Physik-Technik-Reaktoren. 1002 S. Stuttgart 1958.

Sanderson, J. C.: The probable influence of the soil on local atmospheric radioactivity. Amer. J. Sci. **32**, 169—184 (1911). Deutsche Übersetzung s. Phys. Z. **13**, 142ff. (1912).

Satterly, J.: A Study of radium emanation contained in the air of various soils. Proc. Cambridge Phil. Soc. **16**, 336—355, 1910—1912 (1911a).

— The quantities of radium and thorium emanations contained in the air of certain soils. Proc. Cambridge Phil. Soc. **16**, 514—533, 1910—1912 (1911b).

Schmidt, W.: Zur Verteilung radioaktiver Stoffe in der freien Luft. Phys. Z. **27**, 371—378 (1926).

Schumann, G.: Nach freundlicher persönlicher Mitteilung; unveröffentlicht, 1960.

Schwalb, K.: Beiträge zur Kenntnis der Radium-Emanation in der Atmosphäre. Bioklim. Bibl. meteorol. Z. **8**, 82—90 (1941).

Smyth, L. B.: On the supply of radium emanation from the soil to the atmosphere. Phil. Mag. (6) **24**, 632, 637 (1912).

Stewart, N. G., R. N. Crooks and E. M. R. Fisher: The radiological dose to persons in the U.H. due to depris from nuclear test explosions prior to January 1956. A.E.R.E. (Atomic Energy Research Establishment), Harwell, Berksh. H.P./R. 2017 (1957).

Sutton, O. G.: Atmospheric turbulence and diffusion. Compendium of Meteorology, S. 492—509. Boston, Mass. 1951.

Vancour, R. P.: Thesis, Fordham University, New York 1950.

Vries, D. A. de: Some remarks on gaseous diffusion in soils. Trans. 4th Intern. Congr. Soil Sci. **2**, 41—45 (1950).

Way, K., and E. P. Wigner: The rate of decay of fission products. Phys. Rev. **73**, 1318—1330 (1948).

Wilkening, M. H., and J. E. Hand: Radon flux at the earth-air interface. J. Geophys. Res. **65**, 3367—3370 (1960).

Wilson, J. T., R. D. Russel and R. M. Farquhar: Radioactivity and age of minerals. In Handbuch der Physik (S. Flügge), Bd. 47, S. 288—363 (1956).

Wright, J. R., and O. F. Smith: The variation with meteorological conditions of the amount of radium emanation in the atmosphere, in the soil, and in the air exhalated from the surface of the ground at Manila. Phys. Rev. **5**, 459—482 (1915).

Zeilinger, P. R.: Über die Nachlieferung von Radiumemanation aus dem Erdboden. Terr. Magn. **40**, 281—294 (1935).

Zupancic, P. R.: Messungen der Exhalation von Radiumemanation aus dem Erdboden. Terr. Magn. **39**, 33—46 (1934).

Erzeugung radioaktiver Kernarten durch die kosmische Strahlung

von

O. HAXEL und G. SCHUMANN

Mit 10 Figuren

Summary

Radioactive nuclides are produced by cosmic radiation in the atmosphere, the hydrosphere and the lithosphere. The present knowledge mainly refers to production in the atmosphere and in meteorites. The most efficient nuclear reactions inducing radioactivity are the spallations of the constituents of the atmosphere or of iron in the case of meteorites. There is only one important exception: thermal neutron capture by nitrogen bringing forth C^{14}.—The production of radioactive nuclides strongly depends on altitude and geomagnetic latitude according to the behaviour of the nucleonic or star-producing component of cosmic radiation. The intensity of the cosmic radiation may vary considerably during short periods. Its mean value appears to have been constant within a factor of 2 for millions of years and to have altered not much more during the last 10^9 years as indicated by observations on long-lived radioactive species in meteorites.—The C^{14} produced in the atmosphere is distributed over the biosphere and the ocean also. A steady-state model gives a good approximation for the reservoirs and the exchange times. The natural concentration of C^{14} is diluted by "old" CO_2 due to industrial combustion and enlarged by additional production due to nuclear weapon tests. C^{14} measurements have proved to be a valuable tool for studying atmospheric and oceanic transport phenomena and ground water problems.—Be^7 and Be^{10} are spallation products of nitrogen. Be^7 is easily detected in atmospheric air and precipitation. Be^{10} was observed in deep-sea cores. Numerous nuclides are produced by spallation of argon. Up to now Na^{22}, Si^{32}, P^{32}, P^{33}, S^{35}, Cl^{36}, Cl^{39} were detected. There is a certain contribution from weapon tests to S^{35} and a very large one to Cl^{36}. Short-lived nuclides reaching ground-level are mainly produced in the troposphere. Their concentrations can be used to calculate storage times.—The natural tritium production is not known too well. Only very few measurements were made before the start of artificial production by weapon tests which was higher by orders of magnitude. Some investigations with Greenland ice were carried out back to 1944. The fluctuations of the atmospheric tritium content in recent years can not be explained without doubt as yet. Tritium has been used for hydrological studies. To construct a model for the world-wide distribution similar to that for C^{14} is still unfeasible.—Production of radioactive nuclides in the terrestrial lithosphere was detected only in high altitude. Many radioactive substances have been found in the meteorites which were exposed to a rather intense and energetic cosmic radiation for a long time. With recent falls short-lived activities down to 16 days half-life have been observed. The investigations are useful to study possible variations of cosmic radiation in time and space. Regarding space too few measurements are available at present.

A. Entstehungs-Prozesse

I. Kosmische Strahlung in der Atmosphäre

Eine von den mannigfachen Sekundärwirkungen der aus dem Weltraum in die Erdatmosphäre einfallenden kosmischen Strahlung ist die Erzeugung verschiedener Arten radioaktiver Nuklide. Die Primärteilchen, d.h. die von außen in die Atmosphäre eintretenden Teilchen, setzen sich zusammen zu 91,5% aus Protonen, 7,8% aus α-Teilchen und 0,7% aus schweren Kernen, wobei solche mit Kernladungszahlen über 30 äußerst selten sind [Peters (1)]. Freie Neutronen sind in der Primärstrahlung praktisch nicht enthalten, da ihre mittlere Lebensdauer nur 17 min beträgt und sie daher nur kurze Strecken zurücklegen können. Die Energie der Primärteilchen kann außerordentlich hohe Beträge erreichen (Fig. 1). Bei ihren Zusammenstößen mit Atomkernen der Atmosphären-Bestandteile werden Nukleonen frei, die ihrerseits mit weiteren Kernen in Wechselwirkung treten. Mit zunehmendem Eindringen in die Atmosphäre bildet sich so die Nukleonenkaskade. Die Nukleonen-Komponente der kosmischen Strahlung ist hauptsächlich für die Erzeugung radioaktiver Nuklide verantwortlich.

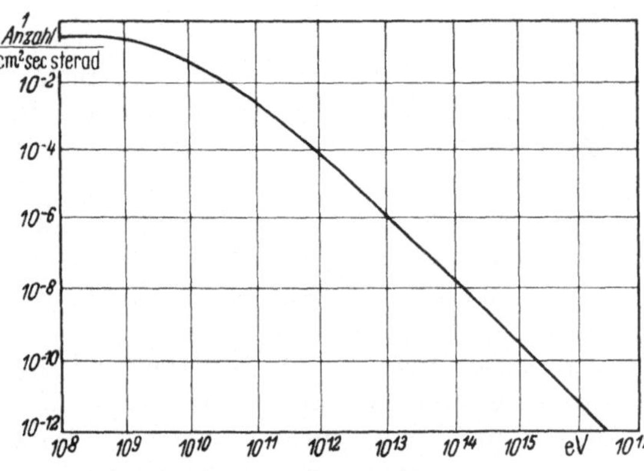

Fig. 1. Integrales Primärspektrum der Protonen der kosmischen Strahlung

Bei den hochenergetischen Wechselwirkungen werden ferner π-Mesonen erzeugt. Beim Zerfall der neutralen π^0-Mesonen entstehen energiereiche γ-Quanten. Aus ihnen bildet sich im weiteren Verlauf durch Paarerzeugung und Vernichtung sowie andere elektromagnetische Wechselwirkungen im Feld von Atomkernen die Elektronen-Photonen-Kaskade. An Reaktionen mit den Kernen der Atmosphären-Bestandteile kommen hier nur Kernphotoeffekte in Betracht.

Dabei könnten nur γ-Quanten mit Energien in der Gegend der Riesenresonanz einen wesentlichen Beitrag liefern. Doch die Wirkungsquerschnitte sind selbst in diesem Bereich so klein, daß sie hinter denen der durch Nukleonen ausgelösten Reaktionen weit zurückbleiben (für N^{14}, O^{16}, A^{40} maximal 15,3; 7,7; 31,2 mbn, integral 0,061; 0,031; 0,23 MeV·bn nach G. A. Fergusson u. a.). Hinzu kommt, daß die entstehenden radioaktiven Kerne für die Messung ungünstige Halbwertszeiten (sehr kurz oder sehr lang) haben, so daß Erzeugung durch Kernphotoeffekt in der Atmosphäre bisher nicht nachgewiesen wurde.

Von den geladenen π-Mesonen sind die positiven einschließlich ihrer Zerfallsprodukte ohne Bedeutung für die Erzeugung radioaktiver Nuklide. Ihre Lebensdauer ist zu kurz und die Materiedichte in den höheren Atmosphärenschichten, wo sie hauptsächlich zu finden sind, zu klein, als daß sie wesentliche Beiträge zur Erzeugung radioaktiver Nuklide liefern könnten. Anders die aus den negativen π-Mesonen entstehenden μ^--Mesonen. Diese können wie die μ^+-Mesonen im Flug zerfallen oder, falls ihre Energie hoch genug ist, durch die Atmosphäre bis zum Erdboden dringen. Außerdem aber besteht eine endliche Wahrscheinlichkeit, daß ein solches μ^--Meson nach weitgehender Abbremsung durch Ionisationsvorgänge von einem Atomkern unter Bildung eines Mesonenatoms eingefangen wird (WHEELER). Die Lebensdauer eines solchen ist kurz und sinkt bei Kernen mit $Z > 10$ unter die normale Lebensdauer eines μ-Mesons, die 2,15 μsec beträgt. Die Abnahme erfolgt proportional Z^{-4}. Für $Z \approx 10$ ist der Zerfall des μ^--Mesons in ein Elektron und ein Neutrino gleich wahrscheinlich wie die Wechselwirkung des Mesons mit dem Kern nach dem Schema $\mu^- + p \to n + \nu$. Die Wahrscheinlichkeit der letzteren steigt dann mit zunehmendem Z stark an. Durch die Absorption des μ-Mesons wird der Kern hoch angeregt, und ein oder mehrere Neutronen verdampfen. Prozessse dieser Art liefern für einen kleinen Teil der von der kosmischen Strahlung erzeugten radioaktiven Stoffe Beiträge.

II. Nukleonen-Kaskade, Neutronen-Einfang und Spallation

Die weitaus größte Rolle für die Erzeugung aktiver Kerne in der Atmosphäre wie auch in der Lithosphäre spielt die Nukleonen-Komponente der kosmischen Strahlung. Sie hat für Energien oberhalb der Größenordnung GeV praktisch das gleiche Energiespektrum wie die Primärstrahlung. In der Nukleonen-Kaskade wird die Energie der einzelnen Nukleonen immer geringer. Bis etwa 500 MeV herunter

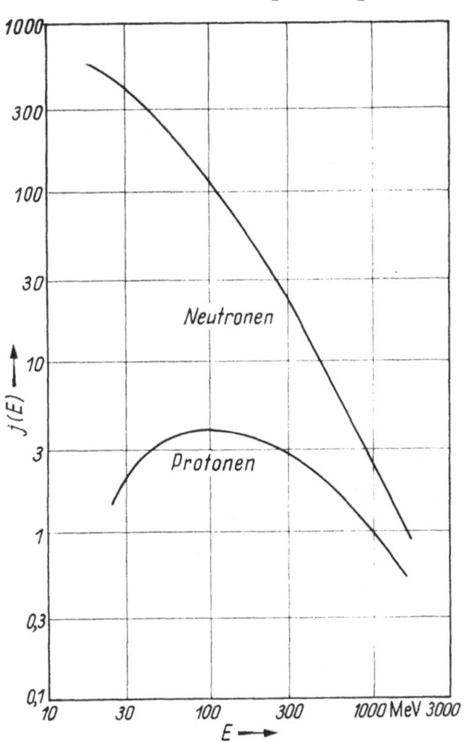

Fig. 2. Protonen- und Neutronen-Intensität in relativen Einheiten als Funktion der Energie für 700 g/cm² (etwa 3 km Höhe) nach ROSSI

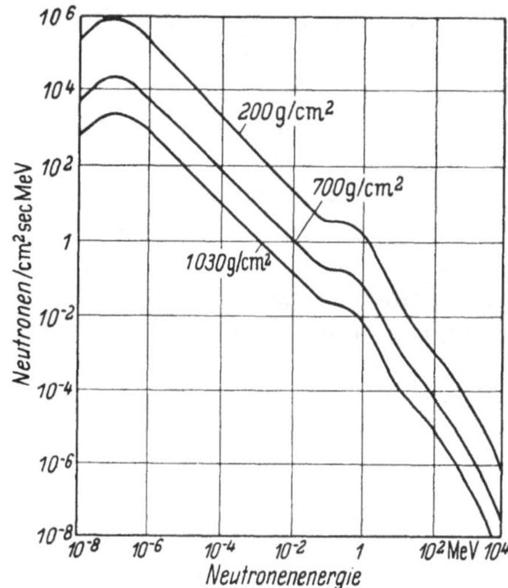

Fig. 3. Neutronenspektrum der kosmischen Strahlung in der Atmosphäre für 1030 g/cm² (Boden), 700 g/cm² (etwa 3 km Höhe) und 200 g/cm² (etwa 12 km Höhe) nach W.N. HESS u. a.

sind Protonen und Neutronen an den Kernwechselwirkungen praktisch gleichmäßig beteiligt. Bei tieferen Energien jedoch verlieren die Protonen ihre Energie nahezu ausschließlich durch Ionisation. Das kommt auch in dem mit abnehmender Energie stark zunehmenden Verhältnis der Neutronen- zur Protonendichte zum Ausdruck. ROSSI hat für eine Höhe von 700 g/cm² (etwa 3 km) die differentiellen Energiespektren der Protonen und Neutronen berechnet. Das Ergebnis gibt die Verhältnisse mindestens qualitativ gut wieder (Fig. 2). Die Neutronendichte nimmt nach kleineren Energien hin immer mehr zu und erreicht für thermische Energien ein Maximum (Fig. 3). In diesem Energie-Gebiet werden die Neutronen schließlich von den Stickstoff-Kernen der Atmosphäre eingefangen. Dabei entsteht C^{14}.

Für die Bildung aller übrigen radioaktiven Nuklide ist nahezu ausschließlich die hochenergetische Wechselwirkung der Nukleonen-Komponente mit den Atomkernen der Atmosphären-Bestandteile verantwortlich. Bei diesen Kernreaktionen handelt es sich um einen schwierig zu beschreibenden Typ, der bei Laborversuchen den Namen Spallation erhielt. Dabei haben wir es im vorliegenden Fall in erster Linie mit einem speziellen Bereich dieser Art von Reaktionen zu tun. Einerseits sind die Targetkerne auf diejenigen beschränkt, die Bestandteile der Atmosphäre sind, also insbesondere N^{14}, O^{16}, A^{40}. Andererseits kommen in erster Linie Energien unterhalb etwa 200 MeV in Betracht. Die Beteiligung extrem energiereicher Teilchen ist gering, weil in der unteren Stratosphäre und der Troposphäre der Anteil hochenergetischer Nukleonen um mindestens eine Größenordnung kleiner ist als der der niederenergetischen. Die Schwellenenergien für die in Frage kommenden Reaktionen liegen ungefähr zwischen 30 und 80 MeV.

Spallationen an anderen Atomkernen als den oben genannten kommen bei der Einwirkung der kosmischen Strahlung auf die Lithosphäre in Betracht. Allerdings ist die Intensität der Nukleonenkomponente der kosmischen Strahlung in Seehöhe schon so niedrig, daß es bisher nicht gelungen ist, solche Reaktionen in diesem Niveau sicher nachzuweisen. Dagegen sind Einwirkungen der kosmischen Strahlung auf Gesteine in Hochgebirgen festgestellt worden. Ein besonders geeignetes Studienobjekt sind die Meteorite, die vor dem Eindringen in die Erdatmosphäre einer sehr intensiven Bestrahlung ausgesetzt waren. Zwar werden die äußeren Schichten der Meteorite beim Eindringen in die Erdatmosphäre auf sehr hohe Temperaturen erhitzt und schmelzen weg, doch geht das so schnell vor sich, daß die hohe Temperatur nicht in den inneren Kern des Meteoriten eindringt, der den Erdboden erreicht.

Der Begriff Spallation ist phänomenologisch. Man versteht darunter Kernreaktionen, bei denen mehrere Nukleonen frei werden. Der Nachweis im Labor erfolgt üblicherweise durch Identifizierung der Endkerne. Das entsprechende Verfahren in der Geophysik ist das Aufsuchen in der Atmosphäre vorkommender radioaktiver Isotope. Diese Methode schließt eine genaue Information über die Reaktion selbst und die freigewordenen Nukleonen oder größeren Bruchstücke aus. Andererseits hat man die Möglichkeit der Beobachtung mit der Nebelkammer oder der Photoemulsion, ein gerade beim Studium der kosmischen Strahlung in der Atmosphäre oft angewendetes Mittel. Hier manifestiert sich die Reaktion als Stern, und auch dieser Begriff hat phänomenologischen Charakter. Dabei ist die Information nach der anderen Seite beschränkt. Sie erstreckt sich nur auf die Zahl der freigewordenen geladenen Teilchen, ihre Energie und Winkelverteilung. Man erhält aber keine Aussage über die Zahl der freigewordenen Neutronen oder die entstehenden Endkerne.

Man deutet solche Reaktionen qualitativ nach Serber mittels zweier aufeinander folgender Vorgänge. Zunächst erzeugt das einfallende Teilchen bei ausreichender Energie in der Kernmaterie durch direkte Stöße mit einzelnen Nukleonen des Kerns eine Nukleonenkaskade ganz analog zu der in der Atmosphäre. Dabei können Nukleonen, denen eine ihre Bindungsenergie übersteigende Energie zugeführt wird, den Kern verlassen. In der Kernmaterie gilt allerdings das Pauli-Prinzip, und es sind nur Stöße erlaubt, bei denen beide beteiligte Nukleonen in einen unbesetzten Zustand gelangen. Nukleonen, die nicht genügend Energie zum Verlassen des Kerns erhalten, übertragen ihre Energie in wesentlich längeren Zeiten auf den gesamten Kern. Dieser gelangt dadurch in einen hoch angeregten Zustand, der zur Nukleonenverdampfung führt, wie es von niederenergetischen Kernreaktionen her bekannt ist.

Die Ausbeuten der Spallation sind am höchsten für kleine Unterschiede in Masse ΔA und Kernladungszahl ΔZ zwischen Anfangs- und Endkern (Fig. 4). Sie nehmen bei kleinen Stoßenergien mit Anwachsen dieser Differenz sehr schnell, bei höheren Stoßenergien langsamer ab. Diese Abnahme erfolgt nach allen Seiten monoton und am langsamsten über dem Energietal der Kernmassen-Verteilung. Die Beobachtung zeigt ferner, daß bei kleinem ΔZ neutronenarme Kerne bevorzugt werden. Die höchsten Wirkungsquerschnitte findet man für $3 \leq \Delta Z + \Delta N \leq 7$. Es besteht weitgehende Unabhängigkeit von der Art der Beschußteilchen, ab etwa $A = 40$ auch vom beschossenen Teilchen. Die in Fig. 4 eingezeichneten Linien gleicher Ausbeute unterscheiden sich jeweils um einen Faktor 10.

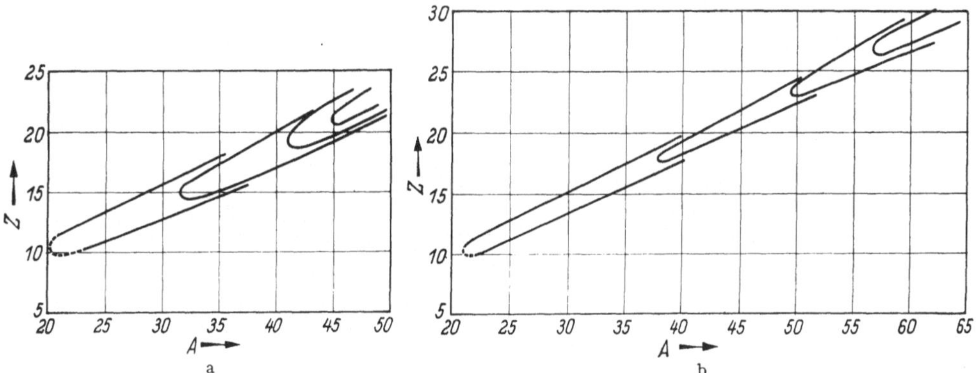

Fig. 4a u. b. Spallation. Linien gleicher Erzeugungs-Querschnitte (20; 2; p,2; 0,02 mbarn). a Spallation von V mit 180 MeV-Protonen [Rudstam (1)]. b Spallation von Cu mit 340 MeV-Protonen (Batzel, Miller und Seaborg)

Eine quantitative numerische Auswertung der Spallations-Theorie für spezielle Fälle ist nur halbempirisch unter Benutzung bestimmter experimenteller Daten möglich. Man kann die Monte Carlo-Methode zur Berechnung des Kaskaden-Prozesses benutzen. Dabei macht man gewöhnlich zwei Annahmen. Einmal werden nur Stöße einbezogen, bei denen beide Partner Nukleonen sind, d.h. andere Komplexe wie z.B. α-Teilchen werden nicht berücksichtigt. Ferner soll die Kernstruktur im wesentlichen ungeändert bleiben, d.h. die Zustände etwa emittierter Nukleonen bleiben unbesetzt, was zu einer Erhöhung der Zahl der erlaubten Stöße auf Grund des Pauli-Prinzips führen kann. Die erste Annahme kann besonders bei leichten Kernen wie N und O zu beträchtlichen Fehlern führen. Abschätzungsversuche auf rein empirischer Grundlage sind von Rudstam und von Kurtschatov u. Mitarb. gemacht worden. Dabei sind die Formeln der letzteren Autoren für den Wirkungsquerschnitt σ eleganter, aber weniger leicht von experimentell untersuchten Kernen auf andere zu übertragen (a_i, b_i Konstanten, N Neutronenzahl, Z Protonenzahl):

$$\log \sigma_0 - \log \sigma = a_N (N_0 - N)^{b_N}, \quad \log \sigma_0 - \log \sigma = a_Z (Z_0 - Z)^{b_Z}. \tag{1}$$

Die Formel von Rudstam

$$\sigma(A, Z) = \exp\left[p A - q - r(Z - s A)^2\right] \tag{2}$$

enthält vier Parameter p, q, r, s. p bestimmt die Konstante in der als exponentiell vorausgesetzten Näherungsbeziehung zwischen $\sigma(A, Z_0)$ und A, wo Z_0 die Kernladungszahl ist, bei der σ für konstantes A ein Maximum erreicht. q dient zur

Normierung von σ auf die Einheit mbarn. r ist ein Maß für den Scheitelparameter bei parabolischer Näherung für die Abhängigkeit des $\log \sigma$ von Z, also die Breite der Gauß-Verteilung der relativen isobaren Wirkungsquerschnitte. s ist der Proportionalitätsfaktor in der näherungsweisen Proportionalität zwischen Z_0 und A. Anwendung auf die Spallation von Argon und Eisen (Atmosphäre bzw. Meteorite) ist möglich, aber nur als Aushilfe zweckmäßig, wo keine besser fundierten experimentellen Bestimmungen für die betreffenden Wirkungsquerschnitte vorliegen.

Die Methode gewinnt an Zuverlässigkeit bei der Bestimmung des Verhältnisses der Wirkungsquerschnitte von Isobaren. Dann fallen zwei von den vier Parametern heraus:

$$\frac{\sigma(A, Z)}{\sigma(A, Z')} = \frac{\exp\left[-r(Z - sA)^2\right]}{\exp\left[-r(Z' - sA)^2\right]} = \exp\left[-r(Z + Z' - 2sA)(Z - Z')\right]. \tag{3}$$

III. Einfluß der Höhe und der geomagnetischen Breite

Um eine globale Übersicht über die Erzeugung von radioaktiven Substanzen zu erhalten, muß man die Abhängigkeit der Erzeugungs-Reaktionen von der

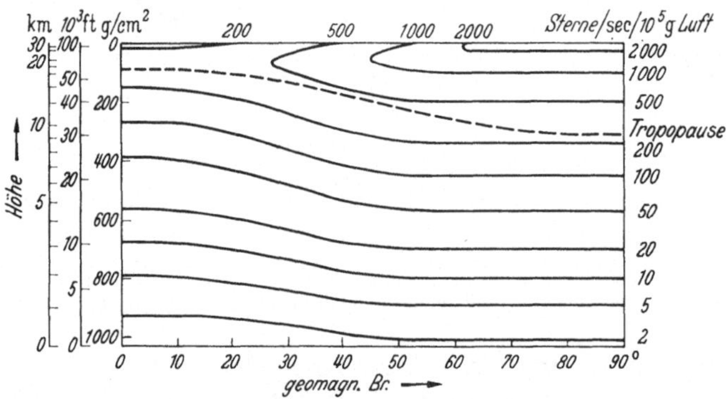

Fig. 5. Zahl der Sterne pro Gramm Luft und Sekunde als Funktion der Höhe und der geomagnetischen Breite [nach Rechnungen von Peters (4)]

geomagnetischen Breite und der Höhe kennen. Messungen der Sternerzeugung in Stickstoff- und Argon-Nebelkammern wurden von W. W. Brown 1954 und Bullock 1957 vorgenommen. Diese Untersuchungen beziehen sich auf Seehöhe und Höhen von 2000 und 3200 m ü. M. für A, für N nur auf die Höhe 3200 m. Für große Höhen existieren nur Messungen mit Photoemulsionen. Diese sind zwar nicht direkt für die Sternerzeugung in der Atmosphäre verwertbar, weil nur ein kleiner Teil der Reaktionen an N- und O-Kernen vor sich geht. Doch ermöglichen sie eine relative Abschätzung der Höhenabhängigkeit, insbesondere für die großen Höhen, für die keine anderen Daten vorliegen. Derartige Messungen gehen hinauf bis 14 g/cm² (29 km Höhe) (Lord).

Auch Messungen an dünnwandigen Hochdruck-Ionisationskammern mit Argon-Füllung (Simpson, Baldwin und Uretz) geben zwar überwiegend Sternbildung in den Kammerwänden gegenüber Sternbildung im Gas. Sie liefern aber trotzdem brauchbare zusätzliche Informationen über die Breiten- und Höhenabhängigkeit der Sterne, desgleichen Schauer-Registrierungen mit Impulskammern (Vernov und Grigorov).

Abgesehen von der alleruntersten und der allerobersten Atmosphäre, wo die Neutronendichte durch die Anwesenheit von Wasserdampf bzw. durch die Möglichkeit des Entweichens aus der Atmosphäre beeinflußt wird, besteht Gleichgewicht zwischen Neutronen und Sternen. Wegen dieses Gleichgewichts kann man die Neutronendaten [SIMPSON (1), SIMPSON und FAGOT, ROSE u. a., SOBERMAN] zur Abschätzung der Breiten- und Höhen-Abhängigkeit der Sterne mit heranziehen. Auf diese Weise haben LAL, MALHOTRA und PETERS die Stern-erzeugungs-Rate als Funktion der Höhe und der geomagnetischen Breite errechnet (Fig. 5). Für Höhen über 12 km (<200 g/cm²) wurde die Höhen-abhängigkeit bei konstanter geo-magnetischer Breite den Emulsions-Messungen von LORD entnommen und dann mit Neutronendaten die Breiten-abhängigkeit bestimmt.

Entsprechend der Zusammensetzung der Atmosphäre verteilen sich die erzeugten Sterne auf N, O und A zu etwa 76,5; 22,5; 1%. Die Anteile, die auf Stratosphäre und Troposphäre entfallen, sind aus Fig. 6 ersichtlich. Als globaler Mittelwert ergibt sich 1,7/cm² sec.

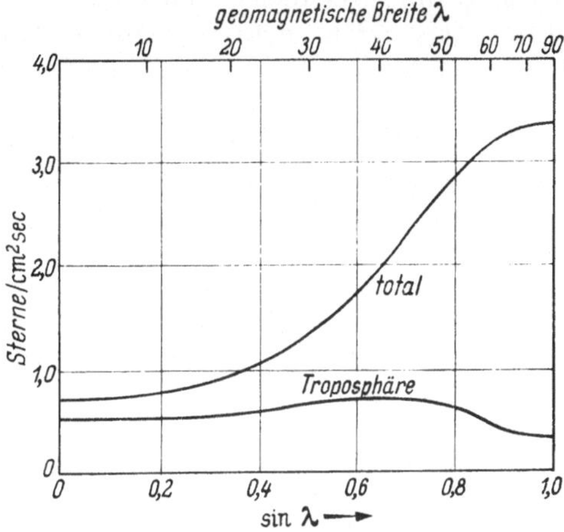

Fig. 6. Zahl der Sterne pro cm² Erdoberfläche und Sekunde als Funktion der geomagnetischen Breite [PETERS (4)]

IV. Zeitliche Konstanz der Intensität der kosmischen Strahlung

Zur Frage zeitlicher Änderungen der kosmischen Strahlung liegen direkte Untersuchungen vor bezüglich der Abhängigkeit von der Sonnenfleckenperiode und bezüglich des Einflusses größerer Sonneneruptionen. Die integrierte Primär-intensität variiert im Laufe eines Sonnenfleckencyclus um mehr als einen Faktor 2, möglicherweise bis zu einem Faktor 4 [SIMPSON (2), WINCKLER]. Jedoch werden davon überwiegend energiearme Teilchen betroffen, die auf die Umgebung der Pole beschränkt sind. PETERS (4) hat geschätzt, daß die entsprechende Änderung in der Erzeugungsrate radioaktiver Nuklide 5% nicht überschreiten dürfte.

Bei Sonneneruptionen ist ein starkes Anwachsen der Nukleonenkomponente der kosmischen Strahlung beobachtet worden. Ereignisse dieser Art, von denen etwa 30 während des letzten Fleckencyclus auftraten, erhöhen den Protonenfluß über den Polen bis herunter zu Breiten von etwa 65° für mehr als 48 Std. Eine Abschätzung des Effekts für die Tritium-Produktion [SIMPSON (2)] gibt folgende Zahlenwerte: Die Erzeugung von T-Atomen durch 100 MeV-Protonen relativ zu der durch 4 GeV-Protonen (mittlere Energie in der kosmischen Strahlung) wird zu $5 \cdot 10^{-2}$ angesetzt. Der Anteil an der Gesamtzeit, den Sonneneruptionen der besprochenen Art haben, beträgt $1,4 \cdot 10^{-2}$. Der den energiearmen Protonen im Verhältnis zu den 4 GeV-Protonen zugängliche Teil der Atmosphäre umfaßt mindestens 10%. Das Verhältnis des Protonenflusses einer Eruption zum mitt-leren Fluß für 4 GeV ist $5 \cdot 10^4$. Nach dieser Abschätzung würde also die Tritium-Erzeugung durch Protonen der solaren Komponente von derselben Größenordnung sein wie die durch die übrige kosmische Strahlung. Auch für C¹⁴ müßte sich danach die Erzeugung durch die Sekundärneutronen der solaren Protonen deutlich be-merkbar machen. Ein tatsächlicher Nachweis für einen derartigen Effekt liegt bisher weder für H³ noch für C¹⁴ vor. Die bei der Abschätzung benutzten Zahlen-werte stammen überwiegend von Untersuchungen während des letzten Sonnen-fleckencyclus, bei dem die Sonnenaktivität die stärkste seit mehr als 100 Jahren war.

Man hat ferner versucht, auf indirektem Wege über die Messung einzelner von der kosmischen Strahlung erzeugter radioaktiver Nuklide Anhaltspunkte für etwaige Änderungen der Intensität der kosmischen Strahlung zu erhalten. Die Übereinstimmung von C^{14}-Altersbestimmungen mit archäologischen Datierungen bestätigt die angenäherte Konstanz der C^{14}-Erzeugung für mehrere tausend Jahre [Libby (1)]. Zieht man geologische Altersbestimmungen und solche nach der Ionium-Methode zum Vergleich mit der C^{14}-Methode heran (Kulp und Volchok), so läßt sich dieser Schluß auf mehrere 10^4 Jahre ausdehnen.

Zu weiter zurück liegenden Zeiten kann man gelangen durch Untersuchungen an sehr langlebigen Aktivitäten, wie sie vor allem bei Meteoriten vorliegen. Für Nuklide ähnlicher Masse und sehr verschiedener Lebensdauer haben die Verhältnisse der gefundenen Aktivitäten gute Übereinstimmung mit den Verhältnissen ergeben, die man aus den experimentellen Produktionsquerschnitten unter Annahme konstanter Bestrahlung ermittelt. Das gilt z. B. für Mn^{54} (Halbwertszeit 290 d) und Mn^{53} ($2 \cdot 10^6$ a) oder Na^{22} (2,6 a) und Al^{26} ($7,4 \cdot 10^5$ a) oder A^{39} (320 a) und Cl^{36} ($3 \cdot 10^5$ a) (s. auch F II, III, z. B. Tabelle 13). Derartige Beobachtungen lassen den Schluß zu, daß die Änderungen in der Intensität der kosmischen Strahlung über Zeiten von der Größenordnung dieser Halbwertszeiten, also über Millionen Jahre, nicht größer waren als ein Faktor 2 (Arnold, Honda und Lal). Weitere Ausdehnung auf die Größenordnung 10^9 Jahre mittels Messungen an K^{40} ist mit etwas größerer Unsicherheit behaftet, doch dürfte die Konstanz der kosmischen Strahlung selbst über solche Zeiträume auf einen Faktor von nicht mehr als 5 gesichert sein (Zähringer).

Indirekte Untersuchungen liegen auch vor hinsichtlich kürzerzeitiger Variationen der kosmischen Strahlungsintensität. de Vries hat Schwankungen von C^{14}-Werten bei der Untersuchung von Wachstumsringen an Baumstämmen gefunden, die auch von anderen Autoren im wesentlichen bestätigt und noch weiter zurückverfolgt werden konnten (Willis, Tauber und Münnich). Diese Änderungen lassen sich für die letzten 500 Jahre mit den Änderungen der Sonnenflecken-Relativzahl in Verbindung bringen, wobei eine negative Korrelation festzustellen ist (Stuiver). Eine Intensitätsänderung der kosmischen Strahlung um einen Faktor 2 während eines Sonnenfleckencyclus führt zu einer Änderung der C^{14}-Konzentration in der Atmosphäre um nicht mehr als 1%. Eine ebensolche Konzentrationsänderung kann erzeugt werden durch eine Änderung der Produktionsrate um 12% während einer Zeitdauer von etwa 100 Jahren. Die Werte der Sonnenflecken-Relativzahl haben für Zeitintervalle dieser Größenordnung in den vergangenen Jahrhunderten manchmal bis zu 50% über oder unter dem Mittelwert gelegen. Trotzdem erreichen die beobachteten Abweichungen der C^{14}-Konzentration nur etwa $\pm 1,5$%. Solche Abweichungen könnten aber z. B. auch durch Änderungen in den Austauschraten zwischenden verschiedenen C^{14}-Reservoiren oder durch klimatische Änderungen zustandekommen.

An grönländischen Schneeproben hat Begemann im Tritiumgehalt zeitliche Änderungen gefunden, die für eine negative Korrelation zur Sonnenflecken-Relativzahl sprechen. Dies wäre in Übereinstimmung mit dem Verhalten der Totalintensität der kosmischen Strahlung und läßt ebenso wie die erwähnten C^{14}-Untersuchungen keinen Hinweis auf einen ins Gewicht fallenden Beitrag der solaren Komponente der kosmischen Strahlung erkennen.

Eine besondere Rolle könnte, wenn einmal mehr Meßwerte vorliegen, dem Be^{10} in Tiefseesedimenten zufallen [Peters (2), (3); Geiss, Oeschger und Schwarz]. Die vertikale Wanderung des Be^{10} innerhalb der Sedimente ist offenbar zu vernachlässigen (Merrill, Lyden, Honda und Arnold). Man erhält also eine differentielle Verteilung über Zeiten bis zu mehreren Millionen Jahren, die durch die Sedimentationsgeschwindigkeit und die Produktion bestimmt ist. Wenn man aus anderen Daten hinreichende Unterlagen über die Sedimentationsgeschwindigkeit als Funktion der Zeit zur Verfügung hat, liefert die Verteilung Aufschluß über die Frage nach der Konstanz der Produktion und damit der Konstanz der kosmischen Strahlung in solchen Zeiträumen.

B. Der radioaktive Kohlenstoff C^{14}

I. Natürliche Erzeugung

C^{14} ist ein β-Strahler, der sich durch Emission eines β-Spektrums mit 155 keV Maximal-Energie in das stabile N^{14} umwandelt. Die Halbwertszeit wurde von LIBBY mit 5568 ± 50 Jahren als gewogenes Mittel einer Reihe von Messungen in verschiedenen Laboratorien bestimmt. Der neueste Wert liegt etwas höher mit 5760 Jahren (MANN u. a.). Erstmals in der Natur nachgewiesen wurde C^{14} von ANDERSON, LIBBY u. a. an Methan aus einer Kläranlage.

C^{14} wird durch die exotherme Reaktion $N^{14}(n, p)\, C^{14}$ mit einer Reaktions-Energie von 620 keV erzeugt. Der Wirkungsquerschnitt für thermische Neutronen beträgt 1,75 bn (COON und NOBLES). Zwischen 0,5 und 2 MeV bestehen für die (n, p)-Reaktion an N^{14} noch einige Resonanzen, deren Maxima bis 200 mbn hinauf reichen (120, 200, 20, 195 mbn bei 499, 640, 993, 1415 keV nach JOHNSON und BARSCHALL), aber sehr steil auf etwa 10 mbn abfallen.

Der Wirkungsquerschnitt im thermischen Bereich ist der höchste für alle Reaktionen, die von den Neutronen der kosmischen Strahlung in der Atmosphäre ausgelöst werden. Die Querschnitte für andere Reaktionen an N^{14} liegen mindestens um zwei Größenordnungen niedriger. Auch bei Sauerstoff kommen höchstens Querschnitte von der Größenordnung 10 mbn in Betracht. Alle übrigen Luftbestandteile sind in so geringer Konzentration in der Atmosphäre vorhanden, daß an ihnen stattfindende Reaktionen nur vernachlässigbare Verluste an Neutronen für die $N^{14}(n, p)$-Reaktion verursachen. Man darf in guter Näherung unterstellen, daß praktisch alle von der kosmischen Strahlung erzeugten thermischen Neutronen zur C^{14}-Erzeugung verwendet werden.

Nach den Messungen von SOBERMAN, mit denen die von GABBE auf 5 % übereinstimmen, ergibt sich ein globaler Mittelwert von 1,1 Einfang-Neutronen pro cm² Erdoberfläche und sec. ANDERSON hat den von den Resonanzen zwischen 0,5 und 2 MeV herrührenden Beitrag zur C^{14}-Erzeugung zu 0,35/cm² sec berechnet. Einen quantitativ schwer abzuschätzenden Beitrag hat man von Prozessen zu erwarten, die durch Neutronen mit Energien über 2 MeV ausgelöst werden, darunter auch von Spallationen. Während ältere Untersuchungen auf eine Gesamterzeugung von merklich über 2/cm² sec führten, sieht CRAIG (1) diesen Wert als obere Grenze an. Die niedrigeren Werte schienen mit den Untersuchungen von SOBERMAN, von GABBE u. a. in der Atmosphäre besser in Einklang zu sein. Doch liefern die neuesten sehr gründlichen Neutronen-Messungen von HESS u. a. auch etwa den Wert 2/cm² sec. Da die Berechnungen naturgemäß nicht allzu genau sind, muß man mit mindestens 15 % Fehler rechnen, wahrscheinlich mit noch etwas mehr.

II. C^{14}-Reservoir

Das erzeugte C^{14} wird in der Atmosphäre relativ schnell zu $C^{14}O_2$ oxydiert. Über die Reaktionszeiten für diesen Oxydationsvorgang ist praktisch nichts bekannt. Die Möglichkeit, daß sich ein Monoxyd in höheren Atmosphärenschichten längere Zeit halten könnte, ist diskutiert worden [HARTECK, zit. nach SUESS(2)], bisher liegt jedoch kein Nachweis über seine Existenz vor. Als Dioxyd wird das C^{14} horizontal gleichmäßig in der Atmosphäre verteilt und hat dort eine so lange Aufenthaltsdauer, daß die für die Neutronenverteilung noch bestehende Breitenabhängigkeit praktisch ausgeglichen wird. Aus der Atmosphäre gelangt C^{14} durch die Assimilation in die Pflanzenwelt und von dort mit der Nahrung in die Körper der Tiere und Menschen. Von der Pflanzen- und Tierwelt geht ein Teil bei der Atmung und Verwesung wieder in die Atmosphäre über, während ein anderer Teil in den Humus des Bodens gelangt und auch in das Grundwasser. Mehr aber noch als in die Biosphäre tritt von der Atmosphäre in das Weltmeer über, in dem sich überhaupt der größte Teil des C^{14} auf der Erde befindet, und zwar in der Hauptsache als gelöstes Bikarbonat. Einen kleineren Teil C^{14} enthält das Meer auch in Form gelöster organischer Substanzen.

Die Aufnahme von C^{14} durch die Biosphäre ist an Lebensvorgänge gebunden, hört also auf, wenn das Lebewesen stirbt. Darauf gründet sich die C^{14}-Altersbestimmung, weil der C^{14}-Gehalt von diesem Zeitpunkt an mit der radioaktiven Halbwertszeit von etwa 5700 Jahren abnimmt. Die so bestimmten Alterswerte für Objekte, die z.B. auch archäologisch oder geologisch datiert werden konnten, stimmen mindestens für die letzten 15000 (Libby) (Archäologie) bzw. 30000 Jahre (Kulp und Volchock) (Ionium) mit solchen andersartigen Altersbestimmungen überein. Daraus ist zu schließen, daß mindestens seit dieser Zeit stationäre Verhältnisse im irdischen C^{14}-Reservoir herrschen und daß insbesondere die Erzeugung von C^{14} und die Intensität der kosmischen Strahlung während dieser Zeit im wesentlichen unverändert geblieben ist. Mann kann also auch hinsichtlich der C^{14}-Reservoire in Atmosphäre, Biosphäre und Hydrosphäre mit stationären Verhältnissen rechnen und insbesondere für die Verteilung des C^{14} auf die einzelnen Teil-Reservoire und für den Austausch zwischen ihnen die Bedingung des stationären Zustandes zugrunde legen. Sie wird charakterisiert durch

$$k_{ij} N_i = N_j k_{ij} \qquad (4)$$

für alle Reservoire i, j; N_i Zahl der C-Atome im Reservoir i, $k_{ij} = 1/\tau_{ij}$ Austauschkonstante für die Reservoire i, j:

$$(d N_i/d t)_j = - k_{ij} N_i. \qquad (5)$$

Bei den Austausch-Reaktionen muß man beachten, daß Isotopen-Trenneffekte auftreten können, weil der relative Massenunterschied beim Kohlenstoff noch ins Gewicht fällt. Eine brauchbare Kontrolle für diesen Effekt bietet das natürliche Vorkommen des stabilen Isotops C^{13}, das 1,1% des natürlichen Kohlenstoffs im Holz ausmacht. Da der Massenunterschied zwischen den Isotopen um eine Größenordnung kleiner als die Gesamtmasse ist, sind in erster Näherung die Trenneffekte bei C^{14} doppelt so groß wie bei C^{13}. Die geochemische Verteilung des Kohlenstoffs auf die Austauschreservoire läßt sich nur mit einer gewissen Ungenauigkeit abschätzen, wie sie bei solchen Angaben unvermeidlich ist. Relativ genau ist noch die Angabe für die Atmosphäre (Hutchinson). Dagegen ist sie für die terrestrische Biosphäre sehr unsicher. Der Wert von Goldschmidt ist auf Grund von Abschätzungen von Schroeder (Assimilation) angegeben worden. Für den Humus liegt der Goldschmidtsche Wert um einen Faktor 2 höher als der von Rubey; so groß ist also die Unsicherheit mindestens. Der Wert für die marine Biosphäre ist ebenso wie der für die Landbiosphäre kaum der Größenordnung nach sicher (Riley, Stommel und Bumpus). Für die Angabe über den im Meer gelösten organischen und anorganischen Kohlenstoff sind die Werte von Revelle und Suess benutzt.

Tabelle 1. *Kohlenstoffgehalt in g/cm² [nach* CRAIG (1)]

Atmosphärisches CO_2	0,126
Terrestrische Biosphäre	0,06
Humus	0,215
Marine Biosphäre	0,002
Gelöstes organisches C im Meer	0,533
Gesamtes anorganisches C im Meer	6,94
Gesamtes Austausch-System	7,88

Der Austausch mit den Sedimenten der Erdkruste und des Meeres erfolgt wahrscheinlich so langsam, daß er für die C^{14}-Bilanz außer Betracht bleiben kann. Die mittlere Sedimentationsrate am Meeresgrund beträgt einige mm/1000 Jahre. Die mittlere Aufenthaltsdauer des Ca im Weltmeer bis zur Abscheidung als $CaCO_3$ wird zu etwa $8 \cdot 10^6$ Jahre veranschlagt (GOLDBERG und ARRHENIUS), die des Kohlenstoffs dürfte von ähnlicher Größenordnung sein. Dem Weltmeer werden durch die Flüsse $2,45 \cdot 10^{-5}$ g „altes" C/cm$_e^2$ * und Jahr zugeführt (H. BROWN), womit man auf eine Austauschzeit von etwa $3 \cdot 10^5$ Jahren kommt.

In der Modell-Darstellung (Fig. 7) bedeuten die N_i den Inhalt des Reservoirs i an C-Atomen. Dabei sind die Werte jeweils auf den Gehalt in der Atmosphäre bezogen. R_i bedeutet das C^{14}:C^{12}-Verhältnis im Reservoir i. Diese Werte sind bezogen auf das Verhältnis R_w bei den üblichen Standard-Proben von rezentem Holz.

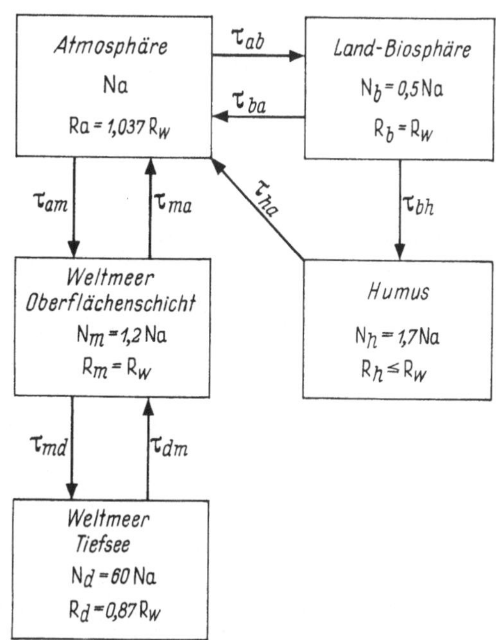

Fig. 7. C^{14}-Reservoir und -Austausch nach CRAIG (1) mit Ergänzungen

Das Verhältnis C^{14}:C^{12} für rezentes Holz hat SUESS (1) zu $1,24 \cdot 10^{-12}$ angegeben. Es kann für den vorliegenden Zweck als repräsentativ für die Land-Biosphäre angenommen werden. Über den Wert des Verhältnisses im Humus liegt keine ausreichende Information vor, es läßt sich daher kaum mehr sagen, als daß es höchstens gleich R_w sein kann. In der Oberflächenschicht des Weltmeeres ist der Wert nicht wesentlich anders als für rezentes Holz. Diese Gleichheit ist zufällig. Die nach dem C^{13}:C^{12}-Verhältnis zu erwartende Anreicherung des C^{14} relativ zum C^{12} infolge der Fraktionierung wird aufgehoben durch den Zerfall. In der Tiefsee beträgt das Verhältnis C^{14}:C^{12} $0,87\,R_w$ (BROECKER, GERARD, EWING und HEEZEN) und liegt damit 13 % unter dem Wert für die Oberflächenschicht.

Da die mittlere Lebensdauer des C^{14} 8000 Jahre beträgt, macht der Zerfall des C^{14} und dementsprechend die Produktion pro Jahr im stationären Zustand 1/8000 des gesamten C^{14} aus. Da die gesamte C^{14}-Produktion in der Atmosphäre

---

* cm$_e^2$ bedeutet cm² Erdoberfläche und wird überall dort so bezeichnet, wo es für das Verständnis nötig ist.

vor sich geht, die nur $^1/_{60}$ des gesamten C^{14} enthält, stellt die jährliche Produktion 0,75% des atmosphärischen C^{14} dar. Praktisch all dieses C^{14} wird später in den Ozean überführt, und wegen der dafür erforderlichen endlichen Zeit kommt es zu einer Stauung des C^{14} in der Atmosphäre, die Differenz zwischen dem $C^{14}:C^{12}$-Verhältnis in der Atmosphäre und im Meer wird um 0,75% erhöht für jedes Jahr mittlere Aufenthaltsdauer eines $C^{14}O_2$-Moleküls in der Atmosphäre.

Das Modell setzt den stationären Zustand voraus und verlangt, daß die Mischung innerhalb eines Reservoirs i in Zeiten bewirkt wird, die klein sind gegen alle τ_{ij}.

Die wichtigste der in dem Schema der Reservoire auftretenden Austauschzeiten ist die zwischen Atmosphäre und Ozean. Denn in der Atmosphäre wird das C^{14} erzeugt, und der Ozean ist der Hauptspeicher. Der Austausch erfolgt dabei mit der Oberflächenschicht des Weltmeeres. Für τ_{am} sind von verschiedenen Autoren sehr unterschiedliche Werte angegeben worden. Rafter und Fergusson (2) fanden 2 Jahre, Craig (1) 5 bis 7, ebenso Bolin (2), Arnold und Anderson 10 bis 20 Jahre.

Kanwisher (1) hat auf Grund von Windblas-Experimenten gezeigt, daß die Luftbewegung und vor allem die Wellenbildung der Wasseroberfläche großen Einfluß auf den Austausch hat. Man wird also kleine Werte für τ_{am} dort erhalten, wo häufig Stürme auftreten, große, wo solche selten sind. Unter diesen Gesichtspunkten erscheint z.B. der Wert von Fergusson für bestimmte Gebiete auf der Südhalbkugel durchaus verständlich. Dagegen ist der globale Mittelwert sicher wesentlich höher, wahrscheinlich in der Größenordnung 7—10 Jahre.

Ziemlich unsicher ist der Wert τ_{md} bzw. der durch die Bedingung des stationären Zustandes damit verbundene Wert von τ_{dm}. Er hängt erheblich davon ab, welches Modell man für das Weltmeer zugrunde legt. Zum Beispiel ist die Oberflächenschicht in verschiedenen Teilen des Weltmeeres verschieden dick, in der Umgebung der Pole existiert sie gar nicht. Unter Berücksichtigung solcher Einzelheiten sind detaillierte Modelle entwickelt worden (Broecker, Gerard, Ewing und Heezen). Für das globale Mittel von τ_{dm} kann man nur die Größenordnung angeben, sie liegt bei etwa 200 bis 1000 Jahre (s. z.B. Bolin). Messungen an Proben aus größeren Tiefen liegen insbesondere vor von Broecker, Gerard, Ewing und Heezen, von Bien, Rakestraw und Suess sowie von Rafter und Fergusson. Wegen der Einzelheiten darf auf den Beitrag von F.F. Koczy und J.N. Rosholt in diesem Buche verwiesen werden.

Die Aufnahme von CO_2 durch die Land-Biosphäre aus der Atmosphäre beträgt 0,004 g/cm^2_e Jahr. Daraus berechnet sich die mittlere Lebensdauer eines CO_2-Moleküls in der Atmosphäre vor Übergang in die Biosphäre zu etwa 30 Jahren.

Eine geophysikalisch besonders interessante Abart der C^{14}-Altersbestimmung ist die an Grundwasservorkommen in humidem Klima [Münnich und Vogel (4); Brinkmann, Münnich und Vogel]. Auf Grund der chemischen Reaktionen bei der Lösung von $CaCO_3$ in CO_2-haltigen Wässern würde man für Grundwasser einen C^{14}-Gehalt von 0,5 R_w erwarten. Doch ist im Boden ein Überschuß an biogenem CO_2 mit 1,0 R_w vorhanden, das im Austausch mit dem Wasser steht. Dabei resultiert in rezentem Grundwasser ein im allgemeinen recht gut definierter C^{14}-Gehalt von etwa 0,85 R_w (Münnich). Dieser Wert ändert sich auch im weiteren Verlauf der Zeit nicht wesentlich, so daß man das Alter von Grundwasservorkommen auf dieser Basis bestimmen kann.

III. Industrie-Effekt

In neuester Zeit sind Abweichungen von der natürlichen Bilanz des C^{14} aufgetreten, die anthropogener Natur sind. Der eine Effekt rührt von der Industrialisierung in der Zivilisation seit Mitte des 19. Jahrhunderts her. Im Zuge der Industrie-Entwicklung wurde durch die Verbrennung von Kohle, insbesondere Steinkohle und neuerdings von Erdöl die wegen ihres Alters von Millionen Jahren keine C^{14}-Aktivität mehr besitzen, in wachsendem Ausmaß zusätzlich C^{14}-freies CO_2 der Atmosphäre zugeführt. Die ersten Angaben über diesen Effekt hat Suess veröffentlicht. Besonders starke Verminderung der C^{14}-Aktivität wurde an Hölzern aus stark industrialisierten Gebieten gefunden. In Mitteleuropa lagen die Werte zwischen 1930 und 1953 um 4% tiefer als um die Mitte des 19. Jahrhunderts. Heidelberger Messungen an südafrikanischen Proben zeigten eine Erniedrigung um 2,7% gegenüber dem Heidelberger Standard [Münnich und Vogel (1)]. Craig (1) schätzt den Mittelwert für die ganze Atmosphäre 1957 auf 1,75%. Lokal können höhere Abweichungen auftreten, wenn die Proben in der Nähe einer Quelle von C^{14}-freiem CO_2 genommen werden. Zum Beispiel kann schon die Nähe einer viel befahrenen Autostraße die C^{14}-Konzentration in Pflanzen spürbar herabsetzen [Münnich und Vogel (1)]. Bolin und Eriksson haben die Größe des Industrie-Effekts unter Berücksichtigung der Änderung des Dissoziations-Gleichgewichts für CO_2 im Ozeanwasser abgeschätzt. Dabei ergaben sich gegenüber den Messungen etwas zu hohe Werte, wenn man mittlere Werte für die Austauschzeiten zwischen Atmosphäre und Ozean bzw. zwischen Ozean-Oberflächenschicht und Tiefsee annimmt. Bringt man andererseits Rechnung und Messung in Übereinstimmung, so folgen für die Austauschzeiten Werte, die nach derzeitiger Kenntnis an der unteren Grenze liegen: τ_{am} etwa 2 Jahre bzw. τ_{dm} 200 bis 300 Jahre. Der errechnete Wert des Industrie-Effekts von 3 bis 4% entspricht einer CO_2-Zunahme von 8 bis 10% in der Atmosphäre. Kanwisher (2) hat auf Grund von Laborexperimenten über die Löslichkeit von CO_2 in Meerwasser ebenfalls nachweisen können, daß das industriell erzeugte CO_2 zwar nach relativ kurzer Verweilzeit in das Meer übergegangen ist, daß aber wegen des schlechten Pufferungsvermögens des Ozeanwassers ein starker Rückfluß auftrat, so daß der Anstieg des CO_2-Partialdruckes in der Atmosphäre durch die Wirkung des ozeanischen Reservoirs wahrscheinlich nicht wesentlich gehemmt wurde.

IV. C¹⁴ aus Kernwaffen-Explosionen

Außer dieser seit längerer Zeit in Gang befindlichen industriellen Injektion von C^{14}-freiem CO_2 in die Atmosphäre sind im letzten Jahrzehnt bei Atombomben-Explosionen, insbesondere bei Versuchen mit Wasserstoffbomben, ins Gewicht fallende Mengen C^{14} zusätzlich erzeugt worden. Der Erzeugungsmechanismus ist der gleiche wie in der ungestörten Natur, nur werden die Neutronen, welche die (n, p)-Reaktion am N^{14} auslösen, nicht von der kosmischen Strahlung geliefert, sondern bei den Bomben-Explosionen freigemacht. Libby (2) hat geschätzt, daß je Megatonne Trinitrotuluol-Äquivalent $3{,}2 \cdot 10^{26}$ C^{14}-Atome erzeugt werden, wobei ein Mittelwert über die verschiedenen Reaktionstypen angenommen ist.

Die ersten Veröffentlichungen betreffs des durch Bomben verursachten Anstiegs der C^{14}-Aktivität in der Atmosphäre stammen von Patterson und Blifford über Messungen in USA, Rafter und Fergusson (1) über Ergebnisse aus Neuseeland und von Münnich und Vogel (1), die Messungen an Proben aus Mitteleuropa und Südafrika vornahmen.

Der Anstieg wird deutlich ab Frühjahr 1954 (Wasserstoffbomben-Versuch der USA, „Castle"-Test). Für die Zeit bis 1958 ergibt sich ein gleichmäßiger Anstieg [Münnich und Vogel (1)]. In den Jahren 1958/59 war der Anstieg wesentlich stärker, bedingt durch die speziellen Bomben-Explosionen in höheren Breiten (Fig. 8). Die Werte von Broecker und Walton bestätigen dieses Bild. Der ursprünglich von diesen Autoren interpolierte mittlere Aktivitätszuwachs pro Jahr wird durch die Zusammenfassung dieser beiden Zeitabschnitte und dadurch, daß ihre Messungen erst seit 1955 laufen, etwas höher.

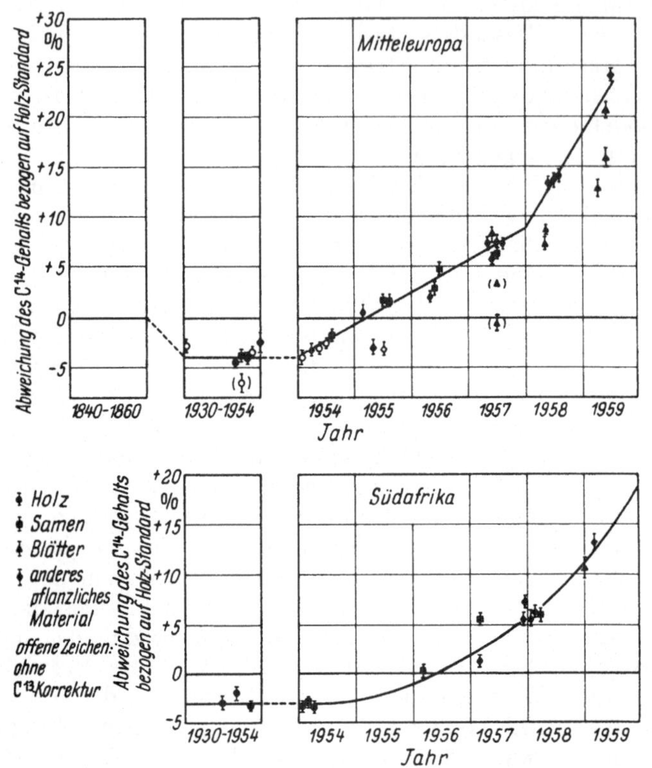

Fig. 8. Anstieg des von Kernwaffen-Explosionen herrührenden C¹⁴ (Münnich und Vogel)

Über den C^{14}-Gehalt der Atmosphäre in größeren Höhen liegen Messungen von Hagemann, Gray, Machta und Turkevich vor, die mit einzelnen Werten bis 30 km Höhe hinaufreichen. Die Konzentration des künstlichen C^{14} in der Stratosphäre betrug danach in den Jahren 1955 bis 1958 im Mittel etwa 125% des Standardwertes. Die zeitlichen Schwankungen waren unregelmäßig und betrugen während der genannten Zeit bei Mittelung über jeweils sechs Monate kaum mehr als ±25%.

Auf eine Komplikation bei der Abschätzung eines Durchschnitts-Wertes für den Anstieg haben Münnich und Vogel (2) hingewiesen. Sie fanden in den Jahren 1958 und 1959 einen ausgeprägten Jahresgang für das Bomben-C^{14} mit einem Maximum im Sommer und einem Minimum im Winter (Fig. 9). Im Maximum 1959 erreichte die Abweichung vom Standardwert 32% [Münnich und Vogel (3)]. Sie deuten diesen Jahresgang als eine Eigenschaft des stratosphärischen C^{14}-Reservoirs, in das bei den hochenergetischen Explosionen praktisch alle C^{14}-Atome injiziert wurden (Patterson und Blifford), und der Art seiner Entleerung. Diese findet bevorzugt im Frühjahr in höheren Breiten statt und steht im Zusammenhang mit dem Frühjahrs-Maximum der Spaltprodukt-Aerosole, z. B. Sr^{90} (STEWART u. a., SCHUMANN und EULITZ). Im Gegensatz zu diesen hat das CO_2 eine viel längere Verweilzeit in der Troposphäre, nämlich Jahre (s. oben τ_{am}) gegenüber wenigen Wochen. Es wird in höheren Breiten aufgestaut und führt zu einem C^{14}-Maximum im Sommer, dessen Abfall der Ausbreitung in Richtung niederer Breiten entspricht. Die Amplitude des Jahresganges ist wesentlich niedriger als bei Sr^{90}. Übrigens ist auch beim Tritium neuerdings ein dem Sr^{90} ähnlicher Jahresgang gefunden worden (vgl. EIII).

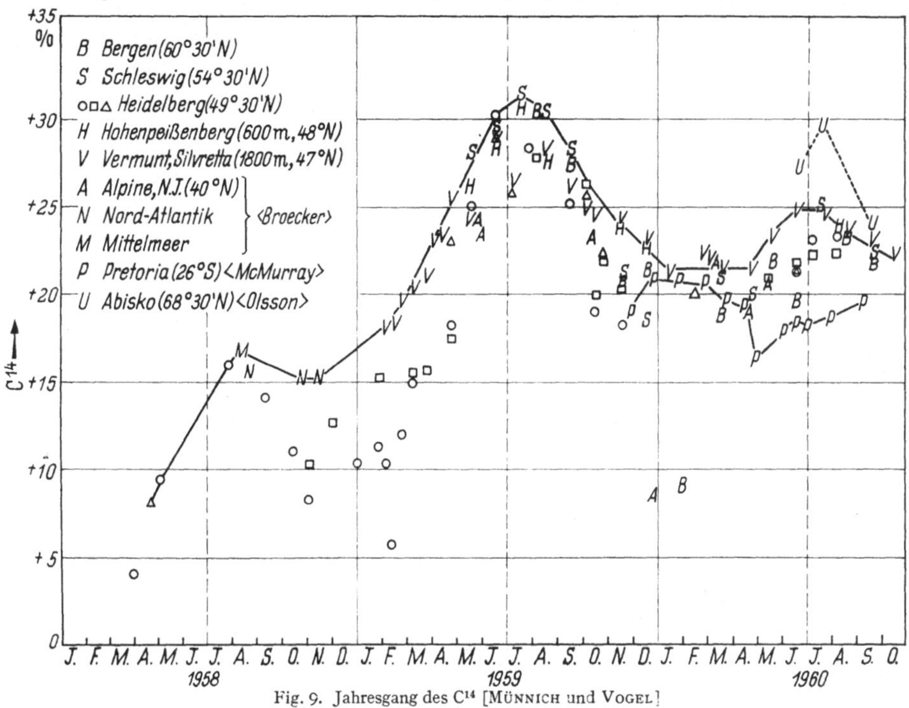

Fig. 9. Jahresgang des C^{14} [MÜNNICH und VOGEL]

C. Beryllium

I. Wirkungsquerschnitte und Erzeugungsraten

Die radioaktiven Be-Isotope, die in der Atmosphäre durch die kosmische Strahlung erzeugt werden, sind Be^7 und Be^{10}. Be^7 wandelt sich durch Elektroneneinfang um in Li^7. Dabei verlaufen 12% der Umwandlung über ein angeregtes Niveau bei 477 keV. Die beim Übergang aus diesem Anregungszustand in den Grundzustand emittierte γ-Strahlung bietet praktisch die einzige Möglichkeit des Nachweises. Die Halbwertszeit des Be^7 beträgt 53 Tage. Be^{10} ist ein reiner β-Strahler, dessen Spektrum eine Maximalenergie von 555 keV besitzt. Seine Halbwertszeit hat einen Wert von $2,5 \cdot 10^6$ Jahren.

Für die Erzeugung der Be-Isotope kommen ausschließlich Spallationsprozesse an N^{14} und O^{16} in Betracht. Dabei müssen drei bzw. vier Protonen emittiert werden. Eine gewisse Wahrscheinlichkeit besteht auch für die Emission von α-Teilchen (HODGSON, BØGGILD und TENNEY).

Eine direkte Messung der Produktion wurde von LAL, ARNOLD und HONDA bei 51 n. Br. und 685 g/cm² an Wasser durchgeführt. Dabei ergab sich ein Wert von $9,0 \cdot 10^{-6}$ Be^7-Atomen pro Gramm Sauerstoff und Sekunde. Messungen der Wirkungsquerschnitte im Labor mit Beschleunigern wurden überwiegend an C

vorgenommen, neuerdings aber auch an N und O. Nach Honda und Lal stimmen
die Querschnitte für Be[7]-Erzeugung aus C, N und O im Energiebereich 225 bis
730 MeV auf etwa 10% überein, nach Benioff (2) bei 5,7 GeV auf etwa 20%.
Daraus ergibt sich eine gewisse Berechtigung, die Messungen an C innerhalb der
sonstigen Meß- und Berechnungs-Fehlergrenzen auf die Erzeugung in der Atmosphäre anzuwenden. Diese Messungen (Fig. 10) erfassen einen weiten Energie-
Bereich von der Schwellenenergie bei etwas unter 30 MeV bis etwa 6 GeV, oberhalb
etwa 500 MeV ändert sich der Querschnitt kaum noch (Baker, Friedlander
und Hudis). Bis dahin von etwa 200 MeV an macht sich ein leichter Anstieg
bemerkbar. Bei 30 MeV liegt der Querschnitt noch unter 0,1 mbarn, bei etwa
55 MeV erreicht er ein Maximum mit etwa 15 mbarn (Dickson und Randle).

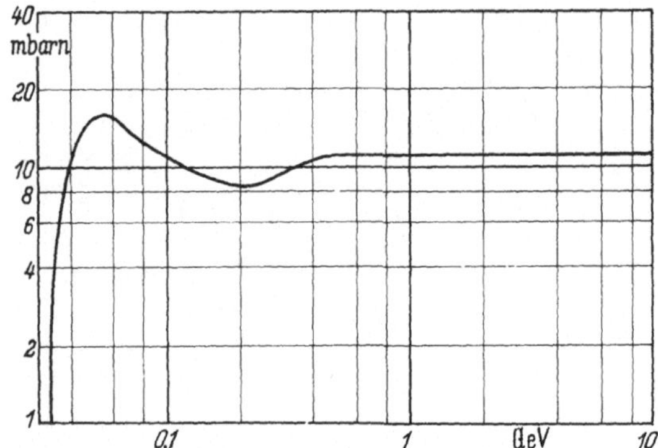

Fig. 10. Erzeugungs-Querschnitt für Be[7] aus C als Funktion der
Beschußenergie

Die Meßergebnisse hinsichtlich der Wirkungsquerschnitte
liefern zuverlässige Absolut-
werte für die Produktion. Es
handelt sich nur noch darum,
die Abhängigkeit von der Höhe
und der geomagnetischen Breite
einzubeziehen und einen Mittelwert für die gesamte Erdoberfläche zu bestimmen.

Lal, Arnold und Honda
geben als Wirkungsquerschnitt
für Sternerzeugung in einer
der Atmosphäre entsprechenden Mischung von Stickstoff

und Sauerstoff 225 mbarn an. Unterstellt man die für höhere Energien bestätigten
geringen Unterschiede zwischen den Wirkungsquerschnitten der Be[7]-Erzeugung an
C einerseits und an N und O andererseits auch für die niedrigen Energien, so ergibt
sich für die Produktion von Be[7] in Luft durch das Neutronenspektrum der kosmischen Strahlung bei 50° geomagnetischer Breite und 680 g/cm² ein Wert von
11 mbarn. Danach wird der Anteil der Sterne in der Atmosphäre, bei denen Be[7]
erzeugt wird, für die genannten Koordinaten 4,9%, im Mittel etwa 4,5%. Dieser
relative Anteil kann als breiten- und höhenunabhängig angesehen werden. Die
Änderung der Be[7]-Produktion mit der Höhe und der geomagnetischen Breite ist
also proportional der entsprechenden Änderung der Sternerzeugung. Man hat,
um die Be[7]-Erzeugung zu ermitteln, jeweils nur die aus Fig. 5 ersichtlichen Werte
mit dem Faktor 0,045 zu multiplizieren. Für das globale Mittel der Erzeugung
von Be[7] erhält man so $7,7 \cdot 10^{-2}/cm^2$ sec.

Dieser durch zahlreiche Meßergebnisse fundierte Wert kommt dem von
Benioff auf einem ganz anderen Weg, nämlich mit Hilfe der Messelschen Theorie
der Nukleonenkaskade ermittelten von $1 \cdot 10^{-1}/cm^2$ sec recht nahe. Man darf
sich allerdings durch diese recht gute Übereinstimmung des Mittelwertes nicht
darüber hinwegtäuschen lassen, daß die Breiten- und Höhenabhängigkeit in
beiden Fällen verschieden herauskommt und Einzelwerte für bestimmte Koordinaten größere Diskrepanzen aufweisen können.

Für Be[10] sind mangels experimenteller Daten bisher nur grobe Abschätzungen
möglich. Lal, Malhotra und Peters haben als Grundlage für solche Abschätzungen Beobachtungen benutzt, die Fuller beim Beschuß einer mit Sauerstoff
gefüllten Nebelkammer mit 300 MeV-Neutronen und Kellogg beim Beschuß
von Polyäthylen und einer mit Methan und Wasserstoff gefüllten Nebelkammer

mit 90 MeV-Neutronen gemacht haben. FULLER fand unter 600 ausgewerteten Kernreaktionen in 34 Fällen Be-Kerne als Bruchstücke, während in 21 Fällen nur eine Identifizierung als Li, Be oder B möglich war. KELLOGG beobachtete unter 1033 Reaktionen in der Nebelkammer 304 Ereignisse, bei denen Be-Kerne als Bruchstücke auftraten (einschließlich $Be^8 \rightarrow 2\alpha$). Der von ihm angegebene Wirkungsquerschnitt für Erzeugung von Be^{10} liegt entgegen der Erwartung unter dem für Be^7, allerdings sind die Fehlergrenzen sehr weit. LAL hat unter Berücksichtigung aller in Frage kommenden Daten das Verhältnis der Erzeugung $Be^{10}:Be^7$ abgeschätzt zu 1,16. Damit würde sich für Be^{10} ein Anteil von 5,2% an den in der Atmosphäre erzeugten Sternen ergeben und eine Produktion von $8,9 \cdot 10^{-2}/cm^2$ sec.

II. Be⁷ in der Atmosphäre

Be^7 wurde erstmals in der Atmosphäre nachgewiesen von ARNOLD und AL SALIH 1955 in Chikago an Regenwasser. Eine große Zahl von Messungen an Regen wurden von PETERS u. Mitarb. in Indien durchgeführt. Speziell Messungen an Schnee erfolgten in Schweden. Eine Sonderstellung nehmen die Messungen von CRUIKSHANK u. a. an Luftfiltern in Kanada ein. Diese systematisch über 1 Jahr hinweg betriebenen Beobachtungen zeigten an zwei Stationen einen ausgeprägten Jahresgang, der von den Verfassern mittels des Jahresganges der Niederschläge erklärt wird. Messungen an Luftfiltern haben gegenüber Niederschlagsmessungen den großen Vorteil wirklicher Kontinuität und sind von den meteorologischen und klimatischen Zufälligkeiten der Niederschlagtätigkeit unabhängig.

GOEL, NARASAPPAYA u. a. haben darauf aufmerksam gemacht, daß zwar die absoluten Konzentrationen des Be^7 im Niederschlag um mehr als eine Größenordnung schwanken. Doch waren die Schwankungen des Verhältnisses der Konzentrationen $Be^7:P^{32}$ kleiner als ein Faktor 4.

In der Äquatorialzone, wo diese Messungen durchgeführt wurden (Indien), ist die Wahrscheinlichkeit, daß in mit den in Frage kommenden Halbwertszeiten vergleichbaren Zeiträumen nennenswerte Mengen stratosphärischer Luft in die Troposphäre gelangen, sehr gering. Das ist auch von den Beobachtungen an Atombomben-Spaltprodukten her bekannt. Wenn keine Zulieferung aus der Stratosphäre erfolgt und die Troposphäre hinreichend gut durchmischt ist, gilt

$$\left.\begin{aligned}
\frac{dC_i}{dt} &= q_i S - \frac{C_i}{\tau_i} - F_i \qquad && F_i = \frac{C_i}{\tau'} = \lambda' C_i \\
&= q_i S - (\lambda_i + \lambda') C_i \qquad && \lambda_i + \lambda' = \frac{1}{\tau_i} + \frac{1}{\tau'} = \frac{\tau_i + \tau'}{\tau_i \tau'} \\
&= q_i S - \frac{\tau_i + \tau'}{\tau_i} F_i .
\end{aligned}\right\} \qquad (6)$$

C_i Konzentration des Nuklids i in der Atmosphäre, q_i Erzeugungsrate bezogen auf S, τ_i Lebensdauer, λ_i Zerfallskonstante, F_i Fallout-Rate, τ' mittlere Aufenthaltsdauer in der Troposphäre, λ' Abfallkonstante für Ausscheidung aus der Atmosphäre, S Sternerzeugungsrate. Stationärer Zustand beim Mitteln über große Zeiten:

$$\left.\begin{aligned}
q_i S &= \frac{\tau_i + \tau'}{\tau_i} F_i \qquad && \frac{F_i}{F_k} = \frac{q_i \tau_i}{q_k \tau_k} \frac{\tau_k + \tau'}{\tau_i + \tau'} = \frac{q_i \tau_i \tau_k + q_i \tau_i \tau'}{q_k \tau_k \tau_i + q_k \tau_k \tau'} \\
\tau' &= \frac{F_k q_i \tau_i \tau_k - F_i q_k \tau_k \tau_i}{F_i q_k \tau_k - F_k q_i \tau_i} = \frac{\tau_i \tau_k (F_k q_i - F_i q_k)}{F_i q_k \tau_k - F_k q_i \tau_i} .
\end{aligned}\right\} \qquad (7)$$

$$\text{Für} \quad \tau' \ll \tau_i, \tau_k \quad \text{wird} \quad \frac{F_i}{F_k} = \frac{q_i}{q_k} .$$

$$\text{Für} \quad \tau' \gg \tau_i, \tau_k \quad \text{wird} \quad \frac{F_i}{F_k} = \frac{q_i \tau_i}{q_k \tau_k} .$$

Voraussetzung ist bei allen diesen Überlegungen, daß der Fallout-Mechanismus keine selektiven Eigenschaften bezüglich einzelner Nuklide aufweist, also alle in gleicher Weise erfaßt.

Für τ' geben die Meßwerte von $Be^7:P^{32}$ in Indien für die Jahre 1957/58 einen Wert von etwa 40 Tagen [Peters (4)]. Selbstverständlich kann man auch die Verhältnisse anderer Nuklide für solche Abschätzungen benutzen, doch ist das vorliegende Beobachtungs-Material bisher zu dürftig.

Die Umrechnung der gemessenen spezifischen Aktivitäten auf eine jährliche Abscheidungsrate erfolgt bei Peters u. Mitarb. über die Mittelwerte für die Niederschlagsmengen in bestimmten Breiten. Diese Methode läßt natürlich lokale Unterschiede unberücksichtigt. In extremen Fällen kann das zu fehlerhaften Ergebnissen führen. Ein Beispiel dafür ist die Station Shillong, die in Indien in dem praktisch regenreichsten Gebiet der ganzen Erde liegt. Goel, Narasappaya u. a. haben daher die dortigen Beobachtungen aus ihren allgemeinen Betrachtungen ausgeschlossen. Hier ist die Konzentration wegen der übergroßen Regenmengen abnorm gering, und das Verfahren der Umrechnung mit der mittleren Niederschlagsmenge der betreffenden geographischen Breite versagt, weil der lokale Wert um mehr als einen Faktor 10 über diesem Mittelwert liegt. Auch aus diesem Sachverhalt geht hervor, daß Messungen an Niederschlägen (s. Tabelle 2) wegen der vielen zeitlichen und räumlichen Zufälligkeiten in deren Verteilung mit Vorsicht gedeutet werden müssen.

Tabelle 2. *Be^7-Gehalt von Niederschlägen*

Ort	Geogr. Br. Grad Nord	Geom. Br. Grad Nord	Jährl. Niederschlag cm	Jahr	Zahl der Proben	Dem Boden pro cm² zugeführte Atomzahl jährlich
Kodaikanal. . . .	10	1	175	1956	47	$5,1 \cdot 10^5$
				1957	16	$3,4 \cdot 10^5$
				1959	16	$3,3 \cdot 10^5$
Bombay	19	14	99	1956	29	$4,5 \cdot 10^5$
				1957	24	$3,4 \cdot 10^5$
				1958	19	$4,4 \cdot 10^5$
				1959	25	$2,8 \cdot 10^5$
Delhi	29		87	1957	3	$3,4 \cdot 10^5$
Mussoori	30			1957	8	$2,7 \cdot 10^5$
Pathankot	32			1957	3	$5,9 \cdot 10^5$
Srinagar	34			1957	5	$3,9 \cdot 10^5$
Chicago und	42	53	89	1954	22	$5,3 \cdot 10^5$
Lafayette . . .	41					
Uppsala	60	56	62	1956	3	$2,7 \cdot 10^5$
				1957	3	$9,3 \cdot 10^4$
Abisko	68	66	31	1957	2	$9,3 \cdot 10^4$

III. Beryllium-10

Die Messung von Be^{10} begegnet beträchtlichen Schwierigkeiten wegen der langen Halbwertszeit und der Tatsache, daß es sich um einen reinen β-Strahler handelt. In Anbetracht der großen Lebensdauer muß man damit rechnen, daß das Nuklid im Weltmeer angereichert ist, und man sah die günstigste Chance in der Untersuchung von Tiefsee-Sedimenten. Tatsächlich wurde Be^{10} in Bohrkernen nachgewiesen, die am Boden des Pazifik entnommen waren. Arnold fand in Tiefsee-Tonen Aktivitäten von etwa $5 \cdot 10^{-3}$ dpm/cm³. Dabei wurde ferner festgestellt, daß die Aktivität mit zunehmender Tiefe in den Bohrkernen allgemein abnahm. Das wurde auch bei den Untersuchungen von GOEL, KHARKAR u. a. bestätigt, die ebenfalls an Proben aus dem Pazifik vorgenommen wurden. Die Aktivitätswerte hatten hier ein Mittel von $8 \cdot 10^{-3}$ dpm/g.

In derselben Größenordnung, nur im Mittel ein wenig niedriger, lagen die Meßwerte von MERILL, LYDEN, HONDA und ARNOLD an zwei weiteren Bohrkernen aus dem Pazifik. Eine Ausnahme machten bei diesen Messungen die obersten 12 cm des einen Kerns, die eine um rund einen Faktor 10 höhere Aktivität zeigten ($61 \cdot 10^{-3}$ dpm/g feuchten Ton). Diese Probe wurde mehrfach chemisch gereinigt, und mit einer Ausnahme stimmten die Meßwerte auf 10% überein. An dieser Probe wurde erstmals bei den geophysikalischen Untersuchungen das Be^{10} physikalisch identifiziert durch Absorptionsmessungen.

D. Argon-Spallationsprodukte
I. Erzeugung

Die Produktion von Argon-Spallationsprodukten durch die kosmische Strahlung ist direkt untersucht worden mittels mehrmonatiger Exposition von Argonflaschen an einer Gebirgsstation (LAL, ARNOLD, HONDA) bei 685 g/cm² (etwa 3 km Höhe) und 51° n. Br. Gegenüber den Verhältnissen in der freien Atmosphäre ist die Produktion durch schnelle π-Mesonen oder Einfang langsamer π^--Mesonen, die in der Flasche oder im Gestein der Umgebung erzeugt werden, etwas bevorzugt. Doch wird der Beitrag zur Gesamtproduktion unter den Bedingungen des Experiments auf weniger als 1% geschätzt. Die Meßergebnisse wurden korrigiert wegen der Absorption der auslösenden Strahlung in der Flaschenwand unter Annahme einer $A^{\frac{1}{3}}$-Abhängigkeit.

Tabelle 3

	Atome/g sec	Atome/cm² Jahr		
		Stratosphäre	Troposphäre	insgesamt
P^{32}	$7,6 \cdot 10^{-6}$	$1,6 \cdot 10^4$	$8,0 \cdot 10^3$	$2,4 \cdot 10^4$
P^{33}	$6,2 \cdot 10^{-6}$	$1,3 \cdot 10^4$	$6,6 \cdot 10^3$	$2,0 \cdot 10^4$
S^{35}	$1,4 \cdot 10^{-5}$	$2,7 \cdot 10^4$	$1,5 \cdot 10^4$	$4,2 \cdot 10^4$

Die Erzeugung pro cm² Erdoberfläche und Jahr ist ermittelt mit Hilfe von Daten für die Höhenabhängigkeit der auslösenden Strahlung in Höhen über 9 km (< 312 g/cm²), die aus dem Jahre 1948/49 stammen. Nach Korrektur wegen des Sonnenfleckencyclus und der dadurch bedingten Änderung der Intensität der kosmischen Strahlung liegen diese Werte um 1,82; 1,37; 1,83 für die drei Nuklide höher als die von LAL, MALHOTRA und PETERS berechneten.

Diese Rechnungen basierten auf Beobachtungen über die relative Häufigkeit einfach und mehrfach geladener Teilchen bei Sternen und auf der empirischen Beziehung von RUDSTAM. Aus der Häufigkeit von Sternen in Argon mit 1, 2, 3 Spuren und von einfach und doppelt geladenen Teilchen wurde die Ausbeute der Elemente S und P bestimmt. Anschließend erfolgte die Ermittlung der Wirkungsquerschnitte für die bereits im Niederschlag nachgewiesenen Kernarten P^{32}, P^{33} und S^{35} mittels der Relation von RUDSTAM.

In der Stratosphäre niedriger Breiten dürfte die Produktion von Argon-Spallationsprodukten etwas vermindert sein. Das Energiespektrum der Nukleonen der kosmischen Strahlung ist in diesem Bereich nach höheren Energien verschoben. Der Anteil von Sternen mit Emission weniger Teilchen ist für solche Energien bei der Sternerzeugung an Kernen mittlerer Masse gegenüber dem mit Emission zahlreicher Teilchen etwas kleiner. Nach den von Lal, Malhotra und Peters berechneten Kurven liegt die Höhe dieses Effektes zwischen 20 und 30% und ist naturgemäß am höchsten für S^{35}. Die Verteilung nach Höhe und geographischer Breite weicht daher dementsprechend von der Verteilung der Sterne nach Fig. 5 etwas ab, wenn auch der allgemeine Verlauf als Funktion dieser Größen der gleiche ist.

Für Si^{32} und Cl^{36} liegen nur Relativ-Abschätzungen der Produktion bezogen auf P^{32} bzw. S^{35} vor. Sie basieren auf den empirischen Formeln von Rudstam. Unter Zugrundelegung der bei der Spallation von V erhaltenen Parameter ergibt sich für den Wirkungsquerschnitt der Produktion von Si^{32} bezogen auf den des P^{32} der Wert 0,2 (Lal, Goldberg und Koide). Daraus folgt eine Produktionsrate von $4,8 \cdot 10^3$ Atomen/cm² Jahr. Die direkte Absolutbestimmung nach der Rudstamschen Formel (vgl. A II) unter Berücksichtigung der Beobachtungen an Sternen wie oben für P^{32}, P^{33}, S^{35} beschrieben, liefert $5,4 \cdot 10^3$ Atome/cm² Jahr. Wie früher erwähnt ist die Anwendung der Formel auf das Verhältnis der Wirkungsquerschnitte von Isobaren zuverlässiger als eine Absolutbestimmung, so daß dem ersteren Wert mehr Gewicht zukommt. Allerdings liegt die Differenz der beiden Abschätzungen ohnehin innerhalb der Fehlergrenze.

Die Cl-Produktion aus A läßt sich relativ zur S-Produktion abschätzen unter Berücksichtigung der Tatsache, daß für Cl nur einspurige Sterne, für S ein- und zweispurige in Betracht kommen (Schaeffer, Thompson und Lark). Geht man davon aus, daß der Anteil einfach geladener Teilchen 0,8 und die Zahl der einspurigen Sterne um einen Faktor 1,7 höher als die der zweispurigen ist (Hodgson), so ergibt sich für Cl eine um 40% höhere Produktion als für S. Aus den Formeln von Rudstam folgt dann, daß 30% des Cl auf das Isotop der Masse 36 entfallen und 34% des S auf das Isotop 35. Die Produktion des Cl^{36} wird damit um 20% höher als die von S^{35}, d.h. man kommt auf einen globalen Mittelwert von $5 \cdot 10^4$ Atomen/cm² Jahr.

Das Isotop Cl^{39} ist wohl das einzige, bei dem die Produktion durch Einfang langsamer μ^--Mesonen einen wesentlichen Anteil an der Erzeugung liefert: $A^{40} + \mu^- \rightarrow Cl^{39} + n + \nu$. Die Bestimmung von Absolutwerten für den μ^--Einfang ist schwierig, und das gleiche gilt für die Erzeugung von Cl^{39} durch andere Kernprozesse, weil die Differenz in Masse und Kernladung gegenüber dem Ausgangskern A^{40} nur jeweils 1 beträgt und experimentelle Anhaltspunkte fehlen. Winsberg ermittelte für die Zahl der Kernreaktionen zwischen langsamen μ^--Mesonen und A-Kernen $2 \cdot 10^{-6}\, e^{-x/247}$ pro g Luft und sec, wo x die atmosphärische Tiefe in g/cm² bedeutet. Durch Integration ergibt sich für die Gesamtzahl der zwischen dem Meeresniveau und der Höhe x stattfindenden Reaktionen je cm² und sec $2,3 \cdot 10^{-4} (e^{-x/247} - 0{,}015)$. Wegen der Unsicherheit in großer Höhe folgt daraus für die Produktion ein Wert von nicht mehr als $7 \cdot 10^3$/cm² Jahr.

Aus ähnlichen Überlegungen wie bei Cl^{36} folgt, daß der Anteil des Cl^{39} an der Gesamterzeugung von Cl-Isotopen durch die Nukleonenkomponente der kosmischen Strahlung jedenfalls von derselben Größenordnung wie der des Cl^{36} ist. Der Beitrag der Nukleonenkomponente zur Erzeugung des Cl^{39} ist also vermutlich wesentlich größer als der des μ^--Einfangs. Doch liefert der letztere einen merklichen Anteil, der jedenfalls nicht vernachlässigbar ist wie wohl bei allen anderen von der kosmischen Strahlung erzeugten Nukliden.

Die Abschätzung der Produktion für Na^{22} ist noch schwieriger als für die übrigen Argon-Spallations-Produkte wegen des großen Massenunterschiedes. Man wird, wie man aus Fig. 4 ersieht, etwa mit einem Faktor 10 weniger zu rechnen haben als bei den S- und P-Isotopen. Damit kommt man in die Größenordnung 10^3 Atome/cm² Jahr.

Tabelle 4. *Argon-Spallationsprodukte*

Kern	Halbwerts-zeit	Zerfall	Teilchen-energie MeV	γ-Strahlung	
				Energie MeV	Intensität % d. Zerfälle
Na^{22}	2,6 a	β^+ 90% E 10%	0,54	} 1,28	100
Si^{32}	710 a	β^-	0,1	—	—
P^{32}	14 d	β^-	1,7	—	—
P^{33}	25 d	β^-	0,25	—	—
S^{35}	87 d	β^-	0,167	—	—
Cl^{36}	$3 \cdot 10^5$ a	β^- 98,3% E 1,7%	0,71	—	—
Cl^{39}	55 m	β^-	1,91 85%	0,246	44
				1,52	41
				1,27	} 52
			2,18 8%	1,27	
			3,45 7%	—	—

II. Meßergebnisse

Von den Argon-Spallations-Produkten wurde als erstes P^{32} im Niederschlag nachgewiesen (MARQUEZ und COSTA). Die meisten Messungen sind in Indien durchgeführt worden (GOEL, NARASAPPAYA u. a.; LAL, RAMA und ZUTSHI; RAMA). Dort wurde auch erstmals P^{33} ebenfalls im Niederschlag gefunden (LAL, NARASAPPAYA und ZUTSHI). Die Messung von P^{33} mit seiner sehr weichen β-Strahlung und dem geringen Unterschied in der Halbwertszeit gegenüber dem gleichzeitig vorhandenen P^{32}, das eine wesentlich energiereichere β-Strahlung besitzt (s. Tabelle 4) bietet gewisse Schwierigkeiten, so daß das bisher vorliegende Meßmaterial nicht sehr umfangreich ist. Dagegen existiert für P^{32} wenigstens für einen Meßort (Bombay) eine verhältnismäßig gute Übersicht über mehrere Jahre. Die in der nachstehenden Tabelle 5 angegebenen Meßergebnisse sind in der gleichen Weise wie bei Be^7 auf die mittlere Niederschlagsmenge der betreffenden geographischen Breite umgerechnet.

Tabelle 5. *Meßergebnisse für P^{32} und P^{33}*

Jahr	Atome/cm² Jahr	Zahl der Messungen	Ort
		P^{32}	
1956	$3,1 \cdot 10^3$	10	Bombay
1957	$2,7 \cdot 10^3$	11	Bombay
1958	$4,3 \cdot 10^3$	19	Bombay
1959	$1,2 \cdot 10^3$	27	Bombay
	$1,6 \cdot 10^3$	16	Kodaikanal
		P^{33}	
1957	$5,2 \cdot 10^3$	12	Bombay

Die meisten Meßergebnisse liegen vor für S^{35}, das zuerst von GOEL beobachtet worden ist. Für dieses Nuklid sind neben zahlreichen Messungen für Niederschläge in Indien auch einige aus Schweden in der Literatur veröffentlicht.

Da die Aufenthaltsdauer in der Stratosphäre groß ist gegen die Halbwertszeiten der vorstehend aufgeführten radioaktiven Kernarten, ist anzunehmen, daß vorwiegend die in der Troposphäre erzeugten Anteile durch die Niederschläge zum Erdboden befördert wurden. Wendet man die früher abgeleitete Gleichung

$$F_i = q_i \, S \, \frac{\tau_i}{\tau_i + \tau'} \tag{8}$$

Tabelle 6. *Meßergebnisse S^{35}*

Jahr	Atome/cm² Jahr	Zahl der Messungen	Ort
1956	$1,8 \cdot 10^4$	1	Kodaikanal
1957	$2,1 \cdot 10^4$	3	Kodaikanal
1956	$1,3 \cdot 10^5$	9	Bombay
1957	$4,9 \cdot 10^4$	4	Bombay
1958	$1,5 \cdot 10^5$	19	Bombay
1959	$3,7 \cdot 10^4$	13	Bombay
1957	$3,6 \cdot 10^4$	2	Mussoori
	$3,6 \cdot 10^4$	2	Delhi
	$1,8 \cdot 10^4$	3	Pathankot
	$2,9 \cdot 10^4$	2	Srinagar
	$1,4 \cdot 10^4$	5	Uppsala
	$5,5 \cdot 10^3$	4	Abisko

mit dem Wert $\tau' = 40$ d auf die Meßwerte der dem Boden zugeführten Aktivität an, so erhält man die nachstehenden Werte für die troposphärische Produktion im Vergleich zu den aus den Daten über die kosmische Strahlung berechneten (s. Tabelle 7):

Tabelle 7

	Aus Beobachtung (Bombay)		Berechnung Atome/cm² Jahr
	Jahr	Atome/cm² Jahr	
P^{32}	1956	$9,3 \cdot 10^3$	$8,0 \cdot 10^3$
	1957	$8,1 \cdot 10^3$	
	1958	$12,9 \cdot 10^3$	
	1959	$3,6 \cdot 10^3$	
P^{33}	1957	$10,9 \cdot 10^3$	$6,6 \cdot 10^3$
S^{35}	1956	$1,7 \cdot 10^5$	$1,5 \cdot 10^4$
	1957	$6,5 \cdot 10^4$	
	1958	$2,0 \cdot 10^5$	
	1959	$4,9 \cdot 10^4$	

In allen Fällen erscheint eine Tendenz zur Überschreitung der berechneten Erzeugungsraten vorhanden zu sein. Bei den Phosphor-Isotopen bleibt der Überschuß mit Ausnahme des Jahres 1958 innerhalb der Fehlergrenze der Abschätzung. Dagegen ist die Abweichung bei S^{35} signifikant. Es handelt sich hier offensichtlich um zusätzliche Erzeugung durch Kernwaffen-Explosionen, vermutlich vorwiegend über die Reaktion $Cl^{35}(n, p)\, S^{35}$, die besonders bei Bombenversuchen in der Nähe der Meeresoberfläche von den entstehenden Neutronen ausgelöst werden kann. Auch P^{32} kann in solchen Fällen durch die Reaktion $S^{32}(n, p)\, P^{32}$ entstehen, doch ist der mittlere Gehalt des Meereswassers an Schwefel um über eine Größenordnung niedriger als der an Chlor. Immerhin scheinen auch bei P^{32} die hohen Werte in den Jahren 1956 bis 1958 und ihre Korrelation zu den gleichzeitigen S^{35}-Werten auf einen Beitrag der Bombenexplosionen hinzudeuten, der aber sicher wesentlich kleiner ist als beim Schwefel. Im Falle des S^{35} überwiegt während der Beobachtungsjahre, aus denen bis jetzt Meßergebnisse vorliegen, offenbar die Erzeugung durch Kernwaffenversuche diejenige durch die kosmische Strahlung.

Ein weiteres Nuklid, bei dem dies offensichtlich noch in hohem Maße der Fall ist, ist das Cl^{36}. Sein Nachweis in Niederschlägen ist erst neuerdings gelungen (SCHAEFFER, THOMPSON und LARK). Die beobachteten Aktivitäten lagen zwischen 0,3 und 12 dpm/g Cl, die Atomkonzentration betrug 10^4 bis 10^6 Cl^{36}-Atome/cm³ Regenwasser. Das sind um drei Zehnerpotenzen mehr als man nach der Überschlagsrechnung für die Produktion durch die kosmische Strahlung erwarten kann. Im Grundwasser, Brunnenwasser und den Oberflächengewässern wurden in den Jahren 1957 und 1960 im Mittel um eine Zehnerpotenz niedrigere spezifische Aktivitäten (bezogen auf Cl) gefunden. Bei Proben aus dem großen Salzsee und aus dem Ozean blieb die Aktivität unter der Nachweisgrenze (DAVIS und SCHAEFFER).

Der Nachweis im Niederschlag gelang auch bei einem kurzlebigen Cl-Isotop, dem Cl^{39} (WINSBERG). Die Probenahme erfolgte dabei in Zeiten zwischen 5 und 30 min von einer Fläche von etwa 300 m². Die Aktivität konnte an mehreren Proben mittels ihrer Halbwertszeit physikalisch identifiziert werden. Ihr Beitrag bleibt wesentlich unterhalb der auf die berechneten Produktionsraten gestützten Erwartung.

Für Na^{22} liegen bisher nur zwei Meßwerte aus Brasilien vor, wo dieses Nuklid erstmals in großen Regenmengen (3 bis 4 m³) nachgewiesen wurde (MARQUEZ COSTA, ALMEIDA). Die Aktivität betrug 0,012 und 0,038 dpm/l.

Bei den langlebigen Kernarten ist der Nachweis in der Atmosphäre bzw. im Niederschlag wegen der geringen Zerfallswahrscheinlichkeit pro Zeiteinheit besonders schwierig. Si^{32} wurde in Schwämmen gefunden, die das Silicium aus dem Meerwasser stark anreichern (LAL, GOLDBERG, KOIDE). Die Messung erfolgte nach vorhergehender chemischer Trennung über das Folgeprodukt P^{32}. Das Ergebnis war 19,6 dpm/kg Si. Die spezifische Aktivität ist also sehr viel kleiner und der Nachweis im Meerwasser selbst oder in der Atmosphäre würde deshalb außerordentliche Probenmengen erfordern.

E. Tritium

I. Natürliche Erzeugung

Tritium ist ein reiner β-Strahler mit einer Maximalenergie von nur 18 keV. Es findet sich im atmosphärischen Wasserstoff sowie im Wasser aller Aggregatzustände in Atmosphäre, Hydrosphäre, Lithosphäre und Biosphäre. Durch seinen radioaktiven Zerfall wandelt es sich mit einer Halbwertszeit von 12,3 Jahren um in Helium-3. Tritium-Konzentrationen werden meist in Tritium-Einheiten angegeben: 1 Tritium-Einheit (TU = tritium unit) bedeutet ein Verhältnis $H^3 : H^1$ von 10^{-18}.

Für die nachstehend vielfach benutzten Begriffe Produktionsrate und Zerfallsrate werden die Definitionen von BEGEMANN (1959) verwendet, d.h. die Entstehung soll eventuelle Zufuhren von außen in die Erdatmosphäre einschließen, der Zerfall mögliches Entweichen nach außen.

Die Erzeugung von H^3 erfolgt durch relativ energiereiche Nukleonen. Die Reaktion $N^{14} + n = C^{12} + H^3$ hat einen Wirkungsquerschnitt von 11 ± 2 mbn für Verdampfungsneutronen, gemessen oberhalb der Schwellenenergie von 4,4 MeV mittels Spaltungsneutronen einer mittleren Energie von 6 MeV [FIREMAN (1)]. Weiterhin liegen Labormessungen des Wirkungsquerschnitts für die Erzeugung von H^3 an N und O vor für Protonen. In die Tabelle 8 sind auch die Werte für Erzeugung aus C aufgenommen, die hier recht erheblich von denen für N und O abweichen.

Tabelle 8. *Erzeugungs-Querschnitte für Tritium in mbn*

Beschuß-Energie MeV	C	N	O	
Neutronen				
6	11			Fireman
Protonen				
225	7,0	16	13	Honda, Lal
300	7,0		19	Honda, Lal
400	8,6		36	Honda, Lal
450	7,3	25	28	Currie, Libby, Wolfgang
730	7,6			Honda, Lal
2050		25	30	Currie, Libby, Wolfgang
2200		28	33	Fireman, Rowland
6200		35		Currie

Die Berechnung der natürlichen Produktionsrate in der Atmosphäre begegnet erheblichen Schwierigkeiten. Neben Spallationen können auch andere Prozesse, insbesondere (n, t)-Reaktionen, zur Tritium-Erzeugung merkliche Beiträge liefern. Allerdings ist der Anteil, der von Spallationen herrührt, auch hier absolut der größte. Currie, Libby und Wolfgang schätzen den Prozentsatz von Sternen mit Tritium-Erzeugung in Stickstoff und Sauerstoff bei einer der Atmosphäre entsprechenden Zusammensetzung des Gemisches auf 8%. Es besteht die Möglichkeit, daß dieser Wert etwas zu niedrig gegriffen ist. Man erhält für diesen Beitrag mit einer Sternhäufigkeit von 1,7/cm² sec nach Peters (4) anstelle des von den Autoren verwendeten Wertes eine untere Grenze von 0,14 T-Atomen/cm² sec.

Der Wirkungsquerschnitt für die Reaktion $N(n, t)$ oberhalb der Schwelle 4,4 MeV beträgt 11 mbn. Die mittlere Energie der bei der Bestimmung benutzten Spaltungsneutronen von 5,7 MeV ist gleich der des Spektrums, das sich ergibt unter der Annahme, daß alle bei den Messungen mit BF_3-Zählern in der Atmosphäre erfaßten langsamen Neutronen aus Verdampfungsprozessen stammen und mit einem Energiespektrum $E\,e^{-E/3}$ erzeugt werden (Freese und P. Meyer). (n, t)-Reaktionen an O spielen praktisch kaum eine Rolle, da ihre Schwelle bei etwa 15 MeV liegt.

Aus den Messungen der Neutronenintensität in Abhängigkeit von der Höhe (Yuan, s. a. Pfotzer) berechnen Currie, Libby und Wolfgang unter Berücksichtigung des Entweichens aus der Atmosphäre, der Streuung und der Absorption, daß ein Anfangsfluß von 3,8 Neutronen/cm² sec mit der genannten Energieverteilung dem beobachteten Neutronenfluß unter 0,4 eV entspricht. Mit dem oben angegebenen Wirkungsquerschnitt folgt, daß 1,3% der Neutronen Tritium erzeugen. Somit beläuft sich die Produktion durch Verdampfungsneutronen auf 0,05 T-Atome/cm² sec.

Sehr unsicher ist die Abschätzung für die Nukleonen mittlerer Energie im Bereich von etwa 10 bis 100 MeV, für die Currie, Libby und Wolfgang eine Tritium-Produktion von etwa 0,01/cm² sec angeben, wobei jedoch allenfalls die Größenordnung sicher ist. Insgesamt ergibt sich auf dieser Basis eine Erzeugungsrate von 0,2 T-Atomen/cm² sec. Dieser Wert wurde auch von Fireman und Rowland aus ihren Wirkungsquerschnitt-Messungen abgeleitet.

Craig und Lal haben ihre Berechnung der Produktion ganz auf die Sternerzeugungsrate bezogen, und alle entstehenden Tritiumatome in die Ausbeute pro Stern eingeschlossen. Als Grundlage für die Rechnung benutzten sie die experimentellen Werte der Tritiumerzeugung von Honda und Lal, aus denen für ein der atmosphärischen Luft entsprechendes Stickstoff-Sauerstoff-Gemisch eine Anregungsfunktion für Reaktionen errechnet wurde, bei denen Tritium entsteht. Mit dieser Anregungsverteilung und dem wohl dem neuesten Stand der Erfahrung entsprechenden atmosphärischen Neutronenspektrum von HESS u. a. ergibt sich für die Tritium-Ausbeute 0,144 Atome pro Stern. Das gilt für den Höhenbereich bis 200 g/cm². Oberhalb dürfte das Neutronenspektrum flacher werden und daher die Tritium-Ausbeute höher. CRAIG und LAL schätzen diese Erhöhung umgerechnet auf die ganze Atmosphäre auf 8%. Mit der Sternerzeugungsrate 1,7/cm² sec folgt dann eine Tritium-Produktionsrate von 0,26/cm² sec. Berücksichtigt man die Abhängigkeit der kosmischen Strahlungsintensität von der Sonnenflecken-Relativzahl, so erhöht sich dieser Wert noch auf 0,27/cm² sec als Mittel über einen Fleckencyclus.

II. Tritium im atmosphärischen Wasserstoff

Der erste Wert für die relative Konzentration von HT im atmosphärischen Wasserstoff wurde von FALTINGS und HARTECK an Anfang 1949 in der Nähe von Hamburg gewonnenen Proben bestimmt zu $(3,8 \pm 1,2) \cdot 10^{-15} = 3800$ TU. GROSSE u. a. fanden im Frühjahr 1952 in Luftproben, die in der Nähe von Buffalo gesammelt wurden $(1,6 \pm 0,1) \cdot 10^{-14} = 16\,000$ TU. Dies sind die beiden einzigen Werte, die aus der Zeit vor der ersten Wasserstoffbomben-Explosion stammen. Durch die Bombenversuchs-Explosionen wurden die Verhältnisse grundlegend geändert. Die damit verbundene künstliche Erzeugung von Tritium hat die Produktion durch die kosmische Strahlung größenordnungsmäßig übertroffen, so daß Untersuchungen über die natürlichen Verhältnisse an Proben aus der späteren Zeit überhaupt nicht mehr möglich sind. Immerhin hat auch die Untersuchung von Proben aus späteren Zeiten ein geophysikalisches Interesse, da sie insbesondere neue Gesichtspunkte bezüglich der Austausch- und Transport-Probleme ins Spiel bringen. GONSIOR (1) hat zuerst Messungen aus den Jahren 1955 bis 1958 veröffentlicht. Diese wurden in Heidelberg an Proben aus Hamburg und Nürnberg vorgenommen. Die Werte steigen an bis in die Größenordnung $10^{-13} = 10^5$ TU. Wenn man diesen Anstieg von der natürlichen Konzentration bis 1958/59 verfolgt, so kann man einen mittleren Anstieg interpolieren. Es muß dahingestellt bleiben, wieweit diese Mittelung sinnvoll ist, da die Produktion schubweise erfolgte, wobei allerdings die Hauptmenge in die Stratosphäre gelangte. Die hauptsächlichsten Schwankungen werden von GONSIOR (2) mit den Atombomben-Explosionen in höheren Breiten in Verbindung gebracht. Leider standen nicht immer Proben in regelmäßiger Folge zur Verfügung, so daß manche Fragen offen bleiben. Die wenigen Proben aus dem Jahre 1959 zeigen vom Frühjahr an abfallende Tendenz. Die den Atombomben zugeschriebenen Schwankungen waren erwartungsgemäß nicht von Schwankungen des Deuterium-Gehalts der Proben begleitet. Dagegen haben BEGEMANN und FRIEDMAN eine Korrelation zwischen Tritium- und Deuterium-Schwankungen an ihren Proben aus USA und Deutschland in der Zeit von 1954 bis 1956 gefunden und daraus auf lokale Beimischung von tritiumfreiem Wasserstoff anderer Herkunft geschlossen.

III. Tritium in HTO

Weit mehr Messungen sind für den Tritium-Gehalt von Niederschlägen durchgeführt worden. Die ersten stammen von GROSSE u. a. und sind die einzigen, bei denen man es sicher mit dem natürlichen T-Gehalt zu tun hat. Für Chikago und New York liegen von 1952 an zahlreiche Werte vor (BEGEMANN und LIBBY; GILETTI, BAZAN und KULP), doch beginnt diese Meßreihe etwa zur Zeit der ersten Wasserstoffbomben-Explosion, so daß bereits erhöhte Konzentrationen auftreten. Man hat zwar den Eindruck, daß sich die späteren Werte vor allem des Jahres 1953 wieder dem natürlichen Niveau nähern, doch ist eine Beeinflussung durch das Bomben-Tritium nicht auszuschließen. Die weiteren Wasserstoffbomben-Versuche von 1954 an brachten dann nicht nur vorübergehend extrem hohe Werte (bis 450 TU), sondern erhöhten auch den allgemeinen Pegel. In Heidelberg hat Gonsior (2) im Mai/Juni 1959 Werte über 10^3 TU gemessen, d.h. der Konzentrationsanstieg gegenüber den natürlichen Werten betrug ebenso wie beim atmosphärischen Wasserstoff größenordnungsmäßig einen Faktor 10^2. Man findet übrigens auch bei dem von Bombenexplosionen stammenden H^3 einen Jahresgang ähnlich dem des Sr^{90} (G. Israël, Libby, Östlund).

Eine Bestimmung des natürlichen Tritium-Gehaltes in der Atmosphäre bzw. im atmosphärischen Niederschlag ist also derzeit überhaupt nicht mehr möglich. Aus diesem Grunde hat man versucht, ältere Proben zu erhalten, die freilich einmal nicht zu weit zurückliegen dürfen, weil die Halbwertszeit nur 12 Jahre beträgt, außerdem hinsichtlich ihrer Entstehungszeit genau definiert sein müssen. Ferner sollten sie ein möglichst direktes Maß für die mittlere Zerfallsrate sein. Begemann hat zu diesem Zweck Schneeproben aus Grönland benutzt. Die Stratigraphie ermöglicht eine Bestimmung des Alters. Eine dünne Oberflächenschicht schmilzt während des Sommers, und die damit verbundene Struktur-Änderung liefert eine eindeutige Zuordnung. Die Dicke der schmelzenden Oberflächenschicht ist so gering, daß eine Verfälschung der Zuordnung durch das einsickernde Schmelzwasser nicht eintritt. Hinzu kommt, daß die Niederschläge in Grönland sich hauptsächlich aus verdampftem Meerwasser zusammensetzen, dessen T-Gehalt sehr niedrig liegt (etwa 1 TU), so daß der Gehalt der Proben nur von der Produktionsrate und der Niederschlagsmenge abhängt. Diese Messungen liefern unter der Voraussetzung, daß die Aufenthaltsdauer des Tritium in der Atmosphäre klein gegen seine Halbwertszeit ist, eine Zerfallsrate von 1,12 T-Atomen/cm^2 sec. Der globale Mittelwert muß wesentlich niedriger sein, da die Tritium-Produktion durch die kosmische Strahlung in hohen Breiten den Mittelwert erheblich übersteigt und ein meridionaler Austausch wegen der geringen Aufenthaltsdauer des Wassers in der Atmosphäre diese Breitenabhängigkeit nur unwesentlich verändern kann.

Craig und Lal haben die aus Nordamerika vorliegenden Meßergebnisse eingehend analysiert und sich bemüht, einen von der künstlichen Tritiumerzeugung unabhängigen Mittelwert für den Tritiumgehalt der Niederschläge zu finden. Dabei wurden Deuteriummessungen an Niederschlagsproben zu Hilfe genommen. Als Mittel der natürlichen Tritiumkonzentration in Niederschlägen über dem nordamerikanischen Kontinent ergibt sich dann der Wert 6 ± 2 TU.

IV. Tritium-Bilanz

Für die Abschätzung der stratosphärischen Verweilzeit des Tritium hat man in Ermangelung direkter Informationen die Verweilzeiten radioaktiver Aerosole herangezogen. Über diese Verweilzeiten bestanden zeitweise unrichtige Vorstellungen, wobei sie auf gleiche Größenordnung wie die Tritium-Halbwertszeit geschätzt wurden. Nach inzwischen verbesserter Kenntnis ist ihr Wert von verschiedenen Parametern abhängig und in den meisten Fällen wesentlich kleiner als diese Halbwertszeit. Daher sind die Abschätzungen der Tritium-Zerfallsrate, die auf Grund der früheren Vorstellungen vorgenommen wurden, sicher zu hoch. Der Wert von etwa 1/cm^2 sec, der sich auf die oben beschriebene Weise ergab, wurde in Anbetracht der vermuteten langen stratosphärischen Verweilzeit auf

2/cm² sec erhöht. Nach der heutigen Kenntnis über die Aufenthaltswahrschein-
lichkeit würde man eher auf etwa 1/cm² sec kommen. Nun sind aber folgende
Tatsachen zu berücksichtigen: BEGEMANN läßt für seine Messungen an den Grön-
land-Proben selbst bei Einbeziehung der langen stratosphärischen Verweilzeit
eine untere Grenze von 1/cm² sec zu; CRAIG (3) kommt unter Berücksichtigung
der zu langen Verweilzeit überhaupt nur auf 1,4/cm² sec; die Werte für die Grön-
land-Proben liegen über dem globalen Mittel. Daraus folgt, daß dieses sicher
kleiner als 1/cm² sec ist.

CRAIG und LAL haben auf Grund der neuen Erkenntnisse einen Wert von
0,5/cm² sec für die Zerfallsrate berechnet. Dabei ist nicht nur die nach den jetzigen
Erfahrungen wesentlich kürzere Aufenthaltsdauer in der Stratosphäre berücksichtigt, sondern auch die Breitenabhängigkeit des Vertikalaustausches durch die
Tropopause. Für den Wasserkreislauf über dem nordamerikanischen Kontinent
wird für diese Rechnungen ein Modell entwickelt, das in einigen Zügen von früheren Überlegungen abweicht. Die Autoren beschränken sich auf das Gleichgewicht in der Troposphäre, weil der durch den Wasserabfluß vom Kontinent
zum Meer bedingte Tritiumtransport unzureichend bekannt ist. BEGEMANN und
LIBBY haben an Hand ihrer Messungen an Niederschlagsproben und am Flußwasser des Mississippi eine umfassende Bilanz des Wasserkreislaufes aufzustellen
versucht. Sicher lassen sich prinzipiell Tritiummessungen für hydrologische
Untersuchungen verwenden (s. auch V. BUTTLAR u.a.), doch ist dabei die stark
schwankende künstliche Tritiumerzeugung ein großes Hindernis.

Jedenfalls reduziert sich die Diskrepanz zwischen der berechneten Produktionsrate und der Zerfallsrate, die man auf einen Faktor 10 veranschlagt und
deshalb als prinzipielles Problem betrachtet hatte, ganz beträchtlich. Diese
Diskrepanz war der Anlaß zu verschiedenen Hypothesen über etwa bisher unberücksichtigt gebliebene Beiträge zur Tritium-Produktion. CRAIG, ARNOLD
und FELD [CRAIG (2)] haben auf die Möglichkeit aufmerksam gemacht, daß
Tritium direkt von der Sonne in die Erdatmosphäre gelangen könnte, während
BEGEMANN zur Diskussion gestellt hat, ob nicht der Beitrag energiearmer
Komponenten der kosmischen Strahlung zu gering eingeschätzt worden sei.

Die Emission von Korpuskularstrahlen von der Sonne steht in engem Zusammenhang mit chromosphärischen Eruptionen und damit mit der Sonnenaktivität überhaupt und mit den Sonnenflecken. Man sollte danach erwarten,
daß die H³-Produktion eine positive Korrelation zur Sonnenflecken-Relativzahl
aufweist, wenn Tritium unmittelbar von der Sonne in die Erdatmosphäre gelangt.
Andererseits besteht zwischen der Intensität der kosmischen Strahlung und der
Sonnenflecken-Relativzahl eine negative Korrelation (MEYER und SIMPSON).
Die einzigen Messungen, die zu einer solchen Korrelationsuntersuchung herangezogen werden können, sind die an den grönländischen Schneeproben. Sie sprechen für die Erzeugung durch kosmische Strahlung (BEGEMANN).

Beim heutigen Stand der Kenntnis ist die Diskrepanz zwischen Produktion
und Zerfallsrate kaum mehr als prinzipielles Problem zu betrachten, sondern
dürfte eine Frage des Mangels an Genauigkeit bei der Bestimmung dieser beiden
Größen sein.

HARTECK hat zur Erklärung der hohen Tritium-Konzentration im atmosphärischen Wasserstoff detaillierte Vorstellungen über photochemische und andere
molekulare Prozesse in der Atmosphäre entwickelt, welche die Bildung von HT
ermöglichen könnten. Eine Bestätigung dieser theoretischen Überlegungen liegt
bis jetzt nicht vor. Es erscheint auch möglich, den Konzentrationsunterschied
für Tritium in Wasserstoff und Wasserdampf ohne Bezugnahme auf solche Prozesse zu verstehen. Wenn man annimmt, daß die relative Konzentration in der

Stratosphäre für Wasserstoff und Wasserdampf die gleiche ist, so folgt daraus zwanglos, daß beim Übertritt in die Troposphäre infolge der Verdünnung durch den dort herrschenden hohen Wasserdampf-Gehalt Konzentrationsunterschiede der beobachteten Größenordnung entstehen [Craig (3), Gonsior (1)].

F. Radioaktivität in Gesteinen und Meteoriten

I. Terrestrische Gesteine und Tektite

Gegenüber der Erzeugung radioaktiver Substanzen durch die kosmische Strahlung in der Atmosphäre tritt die in der Lithosphäre und in der Hydrosphäre weit zurück. Die dort erzeugten spezifischen Aktivitäten sind außerordentlich klein, weil praktisch alle durch die Einwirkung der kosmischen Strahlung entstehenden Elemente in den Gesteinen durch inaktive Isotope vertreten sind, und entzieht sich daher weitgehend der Nachweismöglichkeit. Außerdem spielen Konkurrenzprozesse eine Rolle, bei denen durch spontane Spaltung oder durch (α, n)-Reaktionen entstandene Neutronen zur Auslösung von Kernreaktionen beitragen. Obgleich die Zahl der möglichen Kernreaktionen wegen des Vorhandenseins von sehr viel mehr Ausgangsnukliden ganz wesentlich größer ist als in der Atmosphäre, ist ein Nachweis bisher nur in ganz wenigen Fällen gelungen. Bei der Hydrosphäre überwiegen die aus anderen Bereichen, vor allem aus der Atmosphäre zugeführten Aktivitäten bei weitem etwa in der Hydrosphäre selbst erzeugte.

Davis und Schaeffer haben auf Grund von Messungen (Nachweisgrenze 0,02 dpm/g) an verschiedenartigen Proben einige Abschätzungen zur Erzeugung von Cl^{36} angegeben, welche die Verhältnisse verdeutlichen:

Tabelle 9. *Erzeugung von Cl^{36}*

Gestein	Syenit	Phonolit	Wasser (Gr. Salzsee)	Meerwasser
Höhe (Meter über dem Meer)	290	3200	1260	0
% Cl	0,35	0,35	11	1,9
Neutronen d. kosm. Str. a. d. Oberfl./g min	$5 \cdot 10^{-3}$	$27 \cdot 10^{-3}$	$12 \cdot 10^{-3}$	$3 \cdot 10^{-3}$
Bruchteil der in Cl absorb. Neutronen . .	0,30	0,30	0,77	0,32
Neutronen aus spontaner U-Spaltung/g min.	$0,54 \cdot 10^{-6}$	$1,1 \cdot 10^{-5}$		
/cm² min			$2 \cdot 10^{-6}$	$9 \cdot 10^{-4}$
Neutronen aus (α, n)-Reakt./g min	$1 \cdot 10^{-6}$	$2 \cdot 10^{-5}$		
/cm² min . . .			$2 \cdot 10^{-6}$	$4 \cdot 10^{-4}$
Durch kosm. Str. erzeugte Sättigungsaktivität Cl^{36} dpm/g	0,44	2,3	0,028	$2,6 \cdot 10^{-5}$
Gemessene Akt. Cl^{36} dpm/g	$< 0,02$	$0,12 \pm 0,02$	$< 0,02$	$< 0,02$

Hier ist der Neutroneneinfang in Cl^{35} die bei weitem wahrscheinlichste Kernreaktion. Ihr Wirkungsquerschnitt von 32 mbarn liegt um Größenordnungen über allen Spallations-Querschnitten. Für ein U-Erz (Pechblende vom Gr. Bärensee) fand Kuroda:

Tabelle 10

% U	% Cl	Gemessene Akt. Cl^{36} dpm/g Cl	Aus Cl^{36} berechn. Fluß therm. Neutronen/cm² sec	Zahl d. neutronen-induz. Spaltungen	
				je g min	% d. spont. Spaltungen
45,3	0,02	4,0	0,12	0,034	22

Untersuchungen an Granit und Obsidian auf Be^{10} und Al^{26} verliefen negativ. Auch die Suche nach Co^{60} in Cobalt-Mineralien lieferte keine mit Sicherheit die durch den Nulleffekt gegebene Fehlergrenze übersteigende Aktivität (KOHMAN und EHMANN).

Dem eventuellen Nachweis durch die kosmische Strahlung erzeugter radioaktiver Kerne hat man auch Bedeutung beigemessen für die Frage nach der Herkunft der Tektite, glasiger, bis faustgroßer, teilweise angeschmolzener Fundstücke, die in einigen Gebieten der Erdoberfläche vorkommen und denen man schon früher einen möglicherweise kosmischen Ursprung zugeschrieben hatte. Es ist äußerst unwahrscheinlich, daß heute an der Erdoberfläche gefundene Gesteinsstücke dort über ähnlich lange Zeiten gelegen haben und der kosmischen Strahlung ausgesetzt waren, wie das bei den Meteoriten der Fall ist. In Meteoriten kommen daher sehr viel eher nachweisbare von der kosmischen Strahlung erzeugte Aktivitäten vor.

Die Suche nach einem Gehalt an radioaktiven Stoffen in den Tektiten hat jedoch zu keinem positiven Ergebnis geführt. KOHMAN und EHMANN fanden mit einem Geiger-Zähler einen Gehalt von $0{,}048 \pm 1{,}013$ dpm/g Al^{26} und vermuteten, daß der untersuchte Australit kosmischer Herkunft sei. ANDERS erhielt ebenfalls für einen Australit mit einem γ-Spektrographen $0{,}022$ dpm/g als obere Grenze für das nicht nachweisbare Al^{26} und hält terrestrische Herkunft für durchaus möglich.

Dafür sprechen auch Messungen des Kalium-Argon-Alters an Tektiten (GENTNER und ZÄHRINGER). Diese lieferten für Bediasite und Moldavite Werte von $29{,}4 \cdot 10^6$ und $8{,}7 \cdot 10^6$ Jahre, für Tektite aus Indochina und Australien innerhalb der experimentellen Fehler übereinstimmend $6{,}1 \cdot 10^5$ Jahre im Einklang mit dem Alter der geologischen Formation des Fundortes. Daraus ist zu schließen, daß die Tektite völlig entgast worden sind. Auch in diesen Fällen wurden keine Spallationsprodukte gefunden. Das deutet wiederum darauf hin, daß die Tektite nicht kosmischer Herkunft sind, sondern auf der Erde (oder allenfalls in ihrer unmittelbaren Nähe, möglicherweise auf dem Mond) beim Auftreffen großer Objekte unter starker Erhitzung gebildet wurden.

II. Meteorite

Die Frage der Erzeugungsrate der einzelnen Nuklide in Meteoriten ist erheblich komplizierter als bei der Erzeugung in der Atmosphäre. Sicher spielt die Spallation in Fe eine wesentliche Rolle nicht nur bei den Eisenmeteoriten, sondern auch bei vielen Steinmeteoriten, den Chondriten. Bei den Steinmeteoriten können freilich auch an einer Reihe anderer Kerne als Fe, Co, Ni Spallationen stattfinden. Hinzu kommen ferner mögliche Neutronen-Einfangprozesse, z.B. $Co^{59}(n,\gamma)Co^{60}$ oder (p, n)- und (p, pn)-Reaktionen wie bei der Erzeugung von Al^{26} aus Mg, Al, Si. Dementsprechend ist das experimentelle Material auf Grund von Laborversuchen über die Wirkungsquerschnitte recht umfangreich. Trotzdem reicht es keinesfalls aus, um die Erzeugungsraten ohne zusätzliche Annahmen zu berechnen. Man ist in dieser Beziehung auch in keiner besseren Lage als bei der Erzeugung in der Atmosphäre. In gewissem Ausmaß kann man sich dadurch helfen, daß man das Meteoritenmaterial selbst im Labor beschießt. Doch braucht man auch zur Auswertung solcher Experimente gewisse Voraussetzungen. Unter anderem setzt sie eine Kenntnis des Energiespektrums der auslösenden Strahlung voraus, die ebenfalls noch Mängel aufweist. Eine wesentliche Rolle spielen dabei die im Meteoritenmaterial erzeugten Sekundärteilchen. Im Gegensatz zur Atmosphäre liefern hier die π-Mesonen einen merklichen Beitrag. Vor allem muß man die Veränderung des Energiespektrums der erzeugenden Strahlung durch die energiearmen Sekundärteilchen in Betracht ziehen. Eine weitere Komplikation bedeutet die im interstellaren Raum wirksame Oberflächenerosion, die zu einer Unsicherheit hinsichtlich des Abstandes der untersuchten Probe von der ursprünglich bestrahlten Oberfläche führen kann.

Fireman und Zähringer haben an den Beispielen des H^3 und des A^{37} die Abhängigkeit der Erzeugungsraten von der Tiefe unter der Oberfläche an einem Fe-Target untersucht. Sie fanden bis 1 GeV eine exponentielle Abnahme mit der Tiefe, von 3 GeV an aufwärts zunehmend geringe Übergangseffekte, also zunächst einen Anstieg und Überschreitung eines Maximums. Arnold, Honda und Lal haben relative Erzeugungsquerschnitte für zahlreiche stabile und radioaktive Kerne im Falle von Eisenmeteoriten berechnet. Sie benutzten dazu Beobachtungen über das Primärspektrum der kosmischen Strahlung (Macdonald) und daraus abgeleitete Sekundärspektren in 10 und 100 g/cm² Tiefe des Meteoritenmaterials sowie Anregungsfunktionen, die aus experimentellen Wirkungsquerschnitten bei

Tabelle 11. *In Meteoriten nachgewiesene, durch kosmische Strahlung erzeugte radioaktive Kerne*

Kern	Halbwertszeit	Zerfallsart		Max. Teilchen-Energie keV	γ-Strahlung	
					Energie keV	Intensität % d. Zerf.
H^3	12,3 a	β^-		18	—	—
Be^{10}	$2,5 \cdot 10^6$ a	β^-		555	—	—
Na^{22}	2,6 a	β^+	90%	540	1280	100
		E	10%			
Al^{26}	$7,4 \cdot 10^5$ a	β^+	84%	1170	2960	0,3
		E	16%		1830	96
					1120	3,7
Si^{32}	710 a	β^-		100	—	—
Cl^{36}	$3,1 \cdot 10^5$ a	β^-	98,3%	714	—	—
		E	1,7%		—	—
A^{37}	35 d	E			—	—
A^{39}	~320 a	β^-		565	—	—
K^{40}	$1,3 \cdot 10^9$ a	β^-	89%	1330	—	—
		E	11%		1460	100
Ca^{45}	164 d	β^-		256	—	—
Sc^{46}	84 d	β^-		360	1119	100
					890	100
Ti^{44}	~200 a	E			160	
					72	
V^{48}	16 d	β^+	56%	692	1314	98
		E	44%		986	100
					2250	2
V^{49}	330 d	E			—	—
Cr^{51}	27,8 d	E			325	9
Mn^{53}	$2 \cdot 10^6$ a	E			—	—
Mn^{54}	290 d	E			840	100
Fe^{55}	2,6 a	E			—	—
Co^{56}	77 d	β^+	20%	1500 (komplex)	1220	
					845	
		E	80%		und schwächere	
Co^{57}	270 d	E			136	
					122	
					14	
Co^{58}	71 d	β^+	15%	470	800	100
		E	85%			
Co^{60}	5,2 a	β^-		312	1330	100
					1170	100
Ni^{59}	$8 \cdot 10^4$ a	E			—	—

Protonenbeschuß von Eisen und, wo solche fehlten, aus der Rudstamschen Formel gewonnen wurden.

Als Spallationsprodukte wurden zuerst stabile Edelgasisotope in Meteoriten nachgewiesen, He (PANETH, REASBECK und MAYNE), Ne (REASBECK und MAYNE) und A (GENTNER und ZÄHRINGER). Später gesellten sich dazu einige andere stabile Nuklide: Sc^{45} (WÄNKE) und K^{41} (VOSHAGE und HINTENBERGER). Einige Angaben über die absoluten Substanzmengen sind aus der nachstehenden Tabelle zu entnehmen:

Tabelle 12

Meteorit	He 10^{-6} cm³/g	Ne 10^{-8} cm³/g	A* 10^{-8} cm³/g	Sc 10^{-9} g/g
Toluca	0,5		0,38	
Henbury	9,6		6,8	
Tamaragul . . .	23,1	14,7		2,0
Carbo	20,9	12,3		1,9

* $A^{36} + A^{38}$.

Die relativen Häufigkeiten gibt für den Meteoriten Carbo die folgende auf $He^3 = 100$ normierte Zusammenstellung (WÄNKE):

He^3	He^4	$Ne^{20} + Ne^{21} + Ne^{22}$	A^{36}	A^{38}	Sc^{45}
100	345	3,6	4,0	6,7	20

Von radioaktiven Produkten der kosmischen Strahlung wurde in Meteoriten als erstes Tritium gefunden (FIREMAN und SCHWARZER). Eine Reihe weiterer Messungen dieses Nuklids sind inzwischen durchgeführt worden [z.B. BEGEMANN, EBERHARDT und HESS; GOEBEL und SCHMIDLIN (1)]. Die meisten Ergebnisse betreffen Steinmeteoriten. Bei Eisenmeteoriten stehen offenbar beträchtliche technische Schwierigkeiten einer zuverlässigen H^3-Bestimmung im Wege. Wegen der kurzen Halbwertzeit des H^3 sind diese Untersuchungen auf in jüngeren Zeiten gefallene Meteoriten beschränkt. Das Folgeprodukt He^3 kann nicht unmittelbar zur Abschätzung des H^3-Gehaltes dienen, weil es seinerseits auch direkt durch die kosmische Strahlung erzeugt wird.

Die langlebigen Kernarten Be^{10} und Al^{26} wurden in Eisen- und Stein-Meteoriten nachgewiesen (KOHMAN und EHMANN; VAN DILLA, ARNOLD und ANDERSON). Allerdings bestehen hinsichtlich der gefundenen Aktivitätsbeträge noch Diskrepanzen (HONDA, SHEDLOVSKY und ARNOLD). Cl^{36} konnte an einer größeren Zahl von Meteoriten nachgewiesen werden [GFELLER u. a.; SPRENKEL, DAVIS und WIIG; VILCSEK und WÄNKE (2)]. Nach den stabilen sind neuerdings auch die radioaktiven A-Isotope A^{37} und A^{39} in Meteoriten nachgewiesen und für Untersuchungen über die zeitliche und räumliche Konstanz der kosmischen Strahlung nutzbar gemacht worden [FIREMAN (3); FIREMAN und DE FELICE (1), (2); STOENNER, SCHAEFFER und DAVIS; WÄNKE und VILCSEK; HEYMANN und SCHAEFFER). Neue Ergebnisse aus Brookhaven zeigen insbesondere eine sehr gute Übereinstimmung in der Erzeugung von A^{39} und Cl^{36} durch die kosmische Strahlung in Meteoriten und durch hochenergetische Protonen in einem entsprechenden Target (Tabelle 13).

K^{40} kommt eine besondere Rolle zu. Ein Meteorit enthält im allgemeinen einen gewissen natürlichen Gehalt an K^{40} entsprechend der normalen Isotopenzusammensetzung dieses Elements. Daneben wird durch die kosmische Strahlung zusätzlich K^{40} erzeugt. In Eisenmeteoriten ist der Kalium-Gehalt sehr niedrig (STOENNER und ZÄHRINGER, VOSHAGE und HINTENBERGER) und die

Bestrahlungszeit lang. Daher ist in diesem Fall der Anteil an natürlichem K^{40} klein gegen den von der kosmischen Strahlung erzeugten Anteil. Umgekehrt ist das Verhältnis bei Steinmeteoriten, die höheren Kalium-Gehalt besitzen, während die Bestrahlungszeit wesentlich niedriger liegt als bei den Eisenmeteoriten. Man findet daher bei Steinmeteoriten annähernd den normalen K^{40}-Anteil an der Isotopenzusammensetzung des Kaliums, während bei Eisenmeteoriten dieser Anteil stark erhöht ist.

Tabelle 13. *A^{39} und Cl^{36} in Meteoriten* (Heymann und Schaeffer)

Meteorit	Aktivität in dpm/kg		Verhältnis A^{39}/Cl^{36}
	A^{39}	Cl^{36}	
Aroos	16,3	15,8	1,03
Norfolk	14,1	14,7	0,96
Sikhote Alin	5,5	6,3	0,87
dgl. nach Sprenkel . . .	6,7	6,9	0,97
Treysa	22,1	20,3	1,09
Estherville	22,1	22,7	0,98
Mittel in Meteoriten			$0,98 \pm 0,05$
Mittel bei Beschuß mit 6,2 GeV-Protonen			$0,96 \pm 0,05$

An dem 1959 gefallenen Eisenmeteoriten Aroos sind zahlreiche kurzlebige radioaktive Nuklide nachgewiesen worden (Honda und Arnold). Na^{22} und Co^{60} wurden auch in einigen anderen Objekten gefunden [Vilcsek und Wänke (1); van Dilla, Arnold und Anderson]. In der Tabelle 14 sind die Werte für Aroos zusammengestellt. Sie enthält neben den beobachteten Aktivitäten von Honda und Lal gemessene Erzeugungsquerschnitte für Beschuß von Fe mit Protonen von 730 MeV. Der Vergleich zwischen Zählraten und Wirkungsquerschnitten gestattet Schlüsse auf die hauptsächlich an der Produktion beteiligten Energien. Für die Erzeugung von Kernen in der Umgebung von Mn liefern niedrige Energien hohe Beiträge. Bei mittleren Kernladungszahlen erfolgt die Erzeugung hauptsächlich in dem durch den gemessenen Wirkungsquerschnitt charakterisierten Energiebereich. Bei den leichteren Kernen fällt der Wirkungsquerschnitt für 730 MeV ab, und die Produktion erfolgt überwiegend bei hohen Energien.

Tabelle 14. *Aktivitäten im Eisenmeteoriten Aroos zur Zeit des Falls und Wirkungsquerschnitte für Beschuß mit 730 MeV-Protonen*

Kern	Be^{10}	Na^{22}	Al^{26}	Si^{32}	Cl^{36}	A^{37}	A^{39}	K^{40}	Ca^{45}	Sc^{46}	Ti^{44}
dpm/kg	4,1	2,1	3,6	0,8	14	22	16	8	5	30	4,4
mbn	0,7	0,36	0,43	0,3	6,8				1,4	6,4	2

Kern	V^{48}	V^{49}	Cr^{51}	Mn^{53}	Mn^{54}	Fe^{55}	$Co^{56}+Co^{58}$	Co^{57}	Co^{60}	Ni^{59}
dpm/kg	90	164	260	530	470	1600	120	89	17	60
mbn		30	27		33					

III. Strahlungsalter

Zur Bestimmung der Bestrahlungszeit T eines Meteoriten muß man den Wirkungsquerschnitt σ_i für die Erzeugung des untersuchten Nuklids i kennen, seine Energieabhängigkeit $\sigma_i(E)$ und das Spektrum der kosmischen Strahlung

$F(E)$. Dann wird die Erzeugungsrate in Atomen pro g und sec

$$q_i = 4\pi\,\bar{\sigma}_i\,J_0\,L/A \quad \text{mit} \quad \bar{\sigma}_i = \int\limits_{E_{\min}}^{\infty}\sigma(E)\,F(E)\,dE \Big/ \int\limits_{E_{\min}}^{\infty}F(E)\,dE. \tag{9}$$

L Loschmidtsche Zahl, A Atomgewicht der beschossenen Substanz, wobei in erster Linie Fe bei den Eisenmeteoriten in Betracht kommt, J_0 Intensität der kosmischen Strahlung pro cm^2 sec sterad. Die Gesamtzahl der erzeugten Atome eines (stabilen) Nuklids je g ist dann

$$N_i = q_i\,T. \tag{10}$$

Im Fall der Erzeugung radioaktiver Nuklide kann man für nicht zu große Halbwertszeiten mit Sättigung, d.h. Gleichheit von Produktion und Zerfall rechnen und erhält dann

$$N_k = q_k\,\tau_k = \frac{q_k}{\lambda_k} = \frac{4\pi\,\bar{\sigma}_k\,J_0\,L}{\lambda_k\,A}. \tag{11}$$

N_k Zahl der beim Eintritt in die Erdatmosphäre vorhandenen Atome pro g, λ_k Zerfallskonstante, τ_k mittlere Lebensdauer.

T kann man direkt aus der obigen Beziehung für ein stabiles Nuklid ausrechnen, wenn man alle anderen Größen kennt. Eine wesentliche Verbesserung ergibt sich durch Bestimmung zweier Nuklide, weil dadurch das schwer abschätzbare J_0 aus der Bestimmungsgleichung eliminiert wird. Benutzt man eine radioaktive Kernart, so wird

$$T = \frac{N_i}{q_i} = \frac{N_i}{\bar{\sigma}_i}\,\frac{A}{4\pi\,J_0\,L} = \frac{N_i}{\bar{\sigma}_i}\,\frac{\bar{\sigma}_k}{N_k\,\lambda_k}. \tag{12}$$

Eine Komplikation ergibt sich, wenn das Nuklid k durch radioaktiven Zerfall in das Nuklid i übergeht. Dann ist die gemessene Konzentration

$$N_i = N_i' + N_i''. \quad N_i' = q_i\,T, \quad N_i'' = \int\limits_0^T N_k\,\lambda_k\,dt = N_k\,\lambda_k\,T. \tag{13}$$

In diesem Fall erhält man

$$T = \frac{N_i' + N_i''}{q_i + N_k\,\lambda_k} = \frac{N_i\,A}{4\pi\,J_0\,L(\bar{\sigma}_i + \bar{\sigma}_k)} = \frac{N_i}{\bar{\sigma}_i + \bar{\sigma}_k}\,\frac{\bar{\sigma}_k}{N_k\,\lambda_k}. \tag{14}$$

Von relativ wenigen Ausnahmen abgesehen liegt das Strahlungsalter für Steinmeteoriten in der Größenordnung 10^7 Jahre, für Eisenmeteoriten in der Größenordnung 10^8 Jahre. Es bezieht sich auf die seit der Entstehung des betreffenden Bruchstücks vergangene Zeit, in der es der intensiven Einwirkung der kosmischen Strahlung ausgesetzt war. Voraussetzung für das Ermittlungsverfahren ist, daß die Lage der untersuchten Probe innerhalb des Meteoriten während der ganzen Zeit unverändert geblieben ist.

Eine Bestimmung des Gesteinsalters für das Meteoritenmaterial ist nach denselben Grundsätzen möglich wie bei terrestrischen Gesteinen, z.B. mittels der Kalium-Argon-Methode (GERLING und PAVLOVA, STOENNER und ZÄHRINGER, GEISS und HESS). Dabei muß man natürlich die durch die Bestrahlung von außen entstandenen Beiträge zu den für die Bestimmung benutzen Isotopen abziehen, etwa auf Grund einer Abschätzung unter Zuhilfenahme eines Isotops, dessen Gesamtmenge durch die kosmische Strahlung erzeugt wurde. Die erhaltenen Werte für das Gesteinsalter liegen fast alle in der Größenordnung 10^9 Jahre, für Eisenmeteoriten wesentlich höher als für Steinmeteoriten.

Den Berechnungen des Strahlungsalters liegen die Annahmen zugrunde, daß die kosmische Strahlung hinsichtlich Intensität und Energieverteilung räumlich und zeitlich in den in Frage kommenden Bereichen konstant geblieben ist und daß keinerlei Verluste der durch sie erzeugten Nuklide aus dem Meteoriten eingetreten sind. Bezüglich des letzten Punktes können vor allem Diffusionsverluste von Edelgasen und von Tritium eine Rolle spielen. Wegen der Abhängigkeit des Energiespektrums der erzeugenden Strahlung von der Eindringtiefe in den Meteoriten kann auch die Oberflächenerosion im interplanetaren Raum die Bestimmung des Strahlungsalters beeinträchtigen. Was die zeitliche Konstanz der kosmischen Strahlung angeht, so lassen sich die Meßergebnisse über das Strahlungsalter, die man mit radioaktiven Nukliden sehr verschiedener Halbwertszeiten erhält, dahingehend deuten, daß diese Konstanz jedenfalls auch über Zeiten von der Größenordnung 10^8 Jahre im wesentlichen ungestört geblieben ist (s. auch A IV). Über die räumliche Konstanz scheinen erst die neuesten Messungen an den Argon-Isotopen 37 und 39 (Halbwertszeiten 35 Tage bzw. 320 Jahre) Möglichkeiten zu Informationen zu eröffnen. Allerdings sind die Ergebnisse noch widerspruchsvoll. Stoenner, Schaeffer und Davis fanden für den Steinmeteoriten Hamlet einen Meßwert des Verhältnisses dieser beiden Isotope, der zwar etwas höher war als der im Cosmotron beim Beschuß mit 3 GeV-Protonen gemessene Wert. Aber die Differenz blieb innerhalb der Fehlergrenzen. Fireman und de Felice fanden für den gleichen Meteoriten einen um einen Faktor 2 höheren Wert für das Verhältnis als im Cosmotron bei Beschuß mit 2 GeV-Protonen. Auch bei dem Eisenmeteoriten Aroos lag ihr Meßwert außerhalb der Fehlergrenze über dem Wert des Verhältnisses der Wirkungsquerschnitte für Argon-Erzeugung an Fe. Daraus wäre zu schließen, daß die Bestrahlung in den letzten 50 Tagen, d.h. in einer Entfernung von 1 astronomischen Einheit (=große Halbachse der Erdbahn) stärker war oder daß die während dieser Zeit auf den Meteoriten einwirkende kosmische Strahlung einen höheren Anteil an hochenergetischen Teilchen aufwies. Gültige Schlüsse lassen sich aus diesen Messungen noch nicht ziehen.

Literaturverzeichnis

Anders, E.: The record in the meteorites. II. On the presence of aluminium-26 in meteorites and tektites. Geochim. et Cosmochim. Acta **19**, 53—62 (1960).

Anderson, E.C.: The production and distribution of natural radiocarbon. Annual Rev. Nucl. Sci. **2**, 63—78 (1953).

— W.F. Libby, S. Weinhouse, A.F. Reid, A.D. Kirshenbaum and A.V. Grosse: Natural radiocarbon from cosmic radiation. Phys. Rev. **72**, 931—936 (1947).

Arnold, J.R.: Beryllium-10 produced by cosmic rays. Science **124**, 584—585 (1956).

—, and H.A. Al-Salih: Beryllium-7 produced by cosmic rays. Science **121**, 451—453 (1955).

—, and E.C. Anderson: The distribution of C-14 in nature. Tellus **9**, 28—32 (1957).

— M. Honda and D. Lal: The record of cosmic ray intensity in the meteorites. (Im Druck.)

Baker, E., G. Friedlander and J. Hudis: Formation of Be[7] in interactions of various nuclei with high-energy protons. Phys. Rev. **112**, 1319—1321 (1958).

Batzel, R.E., D.R. Miller and G.T. Seaborg: The high energy spallation products of copper. Phys. Rev. **84**, 671—683 (1951).

Begemann, F.: Neubestimmung der natürlichen irdischen Tritiumzerfallsrate und die Frage der Herkunft des „natürlichen" Tritium. Z. Naturforsch. **14a**, 334—342 (1959).

— P. Eberhardt u. D.C. Hess: He-3-H-3-Strahlungsalter eines Steinmeteoriten. Z. Naturforsch. **14a**, 500—503 (1959).

—, and J. Friedman: Tritium and deuterium content of atmospheric hydrogen. Z. Naturforsch. **14a**, 1024—1031 (1959).

— J. Geiss and D.C. Hess: Radiation age of a meteorite from cosmic-ray-produced He-3 and H-3. Phys. Rev. **107**, 540—542 (1957).

—, and W.F. Libby: Continental water balance, ground water inventory and storage time, surface ocean mixing rates and world-wide water circulation patterns from cosmic-ray and bomb tritium. Geochim. et Cosmochim. Acta **12**, 277—296 (1957).

BENIOFF, P.A.: (1) Cosmic-ray production rate and mean removal time of beryllium-7 from the atmosphere. Phys. Rev. **104**, 1122—1130 (1956).
— (2) Nuclear reactions of low-Z elements with 5.7 BeV protons. Phys. Rev. **119**, 316—323 (1960).
BIEN, G.S., N.W. RAKESTRAW and H.E. SUESS: Radiocarbon concentration in Pacific Ocean water. Tellus **12**, 436—443 (1960).
BØGGILD, J.K., and F.H. TENNEY: Neutron stars in oxygen, neon and argon. Phys. Rev. **82**, 307 (1951).
BOLIN, B.: (1) On the use of tritium as a tracer for water in nature. 2nd Conf. Atomic Energy Geneva, Proc. 18, pp. 336—343 (1958).
— (2) Atmospheric chemistry and broad geophysical relationships. Proc. nat. Acad. Sci. Wash. **45**, 1663—1672 (1959).
— (3) On the exchange of carbon dioxide between the atmosphere and the sea. Tellus **12**, 274—281 (1960).
—, and E. ERIKSSON: Changes in the carbon dioxide content of the atmosphere and sea due to fossil fuel combustion. Rossby Memorial Volume. New York: Rockefeller Institute Press 1959.
BRINKMANN, R., K.O. MÜNNICH u. J.C. VOGEL: Anwendung der C-14-Methode auf Bodenbildung und Grundwasserkreislauf. Geol. Rdsch. **49** (1), 244—253 (1960).
BROECKER, W.S., R.D. GERARD, M. EWING and B.C. HEEZEN: Natural radiocarbon in the Atlantic Ozean. J. Geophys. Res. **65**, 2903—2931 (1960).
—, and E.A. OLSON: Radiocarbon from nuclear tests. II. Science **132**, 712—721 (1960).
—, and A. WALTON: Radiocarbon from nuclear tests. Science **130**, 309—314 (1959).
BROWN, H.: The carbon cycle in nature. Fortschr. Chemie org. Naturstoffe **14**, 317—333 (1957).
BROWN, R.M., and W.E. GRUMMIT: The determination of tritium in natural waters. Canad. J. Chem. **34**, 220—226 (1956).
BROWN, W.W.: Cosmic-ray nuclear interactions in gases. Phys. Rev. **93**, 528—534 (1954).
BULLOCK, F.W.: Nuclear disintegrations produced by cosmic ray particles in a high pressure cloud chamber. Proc. Phys. Soc. Lond. A **70**, 134—141 (1957).
BUTTLAR, H.v., and W.F. LIBBY: Natural distribution of cosmic-ray produced tritium. II. J. Inorg. and Nucl. Chem. **1**, 75—91 (1955).
—, and J. WENDT: Ground water studies in New Mexico using tritium as a tracer. 2nd Conf. Atomic Energy Geneva, Proc. 18, pp. 591—597 (1958).
COON, J.H., and R.A. NOBLES: Disintegration of He-3 and N-14 by thermal neutrons. Phys. Rev. **75**, 1358—1361 (1949).
CRAIG, H.: (1) The natural distribution of radiocarbon and the exchange time of carbondioxide between the atmosphere and sea. Tellus **9**, 1—17 (1957).
— (2) Distribution, production rate, and possible solar origin of natural tritium. Phys. Rev. **105**, 1125—1127 (1957).
— (3) Isotopic tracer techniques for measurement of physical processes in the sea and the atmosphere. Nat. Acad. Sci. — Nat. Res. Council Publ. No. 551, 103—120 (1957).
— (4) A critical evaluation of mixing rates in ocean and atmosphere by use of radiocarbon techniques. 2nd Conf. Atomic Energy Geneva, Proc. **18**, pp. 358—363 (1958).
—, and D. LAL: The production rate of natural tritium. Tellus **13**, 85—105 (1961).
CRUIKSHANK, A.J., G. COWPER and W.E. GRUMMIT: Production of Be-7 in the atmosphere. Canad. J. Chem. **34**, 214—219 (1956).
CURRIE, L.A.: Tritium production by 6 BeV-protons. Phys. Rev. **114**, 878—880 (1959).
— W.F. LIBBY and R.L. WOLFGANG: Tritium production by high-energy protons. Phys. Rev. **101**, 1557 (1956).
DAVIS, R., and O.A. SCHAEFFER: Chlorine-36 in nature. Brookhaven Nat. Lab. BNL 340 (T-59), June 1955. Ann. New York Acad. Sci. **62**, 105—121 (1955).
DICKSON, J.M., and T.C. RANDLE: The excitation function for the production of Be-7 by the bombardment of C-12 by protons. Proc. Phys. Soc. Lond. A **64**, 902—905 (1951).
DILLA, M.A. VAN, J.R. ARNOLD and E.C. ANDERSON: Spectrometric measurement of natural and cosmic-ray induced radioactivity in meteorites. Geochim. et Cosmochim. Acta **20**, 115—121 (1960).
FALTINGS, V., u. P. HARTECK: Der Tritiumgehalt der Atmosphäre. Z. Naturforsch. **5a**, 438—439 (1950).
FERGUSSON, G.A., J. HALPERN, R. NATHANS and P.F. YERGIN: Photoneutron cross sections in He, N, O, F, Ne, and A. Phys. Rev. **95**, 776—780 (1954).
FERGUSSON, G.J.: Reduction of atmospheric radiocarbon concentration by fossil fuel carbon dioxide and the mean life of carbon dioxide in the atmosphere. Proc. Roy. Soc. Lond. A **243**, 561—574 (1958).

Fireman, E.L.: (1) Measurement on the (n, H-3) cross section in nitrogen and its relationship to the tritium production in the atmosphere. Phys. Rev. **91**, 922—926 (1953).
— (2) Tritium production by 2,2-BeV-protons on iron and its relation to cosmic radiation. Phys. Rev. **97**, 1303—1304 (1955).
— (3) Argon-39 in the Sikhote-Alin meteorite fall. Nature, Lond. **181**, 1613—1614 (1958).
— (4) The distribution of Helium-3 in the grant meteorite and a determination of the original mass. Planet. Space Sci. **1**, 66—70 (1959).
—, and J. de Felice: (1) Argon-37, argon-39 and tritium in meteorites and the spatial constancy of cosmic rays. J. Geophys. Res. **65**, 3035—3041 (1960).
— — (2) Argon-39 and tritium in meteorites. Geochim. et Cosmochim. Acta **18**, 183—192 (1960).
—, and F. S. Rowland: Tritium and neutron production by 2,2 Bev protons on nitrogen and oxygen. Phys. Rev. **97**, 780—782 (1955).
—, and J. Zähringer: Depth variation of tritium and argon-37 produced by high-energy protons in iron. Phys. Rev. **107**, 1695—1698 (1957).
Fonselius, S., and G. Östlund: Natural radiocarbon measurements on surface water from the North Atlantic and Arctic Sea. Tellus **11**, 77—82 (1959).
Freese, E., u. P. Meyer: Neutronen in der Atmosphäre. In: Kosmische Strahlung (ed. W. Heisenberg), 2. Aufl., S. 225—237. Berlin-Göttingen-Heidelberg: Springer 1953.
Fuller, M.O.: Disintegration of oxygen by 300 MeV neutrons. University of California Radiation Laboratory UCRL 2699, 1954.
Gabbe, J.D.: Some measurements of atmospheric neutrons. Phys. Rev. **112**, 497—502 (1958).
Geiss, J., and D.C. Hess: Argon-potassium ages and the isotopic composition of argon from meteorites. Astrophys. J. **127**, 224—235 (1958).
— H. Oeschger and U. Schwarz: The history of cosmic radiation as revealed by isotopic changes in the meteorites and on the earth. Varenna Summer School Course on Cosmic Rays, Solar Particles and Space Research 1961.
— — and P. Signer: Radiation ages of chondrites. Z. Naturforsch. **15**a, 1016—1017 (1960).
Gentner, W., u. J. Zähringer: (1) Argon und Helium als Kernreaktionsprodukte in Meteoriten. Geochim. et Cosmochim. Acta **11**, 60—71 (1957).
— — (2) Das Kalium-Argon-Alter von Tektiten. Z. Naturforsch. **15**a, 93—99 (1960).
Gerling, G.K., u. T.G. Pavlova: Bestimmung des geologischen Alters von zwei Meteoriten nach der Argon-Methode. (Orig. russ.) Dokl. Akad. Nauk SSSR. **77**, 85—86 (1951).
Gfeller, C., W. Heer, F.G. Houtermans and H. Oeschger: Cl-36 in meteorites. Helv. phys. Acta **32**, 277—279 (1959).
Giletti, B. J., F. Bazan and J.L. Kulp: The geochemistry of tritium. Trans. Amer. Geophys. Un. **39**, 807 (1958).
Goebel, K., u. P. Schmidlin: (1) Tritium-Messungen an Steinmeteoriten. Z. Naturforsch. **15**a, 79—82 (1960).
— — (2) Der Meteorit von Breitscheid. IV. Radiochemische Untersuchungen. B.Tritium. Geochim. et Cosmochim. Acta **17**, 342—349 (1959).
— — (3) The radiation age of meteorites. Conference on the Use of Radioisotopes in the Physical Sciences and Industry 1960.
—, u. J. Zähringer: Erzeugung von Tritium und Edelgasisotopen bei Bestrahlung von Fe und Cu mit Protonen von 25 GeV Energie. Z. Naturforsch. **16**a, 231—236 (1961).
Goel, P.S.: Radioactive sulphur produced by cosmic rays in rain water. Nature, Lond. **178**, 1458—1459 (1956).
— S. Iha, D. Lal, P. Ramakrishna and T. Rama: Cosmic-ray produced beryllium isotopes in rain water. Nuclear Phys. **1**, 196—201 (1956).
— D.P. Kharkhar, D. Lal, V. Narasappaya, B. Peters and V. Yatirajam: The beryllium-10 concentration in deep sea sediments. Deep Sea Res. **4**, 202—210 (1957).
— N. Narasappaya, T. Rama, P.K. Zutshi and C. Prabhakara: Study of cosmic ray produced short-lived P-32, P-33, Be-7 and S-35 in tropical latitudes. Tellus **11**, 91—100 (1959).
Goldberg, E.D., and Arrhenius: Chemistry of Pacific pelagic sediments. Geochim. et Cosmochim. Acta **13**, 153—212 (1958).
Goldschmidt, V.M.: Geochemistry. Oxford: Clarendon Press 1954.
Gonsior, B.: (1) Tritium-Anstieg im atmosphärischen Wasserstoff. Naturwissenschaften **46**, 201—202 (1959).
— (2) Die Konzentration des Tritiums in der Atmosphäre. Diss. Heidelberg 1960.
Grosse, A.V., W.H. Johnston, R.L. Wolfgang and W.F. Libby: Tritium in nature. Science **113**, 1—2 (1951).
— A.D. Kirshenbaum, J.L. Kulp and W.S. Broecker: The natural tritium content of atmospheric hydrogen. Phys. Rev. **93**, 250—251 (1954).

HAGEMANN, F., J. GRAY, L. MACHTA and A. TURKEVICH: Stratospheric carbon-14, carbon dioxide, and tritium. Science **130**, 542—552 (1959).

HARTECK, P.: Relative abundance of HT and HTO in the atmosphere. J. Chem. Phys. **22**, 1746—1751 (1954).

HERNEGGER, F., u. H. WÄNKE: Über den Uran-Gehalt der Steinmeteorite und deren „Alter". Z. Naturforsch. **12**a, 759—762 (1957).

HESS, W.N., H.W. PATTERSON, R. WALLACE and E.L. CHUPP: Cosmic-ray neutron energy spectrum. Phys. Rev. **116**, 445—457 (1959).

HEYMANN, D., and O.A. SCHAEFFER: Geochim. et Cosmochim. Acta (im Druck).

HODGSON, P.E.: Nuclear disintegrations caused by 50—125 MeV protons. Phil. Mag. **45**, 190—201 (1954).

HOFFMAN, J.H., and A.O. NIER: Production of helium in iron meteorites by the action of cosmic rays. Phys. Rev. **112**, 2112—2117 (1958).

HONDA, M.: Cosmogenic potassium-40 in iron meteorites. Geochim. et Cosmochim. Acta **17**, 148—150 (1959).

—, and J.R. ARNOLD: Radioactive species produced by cosmic rays in the Aroos iron meteorite. Geochim. et Cosmochim. Acta **23**, 219—232 (1961).

—, and D. LAL: Some cross sections for the production of radio-nuclides in the bombardment of C, N, O, and Fe by medium energy protons. Phys. Rev. **118**, 1618—1625 (1960).

— J.P. SHEDLOVSKY and J.R. ARNOLD: Radioactive species produced by cosmic rays in iron meteorites. Geochim. et Cosmochim. Acta **22**, 133—154 (1961).

HUTCHINSON, G.E.: The biochemistry of the terrestrial atmosphere, carbon dioxide and other compounds. In: The earth as a planet (G. KUIPER, ed.). Chicago: University of Chicago Press 1954.

ISRAËL, G.: Dipl.-Arbeit Heidelberg 1961.

JOHNSON, C.H., and H.H. BARSCHALL: Interaction of fast neutrons with nitrogen. Phys. Rev. **80**, 818—823 (1950).

KANWISHER, J.: (1) General Assembly Intern. Union of Geophysics and Geodesy Helsinki 1960, Symposium on carbon dioxide in the atmosphere and the sea surface.

— (2) pCO_2 in sea water and its effect on the movement of CO_2 in nature. Tellus **12**, 209—215 (1960).

KAUFMAN, S., and W.F. LIBBY: The natural distribution of tritium. Phys. Rev. **93**, 1337—1344 (1954).

KELLOGG, D.A.: Cross sections for products of 90 MeV neutrons on carbon. Phys. Rev. **90**, 224—232 (1953).

KOHMAN, T.P., and W.D. EHMANN: Cosmic-ray-induced radioactivity in meteorites and tektites. Radioisotopes in Scientific Research, UNESCO Conf. Paris, vol. 2, pp. 661—680 (1957).

KULP, J.L., and H.L. VOLCHOK: Constancy of cosmic-ray flux over the past 30000 years. Phys. Rev. **90**, 713—714 (1953).

KURCHATOV, B.V., V.N. MEKHEDOV, N.J. BORISOVA, M.YA. KUZNETSOVA, L.N. KURCHATOVA and L.V. CHISTYAKOV: Radiochemical investigation of the products of the spallation of silver with high-energy particles. Conf. Acad. Sci. USSR Atomic Energy 1955, Div. Chem. Sci., Transl. p. 111.

LAL, D.: Cosmic ray produced radioisotopes for studying the general circulation in the atmosphere. Indian J. Met. Geophys. **10**, 147—154 (1959).

— J.R. ARNOLD and M. HONDA: Cosmic-ray production rates of Be-7 in oxygen, and P-32, P-33, S-35 in argon at mountain altitudes. Phys. Rev. **118**, 1626—1632 (1960).

— E.D. GOLDBERG and M. KOIDE: Cosmic-ray produced silicon-32 in nature. Science **131**, 332—337 (1960).

— P.K. MALHOTRA and B. PETERS: On the production of radioisotopes in the atmosphere by cosmic radiation and their application to meteorology. J. Atmosph. Terr. Phys. **12**, 306—328 (1958).

— K. NARSAPPAYA and P.K. ZUTSHI: Phosphorus isotopes P-32 and P-33 in rain water. Nuclear Phys. **3**, 69—75 (1957).

—, and B. PETERS: Cosmic ray produced isotopes as tracers for studying large scale atmospheric circulation. 2nd Conf. Atomic Energy Geneva 1958, Proc. **18**, pp. 533—544 (1958).

— — Cosmic ray produced isotopes and their application to geophysics. Progr. Cosmic Ray Physics (im Druck).

— T. RAMA and P.K. ZUTSHI: Radioisotopes P-32, Be-7, and S-35 in the atmosphere. J. Geophys. Res. **65**, 669—674 (1960).

LIBBY, W.F.: (1) Radiocarbon dating, 2nd ed. Chicago: Chicago University Press 1955.

— (2) Radioactive fallout. Bull. schweiz. Akad. med. Wiss. **14**, 309—347 (1958).

— (3) Tritium geophysics; recent data and results. May 1961. Preprint.

LORD, J.J.: The altitude and latitude variation in the rate of occurrence of nuclear disintegrations produced in the stratosphere by cosmic rays. Phys. Rev. **81**, 901—909 (1951).

MacDonald, F. B.: Primary cosmic-ray intensity near solar maximum. Phys. Rev. **116**, 462—464 (1959).

Mann, W. B., u. a.: Vgl. News Note: Carbon-14 half-life redetermined. Science **133**, 183 (1961).

Marquez, L., and N. L. Costa: The formation of P-32 from atmospheric argon by cosmic rays. Nuovo Cim. **2**, 1038—1041 (1955).

— — and I. G. Almeida: The formation of Na-22 from atmospheric argon by cosmic rays. Nuovo Cim. **6**, 1292 (1957).

—, and I. Perlman: Observations on lithium and beryllium nuclei ejected from heavy nuclei by high energy particles. Phys. Rev. **81**, 953—957 (1951).

Marshall, R. R.: Calculation of a 'cosmic ray age' for the iron meteorite 'Carbo'. Nature, Lond. **184**, 117—118 (1959).

Merrill, J. R., M. Honda and J. R. Arnold: Beryllium geochemistry and beryllium-10 age determinations. 2nd Conf. Atomic Energy Geneva 1958, Proc. **2**, pp. 251—254 (1958).

— E. F. X. Lyden, M. Monda and J. R. Arnold: The sedimentary geochemistry of the beryllium isotopes. Geochim. et Cosmochim. Acta **18**, 108—129 (1960).

Messel, H.: The development of a nucleon cascade. Progr. Cosmic Ray Physics **2**, 135—216 (1954).

Meyer, P., and J. A. Simpson: Changes in the low-energy particle cutoff and primary spectrum of cosmic rays. Phys. Rev. **106**, 562—571 (1957).

Münnich, K. O.: Messung des C^{14}-Gehaltes von hartem Grundwasser. Naturwissenschaften **44**, 32—33 (1957).

—, u. J. C. Vogel: (1) Durch Atombomben erzeugter Radiokohlenstoff in der Atmosphäre. Naturwissenschaften **14**, 327—329 (1958).

— — (2) Variations in C-14 content during the last years. Internat. C-14-Symposium Groningen 1959.

— — (3) Jahreszeitliche Schwankungen im C-14-Gehalt der Atmosphäre. Phys. Verh. **11**, 62—63 (1960).

— — (4) C-14 age determination of deep ground-waters. Internat. Assoc. Sci. Hydrology. Commission of Subterrenean Waters, Publ. 52, 537—541 (1960).

Nilsson, R., K. Siegbahn, A. Berggren and B. Ingelman: Be-7 content in snow. Ark. Fysik **11**, 445—451 (1957).

— I. Olsson, A. Berggren and K. Siegbahn: S-35 and Be-7 contents in rain and snow. Ark. Geofys. **3**, 111—122 (1959).

Östlund, G.: Intern. Atomic Energy Agency Symposium on Detection and Use of Tritium in the Physical and Biological Science. Wien 1961.

Patterson, R. L., and J. H. Blifford: Atmospheric carbon-14. Science **126**, 26—28 (1957).

Peters, B.: (1) The nature of primary cosmic radiation. Progr. Cosmic Ray Physics **1**, 193—244 (1952).

— (2) Radioactive beryllium in the atmosphere and on the earth. Proc. Indian Acad. Sci. **41**, 67—71 (1955).

— (3) Über die Anwendbarkeit der Be-10-Methode zur Messung kosmischer Strahlungsintensität und der Ablagerungs-Geschwindigkeit von Tiefseesedimenten vor einigen Millionen Jahren. Z. Physik **148**, 93—111 (1957).

— (4) Cosmic-ray produced radioactive isotopes as tracers for studying large-scale atmospheric circulation. J. Atmosph. Terr. Phys. **13**, 351—370 (1959).

Pfotzer, G.: Die Neutronenkomponente der Ultrastrahlung. Naturwissenschaften **39**, 149—158 (1952).

Rafter, T. A., and G. J. Fergusson: (1) 'Atom bomb effect'—recent increase of carbon-14 content of the atmosphere and biosphere. Science **126**, 557—558 (1957).

—, and G. J. Fergusson: (2) Atmospheric radiocarbon as a tracer in geophysical circulation-problems. 2nd Conf. Atomic Energy Geneva 1958, Proc. **18**, pp. 526—532 (1958).

Rama, T.: Investigations of the radioisotopes Be-7, P-32 and S-35 in rain water. J. Geophys. Res. **65**, 3773—3776 (1960).

—, and P. K. Zutshi: Annual deposition of cosmic ray produced Be-7 at equatorial latitudes. Tellus **10**, 99—103 (1958).

Reed, G. W., and A. Turkevich: Uranium, helium and the age of meteorites. Nature, Lond. **180**, 594—596 (1957).

Revelle, R., and H. E. Suess: Carbon dioxide exchange between atmosphere and ocean and the question of an increase of atmospheric CO_2 during the past decades. Tellus **9**, 18—27 (1957).

Riley, G. A., H. Stommel and D. F. Bumpus: Quantitative ecology of the plankton of the western North Atlantic. Bull. Bingham Ocean. Coll. **12**, Art. 3, 169 pp. (1949).

Rose, D. C., K. B. Fenton, J. Katzman and J. A. Simpson: Latitude effect of the cosmic ray nucleon and meson components at sea level from the arctic to the antarctic. Canad. J. Phys. **34**, 968—984 (1956).

Rossi, B.: High-energy particles. New York 1952.

Rubey, W.W.: Geologic history of sea water. Bull. Geol. Soc. Amer. **62**, 1111—1147 (1951).

Rudstam, S.G.: (1) Spallation of vanadium, manganese, and cobalt with 187 MeV protons. Phil. Mag. (7) **44**, 1131—1141 (1953).

— (2) Spallation of elements in the mass range 51—75. Phil. Mag. (7) **46**, 344—356 (1955).

Schaeffer, O.A., and D.E. Fisher: Exposure ages for iron meteorites. Nature, Lond. **186**, 1040—1042 (1960).

— S.O. Thompson and N.L. Lark: Chlorine-36 radioactivity in rain. J. Geophys. Res. **65**, 4013—4016 (1960).

—, u. J. Zähringer: Helium- und Argon-Erzeugung in Eisentargets durch energiereiche Protonen. Z. Naturforsch. **13**a, 346—347 (1958).

Schumann, G., u. G. Eulitz: Die jahreszeitliche Variation der stratosphärischen Fallout-Komponente. Naturwissenschaften **47**, 13—14 (1960).

Serber, R.: Nuclear reactions at high energies. Phys. Rev. **72**, 1114—1115 (1947).

Shedlovsky, J.P.: Cosmic ray produced Mn-53 in iron meteorites. Geochim. et Cosmochim. Acta **21**, 156—158 (1960).

Simpson, J.A.: (1) Neutrons produced in the atmosphere by the cosmic radiations. Phys. Rev. **83**, 1175—1188 (1951).

— (2) The production of tritium and C-14 in the terrestrial atmosphere by solar protons. J. Geophys. Res. **65**, 1615—1616 (1960).

— H.W. Baldwin and R.B. Uretz: Nuclear bursts produced in the low energy nuclear component of the cosmic radiation. Phys. Rev. **84**, 332—339 (1951).

—, and W.C. Fagot: Properties of the low energy nucleonic component at large atmospheric depths. Phys. Rev. **90**, 1068—1072 (1953).

Soberman, R.K.: High-altitude cosmic-ray neutron intensity variations. Phys. Rev. **102**, 1399—1409 (1956).

Sprenkel, E.L., R. Davis and E.O. Wiig: Cosmic-ray produced Cl^{36} and A^{39} in iron meteorites. Bull. Amer. Phys. Soc. **4**, 223 (1959).

Stewart, N.G., R.G.D. Osmond, R.N. Crooks and E.M. Fisher: The world-wide deposition of long-lived fisson products from nuclear test explosions. Atomic Energy Res. Establ. Harwell HP/R. 2354 (1957).

Stoenner, R.W., O.A. Schaeffer and R. Davis: Meteorites as space probes for testing the spatial constancy of cosmic radiation. J. Geophys. Res. **65**, 3025—3034 (1960).

—, and J. Zähringer: Potassium-argon age of iron meteorites. Geochim. et Cosmochim. Acta **15**, 40—50 (1958).

Stuiver, M.: Variations in radiocarbon concentration and sun spot activity. J. Geophys. Res. **66**, 273—276 (1961).

Suess, H.E.: (1) Radiocarbon concentration in modern wood. Science **122**, 415—417 (1955).

— (2) The radioactivity of the atmosphere and hydrosphere. Annual Rev. Nucl. Sci. **8**, 243—256 (1958).

Vernov, S.N., and N.L. Grigorov: Interaction of primary cosmic ray particles of different energies with the nuclei in the atmosphere. Nuovo Cim. (10) Suppl. **4**, 879—889 (1956).

Vilcsek, E., u. H. Wänke: Natrium-22 im Meteorit Breitscheid. Z. Naturforsch. **15**a, 1004—1007 (1960).

Voshage, H., u. H. Hintenberger: Die Kalium-Isotope als Reaktionsprodukte der kosmischen Strahlung im Eisenmeteoriten Carbo. Z. Naturforsch. **14**a, 194—195 (1959).

Vries, Hl. de: Variation in concentration of radiocarbon with time and location on earth. Proc. Kon. Ned. Akad. Wet. B **61**, No. 2, 1—9 (1958).

Wänke, H.: (1) Scandium-45 als Reaktionsprodukt der Höhenstrahlung in Eisenmeteoriten I. Z. Naturforsch. **13**a, 645—649 (1958).

— (2) Über den Kaliumgehalt der Chondrite, Achondrite und Siderite. Z. Naturforsch. **16**a, 127—130 (1961).

— and E. Vilcsek: Argon-39 als Reaktionsprodukt der Höhenstrahlung in Eisenmeteoriten. Z. Naturforsch. **14**a, 929—934 (1959).

Wheeler, J.A.: Some consequences of the electromagnetic interaction between μ-mesons and nuclei. Rev. Mod. Phys. **21**, 133—143 (1949).

Willis, E.H., H. Tauber and K.O. Münnich: Variations in the atmospheric radiocarbon concentration over the past 1300 years. Amer. J. Sci. Radiocarbon Suppl. **2**, 1—4 (1960).

Winckler, J.R.: Balloon study of high-altitude radiations during the International Geophysical Year. J. Geophys. Res. **65**, 1331—1360 (1960).

Winsberg, L.: The production of chlorine-39 in the lower atmosphere by cosmic radiation. Geochim. et Cosmochim. Acta **9**, 183—189 (1956).

Yuan, L.C.L.: Distribution of slow neutrons in free atmosphere up to 100000 ft. Phys. Rev. **81**, 175—184 (1951).

Zähringer, J.: Priv. Mitteilung.

Transfer and Circulation of Radioactivity in the Atmosphere

by

Bert Bolin

With 16 Figures

Zusammenfassung

Die Bewegung der Atmosphäre ist außerordentlich kompliziert. Es gibt permanente Zirkulationsräder und ebenso turbulente Bewegungen von den großen Wirbeln, die von Zyklonen oder Antizyklonen gebildet werden, bis zu den kleinen Windböen von der Dauerhaftigkeit eines Bruchteils einer Sekunde. Es ist wichtig, diese verschiedenen Größenordnungen der atmosphärischen Bewegungen bei einer Diskussion über den Transport und die Zirkulation der Radioaktivität in der Atmosphäre zu unterscheiden. In der folgenden Diskussion wird eine Einteilung durchgeführt, bei der drei verschiedene Bereiche unterschieden werden:

a) Der Bereich bis zu 10 km (z.B. die Grenzschicht der Atmosphäre zur Erde).

b) Der Bereich 10 bis 1000 km (z.B. die Zerstreuung radioaktiver Wolken in der freien Atmosphäre).

c) Der Bereich größer als 1000 km (z.B. die globale Verteilung der Radioaktivität in der Atmosphäre).

Zunächst wird eine kurze Zusammenfassung der modernen Turbulenztheorie gegeben mit Hinweis auf die in diesem Zusammenhang wichtigsten Arbeiten.

Bei der Behandlung der Verbreitung der Radioaktivität in der erdnahen Schicht der Atmosphäre wird die Anwendbarkeit von Suttons Diffusionstheorie diskutiert, weiterhin wird auf die modernen russischen Arbeiten betreffend die Anwendbarkeit der Gleichartigkeitstheorien für die Verbreitung in einer geschichteten Atmosphäre, auf Gesichtspunkte für die Bedeutung des Looping bei der Verbreitung von Rauchpilzen, den Zusammenhang zwischen den Korrelationsfunktionen von Lagrange und Euler sowie eine Diskussion des Begriffes Verbreitungsgeschwindigkeit und dessen Bedeutung, speziell im Zusammenhang mit der Absetzung von I^{131} hingewiesen. Eine zusammenfassende Behandlung der Verteilung von Rn und Tn und deren Tochterprodukte in der erdnahen Schicht wird mit besonderem Hinblick auf die Messungen von RaC und RaD gegeben.

Eine Diskussion über die Verbreitung und Zirkulation von Radioaktivität in der freien Atmosphäre weist vor allem auf die Bedeutung atmosphärischer Deformationsfelder für Strukturveränderungen radioaktiver Wolken über längere Zeitdauer hin. Eine Schätzung des Austausch-Koeffizienten für die Verbreitung von Wolken der Dimension 100 km ergibt $K = 3 \cdot 10^8$ cm^2 sec^{-1}. Weiter wird eine Zusammenfassung gegeben zu dem Versuch, die wahrscheinliche Position einer Wolke in der freien Atmosphäre zu verschiedenen Zeitpunkten zu berechnen, nachdem die Wolke beobachtet wurde.

Für die Diskussion über die globale Zirkulation von Radioaktivität in der Atmosphäre wird zuerst eine Zusammenfassung unserer Auffassung über die Zirkulation der Atmosphäre gegeben, die durch direkte Wind- und Temperatur-beobachtungen gewonnen wurde. Für das Studium der globalen Zirkulation der Atmosphäre werden 1. die natürliche Radioaktivität (Rn und Tn samt deren Tochterprodukte), 2. die durch kosmische Strahlung gebildete Radioaktivität und 3. die von Atombombenexplosionen herrührende künstliche Radioaktivität verwendet.

1. Durch Betrachtung eines vereinfachten Modells der Atmosphäre, in welchem die Auswaschung der Atmosphäre in den unteren 5 km vor sich geht, und turbulenten Transport in höheren Schichten kann man zeigen, daß die Verteilung natürlicher Radioaktivität in der Troposphäre zu einem Austausch-Koeffizient von $K_z = 2 \cdot 10^5$ cm² sec⁻¹ führt.

2. Die Radioaktivität, die durch kosmische Strahlung gebildet wird, ist besonders geeignet zum Studium der Zirkulation der Atmosphäre auf Grund der relativ guten Kenntnis der Verteilung der Produktion von Radioaktivität und weiter dadurch, daß die Produktion in der Stratosphäre pro Masseneinheit Luft 100- bis 1000mal größer ist als die Produktion an der Erdoberfläche. Auf Grund dessen ist besonders ein Studium des vertikalen Luftaustausches möglich. Der C¹⁴-Gehalt und seine Veränderungen durch Verbrennung von Kohle und Öl einerseits und durch Neubildung bei Atombombenexplosionen andererseits geben Möglichkeiten zum Studium des Austausches zwischen nördlicher und südlicher Hemisphäre und des vertikalen Luftaustausches zwischen Stratosphäre und Troposphäre.

3. Es stößt auf große Schwierigkeiten, die Verteilung und den Ausfall von künstlicher Radioaktivität, Sr⁹⁰, Cs¹³⁷ usw. quantitativ auszuwerten, da Menge, Höhe und Zeitpunkt der freigegebenen Radioaktivität nicht bekannt sind. Die umfassenden Meßserien, die existieren, lassen jedoch gewisse allgemeine Schluß-folgerungen zu:

a) Radioaktivität in der polaren Stratosphäre erreicht den Boden bedeutend schneller als Radioaktivität in der äquatoriellen Stratosphäre.

b) Der turbulente Transport von Radioaktivität ist zumindest von gleicher Bedeutung wie der, verursacht von einer zeitlich konstanten meridionalen Zirkulationszelle.

c) Die Aufenthaltszeit für bisher freigegebene Radioaktivität in der Stratosphäre ist wahrscheinlich zwei Jahre oder weniger.

d) Tritium gibt die Möglichkeit zum Studium der globalen Zirkulation von Wasserdampf.

Zusammenfassend wird eine Methode angedeutet, nach der eine gleichzeitige Behandlung der Verteilung verschiedener (radioaktiver oder stabiler) Spurenelemente die Möglichkeit gibt, sowohl den großen turbulenten Austausch zwischen verschiedenen Breiten und Höhen zu bestimmen sowie das eventuelle Vorkommen von permanenten Zirkulationsrädern in der Meridionalfläche zu ermitteln.

A. Introduction

The motion of the atmosphere is indeed very complex with large variations both in time and space. The time fluctuations range from the slow seasonal changes of the mean flow pattern of the atmosphere to rapid wind gusts of a second's duration or less. In space we know of scales of several thousand kilometers as the long waves in the westerlies, but we also notice variations in the wind over distances of a few meters. A general dynamical theory for this whole spectrum of motions and the interaction between various scales is lacking. We do not even always know their main characteristics from ordinary meteorological observations. Undoubtedly radioactive tracers may be of great help is the further exploration of these atmospheric motions.

The motion of the atmosphere is maintained by the non-uniform heating of the earth by the sun. Hereby both small scale convection (*e.g.* cumulus clouds), intermediate scale weather disturbances (*e.g.* cyclones) and large scale circulations (*e.g.* the trade winds) are generated. The energy input thus takes place at various scales of motion. The non-linear character of these motions implies energy transfer from one scale to another. Energy dissipation finally takes place by small scale turbulence transferring its energy into scales where molecular dissipation occurs. To a considerable extent this takes place in the surface layer (about 1 km deep) of the atmosphere, but recent studies of the general circulation of the atmosphere make it plausible that internal friction in the free atmosphere is of considerable importance.

The turbulent nature of atmospheric motions is thus a basic feature which must be recognized in all studies of transfer of matter through the atmosphere. Depending on the scale of the phenomenon to be studied we usually can define a mean motion, constant or slowly varying in time, which includes scales of motion larger than the scale of the phenomenon and call the comparatively rapid small-scale fluctuations turbulent motion. The characteristics of the turbulent motions will, of course, be dependent on the scales of motions they include.

Extensive studies of turbulent transfer in the surface boundary layer of the atmosphere have been conducted during the last thirty years. Here the synoptic flow pattern, with a scale of some hundred kilometers is considered as constant and transfer of matter (and heat and momentum) is studied as dependent on the mean velocity profile and the vertical temperature distribution in this layer. The more large scale synoptic features of the motion (cyclones and anticyclones) will, however, carry matter over great distances in the atmosphere. Over a time scale of a few hours to a day we can recognize a cloud of smoke or radioactive debris as a well-defined entity, but soon these scales of motion act as turbulent motion on a larger scale, the dynamics of which in many respects are quite different from those studied for the boundary layer. The cloud becomes very distorted, while concentrations may remain high at individual points. We do not find a simple more or less symmetrical pattern as obtained from the equation describing molecular diffusion. Our present knowledge about atmospheric motions (in the free atmosphere) of the scales in between the synoptic scale and the presumably more or less isotropic small-scale turbulence is indeed limited and information gained indirectly by studying the deformation and diffusion of clouds of radioactivity has been valuable and most likely will be even more so.

Considering finally the global distribution, transfer and circulation of matter, the synoptic disturbances become turbulent elements. A look at any series of weather maps immediately indicates that any small-scale turbulence must be completely insignificant for the horizontal exchange over such distances. In addition to the transfer caused by synoptic eddies we must also consider the possible existence of more permanent features of the atmospheric circulation as meridional circulation cells or the long waves fixed in their position by the orography of the earth's surface or non-uniform heating from below.

The *vertical* transfer may be due to small-scale turbulence (convection) but undoubtedly the vertical velocity field associated with the synoptic disturbances plays an important role as well as the more permanent circulations that may exist. For the lower parts of the atmosphere in particular scavenging by clouds and transfer by rain is very important (cf. the following chap.).

It is obvious from the previous brief survey that we have to consider different scales of motion in the atmosphere in attempts to use radioactive tracers for atmospheric studies. The following more detailed discussion will therefore naturally be separated into

a) Small-scale atmospheric transfer, essentially dealing with scales less than 10 km and particularly considering the surface boundary layer of the atmosphere.

b) Intermediate scale transfer and circulation, discussing dispersion of radioactive clouds in the free atmosphere on a scale from about 10 to 1000 km.

c) Large scale (global) transfer and circulation of radioactivity considering scales larger than about 1000 km and dealing with problems such as the transfer between the troposphere and the stratosphere and exchange between the northern and southern hemisphere.

The division of the subject in this way is somewhat arbitrary but will serve the purpose of stressing the great differences of atmospheric transfer problems as a function of the scale of the phenomenon.

Before proceeding to a detailed discussion of this kind a brief account of the theory of diffusion will be given. For a more complete treatment the reader is referred to the extensive literature on the subject and some excellent monographs (e.g. SUTTON 1953, PRIESTLEY 1959).

B. Theory of diffusion

The distribution of any compound q (in g per g of air) in the atmosphere and its change in time is given by

$$\frac{dq}{dt} = \frac{\partial q}{\partial t} + \boldsymbol{v} \cdot \nabla q = P(\boldsymbol{r}, t) - \lambda q, \tag{1}$$

where t indicates time, $\boldsymbol{v}$ is the instantaneous three-dimensional wind vector, ∇ the nabla operator, $P(\boldsymbol{r}, t)$ expresses the production and/or removal of q from the point $\boldsymbol{r}$ in space and λ is finally the radioactive decay constant for the element considered. Molecular diffusion has been neglected as being of no importance in the problems discussed here. Furthermore $P(\boldsymbol{r}, t)$ includes the possible effect of settling by gravity. We are, however, usually not interested in the instantaneous variations of q but the average value over a certain time interval or volume in space. This average will be denoted by (‾). We assume that

$$\bar{\bar{q}} = \bar{q}, \qquad \overline{\bar{q}\bar{A}} = \bar{q}\bar{A}. \tag{2}$$

Let q' denote the deviation from the mean value $\bar{q}$ and we have

$$q = \bar{q} + q', \qquad \bar{q}' = 0. \tag{3}$$

Similarly we put (ϱ = air density)

$$\varrho = \bar{\varrho} + \varrho' \tag{4}$$

and introduce the concept of a weighted (^) mean according to

$$\overline{\varrho q} = \bar{\varrho}\hat{q}. \tag{5}$$

Deviations from the weighted mean will be denoted by (″) and thus

$$\left. \begin{array}{l} q = \hat{q} + q'' \\ \boldsymbol{v} = \bar{\boldsymbol{v}} + \boldsymbol{v}''. \end{array} \right\} \tag{6}$$

With the aid of the continuity equation

$$\frac{\partial \varrho}{\partial t} + \nabla \cdot (\varrho \boldsymbol{v}) = 0 \tag{7}$$

we may transform Eq. (1) into

$$\frac{\partial}{\partial t}\,(\varrho\,q) + V\cdot(\varrho\,q\,\boldsymbol{v}) - \varrho\,P(\boldsymbol{r},t) + \lambda\,\varrho\,q = 0. \tag{8}$$

Next we introduce (6) into Eqs. (7) and (8), average and make use of (5). We finally obtain (dropping $\hat{\ }$ for the weighted mean)

$$\frac{\partial\bar{\varrho}}{\partial t} + V\cdot(\bar{\varrho}\,\boldsymbol{v}) = 0 \tag{9}$$

and

$$\frac{\partial q}{\partial t} + \boldsymbol{v}\cdot V q + \frac{1}{\bar{\varrho}}\,V\cdot(\overline{\varrho\,q''\,\boldsymbol{v}''}) - P(\boldsymbol{r},t) + \lambda\,q = 0 \tag{10}$$

where we have assumed $P=\bar{P}$.

Here $\overline{\varrho\,q''\,\boldsymbol{v}''}$ expresses the turbulent transfer and the main difficulty encountered now is to express it in terms of the mean fields q and $\boldsymbol{v}$. Different approaches to this problem have been tried in the past with various degrees of success. So far no completely satisfactory theory has been advanced.

The first attempts were an application of the theory of molecular diffusion by introducing an *eddy* diffusivity (cf. for example Schmidt 1925). We then put (using a cartesian coordinate system $x,\ y,\ z$)

$$\left.\begin{aligned}
-\,\overline{\varrho\,q''\,u''} &= \bar{\varrho}\,K_x\,\partial q/\partial x \\
-\,\overline{\varrho\,q''\,v''} &= \bar{\varrho}\,K_y\,\partial q/\partial y \\
-\,\overline{\varrho\,q''\,w''} &= \bar{\varrho}\,K_z\,\partial q/\partial z.
\end{aligned}\right\} \tag{11}$$

The eddy exchange coefficients $K_x,\ K_y,\ K_z$ may be functions of space. In case of constant $\bar{\varrho}$ and constant values of $K_x,\ K_y$ and K_z we arrive at Fick's diffusion law. The solution of the diffusion problem is thereby reduced to the solution of the heat conduction equation. Of course little has been gained by formally introducing (11) if no simple relation for the exchange coefficients can be derived or hypothesized. The introduction of the mixing length concept by Prandtl (1934) has been of great value and combined with v. Karman's similarity hypothesis led to the result that the eddy coefficient to the first approximation is proportional to the distance from the boundary. On the basis of this result vertical transfer in the surface layers of the atmosphere has been studied assuming horizontal homogeneity. In most applications of the eddy exchange theory (or "K-theory") $\bar{\varrho}$ is assumed to be constant. In the surface layers of the atmosphere, this is certainly justified since the variation of $\bar{\varrho}$ with the vertical is much less than that of K_z. This is not necessarily so when dealing with deep strata of the atmosphere. Our knowledge about turbulent exchange in the free atmosphere is still so limited, however, that we often will be forced to assume constant values for $K_x,\ K_y$ and K_z. It is then often not justified to retain a variability of $\bar{\varrho}$ with height, which expresses the fact that when mixing the much less dense strata above with more dense layers below the influence of the latter will be the dominating one due to its larger content of matter. In the case of a constant transfer (and constant K_z) the gradient of q should be inversely proportional to $\bar{\varrho}$.

There are several obvious shortcomings of the K-theory. The most severe one, already pointed out by Richardson (1926), is the dependence of K on the scale of the phenomenon studied. It may range from the value of the molecular diffusivity of about $0.1\ \mathrm{cm^2\ sec^{-1}}$ to about $10^{11}\ \mathrm{cm^2\ sec^{-1}}$ for the large-scale horizontal eddy diffusivity of importance for the global distribution of heat, humidity, etc. Lettau (1951) has summarized our knowledge about the eddy diffusivity in the atmosphere (Fig. 1). The dependence of K on the scale of the phenomenon means that the K-theory is not suitable for studies of the diffusion of a cloud in course of time. On the other hand, if there are reasons to believe that the intensity of turbulence is constant in space or a simple (e.g. linear) function of distance from the boundary, the K-theory often gives useful information about the distribution in a steady state. Due to our limited knowledge about turbulence in the free atmosphere, we shall frequently make use of this theory in the following.

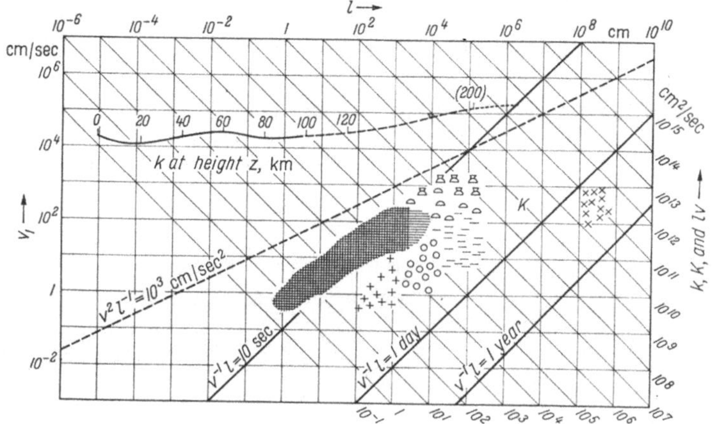

Fig. 1. Diffusion diagram. Each point of the l, v plane (l = scale, v = velocity) determines a diffusion coefficient (cm² sec⁻¹). In molecular diffusion $l \approx$ free path and $v \approx$ mean molecular speed; the height variation of $K = l \cdot v$ is shown. In eddy diffusion $l \approx$ mixing length and $v \approx$ mixing velocity. $K = l \cdot v$ is shown for various layers of the atmosphere: Ordinary turbulent diffusion in the vertical, ▦ 0 to 1 km; ▦ 1 to 10 km; ⧗ ⧗, 10 to 25, 35 to 45 and 80 to 100 km; ‡ ‡, 25 to 35 km; ▤, 45 to 80 km. ◠◠, cumulus convection; ⋈⋈ cumulo-nimbus convection; ⤬ ⤬, large scale horizontal exchange (Lettau 1951)

In his original paper Richardson (l. c.) proposed his "distance neighbor theory" to overcome this difficulty. It has never been developed into practical results, but the basic ideas have been of great importance for the development of the statistical theory of turbulence. Much of the initial work was due to Taylor (1921) and Sutton (see Sutton 1953). A basic theorem derived by Taylor tells that the rate of diffusion depends on the variance of the wind velocity fluctuation distributions according to the formula (in the x-direction)

$$\overline{x^2} = 2\overline{u^2} \int_0^t d\tau \int_0^\tau R(\xi)\, d\xi \tag{12}$$

where $R(\xi)$ is the Lagrangian auto-correlation function

$$R(\xi) = \frac{\overline{u(t)\, u(t+\xi)}}{\overline{u^2}}. \tag{13}$$

Obviously $R(0) = 1$ and goes to zero for large values of ξ, *i.e.* when ξ exceeds the size of the largest turbulent "eddies". Integrating Eq. (12) for different times then yields the well-known facts that the spreading of the plume from a continuous source to begin with is proportional to t^2 but gradually becomes proportional to t. If one assumes that the mean concentration within a diffusion cloud is distributed according to a Gaussian law the distribution originating from an instantaneous point source in an infinite region is given by

$$q(x, y, z, t) = \frac{Q}{(2\pi)^{\frac{3}{2}}(\overline{x^2} \cdot \overline{y^2} \cdot \overline{z^2})^{\frac{1}{2}}}\, e^{-\frac{1}{2}\left(\frac{x^2}{\overline{x^2}} + \frac{y^2}{\overline{y^2}} + \frac{z^2}{\overline{z^2}}\right)}, \tag{14}$$

where the coordinate axes are following the centre of the cloud. Q is the source strength. In this expression $\overline{x^2}$, $\overline{y^2}$ and $\overline{z^2}$ are functions of time (Fleishman and Frenkiel 1955).

On dimensional grounds Sutton proposed the following expression for $R(\xi)$ in the surface layers of the atmosphere

$$R_x(\xi) = \left[\frac{\nu}{\nu + \overline{u'^2}\xi}\right]^n \tag{15}$$

and correspondingly for $R_y(\xi)$ and $R_z(\xi)$. Here ν is the kinematic viscosity of air and n is a number defined by the wind velocity profile law.

$$\bar{u}(z) = \bar{u}_1\left(\frac{z}{z_1}\right)^{\frac{n}{2-n}}, \tag{16}$$

which is derived on the basis of the mixing length hypothesis. We then can derive the following expression for the concentration originating from an instantaneous point source.

$$q(x, y, z, t) = \frac{Q}{\pi^{\frac{3}{2}} C_x C_y C_z (\bar{u}t)^{3(2-n)/2}} \exp\left[(\bar{u}t)^{n-2}\left(\frac{x^2}{C_x^2} + \frac{y^2}{C_y^2} + \frac{z^2}{C_z^2}\right)\right], \tag{17}$$

where

$$C_x^2 = \frac{4\nu^n}{(1-n)(2-n)\,\bar{u}^n}\left[\frac{\overline{(u')^2}}{\overline{u^2}}\right]^{1-n} \tag{18}$$

and similarly for C_y and C_z. This equation can be generalized to continuous point or line sources in the vicinity of the surface of the earth and for elevated sources. They have been extensively applied to the diffusion of smoke from stacks (cf. Sutton 1953 and U.S. Department of Commerce, Weather Bureau 1955). Studies of the spread of radioactivity has been limited by the hazards involved, but the measurements made have been of great interest due to the possibility of detecting very minute quantities and thereby making it possible to test these equations over larger distances.

The atmosphere is usually gravitationally stable. Since the vertical displacement of a particle is greatly influenced by the degree of stability the vertical transfer of matter and radioactivity must be dependent on this stability. Richardson (1925) first fully realized this and derived a criterion for "just-no-turbulence" by balancing the supply of energy made available by the Reynold stresses and the rate at which work has to be done against gravity by the turbulence. He found

$$Ri = \frac{g(\partial T/\partial z + \Gamma)}{T(\partial \bar{u}/\partial z)^2} \begin{cases} > 1 \text{ turbulence decaying} \\ < 1 \text{ turbulence increasing.} \end{cases} \tag{19}$$

Here Γ denotes the dry adiabatic lapse rate (or the moist adiabatic lapse rate in case of saturated air). This criterion applies to turbulent motions which are directly influenced by the gravity field, which is true for small scale turbulence in a stratified atmosphere. When we are concerned with vertical diffusion on a global scale, where the synoptic disturbances probably are the most important "turbulent eddies" Richardson's criterion does not apply. Experimental investigations to verify (19) have yielded a lower value for the critical Richardson number than unity as given in Eq. (19).

The Richardson criterion clearly shows that the transfer of heat and momentum are not independent of each other. On the basis of some similarity hypotheses Monin and Obukhov (1954) deduced a more general theory for the dependence of diffusion on the static stability and conversely how the stability (and thus heat flux) is dependent on the transfer of momentum. Experimental tests of this theory have recently given additional support to these results (Taylor 1960). A more detailed account has been given by Priestley (1959).

In 1941 Kolmogoroff put forward his similarity hypothesis for turbulent motions, emphasizing the importance of the energy transfer from one part of the spectrum to another. For the so-called inertial subrange where no direct energy input takes place Obukhov (1941) deduced that the diffusion coefficient in the classical sense should vary roughly according to

$$K \sim l_0^{\frac{4}{3}}, \tag{20}$$

where l_0 is the characteristic scale. This agrees quite well with the variation of K as shown in Fig. 1, even if it is doubtful whether the similarity theory can be applied to this whole range of motion, since energy input takes place at some intermediate scales. Eq. (20) is still useful to give a first idea of the dependence of K on scale and was already formulated by Richardson (1926). Applications of the similarity hypothesis to atmospheric turbulence are as yet few. Batchelor (1951, 1952) has particularly stressed the difference of diffusion from a continuous source and the dispersion of a smoke puff, an instantaneous source. In the latter case he shows that the initial rate of growth should be proportional to t^2, later to become proportional to t^3. This should be valid as long as the distance between any two particles (i.e. the size of the cloud) is less than the largest scale still governed by the laws of the inertial subrange.

Finally a few words about boundary conditions to be applied in using the equations describing transfer of radioactivity through the atmosphere. At the surface of the earth $(z=0)$ we have

$$(-F_z + \beta q)_{z=0} = 0, \tag{21}$$

where F_z is the vertical flux due to turbulent (or molecular) diffusion, and β is a proportionality factor expressing the amount of q that per second is retained by the boundary surface. We obviously have two extreme cases:

a) Complete absorption at the earth's surface

$$q = 0. \tag{22a}$$

b) No absorption at the earth's surface.

$$F_z = 0. \tag{22b}$$

In the case of exhalation of a radioactive gas (Rn, Tn) from the soil we get

$$F_z = E, \tag{23}$$

where E is the rate of exhalation. Applying the concept of a coefficient of eddy diffusivity we have $F_z = -\bar{\varrho} K_z \, \partial q / \partial z$. At a smooth surface $\lim_{z \to 0} K_z = \varkappa$, where $\varkappa$ is the molecular diffusivity and also in case of a rough surface it decreases as one approaches the surface.

C. Diffusion and transfer of radioactivity in the surface layers of the atmosphere

I. Diffusion from point sources

Some of the main formulae of diffusion from a point source were given in the previous section. A detailed account of their application can be found in the monograph by Sutton (1953) and in Meteorology and Atomic Energy (U.S. Department of Commerce, 1955). Here a few general comments will be made and some more recent work will be described, in which they have been applied to diffusion of radioactive debris in the surface layers of the atmosphere.

Sutton's formulae are valid for neutral stability and have been tested extensively. The original values for the parameters given by Sutton were

$$n = \tfrac{1}{4}, \qquad C_y = 0.21 \text{ cm}^{\frac{1}{8}}, \qquad C_z = 0.12 \text{ cm}^{\frac{1}{8}}. \tag{24}$$

Similar measurements elsewhere have yielded somewhat higher values. This is probably due to differences in the roughness of the terrain. The appropriate values for a given area should therefore be determined experimentally.

Attempts have been made to extend these formulae to cases of unstable and stable stratification. Thus Stewart, Gale and Crooks (1958) have experimentally determined C_y and C_z from a series of measurements of the diffusion of A^{41} released from the reactor BEPO. It was found that the values obtained (being functions of static stability, wind velocity, distance from the source and sampling time) can be used for describing the diffusion from a point source quite well. Of course the coefficients, then, have lost their original significance as generalized coefficients of diffusion in Sutton's theory.

A more direct approach to the problem of diffusion from a point source is suggested by Monin (1959) based on his similarity theory for diffusion in a stratified fluid. No direct measurements to verify this theory are available, but an extensive investigation of the applicability of the original theory to explain the vertical structure of the wind and temperature fields has been made (Taylor 1960). In a stably stratified fluid the agreement is good while the results are less accurate in unstable conditions.

In the case of unstable stratification the discussion of diffusion is complicated by the phenomenon of "looping", i.e. the plume of smoke or radioactivity while being diffused regularly by comparatively small scale wind variations also moves as a whole in the vertical due to convection on a considerably larger scale. A theoretical treatment of this problem has been presented by Gifford (1959).

Another approach to describe the (horizontal) diffusion of a plume is based on Taylor's equation (12). It is, however, very difficult to determine the Lagrangian correlation function $R(\xi)$ experimentally. Various hypotheses have therefore been forwarded of how to relate it to the Eularian correlation function which of course can be directly determined from wind data. Thus Hay and Pasquill (1959) have assumed that the functional form of these two functions is the same, while the scales are different. They deduce a factor four as the most likely ratio between the Lagrangian and Eulerian scales.

From a practical point of view it is particularly interesting to determine the air concentration close to the surface of the earth and the rate of deposition, and how these conditions vary with the distance from the source and its elevation above the ground. This can be determined by a proper integration of Sutton's formulae, but it is then necessary to know the rate of absorption by the earth's surface under various conditions. Studies by Chamberlain and Chadwick (1953) and Chamberlain (1956) are of particular interest here. Define a velocity of deposition v_g as the ratio of the amount deposited per unit area and unit time to the volumetric concentration above the surface (say 2 m, the exact level is not exceedingly important, since the variations in the z-direction already are quite small at these levels). If assuming that the surface of the earth is a perfect sink one can derive the following expression for the deposition velocity (SHEPPARD 1958)

$$v_g = \frac{k\,u^*}{\ln(k\,u^*\,z_1\,\varkappa^{-1})} \,, \tag{25}$$

where u^* is the friction velocity, k is V. KARMAN's constant, $\varkappa$ the molecular diffusivity and z_1 is the height at which the volumetric concentration is measured.

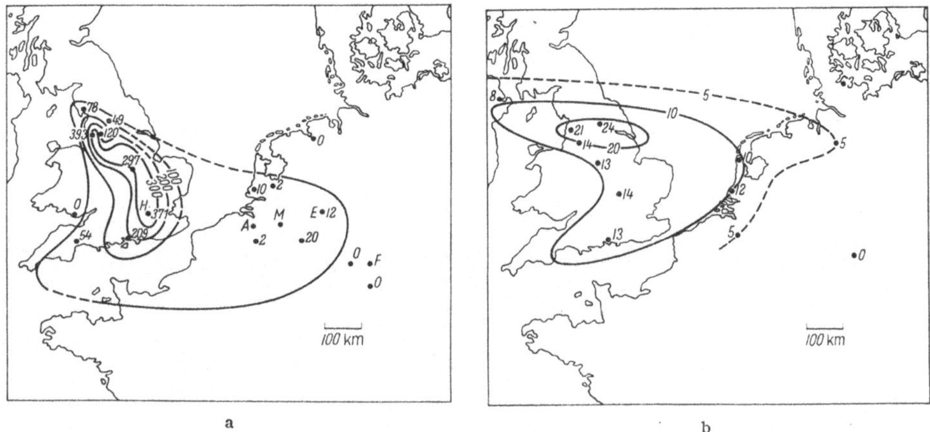

Fig. 2 a and b. Iodine-131 activity in air ($\mu\mu$C m^{-3}). a 0900 GMT, 11 October to 0900 GMT, 12 October 1957   b 0900 GMT, 14 October to 0900 GMT, 15 October 1957 (CRABTREE 1959)

Obviously v_g will be dependent on the intensity of the turbulence and thus parameters such as static stability and surface roughness. For characteristic values of u^*, v_g is 1 to 2 cm sec^{-1}, in quite good agreement with actual measurements of the rate of deposition of radioiodine over a grass field of about 2 cm sec^{-1} (CHAMBERLAIN and CHADWICK 1953). The deposition velocity computed after the Windscale accident in 1957 yielded lower values (0.3 to 0.5 cm sec^{-1}) see CHAMBERLAIN (1959). The special meteorological conditions with light winds and comparatively high static stability may be the reason for this, but it cannot be excluded that the iodine may gradually have been collected on particles and not as easily absorbed by the surface of the earth, due to a small collection efficiency (BOLIN 1959). In general the spread of the radioactive cloud from the Windscale accident (CRABTREE 1959) is a good illustration of the inadequacy of any diffusion theory to describe the travel and diffusion of a cloud for more than a few hours. Figs. 2a and 2b show the iodine-131 activity in air about 36 and 90 hours after the release from Windscale. The changes of the general weather situation with the development of an anticyclone over southern England was of major importance and our capability to foresee such a development in sufficient detail beyond 24 hours is still quite limited.

It should finally be mentioned that SUTTON's theory has been extended by CHAMBERLAIN (1956) to include the effect of wash-out by rain. He considers the efficiency of collection of various sized rain drops and the typical drop spectra for various rain intensities. Fig. 3 shows characteristic changes of the ground level concentration for a ground level source.

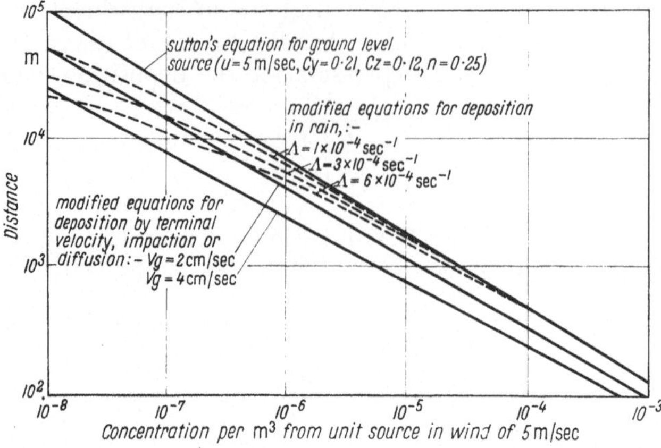

Fig. 3. Air concentration at ground level directly downwind of ground level source. Λ expresses the rate constant for removal by rain, being dependent on size of aerosol particle and intensity of precipitation (Chamberlain 1956)

II. Diffusion from a distributed source at the surface of the earth

Radon (Rn^{222}) and thoron (Tn or Rn^{220}) escape from the surface of the earth at a rate that depends on the character of the ground. Thus very little leaves the ocean surface, rain greatly reduces the transfer of gas up through the soil and a snow cover acts as an effective shield. Consider first an area from which the amount of gas escaping per unit area is constant and which is sufficiently large so that no horizontal gradients exist in the concentration of various radionuclides. We may then disregard all horizontal derivatives in our basic equations (10) and merely consider the vertical exchange. The turbulent problem then becomes almost identical with the problem of evaporation (Sutton 1953). In this way the vertical distribution of radon and thoron was studied early in this century and gave interesting results concerning the vertical exchange mechanisms (Israël 1951). More recently attempts have been made to analyze for the more long-lived isotopes RaD, RaE and RaF. A discussion of the interdependence of the distribution of these elements is therefore of interest.

We denote the concentrations of Rn, RaA, RaB, ..., RaF by $q_R, q_A, q_B, ..., q_F$ and from Eq. (10) we obviously get the following system of equations governing the distribution of this series of radio-isotopes assuming a steady state. A similar treatment of the decay products of thoron can of course be made; no measurements of its daughter products exist, however.

$$\left. \begin{aligned} &\frac{\partial}{\partial z}\left(K_z \frac{\partial q_R}{\partial z}\right) - \lambda_R q_R = 0 \\ &\frac{\partial}{\partial z}\left(K_z \frac{\partial q_A}{\partial z}\right) - (\lambda_A + \lambda)\, q_A = -\lambda_R q_R \\ &\qquad\vdots \\ &\frac{\partial}{\partial z}\left(K_z \frac{\partial q_F}{\partial z}\right) - (\lambda_F + \lambda)\, q_F = -\lambda_E q_E. \end{aligned} \right\} \qquad (26)$$

We have here introduced λ, expressing the rate of removal (wash-out) of particulate matter from the atmosphere due to other processes than turbulence and dry deposition on the surface of the earth. The daughter products of Rn are isotopes of Po, Bi and Pb and are soon attached to particulate matter in the atmosphere. We have furthermore assumed that the variation of $\bar{\varrho}$ with z can be neglected since we deal with the surface layers of the atmosphere within which the variation of K_z is much more important. The use of particularly the long-lived isotopes RaD, RaE and RaF for studies of the circulation of the atmosphere as a whole will be dealt with later (E, II).

The boundary conditions for this set of equations are

$$
\left.
\begin{aligned}
K_z\left(\frac{\partial q_R}{\partial z}\right)_0 &= -E_R/\varrho \\[2mm]
K_z\left(\frac{\partial q_A}{\partial z}\right)_0 + \beta_A (q_A)_0 &= 0 \\
&\ \vdots \\
K_z\left(\frac{\partial q_F}{\partial z}\right)_0 + \beta_F (q_F)_0 &= 0
\end{aligned}
\right\} \quad z = 0 \tag{27}
$$

$$
q_R = q_A = \cdots = q_F = 0 \qquad z \to \infty
$$

which is equivalent of a flux of radon from the interior of the earth $=E_R$. The quantities β_i are not known but depend on the characteristics of the surface and also for example the size distribution of the particulate matter. We shall indirectly get some information of β_i from the observation of natural radioactivity at ground level, which will be of interest with regard to our previous discussion of the deposition velocity. We shall assume that β is the same for all elements $q_A \ldots q_F$.

A large number of measurements of radon in the atmosphere have been made since early this century (cf. ISRAËL 1951). The distribution with the vertical has been tested against our knowledge of the rate of vertical exchange of other substances such as momentum and heat. General agreement is found for an approximately linear increase of K_z as a function of z approaching values of about $K_z = 10^5$ cm² sec⁻¹ in the free atmosphere (above 1 km). During the last decade the concentration of various daughter products (particularly RaC, RaD and RaF) in air and precipitation has been determined (BLIFFORD, LOCKHART, ROSENSTOCK 1952; HAXEL, SCHUMANN 1955; LEHMANN, SITTKUS 1959; BURTON, STEWART 1959). The concentration of these radio nuclides in rain is certainly a result of processes in the free atmosphere and will be dealt with later. We shall here discuss the interpretation of the measurements of air concentration at surface levels.

We may now first ask for the apparent removal time for these constituents by assuming that the removal is simply proportional to the amount present and thus for the moment forget about the effect of turbulent transfer. From Eq. (26) we then simply get

$$
q_F = \frac{E_R}{\lambda_R} \cdot \frac{\lambda_R}{\lambda_A + \lambda} \cdot \frac{\lambda_A}{\lambda_B + \lambda} \cdot \ldots \cdot \frac{\lambda_E}{\lambda_F + \lambda} \tag{28}
$$

and similarly for other compounds. The same expression can also be derived by integration of (26) from $z=0$ to $z=\infty$ and applying the boundary conditions $(\partial q_i/\partial z)=0$ for $z=0$ and $z=\infty$. If knowing the concentration of two elements we can eliminate E_R. We thus for example get

$$
\frac{q_D}{q_C} = \frac{\lambda_C}{\lambda_D + \lambda}, \qquad \frac{q_F}{q_D} = \frac{\lambda_D}{\lambda_E + \lambda} \cdot \frac{\lambda_E}{\lambda_F + \lambda}. \tag{29}
$$

LEHMANN and SITTKUS (1959) derived $\lambda^{-1}=1.5$ and $\lambda^{-1}=14$ days respectively using these two expressions and HAXEL and SCHUMANN (1955) in a similar way determined a residence time of about a few days from the ratio of the RaD activity to that of RaC and RaB (see also correction by Schumann, 1960). The difference between the values indicates that the turbulent transfer cannot be neglected.

Let us therefore next apply Eq. (26) and first assume complete absorption at the earth's surface for all compounds except q_R, i.e. applying the boundary conditions

$$q_A = q_B = \cdots = q_F = 0, \quad z = 0. \tag{30}$$

RaA, RaC' and RaC'' are short-lived compounds and their distribution can be derived without considering the redistribution due to turbulence. The solution of the four differential equations for Rn, RaB, RaC and RaD will of course depend upon K_z and its variation with elevation. It can be shown, however, that the ratios of the Type $\lim_{z \to 0} \lambda_D q_D / \lambda_C q_C$ are independent of K_z. We thus obtain

$$\lim_{z \to 0} \frac{\lambda_D q_D}{\lambda_C q_C} = \lambda_D \frac{\sqrt{\lambda_D + \lambda} + \sqrt{\lambda_C + \lambda} + \sqrt{\lambda_B + \lambda} + \sqrt{\lambda_R}}{(\sqrt{\lambda_D + \lambda} + \sqrt{\lambda_R})(\sqrt{\lambda_D + \lambda} + \sqrt{\lambda_B + \lambda})(\sqrt{\lambda_D + \lambda} + \sqrt{\lambda_C + \lambda})}. \tag{31}$$

For any reasonable values of K_z this will also be approximately true a few meters above the ground. The maximum value of this expression is obtained by putting $\lambda = 0$ (assuming only removal by dry-deposition). One gets $\frac{\lambda_D q_D}{\lambda_C q_C} = 6 \cdot 10^{-5}$ compared with an observed average value of $18 \cdot 10^{-5}$ as determined by Lehmann and Sittkus (1959). This may possibly indicate that the surface of the earth is not a perfect sink. It may also be a result of large scale advection of the more long lived isotopes RaD. A study of the ratio between comparatively short lived isotopes such as RaC and Rn would, however, be of principal interest in this respect.

We may now instead apply the general boundary condition given by Eq. (27), where obviously β has the dimensions of a velocity and is closely related to the deposition velocity defined earlier. It might be of interest to determine the particular value of β that would yield a value for the ratio $\lambda_D q_D / \lambda_C q_C$ at the surface of the earth (say at 2 m elevation) in agreement with the observed value quoted above. It now becomes necessary to specify K_z as a function of distance from the surface of the earth and we must therefore also know average wind conditions and the roughness characteristics of the surface of the earth to treat the problem properly. None of these supplementary observations have been reported in the data published so far. This shows the desirability of collecting ordinary meteorological observations in association with sampling the radioactivity of the atmosphere. A preliminary and very approximate computation according to the lines sketched above indicates that the deposition velocity for RaC and RaD is smaller than the value of 1 to 2 cm sec^{-1} obtained for radio-iodine and also found for oceanic salts in the atmosphere. The fact that the average time between rains is considerably longer than the value of λ^{-1} obtained from the ratio q_D / q_C in Eq. (29) indicates that removal by impaction plays some role. Blifford, Lockhardt and Rosenstock (1952) measured the deposition of RaD by rain and the air concentration and it is easily found that a deposition velocity of $v_g = 0.25$ to 0.5 cm sec^{-1} would mean about equal amounts deposited by rain and by impaction.

Observations of Rn and Tn (and their daughter products) in the atmosphere may also be used for studying the air exchange between land and sea since the exhalation is very different in two such areas. Even more interesting is the use of some of the long-lived isotopes (RaD, RaE and RaF) for large-scale circulation studies in the atmosphere, which will be discussed later as well as the wash-out by rain from higher elevations in the atmosphere.

D. Small- and intermediate-scale diffusion and circulation of radioactivity in the free atmosphere

The turbulence observed in the surface layers of the atmosphere is mainly generated by the frictional stresses acting between the earth and the moving air above. However, also in the free atmosphere we encounter turbulence on this same scale. This is obvious from the diffusion of small clouds, smoke puffs or contrails generated at various elevations. In an unstable atmosphere convection undoubtedly is the main source of energy for this turbulence. In the vicinity of the jet stream, however, strong vertical shear exists and subcritical values of the Richardson number is frequently observed. Our knowledge about turbulence on this scale in the free atmosphere is quite limited. A brief account of some pertinent observations will be given here.

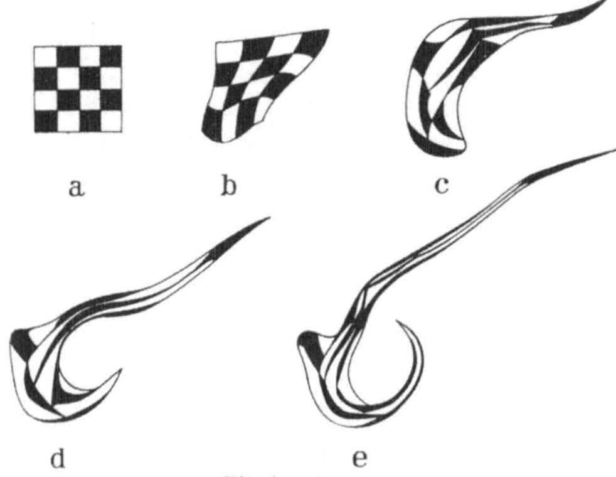

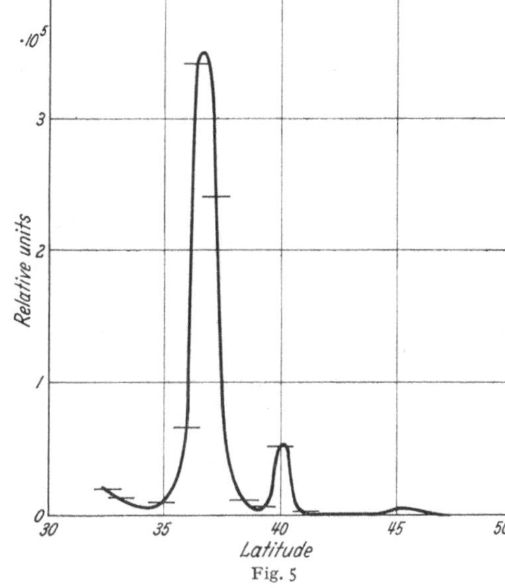

Fig. 4 a—e

Fig. 4 a—e. Deformation of a square (initially 1200×1200 km) by flow at 500 mb. The figures show the development at 12-hour intervals (WELANDER 1955)

Fig. 5. Air concentration (relative units) of radioactivity experienced when flying across an air stream carrying debris from bomb test in Nevada (after MACHTA et al. 1957). The horizontal lines are the actual observations, the length of the line expressing the distance flown while sampling

Studies of the dispersion of smoke puffs in the free atmosphere give some interesting information (see e.g. KELLOGG 1956, GIFFORD 1957). The data support the theory developed by BATCHELOR (1951, 1952) discussed earlier. The intensity of the turbulence in the high troposphere and lower stratosphere was $\sqrt{\overline{u'^2}} = 5$ to 10 cm sec^{-1}, which implies a growth of a 10 m cloud to about 100 m in approximately 5 minutes. It is also possible to estimate the rate of dissipation in the free atmosphere. At these levels one finds about 0.02 cm^2 sec^{-3} as compared to values close to unity in the surface layers of the atmosphere. The intensity of small-scale turbulence in the free atmosphere is thus considerably less than at lower levels in the atmosphere.

If being able to follow a cloud for some time in the atmosphere one observes marked distortions of it in addition to the regular diffusion just discussed (cf. C I). This is obviously the effect of eddies of a scale equal to or larger than that of the cloud itself. Such an effect can be observed for clouds on a scale of a few meters to clouds of atomic debris of 10 to 100 km and also in case of diffusion on an even larger scale. Fig. 4 shows a typical deformation of a square in course of 48 hours by motions on a synoptic scale. (See also Fig. 2 and the deformation of the cloud released from the Windscale accident.) Fig. 5 shows the distribution of radio-activity across such a "band-cloud" originating from a bomb explosion in Nevada about 1500 km away from the observation point and about 36 hours earlier. On this particular occasion as many as four peaks could be found, which is unusual. Machta *et al.* (1957) also computed the effective coefficient of diffusion that is implied by the almost Gaussian distribution of radioactivity found when sampling across the cloud and gives a value of $K_y = 3 \cdot 10^8$ cm² sec⁻¹. It has been stressed before that an eddy exchange coefficient is a very questionable concept when dealing with diffusion from an instantaneous point source. Furthermore the sampling in this case meant an average over a distance comparable to the width of the cloud which certainly influences the estimates of K_y. Still the value quoted should give a rough idea of the effective exchange coefficient in cases of diffusion of clouds of 10 to 100 km size (cf. Fig. 1).

As was mentioned above the larger scale eddies implied by the band-like character of an initially symmetrical cloud can be considered as turbulent eddies themselves. We can obtain an idea about the character of this quasi-two-dimensional turbulence by an analysis of the upper air wind data now available for a number of years from a large part of the northern hemisphere. Thus Solot and Darling (1958) asked, what the probability of displacement of a cloud originally at a certain point would be as a function of time. Using Taylor's theory of continuous movement (Taylor 1921) and available statistics of upper air winds they deduced the probability fields for a cloud initially at 40° N and 120° W as a function of time at an elevation of 50000 feet (about 17 km) for January and July. Similar graphs can be constructed on the basis of the transosonde flights done in recent years (Angell 1959) and Fig. 6 illustrates one interesting result obtained from flights originating in Japan. If the horizontal wind were known at the time of injection considerably reduced areas of probable location would be obtained. By applying quantitative techniques of forecasting the wind field in the atmosphere the areas enclosed by the various probability lines could be further reduced (at least at lower levels in the atmosphere). Here the horizontal trajectories in the atmosphere as deduced from constant level balloons would be of great value (Angell 1959) in assessing the accuracy of forecasting trajectories. The fields shown in Fig. 6 are also approximate due to the actual three dimensional character of atmospheric motions. Present methods of dynamic weather forecasting implies an evaluation of the field of vertical velocities in the course of computations. So far little has been done to study quantitatively three-dimensional diffusion on a synoptic scale making use of the three-dimensional motion of the air. The use of radioactive tracers might here be very useful. Observations from the free atmosphere are very sparse but extensive fall-out measurements give us a picture qualitatively the same as shown in Fig. 6. Thus Fig. 7 illustrates

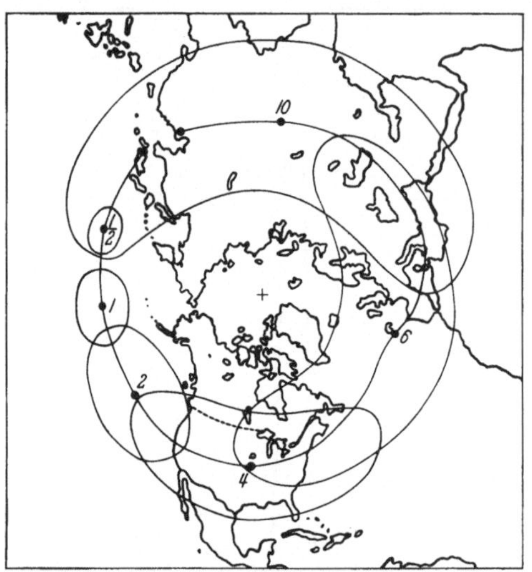

Fig. 6. Average distance travelled by transosonde balloons during the 10 days following release over Japan. Within the closed areas 50% of all sondes released were found (Angell 1959)

the worldwide pattern obtained during the month following the Bravo test in the Pacific 1954 (MACHTA, LIST, HUBERT 1956).

The discussion above clearly illustrates some of the difficulties encountered in forecasting radioactive fall-out from individual radioactive clouds. In addition questions with regard to wash-out, gravitational settling, etc. have to be considered. Present procedures to arrive at the most likely fall-out pattern involve a number of qualitative steps in addition to the more direct numerical computations based on concepts such as those discussed above. For such and other details the reader is referred to existing monographs (e.g. United States Department of Commerce, 1955).

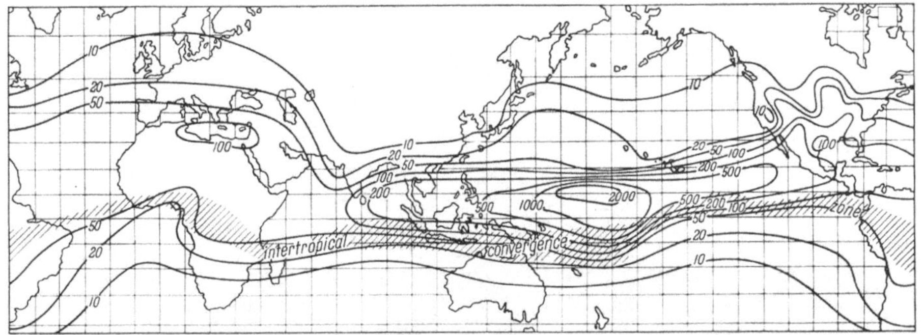

Fig. 7. Total radioactive fall-out from the Bravo cloud in the period 2 to 35 days after detonation (mC per 100 square miles)
Hatching indicates approximate March position of intertropical convergence zone

E. Global transfer and circulation of radioactivity in the atmosphere

There are in principle three different ways in which radio-nuclides are introduced into the atmosphere and are useful for studies of the global circulation of the atmosphere.

a) Exhalation of radon and thoron from the earth's surface and the successive formation of their daughter products. Particularly RaD, RaE and RaF with half-lives of 22 years, 5 days and 140 days are of importance.

b) Formation of radionuclides in the atmosphere by cosmic rays, particularly at upper levels.

c) Introduction of artificial radioactivity by bomb explosions and the indirect generation of other radionuclides due to nuclear reactions with the atmosphere.

The information gained in these various cases is different mainly due to differences in the production function but also due to the different decay rates of the various isotopes formed. Before discussing each series of investigations separately a few general remarks about the large-scale structure of the atmospheric circulation gained by ordinary meteorological observations should be given.

I. The general circulation of the atmosphere

We shall essentially be dealing with the meridional circulation of radioactivity and its possible seasonal variations. Fig. 8 shows a cross-section through the atmosphere from pole to equator in summer and winter. (The figure is based on data in the northern hemisphere, but the differences between the two hemispheres that are known to exist will be of no importance for the following discussion.) We notice here the decreasing temperature in the troposphere up to the tropopause found at 8 and 11 km in the polar regions in winter and summer respectively and at about 16 km in equatorial regions all the year around. Two main breaks in the tropopause are found in middle latitudes, one at about lat. 30° and associated with the subtropical jet-stream, the other one, less pronounced, between lat. 35 and 60°, depending on season and longitude, and associated with the polar front (heavy lines in the diagram) and the polar jet stream.

The surface winds are on an average easterly in the tropics and westerly poleward from lat. 30° (except close to the pole in winter). This implies a flux of westerly angular momentum from the earth to the atmosphere in the tropics and *vice versa* in middle latitudes. The transfer of momentum poleward from

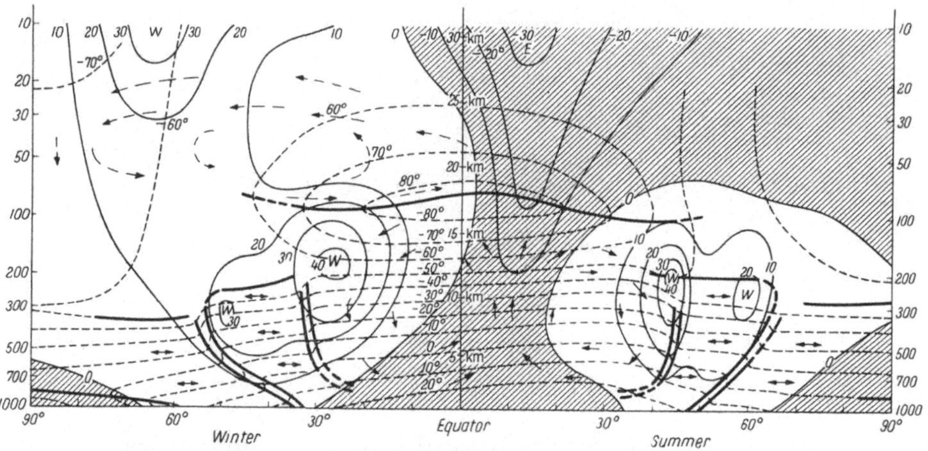

Fig. 8. Meridional cross section from pole to equator (northern hemisphere) in winter and summer. Vertical scale in mb (100 mb at about 16 km, 10 mb at about 30 km). Single heavy lines indicate tropopause, double heavy lines show the position of polar front and subtropical front. Thin solid lines are isotachs (m sec⁻¹) and the hached area indicates easterly winds. Dashed lines are isotherms. Single arrows show probable direction of meridional circulations, double arrows indicate large scale horizontal mixing in troposphere. Similar mixing also exists in the winter stratosphere north of lat 30°

equatorial latitudes that is necessary to maintain a steady state essentially takes place in the upper troposphere (and lower stratosphere). Poleward of lat. 20 to 30° the large horizontal waves and eddies found on the daily synoptic charts are most important for this transfer, while a slow meridional circulation cell is the predominant feature of the transfer in the tropics. This meridional cell is qualitatively indicated in Fig. 8. Palmén (1955) has estimated its intensity and finds an average vertical velocity upwards in the equatorial belt between about 13° N to 13° S of about 0.5 cm sec, and a poleward velocity at upper levels at these two bounding latitudes of about 1 m sec⁻¹. He also showed that the balance of angular momentum requires a downward eddy transfer of momentum in the equatorial belt that implies an eddy exchange coefficient in the mid-troposphere of about $K_z = 4 \cdot 10^5$ cm² sec⁻¹. It should be remembered, however, that a dynamic coupling of the u and w components exists for the large-scale synoptic disturbances giving an upward transfer of momentum *against* the gradient of $\bar{u}$. When discussing the vertical transfer of *other* properties, which do not have any dynamic significance, also these eddies most likely transfer along the mean gradient. The value for K_z given above therefore probably represents an underestimate of the intensity of turbulent diffusion in the vertical in tropical latitudes.

Due to the dynamic coupling of the u and v components in the large-scale atmospheric motions it is difficult to estimate the horizontal eddy diffusivity from studies of the angular momentum (and similarly total energy or sensible heat). However studies of the kind mentioned in the previous section (SOLOT, DARLING 1958) and by computing geostrophic trajectories or studying the trajectories of constant level balloons provides methods for estimating the rate of horizontal exchange (DURST, CROSSLEY, DAVIS 1959). These investigations (as well as earlier ones) indicate values of an effective eddy diffusivity K_y in middle latitudes varying from 0.5 to 1×10^{10} cm² sec⁻¹ at the surface of the earth to about 10^{11} cm² sec⁻¹ at the tropopause level. No estimates of the value of horizontal diffusivity in subtropical and tropical latitudes are available, but it is probably less as indicated by the less intense wind fields observed at these latitudes. Most likely, however, K_y is not less than the value given for the surface layers in the temperature latitudes about 10^{10} cm² sec⁻¹.

Some information about atmospheric circulations in the stratosphere are also available. A more extensive summary has recently been published by REED et al. (1960) and only a few important features will be described here. Above the tropopause and the jet streams the mean wind as well as the absolute magnitude of the fluctuations decreases with elevation. However, during the winter the atmosphere above the polar caps being located in the area of the polar night is cooled and a marked latitudinal temperature gradient develops associated with the so-called polar night jet stream. The maximum velocities approach 100 m/sec⁻¹ at 30 km. Intense synoptic disturbances here move from the west towards the east but also sometimes show retrogression in a manner similar to what may be found in the troposphere. The effect of these disturbances is noticed down to about lat. 30° and implies horizontal exchange between very high latitudes and middle latitudes probably of the same intensity as at the tropopause level, i.e. an effective exchange coefficient of 10^{11} cm² sec. Large vertical velocities also occur implying displacements of more than 5 km in a few days, particularly in February and March (October and November in the southern hemisphere) when the intense vortex is shifted away from the pole and the winter regime breaks down. Generally speaking the winter stratosphere around each pole gives the impression of quite intense horizontal and vertical exchange. In the summer, on the other hand, quite weak easterlies prevail at middle and high latitudes and in general the variations of the wind in space and time are remarkably small. The vertical stratification is also considerably more stable particularly in the polar regions. In equatorial regions we finally find a weak westerly jet-stream (Berson westerlies) at about 22 km and easterly winds above. The rate of horizontal exchange due to synoptic disturbances at these latitudes has not been determined from existing winds, but in view of the observed intensities of the circulation it is probably as intense as at the surface of the earth, i.e. an apparent eddy diffusivity of $K_y = 10^{10}$ cm² sec⁻¹. Finally permanent meridional circulation cells may exist at these levels. Some attempts to compute these have been done (LIBBY, PALMER 1960) and yield an average equatorward flow between 16 and 24 km associated with an upward velocity in the tropics. The intensity of this flow is not very well known, nor how high up the circulation cell implied by the upward velocity at the equator penetrates. Generally more detailed studies are needed before possible meridional circulation cells in the stratosphere have been established conclusively.

Indications about the general circulation of the atmosphere can also be obtained from the distribution of ozone (RAMANATHAN 1956) and water vapour (BREWER 1953) in the upper troposphere and lower stratosphere. In no case is it possible to deduce the character of the necessary flow uniquely, but some constraints are put on the possible circulations which will have to be considered when discussing the distribution of radioactivity in the atmosphere. A marked seasonal variation in the total amount of ozone in a column of air takes place in polar regions. A minimum is reached in the fall with a very rapid increase in the early spring reaching a maximum in March and April in the northern hemisphere. A general decrease takes place through the summer. In tropical latitudes the total content is first of all much lower with quite a sharp transition zone at lat. 30°. The seasonal variations are much smaller with maxima and minima delayed one to three months. The vertical distribution of ozone is characterized by a major maximum at about 25 km and the variations of total amount of ozone referred to earlier are essentially associated with changes below that level. The observations of water vapour reveal very low concentrations above the tropopause sometimes with a frost point depression of as much as 30 to 40° C. This fact is interpreted by Brewer as an indication that much of the air exchange between the troposphere and stratosphere must take place through the "cold trap" at the tropical tropopause with temperatures of -80 to $-90°$ C. We shall return to a discussion of possible circulations later.

II. Daughter products of radon in the upper troposphere and stratosphere

Very few measurements of RaD, RaE and RaF in the free atmosphere exist so far (Burton and Stewart 1960). Some indirect evidence has been obtained from analysis of rain-water for these radio-nuclides (Blifford, Lockhart, Rosenstock 1952; Lehmann and Sittkus 1959). Somewhat different ways of interpreting these data can be followed essentially due to the fact that so far very few data are available.

1. We may assume a steady state and neglect turbulent transfer and compute the removal time (λ^{-1}) due to other processes than radioactive decay [Eq. (29)]. The value thus obtained will of course vary from one part of the atmosphere to another; it will depend upon the particular radio-nuclides studied (RaC, RaD, RaE or RaF) and also be different if studying an air sample or rain water (cf. Sect. C II). Thus λ will not immediately be interpretable in terms of atmospheric circulations, but still be of interest as a general indication of the meteorological half-life of these different constituents. Lehmann and Sittkus (1959) in this way found $\lambda^{-1} = 33$ days (extremes >100 and 10 days) from the ratio $q_F \lambda_F / q_D \lambda_D$ for rain-water, while Blifford, Lockhart and Rosenstock (1952) give an average removal time of 10 days based on the ratio of RaD to RaB and RaC in rain water. These latter authors also find an overall agreement between the absolute amounts of these radio-nuclides in the rain, the production rate and this removal time.

2. Since presumably most of the long-lived decay products of Rn^{222} are removed by condensation and precipitation, we may define the "age" of an air mass as the time elapsed since this occurred last and estimate this time from the gradual build-up of these constituents from the Rn left in the air. Neglecting turbulent mixing we can apply Eq. (10) to these elements and obtain

$$\left. \begin{aligned} \frac{dq_A}{dt} + (\lambda_A + \lambda)\, q_A &= \lambda_R\, q_R \\[1em] \frac{dq_B}{dt} + (\lambda_B + \lambda)\, q_B &= \lambda_A\, q_A \\ &\;\vdots \\ \frac{dq_F}{dt} + (\lambda_F + \lambda)\, q_F &= \lambda_E\, q_E. \end{aligned} \right\} \tag{32}$$

With the initial conditions $q_A = q_B = \cdots = q_F = 0$ we can determine the concentrations of all constituents as a function of time and the ratio between any two elements will be dependent on the time elapsed since the last cleansing took place, but independent of $(q_R)_{t=0}$. BURTON and STEWART (1960) particularly stress the importance of the ratio of the RaF and RaD activities $(\lambda_F\, q_F / \lambda_D\, q_D)$ since both elements are comparatively long-lived. This ratio will thus be increasing from 0 to about 0.5 in 140 days to approach unity as time goes to infinity. Since all other daughter products of Rn^{222} are comparatively short-lived it is not possible to compare the "ages" obtained by using two different ratios and thereby check the assumptions implicit in Eq. (32). Even if the relative concentrations are independent of $(q_R)_{t=0}$ the absolute ones are not, but will be large or small depending on whether $(q_R)_{t=0}$ is large or small. Since Rn^{222} has a half-life of about 3.8 days its concentration normally decreases quite rapidly with elevation (cf. Sect. C II). Large differences also exist between oceanic and continental areas. From measurements of the absolute concentrations of two elements $(q_R)_{t=0}$ can be computed, which will yield some information about the origin of the air sample under study, particularly if the facts deduced from radioactivity measurements in this way are combined with other meteorological observations. In reality of course mixing between different air masses constantly occurs. It is of particular interest to note that the "age" deduced in such a case naturally will be much more determined by the age of the air mass with originally high Rn^{222} concentration.

Using a line of reasoning as sketched above BURTON and STEWART (1960) find that air descending from the stratosphere to the troposphere in middle latitudes has an age of about six months under the assumption that the source of the stratospheric RaD is Rn decay products from the ascending air in the tropics.

3. The meteorological half-life or "age" deduced above generally increases with elevation. This is partly due to the less frequent condensation and removal by precipitation at upper levels and partly due to the time required for Rn^{222} and its daughter products to diffuse or be transferred vertically through the atmosphere. It is of some interest to ask the following question: What is the expected $q_F\, \lambda_F / q_D\, \lambda_D$ ratio (or "age" in the terminology of the previous paragraph) of an air parcel as a function of elevation in the atmosphere under the influence of:

a) radioactive decay,

b) removal by condensation and precipitation in the lower troposphere (rain-bearing layer),

c) vertical transfer by turbulence?

Since comparatively few data are available no detailed theoretical investigation is warranted, but a few general results of interest can be obtained by considering the following model: In the lower part of the atmosphere (say 5 km deep) Rn^{222} is present and decays into RaA ... RaF. We shall here assume that all compounds except RaD, RaE and RaF decay rapidly enough to be in approximate equilibrium with its parent nuclide. The lower layer is well mixed with a removal rate given by λ (an assumption which in principle can easily be removed but which is fairly insignificant for the rough computation made here). In the air above the lowest 5 km no wash-out of radioactivity occurs but vertical transfer by turbulence takes place, characterized by an eddy diffusivity, K_z. We shall finally neglect the variation of $\bar{\varrho}$ as a function of altitude. Applying now equations of the type given in (26) to q_D, q_E and q_F for the upper and lower troposphere with appropriate boundary conditions at the surface of the earth (no dry deposition), the interphase (continuity of q_D, q_E, and q_F) and at the top of the atmosphere ($q_D = q_E = q_F = 0$) we can compute the ratio $q_F \lambda_F / q_D \lambda_D$ for the bottom layer and thus for the precipitation. It is interesting to note that the value 0.17 for this ratio as obtained by Lehmann and Sittkus (1959) is the expected value if the residence time for water in the lower atmosphere is assumed to be 10 days (and thus also for radioactivity in these layers) and $K_z = 2 \cdot 10^5$ cm^2 sec^{-1}. We furthermore find that q_D decreases slowly upwards since it for all practical purposes can be considered as directly formed from Rn (since the life-time of all intermediate elements is short) which is present essentially at low levels. A steady diffusion upward therefore takes place. RaF, however, will have higher concentration at upper levels due to relatively rapid removal below as compared with the rate of decay and diffusion. A downward turbulent transfer of RaF therefore occurs. This shows that neglect of the turbulent transfer and thus use of the Eqs. (29) will yield values that are difficult to interpret in terms of the physical processes taking place in the atmosphere.

The few examples of using naturally radioactive isotopes in the atmosphere for atmospheric circulation studies given above shows the great potentialities of this technique. Many more observations, particularly from the free atmosphere are, however, desirable.

III. Radioactivity formed by cosmic radiation in the atmosphere

As was pointed out in the foregoing Chap. a series of radioactive nuclides are formed in the atmosphere by cosmic radiation. Some of them are of particular interest to meteorology, namely Be^7 (half-life 53 days), Na^{22} (2.5 years), P^{32} (14 days), P^{33} (25 days) and S^{35} (87 days) since their half-lives are of the same order as the characteristic circulation time of the atmosphere (Lal, Malhotra and Peters 1958; Lal, Peters 1959). In addition H^3 (12.5 years) and C^{14} (5.600 years) are also of interest since they enter the two important atmospheric constituents water and carbon dioxide. However S^{35}, H^3 and C^{14} also have been produced by bomb-explosions, which means that special care must be taken when using them for meteorological studies (see below). The natural production of these nuclides is a function of cosmic ray intensity, air density and the yield factor in the nuclear reaction between the bombarding particle and the air molecules nitrogen, oxygen and argon. These are all practically constant in time and fairly well known. We can therefore quite accurately compute the production as a function of altitude and latitude (Fig. 9). To obtain the production of the various isotopes the values in Fig. 9 should be multiplied by the yield factors (Lal 1959)

$$\sigma_{\text{Be}^7} = 4.5 \cdot 10^{-2} \qquad \sigma_{\text{S}^{35}} = 7.6 \cdot 10^{-4}$$

$$\sigma_{\text{P}^{32}} = 4.5 \cdot 10^{-4} \qquad \sigma_{\text{P}^{33}} = 3.9 \cdot 10^{-4}.$$

Fig. 9 shows that the production is much greater at high elevation and high latitudes than at low elevation and low latitudes. It is also clear that the integrated production from the top to the bottom of the atmosphere is much greater at the poles than near the equator, but the production within the troposphere is almost independent of latitude because of the variation of the height of the tropopause.

In making use of these radio-nuclides for atmospheric circulation studies we return to Eq. (10). We again encounter difficulties in trying to apply it in its complete form, in which case the concentration at any one time and any one point is dependent on the production, advection, turbulent transfer, radioactive decay and removal by condensation and precipitation. When more data for several isotopes are available a more thorough treatment may be attempted (see E V). We shall here present some simple consideration that, however, already reveal interesting aspects of the circulation of the atmosphere.

1. Follow an air mass during the interval between wash-out by rain and assume that no exchange with surrounding air masses takes place by turbulent transfer (LAL and PETERS 1959). For a nuclide i we may then write Eq. (10)

$$\frac{dq_i}{dt} = \sigma_i\, S(t) - \lambda_i\, q_i \tag{33}$$

where $\sigma_i S(t)$ is the production, $S(t)$ being the rate of star production per gram of air. If at one instant ($t=0$) the concentration of q_i were $q_i(0)$ we can solve (33) and obtain the following expression

$$\left.\begin{aligned} &q_i(t) - q_i(0) \\ &= \sigma_i\, e^{-\lambda_i t} \int_0^t e^{\lambda_i t} \times \\ &\times S[h(\tau),\varphi(\tau)]\, d\tau \end{aligned}\right\} \tag{34}$$

where $h(\tau)$ is the elevation and $\varphi(\tau)$ is the latitude of the air mass in course of time. This expression shows how the concentration $q_i(t)$ depends on the "history" of the air mass. A few special examples are of interest.

If $S=\mathrm{const}$ we obtain from Eq. (34) (putting $q_i(0)=0$)

$$q_i(t) = \frac{\sigma_i}{\lambda_i}\, S(1 - e^{-\lambda_i t}). \tag{35}$$

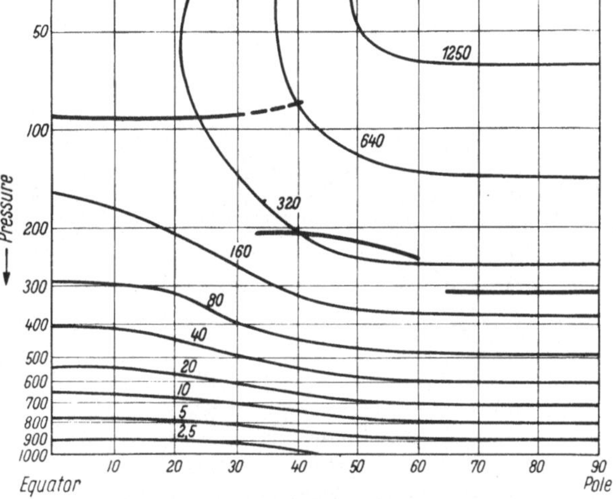

Fig. 9. The number of nuclear disintegrations per gram of air and second as a function of pressure and latitude

By measuring the activity in rain we cannot determine the absolute concentration of an isotope in the air mass from which the rain falls, but assuming the same efficiency in wash-out for the various nuclides, the isotope ratios in the air and in the rain water should be the same. The ratio of two different components becomes

$$\frac{q_i(t)}{q_j(t)} = \frac{\sigma_i\,\lambda_j}{\sigma_j\,\lambda_i}\,\frac{1 - e^{-\lambda_i t}}{1 - e^{-\lambda_j t}} \tag{36}$$

which furthermore is independent of the particular value of S experienced. When t becomes large compared to λ_i^{-1} and λ_j^{-1} the ratio of the concentrations approaches a limiting value. Fig. 10, lower curves shows in which way $q_i(t)$ varies in course of time where $S=6.7\cdot10^{-4}$ stars g^{-1} sec^{-1} corresponding to an average value for the troposphere, and Fig. 11 illustrates the behaviour of some ratios that can be formed. It is seen that constant values are approached in 6 to 8 months.

2. Assume that q_i has attained its equilibrium value for $S=S_1$ and then is moved to another altitude and/or latitude where $S=S_2$. Again integrating (33) yields

$$q_i = \frac{\sigma_i}{\lambda_i}\,[S_2 + (S_1 - S_2)\, e^{-\lambda_i t}]. \tag{37}$$

Here the ratio between concentrations of two different nuclides is not independent of the particular values for S experienced by the air mass. Fig. 10, upper curves shows the variation of q_i according to (37) for some nuclides and Fig. 11 gives the corresponding ratios. It has been assumed that $S_2 = 6.7 \cdot 10^{-4}$ stars g^{-1} sec^{-1} and that $S_1 = 6 \cdot S_2$ approximately corresponding to conditions in the lower stratosphere.

We see that quite different values for the ratios between different nuclides can be expected depending upon the "history" of the air mass. If we accept the assumptions that the troposphere is comparatively well mixed, that complete removal of all radioactivity from an air mass occurs at each rain and that an air mass in the lower stratosphere is brought to equilibrium with

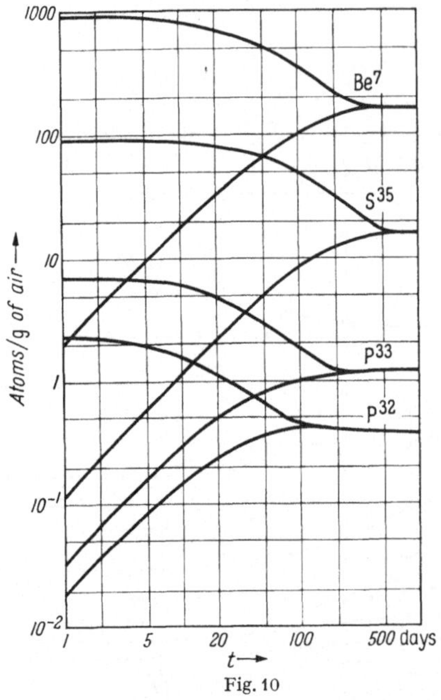

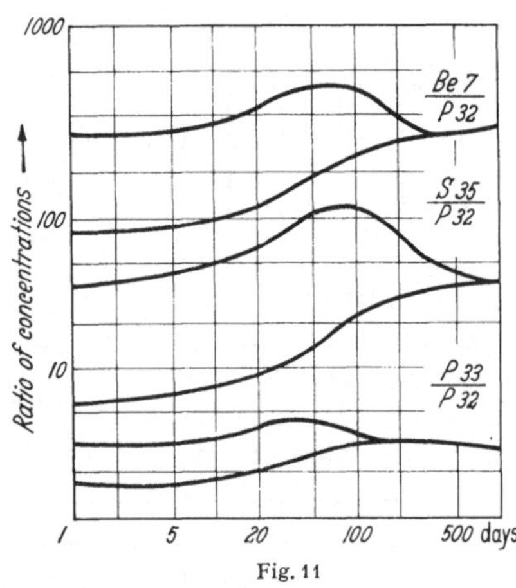

Fig. 10. The number of atoms of various radioisotopes per gram of air as a function of time. a) The atmosphere is completely free of radioactivity at $t = 0$ (lower curves). b) The air mass contains the concentrations in equilibrium with cosmic ray production immediately above the tropopause at $t = 0$ and is moved into the troposphere

Fig. 11. The ratio of the concentrations of different isotopes as a function of time under the conditions given in Fig. 10

regard to its content of these radioactive nuclides before descending into the troposphere, available measurements of Be⁷, P³², P³³ and S³⁵ in rain may be interpreted in the following way (Lal, Peters 1959, where also further references are given).

i. In the latitudes from the equator to lat. 30° N no rain has fallen from an air mass, the major part of which has descended from the stratosphere during the six months prior to the rainfall.

ii. The Be⁷/P³² ratios indicate an average time between rain fall from one and the same air mass ranging between 0 and more than 200 days. An average residence time of about 40 days is obtained.

iii. By comparing the average global fall-out of Be⁷ (based on measurements in India and North America; Arnold, Al-Salih 1955; Cruikshank, Cowper, Grumitt 1956) and the computed production in the troposphere and assuming the stratospheric contribution to the fall-out is negligible again a residence time for Be⁷ in the troposphere of about 40 days is obtained.

It is interesting to note the over-all agreement between these results and the results obtained using naturally radioactive elements (E, I) and artificially introduced radioactivity (E, IV) as tracer elements. Still the assumptions made above are only very approximately fulfilled in the atmosphere. For a more detailed discussion of the circulation of the atmosphere on the basis of radioactive isotopes air samples from different parts of the atmosphere are very desirable. The following simple considerations illustrate some applications.

An air sample with an isotope ratio close to the value of equilibrium [Eq. (35)] must have been close to one and the same isoline in Fig. 9 for about six to twelve months (depending on which elements used, cf. Fig. 11), with no appreciable exchange taking place with other air masses. In regions where these lines are practically horizontal, we may therefore deduce vertical motions from a deviation from the equilibrium value and in areas where these isolines are approximately vertical we can conclude about the rate of horizontal exchange. Since the production of all cosmic ray produced isotopes is given by Fig. 9 except for a constant factor (the yield) we cannot obtain information about the rate of exchange or circulation along the isoline by any observations of cosmic ray produced isotopes only.

A few recent observations of the activity of Be^7, P^{32}, P^{33} and S^{35} in the stratosphere (60000 feet or about 75 mb) indicate that an approximate equilibrium between the different isotopes exist in middle and low latitudes, while this is not the case in polar regions (ARNOLD, personal communication). Values in excess of those in case of equilibrium were recorded indicating a downward motion. To deduce the magnitude of the vertical velocity observations at two levels above each other are required. Put $t = \Delta z \cdot w^{-1}$, where Δz is the height difference between the two points. Eq. (34) then yields (assuming steady state)

$$q_i(z + \Delta z) - q_i(z) = \sigma_i \, e^{-\frac{\lambda_i \Delta z}{w}} \int_0^{\Delta z/w} e^{\lambda_i \tau} \cdot S\left[z + \frac{\tau}{w}\right] d\tau \qquad (38)$$

which can be solved for w numerically. If no absolute measurements are available the use of two different isotopes still permits an evaluation of w.

It was shown in the previous section that the RaD/RaC ratio in precipitation [corresponding to a residence time of about 40 days according to Eq. (29)] is consistent with an eddy diffusivity in the vertical for the troposphere, $K_z = 2 \cdot 10^5$ cm² sec⁻¹. Qualitatively the data on cosmic ray produced isotopes support this conclusion. Further measurements of the concentrations of these various isotopes in the troposphere combined with observations of other radioactive (or stable) isotopes in the atmosphere and various meteorological parameters would make it possible to deduce further details about such features of atmospheric motions (see E, V).

The use of radiocarbon for studies of atmospheric circulations is different due to its long half-life. The C^{14}/C^{12} ratio, which is measured, of course is dependent on changes of the radioactive nuclide C^{14} and variations of the amount of inactive carbon, C^{12}. Fossil fuel combustion has meant a considerable output of CO_2 into the atmosphere (15 % of the total amount in the atmosphere had been added by 1960). The carbon dioxide in the atmosphere is in exchange with the much larger CO_2 reservoir of the oceans and also with the biosphere. A net flux of CO_2 takes place from the atmosphere to the ocean, whereby the chemical equilibrium in the sea is shifted to lower pH and less solubility of CO_2. Hereby a return flow of C^{14} from the sea to the atmosphere can take place. A more complete treatment of these exchange processes yields a residence time for CO_2 in the atmosphere of 3 to 5 years and that the net increase of CO_2 in the atmosphere since the middle of last century is equal to $\frac{1}{2}$ to $\frac{3}{4}$ of the total output due to fossil fuel combustion (Revelle, Suess 1957; Craig 1957; Rafter and Furgusson 1959; Bolin and Eriksson 1959). It is interesting to note that this implies a deposition velocity for CO_2 more than two orders of magnitude less than those previously discussed (Chamberlain and Chadwick 1953). Obviously the limiting factor for CO_2 transfer into the sea is to be found in the sea (Bolin 1960). Finally the difference of the fossil fuel combustion in the northern and southern hemisphere is noticeable in slightly different C^{14}/C^{12} values for the two hemispheres. Furgusson (1958) deduces a residence time of a few years in one hemisphere before transfer to the other. It is interesting to note that this is equivalent to a horizontal exchange coefficient $K_y = 10^{10}\,\mathrm{cm^2\,sec^{-1}}$ somewhat smaller than indicated by meteorological observations for middle latitudes (Durst, Crossley and Davis 1959).

Some aspects of the use of tritium for atmospheric circulation studies will be dealt with in the following section, since bomb explosions are by far the dominating source for this nuclide at present.

IV. The global circulation of the atmosphere as deduced from the distribution and fall-out of artificial radioactivity

In the case of cosmic ray produced radioactivity the production function $P(\mathbf{r}, t)$ is quite well known; also for natural radioactivity it can be deduced with some accuracy. This is not the case for artificial radioactivity and a particular difficulty in making use of these data is the irregularity in time of these injections. It is therefore often quite hard to interpret the observations and deduce the atmospheric motions that have brought about the observed distributions in the atmosphere or the fall-out pattern. Some more precise conclusions can be drawn from the variations of the ratio of various isotopes and from budget computations but otherwise the discussion necessarily has to be qualitative. The general equations for diffusion developed previously can hardly be used.

Quite extensive measurements from all over the world exist particularly at the earth's surface (see e.g.: Joint Committee on Atomic Energy, 1959). The following observed facts most likely depend on the circulation of the atmosphere and will be discussed here.

1. The north-south distribution of the fall-out (Sr^{90}) has a maximum at about lat. 40° in both hemispheres. A slight secondary maximum also shows up at lat. 11° N, the latitude of the American test-site. The Russian test explosions have been made at middle and high latitude in the northern hemisphere, while no tests of significance in this connection have been conducted in the southern hemisphere (cf. Fig. 12, Machta 1959).

2. A seasonal variation of the fall-out has been observed with a maximum in March and April (northern hemisphere). See Fig. 13 (Hardy and Klein 1960).

3. Particularly intense fall-out was recorded in the spring of 1959 about six months after the last Russian test series in October 1958 in northern Siberia. See Fig. 13, and Libby (1959).

4. The concentration of Sr^{90} in soils is closely correlated with the amount of rainfall and its concentration of Sr^{90}, and dry-deposition therefore seems to be of minor significance (Joint Committee on Atomic Energy, 1959).

5. The concentration of radioactivity in the air shows a marked difference between the northern and southern hemisphere, particularly in 1959 (Fig. 14, cf. Joint Committee on Atomic Energy, 1959).

6. The vertical distribution of radioactivity increases rapidly when proceeding from the troposphere up into the stratosphere (Fig. 15, cf. Joint Committee on Atomic Energy, 1959).

7. In early summer 1958 tungsten (W^{185}) was injected into the stratosphere at lat. 11° N and was recorded in the air and precipitation during the following year over large parts of the world (FEELY and SPAR 1960, see Fig. 16).

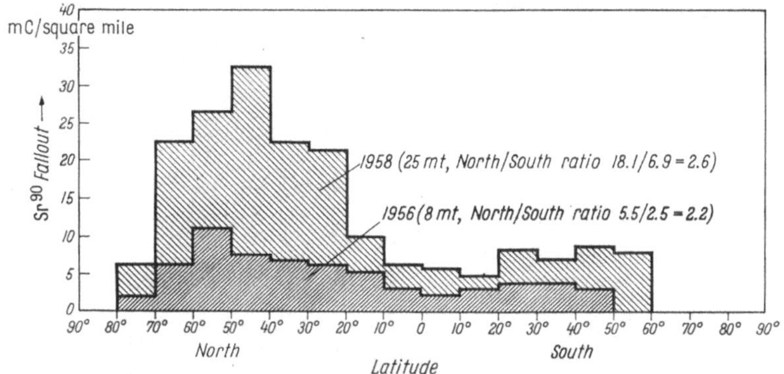

Fig. 12. Distribution of accumulated fallout 1956 and 1958 as a function of latitude (LIBBY 1959)

The input of radioactivity into the atmosphere is essentially due to hydrogen bombs which produce a radioactive cloud rising well above the tropopause. Stabilization occurs at 25 to 50 km elevation and then meteorological factors take over in distributing the debris.

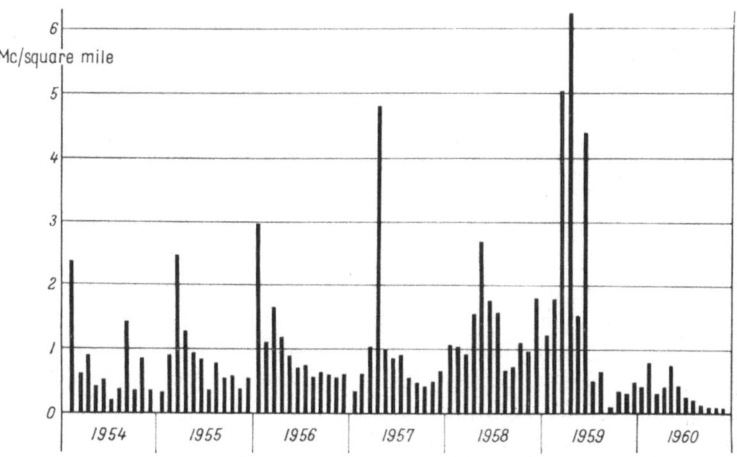

Fig. 13. Monthly fallout at New York during the period 1954 to fall 1960 (HARDY, KLEIN 1960)

When discussing the mechanisms of exchange between different latitudes or levels in the atmosphere some facts deduced from ordinary meteorological observations in the troposphere should be kept in mind. PALMÉN (1955) has shown that a meridional circulation cell is of importance in tropical latitudes but plays a fairly insignificant rôle in middle and high latitudes in transferring heat and momentum. Instead large-scale turbulence is of great importance. This is due to the effect of the rotation of the earth, whereby axial-symmetric meridional motions are surpressed at high latitudes. There are no reasons for this being different in the stratosphere. Support for this is given by the distribution of tungsten (W^{185}) and its change in course of time after the injection in late spring 1958 (FEELY and SPAR 1960).

It was shown in previous sections (D and E, I) that horizontal mixing at high levels (25 km) in winter (October to March in the northern hemisphere) poleward of lat. 30° very likely may approach the same intensity as in the middle and upper troposphere. This is associated with the development of the polar night jet stream in the stratosphere. This implies a horizontal mixing time for air poleward of lat. 30° considerably less than a year. The synoptic disturbances responsible

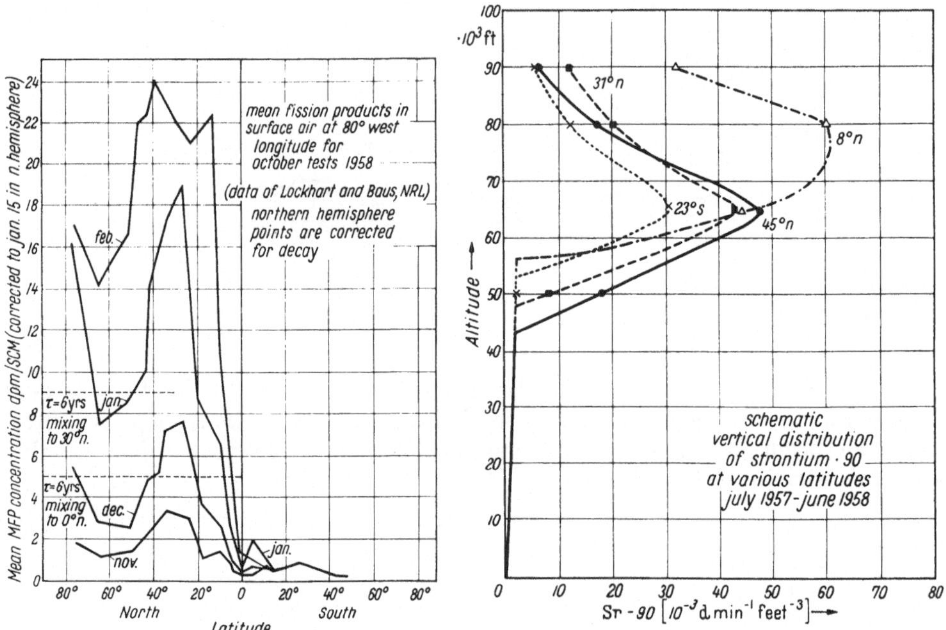

Fig. 14. The concentration of radioactivity in surface air as a function of latitude in late 1958 and early 1959 (Joint Committee on Atomic Energy, 1959)

Fig. 15. The vertical distribution of radioactivity in the troposphere and stratosphere up to about 30 km elevation at different latitudes (Joint Committee on Atomic Energy, 1959)

for this also bring about a vertical mixing. In addition the final decay of the polar night jet stream in spring is associated with a marked subsidence in polar regions whereby transfer down towards the tropopause presumably takes place. In summer conditions should be quite different due to the comparatively weak and regular easterly flow in the stratosphere (above 20 km). The residence time for radioactivity in the stratosphere therefore should very much depend on the *latitude*, *altitude* and *time* of injection. The general picture outlined above is supported by the very rapid fall-out of the debris from polar tests in the fall of 1958 (compare the fall-out in spring 1959 and 1960). A residence time of 3 to 6 months is obtained (MARTELL 1959). A somewhat larger value of about a year is obtained for injections in middle latitudes in summer (MACHTA 1959). These results are deduced assuming a residence time for debris in the troposphere of a month or less.

Turning to equatorial injections into the stratosphere conditions are different. LIBBY (1957) early deduced a characteristic residence time for this debris of about ten years, which later has been modified (LIBBY 1959, MACHTA 1959) but the best estimates at present are of the order of a few years. It was pointed out previously (E, I) that quite a marked change of total ozon content of the atmosphere takes place at lat. 30°, the lowest latitude of the more intense meridional mixing in winter. Probably also a mean meridional flow towards the equator takes place at levels between 20 and 25 km, being associated with upward motion in the vicinity of the equator (LIBBY and PALMER 1960). This would be consistent with this longer residence time and would also explain the comparatively low values of radioactivity above the tropopause at low latitudes (HAGEMANN, GRAY, MACHTA, TURKEVICH 1959).

In tropical latitudes the mechanism of exchange between the stratosphere and troposphere and exchange within the troposphere may be of importance. No large differences in the fall-out of cosmic ray produced radioactivity were found between middle and low latitude (CRUCKSHANK, COWPER, GRUMMITT 1956; LAL and PETERS 1959). Since this fall-out is essentially due to production in

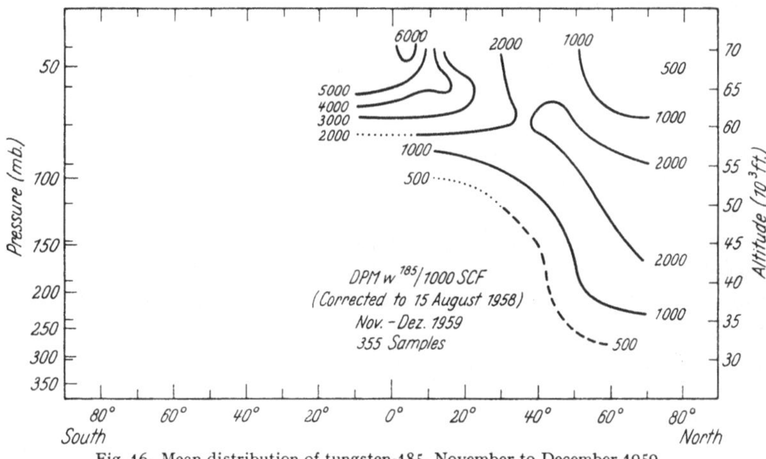

Fig. 16. Mean distribution of tungsten-185, November to December 1959

the upper troposphere, the mean meridional circulation in the tropical troposphere (see E, I) apparently is of minor importance for the fall-out at these latitudes. This fact provides means of computing the relative intensity of the meridional circulation cell and the vertical exchange by turbulence. From this fact we may also conclude that the differences in Sr^{90} fall-out between middle and low latitudes is due to either direct tropospheric injections of radioactivity in middle latitudes (Nevada tests) or the transfer from the stratosphere to the troposphere, being different in the tropics and at higher latitudes. Studies of the ratios Sr^{89}/Sr^{90} or Ba^{140}/Sr^{90} provides a method of approximately dating the debris collected and STEWART et al. (1957) and MARTELL (1959) have shown that the contribution of tropospheric injection to the middle latitude fall-out is insignificant. The conclusion that therefore the rate of the vertical transfer within the stratosphere or from the stratosphere to the troposphere is different in middle and low latitude is inescapable. If the vertical transfer between 16 and 25 km were the same at all latitudes the much more rapid vertical transfer in the high tropical troposphere due to cumulo-nimbus convection than in the low polar stratosphere (10 to 16 km) would necessarily lead to higher fall-out in the tropics than further poleward. We may therefore conclude that vertical transfer in the lower stratosphere in equatorial latitudes is considerably less than further north, which is consistent with the Richardson number being much larger in the stable stratosphere of the tropics.

In explaining the distribution of water vapour in the vicinity of the tropopause BREWER (1953) adopts a circulation model of the atmosphere implying a meridional circulation cell extending from the tropical troposphere into the stratosphere, thereby explaining the very low humidities observed in the low stratosphere in middle latitudes (frost point down to values corresponding to the low temperatures at the tropical tropopause). MACHTA (1959) has discussed this model in connection with radioactive fall-out. The tropical tropopause is usually very sharp and the maintenance of this inversion still having flow across it has not been explained from the point of view of the heat balance. Very low values of the humidities in the low polar stratosphere of course can be explained by merely having an average upward motion *to* the tropopause in the tropics but not through it and then exchange with the polar stratosphere at 12 to 16 km level. Therefore the proposal of having a mass transfer through the tropical tropopause seems less attractive until it has been shown to be possible from the point of view of the heat budget.

In addition to producing solid substances a few radioactive gases are formed due to bomb explosions. Of particular interest are water vapour containing tritium and carbon dioxide containing C^{14}. Since the fall of 1954 regular measurements from balloons have been made up to about 30 km (HAGEMANN, GRAY, MACHTA, TURKEVICH 1959) and together with the C^{14} determinations in surface air and top layers of the ocean (BROECKER, WALTON 1959) some overall budget estimates can be made. On the basis of measurements in 1954 and 1955 a stratospheric residence time (equatorial injection) somewhat less than 5 years is deduced in agreement with the previous discussions. In case of carbon dioxide the tropospheric residence time before transfer into the oceans is about five years (CRAIG 1957; BOLIN, ERIKSSON 1959) as compared with a 30 to 40 days residence time for solid particles. This fact offers a possibility of using differences in the C^{14} activity between the hemispheres to compute the overall horizontal exchange rate in the troposphere and the stratosphere (FERGUSSON 1958); a characteristic value of one to two years is obtained. From 1955 to 1959 the total stratospheric reservoir of C^{14} remained approximately constant indicating an approximate steady state. Data about the meridional and vertical distribution of C^{14} in combination with ozone, water vapour and other tracers can be used to deduce the general circulation and transfer in the stratosphere (see V).

Quite different information is obtained from the distribution and fall-out of tritium. At the very low temperatures in the stratosphere the water vapour brought up from lower layers by the rising atomic cloud and also the vapour formed by the tritium molecules generated soon condenses on solid particles and later form ice crystals. The settling of these crystals may reduce the residence time of tritium in the stratosphere considerably as compared with other radioactive particles (BEGEMANN, LIBBY 1957). If, however, large amount of the water vapour containing tritium were condensed on particles also other radioactive substances would have a similar opportunity. It is not *a priori* clear how much more rapidly water vapour would be removed from the stratosphere due to such an effect. BOLIN (1959a) also has shown that most likely a gradual transfer from the stratosphere to the troposphere occurred extending over a year or even more after the Castle test in 1954. Probably conditions were similar at later tests.

The residence time for water in the troposphere is about 10 days. BEGEMANN and LIBBY (1957) indicate that the corresponding value for tritium is about 40 days similar to what was found for cosmic ray produced isotopes. This difference shows that the latter time to a major extent is determined by the turbulent transfer through the upper parts of the troposphere, where the rain-out mechanism is much less effective.

The circulation of tritium in the lower troposphere is very much complicated by repeated evaporation and condensation (cf. BOLIN 1959a CRAIG and LAL 1961). It is thus quite clear that over the oceans this mechanism of direct molecular exchange between the atmosphere and the sea is more important than the fall-out by rain. It also plays a significant rôle over land, where evapotranspiration by the vegetation also must be considered. The present organization of a world-wide net-work of sampling stations for collection of rain water, ocean water and continental water and analysis for tritium will undoubtedly greatly aid us in obtaining a better understanding of the circulation of water in nature (see above).

V. Final remarks

The previous discussion has quite conclusively shown the usefulness of radioactive tracers for studies of the general circulation of the atmosphere, but also the limitations imposed by for example lack of knowledge of the production or the irregular character of the injections due to bomb tests. Some general results have been deduced but we still do not know many details of the circulation of the stratosphere. One might ask what a more systematic approach to the study of atmospheric motions with the aid of radioactive tracers would involve. It is first of all necessary to know the production function with some accuracy. A first approximate picture of the atmospheric circulation could be obtained by asking for the average steady circulation compatible with the observed distribution of some radioactive (or stable) isotopes. Considering again a meridional plane from the pole to the equator and assuming steady state we may write Eq. (10) as

$$\left.\begin{array}{l} v\,\dfrac{\partial q_i}{r\,\partial \varphi} + w\,\dfrac{\partial q_i}{\partial z} - \dfrac{1}{r\cos\varphi}\,\dfrac{\partial}{\partial\varphi}\left[K_y\cos\varphi\,\dfrac{\partial q_i}{r\,\partial\varphi}\right] - \\[2ex] \qquad - \dfrac{1}{\varrho}\,\dfrac{\partial}{\partial z}\left[K_z\bar{\varrho}\,\dfrac{\partial q_i}{\partial z}\right] - P_i\,(\varphi,\,z) + \lambda\,q_i = 0. \end{array}\right\} \tag{38}$$

We have here assumed that the horizontal density variations are small and that the thickness of the atmosphere is small compared with the radius of the earth r. φ is the latitude. In addition to (38) the continuity equation (11) applied to a meridional plane is valid, whereby a mass stream function can be introduced. Permitting K_y and K_z to be functions of z and φ means that we have three unknown functions of z and φ and they cannot be determined unless we know the steady state distribution of three elements (radioactive or stable). It is only in the case of the cosmic ray produced isotopes that we accurately know the production function, but here the distribution of the stable compounds, ozone and water vapour also should be useful. The production (or destruction) of ozone in the lower stratosphere and upper troposphere is negligible and we may also to a first approximation disregard the condensation and precipitation in the upper troposphere and lower stratosphere. It would of course always be possible to check whether the circulations deduced in this way would be consistent with this latter assumption. Eq. (38) would have to be evaluated by numerical methods (finite difference methods) whereby the system of differential equations would be reduced to $3\,m$ linear algebraic equations, where m is the number of grid points in the finite difference representation. Certain precautions would have to be taken to ascertain that this system of equations has a solution (determinant different from zero) and it would also be necessary to choose isotopes the production functions and half-lives of which were not identical or almost the same, since the accuracy in such a computation then would be very limited (the distributions of these two elements would be almost redundant). This is the reason why the different cosmic ray produced isotopes never would be sufficient for a complete deduction of the atmospheric circulation. Data on ozone and water vapour are accumulating and measurements of cosmic ray produced isotopes in the free atmosphere are under way. Possibly the C^{14} distribution due to bomb explosions also could be used for this purpose. It will indeed be most interesting to see what results a simultaneous and systematic discussion of such data along lines given above will yield. The complications of seasonal variations, differences from one longitude to another and from one year to another would also have to be considered. In combination with ordinary meteorological data, the distribution of radioactive and stable tracers in the atmosphere should gradually help us to understand the complicated machinery of the atmosphere.

References

(It has not been possible to include a complete bibliography on the subject but in the more recent articles referred to further references usually may be found.)

ANGELL, J.K.: A climatological analysis of two years of routine transosonde flights from Japan. Month. Weath. Rev. **87**, 427 (1959).

ARNOLD, J.R., and H.A. AL-SALIH: Beryllium-7 produced by cosmic rays. Science **121**, 451 (1955).

BATCHELOR, G.K.: The application of the similarity theory of turbulence to atmospheric diffusion. Quart. J. Roy. Meteorol. Soc. **76**, 133 (1951); **77**, 315 (1952).

BEGEMANN, F., and W.F. LIBBY: Continental water balance, ground water inventory and storage times, surface ocean mixing rates and world-wide water circulation patterns from cosmic ray and bomb tritium. Geochim. et Cosmochim. Acta **12**, 277 (1957).

BLIFFORD, I.H., B. LOCKHART jr. and H.B. ROSENSTOCK: On the natural radioactivity in the air. J. Geophys. Res. **57**, 499 (1952).

BOLIN, B.: Note on the exchange of iodine between the atmosphere, land and sea. Int. J. Air Poll. **2**, 127 (1959).

— On the use of tritium as a tracer for water in nature. Proc. Sec. Int. Conf. Peaceful Uses of Atomic Energy, Geneva, 18 (1959, a).

— On the exchange of carbon dioxide between the atmosphere and the sea. Tellus **12**, 274 (1960).

—, and E. ERIKSSON: Changes in the carbon dioxide content of the atmosphere and sea due to fossil fuel combustion. In: Rossby Memorial Volume, p. 130. New York: Rockefeller Inst. Press 1959.

BREWER, A.W.: The variations of ozone and water vapour concentrations. Int. Union Geod. Geoph. Ass. Met., Brussels, 1951, 338 (1953).

BROECKER, W.S., and A. WALTON: Radiocarbon from nuclear tests. Science **130**, 309 (1959).

BURTON, W.M., and N.G. STEWART: Use of long-lived natural radioactivity as an atmospheric tracer. Nature, Lond. **186**, 584 (1960).

CHAMBERLAIN, A.C.: Aspects of travel and deposition of aerosol and vapour clouds. AERE. Harwell Rep. HP/R 1261, England 1956.

— Deposition of iodine-131 in northern England in October, 1957. Quart. J. Roy. Meteorol. Soc. **85**, 350 (1959).

—, and R. CHADWICK: Deposition of air-borne radioactive iodine vapour. Nucleonics **11** (8), 22 (1953).

CRABTREE, J.: The travel and diffusion of the radioactive material emitted during the Windscale accident. Quart. J. Roy. Meteorol. Soc. **85**, 362 (1959).

CRAIG, H.: The natural distribution of radiocarbon and the exchange time of carbon dioxide between atmosphere and sea. Tellus **9**, 1 (1957).

—, and D. LAL: The production rate of natural tritium. Tellus **13**, 85 (1961).

CRUIKSHANK, A. I., G. COWPER and W.E. GRUMMITT: Production of Be^7 in the atmosphere. Canad. J. Chem. **34**, 214 (1956).

DURST, C.S., A.F. CROSSLEY and N.E. DAVIS: Horizontal diffusion in the atmosphere as determined by geostrophic trajectories. J. Fluid Mech. **6**, 401 (1959).

FEELY, H.W., and J. SPAR: Tungsten-185 from nuclear bomb tests as a tracer for stratospheric meteorology. Nature, Lond. **188**, 1062 (1960).

FERGUSSON, G. J.: Reduction of atmospheric radiocarbon concentration by fossil fuel carbon dioxide and the mean life of carbon dioxide in the atmosphere. Proc. Roy. Soc. Lond. A **243**, 561 (1958).

FLEISHMAN, B. A., and F. N. FRENKIEL: Diffusion of matter emitted from a line source in a non-isotropic turbulent flow. J. Meteorology **12**, 141 (1955).

GIFFORD, F.: Relative atmospheric diffusion of smoke puffs. J. Meteorology **14**, 410 (1957).

— Statistical properties of a fluctuating plume dispersion model. In: Advances in geophysics, vol. 6, p. 117. New York: Academic Press 1959.

HAGEMANN, F., J. GRAY jr., L. MACHTA and A. TURKEVICH: Stratospheric carbon 14, carbon dioxide and tritium. Science **130**, 542 (1959).

HARDY, E., and S. KLEIN: Strontium program. Quarterly summary report. Health and Safety Laboratory (HASL). Report No. 77. United States Atomic Energy Commission 1960.

HAXEL, O., u. G. SCHUMANN: Selbstreinigung der Atmosphäre. Z. Physik **142**, 127 (1955).

HAY, J. S., and F. PASQUILL: Diffusion from a continuous source in relation to the spectrum and scale of turbulence. In: Advances in geophysics, vol. 6, p. 345. New York: Academic Press 1959.

ISRAËL, H.: Radioactivity in the atmosphere. In: Compendium of meteorology, p. 155. AMS, Boston 1951.

Joint Committee on Atomic Energy, U.S.: Hearings on fallout from nuclear veapons, May 5—8, 1959. 261—948: U.S. Government Printing Office 1959.

KELLOGG, W. W.: Diffusion of smoke in the stratosphere. J. Meteorology **13**, 241 (1956).

KOLMOGOROFF, A. N.: Dissipation of energy in locally isotropic turbulence. C.R. Acad. Sci. USSR. **31**, 538; **32**, 16 (1941).

LAL, D.: Cosmic ray produced radioisotopes for studying the general circulation of the atmosphere. Indian J. Meteor. and Geophys. **10**, 147 (1959).

— P. K. MALHOTRA and B. PETERS: On the production of radioisotopes in the atmosphere by cosmic radiation and their application to meteorology. J. Atmosph. Terr. Phys. **12**, 306 (1958).

—, and B. PETERS: Cosmic ray produced isotopes as tracers for studying large scale atmospheric circulation. Proc. Sec. Int. Conf. Peaceful Uses of Atomic Energy, Geneva, vol.18, p. 533 (1959).

LEHMANN, L., u. A. SITTKUS: Bestimmung von Aerosolverweilzeiten aus dem RaD und RaF-Gehalt der atmosphärischen Luft und Niederschlags. Naturwissenschaften **46**, 9 (1959).

LETTAU, H.: Diffusion in the upper atmosphere. In: Compendium of meteorology, p. 320. AMS, Boston, Mass. 1951.

LIBBY, W. F.: Radioactive fallout. Proc. Nat. Acad. Sci. (Wash.) **43**, 758 (1957); **44**, 800 (1958).

— Radioactive fallout from the Russian October series. Proc. Nat. Acad. Sci. (Wash.) **45**, 959 (1959).

—, and C. E. PALMER: Stratospheric mixing from radioactive fallout. J. Geophys. Res. **65**, 3307 (1960).

MACHTA, L.: Transport in the stratosphere and through the tropopause. In: Advance in geophysics, vol. 6, p. 273. New York: Academic Press 1959.

— H. L. HAMILTON jr., L. F. HUBERT, R. LIST and K. M. NAGLER: Airborne measurements of atomic debris. J. Meteorology **14**, 165 (1957).

— R. J. LIST and L. F. HUBERT: Worldwide travel of atomic debris. Science **124**, 474 (1956).

MARTELL, E. A.: Atmospheric aspects of strontium-90 fallout. Science **129**, 1197 (1959).

— Global fallout and its variability. Geophysical research papers No. 65. GRD, AFCRC, Bedford, Mass. 1959.

MONIN, A. S.: Diffusion in the surface layer under stable stratification. In: Advances in geophysics, vol. 6, p. 429. New York: Academic Press 1959.

—, and A. M. OBUKHOV: The basic laws of turbulent mixing in the atmospheric surface layer. Trud. Geoph. Inst. Akad. Nauk SSSR. No. 24, 163 (1954).

OBUKHOV, A. M.: On the distribution of energy in the spectrum of turbulent motion. Bull. Acad. Sci. USSR., Ser. geogr. and geoph. No. 4, 5 (1941)..

PALMÉN, E.: On the meridional circulation in low latitudes of the northern hemisphere in winter and the associated meridional and vertical flux of angular momentum. Final Rep. Gen. Circ. Project. GRD, Cambridge, Mass. 1955.

PRANDTL, L.: The mechanics of viscous fluids. In: W. F. DURAND (ed.), Aerodynamic Theory, Vol. III Div. G. Berlin 1934.

PRIESTLEY, C. H. B.: Turbulent transfer in the lower atmosphere. Chicago, Ill.: Chicago University Press 1959.

RAFTER, T. A., and G. J. FURGUSSON: Atmospheric radiocarbon as a tracer in geophysical circulation problems. Proc. Sec. Int. Conf. Peaceful Uses of Atomic Energy, Geneva, vol. 18, p. 526 (1959).

Ramanathan, K. R.: Atmospheric ozone and the general circulation of the atmosphere. Sc. Proc. Int. Assoc. Met. UGGI, Rome, Sept. 1954, Publication AIM No. 101c, p. 3. London: Butterworth Sci. Publ. 1956.

Reed, R. J., W. J. Campbell, L. A. Rasmussen and D. G. Rogers: Evidence of a downward propagating annual wind reversal in the equatorial stratosphere. J. Geophys. Res. **66**, 813 (1961).

Revelle, R., and H. Suess: Carbon dioxide exchange between atmosphere and ocean and the question of an increase of atmospheric CO_2 during past decades. Tellus **9**, 18 (1957).

Richardson, L. F.: Turbulence and the vertical temperature difference near trees. Phil. Mag. **49**, 81 (1925).

— Atmospheric diffusion on a distance-neighbor graph. Proc. Roy. Soc. Lond. A **107**, 709 (1926).

Schmidt, W.: Der Massenaustausch in freier Luft und verwandte Erscheinungen. Probleme der Kosmischen Physik, Bd. 7. Hamburg 1925.

Schumann, G.: Correction for tropospheric storage times obtained from Pb 210. Symposium on Atmospheric Chemistry and Radioactivity, Helsinki 1960 (U. G. G. I.).

Sheppard, P. A.: Transfer across the earth's surface and through the air above. Quart. J. Roy. Meteorol. Soc. **84**, 205 (1958).

Solot, S. B., and E. M. Darling jr.: Theory of large scale atmospheric diffusion and its application to air trajectories. Geoph. Res. Papers No. 58. GRD, AFCRC, Bedford, Mass. 1958.

Stewart, N. G., H. J. Gale and R. N. Crooks: The atmospheric diffusion of gases discharged from the chimney of the Harwell reactor BEPO. Int. J. Air Poll. **1**, 87 (1958).

— R. G. D. Osmond, R. N. Crooks and E. M. R. Fischer: The world-wide deposition of long-lived fission products from nuclear test explosions. AERE, HP/R 2354. Harwell, England 1957.

Sutton, O. G.: Micrometeorology. New York-Toronto-London: McGraw Hill, Inc. 1953.

Taylor, G. I.: Diffusion by continuous movements. Proc. Lond. Math. Soc. **20**, 196 (1921).

Taylor, R. J.: Similarity theory in the relation between fluxes and gradients in the lower atmosphere. Quart. J. Roy. Meteorol. Soc. **86**, 67 (1960).

United States Department of Commerce, Weather Bureau: Meteorology and atomic energy. 1955.

Welander, P.: Studies on the general development of motion in a two-dimensional, ideal fluid. Tellus **7**, 141 (1955).

World Meteorological Organization. Meteorological aspects of the peaceful uses of atomic energy. Proc. Sec. Int. Conf. Peaceful Uses of Atomic Energy, Geneve, vol. 18, p. 245. 1959.

Radioaktive Aerosole

von

CHRISTIAN E. JUNGE

Mit 5 Figuren

Summary

A I. For the radon and thoron decay products, the main sources of natural radioactivity in the atmosphere, natural aerosols act as carriers. The main features of these aerosols are discussed, especially the size distribution and concentration under various geographical conditions, including some data on chemical composition and distribution with altitude. Several processes, predominantly those connected with the formation and evaporation of cloud droplets, will modify the size distributions by growth of individual particles.

A II. The decay products of radon and thoron, RaA and ThA are atoms, which readily become molecular clusters (primary particles) similar to small ions and then in turn become attached to the aerosols (secondary particles). The theory of SMOLUCHOWSKY can be applied to this process of attachment and various parameters, such as the half live time of primary particles, the activity ratio of primary to secondary particles and the activity distribution with particle size can be calculated. The agreement with the data available in literature is fair and indicates that the presented model is a good approximation. For longer lived decay products, the removal of aerosols from the atmosphere has to be incorporated in the calculations. Data on the atmospheric residence time of aerosols obtained by the use of radon decay products are critically reviewed.

B. Similar to the decay products of the emanations, the isotopes induced by cosmic radiation can only exist as particles. New data on stratospheric aerosols allows similar calculations as for the radon decay products. It is concluded that most of the isotopes must be attached to particles between 0.02 and 0.2 microns and that the half live of the primary particles is of the order of hours as compared to fractions of a minute in the troposphere.

C I. The structure and chemical composition of close in fallout particles of atomic tests depends considerably on the type of test, i.e., if it is a ground, tower, water or air burst. If no surface material enters the fireball, the particles are formed by condensation and consist of iron oxide with the activity uniformly distributed within them. Surface or tower material may only partly be evaporated or melted, and the result is a variety of particles with the activity primarily deposited on the surface.

C II. The particles of medium range fallout can be characterized by the fact that they are small enough to be carried over large parts of the globe in the first few months after tests, but still large enough to fall out primarily by sedimentation. Activities up to 10^{-9} C are observed in these "hot" particles, which at times may represent a considerable fraction of the activity in air. Their activity is roughly proportional to their volume and they form only the tail of a size distribution which extends to much smaller particles and represents already part of the long range fallout.

CIII. The long range fallout particles are so small that they behave like a gas with respect to meteorological processes. They have residence times in the stratosphere of about a year and are removed from the troposphere primarily by precipitation. The few measurements available from the stratosphere indicate average sizes of a few hundred microns to the tenth of a micron range. After penetration into the troposphere these particles are modified in their size, composition and structure by the same processes which influence the natural aerosols.

CIV. A problem of special interest in the area of reactor aerosols is the absorption of radioactive gases on aerosols. This phenomenon was studied in more detail for Iodine 131, which was released in considerable quantities at the Windscale accident. It appears that absorption of Iodine on aerosols becomes important, when its concentration by weight is about 10^{-3} the concentration of the aerosol.

Einleitung

Die Anwesenheit radioaktiver Substanzen in der Atmosphäre ist von Bedeutung für viele Gebiete, wie z.B. Biologie, Medizin sowie für verschiedene Industrien. Aber die Kenntnis von vorhandenen Konzentrationen der verschiedenen aktiven Isotope genügt nicht. Es ist auch notwendig zu wissen, in welcher Form sie anwesend sind, ob gasförmig oder fest, und wenn fest, in welcher Unterteilung.

Dieser Abschnitt ist der Frage der radioaktiven Aerosole in unserer Atmosphäre gewidmet. Er versucht eine Brücke zu schlagen zwischen zwei Fachgebieten, die, wie Beispiele in der Literatur zeigen, oft nicht viel voneinander wissen, nämlich die Radiochemie und die physikalische Meteorologie. Die folgenden Seiten stellen im wesentlichen eine kritische Zusammenfassung von Arbeiten auf diesen Gebieten dar, aber an einigen Stellen erwies es sich als möglich, darüber hinaus die Analyse noch zu vertiefen.

Drei Gruppen von Radioaktivitäten in der Atmosphäre lassen sich unterscheiden: Die natürlichen Radioaktivitäten, die von den Emanationen herrühren und im wesentlichen auf die Troposphäre beschränkt bleiben. Die natürlichen Radioaktivitäten, die in höheren Schichten durch die kosmische Strahlung erzeugt werden und drittens die künstlichen Radioaktivitäten, die uns das Zeitalter der Atombomben und Reaktoren beschert. Die Diskussion des Stoffes folgt dieser Einleitung. Dabei wurde versucht, besonders unsere Kenntnisse von atmosphärischen Aerosolen, die ja häufig als die Träger der Aktivitäten fungieren, so weit wie möglich zu behandeln und anzuwenden.

A. Die Radioaktivität der Radon- und Thoronzerfallprodukte und ihre Anlagerung an troposphärische Aerosole

I. Die troposphärischen Aerosole

Die meisten radioaktiven Isotope sind unter Normalbedingungen nicht gasförmig. Sie können in der Atmosphäre nur in Form von Aerosolteilchen auftreten, entweder in ihrer ursprünglichen Form als Primärteilchen, oder nach Anlagerung an vorhandene neutrale Aerosole als Sekundärteilchen. Die letztere Möglichkeit ist bei weitem die häufigere. Eine gründliche Kenntnis der natürlichen Aerosole ist deshalb für ein Verständnis der Eigenschaften radioaktiver Aerosole eine Notwendigkeit. Dieser Abschnitt soll einen Überblick über unsere Kenntnisse auf diesem Gebiet vermitteln, wobei wir nacheinander die Größenverteilung, die physikalische Struktur, die chemische Zusammensetzung und die Höhenverteilung der troposphärischen Aerosole besprechen wollen.

Von grundlegender Bedeutung für Eigenart und Verhalten von Aerosolen ist ihre Größenverteilung. Unsere Kenntnisse auf diesem Gebiet sind immer noch recht lückenhaft und erst in jüngster Zeit wurden die Hauptzüge richtig erkannt. Die Schwierigkeiten dabei sind methodischer Natur, denn es sind erhebliche Größen- und Konzentrationsbereiche zu überbrücken, was nur durch eine simultane Anwendung verschiedener Meßmethoden erreicht werden kann.

Fig. 1 veranschaulicht die Verhältnisse in einem Übersichtsbild. Man kann zwischen einer kontinentalen und maritimen Verteilung unterscheiden. Diese Unterscheidung ist ganz allgemein in der Luftchemie von Nutzen, ist allerdings nur im großen gesehen gültig, da die relative Langlebigkeit der Aerosole und die Intensität der atmosphärischen Zirkulation weite Übergangsgebiete zur Folge haben.

Im folgenden werden wir alle Größen von Teilchen in Micron (μ) Radius angeben und für einige wichtige Größenbereiche die folgende, in der Meteorologie vielfach gebrauchte Bezeichnungsweise benutzen:

$$< 0,1 \; \mu \; \text{Aitken-Teilchen}$$

$$0,1 \text{ bis } 1,0 \; \mu \; \text{Große Teilchen}$$

$$> 1,0 \; \mu \; \text{Riesenteilchen.}$$

Der Ausdruck Kondensationskerne oder kurz Kerne wird manchmal gebraucht, wenn die Bedeutung dieser Teilchen für die Wasserdampfkondensation in der Atmosphäre betont werden soll.

Größenverteilungen der reinen Seesalzteilchen liegen meist nur oberhalb $1 \; \mu$ vor (WOODCOCK 1953), und erst neuerdings wurden verläßliche Werte bis zu etwa $0,3 \; \mu$ erhalten (METNIEKS 1958). Verschiedene Anzeichen deuten darauf hin, daß der Anteil an Seesalzteilchen unterhalb $0,1 \; \mu$ gering ist, selbst im Zentrum großer Ozeane und bei starken Winden (JUNGE 1956, MASON 1957), und daß die Mehrzahl dieser Teilchen Reste kontinentaler Aerosole darstellen. Dies führt zu der Vermutung, daß die Seesalzkerne ein Maximum etwas oberhalb $0,1 \; \mu$ haben. Bekannt ist jedoch wirklich nur, daß die Konzentrationen der Aitken-Teilchen über See meist einige hundert beträgt, manchmal bis auf etwa 50 herabsinkt, daß die Gesamtkonzentration der reinen Seesalzkerne nicht viel größer ist als $10/\text{cm}^3$, selbst bei stärkerem Wind, und daß unterhalb $0,01 \; \mu$ vermutlich keine Teilchen vorhanden sind. Der auf Grund dieser Daten vermutete Verlauf der Kurven unterhalb $0,3 \; \mu$ ist in Fig. 1 b angedeutet. Die Verhältnisse bedürfen weiterer Klärung. Die Seesalzkerne entstehen, wie KIENTZLER u.a. (1954) gezeigt haben, wenn Luftblasen im Seewasser die Oberfläche erreichen und zerplatzen. Dadurch entsteht ein feiner Strahl, der in einzelne Tröpfchen aufbricht. Es sind aber Anzeichen vorhanden, daß auch wesentlich feinere Teilchen durch das Zerplatzen der Blasenoberfläche selbst entstehen (MASON 1954), in welchem Falle die Annahmen in Fig. 1 b etwas abgeändert werden müßten.

Der wesentliche Unterschied zwischen maritimen und kontinentalen Größenverteilungen ist in den höheren Konzentrationen der letzteren an großen und vor allem Aitken-Teilchen. Die stetig wachsende Zahl der Aitken-Teilchen, wenn man von abgelegenen ozeanischen Gebieten zu dicht besiedelten oder industriellen fortschreitet, beweist, daß die überwiegende Zahl der Aitken-Teilchen direkt oder indirekt durch menschliche Tätigkeit entsteht. Ein kleiner Anteil mag auch durch Reaktionen verschiedener natürlicher Spurengase entstehen, wie sie vom Boden (z.B. H_2S, NH_3) oder von Pflanzen (z.B. Terpentin) exhaliert werden, oder direkt vom Boden kommen. Im Gegensatz dazu ist der Unterschied in der Konzentration (nicht in der Zusammensetzung!) der Riesenteilchen zwischen Meer und Land im Mittel nicht sehr groß. Dies erscheint bei dem großen Unterschied in den Entstehungsbedingungen rein zufällig. Typische Merkmale der mittleren kontinentalen Größenverteilung sind ein Maximum zwischen 0,1 und 0,01 μ, meist bei 0,03 μ, und ein Abfall, der angenähert nach der dritten Potenz des Radius zwischen 0,1 und 10 μ verläuft (JUNGE 1958). Dieser Exponent, der zwischen 2,5 und 3,5 etwa schwanken kann, wurde durch direkte Messungen sowie indirekt durch

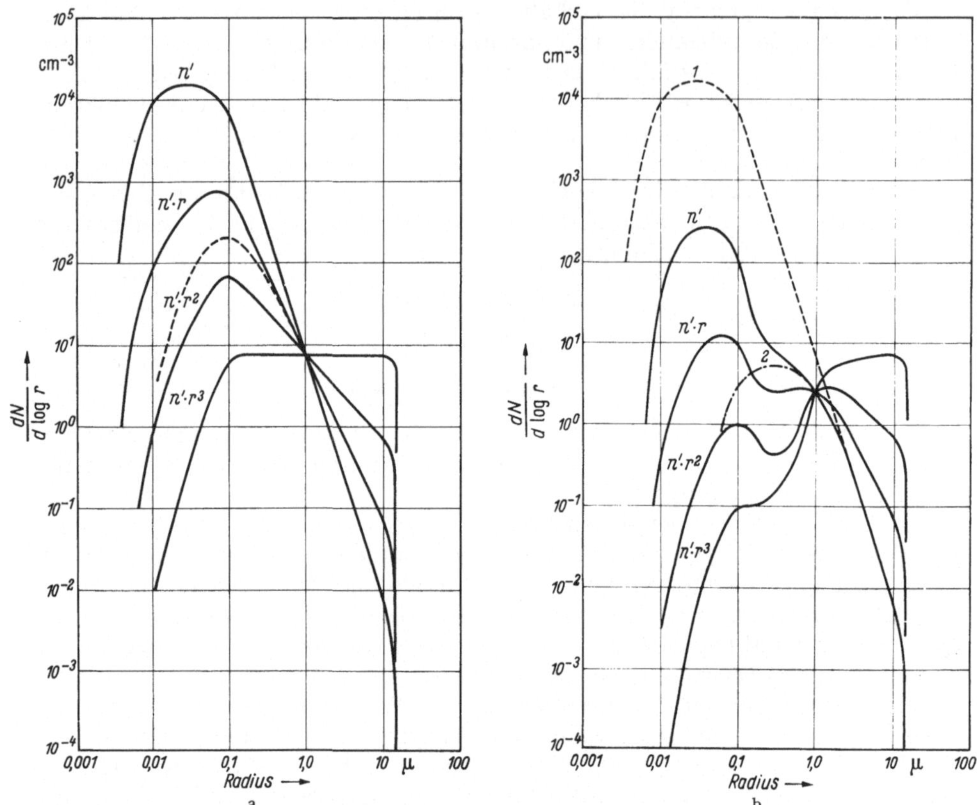

Fig. 1. a Größenverteilung von kontinentalem Aerosol in Bodennähe und die Verteilung der Produkte $n'r$, $n'r^2$ und $n'r^3$ mit $n' = dN/d \log r$. Die Ordinatenskala gilt für n'. Für die anderen Verteilungen ist die Ordinate mit 10^{-4}, 10^{-8} und 10^{-12} zu multiplizieren. b Größenverteilung von maritimem Aerosol in Bodennähe und die Verteilung der Produkte $n'r$, $n'r^2$ und $n'r^3$. Die Ordinatenskala gilt für n'. Für die anderen Verteilungen ist die Ordinate mit 10^{-4}, 10^{-8} und 10^{-12} zu multiplizieren. Kurve 1 gibt zum Vergleich die kontinentale Verteilung. Kurve 2 ist die Seesalzteilchenverteilung. Der Unterschied zwischen n' und 2 repräsentiert den Anteil kontinentaler Restaerosole über dem Ozean

luftoptische Effekte, wie die Wellenlängenabhängigkeit der Dunsttrübung, mehrfach bestätigt (VOLZ 1954, JUNGE 1955, GEORGII 1958, BULLRICH 1960). Er kann am gleichen Ort von erstaunlicher Konstanz sein. Man hat den Eindruck, daß durch Wetteränderungen bedingte Unregelmäßigkeiten rasch wieder ausgeglichen werden, als ob dauernd ein Prozeß wirksam ist, der stets einen Gleichgewichtszustand herzustellen bestrebt ist.

Eine Erklärung für diese Größenverteilung ist bisher nur teilweise gelungen. Sicher ist, daß die untere Grenze dadurch bedingt ist, daß Teilchen mit Radien kleiner als 0,01 μ sich in wenigen Stunden an die größeren angelagert haben. Die untere Grenzgröße hängt deshalb weitgehend von der jüngsten Lebensgeschichte der aerosoltragenden Luftmasse ab. In der Nähe von starken Quellen von Aitken-Teilchen wird man hohe Konzentrationen kleiner Teilchen erwarten dürfen, und mit zunehmender Alterung der Aerosole wird das Maximum der Verteilung zu größeren Radien hin verschoben (JUNGE 1955).

Die obere Grenze ist bedingt durch die Sedimentation der Teilchen im Schwerefeld. Es kann gezeigt werden, daß z. B. für die Meeresoberfläche bei Gleichgewicht zwischen dem durch Turbulenz aufwärtsgerichteten Teilchenstrom und dem abwärtsgerichteten Sedimentationsstrom sich eine obere Grenze der in der Atmosphäre anwesenden Seesalzteilchen von 10 bis 20 μ ergibt. Der Wert der mittleren oberen Grenze wird noch durch zwei Beobachtungen aus ganz anderen Gebieten bestätigt. Diese Grenzgröße wird z. B. sehr genau durch die Größe von Pollenkörnern von Bäumen und Sträuchern repräsentiert, die sich alle in engen Grenzen um 10 μ halten. Pollenkörner müssen für maximale Wirksamkeit bei der Bestäubung klein genug sein, um eine zeitlang in der Luft zu schweben und groß genug, um nicht in zu große Höhen und Weiten entführt zu werden. Ein anderes Beispiel ist die mittlere Größe von Lößteilchen, die ähnlichen Bedingungen wie die der Pollenkörner genügen muß.

Die Ursache für den Abfall der Größenverteilung nach dem Gesetz $dn/d \log r = c/r^3$ ist noch ungeklärt. Er resultiert, wie Fig. 1a zeigt, in einer ungefähr konstanten Massenverteilung zwischen 0,1 und 10 μ und dies ist vielleicht ein Hinweis, in welcher Richtung die Erklärung zu suchen ist. Es ist möglich, daß der Cyclus von Kondensation und Verdampfen von Wolkenteilchen, der weiter unten noch behandelt wird, und der immer mit einer Änderung der Masse des Kondensationskerns einhergeht, eine Rolle spielt. Vielleicht ist der durch die Koagulation bedingte Massenstrom von kleineren zu größeren Teilchen (FRIEDLÄNDER 1960) verantwortlich.

Für Fragen der Anlagerung radioaktiven Materials an die natürlichen Aerosole sind auch die Verteilungen der Produkte $n' \cdot r$, $n' \cdot r^2$ und $n' \cdot r^3$ von Interesse, wobei $n' \equiv dN/d \log r$ ist; sie sind für die maritimen und kontinentalen Modellverteilungen in Fig. 1b dargestellt. Es sei hier betont, daß diese Modelle zum Gebrauch für grundsätzliche Erwägungen gedacht sind, und daß bei Anwendung auf spezielle Fälle Vorsicht geboten ist, da die Abweichungen im Einzelfall merklich sein können.

Ein wichtiges Charakteristikum atmosphärischer Aerosolteilchen ist ihre meist komplexe chemische Zusammensetzung aus verschiedenen Komponenten. Man kann sie am besten als Mischteilchen oder Mischkerne bezeichnen. Viele Teilchen sind zur Zeit ihrer Entstehung recht einheitlich zusammengesetzt wie z. B. die Seesalzkerne, oder Teilchen in gewissen Industrierauchen.

Bei Verweilzeiten in der Atmosphäre von vielen Tagen bis zu mehreren Wochen werden die Teilchen jedoch mancherlei meteorologischen Prozessen unterworfen, die fast alle die Anlagerung weiterer Substanz zur Folge haben. Ein Teil dieser Prozesse hat die Vereinigung von schon vorhandenen Teilchen zur Folge, bewirkt also eine Neuverteilung der Aerosolsubstanz. Diese Prozesse sollen hier nur kurz im Hinblick auf die Zusammensetzung der neutralen Teilchen erläutert werden, aber sie sind auch für die Anlagerung radioaktiven Materials von besonderem Interesse und sollen später noch eingehender besprochen werden. Wir können folgende Zusammenstellung dieser Prozesse geben:

1. Prozesse, die zur Vereinigung von Aerosolteilchen führen.

a) Coagulation von Teilchen in trockener Luft.

b) Verstärkte Coagulation von Teilchen mit Wolken- und Nebelteilchen.

c) Anlagerung von Teilchen an Wolkentröpfchen durch den Facy-Kumai-Effekt.

d) Coagulation von Wolkenelementen.

e) Auffangen von Teilchen durch fallende Wolken- und Regentropfen.

2. Prozesse, die zur Materialvermehrung der Teilchen durch Spurengase führen.

a) Anlagerung von Material, das sich durch photochemische oder andere Reaktionen von Gasen bildet.

b) Anlagerung von Material, das sich bei Reaktionen von Spurengasen an Teilchen oder in Wolkentröpfchen bildet.

3. Prozesse, die zur Aufspaltung von Aerosolteilchen führen können.

a) Zerplatzen von großen Regentropfen.

b) Zersplittern von Eiskristallen und Schneeflocken.

c) Kristallisation und Zerspringen beim Eintrocknen von wasserlöslichen Teilchen.

Die Prozesse 1a und 1b sind der Rechnung leicht zugänglich und resultieren bevorzugt in der Anlagerung der kleinsten Teilchen an die größeren.

Im Facy-Kumai-Effekt werden Teilchen an Wolkenelemente während deren Wachstum durch Kondensation oder Sublimation angelagert. Der Diffusionsstrom des Wasserdampfes bewegt Teilchen verschiedener Größe mit annähernd gleicher Geschwindigkeit gegen wachsende Tropfen. Dieser Vorgang ist wahrscheinlich von sehr allgemeiner Bedeutung (Kumai 1951, Facy 1955). Prozeß 1d und 1e sind in wolkenreichen Klimaten und für größere Teilchen wichtig (Turner 1955), allerdings nur deshalb, weil ein großer Teil der gebildeten Wolkenelemente und zum Teil auch der Regentropfen nicht als Niederschlag zur Erde gelangt, sondern in der Atmosphäre wieder verdampft. In jedem Kreislauf von Kondensation und Verdampfung, den ein Kondensationskern durchläuft, ohne aus der Atmosphäre entfernt zu werden, wird durch diese Vorgänge seine Masse vermehrt.

Neben diesen verschiedenen mechanischen Prozessen der Vereinigung von Aerosolteilchen außer-, inner- und unterhalb von Wolken spielen verschiedene chemische Reaktionen von Spurengasen zweifellos eine Rolle. SO_2 z.B. wird mit einer Halbwertszeit von etwa 20 Tagen durch helles Sonnenlicht zu SO_3 oxidiert, aber auch in Wolkentröpfchen bei Anwesenheit von Spuren von Metalloxyden. Die Bildung von Nitrat in Seesalzkernen (Robbins u. a. 1958) und vermutlich von organischen Bestandteilen in Aerosolen sind weitere Beispiele.

Neben Prozessen 1 und 2, die alle eine Substanzvermehrung der Teilchen bewirken, haben die unter 3 aufgeführten eine Aufspaltung zur Folge. Der Vorgang 3a ist einwandfrei und oft nachgewiesen, für 3b sind starke Hinweise vorhanden, während 3c behauptet aber bisher nicht einwandfrei bewiesen werden konnte (Lodge u. a. 1954, Twomey u. a. 1955).

Das Endergebnis all dieser Vorgänge ist eine ziemlich komplexe Zusammensetzung der Teilchen über weite Gebiete der Erde. Nur dort, wo die Entstehung *einer* Teilchenart stark überwiegt, wie z.B. über See, tritt der Mischkerncharakter zurück. Infolge der verhältnismäßig langen Lebenszeiten von Aerosolen verglichen mit der mittleren Geschwindigkeit der zonalen Zirkulation sind solche Gebiete nur im großen abgrenzbar und die Überlappungen erheblich. Man findet deshalb alle Übergänge zwischen völlig löslichen und unlöslichen Teilchen, vor allem über dichtbesiedelten Kontinenten, von wo die meisten Beobachtungen vorliegen.

Dies kann gut demonstriert werden durch das Wachstum der Teilchen mit der relativen Feuchte. Das beobachtete mittlere Wachstum der Teilchen in verschiedenen Größenbereichen über besiedelten Gebieten deutet auf einen Anteil an löslicher Substanz an der Gesamtmasse von etwa 20% hin. Der Durchmesser solcher Kerne ändert sich unterhalb 60 bis 70% Feuchte kaum, wächst aber dann bis 95% um etwa 30%. Reine Seesalzkerne wachsen schneller und zeigen noch eine zusätzliche Größenänderung, wenn bei Feuchtezunahme der Kristall in Lösung geht.

Die chemische Zusammensetzung der Aerosole soll hier nur kurz gestreift werden, denn sie ist von untergeordneter Bedeutung für unser Thema und insbesondere für den Anlagerungsmechanismus von radioaktivem Material an die Aerosole, der hier im Vordergrund des Interesses steht. Die Frage wurde kürzlich an anderer Stelle ausführlich behandelt (JUNGE 1958).

Wie aus Fig. 1 b hervorgeht, ist fast die gesamte Masse des *maritimen Aerosols* auf die Teilchen mit Radien größer als $0,8\ \mu$ konzentriert, wobei diese Masse recht eng mit der Windstärke korreliert ist (WOODCOCK 1953). Unsere Kenntnisse über das maritime Aerosol basieren fast ausschließlich auf Chloridanalysen unter der Annahme, daß die Seesalzteilchen die gleiche Zusammensetzung zeigen wie das Meerwasser. Diese Annahme scheint für frischentstandene Teilchen zuzutreffen, aber es ist nicht ausgeschlossen, daß beim Verweilen in der Atmosphäre durch die weiter oben genannten Prozesse eine Verschiebung in der Zusammensetzung der Mengenverhältnisse eintritt. Die Interpretation ist deshalb schwierig, weil Chlorid und andere Bestandteile des Meereswassers auch aus kontinentalen Quellen stammen können.

Ein regelmäßiger Bestandteil der Aerosole, vor allem der *großen* Teilchen, ist Sulfat. Dies erscheint teilweise in Verbindung mit NH_3 und teilweise auch mit Ca (SUMI u. a. 1959). Der Schwefelhaushalt der Atmosphäre ist kompliziert und von besonderem geochemischen Interesse, sowohl für die kontinentalen als auch für die sehr reinen Luftmassen polarer Gebiete (JUNGE 1960). Das meiste Sulfat bildet sich in der Atmosphäre durch Oxidation von SO_2 und H_2S.

Analysen von Aerosolen in vorzugsweise industriellen aber auch in ländlichen Gebieten zeigen SiO_2, $CaCo_3$ und Eisenoxyde in merklichen Mengen, sowie erhebliche Anteile von organischen, azetonlöslichen Substanzen. Die Konzentration von Pb, F, Mn und Cu ist nur meßbar in größeren Städten (CHAMBERS u. a. 1955).

Die bisher erwähnten Angaben wurden durch direkte Analysen von Aerosolen gewonnen. Weitere, wenn auch nicht immer eindeutige Rückschlüsse können von Regenwasseranalysen gewonnen werden. Diese zeigen, daß K, Ca, Mg und auch Na über Land in Mengen vorkommen, die weit größer sind als nach der Zusammensetzung des Meerwassers zu erwarten ist. Sie sind vermutlich mineralische Bestandteile des Bodens.

Die Gesamtkonzentration der Aerosole liegt bei etwa 10 $\mu g/m^3$ über dem Meer, 50 bis 100 $\mu g/m^3$ über Kontinenten und mehrere hundert $\mu g/m^3$ über dichter besiedelten und industriellen Gebieten. Diese Werte sind natürlich örtlichen Schwankungen unterworfen.

Die Vertikalverteilung der natürlichen Aerosole in der unteren Troposphäre ist durch verschiedene Messungen belegt. Fig. 2 gibt eine Übersicht. In Einzelprofilen nimmt die Konzentration mit der Höhe zunächst meist langsam ab bis zur oberen Grenze der Durchmischungsschicht, an der dann ein scharfer Abfall einsetzt, oft verbunden mit einer Temperaturinversion. Darüber ist die Abnahme dann wieder gering und oft unregelmäßig mit der Höhe. Dies Verhalten wird deutlich, wenn man die Profile nach der Höhe der Austauschschicht normalisiert, wie es SAGALYN und FAUCHER (1954) im Falle von Großionen (geladene Aitken-Teilchen) taten (Fig. 2a). Mittelt man einfach nach der Höhe, so ergeben Messungen von verschiedenen Autoren angenähert eine exponentielle Abnahme mit der Höhe, jedenfalls schichtweise. Wie Fig. 2b zeigt, erscheint dabei die Abnahme der Aitken-Kerne und der großen

Kerne mit der Höhe ungefähr gleich stark. Dies deutet darauf hin, daß die
Größenverteilung der troposphärischen Aerosole mit der Höhe ungefähr gleich
bleibt, was durch einige wenige direkte Messungen von der Zugspitze auch bestätigt werden konnte (Junge 1958).

Oberhalb 5 km liegen nur einige Konzentrationswerte von Aitken-Kernen vor.
Die Werte von Wigand sind aus methodischen Gründen wohl zu niedrig und
Werte um 50 bis 100/cm³ dürften den mittleren Verhältnissen näher kommen.
Diese noch recht spärlichen Angaben vermitteln den Eindruck, daß die Abnahme
der Aitken-Kerne in der oberen Tropo-
sphäre nicht sehr ausgeprägt ist (Junge
u. a. 1960).

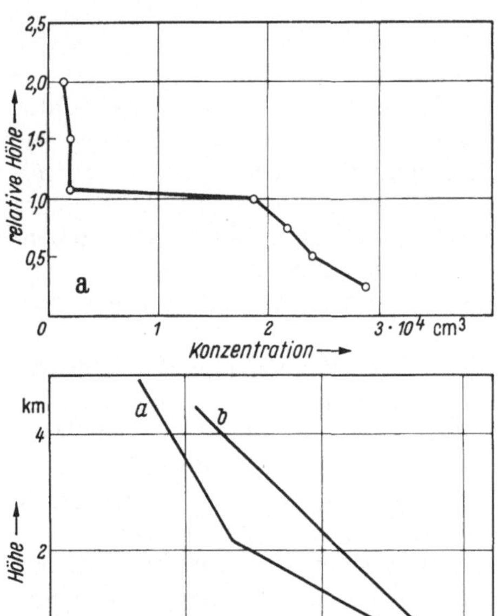

Fig. 2. a Mittelwerte von Ionenzahlen als Funktion der
Höhe nach Sagalyn und Faucher (1954). Höhenwerte
normalisiert nach der Höhe der Durchmischungsschicht,
deren Mittelwert bei 2 km lag. b Mittlere Vertikalprofile
der Konzentration von Aitken-Teilchen (a) und von großen
Teilchen (b) auf Grund von Messungen in Mitteleuropa.
Konzentration normalisiert nach den Bodenwerten. Nach
einer Zusammenstellung von Penndorf (1954)

Die vertikalen Profile stellen das
Ergebnis eines recht komplizierten Zu-
sammenspiels mehrerer Prozesse dar.
Von ausschlaggebender Bedeutung ist
sicherlich der aufwärtsgerichtete Teilchenstrom bedingt durch Turbulenz
und Vertikalgradient auf der einen
Seite und dem Auswaschen der Teilchen durch Niederschläge auf der
anderen Seite, das vorzugsweise im
Niveau um 2 bis 4 km erfolgt. In
trockener Luft nimmt die Zahl der
Aitken-Kerne zusätzlich durch Koagulation ab, wenn die Aerosole langsam
durch Mischung in die Höhe vordringen. Auch geographische Verhältnisse
mögen eine Rolle spielen, wie z.B. in
Westeuropa. Wenn hier die reinen,
an großen und Aitken-Teilchen armen,
maritimen Luftmassen auf den Kontinent übertreten, so werden in den
dicht besiedelten und industriellen
Gebieten erhebliche Aerosolmengen
injiziert, die dann allmählich durch
Mischung nach oben gelangen. Rechnungen zeigen, daß dieses Vordringen
der Aerosole als Funktion der Zeit zu exponentiellen Kurven führt (Junge 1952).
Eine eingehendere Analyse der vertikalen Profile ist schwierig wegen der mannigfachen, nur ungenügend bekannten Parameter, die in diese Rechnungen eingehen.
Ein Sedimentations-Turbulenz-Gleichgewicht kann allerdings nicht vorliegen, da
sich solches viel zu langsam einstellen würde und auch zu einer viel zu geringen
Abnahme mit der Höhe führen würde.

II. Die Anlagerung der Radon- und Thoronzerfallsprodukte
an die troposphärischen Aerosole

Die Quellen der natürlichen Radioaktivität in der Troposphäre sind bekanntlich die Radium-, Thorium- und Aktiniumemanationen und ihre Zerfallsprodukte
(s. z.B. Israël 1951), wobei die Rolle der Aktiniumreihe zu vernachlässigen ist.
Diese Edelgase treten je nach den meteorologischen und geologischen Bedingungen
in verschieden starkem Maße aus dem Boden und werden in der Atmosphäre

durch Mischung horizontal und vertikal verfrachtet. Die bei ihrem Zerfall erzeugten Folgeprodukte gehören zu den Schwermetallen und können deshalb nur in Teilchenform existieren. Die Größenverteilung und die Konzentration dieser Teilchen sind der Gegenstand dieses Abschnittes.

Die Verteilung des Radons in Bodennähe und mit der Höhe zeigt in eindeutiger Weise die Bedeutung der Kontinente als Quellgebiete. Die Radonkonzentration über Land ist um zwei Größenordnungen höher als über abgelegenen Meeresgebieten. Über Land nimmt die Radonkonzentration mit der Höhe ab, während die Vertikalverteilung in maritimen Luftmassen, wie z.B. an der Westküste der USA (WEXLER u. a. 1956) eine Zunahme mit der Höhe aufweist. Die Emanationen und ihre Folgeprodukte sind also eindeutige Tracer für kontinentale Luftmassen.

Zum Verständnis des Folgenden sind die Radon- und Thoronzerfallsreihen hier nochmals aufgeführt (Tabelle 1 und 2). Sie zeigen einmal, daß alle Folgeprodukte Schwermetalle sind und daß weiterhin mit Ausnahme der langlebigen Radonzerfallsprodukte und vielleicht noch des ThB die Halbwertzeiten kurz sind im Vergleich zu meteorologischen Vorgängen. Bekanntlich gilt im radioaktiven Gleichgewicht innerhalb einer Zerfallsreihe $C_i \lambda_i = A$, wo A die Anzahl der pro Zeiteinheit zerfallenden Atome und C_i und λ_i die Atomkonzentrationen und Zerfallskonstanten der einzelnen Glieder bezeichnen. Die Abweichungen von diesem Gleichgewicht, verursacht durch das meteorologische Geschehen, sind es, die das Studium der natürlichen Radioaktivität so anziehend machen. Solche Abweichungen treten vor allem aus zwei Gründen auf: Einmal treten die Emanationen ohne Folgeprodukte aus dem Boden aus und infolge der vertikalen Mischungsprozesse wächst der Anreicherungsgrad der Folgeprodukte von 0 am Boden bis zu 1 in der Höhe, wo Gleichgewicht erreicht wird, an. Untersuchungen dieser Art wurden von JACOBI u. a. (1959) durchgeführt. Dieser Prozeß der Anreicherung ist jedoch unabhängig von der Anwesenheit von Aerosolen und soll deshalb hier nicht weiter behandelt werden. Zum anderen werden Prozesse, die Aerosole aus der Atmosphäre entfernen, zur Störung des Gleichgewichtes führen. Da diese Prozesse der Aerosolvernichtung zu einer Halbwertszeit von etwa 20 Tagen führen, so werden praktisch nur die langlebigen Produkte des Radons beeinflußt, diese allerdings in so starkem Maße, daß die beobachteten Konzentrationen um zwei bis drei Größenordnungen kleiner sind als die nach dem Gleichgewicht zu erwartenden.

Über das Verhältnis der Thoron- zur Radonaktivität in der Atmosphäre liegen nur wenige Schätzungen vor. SCHUMANN (1956) und WILKENING (1952) finden im Mittel ein Verhältnis A_{Tn}/A_{Rn} von etwa 0,05 unter der Voraussetzung,

Tabelle 1. *Radon-Zerfallsreihe und Halbwertszeiten*

Rn^{222}	Radon	Rn	α	3,83 Tage
Po^{218}	Radium A	RaA	99,97% α	3,05 min
Pb^{214}	Radium B	RaB	β	26,8 min
Bi^{214}	Radium C	RaC	99,96% β	19,7 min
Po^{214}	Radium C'	RaC'	99,96% α	$1,5 \times 10^{-4}$ sec
Pb^{210}	Radium D	RaD	β	22 Jahre
Bi^{210}	Radium E	RaE	β	5 Tage
Po^{210}	Polonium	RaF	α	140 Tage
Pb^{206}	Radium G	RaG	stabil	

Tabelle 2. *Thoron-Zerfallsreihe und Halbwertszeiten*

Rn^{220}	Thoron	Tn	α	54,5 sec
Po^{216}	Thorium A	ThA	α	0,158 sec
Pb^{212}	Thorium B	ThB	β	10,6 Std
Bi^{212}	Thorium C	ThC	33,7% α 66,3% β	60,5 min
Tl^{208}	Thorium C''	ThC''	β	3,1 min
Pb^{208}	Thorium D	ThD	stabil	
Po^{212}	Thorium C'	ThC'	α	3×10^{-7} sec
Pb^{208}	Thorium D	ThD	stabil	

daß sich Tn mit ThB im radioaktiven Gleichgewicht befindet, was sicher nicht zutrifft. Das wahre Verhältnis der Aktivitäten ist wohl von der Größenordnung 1 (s. z.B. Beitrag Israël) in Bodennähe bei sehr rascher Abnahme mit der Höhe.

Im Augenblick der Entstehung sind die Zerfallsprodukte naturgemäß atomar dispers, aber sie verwandeln sich sehr rasch in Molekülkomplexe durch Anlagerung von Wasser, Sauerstoff und vielleicht von Spurengasen, ähnlich wie wir es von der Bildung der Kleinionen her kennen. Chamberlain und Dyson (1956) bestimmten die Diffusionskonstante dieser Primärteilchen großer Beweglichkeit im Falle des ThB zu 0,05 cm²/sec. Sie konnten zeigen, daß die Anlagerung der Aktivität in einem zylindrischen Rohr fast ausschließlich durch die Diffusion dieser noch nicht an Aerosolteilchen angelagerten Primärteilchen bedingt ist. Der zum Teil erhebliche Bruchteil der Aktivitäten, der von den Sekundärteilchen repräsentiert wird, nimmt wegen der erheblich kleineren Beweglichkeit dieser Teilchen an dem Anlagerungsprozeß kaum teil. Diese Verhältnisse sind von Bedeutung z.B. bei Retention der Aktivität im Atemtrakt.

Entsprechende Messungen an anderen Zerfallsprodukten liegen nicht vor, aber man kann ganz ähnliche Werte erwarten. Der gemessenen Diffusionskonstanten entspricht ein Teilchen Radius von etwa $5 \times 10^{-4}\,\mu$.

Die Anlagerung dieser Primärteilchen an Aerosole kann mit Hilfe der von Smoluchowsky angegebenen und experimentell im allgemeinen bestätigten Ausdrücke für die Koagulation rechnerisch erfaßt werden (s. Whytlaw-Gray und Patterson 1932). Ist n die Konzentration der sich anlagernden Primärteilchen mit dem Radius ϱ und $n_r\,dr$ die der Aerosolteilchen zwischen r und $r+dr$, so ist die Abnahme der Primärteilchen gegeben durch

$$(dn/dt)_r = 4\pi\,(D+D_r)\cdot(\varrho+r)\,n\,n_r\,dr, \tag{1}$$

wo D die Diffusionskonstante der Primärteilchen und D_r die der Aerosolteilchen mit dem Radius r ist. In unserem Falle ist $D \gg D_r$, $\varrho \ll r$ und folglich

$$(dn/dt)_r = 4\pi\,D\,r\,n\,n_r\,dr. \tag{2}$$

Die totale Anlagerungsgeschwindigkeit ist

$$dn/dt = 4\pi\,D\,n \int_{r_1}^{r_2} n_r\,r\,dr,$$

wo die Grenzen r_1 und r_2 für kontinentales Aerosol bei etwa $3 \times 10^{-3}\,\mu$ und $15\,\mu$ liegen. Die Werte für das Integral über $n_r r$ können leicht von Fig. 1 gewonnen werden und sind, zusammen mit anderen Aerosolparametern in Tabelle 3 zusammengestellt. Wird $\int_{r_1}^{r_2} n_r\,r\,dr = a$ gesetzt, so wird die mittlere Lebensdauer τ und die Halbwertszeit $\tau_{\frac{1}{2}}$ der Primärteilchen

$$\tau = 1/4\pi\,D\,a \quad \text{und} \quad \tau_{\frac{1}{2}} = \tau \times \ln z.$$

Mit den Werten der Tabelle 3 und $D = 0,05$ folgt $\tau = 25$ sec für kontinentales und 2,1 min für maritimes Aerosol.

Chamberlain und Dyson bestimmten diese Lebensdauer experimentell. Sie maßen die Aktivität A' von Radon oder Thoron in $\mu\mu C/cm^3$ sowie gleichzeitig die Konzentration n der in den Primärteilchen enthaltenen RaA- und ThB-Atome, woraus sich dann

$$\tau = n/A'\,\varphi$$

Tabelle 3. *Werte verschiedener Aerosolparameter für logarithmische Größenbereiche*

Radius μ	0,01	0,010 bis 0,032	0,32 bis 0,10	0,10 bis 0,32	1,0 bis 3,2	1,0 bis 3,2	> 3,2	Gesamt
Werte für n cm^{-3}								
Kontinental	1600	6800	5800	940	29	0,94	0,029	15169
Maritim	3	83	105	14	2	0,47	0,029	207
Stratosphäre								
etwa 12 km	2	24	9,1	0,19	0,019	0,001	—	35
etwa 16 km	0,08	3,4	1,7	0,19	0,019	0,001	—	5,3
Werte für $r \cdot n$ cm^{-2}								
Kontinental	$1,2 \times 10^{-3}$	$1,3 \times 10^{-2}$	$3,3 \times 10^{-2}$	$1,3 \times 10^{-2}$	$1,3 \times 10^{-3}$	$1,3 \times 10^{-4}$	$1,3 \times 10^{-5}$	$6,1 \times 10^{-}$
Maritim . .	$3 \ \times 10^{-6}$	$1,8 \times 10^{-4}$	$5,9 \times 10^{-4}$	$2,0 \times 10^{-4}$	$1,3 \times 10^{-4}$	$7,4 \times 10^{-5}$	$1,3 \times 10^{-5}$	$1,2 \times 10^{-}$
Stratosphäre								
etwa 12 km	$2,3 \times 10^{-6}$	$4,6 \times 10^{-5}$	$3,4 \times 10^{-5}$	$3,1 \times 10^{-6}$	$9,5 \times 10^{-7}$	$1,2 \times 10^{-7}$	—	$8,5 \times 10^{-}$
etwa 16 km	$1,7 \times 10^{-7}$	$1,4 \times 10^{-5}$	$8,7 \times 10^{-6}$	$3,1 \times 10^{-6}$	$9,5 \times 10^{-7}$	$1,2 \times 10^{-7}$	—	$2,6 \times 10^{-}$

ergibt, wo $\varphi = 2,20$ sec^{-1} die Zerfallsrate von 1 $\mu\mu$C ist. Tabelle 4 gibt ihre Ergebnisse wieder, die befriedigend mit dem theoretisch gewonnenen Wert von 25 sec übereinstimmen. Die Theorie scheint also, zum mindesten angenähert, die Verhältnisse richtig wiederzugeben. Die Tabelle zeigt auch, daß die Lebenszeit der Kleinionen von ungefähr gleicher Größe ist. Jedoch ist beim Vergleich mit

Tabelle 4. *Messungen der Lebensdauer von Primärteilchen nach Chamberlain und Dyson*

Ort		Radon oder Thoron-Konzentration, C cm^{-3}	Primärteilchen cm^{-3}	Mittlere Lebenszeit sec
Urangrube	A	0,93	1,27	37
100% relative Feuchte mit Tröpfchen	B	0,50	0,81	44
Radium verarbeitende Werkstatt		0,91	0,31	9
Thorium-Fabrik		0,53	0,17	9
Laboratorium unbenutzt		17,7	25,6	39
Laboratorium mit Bunsenflamme		14,1	7,6	15
Kleinionen über Land				50
Kleinionen über Land, aerosolreiche Luft				10

diesem Wert Vorsicht geboten, da bei der Abnahme der Kleinionen auch die Wiedervereinigung entgegengesetzt geladener eine Rolle spielt, ein Prozeß, der bei den Primärteilchen natürlich wegfällt, und überdies die Anlagerung der Kleinionen an die Aerosole noch durch ihre Ladung beeinflußt wird.

Aus den Angaben über die Anlagerungsgeschwindigkeit kann unmittelbar das Verhältnis der Sekundär- zur Primäraktivität n_S/n abgeleitet werden. Es ist

$$d n_S/dt = 4\pi D n a - n_S \lambda$$

und im Gleichgewicht

$$n_S/n = 4\pi D a/\lambda,$$

wo λ die Zerfallskonstante der betreffenden Aktivität ist. Im Falle von RaA und ThB erhalten wir für n_S/n angenähert

10   bzw. 3000 für kontinentales und

1,5 bzw. 300 für maritimes Aerosol.

Die auf RaA bzw. ThB folgenden Zerfallsprodukte sind also praktisch vollständig an das Aerosol gebunden und nur ThA existiert überwiegend als Primärteilchen. Wie weiter unten noch ausgeführt wird, ist vermutlich der überwiegende Teil der Sekundäraktivität an Teilchen um 0,07 μ herum gebunden, deren Diffusionskonstante bei 10^{-5} cm²/sec liegt. Bei Anlagerung an Oberflächen infolge Diffusion wird die Sekundäraktivität daher erst von Bedeutung, wenn das Verhältnis n_S/n von der Größenordnung $5 \times 10^{-2}/10^{-5} = 5 \times 10^3$ wird, in Übereinstimmung mit den oben erwähnten Ergebnissen von CHAMBERLAIN und DYSON.

In den bisherigen Betrachtungen über die Anlagerung wurde vorausgesetzt, daß weder Primär- noch Sekundärteilchen eine elektrische Ladung tragen. Sind die Primärteilchen ungeladen, so ist nur die Ladung der Aerosolteilchen zu berücksichtigen. Der Einfluß dieser Ladung auf den Anlagerungsprozeß ist jedoch zu vernachlässigen. Ist aber ein größerer Anteil der Primärteilchen geladen, so wird der Einfluß merklich. Es ist deshalb notwendig, dieser Frage näher nachzugehen.

Durch die Art der Entstehung der Primärteilchen könnte man vermuten, daß bei α-Zerfall das Teilchen negativ und bei β-Zerfall positiv geladen wird. Die wenigen zufälligen Beobachtungen scheinen diesem Bild nicht zu entsprechen. Es ist möglich, daß die Ionisationswolke hoher Konzentration, in deren Nähe sich das Teilchen direkt nach dem Zerfall befindet, eine rasche Änderung der Teilchenladung bedingt. Wir werden deshalb für unsere Abschätzung zunächst den Extremfall annehmen, daß alle Primärteilchen geladen sind.

Im Aerosol ist bekanntlich ein Teil der Kerne einfach positiv oder negativ geladen, wobei dieser Bruchteil bei Teilchen um 0,1 μ herum etwa $\frac{1}{3}$ ist und mit dem Radius abnimmt. Mehrfachladungen der Teilchen werden erst häufiger bei Radien größer als 0,1 μ und können hier unberücksichtigt bleiben. Die Gesamtzahl der Teilchen mit dem Radius r ist dann

$$n_r = n_{ru} + 2n_{rg}, \qquad (3)$$

wo n_{ru} die Konzentration der ungeladenen und n_{rg} die Konzentration der positiv oder negativ geladenen Teilchen ist, die mit guter Näherung in der Atmosphäre gleichgesetzt werden können. Im Ionisationsgleichgewicht, das in der Atmosphäre stets sehr angenähert erfüllt ist, gilt weiter

$$n_{rg}\,\eta_g = n_{ru}\,\eta, \qquad (4)$$

wo η der Anlagerungskoeffizient zwischen den Kleinionen und den ungeladenen Aerosolteilchen und η_g derjenige zwischen entgegengesetzt geladenen Kleinionen und Aerosolteilchen ist.

Leider sind noch keine befriedigenden Angaben über die Abhängigkeit von η und η_g vom Teilchenradius vorhanden. Die physikalisch bestfundierte Theorie wurde jüngst von KEEFE u. a. (1959) gegeben, jedoch scheint sie vorerst unterhalb von $10^{-2}\,\mu$ zu versagen. Die Ausdrücke anderer Autoren weichen untereinander und von den wenigen vorhandenen Messungen um Beträge ungefähr gleicher Größe ab, die aber unsere Abschätzung kaum ernsthaft in Frage stellen können. Wir benutzen aus diesem Grunde die Beziehung

$$\eta = 4\pi D r \quad \text{und} \quad \eta_g = 4\pi\,(Dr + we) \quad \text{(JUNGE 1955)},$$

wo w die Beweglichkeit der geladenen Primärteilchen im elektrischen Feld und e die Einheitsladung ist. Diese für Kleinionen gültigen Beziehungen wenden wir nun auch für die geladenen Primärteilchen an, wobei wir auch die entsprechenden Zahlenwerte für D und w übernehmen.

Wir haben dann für die Anlagerungsgeschwindigkeit an die ungeladenen Aerosolteilchen

$$(dn_g/dt)_{ru} = \eta\, n_g\, n_{ru}\, dr$$

und für die Anlagerungsgeschwindigkeit an die entgegengesetzt geladenen Aerosolteilchen

$$(dn_g/dt)_{rg} = \eta_g\, n_g\, n_{rg}\, dr,$$

während die Anlagerungsgeschwindigkeit an die gleichgeladenen Aerosolteilchen zu vernachlässigen ist. Mit Hilfe von (4) erhalten wir dann für die gesamte Anlagerungsgeschwindigkeit der geladenen Primärteilchen

$$(dn_g/dt)_r = (dn_g/dt)_{ru} + (dn_g/dt)_{rg} = 2\eta\, n_g\, n_{ru}\, dr.$$

Im Vergleich dazu hatten wir nach (2) für die Anlagerungsgeschwindigkeit der ungeladenen Primärteilchen

$$(dn/dt)_r = \eta\, n\, n_r\, dr.$$

Für den Vergleich setzen wir $n = n_g$ und erhalten mit (3) für das Verhältnis der Anlagerungsgeschwindigkeiten

$$V = (dn_g/dt)_r / (dn/dt)_r = \frac{2(Dr + we)}{3Dr + we}.$$

Mit $D = 0{,}05$ cm²/sec; $w = 1{,}34$ cm²/V sec $= 400$ cm²/ESTV·sec, und $e = 4{,}77 \times 10^{-10}$ ESTL folgen die Werte in Tabelle 5. Bei Integration über den wichtigen

Tabelle 5. *Verhältnis der Anlagerungsgeschwindigkeiten für geladene und ungeladene Primärteilchen*

Radius μ	3×10^{-3}	10^{-2}	3×10^{-2}	10^{-1}	10^{0}	10
Verhältnis V	1,74	1,4	1,06	0,82	0,68	0,67

Größenbereich von 10^{-2} bis $1\ \mu$ kann danach der Fehler von a und τ bei Vernachlässigung der Ladung der Primärteilchen nicht mehr als etwa 30% betragen, fällt also bei dem Stande unseres Wissens auf diesem Gebiet zur Zeit kaum ins Gewicht.

Von besonderem Interesse bei den Sekundärteilchen ist die Verteilung der Aktivität mit der Teilchengröße. Diese ist unmittelbar durch Gl. (2) gegeben. Darin war $n_r = dN/dr$; wird, wie in Fig. 1 usw. die Größe $n_r' = dN/d \log r$ gewählt, so folgt:

$$(dn/dt)_r = 4\pi\, Dn\, r\, n_r\, dr = 4\pi\, Dn\, r\, n_r'\, d \log r.$$

Ist $A_r = dA/d \log r$ die Verteilung der Aktivität auf die Teilchengröße und λ die Zerfallskonstante, so ist im Gleichgewicht

$$A_r = 4\pi\, Dn\, r\, n_r'/\lambda.$$

Der Verlauf von A_r wird also durch die Werte von $r \cdot n_r'$ in Fig. 1 bzw. in Tabelle 3 wiedergegeben. Danach ist zu erwarten, daß für kontinentales bzw. maritimes Aerosol

54 bzw. 48% der Aktivität zwischen 0,032 und 0,1 μ

und

97 bzw. 80% der Aktivität zwischen 0,01 und 0,32 μ

liegt. Der Unterschied zwischen kontinentalem und maritimem Aerosol in dieser Hinsicht ist demnach nicht groß. In jedem Fall trägt die Gesamtheit der Riesenteilchen praktisch keine Aktivität. Die Verteilung der Aktivität auf das *einzelne Teilchen* geht jedoch wie A_r/n'_r, ist also dem Radius proportional, so daß das *einzelne Riesenteilchen* die höchste Aktivität tragen müßte.

Es erscheint unerwartet, daß die Aktivität proportional $r \cdot n'_r$ ist und nicht proportional der Oberfläche, $r^2 \cdot n'_r$. Dies kommt dadurch zustande, daß die Coagulationstheorie von Smoluchowsky auf der Annahme eines Diffusionsfeldes um das Aerosolteilchen herum beruht, dessen Gradient an der Oberfläche bekanntlich umgekehrt proportional dem Radius ist, so daß die Anlagerungsgeschwindigkeit statt mit r^2 nur mit r wächst. Wird die Teilchengröße jedoch klein im Vergleich zur freien Weglänge der Primärteilchen, so wird diese Voraussetzung vermutlich nicht mehr zutreffen, und die Anlagerung mehr nach $r^2 n'_r$ verlaufen. Man kann also für die kleineren Teilchen eine Verschiebung der Anlagerung im Sinne der $r^2 n'_r$-Kurve in Fig. 1 erwarten, wodurch der Schwerpunkt der angelagerten Aktivität zu größeren Teilchen hin verlagert wird.

Ein Versuch, die Theorie der Anlagerung in diesem Sinne zu erweitern, liegt neuerdings von Lassen (1960) vor. Er berechnet den Diffusionsstrom von Primärteilchen zum Aerosolteilchen unter der Bedingung, daß dieser an der Teilchenoberfläche nie größer werden kann als das Produkt aus Teilchenoberfläche und gaskinetischer Bewegungskomponente der Primärteilchen gegen diese. Diese Bedingung ist bei der Smoluchowskischen Lösung nicht erfüllt, wodurch der Strom der Primärteilchen bei kleinen Aerosolradien zu groß wird. Auf diese Weise erhält Lassen einen Ausdruck, der für Teilchen unterhalb $0,01\ \mu$ tatsächlich der Oberfläche der Aerosolteilchen und für Radien oberhalb $1\ \mu$ ihrem Radius proportional ist. Zur experimentellen Prüfung dieser Überlegungen stellt er zwei homogene Aerosole (Dioctylphtalat bzw. Dow Polystyrene Latex) her, deren Radien zwischen 0,044 und $0,586\ \mu$ schwanken und die nach Mischung Thoriumemanation ausgesetzt werden. Mit einem polydispersen Wachsaerosol wurden dann die Experimente auf den Größenbereich $R = 0,7$ bis $5,0\ \mu$ ausgedehnt (1961). Das Verhältnis des pro Teilchen angelagerten ThB der beiden Aerosolgrößen stimmt in allen Fällen befriedigend überein mit der Theorie. Diese wurde auf den Fall elektrisch geladene Aerosolpartikel erweitert (Lassen 1961, II).

Der Frage der Anlagerung natürlicher Radioaktivität an Aerosole als Funktion der Teilchengröße sind Wilkening (1952) und neuerdings Jacobi u. a. (1959) nachgegangen. Wilkening benutzte elektrostalische Abscheider, die alle Teilchen kleiner als $0,015\ \mu$ abschieden, aber deren Abscheidewirksamkeit für größere Radien rasch abnahm.

Wenn er zwei solche Abscheider in Serie schaltete, so enthielt der zweite Abscheider nur sehr wenig Aktivität, und eine ähnliche Erfahrung machte er mit Glaswollfilter. Mit einem Impaktor, dessen untere Abscheidegrenze bei $0,25\ \mu$ lag, erhielt er andererseits nur etwa 10% der Aktivität im elektrischen Abscheider. Er schloß daraus, daß praktisch alle natürliche Aktivität am Teilchen mit Radien kleiner als $0,015\ \mu$ angelagert ist. Dies Ergebnis kann nicht ohne Bedenken angenommen werden. Wie Fig. 1 zeigt, sollte man bei einem kontinentalen Aerosol erwarten, daß die Aktivität

für Radien $\qquad \leqq 0,15\ \mu = 6\%$

für Radien zwischen 0,015 und $0,25\ \mu = 90\%$

für Radien $\qquad \geqq 0,25\ \mu = 4\%$

beträgt. Selbst wenn unser Größenverteilungsmodell von den Bedingungen, unter denen WILKENING maß, abweichen sollte, so bleibt der Tatbestand, daß er in seinen Versuchen einen wesentlichen Teil der Größenverteilung, nämlich den zwischen 0,015 und 0,25 μ, gar nicht erfaßt hat. Es ist überdies nicht unmöglich, daß der Anteil der Teilchen unterhalb 0,015 μ durch die Anlagerung der aktiven Primärteilchen im vorderen Teil des Zylinders erheblich überschätzt wurde in der Weise, wie es weiter oben an Hand der Versuche von CHAMBERLAIN und DYSON erläutert wurde.

Versuche, die eine indirekte Aussage über die Größenabhängigkeit der Anlagerung natürlicher Radioaktivität an Aerosole zulassen, wurden von CHAMBERLAIN u. a. (1957) durchgeführt. Sie bestimmten den mittleren Diffusionskoeffizienten D_1 von künstlichen Aerosolen mittels der dynamischen Methode. Hierbei wird die aerosolhaltige Luft durch ein System enger Kanäle (Diffusionsbatterie) gesaugt und aus dem Verlust der Teilchenzahl kann dann D_1 ermittelt werden. Die Aerosole wurden entweder durch Glühen eines Heizdrahtes oder mit der Bunsenflamme erzeugt. Gleichzeitig wurde der Luft Thoron beigemischt und eine genügende Zeit gewartet, bis die gebildeten Zerfallsprodukte weitgehend in Form von ThB an die Aerosolteilchen angelagert waren. Es wurde dann bei gleichen Bedingungen wie oben der Verlust des ThB in der Diffusionsbatterie gemessen und daraus die mittlere Diffusionskonstante D_2 der radioaktiven Trägeraerosole bestimmt. Das Verhältnis D_1/D_2 in den Versuchsreihen schwankte zwischen 2,5 und 4,2, d. h. als radioaktive Träger fungieren bevorzugt die größeren Teilchen. Wir können die entsprechenden Werte auf Grund der Tabelle 3 leicht berechnen nach

$$D_1 = \frac{\Sigma n D_r}{\Sigma n} \quad \text{und} \quad D_2 = \frac{\Sigma n r D_r}{\Sigma n r}.$$

Das Verhältnis D_1/D_2 hängt natürlich von der Breite der Verteilungskurve ab und ergibt sich für kontinentales Aerosol zu 3,1. Die Größenverteilungen der von CHAMBERLAIN u. a. benutzten Aerosole ist nicht bekannt und die Zahlenwerte für D_1/D_2 sind deshalb nicht direkt vergleichbar. Aber es ist offenbar, daß die Richtung und Größenordnung des Effektes von den theoretischen Betrachtungen richtig wiedergegeben wird.

JACOBI, SCHRAUB u. a. (1959) waren die ersten, die diesen Betrachtungen die Größenverteilungen atmosphärischer Aerosole zugrunde legten und entsprechend ihre Untersuchungen planten. Sie maßen die RaC- bzw. RaC'-Aktivitäten auf Filtern, die in Serie angeordnet waren und verglichen die so erhaltenen Filterwirkungsgrade der natürlichen Aktivität mit den unter gleichen Bedingungen erhaltenen Eichkurven der Filter für verschiedene Teilchengrößen. Diese Eichkurve zeigte ein breites Minimum von etwa 50% zwischen 0,015 und 0,5 μ Radius und wesentlich höhere Abscheidung jenseits dieser Werte. Der gemessene Wirkungsgrad für die kurzlebigen Aktivitäten lag im Mittel um 60%. Dieser Wirkungsgrad blieb ziemlich unverändert, auch wenn mehrere Filter hintereinander geschaltet wurden. Sie schlossen daraus, daß die Abscheidung in den Filtern keine merkliche Verschiebung der Größenverteilung zur Folge haben konnte, und daß deshalb der Bereich des Filterminimums sich ziemlich gut mit dem Bereich maximaler Aktivität decken mußte. Sie konnten zusätzlich durch das Vorschalten eines Impaktors zeigen, daß etwa die halbe Aktivität oberhalb 0,1 μ Radius liegt. Die hier angewandte Filtermethode ließ in dem vorliegenden Falle offenbar deshalb so klare Schlußfolgerungen zu, weil der Bereich maximaler Aktivität sich weitgehend mit dem Bereich minimaler Filterwirksamkeit deckte, was sich durch glückliche Wahl der Versuchsbedingungen ergab.

JACOBI u. a. maßen in Freiluft, in geschlossenen Räumen und in Radiumstollen und fanden ähnliche Ergebnisse. Dies würde mit der Erfahrung übereinstimmen, daß die Größenverteilung des Aerosols im allgemeinen keinen allzu großen Schwankungen unterworfen ist.

Das wesentliche Ergebnis dieser Untersuchungen ist, daß die meiste natürliche Aktivität zwischen 0,015 und 0,5 µ liegt mit einem Maximum bei 0,1 µ in ungefährer Übereinstimmung mit den theoretischen Betrachtungen. Im Hinblick auf die methodischen Mängel der Untersuchungen WILKENINGs und bevor nicht weitere Ergebnisse zu dieser Frage vorliegen, ist es deshalb naheliegend,

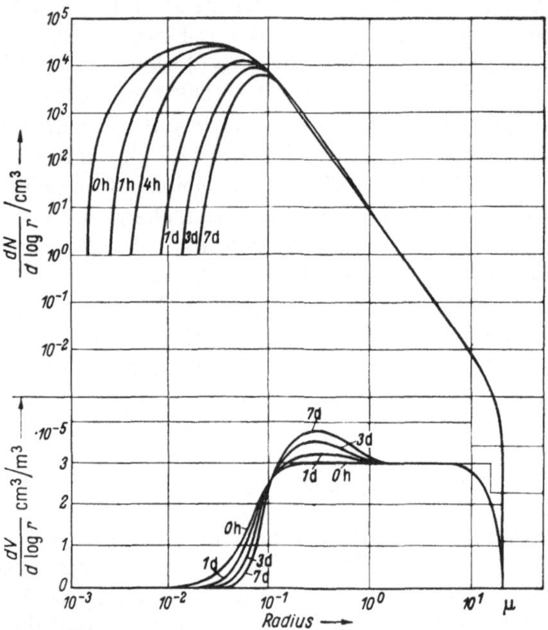

Fig. 3. Berechnete Änderung in der Größenverteilung natürlicher Aerosole über dem Kontinent infolge Coagulation. Oben Teilchenkonzentration, unten Volumenkonzentration. Zeiten in Stunden und Tagen

zunächst das von der Theorie gegebene Bild zu akzeptieren.

Die vorstehenden Betrachtungen über die Größenverteilung gelten selbstverständlich nur für die kurzlebigen Aktivitäten. Für Aktivitäten mit Halbwertszeiten von mehr als etwa einen Tag wird diese Verteilung durch die auf S. 173 aufgeführten Prozesse gestört. Der Koagulationsprozeß (1a) ist der Rechnung verhältnismäßig leicht zugänglich. Quantitative Angaben für atmosphärische Aerosole sind in Fig. 3 dargestellt (JUNGE 1955a); ähnliche Rechnungen wurden von ZEBEL (1958) durchgeführt. Man erkennt, daß dabei Material aus dem Bereich von 0,01 bis 0,1 µ in den Bereich von 0,1 bis 1,0 µ verlagert wird. Bei Aerosolen, die reich an kleinen AitkenTeilchen sind, macht sich diese Verschiebung schon im Verlauf von Stunden bemerkbar, andernfalls sind Tage nötig. Die Riesenkerne werden durch diesen Prozeß kaum beeinflußt.

Dieser letztere Größenbereich wird vor allem durch die Prozesse 1b bis 1e beeinflußt, die alle durch Wolkenbildung ausgelöst werden, und die wir hier etwas näher erläutern wollen.

Beobachtungen und Rechnungen auf dem Gebiet der Tropfenbildung an atmosphärischen Kondensationskernen zeigen, daß die größten Teilchen zuerst aktiviert werden (s. z. B. MORDY 1959). Da in frisch gebildeten Wolken Tropfenkonzentrationen von einigen hundert pro cm³ beobachtet werden, bedeutet dies nach Tabelle 3, daß z. B. in kontinentalen Aerosolen die Teilchen mit Radien unter 0,2 µ als *inaktive* Kerne in der Wolke verbleiben. Diese inaktiven Kerne tragen aber einen erheblichen Anteil der Aktivität und diese wird nun teilweise durch die Prozesse 1b und 1c an *die* Tropfen gelangen, die sich an *Riesenkerne* gebildet hatten. Unterstützt wird dieser Prozeß der Verschiebung der Aktivitäten zu größeren Teilchen hin noch dadurch, daß alle unter 1 angeführten Vorgänge zu einer Vergrößerung der Teilchenmasse führen.

Diese Prozesse sind bisher noch nicht in ihrer gegenseitigen Bedeutung abgeschätzt worden und es ist deshalb hier nicht möglich, quantitative Angaben zu machen. Sie erhalten ihre Bedeutung für unser Problem durch die Tatsache, daß der überwiegende Teil der Wolken wieder verdampft, und die meisten Teilchen den Kreislauf der Tropfenbildung und -verdampfung mehrfach durchlaufen, bevor sie ausgewaschen werden.

Die eben behandelten Prozesse werden gleichsinnig unterstützt durch die unter 2 aufgeführten (S. 174). Ihnen allen ist gemein, daß sie langsam sind, sich erst im Laufe von Tagen auswirken und stark von den Witterungs- und Klimaverhältnissen abhängig sind. Sie alle wirken im Sinne einer Verlagerung der Aktivität zu größeren Teilchen. Man sollte deshalb erwarten, daß RaD bis RaF in stärkerem Maße an größere Teilchen gebunden ist als die kurzlebigen Produkte. Angaben darüber scheinen nicht vorzuliegen.

Die gleichen Prozesse, die zu einer Änderung des Aerosol- und Aktivitätsspektrums führen, entfernen auch stets einen Bruchteil der Aerosole aus der Atmosphäre. Radium D und F wird deshalb in viel geringeren Konzentrationen vorkommen als nach dem radioaktiven Gleichgewicht zu erwarten ist. Das Verhältnis von kurz- zu langlebigen Aktivitäten kann deshalb benutzt werden, um Aussagen über die mittlere Lebensdauer von natürlichen Aerosolen zu erhalten. Betrachtungen und Messungen dieser Art liegen von mehreren Seiten vor (BLIFFORD u. a. 1952, HAXEL und SCHUMANN 1955, SCHUMANN 1956, LEHMANN und SITTKUS 1959).

Es sei C_R die Konzentration von Radon in Atome/cm³, C_1, C_2 usw. diejenige ihrer Folgeprodukte und λ_R, λ_1, λ_2 usw. die entsprechenden Zerfallskonstanten. Nehmen wir ferner an, daß die Geschwindigkeit, mit der die Aerosole aus der Atmosphäre entfernt werden, ihrer Konzentration proportional ist, und daß die Proportionalitätskonstante λ_a zeitlich und räumlich konstant ist, dann gilt:

$$dC_1/dt = C_R \lambda_R - C_1(\lambda_1 + \lambda_a)$$
$$dC_i/dt = C_{i-1} \lambda_{i-1} - C_i(\lambda_i + \lambda_a) \quad i = 2, 3, \ldots .$$

Bei zeitlich und räumlich konstantem C_R erhalten wir für das Gleichgewicht

$$C_i = C_R \lambda_R \cdot \frac{\lambda_1}{\lambda_1 + \lambda_a} \frac{\lambda_2}{\lambda_2 + \lambda_a} \cdots \frac{\lambda_{i-1}}{\lambda_{i-1} + \lambda_a} \cdot \frac{1}{\lambda_i + \lambda_a} .$$

Für die kurzlebigen Produkte ist $\lambda_i \gg \lambda_a$ und die Beziehung vereinfacht sich zu dem bekannten Ausdruck für radioaktives Gleichgewicht in einer Zerfallsreihe

$$C_i \lambda_i = C_R \lambda_R .$$

Für RaD mit $\lambda_{\mathrm{RaD}} \ll \lambda_a$ folgt

$$C_{\mathrm{RaD}} = C_R \lambda_R / \lambda_a ,$$

ein Wert, der um $\lambda_{\mathrm{RaD}}/\lambda_a \cong 10^{-3}$ kleiner ist als die radioaktive Gleichgewichtskonzentration, und

$$C_{\mathrm{RaE}} = \frac{C_R \lambda_R}{\lambda_a} \cdot \frac{\lambda_{\mathrm{RaD}}}{\lambda_{\mathrm{RaE}} + \lambda_a} ,$$

ein Wert, der um etwa $0{,}5 \times 10^{-3}$ kleiner ist als die radioaktive Gleichgewichtskonzentration. Die Beziehung

$$\lambda_a = C_R \lambda_R / C_{\mathrm{RaD}} = C_{\mathrm{RaB}} \lambda_{\mathrm{RaB}} / C_{\mathrm{RaD}}$$

kann zur Bestimmung von λ_a benutzt werden, indem man mittels Filterproben C_{RaB} und C_{RaD} bestimmt. Die auf Filtern mit der Sammelzeit anwachsenden

Aktivitäten der einzelnen Zerfallsprodukte lassen sich berechnen, werden aber durch ziemlich komplizierte Ausdrücke wiedergegeben. Wegen der geringen Konzentration von RaD müssen große Luftvolumina über längere Zeiten gefiltert werden, wodurch sich gleichzeitig die Berechnungsweise vereinfacht. Ist nämlich die Abscheidezeit auf dem Filter lang gegenüber der Einstellzeit des Gleichgewichtes der kurzlebigen Produkte auf dem Filter, so muß gelten

$$A_{\mathrm{RaB}} = \Phi\, C_{\mathrm{RaB}},$$

wo A_{RaB} die Aktivität auf dem Filter, und Φ die Fördermenge durch das Filter bedeutet. Aus meßtechnischen Gründen bestimmt man die kurzlebige Gesamtbetaaktivität

$$A_K = A_{\mathrm{RaC}} + A_{\mathrm{RaB}} = \Phi\, C_{\mathrm{RaB}}\left(1 + \frac{\lambda_{\mathrm{RaB}}}{\lambda_{\mathrm{RaC}}}\right) = 2{,}36\,\Phi\, C_{\mathrm{RaB}}.$$

Da auf der anderen Seite die Halbwertszeit des RaD lang ist gegenüber der Abscheidezeit T, so gilt mit großer Näherung für die langlebige Betaaktivität auf dem Filter

$$A_l = \Phi \cdot T \cdot C_{\mathrm{RaD}} \cdot \lambda_{\mathrm{RaD}}.$$

Damit wird

$$\lambda_a = \lambda_{\mathrm{RaB}} \cdot \lambda_{\mathrm{RaD}} \cdot T \cdot A_K / 2{,}36\, A_l.$$

Die Bestimmung der langlebigen Aktivitäten geschah bei den einzelnen Autoren in verschiedener Weise. Blifford u. a. trennten das RaD chemisch, während Haxel und Schumann die langlebigen Aktivitäten durch Beobachtung der Zerfallskurven vom Hintergrund der verhältnismäßig zahlreichen künstlichen Spaltprodukten zu trennen versuchten. In ähnlicher Weise haben Lehmann und Sittkus Aerosolverweilzeiten aus dem Verhältnis der RaD- zur RaF-Aktivität ermittelt. Die Resultate sind in Tabelle 6 zusammengestellt. Wir werden diese Werte beim Vergleich mit anderen Resultaten noch näher diskutieren. Hier sei nur bemerkt, daß abgesehen von Meßungenauigkeiten, die nicht unerheblich sind,

Tabelle 6. *Aerosolverweilzeiten in der Troposphäre nach verschiedenen Methoden und Autoren*

Beobachter	Ort	Methode	Verweilzeiten in Tagen
Blifford u. a. (1952)	Washington D.C. Französisch Marokko Alaska	Verhältnis von kurzlebigen zu langlebigen Radonzerfallsprodukten	15 ⎫ sehr wahrscheinlich zu hoch 18 ⎭ 4
Haxel und Schumann (1955)	Heidelberg	desgleichen	6
Lehmann und Sittkus (1959)	Freiburg i. Br.	desgleichen Verhältnis RaF/RaD in Luftproben Verhältnis RaF/RaD im Niederschlag	0,8 bis 3,0 8 bis 36 33
Goel u. a. (1959)	Indien	Vergleich von gemessener und berechneter Be⁷-Produktion	30
Lal (1959)	Indien	Verhältnis Be⁷/P³² im Niederschlag	40
Stewart u. a. (1955)	Nord-Atlantik	Spaltprodukte von Nevada-Testen	32

die oben gemachten Voraussetzungen hinsichtlich λ_a und C_R in der Atmosphäre keineswegs erfüllt sind. Diese Voraussetzungen waren

1. Vernichtung des Aerosols proportional der Aerosolkonzentration,

2. λ_a und C_R konstant in Raum und Zeit.

Besonders der letzte Punkt ist keineswegs in Einklang mit den Tatsachen. Die Radonkonzentration am Boden schwankt, wie wir sahen, wie 100:1 zwischen Kontinent und Ozean. Während eines Zeitraumes von der Größenordnung der Aerosollebensdauer können daher die Radonkonzentrationen in Luftmassen ohne weiteres solche Schwankungen durchlaufen. Aber auch λ_a wird keineswegs konstant sein. Da die Vernichtung der Aerosole vornehmlich durch Wolken und Niederschlag erfolgt, wird λ_a in der unteren Troposphäre vermutlich größer sein als in den oberen, bzw. in regnerischen Klimaten größer als in trockenen. Auf Grund dieser Erwägungen glauben wir, daß die Bestimmung der Lebensdauer der Aerosole mittels der Radon-Zerfallsprodukte grundsätzlich nicht viel besser als innerhalb eines Faktors von 3 sein kann.

Messungen des Gehaltes an langlebigen Radon-Zerfallsprodukten in Regen von KING u. a. (1956) bestätigen indirekt die langen Lebensdauern von Aerosolen. Tabelle 7 zeigt die RaE-Konzentration im Regen in verschiedenen Orten. Danach ist auf der Nordhalbkugel der Unterschied zwischen Kontinent und Ozean nicht größer als ein Faktor von 2 bis 3, ganz im Gegensatz zum Radongehalt. Dies bringt unmittelbar zum Ausdruck, daß mit wachsender Lebenszeit der Luftbeimengungen ihre Konzentrationsunterschiede zwischen Quell- und Senkgebieten abnehmen. Erst am Äquator und besonders in südlichen Breiten sinkt der RaE-Gehalt tiefer ab, weil der Meridionalaustausch geringer ist als der Zonalaustausch und weil auf der Südhalbkugel infolge geringerer Landbedeckung weniger Radon in die Atmosphäre gelangt.

Tabelle 7. *RaE-Konzentrationen in Regenwasser nach King u. a. (1956)*

Ort	dpm/Liter
Alaska . .	1,4
Illinois . .	4,6
Hawai . .	2,0
Phillipinen	2,0
Panama .	0,7
Samoa . .	0,4

Abschließend sei hier auf neuere Untersuchungen von BURTON und STEWART (1960) über den Gehalt der Stratosphäre an RaD und dessen Folgeprodukte eingegangen. Sie finden über England eine Zunahme der RaD-Konzentration vom Boden bis zur Tropopause um einen Faktor von 4 und einen weiteren recht steilen Anstieg um einen Faktor 2 in der unteren Stratosphäre von 12 bis 14 km Höhe, der dem bei künstlicher Radioaktivität beobachteten ähnlich ist. Sie finden auch, daß das Verhältnis RaF/RaD in der Stratosphäre größer ist als in der Troposphäre und auf ein Isotopenlebensalter von 6 Monaten hindeutet.

Nimmt man an, daß der Radongehalt der oberen Troposphäre horizontal konstant über die ganze Erde ist, dann käme auf *ein* Radonatom bei *radioaktivem* Zerfallgleichgewicht $\lambda_R/\lambda_{RaD} = 2100$ Atome RaD, jedoch in dem hier herrschenden *Aerosolgleichgewicht* $\lambda_R/\lambda_a = 5,5$ Atome RaD, wenn wir die Aerosolhalbwertszeit in der Troposphäre zu 20 Tagen annehmen. Diese letztere Zahl wird den mittleren Verhältnissen in der Troposphäre wohl annähernd gerecht. Gelangt diese troposphärische Luft in die Stratosphäre, so kann günstigstenfalls die RaD-Konzentration um einen Faktor $(5,5+1)/5,5 = 1,2$ anwachsen. Dabei ist es gleich auf welche Weise der Luftaustausch Stratosphäre—Troposphäre erfolgt. Diese einfache Abschätzung steht im Widerspruch zu den zitierten Beobachtungen. Eine Zunahme des RaD-Gehaltes in der Stratosphäre, wie sie BURTON und STEWART fanden, kann nur auf zweierlei Weise erklärt werden:

1. Anreicherung des RaD-Gehaltes durch meteorisches Material. Abschätzungen von WASSON (1960) zeigen, daß dieser Beitrag vernachlässigbar klein ist.

2. Vordringen von Luft in die Stratosphäre, die im Mittel wesentlich reicher an Radon ist als die hochtroposphärische Luft (s. z.B. Machta und Lucas 1962).

Wir wissen, daß der Radongehalt in *kontinentaler* Luft wesentlich höher ist als in maritimen Luftmassen. Punkt 2 würde dann bedeuten, daß vorzugsweise troposphärische Luft *kontinentalen Ursprungs* in die Stratosphäre gelangt. Ein solcher Prozeß könnte z.B. durch hochreichende Quellbewölkung bedingt sein, die, wie Beobachtungen zeigen, merklich in die untere Stratosphäre vordringen kann. Sollten sich die Messungen von Burton und Stewart bestätigen, so würden sie einen neuen Aspekt in die Betrachtungen über den Luftmassenaustausch Stratosphäre—Troposphäre hineintragen, nämlich die Bedeutung der Injektion von bevorzugt kontinentaler Luft in die Stratosphäre, sei es durch hochreichende Konvektion oder durch großräumige monsulane Zirkulationen. Eine Zunahme der RaD-Konzentration in der Stratosphäre und ein Isotopenalter von etwa 6 Monaten können jedoch kaum durch ein allgemeines stratosphärisches Zirkulationsmodell erklärt werden wie z.B. das von Brewer und Dobson mit generellem Vordringen von troposphärischer Luft in die Stratosphäre über dem Äquator.

Kürzlich gab Machta (1960) neue Daten über die Vertikalverteilung der RaD-Konzentration in der Atmosphäre über den Vereinigten Staaten. Im Gegensatz zu den englischen Messungen zeigen seine Werte einschließlich die der unteren Stratosphäre keine wesentliche Abhängigkeit mit der Höhe. Sollten sich in künftigen Untersuchungen diese Unterschiede bestätigen, so würde eine erhebliche Abhängigkeit der RaD-Profile von der geographischen Länge folgen, und unsere Bemerkungen über Unterschiede zwischen Land und Meer stützen.

B. Die durch kosmische Strahlung erzeugten Radioaktivitäten und ihre Anlagerung an stratosphärische Aerosole

Neben den Emanationen und ihren Folgeprodukten bilden die durch kosmische Strahlung erzeugten radioaktiven Isotope eine zweite, interessante Gruppe natürlicher Aktivitäten. Die Bildung dieser Isotope erfolgt vornehmlich in der unteren Stratosphäre und, vor allem in den Tropen, in der oberen Troposphäre durch Spaltung von Stickstoff-, Sauerstoff- und Argonatomen.

Tabelle 8 gibt nach Lal u. a. (1958) die bisher nachgewiesenen Isotope. Von diesen bleibt T in Form von THO als Wasserdampf und zu einem kleinen Teil als TH als Gas in der Atmosphäre und fällt daher nicht in den Rahmen unserer Diskussion. Das gleiche gilt für C^{14}, das als Kohlendioxyd vorliegt. Die restlichen Isotope können nur als Teilchen existieren und wurden im Regenwasser nachgewiesen. Die letzten vier haben Halbwertszeiten, die sie für das Studium meteorologischer Vorgänge geeignet machen.

Der Grund, weshalb diese durch kosmische Strahlung erzeugten Isotope für die Meteorologie von besonderem Wert erscheinen, ist ein zweifacher: Zunächst wird das Verhältnis der Isotope im Niederschlag das gleiche sein wie in der Luft zu der Zeit, wenn die Kondensation stattfindet. Dies folgt aus der Annahme, daß die Anlagerung der Primärteilchen an die Sekundärteilchen unabhängig

Tabelle 8. *Durch kosmische Strahlung erzeugte radioaktive Elemente nach Lal u. a. (1958)*

Element	Halbwertszeit
Be^{10}	$2{,}7 \times 10^{6}$ Jahre
C^{14}	$5{,}7 \times 10^{3}$ Jahre
H^{3}	12,5 Jahre
Na^{22}	2,6 Jahre
S^{35}	87 Tage
Be^{7}	53 Tage
P^{33}	25 Tage
P^{32}	14,3 Tage

von der chemischen Beschaffenheit der Primärteilchen erfolgt, eine Annahme, die in hohem Grade wahrscheinlich ist. Zum anderen variiert die Intensität der kosmischen Strahlung erheblich innerhalb der Atmosphäre und das Verhältnis von radioaktiven Elementen verschiedener Halbwertszeit wird deshalb eine Funktion der Lebensgeschichte der Luftmassen sein. In den bisherigen Untersuchungen wurden allerdings zunächst hauptsächlich die Grundlagen der vertikalen Verteilung und der Produktionsraten dieser Isotope studiert und die Anwendung auf die Meteorologie beschränkte sich auf Bestimmungen der Lebensdauer der troposphärischen Aerosole.

Ähnlich wie im Falle der Emanationen werden die Isotope zunächst als Atome vorliegen, die sich dann rasch zu Primärteilchen umwandeln, von wahrscheinlich gleicher Größe wie bei ThB. Für die Bildung der Sekundärteilchen ist die Kenntnis der stratosphärischen Aerosole notwendig, die wir hier kurz zusammenfassen:

Die Untersuchung der stratosphärischen Aerosole scheint auf die Existenz von drei Teilchenarten verschiedenen Ursprungs hinzudeuten (JUNGE u. a. 1960) Teilchen unter etwa 0,1 μ Radius zeigen eine Konzentration um 50 bis 100/cm³ in Tropopausenhöhe, die dann oberhalb der Tropopause mit der Höhe rasch abnimmt. Fig. 4 gibt ein Beispiel für ein solches Vertikalprofil, das mit einem

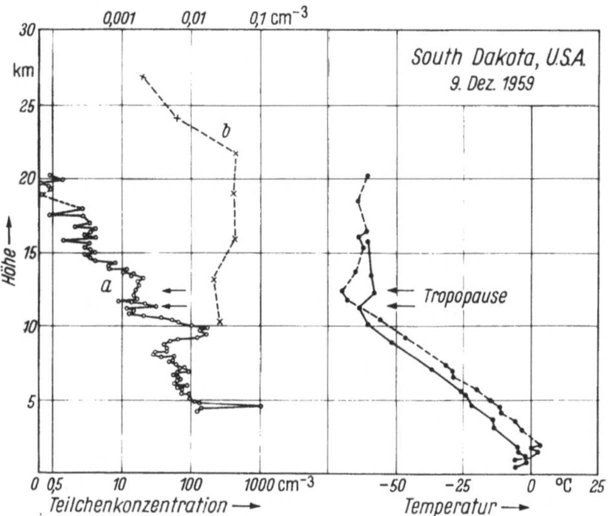

Fig. 4. Beispiel eines Vertikalprofils der Aitken-Teilchen (a) und der großen Teilchen (b) in der Stratosphäre, Süd Dakota, USA. Temperaturangaben nach gleichzeitigen Radiosondeaufstiegen

automatischen Aitken-Kernzähler gewonnen wurde. Diese Vertikalverteilung macht es sehr wahrscheinlich, daß Teilchen in diesem Größenbereich im wesentlichen troposphärischen Ursprungs sind und durch Austauschprozesse in die Stratosphäre gelangen.

Die zweite Teilchenart, die den Bereich von etwa 0,1 μ bis zu 2 μ Radius einnimmt, zeigt in der vertikalen Verteilung ein breites Maximum um etwa 20 km (Fig. 4). Diese Teilchen sind sehr hygroskopisch und enthalten beträchtliche Mengen von Schwefel, wahrscheinlich in Form von Sulfat, aber wenig oder kein Eisen, Calcium und Silicium. Die Vertikalverteilung macht es unwahrscheinlich, daß diese Teilchen von der Troposphäre herauftransportiert werden, und die chemische Zusammensetzung deutet nicht auf extraterrestrischen Ursprung. Sie entstehen möglicherweise in der Stratosphäre durch Oxydation von troposphärischen Spurengasen wie H_2S oder SO_2, die durch Mischung in die Stratosphäre gelangen. Diese Spurengase mit Konzentrationen in der Größenordnung von 1 bis 10 μg/m³ sind in der Troposphäre ein regelmäßiger Bestandteil. Die untere Grenzgröße dieser Teilchen scheint sich um 0,1 μ herum mit der oberen Grenze der troposphärischen Aitken-Teilchen weitgehend zu überlappen. Ihre obere Grenze bei etwa 2 μ erscheint jedoch ziemlich scharf ausgeprägt (JUNGE und MANSON 1960). Oberhalb dieser Grenze sinkt die Konzentration rasch um mehrere Größenordnungen ab, und es ist zu vermuten, daß größere Teilchen extraterrestrischen

Ursprungs sind in Zeiten, die nicht durch Atombomben oder Vulkanausbrüche gestört sind. Wegen ihrer sehr geringen Konzentration liegen aber noch keine sicheren Angaben von dieser dritten Teilchenart vor.

Fig. 5 zeigt, ähnlich wie Fig. 1a und b, eine etwas schematische Größenverteilungskurve der stratosphärischen Aerosole. Der Verlauf der Kurve unterhalb 0,1 µ ist geschätzt, bekannt ist in diesem Größenbereich nur die Gesamtzahl der Teilchen und angenähert die Lage des Maximums. Infolge der raschen Abnahme dieser Aitken-Teilchen mit der Höhe wurden zwei Schätzungen für etwa 12 und etwa 17 km Höhe gegeben. Die Konzentration der zweiten Teilchenart kann für diese Betrachtungen als konstant angesehen werden. Tabelle 3 gibt folgende Werte für a und damit für τ:

$$\text{etwa 12 km Höhe} \quad a = 0{,}8 \times 10^{-4}$$
$$\tau = 1{,}5 \text{ Std,}$$
$$\text{etwa 16 km Höhe} \quad a = 0{,}3 \times 10^{-4}$$
$$\tau = 5 \text{ Std.}$$

Da der Diffusionskoeffizient D der Primärteilchen angenähert proportional der freien Weglänge der Luftmoleküle ist, wurde für D der Wert $4 \times 0{,}05$ $0{,}2$ cm²/sec als für die mittlere Produktionshöhe der Isotope gültig angenommen. Die Lebenszeiten der Primärteilchen in der unteren Stratosphäre, und wahrscheinlich ähnlich in der oberen Troposphäre, wo leider keine Aerosoldaten vorliegen, sind also erheblich länger als in Bodennähe. Dies mag für meßtechnische Betrachtungen von Bedeutung sein. Mit diesen Werten erhalten wir folgende Konzentrationsverhältnisse von Sekundär- zu Primärteilchen bei Annahme von radioaktivem Gleichgewicht und vernachlässigbaren Aerosolverlusten:

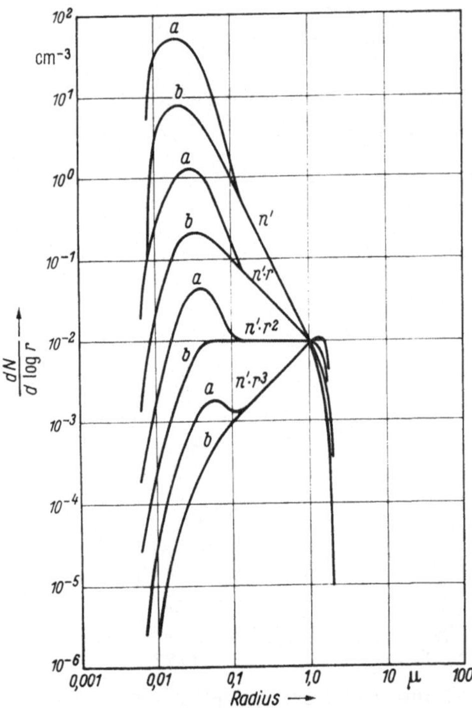

Fig. 5. Größenverteilung stratosphärischer Aerosole in gemäßigten Breiten und die Verteilung der Produkte $n'r$, $n'r^2$ und $n'r^3$. Kurven a und b gelten für 12 und 17 km. Die Ordinatenskala gilt für n'. Für die anderen Verteilungen ist die Ordinate mit 10^{-4}, 10^{-8} und 10^{-12} zu multiplizieren

Isotop:	S³⁵	Be⁷	P³³	P³²
n_S/n_P 12 km	58	35	17	9
n_S/n_P 16 km	17	10	5	3

Mit Aerosollebenszeiten in der Stratosphäre von etwa einem Jahr ist die Annahme verschwindender Aerosolverluste eine gute Annäherung. Der größte Teil der Aktivitäten liegt als Sekundärteilchen vor, aber merkliche Anteile, vor allem der Phosphorisotope, bestehen als Primärteilchen. Die Bedeutung der natürlichen Aerosole als Träger der Radioaktivität ist also auch für die Stratosphäre gegeben.

Die Abschätzungen über den Einfluß der elektrischen Ladung der Aerosole auf den Anlagerungsprozeß von S. 181 gelten unverändert für die Stratosphäre, da im Ausdruck für V das Verhältnis w/D weitgehend vom Luftdruck unabhängig ist.

Die Verteilung der Aktivitäten auf die Teilchengröße wird wiederum durch die Kurve $n' \times r$ in Fig. 5 dargestellt. Die meiste Aktivität sollte danach um $0{,}03\ \mu$ herum konzentriert sein. Da die Größenverteilungskurven unterhalb $0{,}1\ \mu$ aber nur geschätzt sind, ist diese Angabe unsicher. Beobachtungen hierüber, wie über andere Parameter der natürlichen radioaktiven Aerosole in der Stratosphäre, liegen bisher nicht vor.

Von den betrachteten radioaktiven Elementen liegen zur Zeit nur Meßreihen ihrer Konzentration im Regenwasser vor, meist aus Indien (LAL u. a. 1958, LAL u. a. 1959, GOEL u. a. 1959). Mit Hilfe der Niederschlagsmengen lassen sich daraus die pro Zeiteinheit aus der Atmosphäre ausgeschiedenen Mengen berechnen. Beim Vergleich dieser Werte mit der berechneten Produktion durch kosmische Strahlung ist folgendes zu bedenken: Im Falle des Be^7, S^{35}, P^{32} und P^{33} ist die Lebensdauer dieser Isotope kurz im Verhältnis zur Aufenthaltsdauer von Luftmassen in der Stratosphäre, die von der Größenordnung 1 bis 3 Jahre ist. Im Regenwasser kann also praktisch nur der in der Troposphäre erzeugte Anteil dieser Isotope erwartet werden. Außerdem muß in den Regenproben die Abnahme der Aktivität berücksichtigt werden, die während der mittleren Lebensdauer der troposphärischen Aerosole erfolgt. Werden diese Gesichtspunkte berücksichtigt, so finden GOEL u. a. gute Übereinstimmung der gemessenen und berechneten Produktion für Be^7 und P^{32}, wenn die Aerosollebensdauer zu 30 Tagen angenommen wird. Die Genauigkeit dieser Abschätzung ist allerdings nicht sehr groß, vor allem weil im wesentlichen nur Werte aus Indien vorliegen, die dann mit Hilfe mittlerer Niederschlagswerte auf den entsprechenden Breitengürtel umgerechnet werden. Für S^{35} ergeben die Regenmessungen Werte, die um einen Faktor von 3 bis 14 höher liegen als die berechneten, was vermutlich durch zusätzliche Produktion durch Atombombenversuche erklärt werden kann. Unstimmigkeiten beim P^{33} haben vielleicht methodische Gründe.

Genauer ist die Berechnung der Aerosollebensdauer auf Grund des Konzentrationsverhältnisses Be^7/P^{32}. Im Gegensatz zu den Absolutwerten der Konzentration schwankt dieses Verhältnis nur wenig. Sein unterer Grenzwert muß dem Verhältnis der Produktionsraten entsprechen. Infolge des rascheren Zerfalls des P^{32} steigt es dann mit wachsendem Alter der Luftmassen bis auf einen Faktor 4 an. Nimmt man an, daß das im Regenwasser gemessene Verhältnis dem der Luftmassen zur Zeit des Regens entspricht, und daß jeder Regen die Aktivität völlig entfernt, so ergibt sich als mittleres Lebensalter der Aerosole 40 Tage. Die angegebene Genauigkeitsgrenze von ± 5 Tagen ist wohl etwas größer.

Bisher wurden die durch kosmische Strahlung erzeugten Isotope nur im Regenwasser gemessen. Es ist zu erwarten und zu erhoffen, daß künftige Untersuchungen sich auch mit der Konzentration in der Luft selbst als Funktion der Höhe und der geographischen Breite befassen und dadurch interessante Einblicke in die Zirkulation der Atmosphäre ermöglichen.

C. Die Aerosole der künstlichen Radioaktivitäten

Die Aerosole der künstlichen Radioaktivitäten entstehen in grundsätzlich anderer Weise als die der natürlichen Aktivitäten. Wenn wir zunächst die wichtigste Gruppe ins Auge fassen, nämlich die durch Atombombenversuche erzeugte, so läßt sich allgemein folgendes sagen: Im Feuerball entstehen, je nach den Versuchsbedingungen, eine Reihe verschiedenartiger Teilchen, deren Größe einen weiten Bereich von $0{,}01$ bis zu $1000\ \mu$ umfaßt. Die größten dieser Teilchen, mit Radien oberhalb etwa $10\ \mu$, fallen durch Sedimentation ziemlich rasch, d.h. also nahe dem Versuchsort aus. Eine zweite Gruppe von Teilchen, mit Radien zwischen etwa 0,5 und 10 μ, werden durch die Luftströmungen bereits über weite Teile der Erdoberfläche verfrachtet. Diese mittelfristigen Teilchen zeigen noch merkliche individuelle Aktivität und wurden kürzlich unter dem Namen „heiße Teilchen" viel diskutiert. Ihr Ausfall ist noch weitgehend durch Gravition bestimmt, aber auch Turbulenz, Konvektion und Niederschlag werden schon eine merkliche Rolle bei ihrer Entfernung aus der Atmosphäre spielen. Die dritte Gruppe von Teilchen mit Radien kleiner als 0,5 μ werden in ihrer Ausbreitung fast vollständig von den Luftbewegungen beherrscht, besonders wenn sie in die Stratosphäre eindringen. Sie sind die Träger des langfristigen, weltweiten radioaktiven Ausfalls und ihre Entfernung aus der Atmosphäre geschieht vorwiegend durch Niederschlag. Wir werden im folgenden die Diskussion dieser Einteilung entsprechend vornehmen.

I. Die Teilchen im Nahausfall von Atomversuchen

Die Struktur, Zusammensetzung und Größe dieser Teilchen ist vor allem durch die Arbeiten von Adams u. Mitarb. (s. Adams u. a. 1958) bekannt geworden. Diese recht detaillierten Ergebnisse sollen hier nur in großen Zügen behandelt werden, besonders soweit sie Hinweise auf die mittel- und langfristigen Teilchen vermitteln können. Es ergibt sich das folgende Bild: Wird ein Atombombenkörper in genügender Höhe über der Erdoberfläche zur Detonation gebracht, so enthält der Feuerball im wesentlichen verdünnte Gase von Eisen und anderen Konstruktionsmetallen, während der mengenmäßige Anteil der radioaktiven Spaltprodukte klein ist. Wie Stewart (1956) gezeigt hat, entstehen bei der Abkühlung infolge Kondensation und Coagulation Teilchen, die kleiner als 1 μ sind, aber einen merklichen Größenbereich überdecken. Eisendampf in Anwesenheit von Sauerstoff kondensiert als Eisenoxyd aus, da dessen Eisendampfpartialdruck merklich geringer als der von Eisen selbst ist. Der Temperaturbereich, in dem die Kondensation stattfindet, wird in wenigen Sekunden bis zu Bruchteilen von 1 min durchlaufen. Die anwesenden Aktivitäten haben wesentlich geringeren Partialdruck und werden daher später zur Kondensation gelangen als das Eisen, aber nach genügender Abkühlung werden alle Spaltprodukte an die vorwiegend aus Eisenoxyden bestehenden Teilchen angelagert sein. Es sind Anzeichen dafür vorhanden, daß später entstandene Teilchen reicher an solchen Spaltprodukten sind, deren Dampfdrucke gering sind oder die selbst erst durch Zerfall aus gasförmigen oder leicht verdampfbaren Isotopen entstehen. Zum Beispiel entsteht Ba^{140} aus dem gasförmigen Xe^{140} (16 sec Halbwertszeit) und dem Cs^{140} (66 sec Halbwertszeit) und Sr^{89} entsteht aus dem gasförmigen Kr^{89} (3,16 min Halbwertszeit) und Rb^{89} (15,4 min Halbwertszeit).

Im Falle von Luftdetonationen werden also im wesentlichen langfristige und zu einem kleinen Grade mittelfristige Teilchen erzeugt. Diese verhältnismäßig einfachen Entstehungsbedingungen werden bei Testen an der Erdoberfläche oder in deren Nähe, z. B. auf einem Turm, dadurch gestört, daß erhebliche Materialmengen der Erdoberfläche und des Turmes in den Feuerball einbezogen werden. Je nach der Menge des einbezogenen Materials ändert sich Teilchengröße und -art. Ist die Bombe von mittlerer Größe und wird sie von einem Turm zur Explosion gebracht, so wird neben dem Bombenmaterial noch ein Teil des Turmes verdampft. Ein anderer Teil des Turmstahls, sowie das einbezogene Bodenmaterial wird im wesentlichen nur geschmolzen und erscheint als Tröpfchen bis zu 1 mm. Diese bestehen entweder aus Eisenoxyd, meist Magnetit Fe_3O_4, der stabilen Form bei hohen Temperaturen und Sauerstoffüberschuß, aus CaO bei kalkhaltigem

Bodenmaterial (Korallenriffe) oder aus glasartigem Material im Falle von silikatreichem Bodenmaterial. Die Anwesenheit dieser Tröpfchen im Feuerball bewirkt, daß eine erhebliche Menge der radioaktiven Elemente sich an diese anlagern, entweder durch Sekundärkondensation oder durch Vereinigung mit den wesentlich kleineren direkten Kondensationsprodukten. Da die Tropfen während dieses Anlagerungsvorganges überwiegend flüssig sind, verteilt sich das Eisen und die Aktivität ziemlich homogen über ihren ganzen Querschnitt.

Werden die Bomben größer oder direkt an der Erdoberfläche ausgelöst, so gelangen wesentlich größere Mengen von Bodenmaterial in den Feuerball und bei einer größeren Ausdehnung des Feuerballs wird die Verteilung des Materials und der Temperaturen uneinheitlicher. Ein Teil des Materials wird dann nur noch oberflächlich oder überhaupt nicht mehr geschmolzen. Neben Tropfen entstehen unregelmäßige, nur oberflächlich veränderte Teilchen. Bei diesen bildet das angelagerte oder ankondensierte Eisen eine mehr oder minder radioaktive Kruste. Je nach Untergrund findet man unregelmäßige, entweder wieder aus Calciumoxyd oder Glas bestehende Teilchen.

All diese Teilchen mit entsprechenden Übergängen und Kombinationen wurden tatsächlich gefunden und die größeren unter ihnen durch Anfertigung von Dünnschliffen und Radioautogrammen eingehend untersucht. Bei den teilweise oder ganz aus Calciumoxyd bestehenden Teilchen zeigten sich oft Sprünge, die durch Adern von neu gebildeten, relativ inaktivem Material ausgefüllt waren. Diese Strukturänderungen entstanden durch nachträgliche Aufnahme von Wasser und Kohlensäure und die damit verbundene Volumvergrößerung.

Über die wichtige Frage der Verteilung der Radioaktivität mit der Teilchengröße für den Nahausfall liegen Untersuchungen von ANDERSON (1958) und CHAN (1959) vor. Danach scheint z.B. für Nevada-Böden und für Oberflächenexplosionen die ursprünglich im Boden vorhandene Teilchengrößenverteilung von erheblicher Bedeutung zu sein. Die Aktivitätsverteilung der Teilchen ist ihrer Massenverteilung ähnlich, allerdings mit einer verstärkten Bevorzugung der kleineren Teilchen. Dies deutet darauf hin, daß für den Anlagerungsmechanismus der Aktivität die Teilchenoberfläche von gewissem Einfluß ist.

Die von CHAN untersuchten Teilchen stammen von Pazifischen Inseln. Er unterscheidet drei Teilchenklassen, kugelförmige, unregelmäßige und dendritische. Für eine gewisse Größenklasse ergibt sich eine erhebliche Streuung der Aktivitäten über 1 bis 2 Größenordnungen. Die Teilchengrößen variieren zwischen 5 und 300 μ Radius, und die meiste Aktivität entfällt auf die Größenklassen 50 bis 150 μ. Die Verteilung der Aktivität innerhalb der Größenklassen entspricht einer Normalverteilung.

Die mittlere Aktivität variiert in folgender Weise mit der Teilchengröße, ausgedrückt durch den Exponenten des Teilchenradius

alle Teilchen: 2,4

kugelförmige Teilchen: 3,7

unregelmäßige Teilchen: 1,7 bis 2,2

dendritische Teilchen: 2,1.

Danach scheint bei den beiden letzten Klassen die Oberfläche, bei den kugelförmigen eher das Volumen der bestimmende Faktor zu sein. Wahrscheinlich sind die Bedingungen im Einzelfall sehr variabel, so daß allgemeingültige Angaben kaum zu machen sind.

II. Die Teilchen des mittelfristigen Ausfalls

Die wesentlich kleineren Teilchen des mittelfristigen Ausfalls sind wohl überwiegend durch direkte Kondensation und Coagulation im Feuerball entstanden. Neben Resten von Bodenmaterial werden die Konstruktionsmetalle bzw. ihre Oxyde den Hauptanteil in ihrer Zusammensetzung ausmachen. Teilchen dieser Art, kenntlich durch ihre hohe Aktivität, werden weit entfernt vom Entstehungsort gefunden. Aron und Gross (1957) berichteten zuerst über ein solches Teilchen, das sie in Rio de Janeiro etwa 4 Wochen nach einem Atomversuch auf den Christmas-Inseln fanden und das einen Radius von 2 μ hatte. Fallzeiten solcher Teilchen aus Höhen von 20 bis 30 km sind von dieser Größenordnung. Seitdem sind solche Teilchen an vielen Stellen gefunden und untersucht worden, vor allem in Europa. Einzelergebnisse von Bodenbeobachtungen wurden kürzlich in geschlossener Form als Verhandlungen eines Kolloquiums veröffentlicht (Strahlenschutz 12).

Das Interesse an diesen Teilchen relativ hoher Aktivität ist durch ihre möglichen biologischen Auswirkungen bzw. durch methodische Fragen bei der Überwachung der Luft gegeben. Zu gewissen Zeiten war die auf Filtern abgefangene Aktivität bis zu 50% in einem oder wenigen Teilchen konzentriert. Eine solche Verteilung der Aktivität hat natürlich starke Schwankungen der gemessenen Luftaktivitäten mit Ort und Zeit zur Folge und stellt unter Umständen den Sinn von Mittelwerten in Frage. Die Teilchennatur wurde direkt durch Autoradiogramm von Filtern und indirekt dadurch nachgewiesen, daß erhebliche Anteile von Aktivitäten auf kleinsten Stellen der Filter oder in kleinen Bruchteilen von Staubproben oder Niederschlagsrückständen auftraten.

Abgesehen von vereinzelten früheren Beobachtungen traten diese „heißen" Teilchen gehäuft im Frühsommer 1958 und im Spätherbst und Winter 1958/59 auf und konnten Testserien im Pazifik und Sibirien zugeordnet werden. Die höchsten Aktivitäten dieser Teilchen liegen nach verschiedenen Beobachtern zwischen 10^{-10} und 10^{-8} Curie. Ihre Konzentration betrug zeitweise 1 Teilchen auf 100 m^3. Auswertungen der Zahl dieser Teilchen als Funktion der Aktivität von Schumann (Strahlenschutz 12, S. 106—110) lassen einen raschen Anstieg unterhalb 10^{-11} Curie erkennen, der bis zur unteren Meßgrenze von 10^{-13} Curie anhält. Die wenigen sehr „heißen" Teilchen stellen also nur das obere Ende einer breiten Größenverteilung dar. Dadurch erklärt sich, daß Korrelationen der Zahl dieser Teilchen mit der Gesamtaktivität der Luft um so besser werden, je niedriger die untere Grenze dieser Teilchen gesetzt wird, und daß sie ganz verschwinden, wenn nur die wenigen hochaktiven Teilchen erfaßt werden.

Angaben über die Größe dieser Teilchen sind sehr spärlich, da eine Identifizierung solch kleiner Teilchen sehr schwierig ist, wenn nicht ganz spezifische Methoden entwickelt werden. Die aktivsten Teilchen hatten Radien von 0,5 bis 2 μ. Schedling und Müller (Strahlenschutz 12, S. 60—71) finden in einem Einzelfall auch ein Agglomerat von 8 μ Radius, das in mehrere Einzelteilchen von 0,5 μ zerlegt werden konnte. Abschätzungen zeigen, daß die von der Aktivität repräsentierte Substanzmenge wesentlich kleinere Größen ergeben würden; es müssen also noch erhebliche Mengen inaktiven Materials anwesend sein. Vornehmlich mittels γ-Spektrogramm konnte eindeutig nachgewiesen werden, daß diese Teilchen relativ frische Spaltprodukte darstellen, mit Ce^{141}, Ce^{144}, Zr^{95} und Nb^{95} als charakteristische Komponenten. May und Schneider (Strahlenschutz 12, S. 72—83, und 1959) und andere Beobachter finden in vielen einzeln analysierten Teilchen eine relative Anreicherung von Zr^{95} und Nb^{95} gegenüber dem Untergrund auf Filtern und in Regenwasserrückständen, wo auch gleichzeitig Ba^{140}

und andere Spaltprodukte kürzere Lebensdauer nachgewiesen wurden, die in den Teilchen völlig fehlten. Es ist nicht klar, ob diese Anreicherung durch den Entstehungsvorgang bedingt ist oder durch nachträgliche Einwirkung bei Kondensationsvorgängen in der Troposphäre.

Eine sehr eingehende Studie über die Teilchen in Wolken frischer Spaltprodukte liegt von SISEFSKY (1960) vor. Die Proben wurden auf Filtern in 13 km Höhe über Schweden gewonnen, also in der unteren Stratosphäre. Mittels einer genialen Methode (SISEFSKY 1959) wurde für eine große Zahl von Teilchen dieser Proben sowohl die β- und γ-Aktivität als auch die Größe bestimmt. Die Aktivität wurde aus der Größe der Höfe mittels geeichter Beziehungen ermittelt. Die kleinsten noch vermessenen Höfe lagen bei 10 μ. Autoradiographien zur Identifizierung der Teilchen wurden auf die Weise gemacht, daß die Teilchen in eine Kernemulsion eingebettet wurden, die durch Umkehrentwicklung durchsichtige Höfe um die aktiven Teilchen ergab.

Untersucht wurden vor allem zwei Proben vom 25. 2. 58 und 30. 9. 58, die beide Testen in Nordsibirien zuzuordnen sind und außergewöhnlich hohe Aktivität zeigten. Der zeitliche Abfall der Aktivität von einzelnen Teilchen erfolgte nach t^{-K} mit einem mittlerem $K = 1{,}15$ in Übereinstimmung mit der Gesamtheit der Spaltprodukte.

Das wichtigste und interessante Ergebnis ist eine sehr enge Beziehung der Aktivität der einzelnen Teilchen zur dritten Potenz des Radius. Die Streubreite der Aktivität für eine feste Größe ist etwa ein Faktor von 10, der angenähert konstant bleibt über dem ganzen erfaßten Größenbereich von 0,1 bis zu 2,5 μ Radius. Werden die Mittelkurven dieser Punktwolken bestimmt, so ergibt sich die Beziehung

$$\text{Aktivität} = K \cdot R^3 \text{ (Curie)}.$$

Wenn R in μ angegeben wird, hat K für vier untersuchte Proben die Werte 3, 1, 2,5 und 1×10^{-12}. Die Aktivitäten und Teilchengrößen stimmen größenordnungsmäßig mit den oben angeführten Bodenbeobachtungen überein.

Die meisten Teilchen oberhalb 1 μ, für die noch Einzelheiten im Mikroskop erkannt werden konnten, waren kugelförmig oder angenähert so. Der Rest war elliptisch oder polygonal, was auf Kristallstruktur hindeutet. Die Oberfläche war faltig oder körnig und trug manchmal kleinere Teilchen. Die Farben variierten von farblos über gelb bis orange-braun. Die farblosen Teilchen zeigten durchweg eine höhere Aktivität. Die gefundene Proportionalität zwischen Aktivität und Teilchenvolumen ist in Übereinstimmung mit der Vorstellung der Teilchenbildung im Feuerball durch Kondensation und Coagulation. Leider wurden keine Angaben über die Häufigkeit der Teilchen gemacht, so daß Aussagen über die Verteilung der Aktivität mit der Teilchengröße nicht möglich sind.

III. Teilchen des langfristigen Ausfalls

Der Größenbereich der Teilchen in SISEVSKYs Untersuchung greift schon hinüber in denjenigen, den man auf Grund der Fallgeschwindigkeiten den langfristigen Teilchen zuschreiben muß, obwohl eine scharfe Grenze hier kaum zu ziehen ist. Über die Größe der Teilchen unterhalb 0,5 μ liegen nur wenige Daten vor (STERN 1959, JONES 1960). Die Angaben wurden mit Hochballonen auf zwei verschiedene Weisen gewonnen. Einmal wurden zwei Filter gleichzeitig unter solchen Durchflußbedingungen exponiert, daß die Abscheidung der Teilchen an den Filterfasern im Trägheitsbereich stattfand, wodurch die Abscheidewirksamkeit eine bekannte Funktion der Teilchengrößen wird. Bei dem einen Filter war sichergestellt, daß die Abscheidung nahe 100% war. Damit war die mittlere Abscheidewirksamkeit des anderen Filters bestimmt und es konnte der mittlere Radius der Teilchen angenähert bestimmt werden. Im zweiten Fall wurden die Abscheidungen mittels großer Impaktoren vorgenommen mit günstig gewählten Abscheidegrenzen und einem abschließenden Filter. Die Genauigkeit der Daten ist nicht besonders groß. Die Angaben erstrecken sich im wesentlichen über das Jahr 1959. In der ersten Hälfte dieses Jahres wurden die Daten mit der Filtermethode gewonnen und ergaben Radien zwischen 0,02 und 0,05 μ. In der zweiten Hälfte wurden vornehmlich Impaktormessungen gemacht. Dabei zeigte die mittlere Teilchengröße einen erheblichen Anstieg bis auf 0,2 μ Radius. Nach Bestimmungen des Isotopenverhältnisses von Sr^{89} und Sr^{90} u. a. muß geschlossen werden, daß im ersten Halbjahr 1959 vornehmlich Material von Sibirischen Testen zum Ausfall kam, während in der zweiten Hälfte solches von den vorjährigen Pazifischen Testen vorherrschte, was also älter war. Nach dem Alter der Spaltprodukte sollte man also eher den umgekehrten Gang der Teilchengröße erwarten. Es läßt sich noch nicht entscheiden, ob methodische Fehler mitspielen, und weitere Ergebnisse müssen abgewartet werden.

Teilchen mit einer Dichte von 2 g/cm³ und einem Radius von 0,5 μ würden etwa $\frac{1}{2}$ Jahr benötigen, um aus der Stratosphäre auszufallen. Dies dürfte etwa die obere Grenze der langfristigen Spaltproduktteilchen darstellen. Eine untere Grenze ist dadurch gegeben, daß sehr kleine Teilchen sich verhältnismäßig rasch an die oben beschriebenen natürlichen Aerosolen anlagern. Mit Hilfe der jetzt vorliegenden Angaben über die Größenverteilung und Konzentration dieser Teilchen können die Koagulationsgeschwindigkeiten berechnet werden. Es ergibt sich, daß Teilchen unter 0,01 μ Halbwertszeiten haben, die merklich kürzer sind als die Aufenthaltszeiten stratosphärischer Luftmassen, so daß dies die ungefähre untere Teilchengrenze darstellen dürfte, wenn nicht gerade frische Explosionswolken angetroffen werden.

Aus der Troposphäre liegen einige Angaben über die Verteilung der künstlichen Radioaktivität mit der Teilchengröße vor, die in diesem Zusammenhang von Interesse sind. Kalkstein u. a. (1959) schieden mittels eines zweistufigen Impaktors, gestützt durch einen Milliporfilter, während des Zeitraumes Juli 1958 bis Februar 1959 in Bodennähe die Spaltprodukte ab. Auf diese Weise wurde der gesamte Größenbereich der Teilchen in der folgenden Weise aufgeteilt:

Stufe 1   Teilchen $>$0,9 μ Radius,

Stufe 2   Teilchen zwischen 0,1 und 0,9 μ Radius,

Stufe 3   Teilchen $<$0,1 μ Radius.

Die Teilchengrenzen gelten für eine Dichte 2 g/cm³. Die Sammelzeit der einzelnen Proben betrug etwa 1 Woche und gemessen wurde die gesamte langlebige β-Aktivität.

Das Ergebnis war eine mittlere relative Verteilung der β-Aktivität von 57% (1. Stufe), 35% (2. Stufe) und 8% (Milliporfilter). Diese Verteilung hat große Ähnlichkeit mit der mittleren Massenverteilung der natürlichen Aerosole (s. Fig. 1) und würde bedeuten, daß das Verhältnis der Aktivität zur neutralen Aerosolmasse in erster Näherung konstant ist. Die Abweichungen der einzelnen Wochenwerte von diesem Mittelwert waren gering, trotz merklicher Schwankungen der Gesamtaktivität.

Rosinski und Stockham (1960) trennten den Größenbereich oberhalb 0,5 μ Radius in sieben Stufen mittels eines Andersen-Impaktors auf, während der gesamte Bereich unterhalb 0,5 μ von einem Filter abgefangen wurde. Die Messungen wurden in Chicago gemacht. Es wurden von vier Proben mit je etwa drei

Monaten Sammelzeit die Gesamt-β-Aktivität und der Sr^{90}-Gehalt bestimmt. Da die Unterteilung oberhalb 1 μ im Hinblick auf den Gesamtgrößenbereich der natürlichen Aerosole zu fein ist, fassen wir alle Werte oberhalb 1,0 μ in eine Gruppe zusammen und erhalten dann folgende Mittelwerte der relativen Verteilung:

Teilchengröße	Gesamt-β	Sr^{90}
1 μ	19%	30%
0,5 bis 1 μ	25%	24%
0,5 μ	56%	46%

Die Größenbereiche sind nicht genau mit den obigen von KALKSTEIN u. a. benutzten vergleichbar, aber es ist deutlich, daß die Aktivität zu kleineren Teilchen hin verschoben ist. Es kann nicht entschieden werden, ob methodische Unterschiede oder örtliche Aerosol-Verhältnisse (Großstadt gegen offenes Land) diese Abweichungen bedingen. Es ist jedoch offensichtlich, daß in beiden Meßreihen merkliche Anteile der Aktivität an Teilchen oberhalb 0,5 μ Radius gebunden sind, was an sich nicht unmittelbar erwartet werden sollte, da die langfristigen Aktivitäten in der Stratosphäre auf den Größenbereich 0,02 bis 0,3 μ beschränkt sind. Wenn auch im Herbst und Winter 1958/59 zeitweise merkliche Mengen mittelfristiger Teilchen auftraten, so dürfte deren Anteil doch kaum für diesen Effekt verantwortlich sein. Wir müssen also schließen, daß auf dem Weg von der Stratosphäre bis in Bodennähe eine erhebliche Verschiebung in der Größenverteilung der Spaltprodukte zu großen Teilchen hin eintritt. Wir sind davon überzeugt, daß die auf S. 173 und 184 ff. diskutierten Prozesse für diesen Effekt verantwortlich sind, und daß ein weiteres Studium dieser Verschiebung noch interessante Einblicke in die Dynamik der troposphärischen Aerosole liefern kann.

Wie schon oben erwähnt, werden die Teilchen des langfristigen Ausfalls im wesentlichen durch den Niederschlag aus der Troposphäre entfernt. Die Geschwindigkeit, mit der dies geschieht, ist praktisch durch die Lebensdauer der Aerosole bedingt und ist daher von ziemlich grundlegender Bedeutung in der Meteorologie. Sie kann grundsätzlich durch die zeitliche Abnahme von radioaktivem Material in der Troposphäre bestimmt werden, vorausgesetzt, daß solches Material nur in die Troposphäre und nicht auch, wie bei späteren Bombenversuchen, in die Stratosphäre gelangte. In den ersten Jahren der Atomversuche war dies der Fall, und außerdem waren die Pausen zwischen einzelnen Testen noch lang genug, um den Abklingungsvorgang ungestört verfolgen zu können. Es ist das Verdienst von STEWART, CROOKS und FISHER (1955), diese troposphärisch-globalen Tracerexperimente der damaligen Zeit in großzügiger Weise erfaßt zu haben. Seit 1948 verfolgten sie laufend den Gehalt der mittleren und unteren Troposphäre an Spaltprodukten durch Filterproben, die regelmäßig in vielstündigen Flügen über dem Atlantik in Höhen zwischen 600 und 6000 m gewonnen wurden. Die Spaltprodukte von einzelnen Testen oder Testgruppen in Nevada oder in Sibirien konnten so bis zu 20 Wochen lang verfolgt werden, wobei die auf Zerfall korrigierten Werte ziemlich gut einem exponentiellen Verlauf mit einer Halbwertszeit von 22 Tagen (Lebensdauer 32 Tage) entsprachen. Flüge in südlichere und nördlichere Breiten schienen zu zeigen, daß die Spaltprodukte nach dem ersten Umlauf um die Erde im wesentlichen auf den Westwindgürtel der Nordhemisphäre beschränkt blieben. Die beobachtete Abnahme mit der Zeit muß also weitgehend den Prozessen zugeschrieben werden, die die Aerosole aus der Troposphäre entfernen.

Der von Stewart u. a. beobachtete Wert der Aerosollebensdauer ist wohl der zuverlässigste und repräsentativste dieser Art, der bisher vorliegt. Man sieht aus Tabelle 6, daß er recht gut zu den Werten paßt, die bei den Spaltprodukten der kosmischen Strahlung gefunden wurden. Dagegen sind die mittels der Radonzerfallsprodukte gefundenen Werte im allgemeinen kürzer und dieser Unterschied ist wohl reell, selbst wenn man die geringe Genauigkeit der letzteren berücksichtigt. Dieser Unterschied ist verständlich, da die von der Erdoberfläche aufsteigenden Radonzerfallsprodukte sehr bald in die Zone intensivster Niederschlagsbildung von 1 bis 3 km gelangen, wo die mittlere Aerosollebensdauer sicherlich kürzer sein wird als in der oberen Troposphäre. Aerosole, die hier entstehen oder vorzugsweise gespeichert sind, müssen zunächst durch Mischung in die tiefere Troposphäre verfrachtet werden, ehe sie eine gute Chance haben ausgewaschen zu werden. Künstliche Spaltprodukte in der Troposphäre zeigen ähnlich wie die Radonkonzentration über den Ozeanen im allgemeinen auch einen abwärtsgerichteten Gradienten, der diesen Transport ermöglicht. Die mittlere Lebenszeit von Aerosolen in der Troposphäre wird deshalb je nach dem Sitz der Quelle ein anderer sein und vermutlich Werte zwischen 5 und 40 Tagen annehmen können, wobei 30 Tage ein recht zuverlässiger Wert für verhältnismäßig gleichmäßig verteilte Materie darstellen wird.

Es ist noch zu bedenken, daß die Verteilung der Aktivitäten auf die Teilchengrößen bei den natürlichen und künstlichen Aktivitäten verschieden ist. Dies könnte zu weiteren Unterschieden Anlaß geben. Auch wird naturgemäß die Aerosollebenszeit von den klimatischen Bedingungen, insbesondere von der Höhe des Niederschlages abhängen. Diese Fragen gehen jedoch schon in das Gebiet des Auswaschens und Ausregnens über, das von anderer Seite behandelt wird.

IV. Radioaktivitäten von Reaktoren

Über die von Reaktoren erzeugten künstlichen radioaktiven Aerosole ist sehr wenig in der Literatur bekannt geworden. Von allgemeinerem Interesse ist hier ein Problem, das bisher noch nicht berührt wurde, nämlich die Absorption von radioaktiven Gasen an Aerosolen. Wegen ihres Edelgascharakters war sie bei den Emanationen bisher nicht von Bedeutung gewesen.

Chamberlain u. Mitarb. haben die Verhältnisse im Falle des Jod-131 näher studiert (Chamberlain und Wiffen 1959, Chamberlain 1959). Anlaß dazu gab der Unfall des Reaktors in Windscale am 10. Oktober 1957, bei dem während mehrerer Stunden neben Tellurium-132 und Cesium-137 vor allem Jod-131 in die Atmosphäre gelangten. Jod-131 ist ein Glied in der Zerfallskette

$$\text{Tellurium-131 (25 min)} \rightarrow \text{Jod-131 (8,0 d)} \rightarrow \text{Xe-131 (stabil).}$$

Da Jod-131 aus einem anderen Element entsteht, ist es praktisch unverdünnt durch einen inaktiven Träger. Die recht beträchtliche Jod-131-Aktivität in dem genannten Unfall ging z. B. von einer Jodmenge von nur etwa 1 g aus. Jod wird verhältnismäßig leicht an Metall- und anderen Oberflächen adsorbiert, und es lag daher nahe zu untersuchen, in welchem Ausmaß dies auch am natürlichen Aerosol erfolgt. Versuche zeigten, daß bei einer Gesamtkonzentration von Jod-131 von 0,1 μg/m^3 in Anwesenheit natürlicher Aerosole schon ein merklicher Anteil am Aerosol adsorbiert ist, und daß dieser Anteil mit zunehmender Teilchenzahl steigt. Wurde dieses Jod-131 mit 10 μg/m^3 Jod-127 verdünnt, so war ein Einfluß des Aerosols nicht mehr nachweisbar. Der Einfluß erschien aber wieder, wenn durch Zusatz künstlicher Aerosole deren Konzentration auf 10^4 μg/m^3 erhöht wurde. Da die Konzentration natürlicher Aerosole von der Größenordnung

10^2 µg/m³ ist, kann man schließen, daß im Falle des Jod-131 dessen Absorption bei einer tausendfachen Aerosolkonzentration merklich wird.

Die Konzentration des Jod-131 im Windscale-Unfall sank von etwa 1 µg/m³ an der Quelle auf 10^{-4} µg/m³ wenige Kilometer windabwärts. Es kann deshalb angenommen werden, daß anfänglich ein merklicher Anteil des Jods als Gas vorlag, daß aber bei zunehmender Verdünnung mit aerosolhaltiger Frischluft dieser gasförmige Anteil schnell verschwand. Der anfänglich merkliche Anteil von gasförmigem Jod sollte infolge des höheren molekularen Diffusionskoeffizienten ein rascheres Niederschlagen auf den Boden zufolge haben. Die Untersuchung des radioaktiven Ausfalls des Windscale-Unfalls liefert tatsächlich Hinweise, daß das Jod relativ schnell und schneller als z.B. das Ce-137 am Boden zur Ablagerung gelangte.

Schlußfolgerungen

In der vorliegenden Studie wurde der Versuch unternommen, eine kritische Übersicht unserer Kenntnisse von den radioaktiven Aerosolen zu geben. Es zeigt sich, daß in großen Zügen die vorliegenden Beobachtungen erklärt und zum Teil quantitativ auf andere Parameter zurückgeführt werden können. In Einzelfragen ist jedoch noch vieles offen.

Im Falle der natürlichen radioaktiven Aerosole ist es z.B. die Frage nach der Größenverteilung der angelagerten Aktivitäten die weiterer Klärung bedarf. Dieses Problem ist unter anderem meßtechnisch wie auch biologisch von besonderem Interesse. Man könnte zu diesem Zwecke direkte Messungen dieser Verteilungen vornehmen. Ein anderer Weg wäre, zunächst die Theorie der Anlagerung der Primärteilchen weiter zu verbessern und über weite Größenbereiche der Aerosole zu prüfen und dann umfassendere Bestimmungen der Größenverteilungen natürlicher Aerosole vorzunehmen. Zum Beispiel liegen bisher von maritimen Aerosolen unterhalb etwa 0,4 µ überhaupt keine Angaben über die Größenverteilung vor.

Meteorologisch von besonderem Interesse sind die Vorgänge, die die Lebensdauer der Aerosole in der Troposphäre bestimmen. Es ist zu erwarten, daß ausgedehnte und systematische Messungen auf dem Gebiet der natürlichen und künstlichen Radioaktivitäten wesentliche Einsichten in diese Probleme vermitteln können.

Bei der Fortführung von Atomtesten wäre eine wesentlich vollständigere Erfassung der erzeugten Aerosole, vor allem derjenigen des mittel- und langfristigen Ausfalls erwünscht.

Schrifttum

ADAMS, C.E., N.H. FARLOW and W.R. SCHELL: Compositions, structure and origins of radioactive fallout particles. U.S. Naval Radiological Defense Laboratory Technical Report, USNRDL-TR-209 (1958).

ANDERSON, A.D.: A theory for close-in fallout. U.S. Naval Radiological Defense Laboratory Technical Report, USNRDL-TR-249, 1—55 (1958).

ARON, A., u. B. GROSS: Eine Beobachtung über die von Kernbombenversuchen herrührende Radioaktivität der Luft. Z. Naturforsch. **12**a, 944—945 (1957).

BLIFFORD, I.H., L.B. LOCKHART jr. and H.B. ROSENSTOCK: On the natural radioactivity in the air. J. Geophys. Res. **57**, 499—509 (1952).

BREWER, A.W.: Evidence for a world circulation provided by the measurements of helium and water vapor distribution in the stratosphere. Quart. J. Roy. Meteor. Soc. **75**, 351—363 (1949).

BULLRICH, K.: Streulichtmessungen in Dunst und Nebel. Meteor. Rdsch. **13**, 21—29 (1960).

BURTON, W.M., and N.G. STEWART: Use of long-lived natural radioactivity as an atmospheric tracer. Nature, Lond. **186**, 584—589 (1960).

CHAMBERLAIN, A.C.: Deposition of iodine-131 in northern England in October 1957. Quart. J. Roy. Meteor. Soc. **85**, 350—361 (1959).

CHAMBERLAIN, A. C., and M. T. DUNSTER: Deposition of radioactivity in north-west England from the accident at onidscale. Nature, Lond. **182**, 629—630 (1958).
—, and E. D. DYSON: The dose to the trachea and bronchi from the decay products of radon and thoron. Brit. J. Radiol. **29**, 317—325 (1956).
— W. J. MEGAW and R. D. WIFFEN: Role of condensation nuclei as carriers of radioactive particles. Geofis. pura e appl. **36**, 233—242 (1957).
—, and R. D. WIFFEN: Some observations on the behaviour of radio-iodine vapour in the atmosphere. Geofis. pura e appl. **42**, 42—48 (1959).
CHAMBERS, L. A., J. F. MILTON and C. E. CHOLAK: A comparison of particulate loadings in the atmospheres of certain American cities. Paper presented at the Third National Air Pollution Symposium, Pasadena, California 1955.
— E. C. TABOR and M. J. FOTER: The characteristics and distributions of organic substances in the air of some American cities. Paper presented at 48th Annual Meeting of the Air Pollution Control Association, Detroit, Michigan 1955.
CHAN, H. K.: Activity-size relationship of fallout particles from two shots, operation redwing. U.S. Naval Radiological Defense Lab. Technical Report, USNRDL-TR-314, 1—71 (1959).
DOBSON, G. M. B.: Origin and distribution of the polyatomic molecules in the atmosphere. Proc. Roy. Soc. Lond. A **236**, 187—193 (1956).
FACY, L.: La capture des noyaux de condensation par choes moleculaires air cours des processus de condensation. Arch. Meteor. Geophys. u. Bioklim. A **8**, 229—236 (1955).
FRIEDLANDER, S. K.: On the particle size spectrum of atmospheric aerosols. Paper submitted to the J. Meteorology 1960.
GEORGII, H. W.: Ein Beitrag zum Größenverteilungsgesetz des atmosphärischen Aerosols über dem Kontinent. Meteor. Rdsch. **11**, 33—34 (1958).
GOEL, P. S., N. NARASAPPAYA, C. PRABHAKARA, THOR RAMA, and P. K. ZUTSHI: Study of cosmic ray produced short-lived isotopes P^{32}, P^{33}, Be^7, and S^{35}: in tropical latitudes. Tellus **11**, 91—100 (1959).
HAXEL, O., u. G. SCHUMANN: Selbstreinigung der Atmosphäre. Z. Physik **142**, 127—132 (1955).
ISRAËL, H.: Radioactivity of the atmosphere. Compendium of meteorology. Amer. Met. Soc. 1951, p. 155—161.
JACOBI, W., A. SCHRAUB, K. AURAND u. H. MUTH: Über das Verhalten der Zerfallsprodukte des Radons in der Atmosphäre. Beitr. Phys. Atmosph. **31**, 244—257 (1959).
JONES, S.: Persönliche Mitteilung 1960.
JUNGE, C. E.: Austausch und großräumige Vertikalverteilung von Luftbeimengungen. Ann. Meteor. 380—392 (1952).
— The size distribution and aging of natural aerosols as determined from electrical and optical data on the atmosphere. J. Meteorology **12**, 13—25 (1955).
— Remarks about the size distribution of natural aerosols. In: Artificial stimulation of rain. New York: Pergamon Press 1957.
— Recent investigations in air chemistry. Tellus **8**, 127—139 (1956).
— Air chemistry. Adv. Geophys. **4**, 1—109 (1958).
— Sulfur in the atmosphere. J. Geophys. Res. **65**, 227—237 (1960).
— C. W. CHAGNON and J. E. MANSON: Stratospheric aerosols. J. Meteorology (in Press 1960).
—, and J. E. MANSON: Unpublished 1960.
KALKSTEIN, M. I., P. J. DREVINSKY, E. A. MARTELL, C. W. CHAGNON, J. E. MANSON and C. E. JUNGE: Natural aerosols and nuclear debris studies, progress report II. GRD Research Notes No. 24, Geophysics Research Directorate, AF Cambridge Research Center, 1—36, Nov. 1959.
KEEFE, D., P. J. NOLAN and T. A. RICH: Charge equilibrium in aerosols according to the Boltzmann law. Proc. Roy. Ir. Acad., Sect. A **60**, No. 4, 27—450 (1959).
KIENTZLER, C. F., A. B. ARONS, D. C. BLANCHARD and A. H. WOODCOCK: Photographic investigation of the projection of droplets by bubbles bursting at a water surface. Tellus **6**, 1—7 (1954).
KING, P., L. B. LOCKHART, R. A. BAUS, R. L. PATTERSON, H. FRIEDMAN and J. H. BLIFFORD: RaD, RaE, and Po in the atmosphere. Nucleonics **14**, 78—84 (1956).
KUMAI, M.: Electron-microscope study of snow-crystal nuclei. J. Meteorology **8**, 151 (1951).
LAL, D.: Cosmic ray produced radioisotopes for studying the general circulation in the atmosphere. Indian J. Meteor. and Geophys. **10**, 147—154 (1959).
— P. K. MALHORTA and B. PETERS: On the production of radioisotopes in the atmosphere by cosmic radiation and their application to meteorology. J. Atmosph. Terr. Phys. **12**, 306—328 (1958).
LASSEN, L.: Die Anlagerung von Zerfallsprodukten der natürlichen Emanation an elektrisch geladene Aerosole (Schwebstoffe). Z. Physik **163**, 363—376 (1961).
—, u. G. RAU: Die Anlagerung radioaktiver Atome an Aerosole (Schwebstoffe). Z. Physik **160**, 504—519 (1960).

Lassen, L., u. H. Weicksel: Die Anlagerung radioaktiver Atome an Aerosole (Schwebstoffe) im Größenbereich 0,7—5 μ (Radius). Z. Physik **161**, 339—345 (1961).

Lehmann, L., u. A. Sittkus: Bestimmung von Aerosolverweilzeiten aus dem RaD und RaF-Gehalt der atmosphärischen Luft und des Niederschlages. Naturwissenschaften **46**, 9—10 (1959).

Lodge, J.P., J.E. McDonald and F. Baer: An investigation of the Melander effect. J. Meteorology **11**, 318—322 (1954).

Machta, L.: Symposium über Luftchemie und Radioaktivität, Helsinki, Aug. 1960.

—, and H.F. Lucas jr.: Radon in the upper atmosphere. Science **135**, 296—299 (1962).

Mason, B.J.: Bursting of air bubbles at the surface of sea water. Nature, Lond. **174**, 470—471 (1954).

— The oceans as source of cloud-forming nuclei. Geofis. pura e appl. **36**, 148—155 (1957).

May, R., u. H. Schneider: Verteilung der künstlichen Radioaktivität in Staubproben und Regenwasserrückständen. Atomkernenergie **4**, 28—29 (1959).

Metnieks, A.L.: The size spectrum of large and giant sea-salt nuclei under maritime conditions. Geophys. Bull. No. 15, School of Cosmic Physics, Dublin, 1—50 (1958).

Mordy, W.A.: Computations of the growth by condensation of a population of cloud droplets Tellus **11**, 16—44 (1959).

Penndorf, R.: The vertical distribution of Mie particles in the troposphere. Geophysics Research Papers, No. 25, Geophysics Research Directorate, AF Cambridge Research Center 1954.

Robbins, R.C., R.D. Cadle and D.L. Eckhardt: The conversion of sodium chloride to hydrogen chloride in the atmosphere. J. Meteorology **16**, 53—56 (1959).

Rosinski, J., and J. Stockham: Preliminary studies of scavenging systems related to radioactive fallout. Summary Report ARF 3127-12, Armour Research Foundation, Chicago, Ill. 1—51 (1960).

Sagalyn, R.C., and G.A. Faucher: Aircraft investigation of the large ion content and conductivity of the atmosphere and their relation to meteorological factors. J. Atmosph. Terr. Phys. **5**, 253—272 (1954).

Schumann, G.: Untersuchungen der Radioaktivität der Atmosphäre mit der Filtermethode. Arch. Meteor. Geophys. u. Bioklim. A **9**, 204—223 (1956).

Sisefsky, J.: A method for photographic identification of microscopical radioactive particles. Brit. J. Appl. Phys. **10**, 526—529 (1959).

— Autoradiographic and microscopic examination of nuclear-weapon debris particles. Forsvarets Forskningsaustalt, Stockholm, FOA 4 Rapport A 4130-456, 1—37 (1960).

Stern, S.: The sampling of radioactive debris in the stratosphere. Paper presented at the A.M.S. Symposium on Stratospheric Meteorology, Minneapolis, Minnesota, July 1959.

Stewart, K.: The condensation of a vapour to an assembly of droplets or particles (with particular reference to atomic explosion debris). Trans. Faraday Soc. **52**, 161—173 (1956).

Stewart, N.G., R.N. Crooks and E.M.R. Fisher: The radiological dose to persons in the U.K. due to debris from nuclear test explosions. Report for the M.R.P. Committee on the medical aspects of nuclear radiation. A.E.R.E., Harwell, 1—22, June 1955.

Strahlenschutz Nr. 12, Schriftenreihe des Bundesministeriums für Atomkernenergie und Wasserwirtschaft, S. 1—204. Braunschweig: Gersbach & Sohn 1959.

Sumi, L., A. Corkery and J.L. Moukman: Calcium sulfate content of urban air. In: Atmospheric chemistry of chlorine and sulfur compounds. J.P. Lodge ed. American Geophysical Union No. 652, 69—80 (1959).

Turner, J.S.: The salinity of rainfall as a function of drop size. Quart. J. Roy. Meteor. Soc. **81**, 418—429 (1955).

Twomey, S., and K.N. McMaster: The production of condensation nuclei by crystallizing salt particles. Tellus **7**, 458—461 (1955).

Volz, F.: Die Optik und Meteorologie der atmosphärischen Trübung. Ber. dtsch. Wetterd. US-Zone, No. 13, 2, 1—47 (1954).

Wasson, J.: Persönliche Mitteilung 1960.

Wexler, H., L. Macuta, D.H. Pack and F.D. White: Atomic energy and meteorology. Intern. Conference Atomic Energy, vol. 13, p. 333—344 (1956).

Whytlaw-Gray, R., and M.S. Patterson: Smoke: A study of aerial disperse systems. London: E. Arnold & Co. 1932.

Wigand, A.: Die vertikale Verteilung der Kondensationskerne in der freien Atmosphäre. Ann. Phys. **59**, 689—742 (1919).

Wilkening, M.H.: Natural radioactivity as a tracer in the sorting of aerosols according to mobility. Rev. Sci. Instrum. **23**, 13—16 (1952).

Woodcock, A.H.: Salt nuclei in marine air as a function of altitude and wind force. J. Meteorology **10**, 362—371 (1953).

Zebel, G.: Zur Theorie der Koagulation elektrisch ungeladener Teilchen. Kolloid-Z. **156**, 102—107 (1958).

Radioactive precipitations and fall out

by

L. FACY

With 23 Figures

Zusammenfassung

Der Beitrag behandelt die „Physik des Fall-out".

Nach einer kurzen Übersicht über die verschiedenen Fall-out-Arten, die man heute unterscheidet

Nah-Fall-out (Nahbereich des Explosionsortes)
Troposphärischer Fall-out (Regional)
Stratosphärischer Fall-out (weltweit reichend)

werden diese unter meteorologischen Gesichtspunkten zusammengefaßt in die beiden Gruppen des „Trocken-Fall-out" („dry weather removal") und des „Naß-Fall-out" („wet removal due to actions of the atmospheric moisture") und in ihren Einzelheiten behandelt.

Der erste Teil gilt der Betrachtung der mikrophysikalischen Gesetzmäßigkeiten, die den „Trocken-Fall-out" beherrschen: Stokessches Fallgesetz kleiner Teilchen, Cunninghamsche Erweiterung, „Formfaktor", Diffusion, Turbulenz, Koagulation u. a. m.

Im zweiten Teil werden in ähnlicher Weise die beim „Naß-Fall-out" wirksamen Gesetzmäßigkeiten besprochen: Anlagerung von kleinen Teilchen an Wolken- und Niederschlagselemente, „Rain-out", „Wash-out", Regen- und Schnee-Wirksamkeit, Bilanzabschätzung der einzelnen Anteile u. a. m.

Ein letzter Teil behandelt schließlich die bei Ablagerung am Boden zu beobachtenden Einzelvorgänge: Turbulenz, Austausch, Sedimentationsgeschwindigkeit, Koagulation von Fall-out-Partikeln und Boden-Teilchen, orographische und Vegetations-Einflüsse, Reif- und Tau-Wirkung u. a. m.

A. Introduction

I. Definition

The phenomenon of deposition of radioactive particles dispersed by experimental tests on the surface of the earth is commonly known as "fall out".

Generally, in a broad sense, this refers to all processes which are due to the integration of the radioactive dusts into the soil or surface waters. It includes the deposition: during the non-precipitant weather and also the radioactivity collected first by cloud droplets, and then by rain or snow which finally reaches the surface of the earth.

Fall out of radioactive debris from atomic weapons are usually classed as follows:

1. "Close or local fall out" for the debris deposited just below the atomic cloud or within a range of a few ten or a few hundred miles off down wind from the detonation point.

This fall out is generally dry, except in the case of explosions over the ocean or below the sea surface, where heavy rains are produced by the vaporized water. The particles are large, and the radioactivity from the material of the weapon is attached to soil particles or sea salt evaporated from the ocean waves. The settling velocity is high, the concentration is large, and dry scavenging due to the coagulation processes or to the inertia may increase the diameter of particles whatever the initial size of the distribution may be.

Only surface bursts or detonations at very low levels in free atmosphere can be held responsible for a local fall out. Experiments at high levels, even in the megaton range, can scarcely cause fall out of substantial debris in the area below the fire ball.

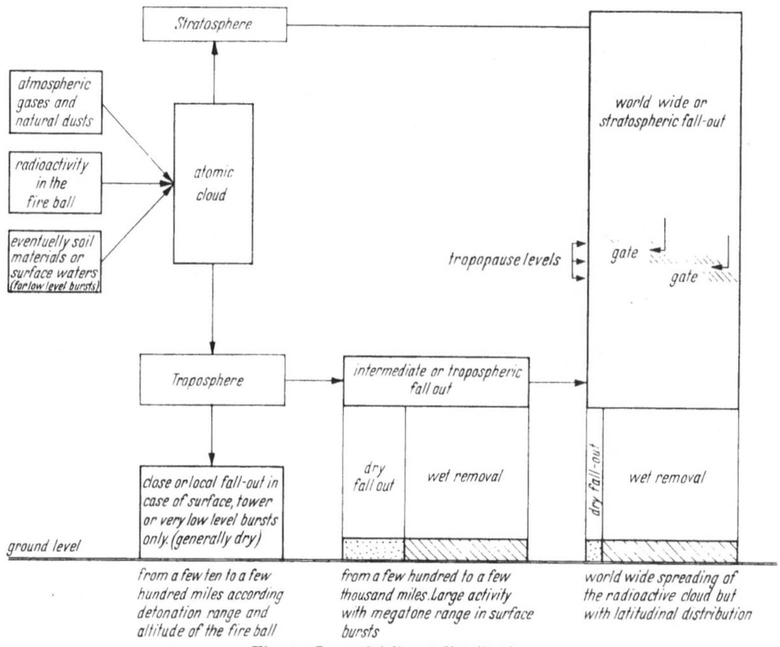

Fig. 1. General fall out distribution

2. "Intermediate or tropospheric fall out" consists of nuclear debris injected at tropospheric levels below the tropopause, and the air mass where it takes place has already performed a complete mixing history with large synoptic scale.

3. "World wide or stratospheric fall out" is mainly composed of atomic debris injected with radioactive clouds which initially rise to high levels above the tropopause in experiments of megaton range and in all high level detonation; more accurately, when the total energy field exceeds 100 kilotons in surface bursts at high latitudes, and 200 kilotons at low latitudes, these values take into account the average height of the polar or tropical tropopause.

The remainder of the intermediate fall out will obviously be mixed with the stratospheric debris introduced into the troposphere, either by the tropopause gate, or by turbulence along the path of the jet streams, or by some other mechanism not yet sufficiently defined.

Concentration of long-life fission products is usually performed at various places of the earth, on ground level, by filtering a reasonable number of cubic metres of air at high flow rate. But meteorological and micro-meteorological conditions are decisive for the deposition on soil surface, for the foliar uptake by plants and for the mean air concentration measured on the filters.

In dry weather conditions there is little fall out, even when the particular activity of surface air is abnormally high. But some very high values of activity have been measured in precipitations; between 95 and 80% of the total amount of radioactivity received on the ground surface is included in rain or snow; obviously the atmospheric moisture is responsible for this tremendous increase in the radioactive fall out for a given activity of the air mass in which condensation and precipitation take place.

For these reasons, it is customary to investigate the removal processes of radioactive dusts in the atmosphere according to conditions prevailing during the deposition.

II. Dry weather deposition

Dry weather deposition implicates the following processes which are not dependent on atmospheric moisture in gaseous or condensed form, viz.:

Sedimentation under gravitational forces; the largest particles are removed from the dust cloud practically by gravity alone; furthermore various processes may increase this action either by repeated coagulations of small particles into larger heavy particles, or by capture of small particles by larger ones due to differential settling velocities.

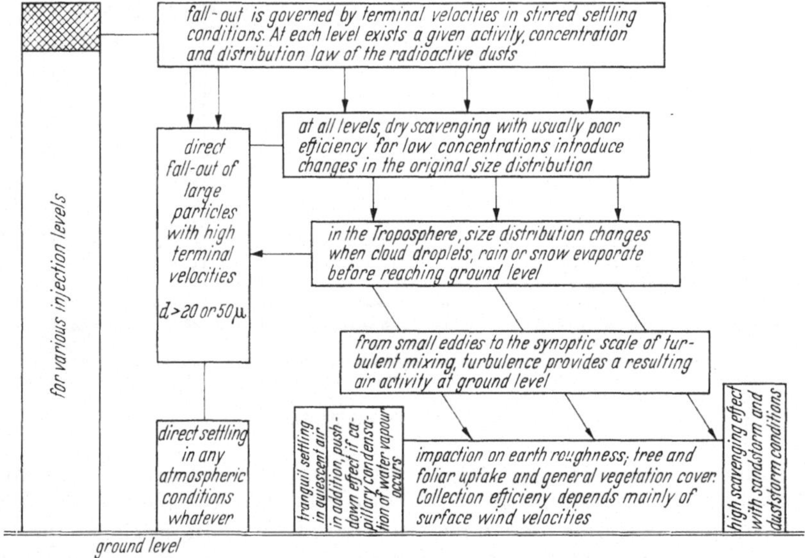

Fig. 2. Dry removal of radioactive particles in free atmosphere

Velocities caused by originated electrostatic fields, diffusive expansion, photophoresis or pure thermal repulsion may act on the radioactive particles and increase the coagulation processes or directly transport the particles to soil surface.

Impaction on the roughness of the earth, and mainly uptake by foils or branches in areas covered by vegetation and forests. These processes are the most effective in the so-called "dry fall out deposition", inasmuch as it is obvious that a smooth polished surface would have a fairly poor collection efficiency.

III. Wet deposition due to atmospheric moisture

Wet deposition includes all processes which are induced by water vapour, by cloud droplets, by mechanical action of rain or snow.

Wet deposition is the most important factor in budgeting world wide fall out, and the cleaning effect of condensation processes or precipitation captures are responsible for at least 80% of the total amount of radioactivity deposited on the earth surface. We may distinguish however the initial stage of capture of dust particles by cloud droplets during all the time when condensation takes place, that is to say inside the cloudy air mass, and the capture by collection efficiency of the falling rain drop or snow flake.

We may call the first mechanism "scavenging effect" similar to the coagulation process inside a dust cloud in dry conditions, but more efficient by several orders of magnitude.

Rain-out or snow-out may design the collection of radioactive particles during the fall of the precipitating element.

All these processes are in fact "wash out" mechanisms even when deposition occurs by water vapour action in rime or frost conditions, or when the soil absorption of water vapours by capillary condensation is responsible for the cleaning of the lower layers of the atmosphere in desert-like areas during the periods of radiative cooling of the earth.

B. Dynamics of small particles in dry weather conditions

I. Velocity of fall or "settling velocity" of a small particle

1. Expression of the drag resistance

The resistance C of a particle in motion can be expressed as a function of various parameters: d diameter of the particle, v the velocity, μ the viscosity of the medium, ϱ_0 the density of the medium.

If the resistance is assumed to vary with the viscosity μ we have the general equation:

$$C = K\, d^a\, v^b\, \mu^c.$$

Dimensional analysis (C being expressed by LMT^{-2}) will give a simple relationship whence the values of the exponents are easily deduced. With $a=b=c=1$, the equation becomes

$$C = K\, d v \mu. \tag{1}$$

This is the STOKES' law, the flow is laminar and the drag force C for a sphere is given by:

$$C = 3\pi\, d v \mu. \tag{2}$$

If we assume now that the resistance C varies according to the density ϱ_0 of the medium instead of the viscosity:

$$C \approx K'\, d^u\, v^v\, \varrho_0^w = K'\, d^2\, v^2\, \varrho_0 \tag{3}$$

which is the NEWTON's law, usually valid at high velocities and turbulent motion. For spheres with $K' = \pi/16$ the drag coefficient is:

$$C = \frac{\pi}{16}\, d^2\, v^2\, \varrho_0. \tag{4}$$

In fact there is no sharp boundary between the streamline and the turbulent flow. The general law of the resistance C can be written as follows:

$$C = K''\, d^n\, v^n\, \mu^{2-n}\, \varrho_0^{n-1} \tag{5}$$

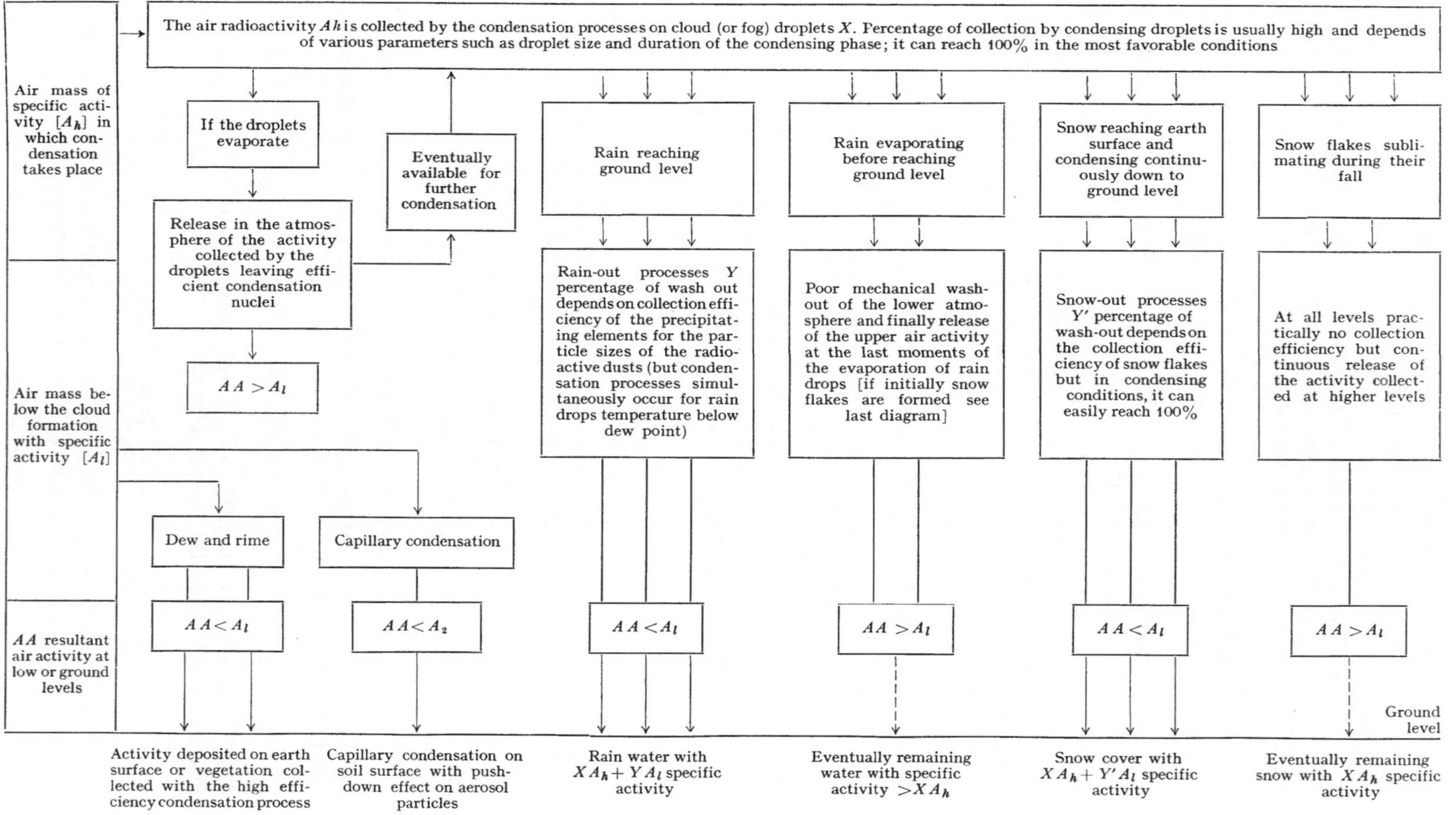

Fig. 3. Wet removal of radioactive particles by atmospheric condensation processes

with $n=1$ we have the STOKES' law, and the NEWTON's law with $n=2$. In the intermediate region between the two flows, the resistance depends upon both the density and the viscosity of the medium, and the value of "n" was found to be $\frac{3}{2}$ according to ALLEN.

2. The terminal velocity

The terminal velocity is obtained by equating the drag force C and the gravitational forces:

$$F = \frac{\pi d^3}{6}\,(\varrho - \varrho_0)\,g \tag{6}$$

for spheres of diameter d and density ϱ and g being the acceleration due to gravitation.

Since the resistance C and μ may vary at the same time, it is more convenient to introduce the dimensionless coefficient Re or REYNOLDS' number which is the quotient of the inertial forces $\left(\dfrac{\varrho\,v^2}{L}\right)$ by the frictional forces $\left(\dfrac{\mu\,v}{L^2}\right)$, where L, the linear dimension, is here the diameter d, and so the REYNOLDS' number is thus written:

$$Re = \frac{dv\,\varrho}{\mu} = \frac{dv}{\eta}$$

if η, the kinematic viscosity, is written for μ/ϱ.

The various equations of motion, (2), (4) and (5) with $n=\frac{3}{2}$, will give, equated with (6):

STOKES' law
$$v_t = K_1\left(\frac{\varrho - \varrho_0}{\varrho_0}\right)d^2\,\eta^{-1}, \tag{7}$$

NEWTON's law
$$v_t = K_2\left(\frac{\varrho - \varrho_0}{\varrho_0}\right)^{\frac{1}{2}} d^{\frac{1}{2}}, \tag{8}$$

Intermediate flow
$$v_t = K_3\left(\frac{\varrho - \varrho_0}{\varrho_0}\right)\eta^{\frac{1}{2}}\,d'. \tag{9}$$

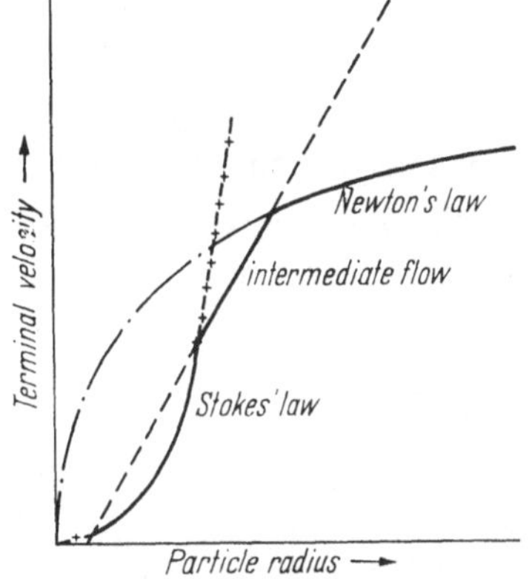

Fig. 4. Flow distribution diagram for free-falling particles

d' of relation (9) is composed of the effective diameter minus a fraction, usually 4/10ths for spheres, of the limiting diameter to which STOKES' equation still applies. The uppermost limit of size for particles corresponds to REYNOLDS' numbers smaller than 2, that is to say to particles less than 50 microns if $\varrho = 1$.

If we admit only 1, 2 for the REYNOLDS' number, the corresponding diameter will only be 25 microns.

Various formulas have been proposed for the uppermost limit of validity of the streamline motion.

According to ALLEN, the value of d can be obtained as follows:

$$d = \sqrt[3]{\frac{36\mu^2}{g\,\varrho_0(\varrho - \varrho_0)}}. \tag{10}$$

GOLSTEIN gives for the resistance C of a particle in motion

$$C = 3\pi\,d\mu\,v\left(1 + \frac{3}{16}\,Re - \frac{19}{1280}\,Re^2 + \cdots\right), \tag{11}$$

a value limited to the second term by OSSEN.

As a matter of fact all these corrections are practically negligible below 10 microns.

a) The Cunningham's correction. The Cunningham's correction must be applied when the diameter of the particle is small compared to the mean free path of the medium. If there is, at the scale of the particle, a vacuum between the molecules of the medium, the falling particle may slip in this vacuum, and the velocity of fall will increase.

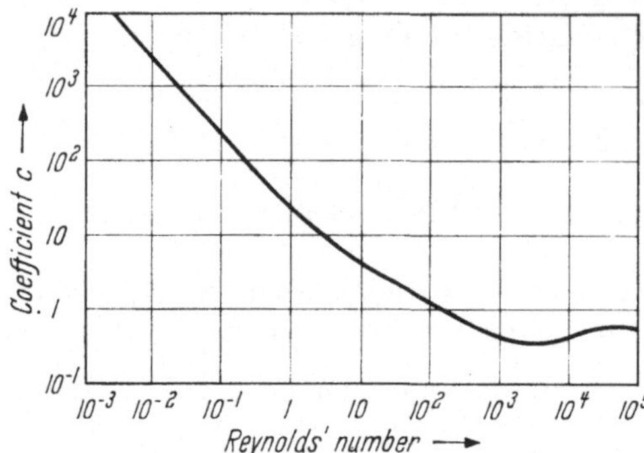

Fig. 5. Coefficient C in terms of Reynold's number

For $d/\lambda = 10$ the terminal velocity is 15% higher than calculated by the Stokes' law.

A complete theory of the resistance experienced by spheres developed by Epstein (1924) introduces two ranges of flow, for $\dfrac{d}{\lambda} \ll 1$ and for $2 \leqq \dfrac{d}{\lambda} \leqq 100$.

The drag coefficient for low values of d/λ can be written in the form:

$$C = \frac{2(A+B)\,\lambda}{2\,\pi\,\mu\,d^2} \qquad (12)$$

and for d of the same order of magnitude as λ inform of the well known Cunningham's correction: $(1 + A\,\lambda/d)$ which fits well with the experience.

For diameters d less than 0, 1 micron at normal P, T conditions we must introduce a correction to V_s, the velocity of fall determinated by Stokes' equation.

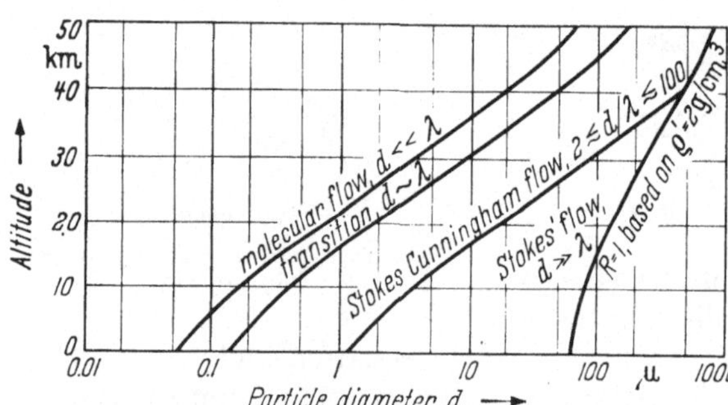

Fig. 6. Free-falling particles for various flow conditions

$$v = v_s\left(1 + \frac{A\,\lambda}{d}\right). \qquad (13)$$

where the mean free path of the molecules is λ and which is expressed according to the kinetic theory of gases by

$$\lambda = \frac{3\,\mu}{\varrho\,\overline{V}}.$$

$\overline{V}$ the average molecular velocity being calculated from:

$$\overline{V} = \sqrt{\frac{8}{\pi}\,\frac{RT}{M}}.$$

Approximately the value of A in relation (13) for air is 1,72 at ground level.

For more accurate developments and to establish the diagrams of the variation of settling velocity with altitude, the value of A must be expressed by

$$A = 1{,}644 + 0{,}522\,e^{-0{,}656\left/\frac{\lambda}{d}\right.}. \qquad (14)$$

If we summarize the condition of validity of equations (7), (8), (9) expressed in Reynolds' number Re with the corresponding coefficient of resistance C, we have

$$
\left.
\begin{array}{lll}
\text{Stokes' law} & 10^{-4} < Re < 2 & C \approx \dfrac{24}{Re} \\[2ex]
\text{Newton's law} & 500 < Re < 10^5 & C \approx 0.44 \\[2ex]
\text{Intermediate flow} & 2 < Re < 500 & C \approx 0.4 + \dfrac{40}{Re}.
\end{array}
\right\} \qquad (15)
$$

Fig. 6 summarizes the various flows of free falling particles in the atmosphere. The particle diameter in microns is plotted on the abscissa, versus the altitudes, expressed in kilometers. The boundaries of the various flows correspond accurately to 10% of the drag formula.

b) The shape factor. In fact, radioactive particles, like other atmospheric dusts, have no regular spherical shape. We may suppose (and this is controlled by electron microscope photographs) that they have either polyhedral or ellipsoidal shape.

In the first case we can take roughly as REYNOLDS' number, the one corresponding to a larger sphere enveloping the particle. The density will consequently be reduced, and the velocity of fall of a particle of irregular shape is smaller than the corresponding velocity of a sphere of the same volume.

When the shape of the particle is more or less ellipsoidal the ratio of the terminal velocity of fall for ellipsoids to the terminal velocity of spheres of the same volume can be computed from theoretical consideration. The curves of Fig. 7 represent the decrease of the terminal velocity for various values of the ellipsoid's parameters, according to computations from OBERBECK's (1876) and LAMB's (1945) formulas.

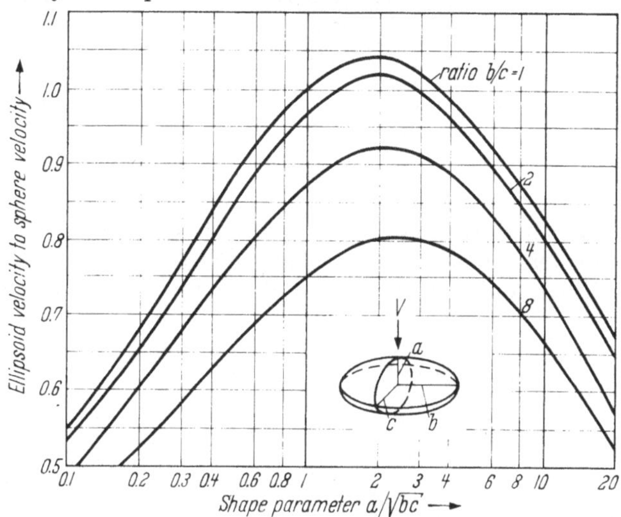

Fig. 7. Ratio of terminal velocity of ellipsoïd to terminal velocity of sphere of same volume

3. Settling of particles under gravity

Ordinary settling of particles under gravity represents only a few per cent of the total amount of radioactive materials released from experimental tests.

As a direct application of the computation of the velocity of fall, radioactive dusts can deposit on soil and vegetation, but a greater percentage of the dry fall-out is due to both gravitational settling and physical interception by inertial deposition from wind effects.

In close fall-out, specially with surface bursts, there is a tremendous quantity of soil or sea salt particles coagulated with radioactive materials.

Fig. 8 represents for various altitudes, that is to say for various λ, η values calculated from the *NACA* standard atmosphere, the terminal velocity of fall of the spherical particles of radius d and of density $\varrho = 2$; velocity in cm per sec in abcissa, versus diameter in microns.

Particles of 10, 50, 100 or a few hundred microns fall out before being distributed at a planetary scale.

Fig. 9 gives a more accurate idea of the variation of the settling velocity with altitude for particle sizes between 0.01 and 10 microns.

The total amount of this so called "close fall out" depends obviously in surface experiments on the nature and size of the outburst. It can contain 50 to 80% of the radioactivity produced, and extend to a range of a few hundred kilometres for the particles of diameters below 100 microns. Small particles, such as produced when the fireball is entirely in the atmosphere, can remain in the troposphere for a sufficient number of months to circle the earth many times.

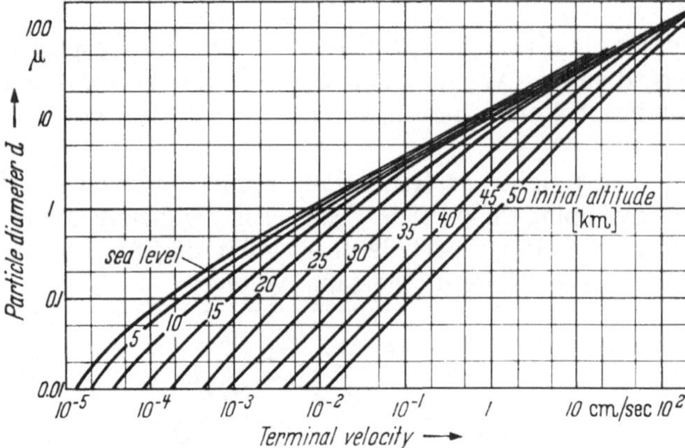

Fig. 8. Terminal velocity of spherical particles of density 2 (Stokes' flow)

Fig. 10 shows the cumulative time required, for particles of a given diameter d in micron and of density 2, to fall from a given altitude to sea level. Additional scales for more accurate computation of time, in days and years, are drawn below the abcissa.

For high level outbursts and shots in the megaton range the stratosphere will store for a longer time the fine debris. The settling of small particles, in dry weather conditions or during precipitation, is called "world-wide" fall out and corresponds generally to debris formed by condensation processes during the cooling of the fire ball, for every range of weapon yields, performed at sufficient high level to avoid the raise of soil material.

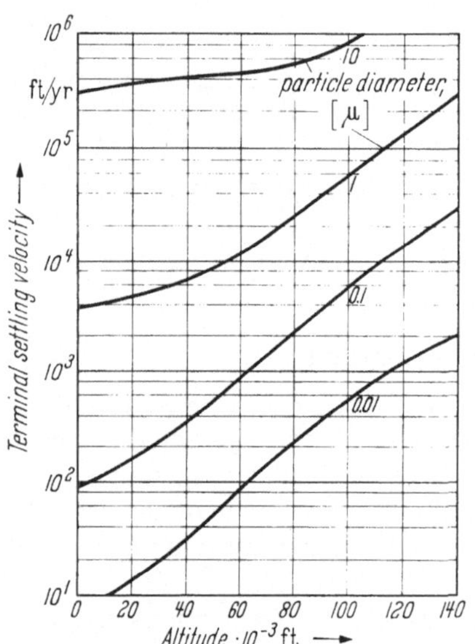

Fig. 9. Change in settling velocity for various particle diameters

a) Tranquil settling. "Tranquil settling" of heterogeneous aerosols composed of particles of many sizes will decrease the concentration of larger particles with high velocity of fall. The differential settling velocities will reduce both the number and the average particle size. But the conditions of a tranquil settling are not usually encountered in the free atmosphere, even not in the lowerlayers during stable meteorological inversion conditions.

b) Stirred settling. The stirred settling is the motion of the particles downward under gravity forces, complicated by random motions of convection currents. The horizontal components of the turbulent mixing due to convection currents will be considered sufficiently low to remain without any effect on the capture by impingement of foliar or ground extensions.

As the upward components of the convection currents will, at a convenient scale, statistically compensate the downward convection currents, the velocity of fall is not modified on the average. But the main effect of the convection is to avoid the differential deposition of the tranquil settling and to keep the concentration uniform throughout a given layer of the atmosphere.

For an homogeneous cloud of particles of diameter d, velocity of fall V, concentration n by cubic centimetre, the number dn of particles settling during the time dt from a level H is given by

$$\frac{V}{H}\,dt = \frac{-\,dn}{n}\tag{16}$$

or by integrating, n_0 being the initial concentration:

$$n = n_0\, e^{-Vt/H}\tag{17}$$

concentration, and rate of settling, will decrease exponentially with time.

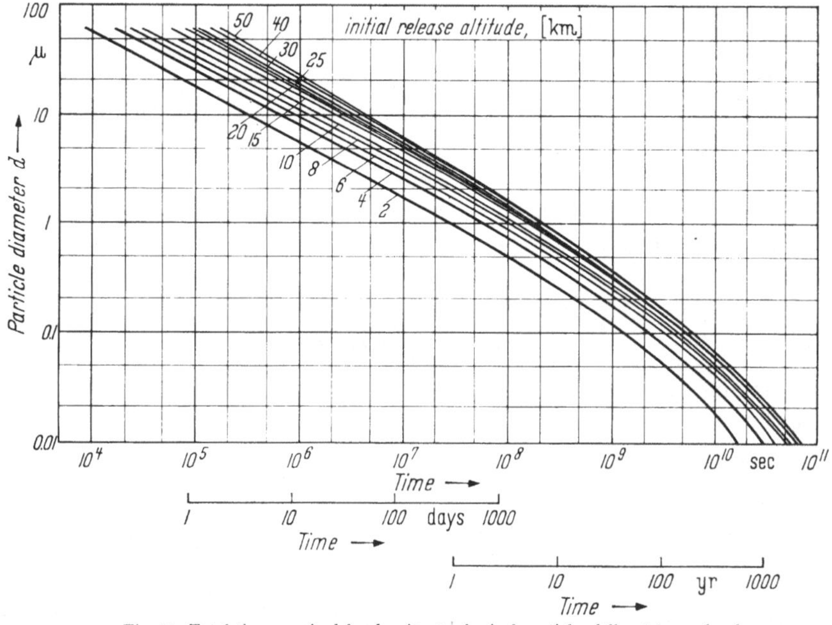

Fig. 10. Total time required for density 2 spherical particles fall out to sea level

For an heterogeneous aerosol the mathematical analysis of stirred settling is rather complicated; an assumption must be made on the probability of a distribution of the particle sizes, and, when the particle density is known, the final distribution can proceed.

II. Diffusion processes and non-gravitational forces

In the preceding paragraphs, gravity forces were only taken in account for the computation of the velocity of fall and the settling rate was discussed as a function of the initial diameter of the particle.

At the physical scale of an aerosol various forces are acting on a particle whose order of magnitude is one micron or less. All physical processes responsible for the coagulation of the primary elements in the radioactive cloud will increase the mean diameter of the particles, consequently the velocity of fall and at least a decrease in the concentration n of the cloud.

Brownian diffusion due to the random molecular motion, thermal diffusion or photophoresis action, electrostatic attraction by charged particles, interception due to turbulence, are the mechanisms which, for dry weather conditions, can directly, but with very poor efficiency for small concentrations, increase the coagulation or the "dry scavenging" effect in a cloud of radioactive particles.

14*

1. Brownian movement

Brownian movement is the sort of random walk performed by a particle of a mass much greater than that of the molecules of the medium in the collision of the molecules in motion with their usual high velocities. The well known Einstein's equation of Brownian movement, viz.:

$$X^2 = \frac{RT}{N} \frac{1}{3\pi\mu r} t \qquad (18)$$

or

$$X \approx 4.8 \cdot 10^{-6} \sqrt{\frac{t}{r}}$$

gives the square average displacement X^2 in the time t for a spherical particle of radius r.

Table 1. *Compared values of terminal velocity in air for particles of density 1.4 to the mean displacement in Brownian movement*

Radius in microns	Terminal velocity cm/s	Brownian movement cm
1	$31.4 \cdot 10^{-3}$	$4.5 \cdot 10^{-4}$
0.5	$114.7 \cdot 10^{-4}$	$6.3 \cdot 10^{-4}$
0.1	$16.2 \cdot 10^{-4}$	$14.1 \cdot 10^{-4}$
0.05	$76.7 \cdot 10^{-5}$	$2 \cdot 10^{-5}$
0.01	$14.7 \cdot 10^{-5}$	$4.5 \cdot 10^{-5}$
0.005	$73.5 \cdot 10^{-6}$	$6.3 \cdot 10^{-3}$

At low pressure there is a considerable increase of the velocity and mean displacement of a particle. The above practical value of X is obviously given for ground level conditions.

2. Diffusion coefficient

If we assume that the behaviour of a dust cloud is similar to the behaviour of gases, the osmotic pressure of the particulate atmosphere is, n being the concentration by cubic centimetre:

$$p = \frac{nRT}{N} \qquad (19)$$

and the force acting on a single particle, in a concentration gradient dn/dx, in a given direction x of displacement of the particle, at a given instant of its random Brownian movement is:

$$f = -\frac{RT}{N} \frac{1}{n} \frac{dn}{dx} \qquad (20)$$

from the relation $F = Cv$ in Stokes' motion applied here with the above value of f and the use of the diffusion coefficient D expressed by the equation:

$$v = -\frac{D}{n} \frac{dn}{dx} \qquad (21)$$

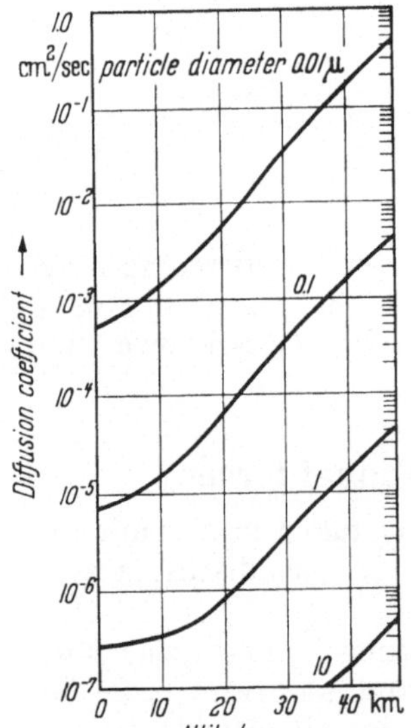

Fig. 11. Diffusion coefficient of small particles in terms of altitude

Table 2. *Some values of the diffusion constant for various sizes of particles*

Diameter in micron	D: cm²/sec	Diameter in micron	D: cm²/sec
0.001	$4.1 \cdot 10^{-2}$	0.1	$6.5 \cdot 10^{-6}$
0.01	$4.4 \cdot 10^{-4}$	0.5	$6.4 \cdot 10^{-7}$

we obtain for the diffusion coefficient of a particulate matter

$$D = \frac{RT}{NC} \qquad (22)$$

where $C = 6\pi\mu r$, for particles of radius r. It is easy to demonstrate that the diffusion coefficient can also be expressed by the mean quadratic displacement X^2 in a given direction for the corresponding time t:

$$D = \frac{X^2}{r\,t}.$$ (23)

For comparison purpose the diffusion coefficients for SO_2 and CO_2 are respectively $11.8 \cdot 10^{-2}$ and $16.4 \cdot 10^{-2}$.

Fig. 11 gives for various particle sizes, the variation of the diffusion coefficient with the altitude. It can be seen from this graph that at a given level (between 10 to 50 km) the mobility of a particle of range 0.01 micron is greater by at least 2 orders of magnitude than the corresponding mobility of a particle with 1 micron, the mean linear displacement being proportional to $\sqrt{D}$. The 1 micron particle must be considered in this occurrence as perfectly motionless in regards to the 0.01 micron particle.

3. Coagulation of particles

Coagulation of particles in an homogeneous aerosol has been found experimentally and varies inversely with the time, n_0 being the initial concentration. At time t the concentration is n, according to the relation:

$$\frac{1}{n} - \frac{1}{n_0} = K t$$ (24)

or in differential form:

$$-\frac{dn}{dt} = K n^2$$ (25)

where K is the coagulation constant, theoretically determined by SMOLUCHOWSKI.

The coagulation depending on pure diffusion processes, is quite low at the ordinary concentrations. The above equation written in terms of percent coagulation per hour for n expressed in cubic centimetre is:

$$-100\,\frac{dn}{n} = 1.08 \cdot 10^{-4}\,n.$$

The following table gives the rate of decrease of an aerosol of radius r for a given concentration.

Table 3

Radius r (microns)	Time (in sec) for a reduction to $n/2$ of a particulate cloud of density n by cc			
	$n = 10^5$ sec	$n = 10^6$ sec	$n = 10^7$ sec	$n = 10^8$ sec
0.01	$3 \cdot 10^3$	$3 \cdot 10^2$	$3 \cdot 10^1$	3
0.05	10^4	10^3	10^2	10
0.1	$1.8 \cdot 10^4$	$1.8 \cdot 10^3$	$1.8 \cdot 10^2$	$1.8 \cdot 10^1$
1.0	$3 \cdot 10^4$	$3 \cdot 10^3$	$3 \cdot 10^2$	$3 \cdot 10^1$

WHYTLAW-GRAY and PATTERSON have developed a more elaborate equation for the coagulation of particles of different sizes.

If we have $n_1 + n_2 = n$ particles per cubic centimeter, n_1 particles of radius r_1 and diffusion coefficient D_1, n_2 particles of radius r_2 and diffusion coefficient D_2, we may write for $n_1 \neq n_2$, if we assume that all n_1 small particles adhere upon touching the collecting particle of radius $r_2 > r_1$:

$$- \frac{dn}{dt} = 2\pi (D_1 + D_2)(r_1 + r_2) n^2. \tag{26}$$

As we only have destruction of the smaller particles:

$$- \frac{dn_1}{dt} = 2\pi (D_1 + D_2)(r_1 + r_2) n_1 n_2 \tag{27}$$

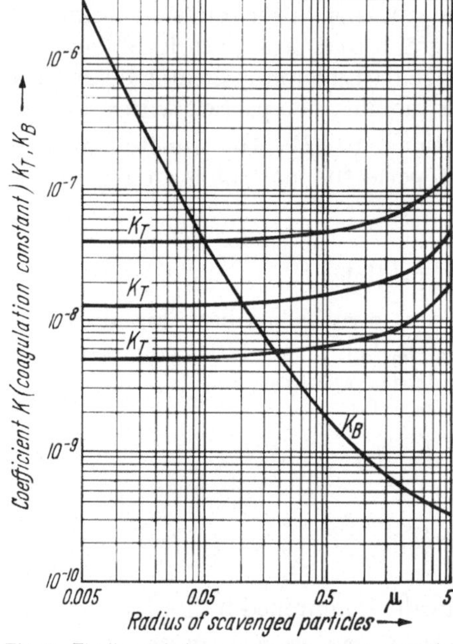

Fig. 12. The "coagulation constant" in terms of particle radius

Fig. 13. Half life of various diameters particles scavenged by 100 scavenging particles, 10 microns radius per cc

$D_1 + D_2$ are diffusion coefficients due to Brownian movement, corrected by the Cunningham's formula to Stokes' law.

Eq. (27) gives

$$- \frac{dn_1}{dt} = K n_1 n_2 \tag{28}$$

where the coagulation constant K is in fact an addition of K_B taking in account Brownian motion and K_T for turbulent motion, which is, according to Smoluchowski, of the form

$$K_T = \tfrac{4}{3} w (r_1 + r_2)^3. \tag{29}$$

w being the velocity gradient normal to streamline, with 4 cm/sec/cm for the lower value according to Whytlaw-Gray and Patterson until 30 cm/sec/cm for a possible upper limit. In graph (12) according to Greenfield (1956) who made the calculations for submicronic particles scavenged by rain droplets, the values of K_T are calculated for three values of w (4, 10, and 30 cm/sec/cm) for scavenging particles of 10 microns radius collector and for various sizes, up to 5 microns radius, for the scavenged particles.

The Brownian coagulation constant K_B is also plotted for a 10 micron radius collector against radius of scavenged particles.

For particles below 1/10th micron in diameter Brownian's motion is more effective in coagulation by diffusional processes rather than turbulence, but it is of the same order of magnitude in scavenging efficiency for particles between 0.1 to 0.4 micron.

The efficacity of such coagulation processes is very small except for close fall out where we may have a high density of large particles per cubic centimeter. For example, the time for the number of scavenged particles (radius 1/10th micron) decreased to $1/e$ of the initial concentration against large particles of "collectors" of 10 microns radius, is about:

10000 hours for   1 collector per cubic centimetre,

1000 hours for 10 collectors per cubic centimetre,

100 hours for the tremendous concentration of 100 large collecting particles per cubic centimetre.

These numbers of hours can be reduced by 6/10ths if we take the largest value for the turbulent coagulation constant corresponding to a velocity gradient of 30 cm/sec/cm, such as that associated with moderate scale turbulent eddies in a dust cloud. There is an increase by a factor 2 or 3 for these times to get the half life corresponding to scavenged particles of one micron radius. But for very small particles of 1/100th micron radius the hours given above must be divided by 20.

4. Thermal diffusion and photophoresis

Thermal diffusion was observed for the first time by TYNDALL (1870), RAYLEIGH (1882), AITKEN (1884). It is easily observed on the boundary of a hot body in a suitable illuminated dust-filled chamber.

The dark space or dust free region surrounding the hot body results from a steady state in which Brownian motion (the effect of osmotic pressure) of the particulate cloud and convection currents are balanced with the force of thermal repulsion.

EINSTEIN (1924), PARANJPE (1936) gave the following values for the force f acting on a particle of radius r in a thermal gradient dT/dx

$$\left. \begin{aligned} f &= K\,\lambda^2\,r\,\frac{P}{T}\,\frac{dT}{dx} \quad &\text{for} \quad r \gg \lambda \\[2mm] f &= K'\,\lambda\,r^2\,\frac{P}{T}\,\frac{dT}{dx} \quad &\text{for} \quad r \lesseqgtr \lambda. \end{aligned} \right\} \tag{30}$$

$\lambda,\ P,\ T$ being respectively the mean free path, the pressure and temperature, $K,\ K'$ constants. According to the values of the drag coefficients for spheres in the ranges $r > \lambda$ and $r \ll \lambda$ the velocities of displacement versus the cold regions in a thermal gradient are independent of the radius of the particle in each range [see Eq. (12)]. These forces of thermal repulsion depend in fact upon numerous parameters. The phenomenon is quite different at low and normal pressures, but they are identical with those observed in the radiometer effect. The most comprehensive development of the radiometer theory is due to EPSTEIN (1929) who took account of the behaviour of a particle in a gaseous medium depending on the ratio of heat received from molecular impacts of gas molecules upon the heat transported through the particle. The final expression for the thermal force derived by EPSTEIN is

$$F = -\,9\pi\,\frac{\mu^2}{\varrho}\cdot\frac{1}{r}\left(\frac{K_a}{2K_a + K_i}\right)\frac{1}{T}\,\frac{dT}{dx} \tag{31}$$

where μ, ϱ are the viscosity and density gases, r the radius of the particle, K_a is the thermal conductivity in the gas (cal/sec-cm-degree K), K_i the thermal conductivity of the sphere (idem). Photophoresis is a phenomenon very similar to the thermal repulsion. Particles of the order of one micron to a few hundredths of micron in diameter, floating in a gaseous medium, are set in motion when illuminated by an intense beam of light.

The displacement is either towards the light source (negative photophoresis) or away from the light source (positive photophoresis).

The calculation of the force exercised by the light source on the particle supposed spherical is based on the assumption that there is a difference ΔT in temperature between the poles of the sphere. There is theoretically no motion if the particle cannot act as a thermal dipole. The equations of motion are very similar to the thermal repulsion or radiometer effect, and there are two different approaches for low or high pressure mediums.

Surface absorption, reflection coefficient, thermal conductivity of the particle are determinant for the ΔT computation, and make clear the positive or negative photophoresis or no photophoresis at all, if there is a poor absorption coefficient or a high thermal conductivity.

Since radioactive materials are usually in form of minute particles of metallic oxydes, or attached to natural nuclei, the behaviour of such particles must be studied with high solar illumination at low pressures corresponding to various levels of the stratosphere. All discrepancy of the particle's velocities in a dust cloud can introduce captures by impact and the numerous physical processes responsible for such motions must be described as factors responsible for the coagulation of particles.

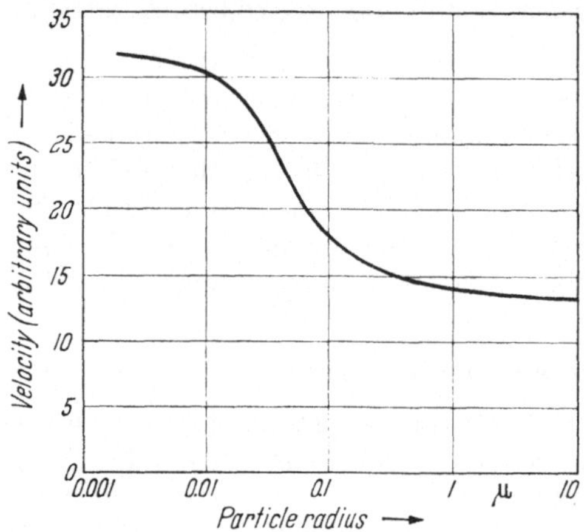

Fig. 14. Thermal gradient velocities in terms of particle radius

5. Electrostatic and sound vibration actions

a) Particles with charges q (esu) placed in an electrostatic field with a voltage gradient E (practical volts/cm) are submitted to a force:

$$F = \frac{qE}{300}$$

and consequently the velocity is given by

$$V = \frac{qE}{300} \frac{1}{3\pi d\mu} \tag{32}$$

for particles of diameter d larger than the mean free path.

The radioactive decay processes are responsible for the formation of positive charges on the particles, but Greenfield (1956) pointed out that because of the creation of high molecular ion densities, there was finally a very small percentage of particles with a single charge, either positive, or negative at any one time. Besides, with the very short life of charged particles, the resultant net charge is practically equal to zero.

But in the free atmosphere radioactive particles attached to natural atmospheric dust or condensation nuclei, are in fact small or large ions, with the various natural processes which may build up positive or negative charges on their surface, and they do not expect their charges only from the radioactive decay processes.

If we admit only one single elementary electrostatic charge on a particle the relation (32) gives, for the velocities in a field of 100 volts/cm:

Particle diameter in micron	Velocity cm/sec	Particle diameter in micron	Velocity cm/sec
0.1	$92 \cdot 10^{-4}$	0.5	$18 \cdot 10^{-4}$
0.2	$46 \cdot 10^{-4}$	1	$9 \cdot 10^{-4}$

In fair weather conditions values of the terrestrial field such as 10000 volts per metre never occur; but values of a few thousand volts per metre are frequently encountered in thunderstorm conditions outside the cloud itself.

For 0.1 micron single charged particle, the velocity of displacement in a reasonable field of 1000 volts/metre is of the same order of magnitude as that of settling velocity. As the motions are opposite for positively or negatively charged particles, relative velocities between two particles may be double of the calculated velocity. So the coagulation between particles of opposite charges is more frequent, partly due to the displacement of particles, partly due to the apparent increase in the collision cross section of particles of opposite charges.

But if it is difficult to give an order of magnitude of the increase in dry scavenging processes due to electrostatic charges, it is nevertheless obvious that all secondary mechanisms such as thermal gradient, photophoresis, are on the same order of magnitude as the Brownian motion for particles below one micron. So much the more since Brownian displacement is a random motion and other mechanisms are, like gravitationnal settling, organized motions with well defined directions.

b) The coagulation of smoke by means of sound generators in the supersonic frequency range is well known. Even frequency as low as 4000 cycles p/s will cause a rapid coagulation of particles of one micron radius if a sufficient intensity is available. The larger the particle, the lower is the frequency needed for these coagulation mechanisms. Like all other mechanisms, coagulation is due mainly to the relative motion of different sized particles which vibrate with different amplitudes.

In free atmosphere coagulation by sonic vibrations has not to be taken into consideration. Natural turbulence, which reduces the energy distribution of the sound waves; the absorption of the sound in the air, not negligible except in the case of very low frequencies; and the very few continuous loud sound sources at the earth surface from human activity, are the main factors for the inefficiency of this mechanism in dry scavenging of fall out.

These considerations do not mean that sonic coagulation occurs only in the man made "sonic flocculators"; thunder claps and explosions are sufficiently known to precipitate rain and dust.

6. Turbulence as a coagulation process

It seems impossible to discuss the exact trajectory of a particle in a turbulent field. But for every change in velocity and direction we can calculate the split or path difference between the medium and the particle. If we consider that turbulence is the normal state of the atmosphere, that even in quiet conditions small scale eddies are always present, it is difficult to deny any possible action of inertia coagulation in the flow pattern of the sustaining medium.

Visual observation of dusts in a light beam gives evidence of small eddies and differential velocities on particles of different sizes. Arenberg (1929) suggested that a similar mechanism was responsible for the coagulation of heterogeneous cloud droplets in a turbulent medium and made his calculations with simplified assumptions for the turbulent flow.

We can assume that a particle, at the edge of an eddy, can suddenly be caught by a different flow and loose its kinetic energy during the path difference S.

The loss of energy of the particle equalizes the work by viscous drag along S.

$$\frac{1}{2} m v^2 = \frac{\pi d^3 v^2}{6} = \int_0^s F\, ds \qquad (33)$$

with

$$F = -C v = -3\pi \mu\, dV,$$

and

$$v = \frac{ds}{dt} \qquad m = \frac{\pi d^3}{6}\varrho.$$

If we differentiate with regard to time:

$$m v \frac{dv}{dt} = -c \cdot v \cdot v \cdot dt \qquad (34)$$

$$\frac{dv}{v} = -\frac{c}{m} dt \qquad (35)$$

hence

$$v = v_0 \exp\left(-\frac{c}{m} t\right) \qquad (36)$$

with v_0 initial velocity.

Integrating again we have the length of the split

$$S = \frac{m v_0}{c}\left[1 - \exp\left(-\frac{c}{m} t\right)\right]. \qquad (37)$$

As v_0 will decrease to $v = v_0/e$ in a time $t = m/c \approx 300\, d^2$ that is to say approximately $3 \cdot 10^{-4}$ sec for a particle 10 microns diameter and density 1, we can admit that it is always large and Eq. (37) becomes

$$S = \frac{m v_0}{c} = 300\, d^2 \varrho\, v_0. \qquad (38)$$

For an initial velocity of 10 m/s particles of density 2 with diameter 1 micron will traverse the medium of 0.6 mm; for 10 microns 6 mm and for 20 microns 24 mm.

Relative velocities between particles of different sizes in high turbulent fields may increase the scavenging effect.

Practically the coefficient K_T for coagulation with turbulent motion introduced by Smoluchowski in Eqs. (28) and (29) gives results of the same order of magnitude.

III. Scavenging due to particle inertia

Diffusional growth either in homogeneous aerosols or in heterogeneous clouds with larger particles or "collectors" neglects the gravitationnal forces and is a more or less valid extrapolation of the diffusion coefficient in kinetic theory of gases.

If we introduce forces acting in prevailing directions such as electrostatic fields, photophoresis and chiefly gravity on the path of a scavenging particle of diameter d and relative velocity v with regard to scavenged particle, we may assume that in dry removing conditions all other particles will be collected and will adhere upon touching the collector.

To simplify the computation of the particles removed we suppose that there is only one size of collecting particle of diameter D and one size of scavenged particle of diameter d

$$n_d = n_{0d} \left[1 - E_{D,d}\right]. \tag{39}$$

Which means that in the cylinder with a cross section of diameter D, n_d the number of particles with diameter d remaining after the passage, or "fall", in case of gravity forces, is expressed as function of n_{0d} the initial number of particles in the cylinder, $E_{D,d}$ being the "collection efficiency" of a sphere of a given density with diameter D, versus spheres of diameter d. This collection efficiency takes into account the departure of the particle trajectories from the air flow lines. ALBRECHT (1931), BRUN (1953), LANGMUIR (1948), and many other authors have studied the problem of the inertia of a particle in the streamline on the air flow before a sphere or a cylinder. More recently TELFORD and Alii (1950 to 1954) have shown the possibilities of capture of water droplets of identical sizes behind the leading drop in the turbulent path, as far as 40 diameters and for as much as 10 diameters radially.

Similar investigations have to be performed to make sure that the same mechanism of coalescence can occur with particles of densities higher than 1.

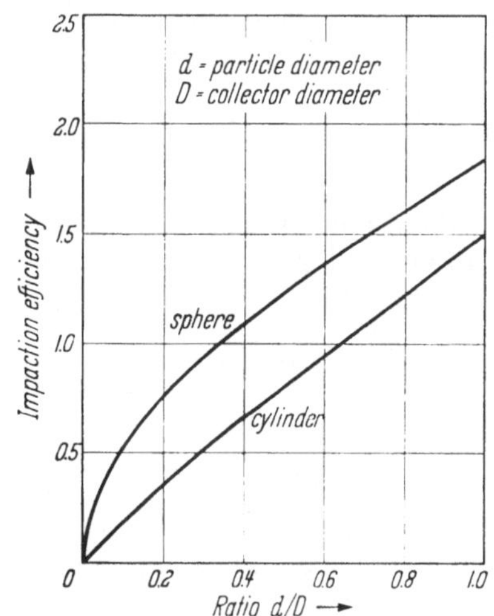

Fig. 15. Collection or impaction efficiency

In Fig. 15 the efficiency of collection or impaction for sphere or cylinder is given. The ratio d/D, diameter of particle/diameter of collector is on the abscissa.

For minute particles in the micron range and less, the efficiency of dry scavenging due to the particle inertia with differential settling velocities is very small as expressed in the Table 4 or in Fig. 19. In case of gravity, the large particles with relative high velocity act as a collector only during their descent and are definitely lost for further scavenging effect. For other mechanisms involving repeated and alternate displacements, the same particle can act as collector many times and the scavenging efficiency repeated. For this reason it is difficult to deny without discussion any valuable effect of non gravitationnal forces in dry scavenging processes in free atmosphere.

For an appreciable scavenging effect with particles of 100 microns diameter under gravitationnal settling, the total amount of collector particles per square centimetre is in the order of $5 \cdot 10^4$.

The order of magnitude for the weight of scavenging material required for cleaning with 95 % scavenging efficiency radioactive particles of 0.1 to 1 micron diameter is about 700 tons per square kilometre in the case of 100 microns sand particles ($\varrho = 2.6$).

Table 4. *Collection efficiency E for drops of diameter D falling through a cloud of smaller drops of diameter d* (Langmuir)

D, μ	d, μ							
	4	6	8	12	16	20	30	40
30					0.092	0.269	0.500	0.643
50				0.050	0.277	0.411	0.613	0.724
80				0.205	0.394	0.510	0.690	0.782
140			0.035	0.340	0.500	0.608	0.750	0.834
200			0.133	0.418	0.564	0.660	0.793	0.862
300		0.010	0.245	0.498	0.631	0.713	0.829	0.887
400		0.085	0.326	0.564	0.684	0.756	0.859	0.908
600		0.213	0.425	0.643	0.749	0.810	0.892	0.929
800	0.040	0.303	0.500	0.698	0.793	0.849	0.919	0.950
1200	0.121	0.355	0.530	0.731	0.827	0.876	0.939	0.963
2000	0.140	0.358	0.535	0.738	0.834	0.886	0.944	0.966
2800	0.168	0.360	0.534	0.735	0.840	0.890	0.950	0.970
3600	0.117	0.288	0.456	0.680	0.800	0.865	0.935	0.965
4800	0.075	0.220	0.372	0.606	0.743	0.823	0.920	0.950
6000	0.050	0.170	0.306	0.546	0.690	0.785	0.900	0.940

Even for surface bursts of few kilotons such conditions can be practically realized and the dry scavenging by soil particles may be responsible for the close fall out of high activity, but only in limited areas downwind the test.

C. Wash out processes or wet removal of radioactive particles

The residence time of a radioactive cloud in the atmosphere depends principally on the turbulence and stability conditions of the lower layers. For given meteorological conditions a particular relatively coarse cloud will remain only a limited time in free atmosphere. This is not the case for the submicronic particles, particularly for those below 1/100th micron. The phenomena are based in this case on a new scale, the sedimentation velocities become of the order of 1 centimetre per day, and the diffusion coefficient becomes of the same order of magnitude as that of the heavy hydrocarbon vapours. Under these conditions the submicronic aerosol residence time may become extremely large, if no other mechanism does at this moment interfere to modify the action of the particles solely under the influence of gravity.

Coagulation processes, electrostatic attractions, photophoresis phenomena may in a limited part modify the titration and the dimensional repartition of the submicronic aerosols, but, in fact, as long as the water does not occur, in liquid or solid form, there is no reason to assume that the general turbidity of our atmosphere reaches, for this scale of aerosols, concentrations several thousand times greater than those corresponding to the most severe pollutions, by particles equal to or in excess of a micron.

While the ponderable absolute value of a particulate cloud of submicronic dimensions with stable characteristics is in itself extremely weak, its toxicity may however be very high as a result of its state of dispersion and through the facility with which it may penetrate, via the respiratory system, into the blood circulation. Such is the case for the radioactive particles resulting from nuclear experiments, whose primitive dimensions are on the average of about 1/100th micron.

Fortunately the atmospheric water is in fact the main purification agent, and it is thanks to the precipitations, to the phenomena of the phase change of the water, that we can yet breath an adequately pure air to allow us to live, notwithstanding the thousands of tons of toxic products produced each year by our industrial activity, notwithstanding the megacuries of radioactive, injurious elements already contained in the atmosphere. As long as the water contained in the atmosphere remains in the shape of vapour, its molecules behaving like those of a gas, no modification is required in the method of transport by means of dry removal. On the other hand, as soon as condensation occurs, as soon as cloud droplets are formed, new mechanisms come into action, which effect a considerably increase in the possibility of the capture of aerosols, radioactive or not, present in the atmosphere.

We must, however, distinguish two different essential phases in the processes of the purification of the atmosphere, with the assistance of the presence of the liquid phase.

In the first stage, several mechanisms, where the molecular scale predominates, are produced exclusively in the cloud itself, in the course of its formation and of its evolution, whatever may be the final outcome of the cloud so formed. These mechanisms are capable of conveying upon the droplets the aerosols present in the air mass where the condensation took place. According to the thermodynamic evolution conditions of this air mass, according to the age of the cloud, a portion more or less important of the specific activity of the air will thus be transferred and accumulated upon the liquid droplets. These droplets, even if they do not contribute in forming the precipitations intended to reach the soil, will nevertheless allow aerosol particles larger than the primitive condensation nuclei to subsist afterwards, after the evaporation: particles enriched with all the impurities present in the cloud inner droplet space, and which will again be able to play the part of a condensation nucleus in the course of the thermodynamic modifications which using the ambient air mass as vehicle, may be able to sustain.

This is the whole of the mechanisms acting in the air mass where the cloud was formed, at the scale of the cloud droplets which we shall call "wet scavenging" or "condensation scavenging".

On the other hand, within a second stage, the following will occur, viz.: all the mechanisms likely to bring back upon the precipitations, either in the liquid state (rain out), or in the solid state (snow out), will bring the aerosol particles present in the atmosphere upon the trajectory of the rain drops, of the crystals or of the snow flakes as much within the clouds, where the precipitating elements are given birth, as below the clouds, between the clouds' base and the ground.

It is quite evident that, according to the importance of the captures which will take place in the case of rain drops or snow flakes and to the activity itself of the air masses in which the cloud was formed, or of the activity encountered between the soil and the cloud, considerable specific activity variations may be noticed in the precipitations. The connection between the activity of the air collected at the ground level during the precipitations, and the activity of the latter depends upon so many meteorological or microphysic factors, that its signification becomes really extremely uncertain.

All the activity of the air volume where the condensation has given birth to a cloud may, eventually, be taken up as a whole by the droplets composing the cloud; but usually only a part of the radioactivity of the air can be collected by the rain drops in the course of their fall. As the specific activity of the air is rarely constant between the ground and the upper level of the clouds, it is difficult to draw practical conclusions as to the comparison of the activity of precipitations reaching the ground and of the activity of the dust contained in the air at this same level.

The mechanisms of condensation within the cloud, the capture of the aerosols by the rain drops in the course of their fall towards the ground, the diffusion effects towards the ice crystal or, at certain levels, towards the snow flake, are processes likely to energetically purify the atmosphere, and to take away rapidly from it the activity carried by the aerosols which it contains. But without an accurate knowledge of the values of the existing radioactivity at all levels, of the determinating meteorological and thermodynamic conditions of the formation of the cloud masses, of the releasing mechanism of the precipitations, it is impossible to calculate the final activity of the snow or of the rain in connection with the individual radioactivity values of the air, even supposed to be known at all levels. It is beyond question, except in the case of certain, especially simple meteorological situations, to connect directly the activity of the precipitations with that of the air taken at the level of the soil.

These privileged meteorological situations will be met for example in the case of orographic clouds causing precipitations, and for which the mass of air insuring the feeding of the cloud is perfectly homogeneous, both from the thermodynamic point of view as well as in connection with the radioactive concentrations. Likewise, in the case of the statistic scale, for a given type of weather, and the precipitations connected with this type of weather, comparisons between the activity of the air and those of the precipitations may again become significant; but nevertheless the precise knowledge of the activities at the levels, where these sundry mechanisms of capture are produced, remains the determinant factor for the knowledge "a priori" of the total activity of the precipitations gathered on the soil.

I. Scavenging due to coagulation particle droplet

We should at the outset consider that all the coalescent mechanisms may interfere within a cloud between droplets and radioactive aerosol, exactly as if these droplets were solid volumes without any particular thermodynamic properties.

If by purely mechanical effects, the aerosol particle and the droplet should effectively come in contact, the Van der Waals forces will then suffice to maintain this contact and to bring the coalescence to a definitive state, even if the particle cannot become wet. Two particles of one micron require 10^{-10} erg to be separated; for particles with $0.01\,\mu$ the necessary work becomes 10^{-13} erg. However from the fact, that only 4×10^{-14} erg are needed, when the energy resulting from the thermic motion reaches ambient temperatures, it can be seen that as long as the gravity forces are negligible—i.e. for spheres of about $30\,\mu$—the Van der Waals forces are adequate to assure the cohesion of the packing.

In addition the discontinuous structure of an aerosol agglomerate, while facilitating even with low relative humidity the capillary condensation phenomena, affords the cohesion of its elements, depending particularly on the wetness of the latter, and, generally, on the value of the superficial tension of the water.

However, it is only with a certain restraint that we may admit that, for example, the application of the Brownian diffusion equation will give for a particulate aerosol and for a cloud of droplets of given characteristics a rate of coagulation in conformity with the statistical values of the kinetic theory. In fact, upon examining with a microscope the behaviour of a water droplet within an aerosol atmosphere, no capture is ever detected, even if the water droplet is in perfect vapour tension equilibrium with the prevailing atmosphere. One can notice on the contrary a real dust free space protecting the droplet from the dusty prevailing atmosphere whenever evaporation occurs. It is only whenever there is a condensation on the droplet that the captures then interfere, the particles being precipitated upon the surface of the droplet with a speed proportional to the gradient of the vapour tension existing in the neighborhood of the droplet.

This total absence of captures between the aerosol and the droplet even in a case of thermodynamic equilibrium of the latter can be attributed to the impossibility to experimentally observe the coalescences between droplets in the midst of a homogeneous artificial water or oil cloud, or any other fluid cloud. This phenomenon can, no doubt, be explained if it is admitted, by a kind of calefaction at low temperature, that a limiting layer forming a mattress of vapour surrounds the droplet in course of evaporation or in a state of equilibrium, and that furthermore, a certain energy is indispensable, if there is no real previous contact, to distort the minimal surface volume such as that of a droplet. This energy has the value $\sigma\,ds$, where σ is the superficial energy and ds the growth of the surface corresponding with the formation of the umbillic.

Particular thermodynamical conditions are necessary, those of condensation, corresponding this time with the deficit in the accumulation of vapour upon the surface of the droplet, so that the real contact between the aerosol and the droplet may be able to act. According to the wetness of the aerosol, the fall of the particle upon the surface of the droplet can clearly be seen with the optical microscope, and afterwards its disappearance, within a more or less long time, within the liquid.

But it is however not impossible, or even certain that all the capture mechanisms by dry removal, likely to bring to bear impact energies superior to $\sigma\,ds$, may be able to cause coalescences between droplets and aerosols in the midst of a cloudy atmosphere. The differential sedimentation under the action of the weight, can alone, in view of the difference of scale between the aerosols and the droplets, bring about captures among the latter, even in case of vapour tension equilibrium. The same applies to the displacements due to electro-static interactions between the droplets or the crystal and aerosol particle electrically charged.

In any case, the importance of these coalescences can only be inferior to those which would take place between dry aerosols of dimensional distribution similar to the radioactive particles and the water cloud, since the latter requires, except in case of condensing phase, an energetic threshold to secure the real contact particle/droplet.

II. Scavenging due to condensation processes on cloud droplets

A phenomenon quite similar to the "dust free space" existing in the neighborhood of heated bodies, or at the photophoresis of the particles, submitted to radiation, can be noticed whenever we examine on a dark field with lateral illumination, the behaviour of a particular cloud in an isotherm enclosure, or, through a suitable evaporation-condensation process, when a tension gradient of partial vapour is created, at the rate of pure molecular diffusion and with a constant global pressure.

The particles in suspension in the atmosphere act as very precise indicators and materialize in a simple kinetic report the flow of vapour molecules moving from the source to the well. The transfer of vapour is thus made directly visible through the continuous displacement of the particles, the displacement of which is easily measurable, being superposed to the usual Brownian motion, whose fluctuations remain perfectly perceptible, particularly in the case of the small displacement velocities corresponding with low gradients.

The source zone, where the evaporation takes place, remains dark at a certain distance and presents exactly the aspect of the dark space existing around a heated body. The well zone, on the other hand, receives the continual flow of the particles impinging its surface, and thus plays the part of cold partition wall of the heat precipitator. The equation of motion of one submicroscopic particle, controlled by a vapour tension gradient can easily be deducted from the mathematical approach already given by EINSTEIN in his study of the radiometer effect, and without introducing here the heat conductibility parameters, as EPSTEIN had been prompted to do to demonstrate the order of magnitude of the effects realized.

Whenever an aerosol particle is located in a field of vapour pressure of a diffusing vapour in the atmosphere everything appears as if the particle was "pushed" by the vapour with a force proportional to the gradient of this partial vapour pressure. The force exerted on the particle, directed in the line of the vapour diffusion, is expressed as follows (FACY 1958a):

$$\left.\begin{aligned} F = k\,\pi\,v^2\,D\,\frac{dp}{dx} \quad \text{for} \quad v \ll \lambda \\[2mm] F = k'\,\pi\,v\,D\,\frac{dp}{dx} \quad \text{for} \quad v \gg \lambda \end{aligned}\right\} \tag{40}$$

where k and k' are constants, v is the radius of the particle, D the vapour diffusion coefficient and dp/dx its gradient. In both cases, for $v \ll \lambda$ or $v \gg \lambda$ the displacement of the particle is constant, independent of the radius and of the shape

$$V = C \cdot D \cdot \frac{dp}{dx} \tag{41}$$

where C is a constant.

In fact, for a same value of dp/dx, there are two systems with uniform velocities, one for the smaller particles, the other for the particles larger than the mean free path; the two constant velocity systems are connected by a variable velocity range, practically between 1/100 th of a micron and 0.5 micron. The speed being recorded then by means of an expression connecting the high speed domain for v small to that of the low speed for v large:

$$V = \left(C_1 + C_2\,\frac{\lambda}{v}\right) \cdot D \cdot \frac{dp}{dx}. \tag{42}$$

Of course these forces applied on the aerosol particles in a vapour pressure field are to be compounded with all those which may exist outside of this pressure field.

The vapour transfers around a droplet, always under the control of the pure molecular diffusion, comprise a distribution of vapour densities, designated $r_0\,\delta_0 = \varrho\,\delta_\varrho = R\,\delta_\infty$ where r_0, R, ϱ apply respectively to the radius of the droplet, a radius at a great distance from the droplet, an intermediary radius between r_0 and R, and δ_0, δ_ϱ, δ_∞ the vapour densities ruling at these distances. Starting from these relations between the radii and densities we get for the δ_ϱ density at the variable, contained between the surface of the droplet and the distance R, the values

$$\delta_\varrho = [(\delta_0 - \delta_\infty)/(R - r_0)]/(R\,r_0)\,(1/\varrho) \tag{43}$$

and the vapour tension gradient for the radius ϱ is then written

$$\frac{d\delta_\varrho}{d\varrho} = \frac{(\delta_0 - \delta_\infty)}{R - r_0}\,R\,r_0\,\frac{1}{\varrho^2}. \tag{44}$$

At the ϱ distance from the centre of the droplet, where the particulate cloud is stopped, we may write that the osmotic pressure of the particulate atmosphere acting upon a particle of v radius [the force $F = n(RT/N)\,\pi v^2$, n designating the concentration of the particulate cloud] should equilibrate the force F, due to the vapour tension gradient $d\delta_\varrho/d\varrho$.

For a given surrounding pressure the relation determinating the radius ϱ of the dust free space limiting for example a particulate cloud of radius v small against λ (case of Fig. 16) we will write as function of concentration N of the cloud

$$k\,\pi v^2 D\,\frac{RT}{N}\,\frac{(\delta_\infty - \delta_0)}{R - r_0}\,R r_0\,\frac{1}{\varrho^2} = n\,\frac{RT}{N}\,\pi v^2 \tag{45}$$

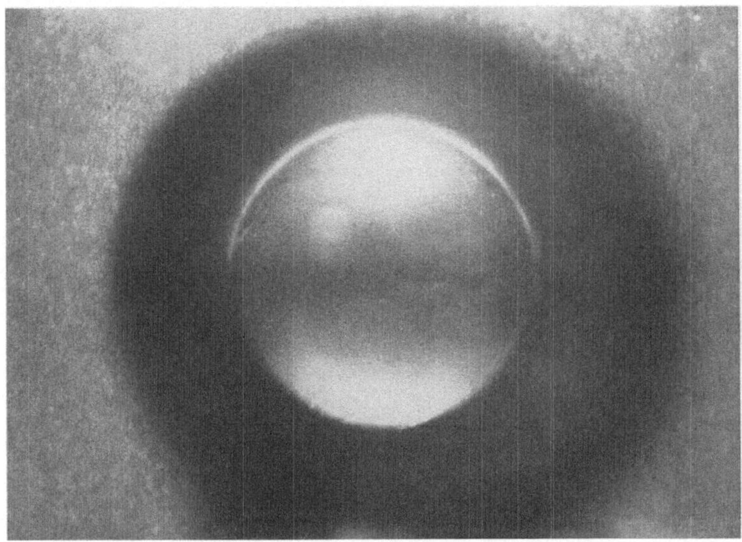

Fig. 16. Dust free space around an evaporating droplet (particle radius 0.5 micron)

and can obtain for ϱ the simple expression

$$\varrho = \sqrt{\frac{k\,(\delta_\infty - \delta_0)\,R r_0}{R - r_0}\cdot\frac{D}{n}}. \tag{46}$$

In the free atmosphere we may consider that n varies very little with the time.

In case the droplet should be in the process of condensation, we can notice a continuous flow of aerosol particles impinging upon the droplet (Fig. 17), striking the surface and then disappear in the interior of the latter, if the particulate cloud is soaked by the solution utilized. Here again, although the temperature should slightly rise on account of the latent condensation heat, the temperature gradients are inadequate to cause a heat repulsion effect of the aerosol particles.

The velocity of the flow of the particles impinging upon the droplet, on account of the osmotic pressure of the particulate atmosphere, is higher than provided for by the expression of the force due to the diffusion of the vapour only. By neglecting gravity in order to simplify, the disappearance of the particles as a result of the capture by the droplet creates here a concentration gradient of the particulate cloud $dn/d\varrho$ at the distance ϱ from the centre of the droplet. From there results a force $F = (RT/N)\cdot(1/n)\cdot(dn/d\varrho)$, which, this time, will be added to the force due to the vapour tension gradients. As for the latter, it can be noted that, in the vicinity of the drop, for ϱ close to r_0, the gradient becomes:

$$\frac{d\delta_0}{d\varrho} \approx - (\delta_0 - \delta_\infty)/r_0 \tag{47}$$

being so much the greater the smaller the radius of the droplet. The capture forces therefore become extremely active (Facy 1958b).

The total force applied on the particle of radius ν (for $\nu \ll \lambda$) in the vicinity of the surface of the drop of radius r_0, taking R great against r_0, is so expressed:

$$F = k\,\pi\,\nu^2\,\frac{\delta_0 - \delta_\infty}{r_0}\,D + \frac{n\,RT}{N}\cdot\frac{1}{D\,k\,(\delta_\infty - \delta_0)}\,. \tag{48}$$

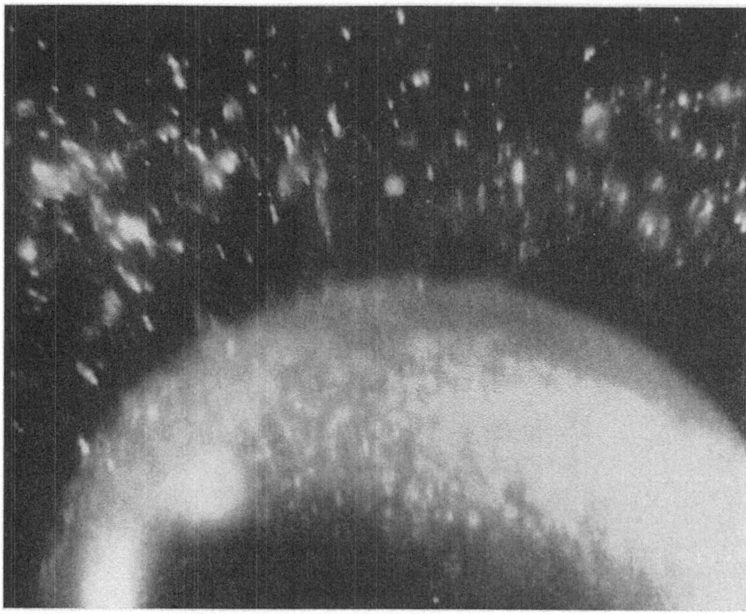

Fig. 17. Impaction of particles on a condensing droplet (particle radius of a few microns)

The repulsion forces of heat motion remain at a much lower scale: in the normal conditions the impact velocity of a particle in the micron range upon a 2 mm droplet of supersatured solution of NaCl within a saturated space with respect to water, reaches $10 \cdot 10^{-3}$ cm/sec while heat gradients of $10°$/cm would only involve displacement velocities 10 times less.

For particles in the micron range the forces attributable to the diffusion of water vapour in the immediate vicinity of a cloud droplet in process of condensation may reach $\frac{1}{3}$ of the gravity forces. For dimensions lower than the micron the kinetic forces due to the diffusion of vapour become superior to those of the gravity; they remain, for the very small particles, much above the effects resulting from the Brownian motion and as they are not random motion but for all the neighbouring particles, cooperating towards the condensing droplet, their importance becomes predominating in comparison to the other diffusion mechanisms.

Of importance in the purification of the atmosphere are the possibilities of repeated coalescences having as final result to bring together by capture, in a single particle of notable size, all those particles which on account of their individual dimension have such low sedimentation velocities that their transit time in the atmosphere becomes considerable.

At the time of the formation even of the cloudy air masses, in the first stages of the condensation, the particles present in the immediate environment of the droplet of the initial cloud have been captured by the inflow of water vapour molecules precipitating towards the condensation nucleus having crossed the critical threshold of growth.

On the other hand, the cloud droplets being in continuous evolution, passing often through alternated cycles of condensation and of evaporation, we can admit that during the fraction T_c of the total life time T of a cloud, there has been condensation, and consequently a period during which the capture mechanisms by molecular kinetic have been rendered possible.

If we indicate by n the number of droplets of the radius r (of falling speed v cm/sec) per cubic centimetre of cloudy atmosphere, the time τ after which the inter-droplet space will be entirely purified will be given by the expression (FACY 1955)

$$\tau = \frac{1}{\dfrac{T_c}{T}\,\chi\,n\,\pi\,r^2\,v}. \tag{49}$$

χ designating the collection efficiency of a condensing droplet. This coefficient is comprised between 1 (even for low values of dp/dx and of the particles of large diameter captured) and a value which may reach several units in the case of high gradients and of very small particles dimensions.

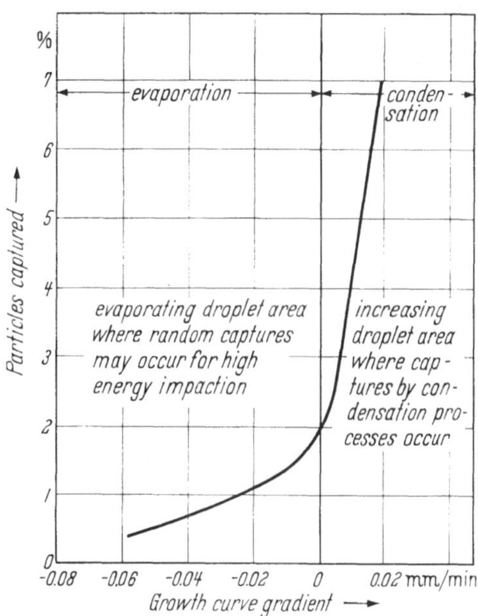

Fig. 18. Effect of vapour gradient on particle capture

For example by taking for χ the unit value, and taking a cloud of 100 droplets (10 microns of radius) per cubic centimetre, a time τ of about 1.40 hours would be sufficient to purify the inter-droplet space of a cloud having devoted only half that time for the growth of its elements by condensation.

Of course in any case, the classical mechanisms of capture by inertia, or more exactly by differential sedimentation between the droplets and the aerosol particles can always interfere, either alone, in the case where the droplet is in thermodynamic equilibrium or during evaporation, either by joining the capture by vapour diffusion, particularly in the case of large particles for which the gravity forces are predominant, and where, consequently, a lesser efficiency of the capture mechanisms during condensation exists.

In the first case it is quite understood that there will be an effective droplet-aerosol capture only if the latter possesses as regards to the drop, a kinetic energy adequate to overcome the $\sigma\,ds$ barier preventing the real liquid-particle contact.

It can thus be seen—providing one is certain to have reached the limit time τ determining the conditions of absolute purification of an interdroplet space— that the value of the activity given in curie per cubic centimetre of condensed water will be identical with the specific activity of the air, given in curie per cubic metre of air, divided by the water content of the cloud, conventionally expressed in grams per cubic metre. In so much as no other scavenging process at the stage of the rain drops or of the snow cristals, should act to modify the final activity of the precipitation collected, we could perhaps—based on the prescriptions of the radio-sounding and with a simple assumption as to the air activity distribution and upon the precipitation mechanism—finally determine by calculations the absolute values of the activity of the latter.

The ice crystal, in the capacity of constituting element of the cloud, plays exactly the same part as a droplet in the process of condensation, but this time with an increased efficiency, at least inasmuch as the crystal is co-existing in the midst of an atmosphere of supercooled water droplets or sursaturated with respect to ice.

In fact, with these conditions which otherwise are thermodynamically unstable, there is a continuous transfer of water molecules resulting from the droplets in process of evaporation upon the increasing crystal since, in negative temperatures, the vapour tension of the ice is lower than that of the supercooled water. The rainfalls in agreement with the BERGERON mechanism (1935) being in fact initiated by an ice phase as an embryo of precipitating element, will accomplish a much better scavenging effect than the rainfalls of the tropical regions where, except in the case of an aged cloud, the mechanisms of capture depends mainly upon the first stages of the increase of the cloudy droplets after the crossing of the condensation level. Likewise a high radioactivity can be collected in the Cirrus clouds whose iced elements perform for the aerosol particles the role of "precipitator by condensation", and this as long as the vapour tension of the ambient atmosphere enables the growing of the crystal already formed.

Naturally, both the crystal and the droplet, after having played the part of captors, may evaporate and leave the cloud before being collected, or become transformed into a precipitating element.

In the case of the droplet, there is a retention with the soluble salts of the primitive condensation nucleus, of the particles captured in the course of the condensing phases. At the last stage of the evaporation a condensation nucleus is re-formed, enhanced with radioactive particles, and available for new cloudy formations.

This mechanism of concentration and of enlargement of the radioactive or atmospheric aerosols may be repeated numerous times in the same air mass and cause the formation of particles having then an appreciable velocity of fall. Such is the case whenever there are fine weather cumulus cloudy formations, alto-cumulus and strato-cumulus formations, and wave clouds associated with the orography.

On the other hand, the case of the ice crystal implicates vise versa, at the time of its sublimation, an individual dismissal of the particles collected in the course of the condensation. This same problem will again be found in the case of the precipitations in the shape of snow falling in an unsaturated atmosphere in relation to the ice.

III. The rain-out of radioactive particles

As to the change of scale which the precipitations involve, other possibilities of capture of atmospheric aerosols present in the trajectories—which the drops of rain or the snow crystals must assume before reaching the ground—must be taken into account.

For the purpose of simplifying, we may estimate for example the scavenging capacity of a rainfall of determined intensity, characterized by the number N

of rain drops, of radius a falling per unit of time upon the unit surface, and by applying our suppositions on the space comprised only between the ground and h, level of the basis of the cloud, the specific activity of the air being homogeneous at the beginning of the precipitation and equal to A_l.

If E_c, the rate of collection efficiency for a rain drop of a radius and a given v radius radioactive aerosol, is:

$$E_c \approx \left[1 + 1.06 \, \frac{9\eta\,a}{2\,dv^2 v} \right]^{-1} \qquad (50)$$

where d is the density, v the radius of the particle collected, v the velocity of fall of the drop, η the coefficient of kinematic viscosity [taking here the approximate comparable value given by ALBRECHT (1931) for one cylinder] we obtain for the decreasing of the radioactivity at the end of a time t

$$A_l(1 - N\,\pi\,a^2\,E_c)^t = A_l\,e^{-N_s E_c t}. \qquad (51)$$

If N_s is written for the product $N \cdot \pi a^2$; the radioactivity a_c which the precipitations will eliminate from the air and will bring to the ground, by surface unity during ϑ seconds, between the time t and the time $t + \vartheta$ can thus be written:

$$a_c = \int_0^h \int_t^{t+\vartheta} A_{l,t_0}\,N_s\,E_c\,dh \cdot dt \qquad (52)$$

which can be expressed in the following simple form:

$$a_c = S\,e^{-\varepsilon t}(1 - e^{\varepsilon\,\vartheta}) \qquad (53)$$

with $S = \int_0^h A_{l,t_0}\,dh$ as sum of the radioactivity comprised between

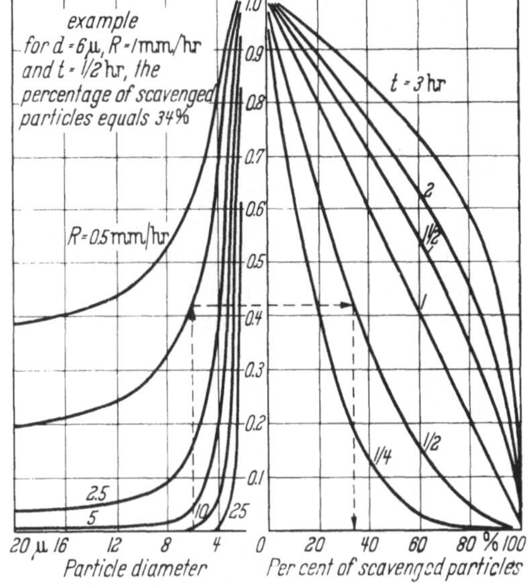

Fig. 19. Percentage of a given diameter particles scavenged from a cloud by raindrops, in terms of rainfall rate R and t, time duration of R rate

the ground and altitude h as the surface of unity, simplified $\varepsilon = N_s E_c$ and admitting A_{l,t_0} as a simple function of h and $\exp(-\varepsilon t)$.

This simple expression of a_c in relation to the parametres of precipitation and of the dimensional characteristics of the radioactive aerosol present under the cloud enables theoretically to calculate the efficiency of the scavenging possibilities of the rain.

GREENFIELD (1956) has planned simple charts enabling to determine with rapidity the percentage of capture by rain-out of aerosol particles of a given diametre, for given intensities of rain (Fig. 19).

The calculations have been made while taking in account the high density of the particles. Likewise the precipitations are expressed at the hour rate, and the distribution law of the rain drops adopted takes account of the observations made by BEST (1950) in this field.

Of course we should in addition take account of the activity already collected by the cloud droplets and add to this value the activity collected by rain-out processes.

All diagrams or charts presented in an attempt to evaluate the possibilities of capture by precipitations figure rather on an order of magnitude. In general they cannot represent the real complexity of the natural phenomena. As a matter of fact the problem is far from being simple. The activity of the air under the cloud might vary by the intervention of air coming either from the return of a convective cell in the vicinity, or from a frontal progression. Furthermore, the possibilities of scavenging of this air are not only connected with the global specific activity, but they depend also closely on the spectrum of the distribution of the particles. Two air masses of the same global activity per cubic metre, one of which would only consist of particles of a few microns, and the other of particles of 1/100th a micron would react in a quite different manner in a scavenging by a rainfall of identical characteristics. Moreover we can see on Fig. 19 that for the smallest particles—those in fact represent almost the total specific activity of the air—the percentage of capture becomes extremely reduced, even in the case of long duration or heavy rainfalls. But this is not entirely true for condensation captures are effective, when the temperature of precipitations are below dew point.

Furthermore, it is quite evident that the outline of a uniform rainfall of N droplets of the "a" radius, per unit surface, or even the adoption of a distribution law is far from corresponding always to the reality; the capture between drops, the ruptures of large drops are normal phenomena in the precipitations exceeding a certain intensity, the coefficient will therefore vary and with it the value of the activity collected by the rain. Finally, in the case of large drops we must not forget that they may vibrate, and at the end of their trajectory they present a face flattened towards the up-stream for which the coefficient E_c will then take very high values, even for aerosols of the micron order. Finally the distribution law and the intensity of the rainfall vary at all levels, principally in the first stage of formation of the precipitation.

IV. Peculiarity of the snow-out processes

The solid form precipitations, from the point of view of the capture of aerosols and scavenging of the atmosphere, proceed from the two previous mechanisms. There is the part played by the elements of large dimensions, the elementary crystal of ice or the snow flake, which in the course of their fall, present on account of their geometry, a given collection efficiency for aerosols of a given size and distribution.

In addition to this mechanism there are also capture effects through diffusion of vapour, since as much in the interior of the cloud as most of the time in the course of the fall of the crystal, the vapour tension of the ice remains below that of the supercooled water, or that of the ambient atmosphere, in the field of the negative temperatures.

In the interior of the cloud, even if the activity of the air has not yet had the time to be transferred in its entirety upon the droplets, the formation of the ice phase causes a continuous flow of aerosols present in the atmosphere around the crystals in formation, and upon the whole of the wake of the trajectory. In fact, the ice phase in the presence of liquid water being in course of continuous and rapid condensation, the vapour tension gradients are of great importance in the neighborhood of the crystal, and we may practically consider that the collection efficiency of a crystal in the course of growth becomes at least equal to the cross section in the course of its fall, taking into account the vapour diffusion effects intervening at the crystal boundaries.

But this only remains true inasmuch as the crystal is in the condensing phase. If, for example, between the base of the cloud and the ground, the snow flake goes through atmospheric layers where should again the temperature be negative, the surrounding vapour tension should become inferior as compared with that of the ice, the snow flake will begin to sublimate and its collection efficiency will instantly drop below the value of the collection efficiency depending upon the geometry of the flake. The effects, which are now repulsive, of the diffusion of vapour place then the ice crystal in the same situation as a rain drop in course of evaporation.

In certain cases it is therefore possible to notice below the snow cloud—and if the temperatures remain negative down to the ground—situations wherein the ice crystal loses all particular efficiency to capture the activity of the air which it traverses.

Inasmuch it is no longer possible to assimilate an ice crystal in process of sublimation to a liquid droplet in course of evaporation. If on one side an energy in excess of $\sigma\, ds$ was at times required for the integration to the drop, on the other hand the aerosol particle once captured would remain acquired, and, even in the case of total evaporation of the droplet, there would remain a residual magma retracing the history of all the captures of the drop.

On the other hand a crystal in course of sublimation resists to any permanent contact between the captured particle by inertia and the crystal itself. The forces of VAN DER WAALS disappear as and when, by sublimation, the crystalline layers disappear under the particle recently captured. The particle is then taken up by the discharge of the air and thrown back into the atmosphere after a very short incorporation in the snow flake.

Furthermore, all the activity picked up by the crystal in the course of its period of growth could then be thrown back into the atmosphere, between the base of the cloud and the ground. This discharge of radioactivity will last as long as the conditions favourable to the sublimation of the crystal will allow the dislocation of its superficial structures and, with them, the particles captured by diffusion during the condensation phases. We will then be able to detect on the contrary an increase in the radioactivity of the air, but solely in the cases where the ambient vapour tension is below that of the ice, taking into account the equilibrium temperature of the precipitation.

It can be seen therefore that there are many various processes, at first sight unaccountable, located most of the time in connection with scales approaching the molecular kinetic or reticular dimensions of the crystal structures, which justify in fact, particularly in the case of solid precipitations the extreme variety of the available data.

V. Total radioactive balance of wet removal processes

The radioactivity brought down finally to the ground by the precipitations thus comprises two distinct contributions successively added.

1. The specific activity of cloud droplets

The first contribution is an activity carried by the droplets of the cloud itself, and which is collected in the air volume where the condensation was performed. If we designate by:

P_a Cloud droplet activity/g of condensed water.

A_h Air activity/standard cubic centimetre.

W Water content of the cloud g/st.c.c.

K Per cent of radioactive particles scavenged by cloud droplets of radius r during condensation processes. We may write

$$P_a = \frac{K \cdot A_h}{W} \ . \tag{54}$$

The volume of cloudy air scavenged after t seconds is according to relation (49)

$$t \frac{T_c}{T} \chi \, n \, \pi \, r^2 \, v \tag{54'}$$

and the value of K indicating the percentage of scavenging processes after these t seconds is:

$$K = \frac{t}{\tau} \quad 0 < t < \tau.$$

τ being the total time in seconds to reach the complete scavenging effect in one cubic centimetre.

T_c/T ratio of time where condensation takes place to total life time of the cloud (with τ for significant upper limit). This ratio can reach 1 for layer cloud continuously rising and remains probably at value $\frac{1}{2}$ for convective clouds with turbulent motions and various scales of eddies.

χ, collection efficiency of droplet in condensation process will always be $\geqq 1$ for large droplets and $\gg 1$ for the smaller ones.

Thus taking 1 as average for T_c/T χ relation (54') now becomes—replacing v by the velocity of fall in STOKES' flow as function of the radius r for a density of 1:

$$K = \frac{t}{\tau} = \frac{4 \pi g}{18 \mu} n \, r^4 \, t \tag{55}$$

and for (54) replacing $W \log n \frac{4}{3} \pi r^3$:

$$P_a = \frac{g}{6\mu} r \cdot A_h \cdot t \tag{56}$$

expressing the value of the specific activity P_a (per example in p.c.) collected by the cloud droplets after t seconds.

After a time $\tau = \frac{18\mu}{4\pi g} \cdot \frac{1}{n \, r^4}$ corresponding to a complete scavenging we obviously have

$$P_a = \frac{3}{4\pi n \, r^3} A_h = \frac{A_h}{W} \tag{57}$$

maximum value for $K = 1$.

This relation is a crude evaluation of radioactivity scavenged by the droplets. For a more accurate computation we have to introduce the law governing the growth of a cloud droplet with time, that is to say from the condensation level H_0 to a level H for a given vertical ascent and moisture of the air mass. P_a will be of the form

$$P_a = = \int_{H_0}^{H} \int_{0}^{t} \Sigma f(r_{h,t}) \, dh \cdot dt. \tag{58}$$

$\Sigma f(r_{h,t})$ indicating that we take into account a distribution law on the cloud droplets whose radii are increasing in function of H and t, the initial instant being that where the cloud droplets started to condensate. For a cloud relatively old—if we can admit that all the radioactivity has already been transferred to the droplets—we may take the relation (57) as an expression of its specific activity.

2. Relation between specific activity and the amount of rain fall

It had already been observed (SUGAWARA 1949) that in the case of rainfalls occurring under similar meteorological conditions, for the same place, there existed an hyperbolic relation, at least approximative between the chlorinity of each rainfall, and the total amount of the precipitation (Fig. 20).

It is similar for the radioactivity, since the distribution law of the radioactive aerosols applies after sedimentation of very large particles, comparable to those of the natural aerosols acting as condensation nuclei. We can therefore say that the specific activity of each rainfall quickly decreases according to the total amount of rainfall. This in fact has been pointed out by numerous authors, and by example on Fig. 21, we have the experimental curves, according to DOPORTO (1960).

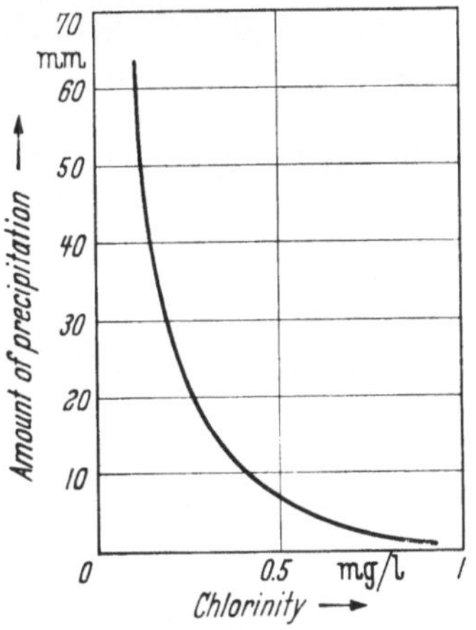

The law giving the activity a_c collected by the rain drops or snow flakes during their fall to the ground level is expressed by relation (53) a_c being the fraction of activity collected between t and $t+\vartheta$. Since we can admit that

$$a'_c \cdot P = C^{te} \qquad (59)$$

Fig. 20. Relation between chlorinity in rain water and amount of precipitation

where a'_c is the specific activity of rain and P the amount of rainfall, we can mention relation (53) to get the activity from the beginning of the rainfall ($t=0$) to the end ($\vartheta = T$)

$$a_c = a'_c \cdot P = S \cdot e\,(1 - e^{\varepsilon\,T}) \qquad (60)$$

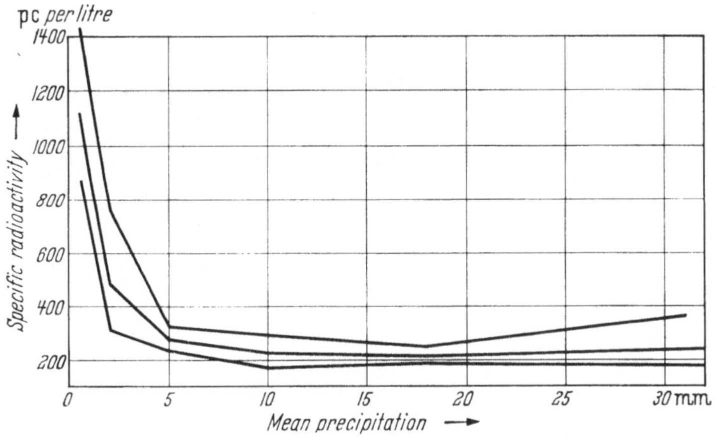

Fig. 21. Mean specific radioactivity for various precipitations

and recalling $\varepsilon = N\pi a^2 E_c$ (where E_c is the collection efficiency), $S = \int_0^{H_0} A_{e,t_0}\,dh$ representing for the surface unit the sum of radioactivity from the ground to the cloud base H_0 with the simplified assumption of N rain drops of radius a per second. We have for P

$$P = \frac{4\pi}{3}\,N\,a^3\,T. \qquad (61)$$

We can write (60) as follows:

$$a_c' = \frac{S}{P}\left(1 - e^{-\frac{3PE_c}{4a}}\right).$$

(62)

3. The total specific activity of precipitations

The final specific activity which will be measured in the precipitation water can thus be given approximately for a frontal rainfall of alto-stratus clouds formed already several hours ago, by completing the relation (57) (given in coherent units) with (62):

$$a_c'' = \frac{A_h}{W} + \frac{S}{P}\left(1 - e^{-\frac{3PE_c}{4a}}\right)$$

(63)

without introducing a distribution law for the droplets or raindrops, and with values of E_c for a definite size of raindrops and radioactive particles in air mass of activity A_l.

If we may express a_c' with more accuracy we should add to Eq. (58) an expression of the form:

$$a_c' = \frac{S}{P}\left(1 - e^{-\Sigma\Sigma\frac{P\cdot E_c}{a}}\right).$$

(64)

where $\Sigma\Sigma\frac{P\cdot E_c}{a}$ would take into consideration a distribution law for the precipitation element as well as a distribution law concerning the sedimentation velocity of the radioactive aerosols, and thereby the coefficient E_c. In any case, it is seen from the number of basic parameters in the equations namely A_h, A_l, a, r, v, n, N, E_c how difficult it is to establish a simple relation between the activity of the precipitations—and that of the air—at the level of the ground, except in the case where $A_h = A_L$ and where the larger part of the radioactivity is due to the scavenging by condensation process in the cloud itself.

D. General considerations on the turbulent mixing

In principle the dry fall-out consists of radioactive particles brought in contact with the ground, or its prolongations, by their own sedimentation velocity. As these velocities are always very small for the aerosols resulting from delayed fall-out it is interesting to review the physical conditions liable to aid the capture of submicronic particles.

A few years ago, only the sedimentation velocities of the sub-micronic particles were taken in account to establish the prevision of residence time in the stratosphere, or the time of residence in the troposphere. In reality, these data, at the particle scale, are of some interest only in certain conditions very rarely realized in free atmosphere, and the gravity forces are determining the concentrations or captures of an aerosol only in the immediate neighborhood of the surfaces where the exchanges are operated.

The determinating factor in the distribution, at the synoptic scale, of the radioactive concentrations is solely the turbulence which, at very variable scales up to 10^{10} orders of magnitude of the value of the diffusion coefficient of the particulate clouds, moves and stirs in a homogeneous manner huge volumes of polluted air.

The direct transport, in a flow, at the hemispheric scale, such as it predominates in the stratosphere, is the first one responsible for the circulation of the atomic clouds originating from the explosions, but the extremely rapid diffusion of the radioactivity towards the lower levels is due to turbulent exchanges between the several layers of the atmosphere finally bringing to the ground—and with a minimum delay—a portion, sometimes important, of the cloud circulating at high altitude.

This accelerated vertical diffusion, which is sometimes established for an isolated level, by the formation of cumuliform clouds, joins in the lower layers the convection rising from the ground. In a direct manner it is still the water vapour which plays the determinating part in these vertical diffusion processes at a very large scale, as it is the stability of the air mass which governs the vertical movements. It is also the struggle of the air masses of different thermodynamic properties which determine the birth and the direction of the low pressure areas responsible for the latitudinal and meridian diffusion in the troposphere.

But it is possible that in the very high levels, and for different causes, these exchanges by vertical diffusion are not solely limited to transfers towards the lower layers. The experiments made above altitudes of from 20 to 30 km are very difficult, and our knowledge in the matter of stratospheric circulation is yet too limited to enable us to affirm, that part of the radioactivity already issued could not be diffused to a height, which, for the finest particles, could reach 1/10th or 1/100th of a millibar. The diffusion coefficient of the very fine particles increases with the altitude, and for explosions made at high levels the radioactive cloud entails almost exclusively particles of the order of 1/100th of a micron.

It is not impossible that for powerful explosions at high levels a part of the radioactivity remains locked up above 40 km during a period liable to reach several half-periods of the most dangerous products such as the strontium 90 or the cesium 137.

Our knowledge of the circulation in the upper atmosphere is yet too limited to enable us to invalidate this assumption, which would give a simple explanation of the deficits noticed in the total amount of the radioactivity spread around since the first atomic bomb. Radiation measurements with rockets might perhaps be interpreted in this sense, and give us the partial explanation of the lack of details experienced up to now in the fall out on the globe.

In each layer of atmosphere, where the vertical diffusion has thus transferred the radioactivity from the atmospheric circulation, the diffusions intervene afterwards according to the meridian or zonal components. If, at the limits of the troposphere, the vertical diffusion was at a maximum in the discontinuous zone existing between the equatorial tropopause and the polar tropopause, there exist likewise privileged zones for the horizontal diffusion. The latitudes swept by the perturbations of the polar front and by the trajectories more or less privileged of the jet-streams, then correspond to the turbulence zones with accelerated diffusion, where the intervention of the wet removal processes is the most effective. At a lower scale, relief accidents can cause on account of certain prevailing winds, high concentrations in the precipitation water, on account of the formation of orographic clouds, and of the permanency of the scavenging mechanisms.

E. The direct ground capture of air activity

I. Coagulation with soil particles and foliar uptake

If in the layers next to the ground abstraction is made of the vertical turbulent exchanges, the number of radioactive particles of dimensions less than the micron should follow a constant progression in the low layers next to the ground, and the balance of the supply should correspond with the list of the advantages coming from the stratosphere and of the captures by the ground. If the latter was comparable to a smooth surface, without roughness, and if no wash-out mechanism should interfere, we would notice in the low layers of the atmosphere a concentration of radioactive aerosols, which quite before the stop of the explosions would have reached dangerous values for the respiratory organs of living beings.

The integration of the radioactive fall-out into the lithosphere—notwithstanding its inevitable consequence of selective fixation and of the introduction of certain elements into the food chain—offers, however, the advantage of avoiding an excessive pollution of the air through an atmospheric radioactive mud, which through turbulent diffusion would very soon have been able to spread over the whole surface of the earth. It is quite evident that if the megacuries of Sr-90 and Cs-137, thrown up so far into the atmosphere, had been in the form of a heavy gas, chemically stable and of reduced solubility in water, there would not have been any problem of the fall-out, but there would exist a generalized radioactive polution of the air of the low layers, a pollution whose decrease would only be effective within the proper life time of the elements.

The integration of the aerosols resulting from the fall-out on and into the ground surface may be considered as established in an irreversible manner as soon as there has been contact between the radioactive particle and the ground. Even if the ground is dry and pulverulent, the granularity of its elements is always much greater than that of the aerosols composing the fall-out. The activity is deposited upon elements adequately coarse, so that even in the case of sand or dust winds can only give afterwards relatively limited displacements in time and in space. This fixation is still increased, for the most soluble salts, through their burying in the deeper layers of the ground.

The conditions favourable for the capture by the ground, without any intervention of the atmospheric water, depend upon both the structure and the geometry of the world surface or the obstacles with which it is connected, and mainly on the intervention of determinating meteorological conditions.

A naked ground, rough, with a profile encouraging the formation of eddies or stationnary ascents due to the orography, or whatever the scale may be: sand ripples, ridges, orographic irregularities, a superficial discontinuing or porous structure, can also hold back and absorb definitely the aerosols which on polished or smooth surfaces, could otherwise be picked up again by the atmosphere.

As to the meteorological conditions, a stability of the lower layers of the atmosphere as well as a weak and localized turbulence, will aid the captures by dry sedimentation.

But it is particularly the vegetation which plays the most important part in the capture of aerosols in suspension, by both the dry and the wet removal. Each stem, each leaf will take up a part of the content of aerosol from the ambient air with an efficiency which, roughly speaking, is proportional to the wind speed, the density, and to the square of the radius of the aerosol particle, and inversely proportional to the diameter of the capturing obstacle.

While a tree trunk, or vegetables of large section, only have a negligible collection efficiency towards the submicronic aerosols, this is not the case with the cereals, in general the graminaceous plants, and the most of the species of the vegetal cover.

The collection efficiency becomes also important for the coniferous trees whose needles make excellent captors. Those trees are the vegetal species first deteriorated in the neighborhood of factories emitting toxic products in the atmosphere. In the neighborhood of aluminum factories for instance, whenever the meteorological conditions allow the formation of fogs at ground level the hydro fluoric acid which is dissolved destroys in less than 24 hours the coniferous needles, whereas the leafy tress only show occasional attacks not imperiling the life of the tree.

It is therefore difficult to compare, within a given region, the quantities of Sr-90 sampled on the foliage parts of vegetation, if, at the same time, we do not mention the nature, or rather the geometry of the growth having been selected for the sampling. Likewise, for a given cereal for instance, at a stage of identical development, and for a same concentration in aerosol, the captures will be directly proportional to the force of the prevailing winds.

II. Water vapour and ground captures

The complexity of the factors increasing the dry captures by the ground or the vegetation is yet increased if we take into account the constant exchanges of vapour water between the ground and the lower layers of the atmosphere.

1. Dew and rime formation

As soon as the condensation occurs, the surface of the ground plays the part of a vapour pressure well, and it then possesses a collection efficiency very high due to the action of the molecular kinetic mechanisms.

There will therefore be an efficient capture of aerosols present in the vicinity of the ground, each time that the formation conditions of dew or of rime will be reached upon the ground or upon the vegetation. The rime will be particularly efficient in the presence of supercooled water fogs, but the definitive integration to the ground will take place only if there is subsequent fusion of the ice deposited; all sublimation having as result to restitute to the atmosphere the aerosols captured by the crystal.

This same mechanism of condensation will facilitate the capture of the aerosols by the water plants, seas, lakes, or rivers, whenever masses of hot air, saturated, will come and will run above cooler water surfaces. Most of the time these exchanges will be made through the relay of fog whose droplets are evaporating in the boundary layer of the intersurface.

2. Capillary condensation on dry surfaces

But the contributions of water to the terrestrial surface are not limited to these examples, the affinity of the ground for atmospheric water vapour is an important phenomenon, continuous at certain hours of the day, and which under the name of "occult precipitations" allows to set the fixation of numerous vegetal species in the arid zones.

These secondary damp ground sources, inasmuch as it may appear paradoxical, constitute the equivalent "in transpiration" of the world surface, of several metres of annual rainfall.

For example, in the 4 or 5 superficial centimetres of a dried up muddy earth, the quantity of water brought by capillary condensation may reach 250 g per square metre, during the summer night, in regions with a climate of the mediterranean type. These contributions are all the more important if the dessication state is more advanced. Even for a relative humidity of the air, of about 70%, these nightly contributions may reach 70 g per square metre.

This night fixation of the water vapour by the capillary elements of the ground applies to superficial temperatures higher than the dew point. It is a true adsorption and, when the dew point is reached, condensation in form of a superficial dew, often not visually noticeable on the bare soil, may be observed. Very high proportions of aqueous depositions can thus be reached, even on the bare ground; about 400 g per square metre and per night (Figs. 22 and 23).

Of course, in the daytime, the sun radiation will recuperate from the ground the greatest portion, or even the whole of the water incorporated during the night, but from the point of view of the capture of the aerosols, these latter nevertheless remain incorporated on the earth surface. This is furthermore the explanation of the effect noticed currently in the regions where the low specific humidity of the ground allows these fixation phenomena; in stable conditions and with little wind the purity of the air of the lower layers is at a maximum at sunrise. In addition we have to the settling of particles under gravity the action of the vapour pressure gradient of the water molecules impinging upon the ground. In view of

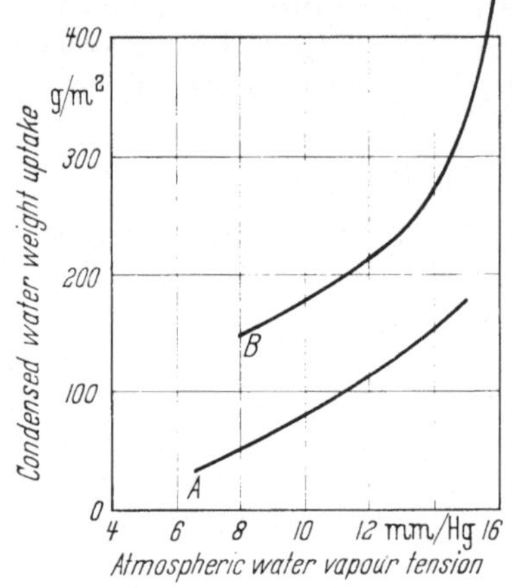

<table>
<tr><td>Fig. 22. Soil adsorption and condensation in terms of atmospheric water vapour tension</td><td>Fig. 23. Condensed water weight (adsorption and condensation phenomenae) in terms of soil moisture</td></tr>
</table>

the quantity of water adsorbed per square metre of the ground, these actions may be extended to a layer of a few hundred metres of atmosphere, and thus increase, sometimes in a spectacular manner, the horizontal visibility in the lower layers.

F. Near future researchs in the fall out field

In fact fall out studies are not peculiar topics in the field of radioactivity, they are mainly geophysical, meteorological problems relative to various branches of physics: dynamic of small particles, micro processes in the atmospheric condensation phasis, and with minor importance, the actions of various forces originated at the particle's scale by electrostatic, gravitational, heat, and radioactive processes.

This is for the microphysical scale. But the macrometeorological scale is obviously determinant when speaking of world wide or close-in fall out.

With the hope that God, and mankind's wisdom, will for ever, keep away the unthinkable use of atomic weapons, the problem concerned with fall out is not completely avoided so far. The peaceful use of atomic energy with the dangers of routine or accidental release from reactor plants, the expected development in aircraft and ship building using atomic power, will induce problems very similar to those of fall out from experimental tests of these last years.

No doubt that an accurate knowledge of all the mechanisms which are able to increase the settling velocities of radioactive submicronic particles will be the best way to lead to an efficient fall out control in the atmosphere.

But a complete knowledge of the meteorological factors is also necessary for both microphysical and meteorological processes responsible of the final accretion to ground level of the dangerous dusts. By example it is not unbelievable that powerful heat sources may be conveniently disposed around or down wind the stacks to introduce strong artificial convection in the areas where accidental releases are to be expected.

On the other hand, all microphysical processes which, by example, are presumed to increase coagulation must be studied separatly with the practical background of a possible use at an industrial scale.

At this same microphysical scale, condensation must provide the most tremendous scavenging effect at the cheapest conditions. Powerful sprays at adequate level in the dangerous area, of both sursaturated salt solutions and pure water are to be experimented whenever the possibility of heavy but local wash-out are supposed to be preferable to natural dispersion of dust clouds; with low wind velocities, and in non desertic areas it will be perhaps one of the most efficient fall out control techniques in emergency situations.

All these studies, eventually all these fall out instrumental modification at an industrial scale, are likely cloud physicists techniques and will progress in the same way. As to the peaceful use of atomic energy, the possibility of an effective interfering of man in this field is surely more expectable than in artificial weather modification. Even in consideration of the huge volumes of polluted air expected from accidental release, they are far beyond the meteorological scale, and reasonably accessible to man prior to the ruling of rain and storms.

General bibliography

High altitude sampling program, DASA 532, Washington, Defense Atomic Support Agency, June 1, 1960.
Micromeretics, J.M. DALAVALLE. New York and London: Pitman Publishing Corp. 1949.
Publications of the A.E.C.:
    Fall-out Control, Final report, SRIA-3, C.E. LAPPLE, August 1, 1958.
    Meteorology and atomic energy, July, 1955.
    Worldwide effects of atomic weapons, Project Sunshine, AECU 3488, Rand Corp., August 6, 1953.

Annoted bibliography

AITKEN, J.: On the formation of small clear spaces in dusty air. Nature, Lond. 29, 322—324 (1884).
ALBRECHT, F.: Theoretische Untersuchungen über die Ablagerung von Staub aus strömender Luft und ihre Anwendung auf die Theorie der Staubfilter. Z. Physik 32, 48—56 (1931).
ARENBERG, D.: Turbulence as the major factor in the growth of cloud drops. Bull. Amer. Meteor. Soc. 20, 444—448 (1939).
BERGERON, T.: On the physics of clouds and precipitations. Proc. 5th Assembly UGGI, Lisbon, vol. 2, 1933, Ed. P. Dupont, Paris, 1935.
BEST, A.C.: The size distribution of rain drops. Quart. J. Roy. Meteor. Soc. 76, 16—36 (1950).
BRUN, E.A.: Impigement of water droplets on a cylinder in an incompressible flow field. NACA/TN 2.904, 1953.
DOPORTO, M.: The deposition of airborne radioactive particles and the cleaning effect of precipitation at Valentia. Met. Serv., Technical note No. 27, Dublin, April 1960, Fig. 2—3.
EINSTEIN, A.: Zur Theorie der Radiometerkräfte. Z. Physik 27, 1—6 (1924).
EPSTEIN, P.: Zur Theorie des Radiometers. Z. Physik 54, 537—563 (1929).
—' On resistance experienced by spheres in their motion through gazes. Phys. Rev. 23, 710 (1924).
FACY, L.: La capture des noyaux de condensation par chocs moléculaires au cours des processus de condensation. Arch. Meteor. Geophys. u. Bioklim., Sér. A 8, 3 (1955).
— Sur le déplacement des particules d'aérosols au cours des processus de diffusion moléculaire. C. R. Acad. Sci. Paris 246, 102—104 (1958).

Facy, L.: Sur un mécanisme de capture des particules d'aérosols par une gouttelette en voie de condensation-évaporation. C. R. Acad. Sci. Paris **246**, 3161—3164 (1958).

Greenfield, S. M.: Ionization of radioactive particles in the free air. J. Geophys. Res. **61**, I, 27—33 (1956).

— Rain scavenging of radioactive particulate matter from the atmosphere, P. 883 AEC, At. Energ. Com., July 16, 1956; s. auch J. Meteorology **14**, 115—125 (1957).

Lamb, H.: Hydrodynamics, p. 604. New York: Dover Publ. 1945.

Langmuir, Irving: The production of rain by chain reaction in Cumulus clouds at temperature above freezing. J. Meteorology **5**, 175—192 (1948).

Oberbeck, A.: Über stationäre Flüssigkeitsbewegungen mit Berücksichtigung der inneren Reibung. J. reine angew. Math. **81**, 62 (1876).

Paranjpe, M. K.: The convection and variation of temperature near a hot surface. Part I. The dust-free or dark layer in relation to surface convection. Proc. Ind. Acad. Sci. A **4**, 4, 423 (1936).

Rayleigh: On the dark plane which is formed over a heated wire in dusty air. Proc. Roy. Soc. Lond. **34**, 414 (1882).

Smoluchowski, M.v.: Versuch einer mathematischen Theorie der Koagulationskinetik kolloider Lösungen. Z. phys. Chem. **92**, 129—169 (1917).

Sugarawa, K., et al.: Separation of the components of atmospheric salt and their distribution. Bull. Chem. Soc. Japan **22**, 47 (1949).

Telford, S.W., N.S. Thorndike and E.G. Bowen: The coalescence between small water drops. Quart. J. Roy. Meteor. Soc. **81**, 241—250 (1955).

Tyndall, S.: On the action of rays of high refrangibility upon gaseous matter. Proc. Roy. Inst. **18** (1870).

Whitlaw-Gray, R., and H.S. Patterson: A study of aerial disperse systems, Smokes, p. 57. London: E. Arnold 1932.

Biological aspects

by

A. T. Krebs and N. G. Stewart

With 14 Figures

Zusammenfassung

Einer kurzen Darlegung der allgemeinen strahlenbiologischen Prinzipien über die Wirkung kleiner und kleinster, über längere Zeiträume wirksamer Dosen, folgt eine Abschätzung der von den Kernstrahlungen im geophysikalischen Raum zu erwartenden biologischen Effekte (somatische Effekte und genetische Effekte). Es wird gezeigt, daß diese Abschätzung nur auf Grund von Extrapolationen der mit viel höheren Dosen gewonnenen experimentellen Befunde möglich ist, und es wird gefolgert, daß bisher, selbst bei Beachtung des Schwellenwertproblem-Komplexes, keine eindeutige Beziehung zwischen biologischen Effekten, wie Krebshäufigkeit, Leukemiafällen, Lebenszeitverkürzung z.B., und den im erdnahen Raum anzutreffenden Strahlendosen nachgewiesen werden konnte. Selbst die spontane Mutationsrate beim Menschen kann nach den besten zur Zeit vorliegenden Schätzungen nur etwa zu 6% auf einen möglichen Einfluß der Umgebungsstrahlung zurückgeführt werden. Für die im Weltenraum anzutreffenden hohen Strahlendosen (van Allen-Gürtel, partikulare Höhenstrahlung, solare Eruptionen) gelten die strahlenbiologisch gut studierten und gesicherten Beziehungen zwischen Strahlendosis und Strahleneffekt im Dosenbereich von 25 r bis zu vielen Tausenden von r.

In bezug auf die vom Menschen geschaffene radioaktive Umweltverseuchung (man-made environmental contamination) liegen besondere Verhältnisse vor, die vorwiegend Fallout-Effekte, Beseitigung und Unschädlichmachung von radioaktiven Abfallprodukten und die mit dem Betrieb von Reaktoren und sonstigen Kernanlagen verbundenen Gefahren und Verseuchungsmöglichkeiten betreffen. Der größte Teil der Strahlungen von Fallout stammt von den Fissionprodukten, die bei und in thermonuklearen Großversuchen erzeugt und mit der Explosionswolke hoch in die Stratosphäre getragen werden. Der „Ausfall" dieser Produkte, die mehr oder weniger schnelle Rückkehr zur Erde, erfolgt nach sehr komplizierten Gesetzen und hängt von mehreren Faktoren ab, so z.B. von der geographischen Breite, der Höhe der ursprünglichen Explosionswolke, Regenverhältnissen, Jahreszeiten und thermodynamischen Bedingungen. Im allgemeinen werden die geringsten Fallout-Mengen in der äquatorialen Zone gefunden; in höheren Breiten scheinen die ausfallenden Mengen in den westlichen Ländern etwas höher zu sein als in den nicht-westlichen Zonen. Die Fallout-Produkte verseuchen die Nahrungsstoffe von Mensch und Tier, und biologisch wichtige Isotope werden vorwiegend über die Kette — Pflanze, Tier (Kuh, Milch), Mensch — in den Körper eingeführt und teilweise in lebenswichtigen Organen gespeichert: I-131 in der Schilddrüse, Cs-137 im Gewebe, Sr-90 im Knochen. Diät und Menge der täglich inkorporierten Nahrungsmittel sind ausschlaggebend für Art und Menge der im Körper verweilenden radioaktiven Isotope. Für Sr-90 scheinen nach den bisher vorliegenden Messungen des Strontiumgehaltes menschlicher Knochen keine großen Unterschiede für die verschiedenen Teile der Erde und große Bevölkerungsgruppen zu bestehen.

Werden die im U.K. gemessenen Gesamtdosen von Fallout als Grundlagen und Richtlinien für die gesamte Erde genommen, so kann gefolgert werden: die von äußerer Fallout-Strahlung und von im Körper abgelagerten Cs-137 an das Knochenmark und die Geschlechtsorgane gelieferten Dosen betragen nur einen kleinen Bruchteil der natürlichen Umgebungsstrahlungsdosen, die hinwiederum viel geringer sind als die von der I.C.R.P. für die Gesamtbevölkerung empfohlenen maximal zulässigen Dosen.

Die bei der technischen Verwertung der Atomenergie entstehenden Abfallprodukte sind so verschieden in ihrer Aktivität und chemischen Form, daß keine für alle Fälle brauchbare Methode zu ihrer „Unschädlichmachung" und Beseitigung empfohlen werden kann. Hochaktive Abfälle werden im allgemeinen — zur Volumverminderung — konzentriert und in auf Dichtigkeit geprüften Spezialbehältern „gelagert"; mittel- und wenig aktive Abfälle werden, da Konzentration zu kostspielig würde, unter sorgfältig kontrollierten Bedingungen in Abfallgruben, Bergstollen oder Höhlen deponiert, in die See versenkt oder fein in die Luft verteilt. Dies erfordert absolut zuverlässige Überwachungsmethoden, die, strikt eingehalten, es möglich gemacht haben, daß bisher die gesamte genetische Dosis von allen Abfallbeseitigungen, Abwässern und Abgasen klein ist im Vergleich zu den von Atomwaffenversuchen gelieferten Dosen, die ihrerseits klein sind im Vergleich zur Strahlenbelastung durch die natürlichen Umgebungsstrahlungen.

Die Gefahren, die durch Freiwerden radioaktiver Isotope in Reaktorbetrieben und anderen Kernanlagen entstehen können, werden in Einzelheiten diskutiert, und Beispiele für die maximal zulässigen Mengen spezieller Isotope die durch Fabrikschornsteine verschiedener Höhen in die Atmosphäre abgeblasen werden können, werden gegeben. Bei dieser Gelegenheit wird die Ergänzung der I.C.R.P.-Empfehlungen durch weitere experimentelle Untersuchungen über atmosphärische Diffusion, Verhalten kleiner Kerne und verwandte Probleme erörtert. Beobachtungen der tatsächlich in die Atmosphäre entweichenden Radioaktivitäten unter normalen Betriebsbedingungen sowohl als auch in Unglücksfällen, führen ebenfalls zur weiteren Untermauerung der Schutzbestimmungen, wie am Windscale-Reaktor-Unfall, im Oktober 1957, diskutiert und dargelegt wird. Auf Grund dieser und anderer Unfallerfahrungen werden Richtlinien für die Wahl des Standortes und der Lage künftiger Reaktoren und anderer Kern-Großanlagen entwickelt. Die ständige Überwachung und Revision dieser Richtlinien an Hand neu gewonnener Erkenntnisse und Erfahrungen ist eine unumgängliche Forderung.

A. Introduction

The nuclear radiations in geophysics present—with a few exceptions [1]—typical low-dose long-term irradiation phenomena. The study of such chronic irradiation phenomena is quite difficult and systematic approaches have been started. The symposia reports on "Low-Level Irradiation", edited by A.M. Brues (1959); on "Immediate and Low-Level Effects of Ionizing Radiations", edited by A.A. Buzati-Traverso (1960); and on "Radioisotopes in the Biosphere", edited by R.S. Caldecott and L.A. Snyder (1960), give a good picture of the situation in the field, especially on the problematic encountered in the range of small and smallest doses.

---

[1] e.g. van Allen belts, some man-made conditions.

Present discussions on the biological significance of high energy radiations in life's radioactive environment have to be based on extrapolations of animal experiments, clinical observations, and radiation accidents in the higher dose ranges, mainly on acute radiation exposures with doses higher than 25 r.[1]

No experimental studies have been done so far in dose ranges comparable with the nature-given conditions.

Of interest in the discussion of the biological aspects are the distinction between actue effects and chronic effects, total-body irradiation effects and partial-body irradiation effects, somatic effects and genetic effects. The response of an irradiated system also depends upon the way in which the radiation is administered; if prompt, fractionated, protracted or chronic. The effects of greatest interest in low-dose long-term exposures are the somatic effects, especially production of cataracts, cancer, leukemia, and life shortening and the genetic effects, primarily the radiation induced mutations. Goal of the studies is, to find the dose-effect relationship in the very low-dose range, down to "zero dose and zero effect".

B. Nature-given radiations

These radiations influence life in two ways, as external irradiations and as internal irradiations. External radiation sources are the radioactive elements in soil, water, and air, in building and construction materials, as well as the cosmic radiations in their different forms. Internal radiation sources are the radioactive substances, living systems incorporate from their environment with food, water, and air.

I. External radiation sources

1. Natural background radiations

The external irradiation of living systems varies with the geographical and geophysical conditions. The range is very broad, covering several magnitudes. Exceptionally high radioactive background doses are found in some monazite sand regions, so in the Kerala region in India and in some Brazil States. Some tables may illustrate the situation and give a basis for the radiobiological evaluation.

External doses over rock formations vary according to Table 1:

Table 1. *Dose rates from external gamma radiation from different rock formations*

Kind of rocks	Dose rates in mrad* per year for			
	Ra^{226}	U^{238}	Th^{232}	K^{40}
Igneous rocks	24	25.8	36.8	34.6
Sedimentary rocks				
Sandstone	13	7.7	18.4	14.6
Shales	20	7.7	30.6	30
Limestone	7.7	8.4	4	—

(Report United Nations)

* rad: absorbed dose. 1 rad equivalent to dissipation of 100 ergs of energy per gram of matter. mrad = 0.001 rad.

---

[1] (r) roentgen: exposure dose. 1 r produces 1.6×10^{12} ion pairs per gram of air.

Table 2 presents some dose rates for different countries:

Table 2. *Dose rates from external gamma radiations out-of-doors in different countries*

Country	Dose rate in mrad per year	Country	Dose rate in mrad per year
France	45 to 350	Austria	58
United States .	50 to 160	Sweden	50 to 120

(Report United Nations)

External gamma radiation dose rates inside buildings have been measured in many locations. Table 3 presents data for Sweden, Great Britain, and Austria.

Table 3. *Doses from external gamma radiation inside buildings*

Location	Building material	Dose rates in mrad per year
Sweden	Wood	48 to 57
	brick	99 to 112
	concrete	158 to 202
Great Britain	all granite	99 to 107
	concretebrick	73 to 81
Austria	wood	54 to 64
	all granite	85 to 128
	brick	75 to 86

(Report United Nations)

For comparison Table 4 brings dose rates from the Kerala region in India and from the Brazil States.

Table 4. *External irradiation dose rates in special areas*

Kerala region (India) (in doors)		*Brazil States* (out doors)		
Location	Mean dose rates (mrad/year)	Location	Dose rates (mrad/year)	
		States of	average	peak val.
Kadiapattam	2814	Rio de Janeiro	500	1000
Midalam	1573	Espirito Santo		
Karamanal	1283	Minas Gerais	1600	12000
Varkala	1376	Goias		

(Report United Nations)

Average dose rates for different dwelling areas were reported by L.R. Solon, W.M. Lowder, A. Shambon, and H. Blatz (1960). They found for the Metropolitan New York area (Bronx, Manhattan) indoor values from 9 to 13.5 µrad per hour and outdoor values of up to 12.4 µrad/hour. The mean annual dose rates for 38 principal US cities varied from 73 to 197 mrad per year. Some cities with their activities are quoted in Table 5 (B. Wallace and Th. Dobzhansky 1959).

The variation of terrestrial dose rate with altitude has been calculated for igneous rock, sedimentary sandstone, sedimentary limestone and intermediate stone by O'Brien, Lowder and Solon (1958).

The radioactivity of fresh surface water is often due to the presence of radon in higher concentrations than corresponding to the radium content of the water.

This holds especially for the radioactive springs of the radium-spas in Europe, the strongest being Radiumspa Oberschlema in Saxony and St. Joachimsthal in Bohemia. Here the geological conditions favor the uptake of high amounts of radon by the water, while penetrating from the depths through the dislocations

Table 5. *Background radiation levels (incl. cosmic radiation) in mrad per 30 years, in some cities of the United States*

City	mrad per 30 years	City	mrad per 30 years
Harrisburg, Pa. . .	2640	Denver, Colo.	4410
Pittsburgh, Pa. . .	2880	Colorado Springs, Colo.	5040
Cleveland, Ohio . .	2730	Grand Junction, Colo.	4140
Toledo, Ohio . . .	2280	Albuquerque, N. M. .	3480
Chicago, Ill. . . .	2640	Amarillo, Tex. . . .	3240
Madison, Wis. . .	2520	Oklahoma City, Okla.	2520
Minneapolis, Minn.	2760	Tulsa, Okla.	2760
Sioux Falls, S.D. .	2850	Little Rock, Ark. . .	3180
Cheyenne, Wyo.. .	4260	Memphis, Tenn. . . .	2850

and the crevasses in the granite massives with their high radium content. The springs consequently show a high radon content, Radiumspa Oberschlema 13 500 Mache units per liter and more.

The radium content of the different waters varies with origin and the kind of water. Ocean water contains $(0.7 \text{ to } 7) \times 10^{-17}$ g/cm^3, rivers show activities from 10^{-15} to 10^{-17} g/cm^3, public water supplies vary usually between 7×10^{-15} and 10^{-16} g/cm^3 while springs can show concentrations in the order of 10^{-10} g/cm^3 (United Nations Report, 1956). The doses delivered by the different water supplies will be considered in the discussion on the role of the "maximum permissible intake" in a later chapter.

The average concentration of radon in soil is about 10^{-13} curie/g. It is injected into the atmosphere at a rate of about 4.3×10^{-10} curie/hour/m^2. The average "equivalent" concentration of radon with its decay products in the air is approximately 1 to 3×10^{-13} curie/liter. The average concentration of thoron in the air is about 0.5×10^{-13} curie/liter, varying with geological and meteorological conditions (Table 6).

Table 6. *Concentrations of radon and thoron in equilibrium with their decay products present in the air in various regions, and corresponding calculated doses*

Country	Average concentrations in curie/liter $\times 10^{13}$		Dose in mrad/year	
	Rn	Tn	Rn	Tn
Czechoslovakia .	8.0	—	11	—
Great Britain . .	3.0	—	4.3	—
France	2.0	0.6	2.8	0.8
Austria	1 to 3.0	—	0.5 to 4.3	—
Sweden	1.0	—	1.4	—
USSR	1.0	—	1.4	0.7

Interesting facts on the specific distribution of the different decay products in the air are given by H. ISRAEL (see above). The natural radioactivity of air, water and soil, including tectonical problems, has been discussed by W. STRATHMANN (1957).

2. Cosmic radiation

The doses delivered by the different cosmic radiations (H.E. Newell and J.E. Naugle 1960) depend on location in space, location on earth and they change with latitude, altitude, and time. Three main radiations have to be distinguished, the cosmic rays, particulate radiation in interplenetary space, and trapped radiations. They contribute different doses to the over-all radiation scheme.

The dose rates arising from "classic" cosmic radiation, taken from UN-Report, are (Table 7):

Table 7. *Cosmic ray dose rates at different altitudes and latitudes*

Altitude m	Dose rates in mrad per year	
	at 50° latitude	near equator
0	41	35
1500	66	44
3050	128	89
4580	263	175
6100	500	340

Some time variations of cosmic ray intensities are correlated with the eleven year solar cycle and solar flares. These two processes influence mostly the components of low penetration power, so that the changes become most apparent at higher altitudes. While the more regular changes, connected with the solar cycle, occur in reasonable limits (S.E. Forbush 1958), the irregular variations associated with solar flares, can be tremendous according to H.J. Schaefer (1959). An example is given by R.I. Allen, A.J. Dessler, J.F. Perkins and H.C. Price (1960), who quote an increase of 10^4 times of the normal cosmic ray intensity for one of the most outstanding solar flare proton events.

The integral doses from particulate radiations mainly from the heavy primaries above 70000 feet with energies up to 10^{18} to 10^{19} eV — are small. These particles produce intensive exposures to a very small amount of material along the track of the individual particle and/or in star formation. The frequency of these events, however, is relatively low, about 28 medium and heavy primaries per cm^3 of tissue per day at the top of the atmosphere, according to H. Yagoda (1957).

Trapped radiations are found in the van Allen belts. These radiations consist of electrons with energies from several eV up to 1 Mev, and of protons with energies of a few thousand eV up to 700 Mev, trapped along the magnetic field lines of the earth. They form the belts, one containing the hard component, the "proton belt", centered at about 11×10^3 km from the earth's magnetic axis, and another containing the electron component, the "electron belt", centered at about 22×10^3 km from the earth's magnetic axis, J.A. van Allen, C.E. McIlwain and G.H. Ludwig (1959); J.A. van Allen (1961). Some estimates on the dose rates encountered in the belts have been made. While the dosage from cosmic radiation at sea level is about 35 to 50 mrad/year and the dosage in interplanetary space about 5 to 12 r/year, the dose rates in the belts are, according to Newell and Naugle about 24 r/hr in the center of the proton belt and up to 200 r/hr in the heart of the electron belt. Characteristic variations of the dose levels along the trajectory of Explorer III and Explorer IV are given by R.I. Allen et al. (1960).

The total external radiation burden from nuclear radiation in the geophysical sphere shows, in a rough outline, the following pattern (Table 8).

Table 8. *External radiation doses from the various natural sources in geophysics*

Radiation sources	Dose rates
A. At sea level	
1. Cosmic radiation	35 mrad/year
2. Ordinary regions	
γ-rays over rocks	25 to 120 mrad/year
γ-rays outdoors	48 to 160 mrad/year
γ-rays aerial sources	1 to 10 mrad/year
3. Active regions	
γ-rays granitic regions France	180 to 350 mrad/year
γ-rays Kerala region India . .	131 to 2800 mrad/year
γ-rays Brazil States	1600 to 12000 mrad/year
B. In space	
Interplanetary space	5 to 12 rad/year
Inner van Allen belt	24 rad/hour
Outer van Allen belt	200 rad/hour
Solar proton event	10 to 1000 rad/hour

II. Internal radiation sources

The doses from incorporated radioactive substances depend on the kind of the incorporated nuclides, their chemical properties, and their distribution in the body. Some organisms are able to cummulate natural radioactive materials from their environment, other organisms discriminate in favor of chemically similar elements and store selectively limited amounts in certain organs (W. VERNADSKY 1930; R.D. EVANS, A.F. KIP, and E.G. MOBERG 1938).

The hazards and dangers, connected with the incorporation of radioactive substances were recognized on a broader basis in the early twenties, when the first radium poisoning cases in USA occurred (radium-dial painters, radium-water drinking cures) and in the mid-thirties, when in some European radium and thorium refining and encapsuling factories several chemists and technicians died from radium and thorium poisoning. The experience gained in the study of these cases gave new impulses to the question into the natural body radio-activity of man, plant, and animal, the "bio-radioactivity" according to M. TOMMASINA (1904). Asking this question again in the thirties with emphasis on the radium content of the human, it was understood that the above experiences would present the best basis for establishing so-called "tolerance doses", suggested by A. MUTSCHELLER (1923) and others.

The first systematic approach to the problem was made by R.D. EVANS (1937), who determined the radium content of living persons. Measurements of the normal and abnormal radioactivity of human and mammalian tissues were made by A. KREBS (1939) followed by measurements on crematory ashes by A. KREBS in 1942 and by J.B. HURSH and A.A. GATES in 1950.

Today a wealth of information is available on the different radioisotopes which constitute the "radioactivity" of the human. The development of modern nuclear equipment, especially of the scintillation counter from the simple original devices to the modern whole-body scintillation counters (A. KREBS 1955; E.C. ANDERSON, R.L. SCHUCH, J.D. PERRINGS, and W.H. LANGHAM 1956; M. BRUCER

1960; H.E. Voress 1960) has contributed decisively to the present knowledge in the field.

The newest data on the amounts of the different radio nuclides deposited in the human skeleton, compiled by A.F. Stehney (1960), are given in Table 9.

Probably the most dangerous of the above mentioned isotopes is radium-226 on account of its affinity with calcium (bone seeker) and its long biological half-life of about 1620 years. Therefore, much attention and much time have been spent to determine the amounts of radium incorporated and cummulated by the human in dependence of environmental factors and dietary habits. While however the study of radium poisoning cases (R.D. Evans and others) had led to the figure of 10^{-7} g (0.1 microcurie) of radium-element as "maximum permissible amount" in the total body (NBS-Handbook, No. 69), the determinations of the radium content of the normal human have given rather a broad range than a single specific value.

It was with this fact in mind that the recommendation was made earlier to express the body activity of the human with its upper and lower limits as this is done for other biological-medical parameters (Krebs 1954).

Recent measurements of the radium content of soft tissues of the human support this suggestion. J. B. Hursh, A. Lovaas, and E. Biltz (1960) attempting to confirm values reported by H. Muth, B. Rajewsky, J. Hantke, and K. Aurand (1960) for the radium content of soft tissues of the human body found much lower activities in soft tissues: in the case of liver tissue up to $1/_{23}$rd and in the case of muscle tissue $1/_{50}$th of the values reported by H. Muth, B. Rajewsky, J. Hantke, and K. Aurand (1960).

Table 9. *Natural radioisotopes in the human skeleton*

Radioisotope	Skeletal amount (units of 10^{-10} Curie)	Average dose rate mrad/year
K^{40}	50	7.0
C^{14}	40	0.5
Ra^{226} . . .	0.4	1.2
Ra^{228} . . .	0.2	1.6
Pb^{210} . . .	1.0	1.4
$U^{238} - U^{235}$.	0.05	0.1

The extremes of the skeletal radium content as a function of geographical and other factors are not yet established. The average amount of radium found in human bones—about 4×10^{-14} curie/g ash according to R.A. Dudley (1959)—varies by a factor of 3 to 5 and a possible variance of up to 10 is seriously considered.

An important step forward in this field was the construction of large-scale, high-pressure ionization chambers by F.W. Spiers and P.R.J. Burch (1950), which led to the development of the modern whole-body counter devices. With aid of these instruments it was possible to get detailed information about the potassium content of the human and its dependence on environmental factors (E.C. Anderson and W.H. Langham 1959). Similar information is lacking for the radium content of the living human; with an increase in sensitivity of the whole-body counters by one magnitude, however, it should be possible to also get such detailed information for radium and other interesting radio nuclides.

The total skeletal dose rates from incorporated radioactive elements (including also some fission products), estimated by Dudley on the basis of an RBE-factor of 4 for α-rays are presented in Table 10.

The body radioactivity is a function of the amounts of food, water, and air taken up by the body, day by day, and by the radioactivity of these substances. Until recently, water was believed to be the main source of body activity. This concept is changing at present, not only on account of the new concept of "maximum permissible intake", but also on account of newer measurements on the

Table 10. *Estimated skeletal dose rates of man*

Source of radiation	Average skeletal dose rate	Comment
K^{40} (in body)	10 mrem/yr	
Ra^{226} series (in body) . . .	12	
Ra^{228} (MsTh) series (in body)	12	Value not very accurately known
Pb^{210} (RaD) series (in body)	8	Value not accurately known
Sr^{90} (in body from fallout) .	3 (1957) 12 (1975)	Dose rate from past explosions (reaching maximum in about 1975)
Cs^{137} (in body from fallout)	0.5 (1957)	
Cosmic rays	30	Value near sea level at higher geomagnetic latitudes
γ-rays from natural radio-activity in environment	45	Typical value for sedimentary rock area, as-suming building materials elevate dose rate by factor of 1.5

γ-rays from fallout . . .	0.5	Average value in U.S. during 1951 to 1956,
		    with reasonable allowance for weathering
		    of fallout and shielding of radiation

Medical x-rays	75	Value very inaccurately known
Total	196 (1957)	

rem (roentgen equivalent man): quantity of ionizing radiation, which produces, when absorbed by man, an effect equivalent to the absorption of one roentgen of X-rays or γ-radiation.

RBE (relative biological affectiveness); ratio of the X-ray or γ-exposure to the exposure of some other radiation needed to produce the same biological affect. RBE is a function of kind of radiation, object, and chosen test reaction.

alpha activity of food. P. SHANDLEY (1953), after measurement of the radioactivity of some foodstuffs and evaluating the dietary habits of the US population, concluded that man incorporates daily an amount of radium in the order of 3.5×10^{-12} g. This is in good agreement with estimates by J.B. HURSH (1957) and with findings by A. F. STEHNEY and H. F. LUCAS (1956). However, recent publications by W.V. MAYNEORD, J.M. RADLEY and R.C. TURNER (1959) based on measurement of alpha activity of food, present different figures. According to their studies the daily intake can be more than 10^{-9} curie.

Since the question into the alpha-activity role in the measurements of the body-activity has played such an important role in previous discussions (A. KREBS 1954, J.B. HURSH 1956), it may be worth while to bring the recent data on body-alpha activities and on food alpha activities as reported by MAYNEORD, RADLEY, and TURNER (1959).

Table 11. *Alpha activities of animals*

Kind	Activity in c/g bone ash
Cattle . .	2 to 6×10^{-12}
Sheep . .	2 to 6×10^{-12}
Horses . .	3.8×10^{-13}

Measurements on 60 persons from the Cornwall, London and Cumberland areas in England and on 11 stillborn children gave an average alpha activity of 5.0×10^{-13} c/g bone ash for the Cornwall area, an average alpha activity of 1.7×10^{-13} c/g bone ash for the London and Cumberland regions, and an average alpha activity of 4.5×10^{-13} c/g bone ash for stillborn children. The activities in foodstuffs vary by a factor of up to 14000, ranging from less than 0.1×10^{-12} c per 100 g of fruits to 1400×10^{-12} c/100 g of nuts. Figs. 1 and 2 give some details as well as some activity values for different materials.

Since little is known about the radioactivity of animals (MAYNEORD and co-workers), values for cattle, sheep, and horses (Table 11) may be of interest.

The conclusions drawn by MAYNEORD and co-workers from their findings are, among others: Solid food is the main source of the daily intake of radioactive material; milk and water, contrary to existing assumptions, contribute only a certain part; the daily activity-intake by food can equal the total-body content; adults can incorporate daily as much as 10^{-9} curie alpha activity and more; two persons, with apparently normal diet can nevertheless differ in activity intake by a factor of several 100; the variations of alpha activity found in investigations of the human are thus not longer surprising.

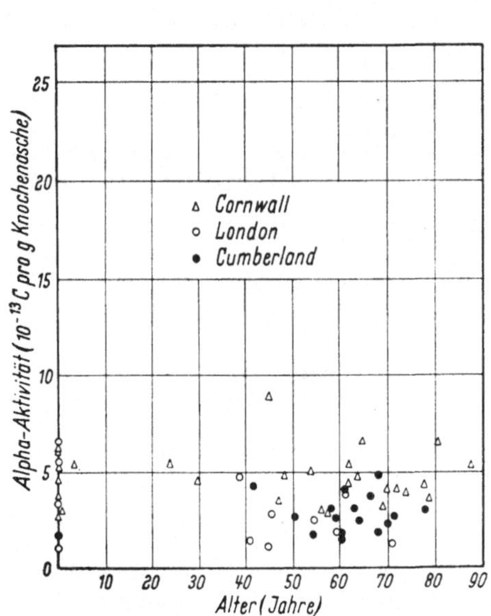

Fig. 1. Total alpha activity of human bone ashes of persons from Cornwall, London and Cumberland, Great Britain (Mayneord et al. 1959)

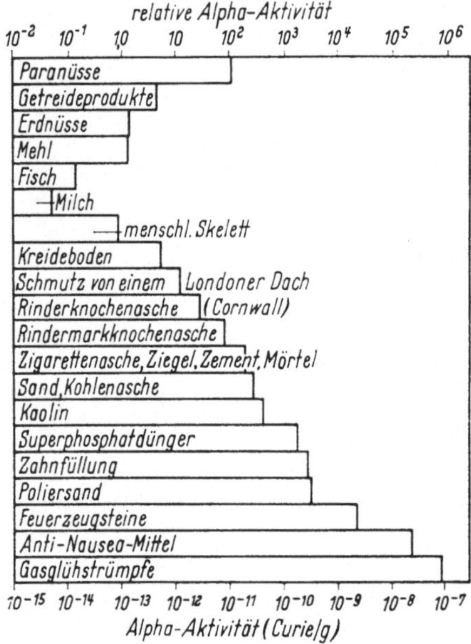

Fig. 2. Alpha activities of food and other materials (Mayneord et al. 1959)

The role of the radioactivity of the consumed water was discussed by J.B. Hursh (1957), L.D. Marinelli (1958) and others. Special studies on the number of people consuming high radioactive waters were made by H.F. Lucas (1959) and are represented in Table 12.

Table 12. *Populations consuming water at various radium 226 levels*

State	Max. concentration consumed ($\mu\mu$g/liter Ra[226])	Number of cities	Population
Illinois	0 to 0.99	134	1 068 610
	1 to 3.9	42	213 532
	4 to 9.9	34	310 588
	10 to 37	19	30 261
Wisconsin	0 to 0.99	10	113 394
	1 to 3.9	10	117 425
	4 to 7	4	51 099

The radiation doses delivered to the lungs by inhaled radon and thoron under different conditions have been estimated from activity measurements in the air, some data for Sweden, are quoted in Table 13.

The importance of the radon daughter products in radon inhalations was emphasized by J. Shapiro (1956) and by K. Aurand, W. Jacobi, H. Muth, and A. Schraub (1960). According to the studies by Aurand et al., breathing

Table 13. *Doses to lungs from radon and thoron in the air (in Sweden)*

Type of building	Dose rates in mrem/year			
	radon		thoron	
	in equilibrium	with ventilation	in equilibrium	with ventilation
Wood	263	73	185	52
Brick	453	128	605	173
Light-weight concrete . .	930	262	640	178

(Report United Nations)

of radon and its decay products outdoors in Bad Gastein delivers a radiation dose to the lungs of maximal 30 mrem/week with an average amount of 3 mrem per week. C. P. STRAUB (1959) reports for the Cincinnati area on intake of radon and thoron daughter activity of 166 to 42000 $\mu\mu$c per day for radon and of 80 to 200 $\mu\mu$c per day for thoron decay products.

A rough estimate of the mean annual doses to gonads, bone and marrow tissue from the different radiation sources is presented in Table 14.

Table 14. *Annual doses from natural radiation sources to different tissues*

Source	Gonad dose (mrem/year)	Osteocyte dose (mrem/year)	Mean marrow dose (mrem/year)
External			
Cosmic rays	28	28	28
Terrestrial radiation .	47	47	47
Atmospheric radiation	2	2	2
Internal			
K^{40}	19	11	11
C^{14}	1.6	1.6	1.6
Rn—Tn	2	—	2
Ra	—	38	0.5
Approximate totals . .	100	130	95

(Report United Nations)

C. Biological evaluation of the nature-given radiation doses

I. General principles

Interest in the biological effects of low-level irradiations started with the studies by M. TOMMASINA (1904) on bio-radioactivity and the work by J. STOKLASA and J. PENKAVA (1912). It prevailed through the following decades, especially in connection with the radium poisoning cases and therapeutical application of radioactive emanations and radium waters in radiumbalneology. It became an important domain of research and a main discussion topic with the first observations of fallout in 1954. Since then, many committees have been formed, many conferences have been held and "the question has attracted more general attention and given rise to more widespread public debate than any other question in which the scientific facts are so essential but so few" (A. BRUES 1959).

Statements on the biological effects of very low dose—chronic irradiations, as encountered in the geophysical and biosphere, are based on extrapolations of experiments and observations in higher dose ranges. However, there are at least two ways to extrapolate the available dose-effect curves. Depending on the premises and assumptions, one way of extrapolation leads to a linear dose-effect relationship down to the lowest doses, the "non-threshold" concept; the other method of extrapolation establishes the "threshold" concept, according to which a certain minimum dose has to be exceeded in order to produce an effect (Fig. 3).

This "threshold or non-threshold" question is the crucial corner-stone in the discussions on the effects of low-dose irradiations, including the nature-given radiations, as well as the man-made radioactive environment of life. Both, the linear theory, as well as the threshold theory, have reasonable theoretical support, but a definite and final answer has not yet been found or agreed upon (Summary-Analysis of Hearings, US Congress, 1959). While in the case of somatic effects some mutual agreements appear possible, concessions are difficult to make in the case of genetic effects.

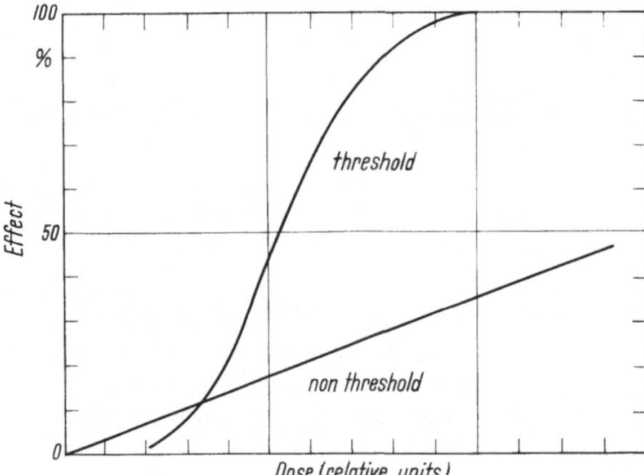

Fig. 3. Threshold and non-threshold response to ionizing radiation (W. H. Langham 1959)

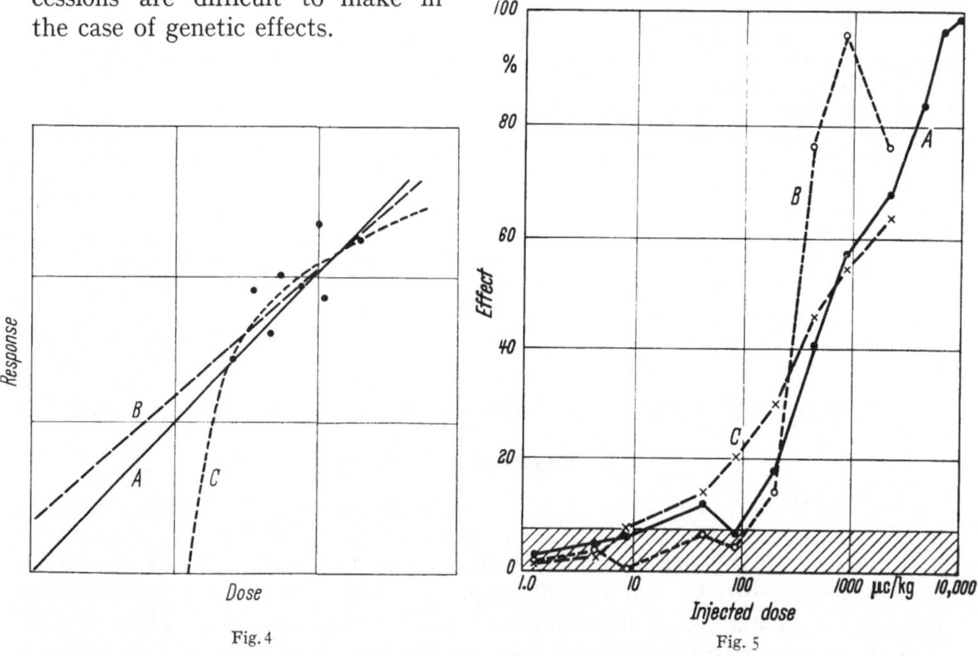

Fig. 4. Possible extrapolation from hypothetical data (M. P. Finkel 1958)

Fig. 5. Effect of strontium-90 on life expentancy (A), on incidence of tumors of bone (B) and bloodforming tissues (C) in mice. Shaded area represents non-significance at the 10% probability level by the t test (M. P. Finkel 1958)

This situation is reflected in the current discussions on the leukemogenic and sarcogenetic responses to chronic and/or low-level irradiations. While E. B. Lewis (1957) on the basis of human data, including the Japanese atomic bomb survivors, X-ray treated spondylitis cases, X-ray treated cases of thymic enlargement, practicing radiologists, and spontaneous incidence of leukemia in Brooklyn, New York, argues in favor of a linear relationship between incidence of leukemia and radiation dose, A. M. BRUES (1958) in his article on "Critique of the Linear Theory of Carcinogenesis", using and discussing the same material, like LEWIS, recommends consideration of a possible "threshold" relationship. M. P. FINKEL (1958) in her excellent study "Mice, Man, and Fallout" presents, before discussing her experimental results, an interesting example of possible extrapolations from hypothetical data (Fig. 4).

From her experimental data (Fig. 5) she concludes that none of the experimental curves can be described by a linear function.

V. E. ARCHER and B. E. CARROLL (1960) discussing FINKEL's plotting of effect against injected dose in microcuries per kilogram and suggesting to consider time, by plotting effect against dosage in millicurie-days per kilogram, favor a linear dose effect interpretation of FINKEL's experiments. This would support the views of the Committee of the NCRP on the Effects of Atomic Radiation (1958) and of LEWIS, according to which no threshold for radiation injury may exist and the dose-response curves may be straight, extending to zero dose and zero response. It may be worth while to mention in this connection the statement by M. P. FINKEL, B. O. BISKIS, and P. J. BERGESTRAND (1960) according to which hazards of chronic exposure cannot be predicted directly from experience with acute exposures and that injury from a contaminated environment should be based on information obtained from continuous exposure studies.

II. Specific effects

1. Somatic effects

In some situations, however, the discussions have been pushed forward enough to allow some guesses as to the biological significance of the environmental radiations: in the cases of leukemia incidences, cancer, and of radiation-induced mutations.

Leukemia. In the case of leukemia, E. B. LEWIS (1957) concludes that 10 to 20% of the spontaneous leukemia cases observed in the Brooklyn area may be caused by the natural background radiation, recently carefully measured in this area. This would be in accordance with the Summary Report (Science **128**, 402, 1958) on the United Nations Study stating: "harmful effects attributable to radiation from natural sources are not known with certainty, but it seems likely that some genetic and possible somatic injury is caused in this way". It would also concur with the fact, that all conclusions are based on experimental and clinical data, obtained at higher radiation levels and that "no experiments aimed at observing these biological effects (leukemia, cancer, lowered resistance to stress, premature aging) have ever been conducted at radiation levels very close to the natural background" (Summary-Analysis of Hearings, US Congress, 1959).

A. M. BRUES (1958) in response to LEWIS' statements emphasizes, however, the lack of enough experimental material on human leukemogenesis, to indicate a linear relation between dose and leukemia incidence. Such a hypothesis cannot be ruled out statistically, of course, since data are scanty, but it seems to BRUES less probable than a non-linear or threshold relations.

Cancer. A similar situation holds for the question into the correlation between natural radiation doses and incidence of cancer. Here our knowledge is based on clinical experience, professional accidents, animal experiments and to a high degree on the study of still living radium poisoning victims from the early twenties.

These persons, still alive 30 to 35 years after incorporation of the radioactive material, give the best information about induction of cancer by chronic internal irradiation and about the correlation between frequency of malignancy and delivered dose. R. D. Evans and his group (1937 to 1960); L. D. Marinelli (1959) and co-workers, W. B. Looney et al. (1955) together with others, have collected in the study of these cases valuable, some times surprising data. The picture is quite dynamic, demonstrating a tremendous biological variability. The interpretation of the data is difficult, so much the more since these persons must have been exposed for long periods of their life to high doses from the not yet excreted radioactive material. R. E. Rowland (1960) examining 19 cases who carried radium from a minimum of 10 years to a maximum of 40 years, with body burdens from 20 µc down to about 1 µc found the following facts. The "hot spots" show about 75 times as much activity per unit bone mass as the diffuse distributions. The life time doses delivered to the osteocytes in their lacunae amount in certain cases to several hundreds of thousands of rads, while the doses delivered to all osteocytes in bone range up to some thousand rads. The Argonne team now has about 250 cases, including persons injected with radium, and R. D. Evans and his group have studied and are following at M.I.T. about 300 cases. So the total pool is over 500 persons under study, about one-half at Argonne, about one-have at M.I.T. (R. D. Evans 1961a), with the goal to find clearer indications of the probability between the induction of recognizable radiation damage to the human and the effective dose and/or dose rate. W. S. Snyder (1959) has pointed out one difficulty in interpretating the findings in the dial painters. We do not know whether the malignancies found in these persons after so many years are attributable to the high dose rates the bone received in the early time after incorporation or whether they should be attributed to the continuous irradiation of the bone in excess of a certain level throughout the whole history of the case. In any event, even the lowest doses, producing detectable effects, are much higher than the radiation doses delivered by life's environmental radiation sources to the bones. So far no correlation has been found between the nature-given radiation doses and the incidence of cancer. Some guesses on the spontaneous incidence of radiation cancers have been made by L. D. Marinelli (1959). Valuable information comes from the *Schneeberg miner* disease studies. Here, long-term exposure to low-level radiation caused cancer of the lungs, the latent period being about 17 years (Congressional Hearings II, 1958). In these cases the miners worked in areas with radon concentrations of about 30 times higher than the accepted maximum permissible amounts of 0.0001 microcurie per liter air. During the 17-year period the miners received—even spread of the radiation dose over the lungs assumed— a total dose to the lungs of 1000 r and more. Comparison of these figures with Shapiro's values and with the doses to the lungs when breathing radon and its decay products out-doors in radium-spa Bad Gastein (max. 30 mrem/y) leads to the conclusion, that many years would be required to deliver 1000 r to the lungs when inhaling radon and its decay products from the air. Also, the external background radiation, with roughly 100 mrad/year, summing up to 7 to 9 r for a man, who lives to be 70 years of age (Hearings, US Congress, 1957), is far out of the lowest dose range found to be cancerogenic. This also holds for thoron and its decay products, the toxicity of which was studied by R. D. Evans and C. Goodman (1940) and by A. Krebs (1940).

The importance of naturally occurring highly radioactive springs, containing primarily radon and its decay products, as therapeutical agents becomes evident from the role the radium-spas are playing in Europe. Joachimstal in Czechoslovakia, Bad Gastein in Austria, Radiumbad Oberschlema in Saxony, Bad Kreuznach in West Germany, are famous for their radiumemantion therapy, which has been often discussed in connection with balneological problems (see, e.g., A. KREBS and H. LAMPERT 1948).

There are variations in the doses delivered by natural sources in the order of one or two magnitudes, such as in the Kerala region in India and in some Brazil States. But no detectable effects have been found so far in people living in these areas. This is supported by a statement from W. S. SNYDER (1959): "In spite of this spread of values, I know of no definite variation in any epidemiological parameter which has been shown to be correlated with the difference in background. Studies of this kind are being undertaken by the United Nations, as well as by other groups, and offer hope that we may be able to obtain human data concerning the effects of low-level chronic irradiations over the life span of an individual or over the centuries, which measure the life of people."

Life shortening. Life shortening by low-level chronic irradiation is another controversial problem. Since H. A. BLAIR's (1952 to 1956) approach, many reports have appeared trying to extrapolate from life shortening in animals to life shortening by ionizing radiations in man. According to the guess of H. B. JONES (1956) exposure to 1 r may shorten life span up to 10 to 15 days. G. FAILLA and P. MCCLEMENT (1957) based on LORENZ's chronic irradiation experiments with mice, derived a formula for calculating the life shortening effect of chronic irradiation. According to their formula exposure to 1 r of accumulated dose at a dose rate not in excess of 0.5 r per day will shorten life by one day for the human. Application of this formula to natural background radiation gives a negligible small figure for the shortening of life by these radiations. The difficulty in these conclusions comes from the highly speculative assumptions and from the fact that radiation induced premature aging may be different from normal aging of species (N. L. BERLIN and F. L. DIMAGGIO 1956). Also, here it is argued that shortening of life in man by ionizing radiation, comparable to that observed in animals has yet to be demonstrated in human populations (A. C. UPTON and A. W. KIMBALL 1960) and that extrapolations from animal experiments to man should be accepted with all reservations (R. F. KALLMANN and H. I. KOHN 1958, N. I. BERLIN 1960, G. A. SACHER 1960).

As to the biological aspects of cosmic and space radiations a great problem for space flight arises from the strong radiation belts, girdling the earth (J. A. VAN ALLEN, C. G. MCILWAIN, and G. H. LUDWIG 1959). Here radiations exist —electrons and protons—which are well known in their biological effectiveness. Since the doses in the belts are high, hazardous effects to the space traveler and to the occupants of space platforms, stationed in these areas, have to be anticipated. By shielding, speed, special escape routes and operating of space stations outside the belt zones, the dangers can be minimized, but they definitely exist (R. D. EVANS 1961 b; H. J. CURTIS 1961).

The classical question into biological hazards from heavy primaries, discussed in details by H. J. SCHAEFER (1950), A. KREBS (1950), C. A. TOBIAS (1952), dwarfed with the discovery of the radiation belts. Heavy primaries, still of interest as a kind of radiation not yet completely available in the laboratories, are of little or no danger in space travel (W. H. LANGHAM 1959, L. J. LEBISH, D. G. SIMONS, H. YAGODA, P. JANSSEN, and W. HAYMAKER 1959, H. E. NEWELL, and J. E. NAUGLE 1960). They are single events, which can produce some cellular damage, but their frequency is negligibly small. The best established biological effect produced by heavy primaries is the greying of hair, as demonstrated by H. B.

Chase and J.S. Post (1956) in black C-57 mice, flown for 70 to 90 hours at altitudes of 90000 feet and more (see also W. Zeman, H. J. Curtis and C. P. Baker).

The biological effects of cosmic radiation-produced isotopes are negligible and also the production of cataracts has not to be considered, since the nature-given radiation doses are too low (W.F. Libby 1957; Nat. Acad. Science, Report to the Public, 1960). One interesting facet may be mentioned: according to C.A. Tobias, a dark adapted person may be able to see very heavily ionizing single tracks as a small light flash, when passing through several retinal receptors.

2. Genetic effects

The question into genetic effect caused by natural background radiation and cosmic radiation was raised soon after Mueller's discovery (1927) of radiation induced mutations. From the linearity between dose and effect for Drosophila in the dose range from several thousand roentgen down to a few hundreds of r (Fig. 6) and recently down to 25 r and 10 r, respectively, it was concluded that also in the lowest dose ranges proportionality between the number of radiation induced mutations and the applied dose would exist. In addition, every amount of radiation, however small it may be, should be cummulative and independent of dose rate.

This direct proportionality between dose and effect at higher doses was used to estimate the contribution of nature-given radiations to the spontaneous mutation rate. Knowing the number of mutations produced by 1 r and the doses from background and cosmic radiation, the calculations could be done easily. For Drosophila with a life span of about 22 days, the contribution from the nature-given radiation to the spontaneous mutation rate is, according to B. Wallace and T.H. Dobzhansky (1959) less than 1%, according to M. Westergaard (1957) practically not more than 0.02%.

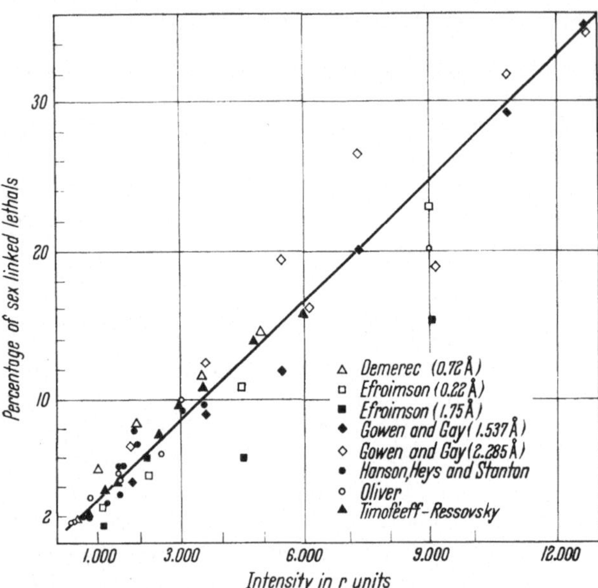

Fig. 6. The percentage of lethals in the X-chromosome of Drosophila melanogaster in relation to the dose of radiation (M. R. Zelle 1960)

Another approach, often used to estimate the genetic hazards from low-level radiations, introduces the "doubling dose" concept, an amount of radiation required to induce mutations at a rate equal to the spontaneous or natural mutation rate. This dose can be calculated when the spontaneous mutation rate and the induced rate per r are known for the species under consideration. Spontaneous mutation rate and induced mutation rate are well known for *Drosophila*; the first data for the mouse have been given by W.L. Russell (1951); knowledge for man is fragmentary. For *Drosophila* with a spontaneous mutation rate of 1×10^{-5} to 3×10^{-6} per gene locus and an induced mutation rate of about 3×10^{-8} per gene locus/r, M. WESTERGAARD gives a doubling dose rate of 30 to 50 r per generation. For the mouse, which is about 10 to 18 times more mutagenic than Drosophila, this figure is 35 to 80 r per generation. For man with an estimated spontaneous mutation rate of 2 to 3×10^{-5} to 1×10^{-7} or less (M. WESTERGAARD) and an induced mutation rate of 3×10^{-7}/r (highest value given for the mouse by RUSSELL) to 2×10^{-7}/r (lowest value given by RUSSELL), J.V. NEELS and W.J. SHULL (1954) taking into account these and other uncertainties calculate a doubling dose of 3 r to 150 r. J.M. MUELLER (1954) assuming a spontaneous mutation rate for man of about 2×10^{-5} and an induced mutation rate of 2.6×10^{-7}/r (average value of RUSSELL) arrives at a figure of about 80 r for the doubling dose.

With this data, based on the best available information, it is now possible to estimate the biological importance of nuclear radiation in geophysics as to genetic effects. For this purpose one has to compare the above given doubling doses with the total dose a productive human generation receives from natural environmental radiations. This dose has been estimated by M. WESTERGAARD (1957) to be about 3 to 7 r per generation, assuming an average dose from environmental radiations of 0.12 to 0.23 r per year and a span of one productive human generation of 25 to 30 years. Comparing this 3 to 7 r per generation with the doubling dose of 3 r, this would mean all spontaneous mutations are produced by natural background and cosmic radiations; comparing the dose of 3 to 7 r with the highest doubling dose of 150 r would lead to a spontaneous mutation figure of about 2.5% from environmental, nature-given radiations; accepting 80 r as doubling dose, as recommended by MÜLLER, would make the natural environmental radiations responsible for about 5% of the spontaneous mutations.

Current, best estimates accept a mammalian doubling dose of 50 r (Congressional Hearings 1959, Science Report 1959), leading to the conclusion that about 6% of the spontaneous mutations are produced by the natural radiation sources.

Recent findings in the Kerala region (Science **131**, 1722, 1960) and experimental studies into biological effects of cosmic radiation support this conclusion. Such experiments gave in general small yields, leaving open the question into other factors responsible for the bulk of the spontaneous mutations. These experiments also justify the tendency to take the nature-given radiation conditions as a basis and as a guide for estimating dangers to large populations from low-level irradiations, and for establishing reasonable radiation protection guides. Life has been exposed to these nature-given radiations probably throughout its history. It has progressed under these conditions and developed to its present form and shape with or in spite of the nature-given radiation burden. "Good-or-bad" is not the question, decisive is how much more radiation can be added to this natural amount without interfering too much with the delicate equilibrium built up over the eons.

Previously, The National Academy of Science Committee on the Effects of Atomic Radiation and the National Committee on Radiation Protection (NCRP) recommended as limit a maximum average-per-capita-dose to the gonads of roughly 3 times background-radiations[1], that is a dose of 10 r per 30 years of all man-made radiations, including medical and dental radiation. The NAS Committee in 1960 (Report for the Public 1960) holds up this advice and continues to recommend that for the general population the average gonadal dose, accumulated during the first 30 years of life, should not exceed 10 r of man-made radiations and should be kept as far below this as is practicable. This is essentially in agree-

---

[1] About 4.3 to 5.5 r/30 years.

ment with the most recent suggestion of the International Committee on Radiation Protection (ICRP), suggesting for long-range planning purposes a permissible average load for the whole population in the general vicinity of the background dose, a man-made radiation level of 1.7 times background, if background is taken to be 100 millirem per year.

These are conservative, but reasonable decisions, valid until experimental research presents new facts and data. This agrees with the statement by K.Z. MORGAN, at the Summary Hearings, 1959: "for the present, it is safe only to assume that types of radiation damage, such as genetic mutations, leukemic incidence and shortening of life span, increase to some extent with any increase in dose taking place at any dose rate following any accumulated dose, viz., there is no threshold. However, it certainly would be ultraconservative—and at least with respect to genetic mutations it would be incorrect—to assume a linear relationship between dose and effect all the way from high chronic dose rates of 400 rad per 30 years to background dose rates of about 4 rad per 30 years" (Summary Analysis of Hearings, US Congress, 1959).

Thus, recent experimental studies by W.L. RUSSELL, L.B. RUSSELL, and E.M. KELLEY (1958) into the influence of dose rate on the number of radiation-induced mutations become of special value. Irradiating mouse spermatogonia in one case with chronic irradiation (90 r per week), in another case with acute irradiation (80 to 90 r per minute) did not result in the same mutation rate for equal doses. J. COCKCROFT (1959) in his Summary Report on Second Atoms for Peace Conference, mentioning similar experiments by T.C. CARTER, described these discoveries as the beginning of a new era in radiation genetics which will probably necessitate a complete reconsideration of the quantitative predictions of the genetic radiation hazards. Already intensive discussions have been started in connection with RUSSELL's observations. It is, however, too early for definite statements and much more experimental studies are needed to solve this problem of such a basic, fundamental importance. This attitude has clearly been expressed by the National Science Committee on the Biological Effects of Atomic Radiation, 1960 and during the Congressional Hearings, 1959, both recommending wise use of radiation until more information has been obtained.

Other interesting questions concern the production of malformations by irradiation in utero, and the influence of radiation on the sex ratio in man. In both cases little is known about the influence of the environmental background radiations. Malformations can be produced by doses as low as 25 r, as shown by L.B. RUSSELL (1952) and others, when irradiating the mother at certain time intervals of the gestation period.

The relationship between the geological environment and mortality from congenital malformations has been discussed by J. KRATCHMAN and D. GRAHN (1959). The available data do not permit any conclusive statements, but justify further investigations.

As to influence of irradiation on the sex ratio re-analysis of the Japanese A-bomb data by W.J. SCHULL and J.V. NEEL (1958) reveals changes, which should be expected if exposure had resulted in the induction of sex-linked lethal mutations. But also here, it is a long way, to prove that similar effects may be produced by the environmental low-level radiations.

The permissible genetic doses to the population-at-large have been discussed in details by K.Z. MORGAN at the Summary Hearings, US Congress, 1959. The doses recommended by NAS and ICRP can be split up in the way as shown in Table 15.

Table 15. *Permissible genetic doses to the population-at-large suggested by the ICRP to serve as a guide (RBE doses in rem to age 30)*

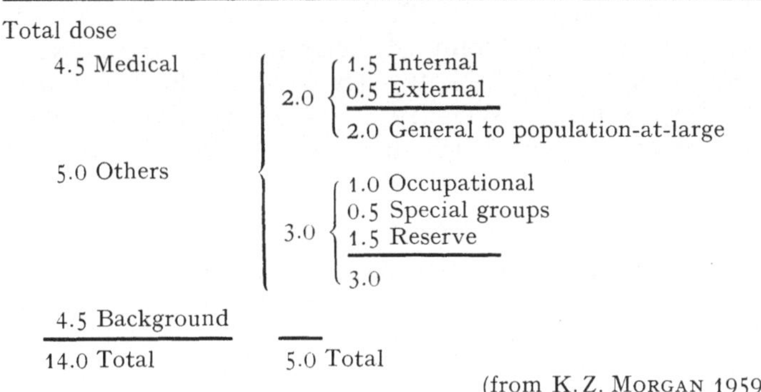

(from K. Z. Morgan 1959)

Systematic remarks on the effects of high energy radiations on man, including data on the maximum permissible concentrations of some practically important isotopes have been made by Catsch (1957).

Table 16. *Background radiation and non-occupational exposures*

Source	Amount of rate
Cosmic rays, sea level	$\sim$0.1 mr/d
Cosmic rays, 15000 ft alt.	0.5 mr/d
Cosmic rays, 55000 ft alt.	7.5 mr/d
Cosmic rays, top of atmosphere.	70 mr/d
Air, radon (incl. thoron)	$\sim 10^{-10}$ μc/cm³
Natural "background"	0.01 to 0.1 mr/hr
Earth's outer crust:	
Uranium, by weight.	6 ppm
Thorium by weight	12 ppm
Radium, by weight	2×10^{-6} ppm
Sea water:	
K^{40}	0.33 μc/m³
U	0.0015 mg/kg
Drinking water	10^{-16} to 10^{-12} curie/cm³
Cow's milk	10^{-14} curies/cm³
C^{14} (isotopic abund. 1.6×10^{-12})	16 d/m per g carbon
Human body, K^{40}, total	0.23 μc
Human body, C^{14}	0.0068 μc
Wrist watch dial, ($\sim$1 μg Ra)	$\sim$1 mr/hr, γ, wrist
Airplane instruments (10 to 100 μg Ra per dial):	
At face of each dial	5 to 10 mr/hr
At pilot's position	< 1 mr/hr
Shoe fitting (20 sec exp)	av 10 to 15 r to feet
Diagnostic X-ray[1]:	
14×17 chest plate	0.05 to 0.25 r
Photofluorographic chest	0.7 to 1.2 r
Extremities	0.25 to 1.0 r
Skull	1.3 r
Abdomen	1.3 r
GI series	0.65 r/plate
Lumbar spine, lat.	5.7 r
Pregnancy, lat.	9.0 r
Fluoroscopy	0.28 r/sec
Dental	0.5 r/film
Spectacle lenses containing U	1 to 8 mr/hr β to eyes
Average background exposure	0.4 mrad/day

[1] There is apt to be considerable variation depending on filtration, distance and techniques.

The radiation burden to man from natural background radiations is far below the dose originating from the radiation-environment man has created by himself. Libby (1957) has given some data on the radiation doses encountered in modern life and Table 16, taken from Atomic Medicine (Ed. C. F. Behrens 1959), gives some more interesting details (Table 16). These include like the tables presented by Franzen, Muszynski and Wiesenack (1957) also data on the man-made-environtal contamination, which will be discussed in the following chapters.

Conclusions

In summarizing the best available data it must be concluded, that so far neither for somatic effects nor for genetic effects a clear relationship between these effects and the doses delivered from nuclear radiation in the geophysical sphere has been established. All elaborations on possible effects in this low and lowest dose range are based on extrapolations of experiments done in higher dose ranges and are wide open for discussion and criticism. Some kinds of nuclear radiations in geophysics, so for instance in the van Allen belts and in interplanetary space, present definite hazards to health and proper measures have to be taken to overcome these dangers. In any case of doubt, the newest radiation-protection guides recommended by the different National and International Committees on Radiation Protection should be advised for possible radiation dangers. The most recently given protection guides are presented in Table 17:

Table 17. *Radiation protection guides for normal peacetime operations (Fed. Rad. Council, No. Rept. 1, 1960)* [1]

Type of exposure	Condition	Dose [2] (rem)
Radiation worker		
a) Whole body, head and trunk, active blood forming organs, gonads, or lens of eye	Accumulated dose	5 times number of years beyond age 18
	13 weeks	3
b) Skin of whole body and thyroid	year	30
	13 weeks	10
c) Hands and forearms, feet and ankles	year	75
	13 weeks	25
d) Bone	body burden	0.1 microgram of radium-226 or its biological equivalent
e) Other organs	year	15
	13 weeks	5
Population		
a) Individual	year	0.5 (whole body)
b) Average	30 years	5 (gonads)

[1] See also table 2 b.

[2] Minor variations here from certain other recommendations are not considered significant in light of present uncertainties.

D. Man-made environmental contamination

I. Dust from nuclear explosions

The particles from a nuclear test explosion consist largely of inert materials from the environment of the explosion but all are impregnated with radioactive fission products more or less uniformly throughout their volume. The larger particles may fall to earth relatively close to the site of the explosion but in every type of burst a significant fraction of the total radioactivity is contained in fine dust particles whose rate of fall under gravity is small and which may therefore remain airborne long enough to diffuse widely throughout the atmosphere and affect large areas of the earth's surface.

The fission products contained in the fine dust are fairly representative of those generated in the parent explosion although, under certain circumstances, some may be preferentially deposited along with the larger particles near the test site. Different types of weapon produce the same assemblage of fission products in slightly different proportions but the rate of decay of the gross activity is almost independent of these differences and varies approximately as the inverse 1.2 power of time. Thus the activity decreases by a factor of 10 for a 7-fold increase of time as measured from the instant of the explosion. Table 18 shows the percentage of the total activity at various times after fission for those fission products which contribute more than 5% of the total activity at any time (BOLLES and BALLOU 1956).

Table 18. *Principal components of fission*

Fission product	Principal radiations emitted	Percentage of total activity at various times after fission				
		10 days	100 days	1 year	5 years	10 years
Sr^{89}	β		10	} 4.5	17	22
Sr^{90}	β					
Y^{90}	β			} 6.2	17	24
Y^{91}	β		13			
Zr^{95}	β, γ		15	7		
Nb^{95}	β, γ		20	14		
Mo^{99}	β, γ	7				
Tc^{99m}	γ, electron, β					
Ru^{103}	β, γ		7			
Rh^{103m}	γ, electron		7			
Te^{132}	β, γ	6				
I^{131}	β, γ	} 12				
I^{132}	β, γ					
Xe^{133m}	γ, electron					
Xe^{133}	β, γ	15				
Cs^{137}	β, γ				11.5	17.5
Ra^{137}	electron, γ				10	16
Ba^{140}	β, γ	10				
La^{140}	β, γ	12				
Ce^{141}	β, γ, electron	} 7	} 16			
Ce^{144}	β, γ			26	6.5	
Pr^{143}	β	} 11	} 9			
Pr^{144}	β			26	6.5	
Pm^{147}	β			6	25	16

1. Internal and external sources of radiation

The fission products from nuclear explosions may enter the human body in several ways: first by direct inhalation; second through deposition on and uptake by vegetation eaten by humans; third by transfer through animals; and fourth by contamination of water supplies. Many of the isotopes produced in fission do not enter into metabolic processes and therefore do not contribute significantly to internal radiation. Of those which do, the most important are Sr^{90} and Cs^{137}, which have physical half-lives of 28 and 30 years respectively and significant biological retention times in the body. The principle source of external radiation is the fission product activity deposited on the ground, since the radiation from fission products suspended in the atmosphere is negligible by comparison except at the immediate site of an explosion (Medical Research Council, 1956).

The relative importance of the internal and external sources of radiation varies over the surface of the earth. The dose from external radiation depends on how much activity is deposited and when, and the distribution of fallout over the earth's surface provides the primary data from which to assess the doses in different countries from this cause. On the other hand, the dose from ingested fission products depends not only on how much is deposited but also on the diet of the people in those countries.

2. Movement and deposition of dust from nuclear explosions

Before the radiation doses associated with fallout can be properly discussed it is necessary to describe some of the physical and meteorological factors which

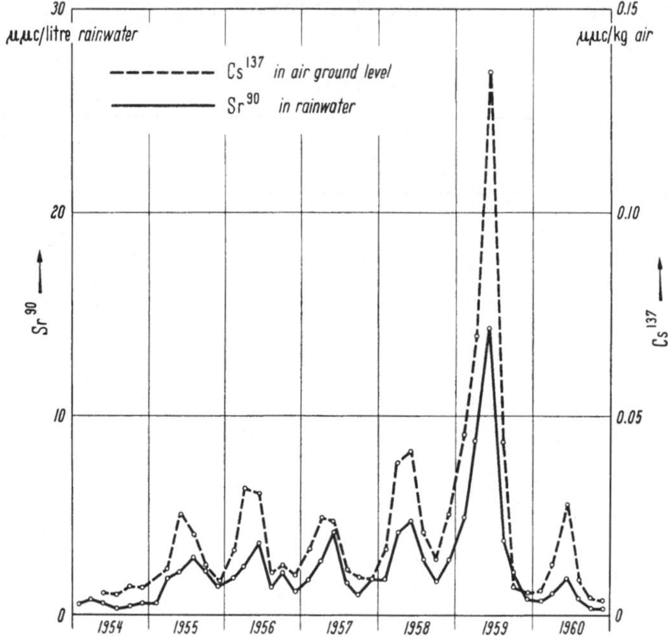

Fig. 7. Cs^{137}(---) at ground level and Sr^{90} in rain water for the years 1954 to 1960

control the transport, diffusion and deposition of the dusts. The main concern will be long-range fallout and local effects will not be treated in detail.

The heat of the fireball from an explosion in the kiloton range is dissipated quite rapidly and the fission products do not normally rise above the tropopause. Under these conditions, half the radioactive material is removed from the atmosphere, mainly by rainfall, in about 22 days (STEWART, CROOKS, FISHER 1956) so that the deposition is effectively complete within 3 months. The latitudes where the tropospheric fallout is deposited are mainly determined by the latitudes of the test sites. The dust from explosions in the megaton range penetrate into the stratosphere and return slowly so that deposition may take place over a period of up to several years (STEWART, CROOKS, FISHER 1956). The pattern of the deposition both in space and in time has some striking features which have been well illustrated by systematic studies carried out in the U.K. and elsewhere, making use of the biologically important isotopes Cs^{137} and Sr^{90} as tracers. The concentration of Cs^{137} in ground-level air and of Sr^{90} in rainwater in the U.K.

(PEIRSON, CROOKS, FISHER 1960) between 1954 and 1960 are shown in Fig. 7. Broadly, the graphs show that the concentration of long-lived radioactivity in rain is proportional to that in the air—fine differences can be attributed at least partially to the fact that air sampling was on a continuous basis whereas rain sampling was naturally confined to periods when rain fell. Another feature of the graphs is the clear seasonal variation in concentration, with peaks in the late spring and troughs in the late autumn of each year. Measurement of the shorter-lived activity—and particularly of the Sr^{89}/Sr^{90} ratio in rain—indicate that the bulk of the Sr^{90} is of fairly old origin and has certainly come from the stratosphere. The peaks in 1955 and 1956 are interesting, because the ratio analysis proves that they both derive primarily from the Pacific explosions in the spring of 1954. These results strongly suggest that the dust stored in the stratosphere is returned to earth periodically and not at a constant rate. The peak in 1960, following a period of months with no major tests, supports the same hypothesis.

An interesting and important feature of stratospheric fallout is that within localised regions of the world the mean concentration of Sr^{90} in rainwater over long periods is virtually constant and independent of local rainfall. Thus Table 19 shows the mean concentration of Sr^{90} in rainfall at 6 measuring stations within the U.K. over the 3 years 1957 to 1959 (CROOKS, OSMOND, FISHER, OWERS and EVETT 1960). The extreme rainfall rates differ by a factor of 6 whereas the concentration figures lie within a very narrow range. This result means that within defined regions, *total* fallout is to a first approximation proportional to rainfall.

Table 19. *Mean concentration of Sr^{90} in rainwater in the U.K.*

Station	Felixstowe	Kinloss	Abingdon	Liverpool	Milford Haven	Snowden
Mean annual rainfall (cm) .	52	62	63	87	104	326
Sr^{90} in rain ($\mu\mu c/l$)	6.33	6.09	5.67	6.80	6.13	6.46

The most striking feature of the pattern of Sr^{90} over the earth's surface is that very little has been deposited in equatorial regions compared with the middle latitudes of the northern hemisphere, despite the fact that many of the test explosions were conducted at low latitudes. The mean Sr^{90} concentrations in rainwater at stations between latitudes 70° N and 50° S between 1957 and 1959 are shown in Fig. 8 (CROOKS, OSMOND, FISHER, OWERS and EVETT 1960). Since a fairly smooth curve can be drawn through the points, despite large differences in the longitudes of the sampling stations, it is reasonable to deduce that the mean concentration of Sr^{90} is a function of latitude only, so that apart from local fallout effects, an approximate figure for the mean rate of deposition of Sr^{90} at any point within the latitudes covered by the survey can be computed from Fig. 8 and the local rainfall.

The seasonal variation of Sr^{90} or Cs^{137} is remarkably similar to the seasonal variation of the total ozone in the atmosphere which has been observed by many workers. Based on measurements of the distribution of ozone and water vapour in the atmosphere, DOBSON (1956) has proposed a model for the general circulation which offers a satisfactory explanation of the observed Sr^{90} data (Fig. 9). In DOBSON's model, a very cold pool of air forms above the winter pole during the late winter months when the air lies in shadow. The ultimate sinking of this pool carrying ozone-rich air to lower levels in the stratosphere, is believed to be the cause of the rapid increase in total ozone in early spring in high latitudes. Since the Sr^{90} concentration in the stratosphere increases rapidly with height

(Stewart, Crooks, Fisher 1956), this subsidence would also bring Sr^{90}-rich air into the lower stratosphere in early spring, leading to a seasonal variation of the concentration in this region of the atmosphere.

The interchange of air between the stratosphere and the troposphere has been discussed by Brewer (1949) who, in order to explain the form of the water-vapour curve in the stratosphere, has suggested a circulation system in which tropospheric air enters the stratosphere at the equator, travels in the stratosphere to temperate and high latitudes and then sinks again into the troposphere. This circulation provides the means for bringing stratospheric Sr^{90} down into the troposphere where the concentration would be expected to show the seasonal features discussed above. The form of the global deposition curve supports the view that the stratospheric air enters the troposphere in the middle latitudes, bringing with it Sr^{90} which is progressively washed out of the troposphere by rainwater as it travels north and south from the region of entry.

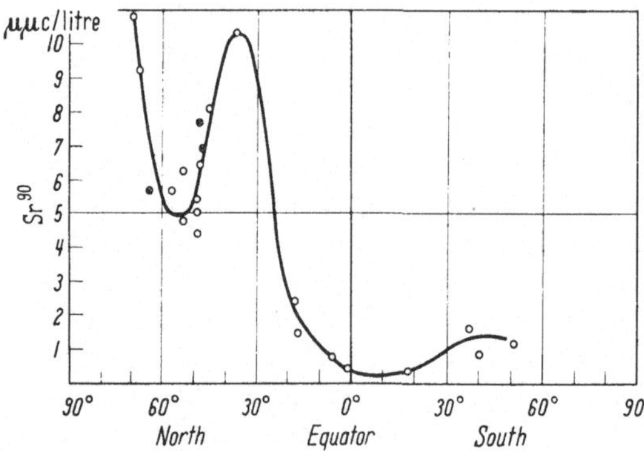

Fig. 8. Mean Sr^{90} content of rainwater at various latitudes. 1957 to 1959. (Points marked $\oplus$ 1958 to 1959 only)

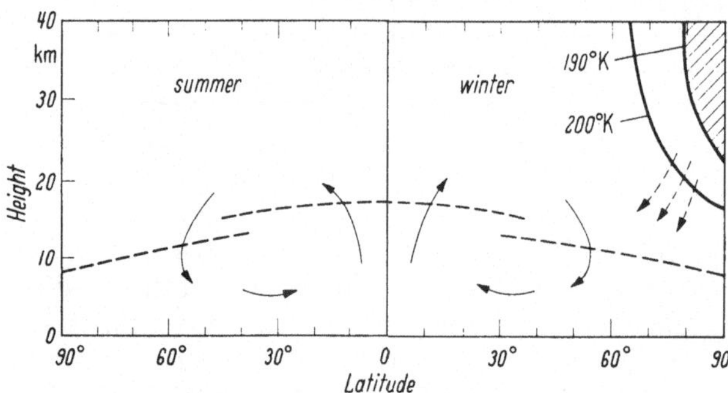

Fig. 9. Atmospheric circulation model (after Dobson and Brewer), showing the cold pool of air above the winter pole. The dashed lines represent the tropopause and the arrows show the sense of the mean air flow

The average length of time required for fission products to return from the stratosphere to the troposphere depends on a number of factors. Thus debris released into the stratosphere at high altitudes in tropical latitudes may take several years before it is in a position to be readily transferred to the lower atmosphere (Stewart, Crooks, Fisher 1956; Libby 1956). On the other hand, Peirson (Peirson, Crooks, Fisher 1960) and Ambrosen (1960) have shown that about 50% of the heavy fallout in early 1959 could be attributed to a number of nuclear explosions held in Arctic latitudes in the autumn of 1958. The mean residence time of this debris has been estimated by Feely (1960) to lie between 6 and 13 months. The shorter residence time is not unexpected from the Dobson-Brewer model since dust injected into the stratosphere at higher latitudes in autumn is particularly well placed for rapid transfer back into the troposphere.

3. Inhalation and ingestion of fission products from nuclear explosions

The daily intake of fission product nuclides by inhalation has been calculated from the concentrations measured in air over several years in the U.K. and found to be extremely small (Medical Research Council, 1960). This is likely to be true for the world in general, except close to the test sites.

The main route by which these radioactive elements gain access to the body is by the ingestion of fission products which are deposited by rain and which contaminate both human and animal foods. Fortunately, most of the fission products are not absorbed to any significant extent by the gut although this is not true for the isotopes of strontium and caesium. Thus Sr^{90}, being chemically similar to calcium, follows calcium through the food chains from environment to man and is eventually incorporated with it in bone. Since Sr^{90} and its daughter product Y^{90} emit only beta particles, which have a limited range, the radiation dose from them is essentially confined to bone and to bone marrow. Cs^{137}, which is responsible for much of the external radiation dose to man, is also taken into the body through food-chains but becomes distributed fairly evenly throughout the tissues and in conjunction with its γ-ray emitting daughter-product Ba^{137} causes uniform radiation of the whole body. I^{131} is also taken up and, like other forms of iodine, is concentrated in the thyroid gland. Finally, the C^{14} which is created in nuclear explosions by the escape of neutrons into the atmosphere (LEIPUNSKI 1957) adds to the amount naturally present in the environment and in human tissues.

a) Strontium-90. In order to assess the hazard from Sr^{90} it is not only necessary to determine current concentrations in human bone but to study the change in these concentrations with time and in particular, the relations between Sr^{90} in food and in bone. The chemical similarities between strontium and calcium make the use of the Sr^{90}/Ca ratio convenient for following the history of Sr^{90} from the environment to human bones. However, the chemical behaviour of the two elements is not identical and their utilisation varies in biological processes. Thus cows utilise calcium about 10 times more efficiently than strontium in producing milk (COMAR, RUSSELL and WASSERMAN 1957). There is also a relatively constant discrimination observed in both the U.K. and U.S., between the two elements in their transfer from human diet to bone, the ratio in bone being approximately one-quarter of the mean ratio in the total diet (BRYANT, CHAMBERLAIN, SPICER, WEBB 1958; KULP, SCHULERT, HODGES 1960). Thus if the ratio of Sr^{90} to calcium in the total diet is known, the level of radiation which the skeleton will receive can be estimated.

Investigations of the extent to which foods were being contaminated with Sr^{90} and Cs^{137} have been carried out since about 1954, particularly in the U.K. and U.S. where broadly similar results have been obtained. The results of a comprehensive food survey carried out by the Agricultural Research Council of the U.K. are summarised in Table 20 in which the estimated total intake of Sr^{90} is derived from the average diet with small added contributions from drinking water and inspired air (Medical Research Council, 1960). The mean daily intake of calcium is estimated as $1.084\,g/day$ so that the Sr^{90} to calcium ratio was $6.1\,\mu\mu c/g$ in 1958 and $9.3\,\mu\mu c/g$ in 1959 as compared with the estimate of $5.5\,\mu\mu c/g$ for 1957 made by BRYANT et al. (BRYANT, CHAMBERLAIN, SPICER, WEBB 1958). It is clear from Table 20 that in the U.K. most of the Sr^{90} ingested is obtained from milk and milk products and a study of the diet of selected countries strongly suggests that this will be true for most of Europe, North America and Oceania. It has also been established by measurement in the U.K. and the U.S. that the average ratio of Sr^{90} to calcium in the average total diet is close to the ratio for milk alone, being only some 20% higher (KULP, SCHULERT, HODGES 1960). It follows that the concentration of Sr^{90} in milk is a valuable criterion for studying the way in which the contamination of diet is related to the fission product deposition. A sequence of milk samples from a large milk depot in an

Table 20. *Estimation of mean daily intake Sr^{90} by members of the population of the United Kingdom, 1958 and 1959*

Source of the Sr^{90}	Estimated Sr^{90} intake ($\mu\mu c$/day)	
	1958	1959
Milk, cream and cheese	4.20	5.57
Root vegetables	0.35	0.58
Leaf and other vegetables, fruit, etc. .	0.54	0.88
Flour and cereals of all types	0.75	1.32
Eggs, meat, fish and other sources . .	0.32	0.55
Diet total	6.16	8.90
Drinking water	0.23	0.43
Tea	—	0.45
Air breathed	0.16	0.30
Total daily intake	6.55	10.08

area of moderate rainfall in the U.K. (BRYANT, CHAMBERLAIN, SPICER, WEBB 1958) showed that during the period 1955 to 1957 the levels in milk rose only slightly and remained in the range of some 4 to 5 $\mu\mu c$/g of calcium despite a steady rise in the accumulated deposition of Sr^{90}. This was the first direct evidence that the rate of fallout as opposed to the accumulated total in the soil might be an important factor in determining the contamination of human diet. More precise information on this topic is provided by experimental studies, and in the U.K. large scale controlled field experiments with either natural fallout or tracers have been carried out as well as experiments on soil and plant in the laboratory. BURTON et al. (BURTON, MILBOURN, RUSSELL 1960) have deduced from these that in 1958 the quantity of Sr^{90} in milk was some 5 times greater than could be attributed to absorption of strontium from the soil into cattle fodder. This implies that about 80% of the Sr^{90} in milk was due to direct contamination of the fodder by rain. A comparable conclusion was reached in 1959. Further information has been obtained from a comparison of the ratios of Sr^{89} to Sr^{90} in milk during 1958 with the ratios simultaneously observed in rain. This comparison indicated that the apparent delay period in the passage of Sr^{90} from rain to milk was about 1 to 2 months during the summer. The situation appeared to be comparable in North America. It is thus apparent that the amount of Sr^{90} from fallout which has entered into human diet in the U.K. has been determined mainly by the quantities deposited in the recent past and not by the accumulated deposition in the soil. Thus the 40% increase in the Sr^{90}/Ca ratio in diet in 1959 above that of 1958 and 1957 is a result of the higher deposition of Sr^{90} in that year. There is, however, no sound basis for predicting how rapidly the decreased fallout level in 1960 will be reflected in diet since a lag period, which cannot be accurately estimated, will be caused by the consumption of stored food. If no further large tests are held, the rate of deposition will continue to decrease while for a time the cumulative deposit will continue to increase slowly. The contribution of the latter to the Sr^{90} in diet will grow in relative importance but its magnitude should always be small compared with that caused by direct contamination in past periods of high fallout.

Few systematic data are available for the Sr^{90}/Ca ratio in the diets of much of the world's population. Some differences from the values observed in Europe and North America are to be expected in some areas due to differences in the level of fallout. This could explain, for example, why the ratio of New Zealand cheese in 1959 was about a quarter of that produced in Europe and Canada. Some differences are more complex. Thus much of the agricultural soil in the world is low in calcium and this not only reduces the amount of calcium in plants but enhances the absorption of Sr^{90} from the soil; both effects will increase the Sr^{90}/Ca ratio in plants. The effect of different kinds of diet are difficult to assess without a comprehensive programme of measurement. Dairy products are less important than leafy vegetables and cereals in some large countries but the data available are inadequate. In the U.K., although flour is less important a factor than milk as a source of Sr^{90}, the Sr^{90}/Ca ratio is 4 to 5 times higher than in milk and for a given level of fallout it might be expected that the ratio in rice-eating countries might be higher than in the U.K. and U.S. Thus the 1958 U.N. report estimated that the ratio in Japanese diet might be about 6 times higher than in countries which derived most of their dietary calcium from milk, although observed Japanese values in 1959 ranged from 11 to 15 $\mu\mu c/g$ in different parts of the country compared with the value of 9.3 for the average diet in the U.K. Undoubtedly some of the controlling factors are self-cancelling and there is no recorded evidence to show that any large group of people in the world have much higher Sr^{90}/Ca ratios in their average diet than those recorded in Europe, America and Japan. Analysis of Sr^{90} in human bones leads to a similar conclusion.

The analysis of human bone for Sr^{90} has been carried out on an increasing scale in recent years and KULP et al. (KULP, SCHULERT, HODGES 1960) have recently reported the results of a study of some 9000 human bone samples collected from several parts of the world since 1953. Since Sr^{90} has been present in the diet of human beings for a relatively short time and the rate of turnover of existing bone is slow, the average Sr^{90} concentration in the adult skeleton is

Table 21. *Average Sr^{90} concentrations (in $\mu\mu c$ per g calcium) in the skeletons of various age groups.* (Number of samples in parentheses)

Population	Age group (years)					
	20 to 29	30 to 39	40 to 49	50 to 59	60 to 69	over 70
New York	0.071	0.098	0.080	0.092	0.097	0.094
Whole skeletons.	± 0.008	± 0.015	± 0.009	± 0.006	± 0.005	± 0.004
Normalised to 1957 . . .	(4)	(9)	(17)	(67)	(94)	(125)
Western culture.	0.14	0.18	0.14	0.12	0.13	0.10
Area, single bones	± 0.03	± 0.03	± 0.02	± 0.02	± 0.02	± 0.02
1957	(26)	(29)	(40)	(46)	(52)	(48)

much lower than in the bones of children, in whom new bone is continually being formed. The highest concentrations of Sr^{90} in bone have in fact been observed in children from a few months to five years old. Table 21 gives comparisons of the average Sr^{90} concentrations in the skeletons of various groups of adults and shows that these are essentially independent of age. It is therefore possible to treat adults as a homogeneous class regardless of age. KULP's values of the average concentration of the Sr^{90} in human bone in 1958 and 1959 are given in Table 22. Included in the averages are the data of BRYANT covering the U.K. and Australia (BRYANT 1958, 1959, 1960). The standard deviation for large sets of single adult bones from an area in which diet is reasonably uniform is about 80%. Therefore since the annual increase in concentration in the diet had been only approximately 50%, useful comparisons between means can be made only between stations with fairly large numbers of samples. If, for example, 70 samples are analysed the standard error of the mean should not exceed 10%,

Table 22. *Average concentrations of Sr^{90} (in $\mu\mu c$ per g calcium) in human bone in 1958 and 1959.* (Number of samples in parentheses)

Station	1958			1959		
	0 to 4 years	5 to 19 years	adult	0 to 4 years	5 to 19 years	adult
30° to 70° N, Western culture area						
New York	2.1 (16)	0.9 (2)	0.13 (28)	2.8 (5)	1.1 (4)	
Boston.	1.5 (3)	0.9 (12)				
Houston	1.6 (8)	1.6 (2)	0.17 (67)			0.23 (46)
Bismarck, N.D.			0.40 (3)			
Vancouver	1.9 (4)	0.8 (21)		2.0 (1)	1.1 (5)	
England	1.4 (50)	0.8 (19)	0.17 (15)	2.7 (27)	1.0 (19)	0.25 (4)
Germany (Bonn) . . ˙ .	1.7 (18)	0.6 (11)	0.34 (61)	2.0 (23)	0.8 (7)	0.38 (55)
Israel	0.2 (6)	0.2 (2)	0.16 (27)	1.6 (3)	0.7 (1)	
30° to 70° N, Orient						
Japan (Tokyo)		0.7 (2)	0.17 (30)		0.6 (1)	
10° to 30° N						
Puerto Rico	0.64 (20)	0.65 (10)	0.21 (137)		0.89 (19)	0.28 (51)
Guatemala	0.1 (1)	0.3 (5)	0.08 (28)			
Taiwan		0.8 (4)	0.20 (11)		1.1 (1)	
Thailand (Bangkok) . . .		0.39 (1)	0.30 (10)			
10° N to 20° S						
Brazil (Recife)	0.72 (3)	0.49 (10)	0.16 (25)	1.96 (8)	0.32 (6)	0.15 (70)
Belgian Congo (Leopoldville).	1.5 (1)	0.27 (1)	0.07 (4)	1.0 (1)		
Colombia (Barranquilla) .		0.2 (2)	0.03 (2)			0.06 (5)
20° S						
South Africa (Durban-Capetown) . .		0.95 (22)	0.22 (2)		0.92 (17)	
Australia.	0.60 (52)	0.33 (25)	0.12 (65)		0.46 (2)	0.23 (22)
Chile (Santiago).		0.46 (8)		0.39 (17)	0.27 (16)	

the value of which is similar to the analytical error. Although data from some parts of the world are at present meagre there is no evidence in Table 22 of higher values than some of those observed in Western countries. Although Puerto Rico lies at latitude 18° N and receives less fallout per inch of rainfall than the average for Western culture areas (Fig. 8) the difference is compensated for by the higher ratio of non-milk sources to milksources in the diet. Similarly the fact that the average for adults in Thailand is 50% higher than that for Western culture areas, although the fallout in Thailand is lower by a factor of 2 to 4, is also probably related to the difference in the diet.

The probable curve for the distribution of the Sr^{90} in Western man in 1959 is shown in Fig. 10 (Kulp, Schulert, Hodgens 1960). The highest values of 2.6 $\mu\mu c$ Sr^{90} per gram of calcium occurs in 1 year old children. The lower value of 1.3 observed in foetal skeletons is due to a discrimination factor of 12 between the mother's diet and the foetus. From published estimates of the amount of Sr^{90} in the stratosphere and its residence time, Kulp has calculated the change in the average concentration of Sr^{90} in skeletons in people of any age as a result of nuclear tests held before the end of 1958 on the assumption that only 20% of the Sr^{90} in diet is due to the cumulative deposits in soil. His results are demonstrated in Fig. 11 which shows curves projected to 1970 of the skeletal burden in individuals who in 1959 were 1, 15 and 26 years old respectively. It may be seen that concentrations in adults will increase slightly and then level off towards the equilibrium level for new bone determined by the diet. Levels in 1-year-olds will rise slightly but then will drop sharply towards the equilibrium level.

In their report of December 1960 (Medical Research Council, 1960), the British Medical Council considered the maximum value of the ratio of Sr^{90}

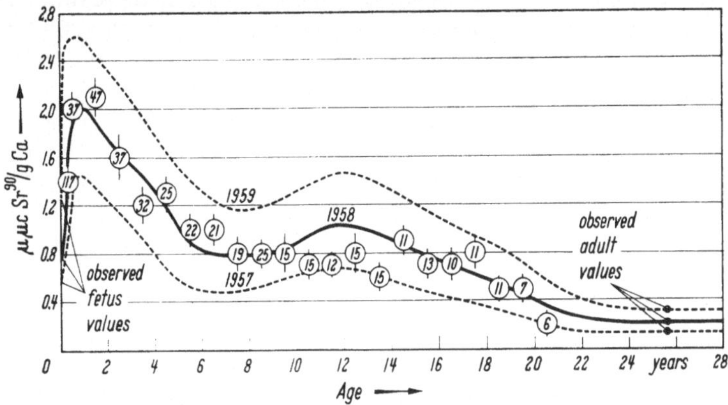

Fig. 10. Calculated curves showing expected Sr^{90} concentration in the bones of children and young people in Western culture areas for 1957, 1958, and 1959; the experimental data include the number of samples in each age group for 1958

to calcium ratio in bone which was theoretically possible on a selected diet. Using the food surveys conducted by the Agricultural Research Council in 1959, it can be calculated that if the diet of any individual had consisted only of those foods with the highest contamination found anywhere in the U., the ratio could have reached 80 $\mu\mu c/g$ in his total diet. The corresponding ratio in new bone laid down by any individual is therefore unlikely to have exceeded 20 $\mu\mu c/g$. Since the assumptions in this calculation are extremely pessimistic it is not surprising that this level is twice as high as the highest value actually observed.

On the basis of the recommendations of the International Commis-

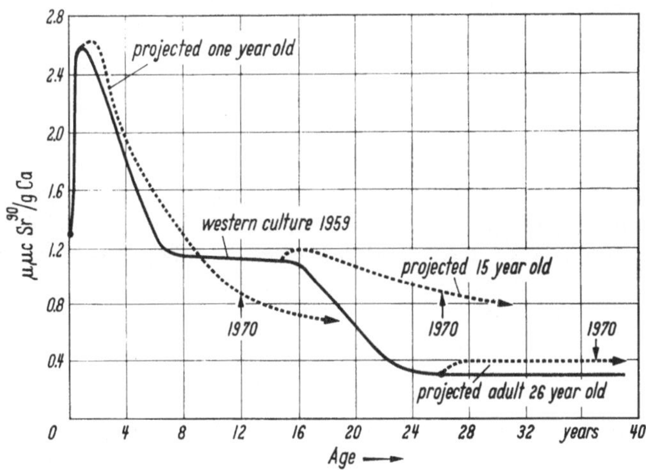

Fig. 11. Concentrations of Sr^{90} in Western culture areas for 1959 and projected curves for individuals aged 1, 15 and 26 years, respectively, in 1959

sion on Radiological Protection (I.C.R.P. 1959), the Council recommended that the maximum permissible concentration of Sr^{90} in bone averaged over the whole population should be 67 $\mu\mu c/g$ of calcium and the maximum for any individual member of the population should be 200 $\mu\mu c/g$. The corresponding dose rates to bone-marrow tissue are 60 and 180 millirems per year respectively. Since these values refer to the permissible level when Sr^{90} is the sole source of radiation in bone additional to that received naturally, a small allow-ance must be made for the irradiation of bone by external radiation.

b) Caesium-137. Cs[137] is a long lived nuclide with a half life of 30 years and its activity in fallout has been found to be on an average 1.7 times that of Sr[90] (Crooks, Osmond, Fisher, Owers and Evett 1960). Unlike Sr[90], Cs[137] is held in the soil in forms which are largely inaccessible to plants and thus its uptake into plants is determined, even more markedly than the uptake of Sr[90], by the rate of fallout and the resulting superficial contamination of plants. Being an alkali metal like sodium and potassium, caesium is freely absorbed from the human intestinal tract and appears to remain in the body for a limited time having an average stay of some four months. The minority of Cs[137] atoms which decay in the body before excretion form Ba[137] which disintegrates almost at once and emits γ-rays. These γ-rays can be detected from outside the body with suitably sensitive equipment and it is therefore possible to measure the degree of internal contamination of an individual with Cs[137]. It is then possible to assess the radiation dose rate both to the whole body and to individual tissues.

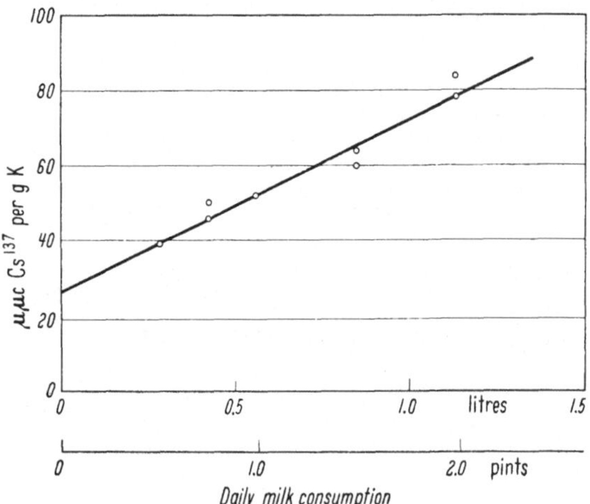

Fig. 12. Correlation of Cs[137]/potassium ratios *in vivo* with daily milk consumption

Analyses for Cs[137] already made in Canada, the Federal Republic of Germany, Norway, Sweden, the United Kingdom and the United States of America lead to the view that the most important dietary sources of this isotope are milk and meat (F.A.O. Report, 1960). Thus in a detailed study of nine subjects in the U.K in late 1958 and early 1959, Rundo (1960) found a clear correlation between the caesium content of the subjects and their average daily milk intake (Fig. 12). However, Rundo has pointed out that variation in the biological half-life among individuals also influence the Cs[137] content in vivo, even if the daily intake is constant, so that a larger number of subjects is required to establish the correlation more firmly. In Fig. 12 the caesium content is expressed by the ratio of Cs[137] to natural potassium content; in this way the Cs[137] content is related to the mass of the muscle, the main site of deposition in the body. There is, however, some evidence that the metabolism and routes of entry into the human body of these two elements are to some degree different and that a biological meaning similar to that of the Sr[90]/Ca ratios should not be implied. Fig. 12 suggests that during the period of measurement, about 26 $\mu\mu c$ Cs[137] per g potassium originated from sources other than milk. The mean value for all the control subjects was 56 $\mu\mu c$ caesium-137 per g potassium so that milk was responsible for one-half the average caesium burden. Calculations have been made which suggest that in the United States of America about 60% of Cs[137] in the human diet comes from milk, 25% from meat, 7% from flour and cereals, 5% from vegetables and 3% from citrus fruits. Similar results have been obtained from surveys in Canada, the Federal Republic of Germany, Norway and Sweden (F.A.O. Report, 1960). On the world scale, available data are unfortunately very limited and no general statements on the effect of diet composition on the intake of Cs[137] can be made until the subject has been more fully examined. There is some evidence that in countries in which the potato is an important article of diet, it may also be a significant contributor of Cs^{137}. No published data are available for countries whose food supplies contain large proportions of cereals or starchy foods other than potatoes. From limited data, it appears that the Cs^{137} concentration in fish from the Pacific Ocean in the period 1954 to 1958 was not more than 0.02 $\mu\mu c$ per g of edible fish and it therefore seems unlikely that fisheries could produce a significant fraction of Cs^{137} in human diet (F.A.O. Report, 1960).

Table 23. *Mean Cs^{137}/potassium ratios (in $\mu\mu c$ per g) in control groups resident in Illinois and in New Mexico (U.S.A.) and in Berkshire and Oxfordshire (U.K.)*

Date	Chicago, Illinois	Los Alamos, New Mexico	Berkshire and Oxfordshire
Jan. to June 1956	32	} ~32	33
July to Dec. 1956	—		30
Jan. to June 1957	35	~32	35
July to Dec. 1957	—	~42	41
Jan. to June 1958	46	58	43
July to Dec. 1958	51	64	53
Jan. to June 1959	63	—	56
July 1959	—	—	63
Sept. 1959	57	—	—
Dec. 1959	55	—	56
March 1960	—	—	58
June 1960	—	—	52
Sept. 1960	—	—	48

The trend in the mean levels of Cs^{137} in human beings in the U.K. and U.S. is shown in Table 23 compiled by RUNDO (1960). This shows that the general levels, which are very similar in both countries despite diet differences, increased between 1956 and 1959 when the rate of fallout reached its peak. There are signs, particularly in the U.K., of a subsequent decrease in sympathy with the reduced rate of fallout in 1960. Since the contamination of animal foods is controlled by fallout rate and since the residence period of Cs^{137} in the human body is only 4 months, a more rapid decrease in human levels might have been expected. However, the relationships are complicated by climatic factors such as local rainfall and, more significantly, by the storage of cattle and human foods and these two factors need to be adequately assessed before reliable predictions can be made.

The radiation doses to the gonads corresponding to measured levels of Cs^{137} in the body have been computed by RUNDO on the assumption that the caesium is distributed throughout the body. For the highest mean value of 63 $\mu\mu c$ Cs^{137} per g potassium observed in the U.K. (July 1959) the dose rate is 1.1 mrad/yr or about 1% of the total natural background radiation. GUSTAFSON (1960) has shown that the maximum whole-body dose rate from fallout materials outside the body occurred in May 1959 in the United States and has calculated that the mean dose rate from internal Cs^{137} at that time was 1.8 mrad/yr.

c) **Iodine-131.** This radioactive isotope has a short half-life of only 8 days and like other forms of iodine it is concentrated in the thyroid gland. No very comprehensive surveys have been made of the distribution of this isotope after test explosions but enough ad hoc measurements have been made to show that the hazard is in general small. Measurements in the United States for the period of 1955 to 1956 show that, excluding areas immediately adjacent to test sites, the annual thyroid dose in man averaged about 5 mrem compared with an average of about 100 mrem from natural sources (U.N. Report, 1958). I^{131} was detected in human thyroids in the U.K. following the large number of weapon tests which took place in the late summer and autumn of 1958 (Robertson, Falconer 1959). The dose calculated from a small number of measurements in a 4-week period at the end of 1958 was 6 mrem, and this was a period of high fallout of fairly recently produced fission products. The British Medical Council have concluded that the radiation dose due to the thyroid gland from I^{131} produced in nuclear weapon explosions is unlikely to have exceeded 24 mrem in any year. The maximum permissible thyroid dose to large populations, according to the recommendations of the International Commission on Radiological Protection is 1000 mrem/yr (I.C.R.P. 1958).

d) Carbon-14. C^{14} is produced naturally by the action of cosmic rays on the earth's atmosphere and the total world inventory has been estimated to be about 2150×10^{27} atoms. About 34×10^{27} of these atoms are present in the atmosphere as CO_2 and take part in the natural carbon cycle. The dose to the human body due to natural radiocarbon in tissue is approximately 1 mrad per year (Medical Research Council, 1960).

Radiocarbon is also produced by the capture in atmospheric nitrogen of neutrons released in nuclear explosions and Hagemann (Hagemann, Machta, Gray, Turkevich 1959) has calculated that 25×10^{27} atoms of C^{14} had been added to those naturally present in the atmosphere by November 1958. This excess C^{14} is expected to move from the stratosphere into the troposphere with a half-period of about 3.5 years (Craig 1958) and ultimately to be transferred into the biosphere and surface waters of the oceans. The future trend of the carbon-14 content of the troposphere due to all explosions prior to 1959 has been described as follows (Medical Research Council, 1960):

(1) A continuing rise, reaching about 40% above normal in about 1962 when all the artificial C^{14} is thoroughly mixed with the natural content of 34×10^{27} atoms.

(2) A fall with a half-period of several years to 25% above normal due to exchange with the carbon reservoir of the biosphere and surface waters of the ocean.

(3) A further fall with a longer half-period of about twenty years (Craig 1958), down to 1% above normal as a result of mixing with the carbon reservoir in deep oceans.

(4) The final fall to the original level existing before nuclear weapon-tests, with a half-period equal to that of the radioactive decay of C^{14} (5570 years).

From these data it has been estimated that the radiation dose to the gonads from artificially-produced C^{14} will reach 10 mrems during the 30 year generation beginning in 1954 and will be 5 mrems in the following 30 year period compared with total natural doses of 3000 mrems in the same periods (Feely 1960). The doses in subsequent generations will rapidly become much lower because of the diffusion of C^{14} into the ocean and into the humus of soil.

The cumulative dose ultimately distributed over succeeding generations is estimated to total 120 mrems which is larger than that which is calculated to result from the decay of all the radiation fission products from test explosions, but the dose will be delivered slowly over tens of thousands of years.

4. External radiation from fission products

External radiation arises primarily from fission products deposited on the ground. The dose rate from the deposit, which contains gamma emitters of various energies and half-lives, can be computed from measurements of the concentration of the important components in rainwater or in soil, using a γ-ray spectrometer, or measured directly by ion chambers. These methods have been used successfully by SPIERS (1959) and PEIRSON and SALMON (1959) but since in general the dose rate increases are small compared with natural background levels the computational methods are generally preferred. Although the dose rate at any time is a useful indicator of the current level of fallout, particularly when compared with natural background, it is of limited biological significance. For the purposes of estimating the possible effects of small doses, the dose accumulated over a number of years is a more-relevant parameter. In the case of genetic effects, for example, the integrated dose for a generation, that is, over a 30 year period is probably the appropriate index (Medical Research Council, 1960). For somatic effects, the appropriate period would depend on the particular effect under consideration.

There are several difficulties associated with calculating either the dose rate or the integrated dose to an individual, even when comprehensive fallout data are available. The first step is to assume that the fission products are uniformly distributed over an infinite plane and to compute the doses to an individual standing on that plane, the so-called "infinite-plane dose". The dose delivered to the gonad and bone marrow of a representative individual must however be computed taking into account shielding, weathering and leaching factors. The shielding factor accounts for the reduced dose rate during the time the individual spends indoors where the dose rate from fallout deposits is reduced, whereas the weathering and leaching factors account for movement of the deposited gamma emitting isotopes from the upper layers of the earth's surface, for example into the lower layers of the soil. These factors have been variously estimated in different countries and composite reduction factors ranging from 3 to 21 have been used in reports submitted to the United Nations Committee on Radiation Hazards (U.N. Report, 1958). A shielding factor of 5 has been used in more recent reports from the United States and the United Kingdom (GUSTAFSON 1960); (Medical Research Council, 1960). A greater uncertainty arises in the computation of the 30-year dose which is made up of dose already received plus dose which is going to be received in the future. The dose to be received in the future is made up of:

(1) Fission product activity still on the ground. Further dose from this source can be computed if the components of the activity in the soil are known.

(2) Fission product activity held in the stratosphere and still to be deposited. This requires a knowledge or estimate of the stratospheric store of fission products and of storage times.

(3) Fission products from weapons still to be exploded. Since no estimate can be made of this, it will be assumed in this chapter that tests ceased in 1958.

Most of the smaller or nominal-sized weapon tests have been staged in the middle latitudes of the Northern hemisphere and the fallout from them has been largely confined to those latitudes. Stratospheric dust is also deposited in the same region of the northern hemisphere (Fig. 8) so that the doses from tropospheric fallout will vary with geographical location in a manner roughly similar to those from stratospheric fallout. The United Kingdom is near the peak of the latitude distribution curve of both sources and is a reasonably representative area in which to compare the relative magnitudes of the doses. It has been shown that the total dose to the average man in the U.K. from some 70 nominal bombs exploded before mid-1955 was 1 mr (shielding factor of 5). The total dose over a period of 50 years due to the 1954 thermonuclear tests in the Pacific has been estimated as 4 mr (STEWART, CROOKS, FISHER 1956). Since that time, the ratio

Table 24. *Dose rate due to fall-out at Argonne National Laboratory ($\mu rad/h$) natural background = 11.12 $\mu rad/h$*

Isotope	1957			1958			1959				
	May	July	Oct.	April	July	Sept.	March	April	May	June	July
Zr^{95}-Nb^{95} .	1.14	1.23	1.46	1.27	0.60	2.00	4.64	6.22	6.62	5.82	4.80
Cs^{137} . . .	0.09	0.08	0.09	0.12	0.15	0.21	0.32	0.34	0.36	0.37	0.38
Ru^{106} . .	0.15	0.15	0.15	0.14	0.13	0.18	0.53	0.69	0.75	1.00	1.03
Ru^{103} . .	0.35	0.41	0.49	0.27	0.07	0.17	0.69	0.48	0.39	0.25	0.18
Ce^{141} . . .	0.01	0.02	0.02	0.01	0.01	0.01	0.13	0.07	0.06	0.04	0.03
Ce^{144} . . .	0.04	0.05	0.05	0.04	0.04	0.05	0.21	0.27	0.28	0.31	0.33
Total	1.78	1.94	2.26	1.85	1.00	2.62	6.52	8.07	8.46	7.79	6.75

of large to small explosions has increased, and the dust from the more northerly large-scale tests has been deposited in a shorter time with the result that the external doses from tropospheric fallout are very small when compared with those from stratospheric fallout.

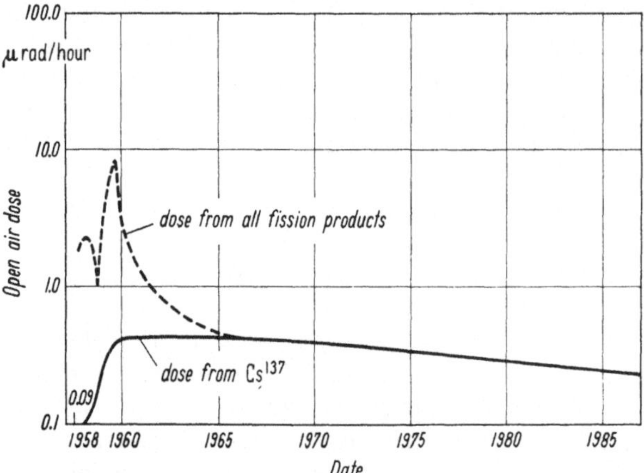

Fig. 13. Calculated γ-ray dose 1 m above the ground from fission products; tests discontinued in 1958

A comprehensive programme of dose-rate measurements due to fallout has been carried out since 1957 by Gustafson at the Argonne National Laboratory, Chicago, U.S.A. (Gustafson 1960). The radioactive content of soil was measured regularly on a γ-ray spectrometer and the dose rate from each component isotope computed by standard methods. The computed "infinite plane" dose rates are shown in Table 24 in units of $\mu rad/h$. During this period, most of the external dose was due to the 65 day Zr^{95}-Nb^{95} chain, whose absolute contribution rose to a peak in May 1959 following the autumn 1958 high-latitude tests and subsequently declined due to radioactive decay. In contrast, the contribution from 30-year Cs^{137} has increased slowly as fresh material from the stratosphere has added to the amount already on the ground. Gustafson has applied these data to estimate future levels of radiation and the integrated dose from material deposited since 1957 on the assumption that testing is not resumed. The results are given in Fig. 13 which shows quite clearly that after a few years the bulk of the external dose is due to Cs^{137}. The underlying assumptions in the calculation are that between 60 and 75% of the fallout in spring 1959 was due to the high-latitude tests in 1958, having a storage time of 6 to 9 months, and the remainder from Pacific tests having a storage time of 4 years. He has calculated that the present situation, in which nuclear weapon testing is no longer being carried out, will lead to a 30-year dose of 52 mrads and a 70 year dose of 62 mrads to roughly one half of the population of the U.S.A. These increases include 10 mr due to whole-body radiation from Cs^{137} absorbed internally in body tissues. The doses represent increases of approximately 1.4 and 0.7% respectively over natural radiation which, in the Chicago area, amounts to 125 mrads/year. The peak dose rate of 8.46 μrad/h unshielded is about 75% of the natural background level. The latitude of Chicago (42° N) is close to the latitude of highest rainwater activity (Fig. 8) so that the Argonne figures provide a reasonable basis on which to assess the importance of gamma radiation from fallout. The world average would be expected to be smaller. The United Nations Scientific Committee considered all data submitted to them in their first report issued in 1958 (U.N. Report 1958), and estimated that if weapon testing ceased at the end of 1958, the population-weighted 30-year dose from fallout gamma radiation would be 10 mr as compared with the Argonne figure of 52 mr. The U.N. estimate is liable to be low as it did not include allowance for the comparatively rapid deposition of stratospheric dust from the high-latitude tests held in 1958.

5. Total doses to tissue from nuclear explosions

It is not possible to assess accurately the doses to which the average citizen of the world has been exposed from radioactive fallout. Substantial measuring programmes have been confined to a few countries and extrapolation is difficult since conditions in other countries differ for a number of reasons. In the first place, the deposition of fallout varies with latitude, and very little is deposited in equatorial regions (Fig. 8). As far as external radiation is concerned, this factor is at least partially cancelled by the fact that the shielding afforded by buildings in north temperate latitudes is greater than in the tropics. Differences in the composition of diet in some countries may cause relatively higher doses to bone marrow from Sr^{90} than would the same level of fallout in western countries, where most of the observations have been made. On the other hand, it has been found that the Sr^{90}/Ca ratio of average diet in Japan in 1959 was not much higher than that in the U.K. (U.N. Report, 1958). The limited data on Sr^{90} in bone obtained by KULP (Table 22) do not indicate any large difference in levels between western culture areas and rice-diet areas. In fact, the levels in Europe and the United States tend to be higher than in any other areas of the world.

With these points in mind, total doses from fallout in the U.K. are put forward as a guide to levels in all countries. The controlling factors must, however, be borne in mind if the data are to be applied to special areas of the world in which these factors are unlikely to be self-cancelling.

The doses appropriate to the U.K. have been summarised by the British Medical Research Council in their 1960 report (Medical Research Council, 1960). The estimated gonad doses over 30 year periods are given in Table 25 on the assumption that nuclear test explosions ceased in November 1958. The integrated gonad doses for the 30 year periods commencing 1954, 1960 and 1984 are about 1.2, 0.75 and 0.2% of the integrated dose from the natural background, which is about 3000 mrads in 30 years.

The I.C.R.P. (1958) have recommended that the genetic dose to large populations from all sources, additional to the natural background, should not exceed 5 rems plus the lowest practical contribution from medical exposure. The commission did not wish to make any firm recommendation as to the apportionment of this genetic dose but as an illustration gave the following:

(A) Occupational exposure . . . . 1000 mrems

(B) Exposure of special groups (e.g. population close to nuclear
    installations) . . . . 500 mrems

(C) Exposure of the population of large . . . . 2000 mrems
             Reserve . . . . 1500 mrems

The total doses in Table 25 are small in comparison with the illustrative figure of 2000 mrems for the maximum permissible genetic dose to the general population.

The Medical Research Council of the U.K. experienced difficulty in defining the significant period over which the doses to bone marrow from Sr^{90} should be integrated, as little relevant information was available. They therefore calculated

Table 25. *Estimated gonad doses from fall-out over 30-year periods nuclear test explosions assumed discontinued from November, 1958*

| | Estimated dose (millirads) | | |
	period commencing 1954	period commencing 1960	period commencing 1984
External radiation from fall-out .	16	8	1.5
Cs^{137} within the body	9	4.5	0
C^{14} within the body.	10	10	5
Total dose from fall-out	35	22.5	6.5

the average dose to bone marrow from new bone in the two year periods 1956 to 1957 and 1958 to 1959. The values were 5 mrems and 12 mrems respectively which are about 3 and 7% of the integrated doses from the natural background in the same periods. The I.C.R.P. make no specific recommendations as to the maximum permissible "somatically" relevant dose to the population. However, by recommending for planning purposes that the maximum permissible level of Sr^{90} in bone applicable to the population at large should not exceed one-thirtieth of the values for individuals who are occupationally exposed, they do in effect imply a maximum dose rate of about 60 mrems per year to the bone marrow of the general population. The observed mean dose rates in the U.K. over the two-year periods stated are small compared with this level.

II. The disposal of radioactive wastes

Many of the processes associated with the development of nuclear power produce radioactive waste products in solid, liquid and gaseous forms. The wastes arise in a number of chemical forms of widely different radioactive content so that no single method of disposal can be universally adopted. It is, however, possible to class the different methods under two broad headings, storage and release to the environment. The latter method has sometimes been regarded as totally undesirable but as in the case of other industrial wastes, a balance has to be struck between the disadvantages to the community of waste disposal and the advantages conferred on society by the industry itself. As a minimum criterion, the dispersion of radioactive waste into the environment should be controlled so that the radiation exposure both of individuals and of the population as a whole remains below the maximum permissible values agreed both nationally and internationally.

The recommendations of the International Commission on Radiological Protection (I.C.R.P., 1958) provide a satisfactory basis for assessing the safety of discharges of radioactivity to the human environment and in most countries legislation exists to control effluent discharge accordingly (BURNS 1961). The recommendations are not enough of themselves, and they need to be supplemented by information on the behaviour of the radioactive material in the environment, and on the use which man makes of his environment.

1. High activity wastes

The amount of radioactive material arising in the form of waste is now so large that dispersal into man's environment cannot be regarded as a feasible solution to the disposal of the principal waste fission products from the generation of nuclear power. These wastes, as well as other small-volume high activity wastes, are concentrated to the minimum practical volume and stored in leak-proof shielded containers until they decay to an insignificant level. This may be several hundred years in the case of Sr^{90} and Cs^{137} which have half-lives of about 30 years. Approximately 65 million gallons of these wastes are stored in underground tanks in several places in the United States. While tank storage represents an interim answer to the problem for the present and the immediate future, it is generally accepted that it is not a practical long term solution from an engineering standpoint. It is expected that waste volumes will be reduced with the development of new and improved chemical processing and waste systems. The following approaches are, for example, being investigated in the waste disposal programme of the U.S. Atomic Energy Commission (National Academy of Sciences, 1960):

1. The fixation of fission products in a solid form, by fusion or by incorporation in glasses or ceramics which are known to be highly resistant to leaching by chemical agents.

2. Storage of this solid material in selected geological formations with current emphasis on salt beds.

3. The direct discharge of liquids to selected geological strata such as salt cavities or deep permeable formations.

2. Low and medium activity wastes

In addition to these high-activity materials there arise substantial quantities of wastes in which the radioactivity is of low or moderate concentration. Treatment of these wastes so as to concentrate the activity in a form suitable for storage is not always an economic proposition and dispersal into the environment under carefully controlled conditions is often a more satisfactory way of disposal. The general aim is to achieve sufficient dilution during and following release that the amount of radioactivity reaching any one person is trivial.

It is clearly necessary to define standards of radiation exposure for the general public and for this purpose the I.C.R.P. have divided exposure of the population into two categories:

(1) Exposure of the public living in the neighbourhood of controlled areas.

(2) Exposure of the population at large.

The former category deals with the exposure of individuals, while the latter is concerned primarily with the genetic problem, and therefore deals with the average exposure of the whole population. The recommended maximum permissible doses vary from organ to organ and include the doses from both external and internal radiation. Table 26 gives the values implied by the 1958 recommendations of the main Commission.

With regard to waste disposal, the relative importance of the two categories of exposure will depend primarily on the remoteness of the site of disposal. For example, waste discharged into sewers or coastal waters is likely to be limited by the possible exposure of special groups such as sewage workers or local fishermen who eat part of their own catch. The contribution to the population average will be trivial because a discharge must be governed by the dose to the most highly

Table 26. *Maximum permissible doses recommended by the I.C.R.P. (1958) for the general public* [1]

Organ	Maximum permissible doses	
	special groups	whole populations
Whole body	0.5 rem/yr	5 rem/30 yr
Gonads	0.5 rem/yr	5 rem/30 yr
Blood-forming organs . . .	0.5 rem/yr	0.5 rem/yr
Lens of eye	0.5 rem/yr	0.5 rem/yr
Thyroid and skin	3.0 rem/yr	1.0 rem/yr
Other single organs	1.5 rem/yr	0.5 rem/yr

[1] See also page 260 (Table 17) and page 285.

exposed individuals of the special group. In contrast, waste discharged into the ocean depths will be very widely distributed before it causes any human exposure and genetic considerations may well provide the limiting factor. In assessing these problems in a practical case, it is clearly necessary to have in mind tentative figures for the permissible genetic dose. The I.C.R.P. (1958) pointed out that their suggested maximum genetic dose of 5 rems per 30 years must not be used up by one single type of exposure and they considered that the apportionment should be made by national authorities. DUNSTER (1960) has suggested that since public exposure to radiation from waste disposal is likely to be due only to the release of wastes of relatively low activity, there seems to be no reason why waste disposal should claim a substantial part of the total permissible genetic dose. He considered that an allowance of 10% of the total should prove adequate for planning purposes.

These maximum permissible doses are rarely capable of direct application to the problems of waste disposal. They have to be supplemented by studies of the processes by which radioactive materials become distributed through the environment and thereby cause radiation exposure of man. In many cases, a review of these processes and of the amount of waste to be released shows the existence of such large safety factors that a release can be declared safe with no further study. In other cases, however, present knowledge proves to be inadequate and no such declaration can be made. The preliminary review will then indicate the lines of investigation which need to be followed and will also show that some aspects of the problem can be neglected as trivial. In this way, a practical programme can be defined for the study of what in most cases is an exceedingly complex problem.

As an example of this approach in the United Kingdom, extensive experience has been gained with discharges of liquid radioactive waste into coastal waters (DUNSTER 1959). The large volume of low activity waste from the Windscale Works, Cumberland, are discharged into the Irish Sea some 3 km beyond the high water line. Preliminary studies showed that only three main potential limitations existed. These were—radioactivity in edible fish, in edible sea-weed and radiation from shore sand. The detailed investigations were therefore concerned with laboratory experiments to relate the radioactivity of fish, sea-weed and shore sand to the radioactivity in sea water predicted by hydrographic studies of the diffusion processes and tidal streams, using dye as a tracer. Other studies were concerned with the distribution and commercial capture of edible fish, with the distribution of edible sea-weed, and with the amounts of fish and sea-weed eaten by the principal consumers. The combination of the results of this work, with the recommendations of the I.C.R.P., provided an estimate of the maximum permissible discharge rates of alpha and beta emitters. Finally, controlled discharges at about one tenth of these rates were made, together with extensive surveys of radioactivity in the marine environment in order to check the predictions and to demonstrate the safety of the proposed rates of discharge. It was found that the contamination of edible sea-weed was the main factor which limited the discharges at Windscale. It is important to stress that although these experiments provide information which can be applied to other coastal sites, the controlling factor in any one area may be a local peculiarity such as extreme stratification of layers of water, or the presence of oyster beds. Each case must therefore be the subject of a local assessment and any necessary investigations must be planned as a result of that assessment. This basic approach to the waste disposal problem has been used at Hanford in the U.S.A. where low-activity liquid wastes are fed directly into the ground (BROWN, PARKER, SMITH 1955). The aim in this method of disposal is to utilise the porosity, hydraulic retention and chemical reactions within the soil to delay the movement of activity sufficiently for natural decay to reduce the levels to negligible values by the time the waste reaches the biological environment. In the first ten years in which this method was used, between 10^5 and 10^6 curies were dispersed into one million cubic metres of soil and considerable field trials were undertaken to demonstrate the safety of the method.

At the Chalk River Plant in Canada, low level solid waste is buried in the ground following careful geological and hydrological studies of the area (MAWSON 1956). Regular samples of leaves, twigs, surface dust, soil, animals, reptiles, fish and insects are taken from the disposal areas and the swamps, streams and lakes into which the disposal areas drain. Auger wells are prepared and periodically sampled. Dry wells are sunk and lined with sealed aluminium casing. Scintillation counters are lowered down these wells to detect and follow the movement of the radioisotopes. In this way, escape of radioactive nuclides can be detected and the use of the disposal area discontinued or remedial action taken before any hazard arises.

Low-activity liquid wastes are sometimes discharged directly into rivers as in the United Kingdom where effluent from the Atomic Energy Research Establishment of Harwell is discharged into the River Thames (BURNS 1960). If the river water is used for drinking, this is likely to be the factor which limits the permissible rate of discharge. The safety assessment is somewhat simpler than in other types of disposal systems since the I.C.R.P. recommendations (I.C.R.P., 1959) define specific values for the maximum permissible concentrations of radioisotopes in drinking water, and since sampling and analysis are easy to carry out.

Deep sea dumping is used as a means of disposal of solid active wastes which contain moderate amounts of radioactivity (BURNS 1960). The waste is generally packed in sealed steel drums, either with or without an inner concrete shell, and with a device for equalising pressure between the interior and exterior of a package. The potential hazard in this case is associated with the concentration of radioactivity in fish and marine organisms. At the present time, it is possible to make only vague estimates of the fate of radioactive isotopes introduced into the deep ocean. Too little is known concerning the circulation and mixing processes in the surface layers and in the deeper, more homogeneous waters of the great ocean basins to evaluate accurately either the rate of dispersal near the surface or how rapidly materials introduced at great depths will be transported by mixing and vertical currents into the surface layers where they will be concentrated by marine organisms. The work of BOGAROV and KREPS (1958) has suggested that the turnover of water may be quite rapid in contrast with widely held opinion. It is not surprising, therefore, that current practice is to restrict dumpings of radioactive materials to amounts which can be shown to be extremely safe on the basis of pessimistic calculations. It is clear that before much fuller use can be made of the ocean's capacity to receive radioactive wastes in this manner, many physical, chemical, biological and geological processes need to be studied in detail. To provide the essential information, comprehensive oceanographic investigations need to be made by all maritime nations.

In the United Kingdom, waste material with any significant associated radioactivity is disposed of in ocean deeps beyond the continental shelf (Burns 1960). The depth of water must be greater than 1500 fathoms and the area must be approved on behalf of the Minister of Agriculture, Fisheries and Food. The amount disposed of in this way is small and the limits set are governed by the radiation level from the containers. These are such as to prevent any hazard to personnel during shipment and dumping operations. Waste materials which are known to be of a very low level of activity are disposed of in a dumping area in the English Channel. The depth of water in this area is 100 fathoms and no fishing takes place in the area. The maximum amount disposed of in this area is 5000 tons per year of low-activity waste estimated not to exceed 200 curies of alpha radioactivity and 4000 curies of beta-gamma activity.

Although no general legislative control exists for the dumping of material outside territorial waters, it is highly desirable that careful records are kept of the total activity content and weight of all consignments. This is the practice in the United Kingdom and for all such disposals sanctioned by the United States Atomic Energy Commission.

3. Gaseous effluent

Many reactors and chemical processing plants produce gaseous effluents which may contain radioactive contamination in either gaseous or particulate form. Efficient methods have been developed for removing much of the contamination from effluent streams (Silverman 1960) but in most cases there remain small amounts of radioactivity to be discharged into the atmosphere, generally through stacks. For example, A^{41} is emitted from reactors in which air is used to cool the core, as in the Bepo reactor at Harwell (Stewart, Gale, Crooks 1958) and a reactor at Brookhaven. Another example of atmospheric release is associated with the handling of slightly active combustible wastes whose bulk must be reduced by incineration. The combustion products are scrubbed and filtered but a small emission of airborne activity cannot be prevented, except at excessive cost. As a final example, I^{131} is emitted from some chemical processing plants such as that at Hanford, U.S.A. (Parker 1955). As in the case of liquid effluents it is necessary to set limits to the amount of radioactivity which may be discharged in this manner and estimates must often be made in advance, while the plants are being designed.

The emission of activity to the atmosphere can give rise to four possible types of hazard:

(a) An inhalation hazard to people breathing the cloud.

(b) Direct external radiation hazard from the radioactive cloud.

(c) Direct external radiation hazard from material deposited on the ground.

(d) Ingestion hazard from material which finds it way into food chains, e.g. through the grass—cow—milk chain.

The type of hazard which predominates will depend on the circumstances of a particular emission. The risk from inhalation is, however, not usually a limiting factor in practice. It is less restrictive than external radiation for nuclides such as A^{41} which are not absorbed by the body and less restrictive than the uptake through food chains of nuclides such as I^{131}.

Since the maximum permissible doses to members of the general public are specified by the I.C.R.P. as average values over annual periods, it is clear that maximum permissible releases must be similarly specified and that the values of ground-level air concentrations and of grass contamination levels from a continuous release must be averaged over all kinds of meteorological conditions. For a given rate of discharge, a fair estimate can be made of ground level concentrations for certain categories of meteorological conditions, using SUTTON's diffusion equations (SUTTON 1947). In this type of calculation, account must be taken of the rise of the effluent plume above its point of discharge due to its initial temperature and momentum (SCORER 1959; BOSANQUET, CAREY, HALTON 1950). The mean concentrations can be computed from a knowledge of the wind-rose and of the frequency of occurrence of the various meteorological conditions considered by SUTTON. By this means, and using the recommendations of the I.C.R.P., permissible discharges of radioactive effluent can be computed for hazards (a) and (b) above. As an example, CHAMBERLAIN (1961) has calculated the permissible emission of A^{41} from stacks of various heights and the results are given in Table 27. The figures are based on a mean dose rate of 0.2 rad/year to

Table 27. *Permissible emission of nuclides from stacks*

Nuclide	Criterion	Permissible emission (μc/sec)		
		$h=20$ m	$h=50$ m	$h=100$ m
A^{41}	external radiation	1×10^4	4×10^4	1×10^5
I^{131}	uptake to thyroid via milk	0.05	0.3	1.3

members of the public who live near the place where the radioactive plume reaches the ground. The I.C.R.P. permissible dose rate is 0.5 rad/year for this case but it is considered inadvisable to design so that this amount of radiation is given to the public from a single installation. The current release from the Bepo reactor at Harwell, which has a stack of effective height 125 m (STEWART, GALE, CROOKS 1958), is about 1.6×10^4 μc/sec of A^{41} and is well within the permissible limit given in Table 27, even for a stack height of 100 m.

Prior assessment is more difficult in the case of hazards (c) and (d) as an estimate must be made of the amount of deposition which will result from a given release to the atmosphere. It is not always possible to predict the size of particles in the effluent gas stream. In addition, where the effluent stream diffuses to ground level, deposition will occur by impaction and diffusion, of vapours and of those particles too small to have appreciable rates of fall under gravity. To describe the deposition of aerosols, CHAMBERLAIN (1953) defined a parameter Vg having the dimensions of a velocity and known as the velocity of deposition:

$$Vg = \frac{\text{Rate of deposition (μc per m}^2 \text{ of ground per sec)}}{\text{Volumetric concentration in air (μc per m}^3)}.$$

For large particles Vg is equal to the terminal velocity, but a very small aerosol may have a high velocity of deposition due to diffusion and impaction. Some experimental values of the velocity of deposition of various aerosols obtained on the airfield at Harwell, England, are given in Table 28. It is seen that there is a variation between them amounting to a factor of 250. The modes of deposition

Table 28

Aerosol or vapour	Surface	Velocity of deposition (cm/sec)
lycopodium spores (32 microns diam.)	grass airfield	1.2
iodine-131 vapour	grass airfield	2.5
fission product aerosol	grass airfield	0.1
strontium and barium fractions of fission product aerosol	grass airfield	0.01

of the four aerosols were very different. The lycopodium spores were deposited primarily by virtue of their terminal velocity (2 cm/sec), and the iodine vapour by diffusion. The fission product aerosol was generated by striking an electric arc between irradiated uranium electrodes and that its terminal velocity was very small was indicated by the fact that deposition on upward and downward facing surfaces was equal. The velocity of deposition of the strontium and barium fractions of the fission product aerosol was determined separately by means of radiochemical analyses of the samples and found to be less by a factor of 10 than for the mixed fission products.

For any particular type of aerosol emitted from a chimney, the rate of deposition in dry weather conditions can be calculated by multiplying the ground-level concentration in air, computed from Sutton's equations, by the appropriate value of Vg. The limit on the amount which may be released can then be assessed by computing the dose rate to a person from the deposited radioactivity or by assessing the amount of the deposited activity which might find its way into the body through food chains. The most important example of the latter case is I^{131} which is absorbed by the human thyroid from the milk produced by cows which graze contaminated pastures. The critical hazard is normally the dose to the thyroid or children, the upper limit of which is given by the I.C.R.P. as 3 rad/year. Using the published data of Garner (1960), Chamberlain (1961) has calculated that the maximum permissible I^{131} activity on grass is 8 mμc per m^2 of ground surface. Using Sutton's equations and a value of $Vg=$ 2.5 cm/sec he has computed the permissible outputs from stacks which ensure that this limit is on average not exceeded, and these are given in Table 27. The value of $Vg=2.5$ cm/sec observed in the field experiments at Harwell (Chamberlain, Chadwick 1953) have been confirmed by measurements at Hanford under average meteorological conditions (Parker 1955). During the Windscale accident in October 1957, measurements were made which led to a value of $Vg=$ 0.3 cm/sec (Stewart and Crooks 1958) but this was probably a low result, to be attributed to low wind velocities at the time and possibly to the non-gaseous form of some of the radio-iodine in that instance. Chamberlain considered it advisable to take the higher values of Vg, to allow for any increased deposition during rainfall (Chamberlain 1959).

4. Genetic dose due to effluent discharges

Because of national legislation and the need to control the dose to members of the public living close to nuclear installations, the total genetic dose from liquid, solid and gaseous effluents discharged locally is extremely small when considered on a world scale. Because of the lack of knowledge of the diffusive properties of the deep oceans and of the relevant biological cycles, the radioactive material being dumped at sea is at present limited to extremely safe quantities and the same conclusion applies. It is certain that the total genetic dose due to all effluent discharges is extremely small compared with that due to the testing of nuclear weapons which in turn is small compared with the dose due to natural background and with the maximum genetic doses suggested by the I.C.R.P. It is accepted by all nations who produce radioactive materials that effluent discharges need to be carefully controlled and that the amounts released can only be increased when the environment into which they are discharged is fully understood physically and biologically. A recognition of this need can be seen in the ever-increasing literature on the subject and on the steps taken by bodies such as the I.A.E.A. to encourage international co-operation and the full exchange of information.

III. Environmental hazards from accidents to reactors or other nuclear plants

1. Accidental release of radioactivity into the atmosphere. The Windscale accident

The methods described for assessing the hazards from the continuous discharge of radioactivity into the atmosphere can be applied to the release of activity from an accident in a reactor or plant in which large quantities of radioactive materials are stored or processed. Particular attention has been paid to the problem of reactor accidents and to the allied problem of reactor siting, but the methods and principles adopted apply equally well to the radiochemical and other plants which are an integral part of a nuclear energy industry. Much of the hazard assessment is based on theory and on small-scale field experiments, but the maximum use is made of the experience gained in actual incidents.

Although several reactor accidents of various degrees of severity have occurred, the only one to release significant quantities of radioactivity into the atmosphere was the accident to the reactor at the Windscale works of the U.K.A.E.A. on October 10th, 1957 (Atomic Energy Office, 1957). A fire occurred during the release of stored Wigner energy from the graphite moderator in the core of the reactor and, before it was extinguished, the following estimated amounts of activity were released into the atmosphere through the filters on the 125 m stack (Medical Research Council, 1960):

I^{131}	20 000 curies
Te^{132}	12 000 curies
Cs^{137}	600 curies
Sr^{89}	80 curies
Sr^{90}	2 curies

Comparison with the relative proportions of these isotopes in the reactor showed that there had been a strongly preferential release of the more volatile elements iodine, tellurium and caesium.

The cloud spread over the southern part of the U.K. and over Europe in accordance with the rather complex meteorological conditions which accompanied the release and which persisted for some days after it. The meteorological analysis of the travel of the cloud has been discussed by CRABTREE (1959) and the spread is indicated on the map in Fig. 14 in which the time-integrals of the iodine-131 concentrations at a number of places in Europe are given in units of $\mu\mu$c-days/m^3.

The external dose to people living in the area around the reactor was due principally to the isotopes I^{131}, Te^{132}, and I^{132}. DUNSTER et al. (1958) have estimated this dose as 30 to 50 millirads while CHAMBERLAIN (1958) gives a total gamma dose of 75 millirads, no allowance being made for the shielding effect of houses. The estimated dose due to radiation from the passing radioactive cloud was small in comparison.

The main hazard to the population arose from the contamination of milk by I^{131} and most of the control measures taken were concerned with limiting doses from this source. The basic criterion adopted was that no child

Fig. 14. Map showing the time integral of the concentration of radioiodine in air (micro-microcurie-days per cubic metre) at various locations in Europe following the Windscale accident in October, 1957

should receive more than 20 rads to its thyroid. From figures available at the time it was calculated that the corresponding limit to the concentration of I^{131} in milk was 100 millimicrocuries per litre. Extensive surveys were carried out as a result of which all milk within an area of some 200 square miles to the south of Windscale was barred for human consumption until the I^{131} concentrations reached acceptable levels.

As a result of the control measures adopted, the estimated highest dose to a child's thyroid was 16 rads, and the corresponding figure for an adult's thyroid was 4 rads. Since the control of milk was based on large numbers of assays there is every reason to believe that the members of the public were satisfactorily protected against excessive doses of internal radiation. The measurements made on thyroid glands gave confirmation to this view. An independent committee set up by the Medical Research Council to report on health and safety aspects of the accident concluded that "It is in the highest degree unlikely that any harm has been done to the health of anybody, whether a worker in the Windscale plant or a member of the general public".

By November 4th the only area remaining restricted was a narrow coastal belt extending some 12 miles south from Windscale. This area remained under restriction until November 23rd, during which time sufficient analyses of strontium in milk had been made for it to be certain that no restrictions were necessary on account of strontium-90.

2. Maximum permissible doses to the public in an emergency

The Windscale accident brought out clearly that the recommendations of the International Commission on Radiological Protection were not sufficient for determining the actions to be taken for the protection of the public in the event of a sudden incident of this sort. As a result, the Medical Research Council's Committee on Protection against Ionising Radiations considered the maximum permissible levels of dietary contamination for the more important isotopes in emergency conditions and their main recommendations are reported in Table 29 (Medical Research Council, 1959).

Table 29. *Maximum permissible dietary intakes of important radioisotopes recommended by the Medical Research Council of the U.K.*

Isotope	Maximum permissible daily intake (mμc/day)	Criteria
I^{131}	60 up to age 6 months, rising to 300 at age 10 and 1300 at ages over 20	thyroid doses less than 25 rads
Sr^{90}	2 at any age	local dose rate in mineral bone not greater than 1.5 rads/yr
Sr^{89}	200 at any age	maximum local dose not greater than 15 rads
Cs^{137}	60 at birth rising to 150 at 6 months and 1150 at ages over 20	maximum dose to whole body not greater than 10 rads

The values take into account the likely sequence of the rise and fall of food contamination and are expressed as those permissible at the time of maximum contamination soon after the accident. It is estimated that these levels of intake if continued during the likely course of the contamination would deliver certain doses of radiation to the tissues most heavily irradiated. These total doses are about equal to the annual doses regarded by the I.C.R.P. as the maximum permissible ones for those occupationally exposed to radiation. Thus the maximum intakes of I^{131} proposed for various ages correspond to a total thyroid irradiation of 25 rads as compared with a maximum annual value of 30 rads during occupational exposure. More recently, in the absence of guidance from the I.C.R.P., the Medical Research Council has also recommended maximum doses of external radiation considered acceptable for members of the general public as a result of an incident at an atomic energy establishment (Medical Research Council, 1960). Some of their recommendations, which are complementary to those made by the I.C.R.P. for long-term chronic exposure (Table 26), are:

(1) The dose of γ-rays giving exposure to the whole body but measured in free air:

Children up to the age of 16 years and pregnant women   . . . 20 roentgens

Other persons . . . . 30 roentgens

(2) Subject to (1) above, the combined dose of β- and γ-rays giving exposure to the whole body surface, and estimated in any superficial tissue:

Children under the age of 16 years . . . . 75 rads

Other persons . . . . 150 rads

An additional allowance of dose is recommended for certain adults who may be required to carry-out essential duties after the accident. The Council considered that few people, and certainly less than one fiftieth of the population, would be involved in the conditions under consideration, so that no further limitations of the doses were necessary on genetic grounds.

3. Reactor siting

Since the escape of even a small fraction of the fission products contained in the fuel of a power reactor could constitute a danger to public health (Kuper and Cowan 1958) elaborate precautions are taken in the design and operation of such reactors to make this exceedingly improbable. Nevertheless, it is usual in assessing the safety of reactors and in selecting reactor sites, to postulate that, in spite of all precautions, a release of fission products has taken place and then to estimate the possible consequences of the release on the health of the general public living in the district.

The types of hazard and the methods of calculation and assessment are identical with those considered in the section on the continuous release of gaseous effluents but because of the possible large amounts of activity involved, the shorter period of release and the possibility of taking emergency action to reduce exposure in the neighbourhood, the relative importance of the hazards may vary. Thus the external radiation dose from the cloud may become more important than that from the deposited material, if the people exposed can be evacuated in time to avoid most of the latter. Similarly, the hazard due to inhalation of I^{131} from the cloud may be greater than the hazard from I^{131} in milk since steps can be taken to restrict milk consumption. A reactor siting policy must take such factors into account and must try to find a balance between the greater hazard to a few people living nearby and the lesser hazard to much larger numbers at a distance. In realising this balance, due regard must be paid to the practicability of remedial action, such as the immediate evacuation of the few and the greater administrative difficulty in dealing with substantial numbers, albeit less urgently.

Marley and Fry, in 1955, were among the first to assess the environmental consequences of a reactor accident (Marley and Fry 1955). They considered the various classes of hazard from an arbitrary release of fission products into the atmosphere and calculated the corresponding ranges within which preventive and remedial measure could be taken to minimise the effects of the release on the public as a whole. Since then, a number of studies have been carried out on the proportions of the various fission products released from reaction fuels under simulated accident conditions (U.S.A.E.C., 1957) and with the results of the Windscale accident in mind, Kuper and Cowan (1958) and Farmer and Fletcher (1959) have reassessed the likely consequences of a nuclear accident, giving due weight to the fact that a release is likely to consist mainly of the volatile fission products. Farmer and Fletcher have also adopted a more realistic approach to the problem of determining the maximum credible release from a given reactor system. They advocate that the safety of a nuclear power plant be studied by considering the consequences to the reactor of various accidents which could be caused by equipment failure or by maloperation. The highest standards of engineering skill must be employed to prevent such accidents but there may still remain some conditions which cannot be proved impossible, though they may be extremely unlikely to occur. In many cases, it can be shown that the reactor has been designed to accommodate the particular faults envisaged and that no significant release of fission products would follow. If not, theoretical studies of the dynamic behaviour of the reactor in conjunction with experimental data on the release of activity from fuel elements will lead to estimated figures for the maximum credible release of activity into the atmosphere. Calculations of this type are generally made using pessimistic assumptions.

Using this pessimistic approach in relation to the Calder Hall gas-cooled reactors in the U.K., FARMER and FLETCHER calculated that in a severe accident 250 curies of total iodine might be released into the atmosphere over a period of a few hours. They then calculated the possible effects of this release on people living in the environment, and using the M.R.C. recommendations on maximum energency doses to the public, drew up a list of actions required to reduce the consequences. Their results are given in Table 30:

Table 30. *Effects of release of 250 curies total iodine (containing 110 curies I^{131}) under fair-weather conditions. Period of release — several hours*

Control level	Limiting exposure	Action	Limiting distance
2×10^{-2} c/m² on ground	25 rem whole-body gamma	evacuate	less than 100 yd
0.01 curie-sec/m³ in air	25 rem integrated dose to child's thyroid	temporary removal during release	440 yd
5×10^{-5} c/m² on exposed food	25 rem integrated dose to child's thyroid	temporary ban on consumption of crops and exposed food	550 yd
0.065 μc/l in milk	25 rem integrated dose to child's thyroid	temporary ban on consumption of milk	4 miles

Members of the public do not live within 100 yards of any Calder Hall reactor so that no evacuation is likely to be required due to radiation from deposited activity. Emergency action would be needed to remove people (particularly children) from within a 440 yard radius during the passage of the cloud to reduce iodine uptake by inhalation. The affected area for exposed food can be determined by ground contamination measurements, and levels of high contamination defined rapidly. The ban on crops would be removed within a few weeks over the greater part of the area. The control of milk consumption would follow the pattern described by DUNSTER, HOWELLS and TEMPLETON (1958) for the Windscale accident. This type of calculation leads to certain desirable features for the population distribution around the site of a Calder Hall type of reactor:

(i) Few people should live within 500 yards of the site. This condition ensures that adequate arrangements can be made to move people away from the drift of the cloud.

(ii) In any 30° sector around the site, not more than 500 people should live within 1 mile.

This is regarded as a secondary restriction which simplifies the emergency action required if, as is even more highly improbable, the release of I^{131} is ten times higher than calculated.

(iii) No large centre of population of 10000 or more should lie within 5 miles.

Although there is no obvious restriction on population distribution at distances beyond 1 mile, nevertheless if an actual emergency situation developed, people at greater distances would require assurance that they and their children are safe and that life can proceed normally. This would require extensive measurements of radiation levels and analysis of foodstuffs. For this reason it is desirable to avoid sites which have large numbers of people living within the near vicinity. This condition remains a matter of judgment, having regard to the available national monitering facilities.

The Farmer-Fletcher siting criteria were based on information available at the time of derivation. This information represented an advance on that available to Marley and Fry, and, as knowledge and experience increase it is clearly desirable that siting policies be kept under continuous review so that sites may be chosen as realistically as possible.

Recommendations

The biological problems encountered in nuclear radiation in geophysics present—with the exception of some space radiations—typical low dose—long time irradiation phenomena. As such, they will profit from every progress in the study of low dose radiation effects, primarily in the dose range from 5 r down to "zero" radiation. Since these low dose effects—somatic effects as well as genetic effects—however depend on so many factors like total dose, dose rate, dose fractionation and require besides proper space and facilities tremendous numbers of animals —in mammalian studies with mice according to H. Nachtsheim (1959) f.i. three quarter of a million of mice for one experimental series—special emphasis should be given to the study of the nature-given radiation conditions. Just to mention a few: The question of the natural radium content of the human presents still a tricky problem. Besides the total amount to be found in the average human, its changes with environmental and dietary factors, recent measurements on the radium content of bone and of soft tissue, do not agree. The amounts reported in the European literature and USA studies differ by factors up to 50, making all previous conclusions and statements questionable. The question of the role of the alpha activity in the average human, so often discussed in the past in USA and European literature, has found, as reported by Mayneord et al. a solution in this direction, that there are definitely high alpha activities in the body. The surprising findings in still-living radium poisoning victims from the twentieths and thirtieths, who still carry high amounts of radium in their bodies, have upset the whole picture of radium poisoning and extensive study of these cases appears highly desirable. It should be kept in mind, that many of the present radiation protection guides are based on these nature-given "constants" and exploration of the situation is mandatory.

As to man made environmental contamination many physical, chemical, meteorological and biologic-physiological factors have to be explored in order to evaluate the overall biological importance of these radiation sources. The present lull in the testing of large nuclear weapons provides an opportunity to study atmospheric phenomena without interference from the injection of fresh debris. Among the problems to be studied, are: the spring maximum in stratospheric fallout, the locations and intensities of the exchange processes, the mechanisms of deposition and washout by rain and the distribution of fallout on the earth's surface. Appropriate dietetic investigations should accompany these studies. The contamination of primary foods such as milk and cereals in western countries has been dependent on the surface contamination of grass and crops by recent fallout and is approximately proportional to the rate of fallout rather than to the accumulated deposit. If no more tests are held in the immediate future, the rate of fallout is likely to decrease progressively and the situation will be reached where uptake through the roots of plants from accumulated fallout is the dominant route for the passage of important isotopes such as Sr^{90} into human food chains. It will therefore become possible to study this factor in isolation, to assess the long-term contamination problem in more detail and to investigate the effects of weathering of the fallout into the soil. Since the composition of total diet, as to kind of food and daily incorporated quantities, is the factor which determines the effective internal dose of radiation from fallout, the comprehensive studies of the type which have been conducted in western countries and in Japan should be extended to the regions where diet is different and where data are at present very sparce.

Research programmes into the possibility of converting high-level radioactive waste products into solid non-leachable forms which can be easely stored in selected geological formations should be intensified in countries with major atomic energy installations. The release of low and medium-level wastes into the environment need to be preceded and accompanied by comprehensive studies into the capacity of the environment to receive them. The object of these studies is to form a link between the amounts discharged and the amounts which man may inhale or ingest. At present time, the knowledge of the metabolism of radioisotopes is not comprehensive, although knowledge is increasing through direct experiments and through fallout studies. The British Medical Research Council has used—where information was found to be inadequate—the more cautious rather than the more likely assumption in deriving maximum permissible dietary contamination levels. There is, therefore, an economic as well as scientific and social reasons for increased research into this topic.

The disposal of radioactive wastes into deep oceans is not a local matter and comprehensive oceanographic investigations involving physical, chemical, biological and geological processes require to be carried out if fuller use is to be made of the ocean's capacity to receive such material.

The assessment of the environmental hazards from accidents to reactors and other nuclear plants raises similar problems to those involved in planned waste disposal, with greater emphasis on the properties of airborne particulates. Further work is desirable on the physical properties of the fission product particles which may escape from reactor fuel elements, on their deposition velocities under different meteorological conditions and on the efficiency with which rain removes them from the atmosphere. The growth of knowledge in this field and in the transfer and metabolism of fission products will help authorities to strike the required balance between safety and economic and social values in the choice of reactor sites.

References

ALLEN, J. A. VAN: Corpuscular Radiation in Space. Radiation Res. **14**, 540 (1961).
— C. E. McILWAIN and G. H. LUDWIG: Radiation observation with satellite 1958 epsilon. J. Geophys. Res. **64**, 271 (1959).
ALLEN, R. I., A. J. DESSLER, J. F. PERKINS and H. C. RICE: Shielding problems in manned space vehicles. Lockheed Nuclear Products Report, NR-104 (1960).
AMBROSEN, J.: Time of residence of radioactive debris in the stratosphere. Nature, Lond. **185**, 301 (1960).
ANDERSON, E. C., and W. H. LANGHAM: Average potassium concentration of the human body as a function of age. Science **130**, 713 (1959).
— R. L. SCHUCH, J. D. PERRINGS and W. H. LANGHAM: The Los Alamos Human Counter. Nucleonics **14** (1), 26 (1956).

ARCHER, V.E., and B.E. CARROLL: Life shortening and tumor production by strontium-90.
Science **131**, 1808 (1960).
Atomic Energy Office (1957): Accident at Windscale No. 1 pile on 10th October, 1957. Cmnd.
302, H.M. Stationary Office, London.
AUB, J.C., R.D. EVANS, L.H. HEMPELMANN and H.S. MARTLAND: Late effects of internally-
deposited radioactive materials in man. Medicine, Baltimore **31**, 221 (1952).
AURAND, K., W. JACOBI, H. MUTH u. A. SCHRAUB: Weitere Untersuchungen zur biologischen
Wirkung des Radons und seiner Folgeprodukte. Strahlentherapie **112**, 262 (1960).
BACQ, A.M., and P. ALEXANDER: Fundamentals of radiobiology. London: Butterworth 1955.
BEHRENS, CH.F.: Atomic medicine, 3rd edit. Baltimore: Williams and Wilkins 1959.
BERLIN, N.I.: Effect on chronic ionizing radiation on the life span of man. Symposium on
the Delayed Effects of Whole-Body Radiation. The Johns Hopkins University, ORO-
SP-127 (1960).
—, and F.L. DiMAGGIO: A survey of theories and experiments on the shortening of life span
by ionizing radiation. AFSWP-608, Washington 1956.
BLAIR, H.A.: A formulation of the injury, life span, dose relations for ionizing radiations.
AEC-Report, UR-206 (1952).
— Data pertaining to shortening of life span by ionizing radiation. AEC-Report UR-442
(1956); Proc. Internat. Conf. Peaceful Uses of Atomic Energy, vol. **11**, p. 118 (1956).
BOGAROV, B.G., and E.M. KREPS: Concerning the possibilities of disposing of radioactive
waste in ocean trenches. Internat. Conf. Peaceful Uses of Atomic Energy, vol. 18, p. 371
(1958).
BOLLES, R.C., and N.E. BALLOU: Calculated activities and abundances of U^{235} fission pro-
ducts. USNRDL-456 (1956).
BOSANQUET, C.H., W.F. CAREY and E.M. HALTON: Dust deposition from chimney stacks.
Proc. Inst. Mech. Engrs. **162**, 355 (1950).
BREWER, A.W.: Evidence for a world circulation provided by measurements of helium and
water vapour distribution in the stratosphere. Quart. J. Roy. Meteor. Soc. **75**, 351 (1949).
BROWN, R.E., H.M. PARKER and J.M. SMITH: Disposal of liquid wastes to the ground.
Internat. Conf. Peaceful Uses of Atomic Energy, vol. 9, p. 669 (1955).
BRUCER, M.: The ORINS human radiation counters. US-AEC Report: ORINS-38 (1960).
BRUES, A.M.: Critique of the linear theory of carcinogenesis. Science **128**, 693 (1958).
— Low-level irradiation. Amer. Assoc. Advanc. Science, Publ. No. 59. Washington, D.C.,
1959.
BRYANT, F.J., J.W. ARDEN, E.H. HENDERSON, G.D. LLOYD and A.G. MORTON: Radio-
active and natural strontium in human bone. U.K. Results for 1959. AERE-R 3246 (1960).
— A.C. CHAMBERLAIN, G.S. SPICER and M.S.W. WEBB: Strontium in diet. Brit. Med. J.
1958, 1371.
— J.C. COTTERILL, E.H. HENDERSON, G.S. SPICER and T.J. WEBBER: Radioactive and
natural strontium in human bone. U.K. Results for mid and late 1958. AERE-R 2988 (1959).
— E.H. HENDERSON, G.S. SPICER and M.S.W. WEBB: Radioactive and natural strontium
in human bone. U.K. Results for 1957. AERE C/R 2583 (1958).
— — — — and T.J. WEBBER: Radioactive and natural strontium in human bones. U.K.
Results for early 1958. AERE C/R 2816 (1959).
BURNS, R.H.: The disposal of radioactive solid wastes, p. 181. Radioactive Wastes (J.C.
COLLINS, Edit.). London: E. & F.N. Spon, Ltd. 1960.
— Radioactive waste control at the United Kingdom. Atomic Energy Research Establish-
ment, Harwell. Disposal of radioactive wastes, vol. 1, Conf. Proceedings, Monaco, 16 to
21 Nov. 1959, p. 411, I.A.E.A. 1960.
— The legal aspects of atomic waste disposal and transport of radioactive materials. Atomic
Energy Waste (E. GLUECKAUF, Edit.), p. 187. London: Butterworths Scientific Publica-
tions 1961.
BURTON, J.D., G.M. MILBOURN and R.S. RUSSELL: Relationship between the rate of
fallout and the concentration of strontium-90 in human diet in the United Kingdom.
Nature, Lond. **185**, 498 (1960).
BUZZATI-TRAVERSO, A.A.: Immediate and low level effects of ionizing radiations. London:
Taylor & Francis, Ltd. 1960.
CALDECOTT, R.S., and L.A. SNYDER: A Symposium on Radioisotopes in the Biosphere.
Minneapolis: Univ. of Minnesota, Center for Continuation Study 1960.
CARTER, T.C.: Radiation-induced gene mutations in adult female and foetal male mice.
Brit. J. Radiol. **31**, 407 (1958).
— M.F. LYON and R.J.S. PHILLIPS: Genetic hazards of ionizing radiations. Nature, Lond.
182, 409 (1958).
CATSCH, A.: Wirkungen energiereicher Strahlen auf den Menschen. Die Atomwirtschaft
März 1957, S. 73ff.

CHAMBERLAIN, A.C.: Relation between measurements of deposited activity after the Windscale accident of October, 1957. AERE HP/R 2606 (1958).
— Deposition of iodine-131 in northern England in October 1957. Quart. J. Roy. Meteor. Soc. **85**, 350 (1959).
— Dispersion of activity from chimney stacks. Atomic Energy Waste. (E. GLUECKAUF, Edit.), p. 308. London: Butterworths Scientific Publications 1961.
—, and R.C. CHADWICK: Deposition of airborne radioiodine vapour. Nucleonics **11**, (8), 22 (1953).
CHASE, E.B., and J.S. POST: Damage and repair in mammalian tissues exposed to cosmic ray heavy nuclei. J. Aviation Med. **27**, 53 (1956).
COCKCROFT, J.: Peaceful uses of atomic energy. Science **129**, 247 (1959).
COGAN, D.G.: Radiation cataracts in man. Operations Research Office, The Johns Hopkins Univ. ORO-SP-127, 59 (1960).
COMAR, C.L., R.S. RUSSELL and R.H. WASSERMAN: Strontium-calcium movement from soil to man. Science **126**, 485 (1957).
CRABTREE, J.: The travel and diffusion of the radioactive material emitted during the Windscale accident. Quart. J. Roy. Meteor. Soc. **85**, 362 (1959).
CRAIG, H.: A critical evaluation of mixing rates in oceans and atmosphere by use of radiocarbon techniques. Proc. II. Internat. Conf. Peaceful Uses of Atomic Energy, vol. 18, p. 358 (1958).
CROOKS, R.N., R.G.D. OSMOND, Miss E.M.R. FISHER, M.J. OWERS and T.W. EVETT: The deposition of fission products from distant nuclear test explosions. Results to the middle of 1960. AERE-R 3349 (1960).
CURTIS, H.J.: Limitations on space flight due to cosmic radiations. Science **133**, 312 (1961).
DOBSON, G.M.B.: Origin and distribution of the polyatomic molecule in the atmosphere. Proc. Roy. Soc. Lond. A **236**, 187 (1956).
DUDLEY, R.A.: Natural and artificial radiation background of man. In: Low-level irradiation. Washington, D.C.: Amer. Assoc. Adv. Science 1959.
—, and R.D. EVANS: Radiation dose to man from natural sources. In: Report of the Congress of the United States, Joint Committee on Atomic Energy, Subcommittee on Radiation, Fallout from Nuclear Weapons Tests. (Washington, D.C.: Government Printing Office 1957) pl. II 1236 (1957).
DUNSTER, H.J.: The public health problems of nuclear waste disposal. Disposal of Radioactive wastes, vol. 1, Conf. Proceedings, Monaco, 16—21 Nov. 1959, p. 543. I.A.E.A., **1960**.
— The disposal of radioactive liquid wastes into coastal waters. Progress in Nuclear Energy, Ser. XII, vol. 1, Health Physics, p. 566. London and New York: Pergamon Press 1959.
— H. HOWELLS and W.L. TEMPLETON: District surveys following the Windscale incident, October, 1957. Proc. II. Internat. Conf. Peaceful Uses of Atomic Energy, vol. 18, p. 296 (1958).
EVANS, R.D.: The quantitative determination of the radium content of living persons. Amer. J. Roentgenol. **37**, 368 (1937).
— Radium and mesothorium poisoning and dosimetry and automative techniques in applied radioactivity. Annual Progr. Report, Contract AT (30-1)-952, Radioactive Center, Massachusetts Inst. Tech. 1957, 1958, 1959, 1960.
— Personal communication 1961 a.
— Principles for the calculation of radiation dose rates in space vehicles. Techn. Report NASA, Contract Nr. NAS 5—664, Report 63270–05–01, July 1961 b.
—, and J.C. AUB: Recent progress in the study of radium poisoning. Cancer Problem Symposium 1937. Amer. Assoc. Adv. Sci. **4** (Suppl. Science) **85**, 227 (1937).
—, and G. GOODMANN: Determination of the thoron content of air and its bearing on lung cancer hazards in industry. J. Indust. Hyg. and Toxicol. **22**, 89 (1940).
— A.F. KIP and E.G. MOBERG: The radium and radon contents of Pacific ocean waters, life and sediments. Amer. J. Sci. **36**, 241 (1938).
FAILLA, G., and P. MCCLEMENT: The shortening of life by chronic whole-body irradiation. Amer. J. Roentgenol. **78**, 946 (1957).
FARMER, F.R., and P.T. FLETCHER: Siting in relation to normal reactor operation and accident conditions. Atti del Congr. Scientifico, VI Rassegna Internaz. Ellettronica e Nucleare, vol. 1, Parte 1, p. 411 (1959).
FEELY, H.W.: Strontium-90 content of the stratosphere. Science **131**, 645 (1960).
FINKEL, M.P.: Mice, men, and fallout. Science **128**, 637, 1580 (1958).
— B.O. BISKIS and P.J. BERGSTRAND: Radioisotope toxicity, significance of chronic administration. In: Radioisotopes in the biosphere. Minneapolis: University of Minnesota 1960.
Food and Agriculture Organisation of the United Nations (1960): Radioactive materials in food and agriculture. Report of Expert Committee.

Forbush, S.E.: Cosmic ray intensity variations during two solar cycles. J. Geophys. Res. **63**. 651, 699 (1958).

Franzen, L.F., G. Muszynski and G. Wiesenack: Natürliche und künstliche Strahlen-belastung. Die Atomwirtschaft, Nov. 1957, S. 362ff.

Garner, R. J.: An assessment of the quantities of fission products likely to be found in milk in the event of aerial contamination of agricultural land. Nature, Lond. **186**, 1063 (1960).

Gustafson, P.F.: Assessment of the radiation dose due to fallout. Radiology **75**, 282 (1960).

Hagemann, F., L. Machta, J. Gray and A. Turkevich: Stratospheric carbon-14, carbon dioxide, and tritium. Science **130**, 542 (1959).

Hevesy, G. v.: Die Bedeutung der Radio-Isotopen-Forschung für Medizin und Biologie. Strahlentherapie **102**, 341 (1957).

Hursh, J.B.: The natural radioactivity of man. AEC-Report, NLCO-595, 111 (1956).
— Natural occurrence of Ra^{226} in human subjects, in water, and in food. Brit. J. Radiol. Suppl. **7**, 45 (1957).
—, and A.A. Gates: Radium content of individuals with no occupational exposure. Nucleonics **7**, 46 (1950).
— A. Lovaas and E. Biltz: Radium in bone and soft tissues of man. AEC-Report, UR-581 (1960).

International Commission on Radiological Protection (ICRP): Recommendations. London and New York: Pergamon Press 1958.

International Commission on Radiological Protection (ICRP): Report of Committee II on Permissible Dose for Internal Radiation. London and New York: Pergamon Press 1959.

Jones, H.B.: Factors in longevity. Kaiser Found. Med. Bull. **4**, 329 (1956).

Kallmann, R.F., and H.I. Kohn: Life-shortening by whole-body and partial-body X-irradiation in mice. Science **128**, 301 (1958).

Kratchman, J., and D. Grahn: Relationship between the geological environment and mortality from congenital malformation. US-AEC, Washington, D.C., Report TID-8204 (1959).

Krebs, A.: Über die normale und anormale Radioaktivität menschlichen und tierischen Gewebes. Fund. Radiol. **5**, 89 (1939).
— Toxicity of thorium emanation. Naturwissenschaften **28**, 266 (1940).
— Radioaktive Spurenelemente der Luft als bioklimatische Faktoren. Bioklim. Beibl. **2**, 64 (1941).
— Untersuchungen zum Problem der Radiumvergiftung. IV. Strahlentherapie **72**, 164 (1942).
— Radiumemanation als Heilmittel. Umschau, Frankfurt a.M. **25**, 57 (1942).
— Possibility of biological effects of cosmic radiation at high altitudes, stratosphere and space. J. Aviation Med. **21**, 481 (1950).
— Szintillationszähler. Ergebn. exakt. Naturw. **27**, 361—409 (1954).
— The radioactivity of the human being. Science **119**, 429 (1954).
— Early history of the scintillation counter. Science **122**, 17 (1955).
—, u. H. Lampert: Radiumemanations-Therapie in Ergebnisse der physikalischen Therapie. Dresden u. Leipzig: Theodor Steinkopff 1948.

Kulp, J.L., A.R. Schulert and Miss E. J. Hodges: Strontium-90 in man. Science **132**, 448 (1960).

Kuper, J.B.H., and F.P. Cowan: Exposure criteria for estimating the consequences of a catastrophe in a nuclear plant. Proc. II. Internat. Conf. Peaceful Uses of Atomic Energy, vol. 18, p. 319 (1958).

Langham, W.H.: Implications of space radiations in maned space flights. Aero Space Med. **30**, 410 (1959).

Lebish, I. J., D.G. Simons, H. Yagoda, P. Janssen and W. Haymaker: Observation on mice exposed to cosmic radiation in the stratosphere. Military Med. **124**, 835 (1959).

Leipunsky, O.I.: Radioactive hazard resulting from the explosions of a "clean" hydrogen bomb and of a conventional fission bomb. Atomnaya Energiya **3**, 530 (1957).

Lewis, E.B.: Leukemia and ionizing radiation. Science **125**, 965 (1957).

Libby, W.F.: Radioactive strontium fallout. Proc. Nat. Acad. Sci. U.S.A. **42**, 365 (1956).
— Dosages from natural radioactivity and cosmic rays. Science **122**, 57 (1957).

Looney, W.B., R. J. Hasterlik, A.M. Brues and E. Skirmont: A clinical investigation of the chronic effects of radium salts administered therapeutically, 1915—1931. Amer. J. Roentgenol. **73**, 1006 (1955).

Lucas, H.F.: Populations consuming water with high natural radium-226 contents. AEC-Report, ANL-6049, 48 (1959).

Marinelli, L.D.: Radioactivity and the human skeleton. Amer. J. Roentgenol. **80**, 729 (1958).

Marinelli, L. D.: Epidemiological studies on radiation cancerogenesis in the human skeleton. AEC-Report, ANL-6049, 30 (1959).

Marley, W. G., and T. M. Fry: Radiological hazards from an escape of fission products and the implications in power reactor location. Proc. Internat. Conf. Peaceful Uses of Atomic Energy, vol. 13, p. 102 (1955).

Mawson, C. A.: Report of the waste disposal system at the Chalk River plant of the Atomic Energy of Canada, Ltd. CP-658 (1956).

Mayneord, W. V., J. M. Radley u. R. C. Turner: Die Alpha-Aktivität des menschlichen Körpers und seiner Umgebung. Strahlentherapie 110, 431 (1959).

Medical Research Council (1956): The hazards to man of nuclear and allied radiations, p. 56. Cmmd. 9780, H.M. Stationery Office, London.

Medical Research Council: Maximum permissible dietary contamination after the accidental release of radioactive material from a nuclear reactor. Brit. Med. J. 1959 (1), 967.

Medical Research Council (1960): The hazards to man of nuclear and allied radiations. Cmnd. 1225, H.M. Stationery Office, London.

Morgan, K. S.: Fallout from nuclear weapons tests. Summary—Analysis of hearings. Joint Committee on Atomic Energy, Congr. USA, p. 19, Washington, D.C., 1959.

Morgan, K. Z.: Human exposure to radiation. Bull. Atomic. Sci. 15 (9), 384 (1959).

Müller, H. J.: The manner of dependence of the "permissible dose" of radiation on the amount of genetic damage. Acta Radiol., Stockh. 41, 5 (1954).

Muth, H., B. Rajewsky, J. Hantke and K. Aurand: The normal radium content and the Ra^{223}/Ca ratio of various foods, drinking water and different organs and tissues of the human body. Health Physics 2, 239 (1960).

Mutscheller, A.: The tolerance dose. Amer. J. Roentgenol. 13, 65 (1923).

National Academy of Sciences—National Research Council. The biological effects of atomic radiation. A report to the public. Washington 1960.

Neel, J. V., and W. J. Schull: Human heredity. Chicago: Chicago University Press 1954.

Newell, H. E., and J. E. Naugle: Radiation environment in space. Science 132, 1465 (1960).

O'Brien, K., W. M. Lowder and L. R. Solon: Beta and gamma dose rates from terrestrially distributed sources. Radiol. Res. 9, 216ff. (1958).

Parker, H. M.: Radiation exposure from environmental hazards. Internat. Conf. Peaceful Uses of Atomic Energy, vol. 13, p. 305 (1955).

Peirson, D. H., R. N. Crooks and Miss E. M. R. Fisher: Radioactive fallout in air and rain. AERE-R 3358 (1960).

—, and L. Salmon: Gamma-radiation from deposited fallout. Nature, Lond. 184, 1678 (1959).

Report of the United Nations Scientific Committee on the Effects of Atomic Radiation. General Assembly, Thirteenth Session Suppl. No. 17 (A/3838), New York, 1958.

Robertson, H. A., and I. R. Falconer: Accumulation of radioactive iodine in thyroid glands subsequent to nuclear weapon tests and the accident at Windscale. Nature, Lond. 184, 1699 (1959).

Rowland, R. E.: Late observations of the distribution of radium in the human skeleton. In: Radioisotopes in the biosphere. Minneapolis, University of Minnesota, 1960.

Rundo, J.: Radiocaesium in human beings. Nature, Lond. 188, 703 (1960).

Russell, W. L.: X-ray induced mutations in mice. Cold Spring Harbor Symp. Quart. Biol. 16, 327 (1951).

— Genetic effects of radiation in mammals. In: Radiation biology, edit. A. Hollaender. New York: McGraw Hill 1954.

— Lack of linearity between mutation rate and dose for X-ray-induced mutations in mice. Genetics 41, 658 (1956).

— Radiation in mice—the genetic effects and their implications for man. Bull. Atomic Scientists 12, No. 1 (1956).

— L. B. Russell and E. M. Kelly: Radiation dose rate and mutation frequency. Science 128, 1546 (1958).

Sacher, G. A.: Problems in the extrapolation of long-term effects from animals to man. Symposium on the Delayed Effects of whole-body radiation. The Johns Hopkins University, ORO-SP-127 (1960).

Schaefer, H. J.: Evaluation of present-day knowledge of cosmic radiation at extreme altitudes in terms of hazards to health. J. Aviation Med. 21, 375ff. (1950).

— Radiation hazards in space exploration. Proc. Lunar and Planetary Exploration Colloquium 1, 23 (1959).

Schull, W. J., and J. V. Neel: Radiation and the sex ratio in man. Science 128, 343 (1958).

Scorer, R. S.: The behaviour of chimney plumes. Int. J. Air Poll. 1, 198 (1959).

Shandley, P. D.: The radium content of some foods. AEC-Report, UR-255 (1953).

Shapiro, J.: Radiation dosage from breathing radon and its daughter products. A.M.A. Arch. Ind. Health 14, 169 (1956).

SILVERMAN, L.: Economic aspects of air and gas cleaning for nuclear energy processes. Disposal of Radioactive Wastes, vol. 1, Conf. Proceedings, Monaco, 16—21 Nov. 1959, p. 139, I.A.E.A. 1960.

SNYDER, W. S.: The scientific basis for maximum permissible limits of internal contamination. US-AEC Report, TID-7577, 26 (1959).

SOLON, L. R., W. M. LOWDER, A. SHAMBON and H. BLATZ: Investigations of natural environmental radiations. Science 131, 903 (1960).

SPIERS, F. W.: Some measurements of background gamma radiation in Leeds during 1955 to 1959. Nature, Lond. 184, 1680 (1959).

— and P. R. J. BURCH: Measurement of the normal radioactivity of the body. In: Biological hazards of atomic energy, edit. A. HADDON. Oxford: Clarendon Press 1952.

STEHNEY, A. F.: Naturally occurring radioisotopes in man. In: Radioisotopes in the biosphere. Minneapolis: Univ. of Minnesota 1960.

—, and H. F. LUCAS jr.: Studies on radium content of humans arising from natural radium of their environment. Proc. Internat. Conference, Geneva, August 1955, vol. 11, p. 49. New York: United Nations 1956.

STEWART, N. G., and R. N. CROOKS: Long-range travel of the radioactive cloud from the accident at Windscale. Nature, Lond. 182, 627 (1958).

— — and Miss E. M. R. FISHER: The radiological dose to persons in the U.K. due to debris from nuclear test explosions prior to January 1956. AERE HP/R 2017 (1956).

— H. J. GALE and R. N. CROOKS: The atmospheric diffusion of gases discharged from the chimney of the Harwell pile (BEPO). Int. J. Air Poll. 1, 87 (1958). AERE HP/R 1452 (1957).

STOKLASA, J., and J. PENKAVA: Biologie des Radiums und der radioaktiven Elemente, vol. I. Berlin: Urban & Schwarzenberg 1952.

STRATHMANN, W.: Radioaktivität in Luft, Wasser und Boden. Die Atomwirtschaft März 1957, S. 85ff.

STRAUB, C. P.: Research in the radioactive contamination of the environment. US Public Health Service, R. A. Taft Sanitary Engineering Center, Congr. on Industrial Health, Cincinnati, 1959.

Summary—Analysis of Hearings, 1959. Fallout from nuclear weapons tests. Joint Committee on Atomic Energy, Congr. of the United States, 1959.

SUTTON, O. G.: The theoretical distribution of airborne pollution from factory chimneys. Quart. J. Roy. Meteorol. Soc. 75, 426 (1947).

TOBIAS, C. A.: Radiation hazards in high altitude aviation. J. Aviation Med. 23, 345 (1952).

— Radiation hazards of primary cosmic particles. Appendix Report AFMDC-TR-59-32, 118 (1959).

TOMMASINA, M.: Bio-radioaktivität. C. R. Acad Sci., Paris 139, 730 (1904).

United Nations Scientific Committee (1958): Report of the United Nations Scientific Committee on the effects of atomic radiation. United Nations, New York, 1958.

United States Atomic Energy Commission: Theoretical possibilities and consequences of major accidents in large nuclear power plants. WASH.-740 (1957).

UPTON, A. C., and A. W. KIMBALL: Shortening of life span by ionizing radiation: Possible relation to accelerated aging. Science 131, 1322 (1960).

VERNADSKII, V. L: Accumulation of radio-isotopes by aquatic organisms. Dokl. Akad. Nauk SSSR., Ser. A, No. 2 (1929); No. 6 (1931).

VERNADSKY, W. C.: Radioactivity of aquatic systems. C. R. Acad. Sci., Paris 191, 421 (1930).

VORESS, H. E.: Whole-body counters. US-AEC: TID-3543 (1960).

WALLACE, B., and TH. DOBZHANSKY: Radiation, genes, and man. New York: Henry Holt & Company 1959.

WESTERGAARD, M.: Die Erbanlagen des Menschen und seine Verantwortung. Med. Klin. 52, 274 (1957).

YAGODA, H.: A study of cosmic ray heavy primary hits in a phantom of a human brain. Acta Medica Belgica, III e Congr. Internat. de Neuropathologie 1957, p. 171.

ZELLE, M. R.: Mutagenic aspects. In: Radioisotopes in the biosphere. Minneapolis: University of Minnesota 1960.

ZEMAN, W., H. J. CURTIS and C. P. BAKER: Histopathologic Effekt of High-Energy-Particle Microbeams on the Visival cortex of the Mouse Brain. Radiation Res. 15, 496 (1961).

Meßmethoden

von

G. Schumann

Mit 9 Figuren

Summary

Principles of measuring radioactivity in geophysics are the same as in nuclear physics generally. But there are some special features: activities are extremely low, and sources have considerable extent in area and thickness. This article therefore deals with the problems of low-level counting many methods of which were developed especially for geophysical applications. Counting statistics lead to special consequences if the effect to be measured is smaller than the background: the figure of merit for a counting device (in general the ratio e/b, e source count, b background count) is now e^2/b. Therefore to increase the efficiency is even more important than to reduce the background. Principles of achieving both are discussed. Measuring low activities takes long times to obtain sufficiently small statistical errors. That requires an extremely good long-time stability of the apparatus, especially of the electronics. Thick sources present particular problems for β-counting because of backscattering, selfabsorption and selfscattering. They are less critical with α-counting, and with γ-counting they only sometimes deteriorate the geometrical efficiency of a device. The discussed principles of low-level counting are then applied to ionization chambers, Geiger counters, proportional counters, scintillation counters and photographic emulsions.

Sampling devices and conditions are discussed in general and for taking gas samples, aerosol samples, water samples and rock samples. The methods of measuring the radioactivity of these different kinds of samples are specified for individual substances of geophysical interest. The samples often contain a mixture of active substances or contributions of unwanted chemical elements. In many cases chemical separations can not be avoided, first of all for pure β-emitters. A short review is given of radiochemical methods used for this purpose. For analyzing mixtures of radioactive substances by purely physical means decay-time and energy discrimination may be used. α- and γ-spectroscopy are readily applicable to geophysical problems. But β-spectroscopy falls short because of the low specific activity of the samples. In some cases β-absorption may be measured.

Einleitung

Bei Untersuchung radioaktiver Substanzen in der Geophysik bedient man sich selbstverständlich der gleichen Meßverfahren, die auch sonst in der Kernphysik üblich sind. Bei dem beschränkten Umfang des vorliegenden Artikels ist es unmöglich, auf die physikalischen Grundlagen dieser Methoden einzugehen. Sie müssen als bekannt vorausgesetzt werden. Zur Orientierung sind im Literaturverzeichnis einige zusammenfassende Darstellungen dieses Gebietes angeführt. Ausgeschlossen bleiben müssen ferner Meßmethoden für Gebiete, die zwar in enger Beziehung zum Thema des Buches stehen, aber nicht direkt dazu gehören, z.B. Untersuchungen der kosmischen Strahlung, des Größenspektrums von Aerosolen, Methoden der angewandten Geophysik u. dgl.

Nachstehend sollen die Besonderheiten behandelt werden, die für die Messungen auf dem Gebiet der Geophysik charakteristisch sind, wo man es fast stets mit sehr niedrigen Aktivitäten zu tun hat.

I. Allgemeine Grundsätze für Messung niedriger Aktivitäten

1. Statistik

Für die Beurteilung der Brauchbarkeit eines Detektors zur Messung sehr geringer Aktivitäten sind etwas andere Gesichtspunkte maßgebend als bei der gewöhnlichen Zählstatistik (Thomas, Loevinger et al.). Allgemein werden, wenn N_0, N_t die gemessenen Impulszahlen für Nulleffekt bzw. Totalregistrierung (Effekt + Nulleffekt) und T_0, T_t die entsprechenden Meßzeiten bedeuten, die Zählraten (Impulse pro Zeiteinheit)

$$n_0 = \left(N_0 \pm \sqrt{N_0}\right)/T_0, \qquad n_t = \left(N_t \pm \sqrt{N_t}\right)/T_t.$$

Als Mittelwert für die gesuchte Zählrate des Präparats ergibt sich

$$n_e = n_t - n_0 = N_t/T_t - N_0/T_0$$

und für den mittleren Fehler

$$\delta \cdot n_e = \sqrt{n_t/T_t + n_0/T_0}.$$

Für die Eignung eines Detektors i ist die zur Erzielung einer bestimmten statistischen Genauigkeit erforderliche Meßzeit

$$T_{ti} = [(n_{ei} + n_{0i})^2 + n_{0i}^2]/(\delta n_{ei})^2 \, (n_{ei} + n_{0i})$$

maßgebend.

Im Fall $n_e \gg n_0$ wird $T_{ti}/T_{tk} \approx n_{ek}/n_{ei}$, d.h. der bessere Detektor ist derjenige mit dem größeren n_e.

Im Fall $n_0 \gg n_e$ wird $T_{ti}/T_{tk} \approx n_{ek}^2 \, n_{0i}/n_{ei}^2 \, n_{0k}$, d.h. der bessere Detektor ist derjenige mit dem größeren n_e^2/n_0.

Wenn es möglich ist, bei einem Detektor die Ansprechwahrscheinlichkeit durch Variation der Betriebsbedingungen zu ändern, z.B. bei einem Szintillationszähler, wählt man den Arbeitspunkt so, daß n_e^2/n_0 ein Maximum wird. Bei weiterer Erhöhung von n_e kommt man in das für kleine Zählraten ungünstige Gebiet.

Bei Detektoren mit Mittelwertmessung (Ionisationskammer mit Integralstrommessung, Ratemeter) gilt für den relativen mittleren Fehler der Einzelablesung und Beobachtungszeiten, die groß sind gegen die Zeitkonstante RC der Apparatur (Schiff und Evans)

$$\sigma_1 = 1/\sqrt{2NRC},$$

wo N die mittlere Zahl der ionisierenden Ereignisse pro sec bedeutet, von denen jedes den gleichen Beitrag liefert. Bei regelmäßigen Ablesungen über eine Zeit t hat der Mittelwert den mittleren Fehler

$$\sigma = \sigma_1 \sqrt{1 + 2t/RC}/(1 + t/RC).$$

2. Ansprechwahrscheinlichkeit

Damit ein möglichst großer Teil der vom Präparat emittierten Strahlung an den Detektor gelangt, ist eine gute Detektor-Geometrie nötig. 4π-Geometrie ist bei Ionisationsdetektoren für Gasproben, die als Füllgas dienen oder ihm beigemischt werden, praktisch gegeben. Sonst kombiniert man 2π-Detektoren. Bei Szintillationszählern kann man von solchen Kombinationen abgesehen Präparat und Szintillator mischen oder das Präparat in ein Bohrloch des Szintillators einführen. Bei der Kernphotomethode läßt sich das Präparat zwischen zwei Emulsionen bringen oder mit flüssiger Emulsion mischen.

Um 2π-Geometrie zu erreichen, kann man der Ionisationskammer leicht ein ausreichendes Volumen geben und das Präparat so anbringen, daß der Halbraum im Innern der Kammer liegt. Großflächige Zählrohre mit ebenen Flächen sind Spezialkonstruktionen, gewöhnlich Durchflußzähler [HEINTZE (2)]. Eine Geometrie, die nur unwesentlich schlechter ist als exakte 2π-Geometrie, läßt sich auch bei normalen Zählrohren leichter realisieren, die Zylinderform, bei der Präparatfläche und Zählrohrmantelfläche gleich sind [ANDERSON et al., SCHUMANN (1)]. Mit Szintillationszählern und Photoemulsionen ist 2π-Geometrie einfach zu verwirklichen, wenn die Präparatfläche nicht zu groß ist.

Anordnungen, bei denen nur ein kleiner Raumwinkel vom Detektor erfaßt wird, besitzen zwar Vorzüge, weil eine ausgeblendete Strahlung sauberere Meßbedingungen bietet. Bei Präparaten geringer Aktivität sind sie jedoch ungünstig, weil sie nur einen kleinen Bruchteil der emittierten Strahlung an den Detektor bringen.

Damit die an den Detektor gelangten Teilchen bzw. Quanten messend erfaßt werden, müssen sie in das empfindliche Volumen des Detektors eindringen. Zwischen diesem und dem Präparat sollte also möglichst wenig absorbierendes Material liegen. Die Bedingung ist ohne weiteres erfüllt, wenn das Präparat in das Detektorinnere gebracht wird. Das ist prinzipiell möglich bei Ionisationsdetektoren wie bei Szintillationszählern, wenn man Präparat und Szintillator mischt, wie auch bei Photoemulsionen, die man mit Präparatlösungen imprägniert. Bleibt das Präparat außerhalb des Detektors, erfolgt bei energiearmer Strahlung unter Umständen merkliche Absorption im Luftraum zwischen Präparat und Detektor und in der Detektorwandung, auch wenn diese ein dünnes Fenster enthält. Größe und Bedeutung der Absorptionseffekte hängen stark von der speziellen Anordnung und der zu messenden Strahlung ab.

Eine Sonderstellung nehmen die Ionisationsdetektoren für γ-Strahlung ein, die wegen ihrer gegenüber den Szintillationszählern sehr geringen Ansprechwahrscheinlichkeit bei schwachen Aktivitäten nur noch selten verwendet werden. Da die Wahrscheinlichkeit der Ionisation im Gasraum überaus gering ist, muß sie in der Detektorwand möglichst hoch gemacht werden. Hilfsmittel dazu sind schweratomige Substanzen und hinreichende Wandstärken. Natürlich hat eine Vergrößerung der Wanddicke nur insoweit Sinn, als die Sekundärelektronen noch in das Gasvolumen des Detektors gelangen.

Bei vielen Szintillatoren und bei Kernemulsionen besteht die Möglichkeit, das Präparat in unmittelbaren Kontakt mit dem Detektor zu bringen. Man schützt diesen gegebenenfalls durch eine dünne Folie gegen Kontamination.

Welche Energie das in den Detektor gelangte Teilchen oder Quant abgeben muß, um registriert zu werden, hängt von der Art der Strahlung und des Detektors ab. Beim Geiger-Zähler genügt ein Ionisationsakt in der Wand oder im Zählerinneren. Bei den meisten anderen Detektoren wird man die Bedingungen so einrichten, daß ein Teilchen oder Quant möglichst viel oder sogar seine gesamte

Energie auf den Detektor überträgt. Auf vollkommene Energieabgabe ist man z.B. angewiesen bei α-Spektroskopie oder Spurenlängen-Messung in Photoemulsionen.

3. Nulleffekt-Reduktion

Die Eigenaktivität des Detektors muß möglichst niedrig sein. Elemente, die natürlich radioaktive Isotope besitzen, auch wenn ihr Anteil noch so gering ist, tragen immer Aktivität bei, z.B. K oder Pb. Alle Materialien, die Zn enthalten, sind aktiv, weil Zn und Pb in der Natur vergesellschaftet vorkommen. Ferner enthält Weichlot außer Sn stets einen Prozentsatz Pb, also Aktivität.

Heute werden aber auch in der Technik radioaktive Stoffe benützt, die manchmal eine sonst unmerkliche, im vorliegenden Zusammenhang aber wesentliche Kontamination verursachen. So sind Al-Chargen auf den Markt gekommen, die Ra-Aktivität enthielten, vermutlich von Leuchtfarben aus Flugzeugschrott. Auch hat man Co^{60} u. a. zur Abbrandkontrolle von Hochöfen verwendet, wobei kleine Mengen in den Stahl gelangen können. Im Zweifelsfall muß man die Materialien genau prüfen.

Unter Umständen hat man mit der Luftaktivität zu rechnen bzw. mit der Aktivität des Staubes, der sich auf Apparaturteilen und eventuell auf dem Detektor absetzt. Wo Verdacht auf solche Kontamination besteht, kann man die Apparatur durch Einhüllen mit Kunststoffolie schützen. Auch die Präparate müssen dann strengstens vor Staub geschützt werden. Ersatz der Normalluft zwischen Präparat und Detektor durch gealterte Preßluft, in der kein Rn und keine kurzlebigen Folgeprodukte mehr enthalten sind, kann in extremen Fällen den Nulleffekt herabsetzen (Boulenger et al.).

Ein weiteres Mittel zur Nulleffekt-Reduktion ist die Abschirmung gegen von außen einfallende Strahlungen. Dazu gehören kosmische Strahlung sowie γ-Strahlung der Umgebung. β-Strahlung wird mehr absorbiert und ist weniger wirksam, doch können Quellen in Detektornähe noch stören. α-Aktivität ist ohne Bedeutung, weil ihre Reichweite in Luft nur wenige cm beträgt und in festen Stoffen nur hundertstel mm.

Pb hat den Vorteil, daß man wegen seines hohen spezifischen Gewichtes mit den kleinsten Mengen auskommt und mit den kleinsten Gesamtgewichten. Es hat den Nachteil, daß seine Eigenaktivität, insbesondere von Pb^{210} und seinen Folgeprodukten, beträchtlich ist. Da der γ-Anteil unbedeutend ist, genügt es oft, die Pb-Abschirmung im Innern mit einem Futter aus Material geringerer Eigenaktivität auszukleiden, z.B. Stahl oder Elektrolytkupfer. Die ganze Abschirmung aus Fe zu bauen, erfordert viel Raum und Gewicht, und auch Eisen und Stahl ist nicht immer ganz aktivitätsfrei. Das Schwermetall mit geringster Eigenaktivität ist Hg, das manchmal als Abschirmung unmittelbar um den Detektor herum verwendet wird [Kulp (1)].

Man kann aber nicht alle von außen kommenden Strahlungen durch solche Abschirmungen ausschalten. Übrig bleibt vor allem die durchdringende Komponente der kosmischen Strahlung. Ein Mittel zu ihrer weitgehenden Ausschaltung bietet ein Antikoinzidenzkranz. Der eigentliche Detektor wird mit einem ringförmigen Detektor oder einem Kranz von Detektoren umgeben und mittels geeigneter elektronischer Schaltungen dafür gesorgt, daß alle Impulse des Meßdetektors, bei denen praktisch gleichzeitig ein oder mehrere Impulse im Kranz ausgelöst werden, nicht zur Registrierung gelangen. Nicht erfaßt werden die Neutronen der kosmischen Strahlung, wohl aber im wesentlichen die Mesonen.

Den Anteil der Nukleonen-Komponente der kosmischen Strahlung kann man herabsetzen, wenn man die Apparatur noch mit neutronenabsorbierenden Substanzen wie Paraffin oder borhaltigem Material abschirmt [DE VRIES (2), (4)] oder im Keller oder einem sonstigen unterirdischen Raum aufstellt.

Sind das Spektrum des Nulleffekts und das des Effekts verschieden, so läßt sich eine weitere Reduktion des Nulleffekts dadurch erzielen, daß man die Impulshöhen diskriminiert. Auf diese Weise kann man sowohl bei Messung einer Teilchenart eine andere ausschalten als auch Teilchen bestimmter Energie gegenüber gleichartigen anderer Energie auswählen.

4. Elektronik

Der Verbesserung der statistischen Genauigkeit durch Verlängerung der Meßzeiten sind Grenzen gesetzt. Die Betriebsdaten einer elektronischen Apparatur können nur innerhalb technisch bedingter Toleranzen konstant gehalten werden, und die Schwierigkeiten wachsen mit zunehmender Dauer. Die Meßzeiten sollen also lang sein, damit der statistische Fehler klein wird, aber nicht zu lang, damit nicht durch Änderungen von Apparatureigenschaften entstehende systematische Fehler die Verbesserung des statistischen Fehlers aufheben.

Die Anforderungen an Spannungskonstanz hängen von Detektor und Meßverfahren ab. Am höchsten sind sie bei Spektroskopie mit Szintillationszähler, Proportionalzähler und Impuls-Ionisationskammer, wo man Schwankungsgrenzen unter 10^{-3}, manchmal bis $2 \cdot 10^{-4}$ (beste kommerzielle Geräte) verlangen muß. Bei bloßer Zählung hängen die Anforderungen von der Plateaugüte des Detektors ab, im allgemeinen verwendet man Geräte mit 10^{-3}, bei sehr guten Plateaus genügen 10^{-2}. In kritischen Fällen empfiehlt sich auch Stabilisierung der Heizspannung. Gute Temperaturkonstanz ist vorteilhaft, besonders empfindlich sind hier Gasentladungsröhren, d.h. auch die Glimmstabilisatoren, die vielfach für die elektronische Spannungsreglung verwendet werden, sowie Photomultiplier.

Unerläßlich ist elektrostatische Abschirmung aller störempfindlichen Teile einschließlich Verbindungsleitungen. Sie muß um so besser sein, je hochohmiger die Spannungsquelle ist, also z.B. bei der Ionisationskammer besonders gut. Für alle Leitungen empfiehlt sich eine niederohmige Spannungsquelle nicht nur der Verluste wegen, sondern auch zur Unterdrückung von Störspannungen. Die Abschirmung muß tadellos geerdet sein, eine schlecht geerdete Abschirmung wirkt als Antenne.

Gute Erdung ist überhaupt ausschlaggebend für Ausmerzung von Störimpulsen. Innerhalb der Apparatur muß man Erdschleifen und Spannungsdifferenzen zwischen verschiedenen Teilen der Erdung vermeiden, also zentral erden und guten Kontakt bei den Zuführungen zu dieser Erdleitung sichern. Spannungsabfälle durch Kontaktwiderstände (z.B. auch an Röhrensockeln) und Selbstinduktion von Leitungen müssen so gering wie möglich bleiben. Die Heizung muß unbedingt auf ein festes Potential gelegt und durch eine Kapazität gegen Erde hochfrequenzmäßig kurzgeschlossen werden, am besten erdet man eine Seite. Für sehr empfindliche Verstärker ist ein getrennter Heiztransformator nötig, eventuell in Verbindung mit einem LC-Siebglied. Bei starken äußeren Störungen kann man einen separaten Störempfänger vorsehen und in Antikoinzidenz zur Meßapparatur schalten.

Falsche Impulse können auch infolge Übersteuerung des Verstärkers durch von der kosmischen Strahlung herrührende übergroße Impulse zustandekommen, die um mehrere Größenordnungen höher sind als die Meßimpulse. Durch Begrenzerschaltungen und geeignete RC-Glieder müssen der Unterschuß solcher Impulse und weitere Reflexionen ausgeschaltet werden, und zwar mindestens auf 10^{-3}, wenn nicht auf 10^{-4} der ursprünglichen Impulshöhe (bei Untersuchungen im niederenergetischen Bereich).

5. Dicke Präparate

Ein weiteres Charakteristikum vieler geophysikalischer Aktivitätsmessungen sind dicke Präparate. Daraus folgt die Notwendigkeit der Korrektur wegen Selbstabsorption und verwandter Effekte. Außerdem haben die Präparate flächenhafte Form. Das erschwert die Bemühungen um eine günstige Geometrie für die Messung. Die Dicke des Präparats und seine Flächenausdehnung stehen natürlich in unmittelbarer Beziehung zueinander. Ob man die Fläche vergrößert und dadurch die Dicke verringert oder umgekehrt, hängt vom Einzelfall ab.

Die Selbstabsorptions-Effekte sind am einfachsten für α-Teilchen der hier interessierenden Energien, weil die Wechselwirkung mit der Materie praktisch ausschließlich in Energieverlusten durch Ionisation besteht. Bedeuten n die gemessene Aktivität, s die spezifische Aktivität der Präparatsubstanz, R die Reichweite der α-Teilchen in dieser Substanz, d die Präparatdicke, so gilt

$$\text{für} \quad d \leqq R \quad n = \frac{s\,d}{2}\left(1 - \frac{d}{2R}\right),$$

$$\text{für} \quad d \geqq R \quad n = s\,R/4.$$

Für die Umrechnung der Reichweite in Luft auf die im Präparat gilt

$$\frac{R_P}{R_L} = \frac{1}{b}\,\frac{\varrho_L}{\varrho_P}\,\frac{A_P}{A_L}$$

ϱ_i Dichte, A_i effektive atomare Massenzahl (Luft 14,4), b Bremsvermögen bezogen auf Luft. Für b gibt es mehrere nahezu gleichwertige Näherungsformeln (Z_i effektive Kernladungszahl):

$$b = \sqrt{A_P/A_L} \;\; (\text{Bragg-Kleeman}), \quad b = (Z_P/Z_L)^{\frac{2}{3}} \;\; (\text{Glasson}),$$

$$b = (Z_P/Z_L)\,\sqrt{(Z_L + 4)/(Z_P + 4)} \;\; (\text{Rosenblum}).$$

Das molekulare Bremsvermögen ergibt sich näherungsweise durch Addition der atomaren Bremsvermögen.

Bei dicken β-Präparaten wird die emittierte Strahlung einmal im Präparat-Material selbst gestreut und zum Teil absorbiert, außerdem wird der in Richtung auf die Unterlage emittierte Anteil teilweise zurückgestreut, so daß er zusätzlich in den Detektor gelangt, soweit er nicht wieder absorbiert wird. Für beide Effekte tritt mit Zunahme der Schichtdicke schließlich Sättigung ein. Man kann übersichtliche Verhältnisse erreichen, wenn man in den Fällen, wo die Präparatschicht nicht dünn gemacht werden kann, mit sättigungsdicker Schicht arbeitet.

Es gibt jedoch Fälle, wo weder das eine noch das andere zu erreichen ist. Dann muß man eine Korrektureichung vornehmen, wobei man zweckmäßig alle Effekte (Selbstabsorption und Selbststreuung im Präparat sowie Rückstreuung) zusammenfaßt. Man bestimmt mit Eichpräparaten bekannter spezifischer Aktivität die Änderung der Zählrate mit der Schichtdicke unter sonst gleichen Bedingungen wie bei der eigentlichen Messung. Bei geringer Schichtdicke kann die Rückstreuung überwiegen, der Meßwert also zu hoch werden, während sonst die Selbstabsorption vorherrscht und der Meßwert zu klein wird.

Will man die Rückstreuung bei dünnen Präparaten möglichst herabsetzen, verwendet man als Unterlage dünne Kunststoffolien. Wichtig ist, daß die Rückstreuung nur von der mittleren Kernladungszahl des streuenden Materials abhängt, nicht von der β-Energie. Wenn bei reinen Rückstreuexperimenten gelegentlich eine leichte Energieabhängigkeit beobachtet wird, ist dies auf Absorption in Fenster oder Wand des Detektors oder auf dem Wege bis zum Eintritt in den Detektor zurückzuführen. Die Absorption ist natürlich wesentlich von der Energie abhängig, dagegen bei monochromatischen Elektronen praktisch unabhängig von der Art der Substanz lediglich eine Funktion der Flächendichte.

Methoden für Absolutzählung an dicken β-Präparaten sind verschiedentlich angegeben worden. Um Geometrie-Korrekturen zu vermeiden, mißt man zweckmäßig in 4π- oder 2π-Geometrie. In 4π-Geometrie hat man nach Baker und Katz mit Variation der Schichtdicke d

$$n = s\,d\,[1 - \exp(-k\,d)]/k\,d, \qquad k = 0{,}0155\,E_{\max}^{-1,41}\ \mathrm{cm^2/mg}.$$

Das Verfahren ist brauchbar, wenn es möglich ist, in so dünner Schicht zu messen, daß der Quotient $[1-\exp(-kd)]/kd$ nur eine kleine Korrektur darstellt. Bei starker Selbstabsorption bestehen Bedenken, da k nicht als Funktion von $E_{\max}$ allein angesehen werden kann [Heintze (2)]. Mit Zählung an sättigungsdicker Schicht verbinden Suttle und

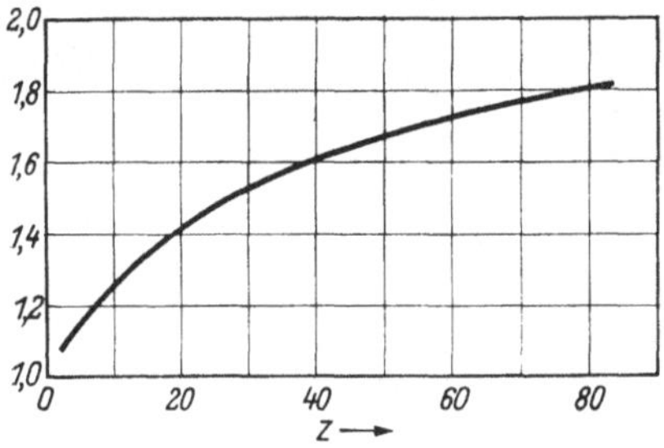

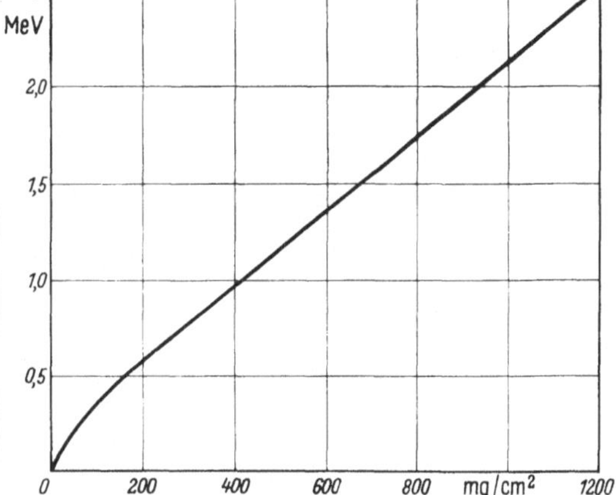

Fig. 1. Rückstreu-Koeffizienten für β-Emission

Fig. 2. Praktische Reichweite monoenergetischer Elektronen

Libby Aufnahme der äußeren Absorption. Ist τ der äußere Absorptionskoeffizient, so hat man ($G=1$ für 4π, $=2$ für 2π-Geometrie)

$$n = s/G\,\tau.$$

Das Verfahren ist auch für β-Spektren unbekannter Maximalenergie anwendbar. Streng genommen müßte die Absorption mit dem Material der Präparatschicht gemessen werden, weil τ von der Kernladung Z nicht unabhängig ist.

Heintze und Fischbeck kommen bei sättigungsdicker Schicht ohne Absorptionsmessungen aus. Aus Analogie zur α-Emission unter Berücksichtigung der Elektroneneffekte folgt

$$n = a\,s\,\bar{R}, \qquad \bar{R} = \int\limits_{0}^{E_{\max}} R_p(E)\,N(E)\,dE \Big/ \int\limits_{0}^{E_{\max}} N(E)\,dE$$

(R_p praktische Reichweite monochromatischer Elektronen). Das Verfahren ist auf alle β-Strahler anwendbar, deren Zerfallsschema und β-Spektrum bekannt ist, und hat eine Genauigkeit von etwa 5%. a besteht aus einem Faktor, dessen Abweichung von $\frac{1}{4}$ (vgl. Formel für α-Emission) ein Maß für die Schwächung durch Streuung in der Präparatschicht ist, und einem zweiten, der durch das Verhältnis der mittleren wahren Elektronenreichweite zu $\bar{R}$ gegeben wird. Der erste Faktor nimmt mit Z ab, der zweite zu, so daß a nur ganz wenig von Z abhängt. Außerdem besteht eine geringe Energieabhängigkeit. Empirische Werte für a: C^{14} 0,24; S^{35} 0,25; K^{40} 0,23; Sr/Y^{90} 0,23; Th/Pa^{234} 0,21.

6. Eichung

Geophysikalische Untersuchungen erfordern fast stets Absolutbestimmung einer Aktivitätskonzentration. Direkte Absolutmessung stellt hohe Anforderungen. Am einfachsten ist sie an α-Strahlern, weil dabei nur Absorptionskorrekturen nötig sind, die man auch ganz vermeiden kann, wenn das Präparat dünn ist und man in 2π-Geometrie mißt. Allerdings muß man bei hoher Genauigkeit eventuell die Rückstreuung berücksichtigen, die einige Prozent betragen kann (Jaffey) und bei anderen Messungen keine Rolle spielt.

Bei β-Strahlern sind im allgemeinen eine größere Zahl von Korrekturen nötig, von denen die energieabhängigen sich noch dazu auf ein ganzes Spektrum beziehen. Die heute wichtigsten Verfahren der direkten Absolutmessung für β-Strahler sind die 4π-Zähler-Methode [Haxel (1)] und die β-γ-Koinzidenz-Methode (Wiedenbeck). Bei der ersten wird das möglichst selbstabsorptions- und rückstreuungsfreie Präparat in einem Geiger-Zähler mit 4π-Geometrie gemessen. Die zweite ist nur auf β-Strahler anwendbar, bei denen ein Teil der Zerfälle oder alle mit einem γ-Übergang gekoppelt und die Einzelheiten des Zerfallsschemas genau bekannt sind. Diese Verfahren fordern für die Präparate Bedingungen, die oft kaum zu erfüllen sind, oder aber Korrekturen, deren Bestimmung beträchtlichen Schwierigkeiten begegnet. Nach Möglichkeit wird man daher der direkten Absolutmessung die Eichung mit Standard-Präparaten vorziehen. Allerdings sind solche nicht immer in der benötigten Größenordnung der Aktivität und in der richtigen Geometrie zu bekommen. Denn die Aktivität sollte die der zu messenden Proben nicht allzusehr übersteigen, weil manche Detektoreigenschaften zählratenabhängig sein können, z.B. der Einfluß von Auflösungszeiten. Und die Geometrie sollte möglichst exakt dieselbe sein, denn auch Geometrie-Korrekturen sind nicht trivial, wie z.B. Diskrepanzen zwischen Rechnungen und Messungen für Endfenster-Zählrohre zeigen. Man muß daher unter Umständen ein Standardpräparat in der für den speziellen Fall nötigen Weise umarbeiten, wobei man sich über die dabei auftretenden Fehlerquellen und Fehlergrenzen genaueste Rechenschaft geben muß.

Selbstverständlich wählt man wenn irgend möglich zur Eichung die gleiche Substanz, die man in den Proben mißt. Das ist jedoch unmöglich, wenn die Proben Gemische aktiver Substanzen enthalten, deren Zusammensetzung von Probe zu Probe wechseln kann, z.B. U und Th mit Folgeprodukten, bei denen die Gleichgewichtsfrage ungeklärt ist, oder Spaltprodukte, deren Zusammensetzung sich mit der seit der Entstehung vergangenen Zeit ändert. Für die Untersuchung von Spaltprodukt-Gesamtaktivitäten ist vielfach K^{40}-Eichung üblich, weil die mittlere Energie des K^{40}-β-Spektrums ganz grob etwa der mittleren Energie eines normalen Spaltprodukt-Gemisches entspricht. Diese Art der Eichung ist schon deshalb nicht ganz korrekt, weil sich das Spektrum des Spaltprodukt-Gemisches mit der Zeit ändert. Auf Grund von experimentellen Ergebnissen an aufgesprühten aktiven Schichten verschiedener β-Strahler, die aus einer vorher geeichten Lösung erhalten wurden (Nervik et al.), wurden Berechnungen über die bei K^{40}-Eichungen nötigen Korrekturen vorgenommen (Schraub). Für 10 Tage alte Spaltprodukte wird der mit K^{40}-Eichung ermittelte Meßwert um 19% zu niedrig, für ein Gemisch von 100 und eines von 1000 Tage um 53%. Dabei ist eine Präparatdicke von 20 mg/cm² zugrunde gelegt. Bei der Rechnung wurden die Maximalenergien der Spaltprodukte entsprechend ihrer Ausbeute beim Spaltprozeß sowie die Halbwertszeiten berücksichtigt, nicht dagegen die Form der β-Spektren.

II. Anwendung auf verschiedene Detektoren

1. Ionisationskammer

Die Ionisationskammer ist eines der ältesten Meßgeräte zum Nachweis radioaktiver Umwandlungen und auch heute noch sehr nützlich, besonders zur α-Messung. Da die Ionisierung durch β- oder gar γ-Strahlen gegenüber der durch α-Teilchen sehr klein ist, kommt es beim Nulleffekt vor allem auf die Eigen-α-Aktivität der Baumaterialien an. Tabelle 1 gibt Übersicht über Meßwerte an verschiedenen

Tabelle 1. *Eigenaktivität verschiedener Substanzen.* α-Emission in cpm/1000 cm^2

	Bearden	Curtiss	Sharpe	McDaniel et al.	Boulenger et al.
Perspex			0		
Stahl	0,5		0,8	0,4	
Korrosionsfreier Stahl				0,5	
Ni			0,5		
Cu	1,5			1,1	
Messing		1,7	3,3		
Al	5		3,3	3,7	
Ag				1,3	0,6
Weichlot	470			430	
Graphit				12	
Aquadag	1			1,7	
Argon techn.			0		
N$_2$ techn.				0	
Literatur	Bearden	Curtiss	Sharpe	McDaniel et al.	Boulenger et al.

Substanzen. Durch geeignete Auswahl der Materialien kann der Nulleffekt auf 0,2 cpm/1000 cm^2 reduziert werden [Bate (1)]. Da ein α-Teilchen pro Stunde im Mittel einen Ionisationsstrom von 10^{-17} A liefert, liegt der vom Kammermaterial herrührende Nulleffekt bei den besten Kammern in der Größenordnung 10^{-16} A. Die kosmische Strahlung erzeugt im Meeresniveau hinter 10 cm Pb $4 \cdot 10^{-19}$ A/cm^3.

Da es prinzipiell nur auf das elektrische Feld ankommt, kann man Flächen-Elektroden durch Drahtkonstruktionen ersetzen, sofern das mit dem sonstigen Aufbau des Detektors vereinbar ist. Diese Möglichkeit zur Nulleffekt-Reduktion ist auch bei den übrigen Ionisations-Detektoren anwendbar. Beim Abätzen zur Reinigung Pb^{210}-haltigen Materials besteht die Gefahr, daß sich Aktivität aus der ganzen von der Säure gelösten Schicht auf der abgeätzten Oberfläche niederschlägt. Das als Folgeprodukt auftretende α-aktive Po ist extrem elektronegativ und neigt dazu, sich in einer Lösung seiner Salze auf jeder Metallfläche niederzuschlagen. Auch das β-aktive Pb^{210} selbst kann unter Umständen abgeschieden werden [S. Deutsch (3)].

Bei Ionisationsstrom-Messung arbeitet man möglichst im Sättigungsgebiet. Sättigung ist bei geringen Drucken leichter zu erreichen als bei hohen. Bei α-Messungen sind wegen der hohen Ionendichten längs der Teilchenbahnen beträchtliche Feldstärken erforderlich. Auch hängt die Sättigungsspannung davon ab, welche Winkel die Teilchenbahnen mit der Feldrichtung bilden. In solchen Fällen müssen Korrekturen angebracht und die Meßwerte auf Sättigung extrapoliert werden. Absolutmessungen sind daher mit erheblichen Schwierigkeiten verbunden, und man eicht zweckmäßig mit Standard-Aktivitäten. Die niedrigsten meßbaren Ströme liegen in der Größenordnung 10^{-16} A, doch ist die Genauigkeit unterhalb 10^{-15} A nicht mehr sehr gut.

Nach Möglichkeit soll die Strahlung in der Kammer völlig absorbiert werden. Das ist bei α-Messungen möglich. Isolierungen müssen sicher sein gegen Unter- und Überdruck. Sie sollen — auch das gilt für alle Ionisationsdetektoren — ebenso wie sonstige Baustoffe, z.B. Kitte, möglichst niedrigen Dampfdruck besitzen, um die Reinheit der Gasfüllung nicht zu beeinträchtigen.

Kompensationsschaltung von zwei Kammern gestattet bei Langzeitmessungen weitgehend Ausschaltung von Nulleffekt-Schwankungen. Der Nulleffekt läßt sich dadurch überhaupt wesentlich herabsetzen, wenn auch die Eigenaktivität zweier Kammern nie genau gleich und ihre räumliche Lage und Exposition gegenüber der Umgebungsstrahlung nicht völlig identisch ist.

Bei Impulsmessung haben langsame Kammern mit Ionensammlung (Zeitkonstante > 10 msec) den Vorteil, daß Verunreinigungen des Füllgases weniger stören und daß die Impulse unabhängig vom Ort ihrer Entstehung sind. Eine unangenehme Eigenschaft ist die Empfindlichkeit gegen mechanische Schwingungen, die zu Kapazitätsänderungen in der Größenordnung der Impulsfrequenzen Anlaß geben (Mikrophonie). Selbst bei umfangreichen Maßnahmen zur Schwingungsdämpfung [Schaumgummi, Filz, Zuleitungen zu empfindlichen Verstärkerteilen mit Spiralfedern aus Litzendraht, zusätzliche Abstützung der Kollektor-Elektrode am Kammerboden (Rajewski et al.)] ist keine absolute Sicherheit gegen falsche Impulse gegeben. Schnelle Kammern mit Elektronensammlung (Zeitkonstante 0,1 bis 0,01 msec) vermeiden diesen Nachteil. Jedoch werden hohe Anforderungen an die Reinheit des Füllgases gestellt. Elektronegative Gase wie O_2, Cl, NH_3, NO, N_2O, H_2S, SO_2, HCl, Wasserdampf dürfen nicht vorhanden sein. Bei Hochdruck-Kammern ist eine spezielle Nachreinigung auch der kommerziellen technisch reinen Gase nötig. Nachteilig ist auch die Abhängigkeit der Impulse vom Ort der Entstehung der Ionen. Der Einfluß der positiven Ionen kann verringert werden, wenn die Kapazität möglichst klein gehalten wird und die Ionen weit entfernt von der Kollektor-Elektrode entstehen. Der zuletzt erwähnte Nachteil wird bei der Gitterionisationskammer nach Frisch vermieden, die besonders geeignet ist zur Messung extrem kleiner α-Aktivitäten. Zwischen den beiden Elektroden befindet sich ein Gitter auf mittlerem Potential. Die Kollektor-Elektrode ist positiv mit Bezug auf die anderen Elektroden. Die Probe ist auf der geerdeten Elektrode angebracht. Die Ionisation wird erzeugt zwischen ihr und dem Gitter. Das Gitter schirmt die Kollektor-Elektrode ab gegen positive Ionen. Durch geeignete Wahl der Gitterdrähte und ihres Abstandes voneinander sowie der Abstände und Spannungen der Elektroden kann man theoretisch erreichen, daß kein Elektron durch das positive Gitter eingefangen wird. So wird die Ladung, die durch Elektronensammlung auf der Kollektor-Elektrode erscheint, gleich der totalen Ionisation, die vom Teilchen erzeugt wird. Der Nulleffekt guter Gitterionisationskammern beträgt nur wenige α-Impulse pro Stunde.

Auch die Gitterionisationskammer ist normalerweise empfindlich gegen elektronegative Verunreinigungen. Diese Empfindlichkeit läßt sich herabsetzen durch Füllung mit Argon und 2 bis 4% N_2. Dann ist einwandfreie Funktion gesichert bei Anwesenheit von O_2 bis zu 0,5% [Facchini (2)]. Solche Mischungen sind auch für Durchflußbetrieb geeignet. Die Gitterionisationskammer eignet sich vorzüglich als Spektrometer. Die Auflösung läßt sich auf 2% für die Po^{210}-α-Linie (5,3 MeV) bringen [Facchini (3)].

2. Geiger-Zähler

Messungen mit dem Geiger-Zähler an einem Präparat oder in einer Umgebung, wo ein Gemisch aktiver Stoffe wirksam ist, sind im allgemeinen nur brauchbar, wenn eine Komponente hinreichend überwiegt oder eine Abfall-analyse oder Energiediskriminierung (integral) vorgenommen werden kann.

Ein in der bodennahen Atmosphäre frei angebrachter Geiger-Zähler registriert z.B. sehr komplexe Ereignisse. Für ein Zählrohr von 0,2 mm Al Wandstärke und 380 cm^2 effektiver Oberfläche wurde in 10 m Höhe über dem Erdboden gefunden (KIEFER et al.): kosmische Strahlung harte Komponente 108 cpm gemessen, 109 cpm berechnet nach GREISEN, weiche Komponente berechnet 59 cpm, γ-Strahlung aus der Luft berechnet nach PLESSET et al. und HIRSCHFELDER et al. 12 cpm, β-Strahlung aus der Luft gemessen 12 cpm, berechnet 10 cpm, γ-Strahlung aus dem Boden 169 cpm umgerechnet aus Untersuchungen der γ-Aktivität von Bodenproben in Abhängigkeit von der Tiefe, γ-Strahlung des Fallout auf der Bodenoberfläche 22 cpm umgerechnet wie vor, Gesamtimpulszahl gemessen 365 cpm, berechnet 381 cpm.

Der Nulleffekt eines Zählers mit möglichst guter Abschirmung im Labor setzt sich zusammen aus der Eigenaktivität des Materials, der Aktivität der Abschirmung, der restlichen Umgebungsstrahlung und verschiedenen Komponenten der kosmischen Strahlung, nämlich den vom Antikoinzidenzkranz durchgelassenen Mesonen, der von den Mesonen in der Abschirmung erzeugten γ-Strahlung und der Nukleonenkomponente (Neutronen). Der Nulleffekt pro Einheit der effektiven Oberfläche beträgt bei Metall-Zählern niedriger Eigenaktivität mit Antikoinzidenzkranz und Abschirmung mittels Fe oder Pb etwa 8 cpm/1000 cm^2 (Tabelle 2) [HEINTZE (1), MÜNNICH (1)], mit Hg-Abschirmung unmittelbar um den Innenzähler 4,5 cpm/1000 cm^2 [KULP (1), FERGUSSON]. Eine weitere Verbesserung ist möglich mit nichtmetallischem Wandmaterial, z.B. Kunststoffolien, die mit möglichst aktivitätsarmen Metall bedampft sind [HOUTERMANS (1), GEISS et al.], oder Quarz [DE VRIES (5)]. Zusätzliche Abschirmung mit Paraffin und borhaltigem Material oder Aufstellung in einem unterirdischen Raum setzt die verbleibenden Effekte der kosmischen Strahlung merklich herab. Die niedrigsten bisher erreichten Werte liegen zwischen 1 und 1,5 cpm/1000 cm^2 [HOUTERMANS (1), DE VRIES (5)].

Tabelle 2. *Nulleffekt von Elektrolytkupfer-Zählrohren mit Pb-Abschirmung und Antikoinzidenz-Kranz in cpm/1000 cm^2* [HEINTZE (1), MÜNNICH (1), SCHOLZ (2)]

β-Eigenaktivität der Zählrohrwand	1,5—3,0
γ-Strahlung von außen	1,8
Neutronen	0,5
Von Mesonen erzeugte γ-Strahlen	1,2
Vom Antikoinzidenzkranz durchgelassene Mesonen	0,5
γ-Strahlung des Aufbaus	0,9
Rest: Pb-Abschirmung	1,3

Verkleinerung des Zählers vermindert zwar den Nulleffekt. Doch machen sich immer mehr die Endeffekte bemerkbar, so daß das Plateau um so schlechter wird, je kleiner die Dimensionen sind. Ein schlechtes Plateau stellt aber bei kleinen Aktivitäten mit langen Meßzeiten hohe Anforderungen an die Konstanz der Apparatur.

Die Geometrie der Geiger-Zähler kann ganz verschieden sein. Die vielfach üblichen Fensterzähler haben für flächenförmige Präparate keinen günstigen Raumwinkel. Sie haben aber einige Vorteile. So sind sie nahezu die einzigen Zähler, die bei mehr oder weniger automatisierten Apparaturen eine einfache Konstruktion ermöglichen. Außerdem kann man bei Flächen bis etwa 5 cm^2 Fensterdicken bis 1,5 mg/cm^2 herunter verwenden und so relativ energiearme β-Strahler messen.

2π-Geometrie läßt sich mit zylindrischen Zählern sehr nahe realisieren. Wenn man einigermaßen energiereiche β-Strahler hat, kann man das Präparat außen anbringen [Schumann (1)]. Wandstärken unter etwa 25 bis 30 mg/cm² sind nicht möglich, weil sie die Druckdifferenz zur Außenluft nicht aushalten, außer bei sehr kleinen Flächen. Man kann allerdings im Durchflußbetrieb messen. Dann sind wegen der kleinen Druckdifferenz Kunststoffolien von 2 bis 1 mg/cm² und darunter verwendbar (Sugihara et al., Martell).

Will man extrem energiearme β-Strahler messen, muß man das Präparat in den Zähler bringen. Jedoch wird dessen einwandfreie Funktion dadurch leicht gestört. Libby hat ein Gitter zwischen der als Präparatauflage benutzten Zählerwand und dem Anodendraht eingeführt (Anderson et al.), um die Homogenität des Feldes in Drahtnähe zu sichern. Ferner ist das Präparat auf einem verschiebbaren Zylinder angebracht, so daß man Effekt und Nulleffekt messen kann, ohne das Zählrohr öffnen zu müssen. Die Ausführung einer Elektrode als Drahtgitter wird auch anderswo verwendet, weil sie den Nulleffekt herabsetzt (McDaniel et al., Nydal et al.).

Nahezu 4π-Geometrie kann man durch Kombination zweier Zähler erzielen, zwischen denen das Präparat angebracht wird. Die beste Annäherung erhält man mit Großflächenzählern. Eine andere Ausführungsform besteht in einem Kranz von Zählrohren, eventuell in Form eines Hohlzylinders mit mehreren Anodendrähten, in den die Probe hineingesteckt wird (Weiner et al.). Die ideale Geometrie liefert der Gaszähler, bei dem das gasförmige Präparat entweder selbst als Zählgas dient oder diesem beigemischt wird. Es ist auch bei einigen normalerweise nicht gasförmig auftretenden Stoffen versucht worden, gasförmige Verbindungen zu finden, die als Zusatz zum Zählgas dienen können, ohne dessen Funktion zu stören. Zum Beispiel wurde Pb^{210} als $Pb(CH_3)_4$ untersucht [Damon (2)], S^{35} als H_2S (Bernstein et al.), P^{32} als PF_3 (Suttle et al.). In Sonderformen für Antikoinzidenz-Anordnungen sind Innenzähler und Antikoinzidenzkranz konstruktiv vereinigt in zwei koaxialen Rohren (Raeth et al.). Alle in der Außenwand des Gesamtrohres erzeugten Elektronen werden durch die Antikoinzidenz-Schaltung eliminiert. Verringerung der Materialdicke zwischen Innenzähler und Kranz ist günstig für niedrigen Nulleffekt. Die optimale Dicke hängt davon ab, welche Energien bei den zu messenden Nukliden auftreten. β-Teilchen, die gezählt werden sollen, dürfen die Zwischenschicht nicht durchdringen, weil sie sonst Antikoinzidenzen erzeugen [Houtermans (1)]. Besonders geeignet ist diese Ausführung für Gaszähler. Sie ist aber auch für feste Präparate verwendbar, wenn diese in den Innenzähler gebracht werden können (Geiss).

Bei einem 4π-Zähler, der aus zwei übereinander angeordneten 2π-Zählern besteht, kann man beide Teile in Antikoinzidenz schalten, so daß nur Einzelimpulse eines Zählers registriert werden. Eine Konstruktion dieser Art für Messungen an Luftfiltern besteht aus zwei Großflächenzählern mit je vier parallelen Anoden und an jeder Längsseite einem zusätzlichen Antikoinzidenzzählrohr. Sie sind alle in einem gemeinsamen Kasten untergebracht, den das Zählgas erfüllt (Barendsen).

Eine ungewöhnliche Form einer Antikoinzidenz-Anordnung haben van Duuren et al. mit Halogenzählern entworfen. Halogen-Zähler (Geiger et al.) mit Füllungen aus Edelgas und einigen Promille Halogen haben die Vorteile niedriger Meßspannungen, hoher Impulse, langer Lebensdauer und, was für einige geophysikalische Anwendungen wichtig ist, geringer Temperaturempfindlichkeit. Sie besitzen nur einen schmalen Proportionalbereich und sind daher lediglich als Geiger-Zähler verwendbar. Der Anode in Halogen-Zählern kann man einen wesentlich größeren Durchmesser geben als in Zählern mit anderen Füllgasen

(HERMSEN et al.). Der Antikoinzidenz-„Kranz" der erwähnten Anordnung ist ein einziger Zähler mit zwei halbkugeligen Elektroden mit anschließendem zylindrischem Teil, die konzentrisch bzw. koaxial sind (Kathode 76 mm, Anode 32 mm Durchmesser), aus korrosionsfreiem Stahl. Den Abschluß bildet ein Glasring, der Anode und Kathode voneinander isoliert. Ein Halogen-Fensterzähler befindet sich innerhalb der Anode des Antikoinzidenz-Zählers zur eigentlichen Präparat-Messung.

3. Proportionalzähler

Der Proportionalzähler hat gegenüber dem Geiger-Zähler Vorteile bezüglich Nachweisempfindlichkeit und Umfang der durch die Messung erhaltenen Information. Erkauft werden sie durch größeren apparativen Aufwand und höhere Anforderungen an die Betriebsbedingungen.

Er ist ein brauchbarer α-Zähler. Daneben ist er besonders geeignet für β-Impulshöhenmessungen, solange die β-Teilchen ihre Energie ganz im Zähler abgeben. Er ist dabei bis unter 1 keV verwendbar, besitzt eine bessere Auflösung als der Szintillationszähler und kann zu Absolutmessungen dienen.

Durch geschickte Diskriminierung kann man das Verhältnis von Effekt zu Nulleffekt bzw. dessen Wurzel unter Umständen verbessern, z. B. bei Messungen an H^3. Verunreinigungen durch natürlich radioaktive Elemente (von K abgesehen) lassen sich auf Grund ihrer α-Emission abschätzen, und man kann in gewissen Grenzen daraus auch auf die damit verbundene β- und γ-Strahlung schließen.

Für laufende Kontrolle auf einwandfreie Funktion hat sich Einstrahlung einer Röntgenlinie (z. B. CuK) bewährt [HEINTZE (1), MÜNNICH (1)], die gegen Unregelmäßigkeiten der Gasverstärkung und elektronegative Verunreinigungen sehr empfindlich ist.

Besonders geeignet für das Arbeiten mit nicht-gasförmigen Präparaten innerhalb des Zählrohrs ist der Durchflußzähler. Man kann zwar auch Proben in Proportionalzähler [HEINTZE (1)] oder Geiger-Zähler (ANDERSON et al.) einschleusen, die anschließend mit dem Zählgas gefüllt und abgeschlossen werden. Der Probenwechsel ist aber wesentlich umständlicher als beim Durchflußzähler, wo die Spülzeit nach Einbringen der Probe je nach den Strömungsverhältnissen bei der speziellen Konstruktion nicht mehr als 10 bis 20 min beträgt.

Bereitet es Schwierigkeiten, die Probe in den Zähler zu bringen, z. B. weil sie durch Form oder Oberflächenbeschaffenheit die elektrische Feldverteilung im Zählerinnern stören würde, so verwendet man einen Zähler mit Fenster.

Beim Durchflußzähler spielen Verunreinigungen des Zählgases keine so große Rolle wie beim abgeschlossenen Zähler. Außerdem kann Gasabsorption oder Gasabgabe am Präparat in gewissen Grenzen in Kauf genommen werden, was beim abgeschlossenen Zählrohr schnell zu Funktionsstörungen führt. Besonders unempfindlich gegen Wasserdampf sind Aethan, Propan, Butan (NILSSON et al.). Methan ist empfindlicher.

Ein gewisser Nachteil ist die wegen des Atmosphärendruckes im Zähler nötige hohe Spannung. Mit dem viel benutzten Methan kommt man bei Abständen zwischen Anode und Kathode um 1 cm auf etwa 5000 V für das β-Plateau. Durch geeignete Zählgasmischungen, z. B. He + Butan, kann man diese Spannungen auf weniger als die Hälfte herabdrücken. Günstige geometrische Verhältnisse lassen sich auch für großflächige Proben leicht erzielen. 2π- und 4π-Geometrie sind mit guter Annäherung zu verwirklichen.

Das Prinzip des Durchflußzählers ist nicht beschränkt auf den Proportionalzähler, doch werden die meisten Durchflußzähler im Proportionalbereich betrieben. Für niedrige Aktivitäten und die damit zusammenhängenden langen

Meßzeiten sind die meist sehr guten Plateaus der Durchflußzähler vorteilhaft. Daher werden keine sehr hohen Ansprüche an die Konstanz der Hochspannung gestellt (Größenordnung 10^{-3} über 24 Std). Im α-Plateau läßt sich der Nulleffekt mit etwa 50 cm² Nutzfläche in die Größenordnung 0,1 cpm bringen, so daß man sehr niedrige Aktivitäten unter einfachen Betriebsbedingungen messen kann. Das β-Plateau ist meist besser als das eines Geiger-Zählers.

Man hat vielfach Durchflußzähler gebaut, deren Anode eine Drahtschleife ist. Eine gute Feldverteilung läßt sich aber mit größerer Sicherheit erreichen, wenn man einen oder eventuell mehrere geradlinig gespannte Drähte benutzt [Heintze (2)]. Insbesondere für Großflächenzähler ist eine solche Anordnung vorzuziehen. Auch kastenförmige Zähler zeigen ausgezeichnete Zähleigenschaften, wenn man für jeden Anodendraht Zylindersymmetrie annähert, d.h. für jeden Teilzähler Breite und Höhe gleich macht. Will man einen Durchflußzähler mit Antikoinzidenzkranz betreiben, kann man auch für diesen den Durchflußbetrieb verwenden. Der Kranz braucht nicht aus Einzelzählrohren zu bestehen, sondern seine Anodendrähte können — durch eine geeignete Wand getrennt — im gleichen Gasraum wie der Innenzähler untergebracht werden.

4. Szintillationszähler

Der Szintillationszähler hat wegen seiner hohen Ansprechwahrscheinlichkeit für γ-Strahlen (Fig. 3) ältere γ-Meßmethoden mindestens im Bereich niedriger Aktivitäten fast völlig verdrängt. NaJ ist sehr hygroskopisch, so daß die Kristalle gegen Luftfeuchtigkeit geschützt werden müssen. Diese Komplikation fällt weg bei CsJ. Doch ist dessen Lichtausbeute geringer, und die Impulse sind länger. Einkristalle kann man bis zu Größen von 20 cm und mehr fertigen. Auch relativ großflächige Präparate lassen sich damit ausmessen.

Tabelle 3. *Nulleffekt von NaJ-Szintillations-Zählern (cpm)* [Miller (1), (2)]

Kristall	$1\frac{1}{2} \times 1''$	$2\frac{1}{2} \times 2\frac{1}{8}''$	$4 \times 1\frac{1}{2}''$		$7 \times 3\frac{1}{2}''$
			Al-Fassung Glas-Fenster	Stahl-Fassung Quarz-Fenster	
MeV	0,07 bis 2,5	0,08 bis 0,8	0,09 bis 2,5		0,08 bis 1,6
Ohne Abschirmung			7000		
8″ Fe	62		480		
8″ Fe $+\frac{1}{8}''$ Pb					535
8″ Fe $+ 2''$ Pb	41				
8″ Fe $+ 1''$ Hg	35	75	360	150	
8″ Fe $+ 2''$ Pb $+ 1''$ Hg	33				
8″ Fe $+ 2''$ Bi					427
8″ Fe $+ 2''$ Bi $+\frac{1}{2}''$ Hg					388
davon:					
Kosmische Strahlung	7	15		40	155
$K^{40}\,\beta$	4	8		15	30
Multiplier	8	25		38	100
Rest	14	27		57	103

Der Nulleffekt von NaJ-Zählern setzt sich zusammen aus: Umgebungsstrahlung einschließlich Eigenaktivität der Abschirmung, kosmischer Strahlung, Eigenaktivität des Kristalls, die hauptsächlich auf dessen K-Gehalt beruht, und Aktivität des Multipliers, die vom K-Gehalt des Glaskolbens herrührt sowie vom K- und Ra-Gehalt des Sockels, während der eigentliche Innenaufbau keine nachweisbare Aktivität besitzt [Miller (1)]. Die Aktivität des Multipliers kann weitgehend ausgeschaltet werden durch Lichtleiter zwischen Kristall und Multiplier und entsprechende Abschirmung. Wo die Ausschaltung wesentlich ist, kann der gewonnene Vorteil wichtiger sein als der durch den Lichtleiter entstehende Lichtverlust. Von der volumenproportionalen Eigenaktivität des Kristalls abgesehen ist der Nulleffekt proportional zur Oberfläche. In Tabelle 4 sind quantitative Angaben über den Nulleffekt eines mit 10 cm Pb und einem Zählrohr-Antikoinzidenzkranz abgeschirmten $1_4''\times 2''$-NaJ-Kristalls zusammengestellt. Der als letzte Position aufgeführte Rest entstammt der Apparatur selbst und der Abschirmung und konnte nicht näher spezifiziert werden. Der als Pb-Aktivität bezeichnete Beitrag ist derjenige, der mit Sicherheit als Eigenaktivität der Abschirmung identifiziert wurde.

Das als α-Szintillator oft benutzte ZnS hat eine geringe Eigen-α-Aktivität. Sie ist oft vernachlässigbar, da wegen der optisch ungünstigen mikrokristallinen Form sehr kleine Mengen verwendet werden. Der Nulleffekt eines kompletten ZnS-Zählers konnte auf 3,5 α-Teilchen pro Stunde heruntergedrückt werden bei Verwendung einer 100 cm²-Präparatschale aus Ag und eines Multipliers EMI 6099 (Boulenger et al.). Die Eigenaktivität von ausgewähltem ZnS (33 Z 20 C RCA) liegt offensichtlich beträchtlich unter dem Durchschnitt (10 α/100 cm² h nach Sharpe).

Für flüssige Szintillatoren ist wichtig, daß Quarzgefäße einen gegen Vycor-Gefäße um einen Faktor 2 und gegenüber anderen Glassorten noch mehr herabgesetzten Nulleffekt besitzen (Agranoff).

Tabelle 4. *Spezifischer Nulleffekt von NaJ-Szintillations-Zählern (cpm/100 cm²)* [Scholz (2)]

	>0,1 MeV	>0,8 MeV
Kosmische Strahlung	28	20
Von Mesonen erzeugte γ-Strahlen	3,5	
γ-Strahlung von außen	2,2	2,7
γ-Strahlung des Zählrohrkanals	2,8	
Kristall (K⁴⁰) $1\frac{3}{4}''\times 2''$	2	0,4
Pb-Aktivität	5,8	
Rest: Aufbau und Abschirmung . . .	13	3

Während man im allgemeinen die Rauschimpulse des Multipliers wegen ihrer geringen Höhe leicht elektronisch wegdiskriminieren kann, machen sie sich bei energiearmer Strahlung als Nulleffekt-Erhöhung bemerkbar. Mittel dagegen sind Kühlung des Multipliers sowie Koinzidenz-Schaltung zweier Multiplier, wobei die Szintillationsimpulse als Koinzidenz gezählt werden, die Rauschimpulse dagegen keine Koinzidenzen erzeugen und eliminiert werden.

Zur Verbesserung der Geometrie kann man flächenhafte Präparate zwischen zwei Szintillatoren bringen und entweder mit einem Multiplier messen, dessen Photokathode senkrecht zum Spalt zwischen den Szintillatoren steht (Ketelle), oder mit zwei Multipliern (Johnson et al.).

Die Ausführung von Bohrloch-Szintillatoren richtet sich nach den zu messenden Präparaten. Man versucht möglichst nahe an die 4π-Geometrie heranzukommen. Szintillatoren mit Bohrloch gibt es für γ-Messung aus NaJ und für β-Messung aus Kunststoff. Zwar sind Plastik-Szintillatoren dem Anthracen in Lichtausbeute und Auflösung unterlegen, doch kann man Anthracen-Kristalle nur in kleinen Abmessungen herstellen, Plastik-Szintillatoren dagegen fast beliebig groß. Bohrlochdurchmesser gibt es von wenigen mm für sehr kleine Präparate bis über 30 cm für sehr voluminöse. Im letzten Fall wird der Detektor von vier Viertel-Zylindern von 54 cm Länge aus 13 cm dickem Kunststoff-Szintillator gebildet, von denen die Impulse mit acht Multipliern mit 12 cm-Photokathoden abgenommen werden (Nuclear Enterprises).

Für Flüssigkeits- (Dratz) und Pulver- [Hurley (2), Adams (2)] Proben kann man Ringgefäße verwenden, die den Szintillator allseitig umschließen mit

Ausnahme der der Photokathode zugekehrten Seite. Die Ansprechwahrscheinlichkeit ist abhängig vom Füllvolumen (Tabelle 5).

Zur Herstellung einer guten Geometrie gehört auch möglichst gute Anpassung des Szintillators an die Photokathode, gegebenenfalls mittels Lichtleiter. Unbedingt benötigt werden Lichtleiter bei dünnen CsJ- oder Kunststoff-Szintillatoren als α-Detektoren. Diese Schichten müssen extrem dünn sein (Größenordnung 0,1 mm), damit der γ- bzw. β-Untergrund niedrig gehalten werden kann. ZnS dagegen läßt sich (gemischt mit einer Haftlösung) direkt auf das Multiplier-Fenster aufbringen. Für gute Reflexion des Szintillationslichtes in Richtung auf den Multiplier werden konische Flächen empfohlen (Souch et al.).

Eine ideale Geometrie erhält man durch Mischung von Probe und Szintillator. Man kann Pulver-Proben mit pulverförmigen Szintillatoren, z.B. ZnS (Rosholt), oder in Form kleiner Kügelchen vorliegendem Kunststoff-Szintillator vermengen (Steinberg).

Bei Szintillationskristallen muß man die Kristalle aus dem pulverförmigen Ausgangsmaterial, dem die Probe beigemischt wird, erst herstellen. Das Probenmaterial muß dabei in Verbindungen vorliegen, die den Einbau in das Kristallgitter

Tabelle 5. *Abhängigkeit der Ansprechwahrscheinlichkeit von der Füllung eines Ringgefäßes mit einer Lösung konstanter spezifischer Aktivität* (Dratz)

Flüssigkeitsmenge	Nulleffekt	Zählrate	Ansprechwahrscheinlichkeit
cm³	cpm	cpm	cpm/cm³
650	286	1964	3,02
1000	268	2711	2,71
1650	244	3343	2,03

ermöglichen. So lassen sich z.B. seltene Erden in NaJTl nicht als Oxyd, dagegen als Chlorid einbauen (Leutz).

Gut entwickelt ist die Mischtechnik für flüssige Szintillatoren. Am besten eignen sich Proben, die in Lösungsmitteln für die Szintillationssubstanz, z.B. Toluol, Xylol, Anisol, löslich sind oder mit einem solchen einwandfrei gemischt werden können (Verdünnungsmittel) [Pringle et al., Hayes (3)]. Kleine Mengen Flüssigkeiten, die diese Bedingungen nicht erfüllen, lassen sich unter Umständen ebenfalls in die Szintillationsflüssigkeit bringen, z.B. Wasserproben für H³-Messung. In diesem Fall sind jedoch mengenmäßig Grenzen gesetzt, weil solche Stoffe die Fluorescenz des Szintillators unterdrücken.

Substanzen, die in Lösungs- oder Verdünnungsmitteln der flüssigen Szintillatoren unlöslich sind, lassen sich in der Szintillationsflüssigkeit [Hayes (3)] oder noch besser in einem Szintillationsgel [Funt (2), White et al.] suspendieren. Bei Suspensionen wäßriger Lösungen wurde für die Zählrate in Abhängigkeit vom H_2O-Gehalt bei niedrigen Konzentrationen ein linearer Anstieg, bei hohen ein konstanter Wert („sättigungsdickes" Präparat) gefunden (Shapira et al.). Bei veraschten Proben konnten bis 10 g Asche in 25 cm³ Szintillationsflüssigkeit suspendiert werden (Rapkin). Die Beeinträchtigung der optischen Verhältnisse ist geringer, als man erwarten würde. Es wird angenommen [Hayes (3)], daß dispergierte farblose Flüssigkeitströpfchen und farblose oder weiße feste Teilchen als Reflektoren wirken. Farbige Substanzen setzen die Lichtausbeute merklich herab. Auf Filterpapier abgeschiedene Proben können nach Tränken des Papiers mit flüssigem Szintillator direkt gemessen werden [Funt (3)]. Das gilt sowohl für Aerosolproben wie für chemische Niederschläge und für Papierchromatographie.

Szintillationszähler sind gut geeignet zur Spektroskopie. Für α-Spektroskopie läßt sich CsJ verwenden. Dünne Kunststoff-Schichten und ZnS geben schlechte Auflösung. Die höchste bei geophysikalischen Untersuchungen vorkommende

α-Energie ist die des Po^{212} mit 8,8 MeV. Man kann daher die Szintillatoren sehr dünn wählen und erhält vollständige Absorption der zu messenden Strahlung bei weitgehender Reduktion unerwünschter Nebeneffekte.

Bei γ-Spektroskopie wird im wesentlichen mit NaJ(Tl)-Kristallen gearbeitet. Für Messungen an schwachen Präparaten kommen nur Photolinien in Betracht. Compton- oder Paar-Spektrometrie sind wegen ihrer geringen Ausbeuten nicht verwendbar. Wenn wie bei vielen geophysikalischen Präparaten mehrere γ-Strahler in einem Präparat vorkommen, stören sich nicht nur die Photolinien verschiedener Nuklide gegenseitig, sondern auch die Compton-Verteilungen energiereicherer Strahlungen die Photolinien energieärmerer. Man kann die Compton-Verteilungen weitgehend unterdrücken, wenn man Szintillator und Multiplier in das Innere eines zweiten Szintillators großen Volumens (Kunststoff-Szintillator-Block oder Gefäß mit flüssigem Szintillator) bringt und durch Antikoinzidenzen alle Impulse eliminiert, bei denen im äußeren Szintillator ein Compton-Photon gezählt wird. Kann man keine ausreichende Analyse der Spektrogramme erreichen, bleibt nur eine radiochemische Trennung der beteiligten Nuklide. Man braucht dabei oft keine vollständige chemische Analyse durchzuführen, sondern nur die gerade störenden γ-Strahler zu eliminieren. Zum Beispiel muß man zur Messung von Be^7 (479 keV) das Ru^{103} (492 keV) quantitativ abtrennen.

Bei β-Spektroskopie mit Szintillationszählern führen Streueffekte oft zu starker Vermehrung der kleinen Impulse, weil die Elektronen nur einen Teil ihrer Energie im Szintillator abgeben. Mit Präparaten zwischen zwei Anthracen-Kristallen erhält man noch die besten Ergebnisse. Bei dickeren Plastik-Szintillatoren wird optische Absorption merklich. Messungen an einzelnen Nukliden können brauchbare Ergebnisse liefern, solche an Gemischen begegnen erheblichen Schwierigkeiten.

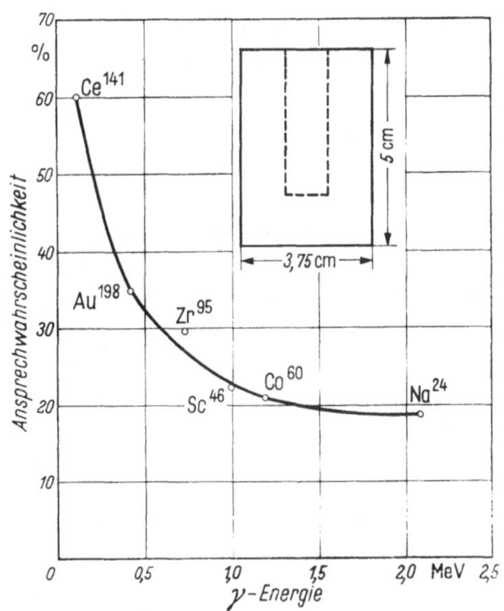

Fig. 3. Ansprechwahrscheinlichkeit eines NaJ-Bohrloch-Kristalls (BORKOWSKI)

Apparativ stellt die Spektroskopie bei schwachen Präparaten weit höhere Anforderungen als sonst. Für die Hochspannung benötigt man eine Konstanz von annähernd 10^{-4} über 24 Std. Temperaturdifferenzen wirken sich nicht nur auf die Spannungsstabilisierung, sondern auch unmittelbar am Kristall und am Multiplier aus. Bei mehreren Grad Temperaturänderung muß man auch bei guter Konstanz der Hochspannung mit Impulshöhen-Änderungen rechnen, die in ungünstigen Fällen die Größenordnung Prozent erreichen können. Man kann Schwankungen in der Elektronik dadurch begegnen, daß man ständig eine Eichlinie unter gleichen Bedingungen registriert und darauf eine Regelung der wesentlichen Betriebsgrößen aufbaut. Im allgemeinen wird es leichter sein, kleine Räume, etwa den Szintillator mit Multiplier und Abschirmung oder besonders empfindliche Teile der Elektronik wie Stabilisatorröhren oder Spannungsteiler innerhalb enger Toleranzen auf konstanter Temperatur zu halten.

5. Photographische Methoden

Für extrem kleine Aktivitäten scheint die photographische Methode besonders geeignet, weil man über sehr lange Meßzeiten jedes Einzelereignis registrieren kann. Aber der Aufwand bei der Auswertung ist unvergleichlich größer als bei allen anderen Verfahren. Besonders schwierig ist das Arbeiten mit β-empfindlichen Emulsionen wegen des starken Untergrundes, der schlechten Identifizierbarkeit energiereicherer β-Einzelspuren und der Grobkörnigkeit der Emulsion.

Tabelle 6. *Abhängigkeit des Fading von Temperatur und Luftfeuchtigkeit* [S. Deutsch (4)]

Aufbewahrungs-Temperatur °C	Relative Luftfeuchtigkeit %	Fading %
+ 15	50	72
+ 4	50	38
− 35	0	0

Emulsionen sind, weil sie jedes Einzelereignis bleibend registrieren, besonders empfindlich gegen Kontamination. Es genügt schon, wenn die Verpackung kontaminiert ist. Für die Handhabung im Labor ist größte Sorgfalt unerläßlich, z.B. auch im Hinblick auf die natürliche Aktivität der Luft und des atmosphärischen Aerosols.

Der Länge der Expositionszeit ist durch das Fading eine Grenze gesetzt. Es ist bei feinkörnigen Emulsionen stärker als bei grobkörnigen. Die Erscheinung ist ein Oxydationsprozeß am Ag der latenten Bildkeime (Albouy et al.). Ihr Ausmaß kann durch Temperaturerniedrigung und Exposition im Vakuum-Exsiccator zur Ausschaltung des Luftsauerstoffs und der Luftfeuchtigkeit herabgesetzt werden (Tabelle 6). Niedrige p_H-Werte fördern das Fading.

Umgekehrt kann man durch entsprechende Maßnahmen vor der Exposition seit Herstellung der Emulsionen entstandene Untergrundschwärzung weitgehend löschen.

Den Nulleffekt während der Exposition muß man durch Vergleichsplatten bestimmen, die von der Herstellung bis zur Auswertung exakt den gleichen Bedingungen unterworfen waren wie die Meßplatten.

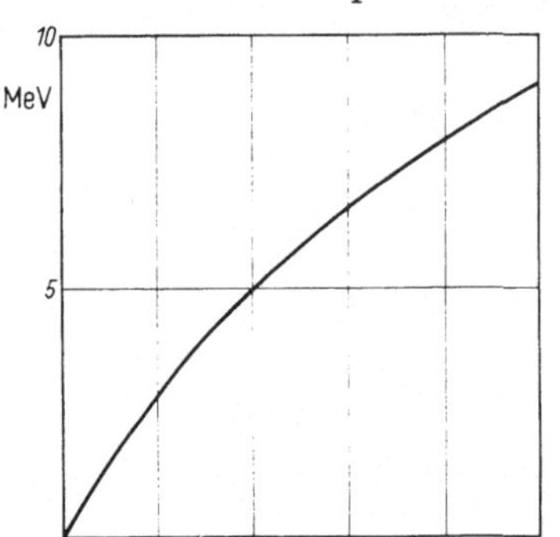

Fig. 4. Reichweite von α-Teilchen in Ilford Bl-Emulsion für geophysikalisch interessierende Energien (Lattes et al.)

Für Ilford-Platten zur α-Messung liegt eine genaue Analyse der Eigenaktivität vor [S. Deutsch (1), (2)]. Einfache Spuren der kosmischen Strahlung mit Restreichweiten $< 60\ \mu$, d.h. im Bereich der geophysikalisch interessierenden α-Reichweiten (hauptsächlich Protonen), sind gegenüber den α-Verunreinigungen zu vernachlässigen. Die Oberflächenaktivität, d.h. die Zahl der α-Spuren, welche die Emulsionsoberfläche durchsetzen, war bei der Standard-Ausführung 0,4, bei speziell gereinigter Gelatine bis herunter zu 0,2 α-Spuren pro cm^2 und Tag. Die Volumenaktivität der Emulsion betrug 42 bzw. 19 Ereignisse pro cm^3 und Tag, darunter überwiegend einfache Spuren (U, Th, Po), 20 bis 35% 3- und 4fach-Sterne (Ra^{226}), sehr wenige 5fach-Sterne (Ra^{228}), die Glasaktivität 3 bis 4 α/cm^2 Tag. Aus dem Vergleich zwischen Oberflächen- und Volumenaktivität wird geschlossen, daß 85% der Oberflächenaktivität von außen in die Emulsion hineinführende α-Spuren von Rn und Folgeprodukten sind. Unter den restlichen 15% sind möglicherweise einige (n, α)-Prozesse. Eine Abschätzung der Volumenaktivität unter der Annahme, daß der aus den 3- und 4fach-Sternen ermittelte Ra^{226}-Wert eine untere Grenze darstellt, weil Rn zum Teil verloren geht, der Ac-Beitrag vernachlässigbar ist, Th sich im Gleichgewicht mit Ra^{228} befindet und die U- und

Tabelle 7. *Radioaktive Verunreinigungen von Kernphotoemulsionen* [S. DEUTSCH (1)]

Verunreinigung	Standard-Ausführung g/cm³	Spez. gereinigte Gelatine g/cm³
Ra²²⁶	$\geqq 3,5 \cdot 10^{-15}$	$\geqq 1,2 \cdot 10^{-15}$
Th	$7 \cdot 10^{-9}$	$0,8 \cdot 10^{-9}$
RdTh(Ra²²⁸)	$9 \cdot 10^{-19}$	$1 \cdot 10^{-19}$
U	$\leqq 1,4 \cdot 10^{-8}$	$\leqq 1 \cdot 10^{-8}$
Po	$\leqq 2 \cdot 10^{-18}$	$\leqq 1,4 \cdot 10^{-18}$
Th im Glas	10^{-6} g/g	

Po-Beiträge sich aus den übrigen Einzelspuren ergeben, also obere Grenzwerte sind, liefert die Werte in Tabelle 7.

Für lösliche Proben gibt es die Imprägniermethode in ihren verschiedenen Abwandlungen. Das Eintauchen der Emulsion in die Lösung ergibt für das Eindringen in die Schicht schwer kontrollierbare Bedingungen. Man muß deshalb mit einer Standard-Lösung unter exakt gleichen Verhältnissen eichen. Das läßt sich vermeiden, wenn man die Lösung in Tropfen auf die Emulsionsschicht bringt. Eine Fehlerquelle ist hier die mögliche Anreicherung der Aktivität an der Oberfläche, was man besonders beachten muß, da es in der Geophysik praktisch immer um Konzentrationsmessungen geht. Konzentrationsunterschiede werden vermieden oder verringert bei hoher Luftfeuchtigkeit, wo Verdunstung des Wassers aus der noch nicht von der Emulsion aufgenommenen Lösung herabgesetzt wird. Übliche Emulsionen nehmen etwa das 3fache ihres Volumens in trockenem Zustand an Wasser auf.

Proben in wäßriger Lösung kann man mit auf 50° C erwärmter flüssiger Emulsion mischen und auf eine mit Gelatine überzogene Glasplatte vergießen.

Bei allen Formen der Imprägniermethode können Emulsionsgifte stören. Schwermetall-Kationen, insbesondere Bi, Pb, Tl, Hg, aber auch Fe führen zu Desensibilisierung. Manchmal kann man dies durch Komplexbildner, z.B. Citronensäure, vermeiden. Der p_H-Wert der Lösung ist am günstigsten im Bereich zwischen 5 und 9. Werte über 9 geben übergroße Sensibilisierung und starke Untergrundschwärzung, Werte unter 5 Desensibilisierung (STOEPPLER). Nach der Exposition wird die Emulsion gründlich gewässert, um das Imprägnierungsmaterial möglichst zu entfernen, und dann erst entwickelt [BRODA (2)].

Es gibt noch eine Form der Imprägniermethode, die aber ebenso für nichtlösliches Material anwendbar ist und damit eine Abwandlung der Auflagemethode darstellt: das 2-Emulsions- oder Sandwich-Verfahren [PICCIOTTO (1)]. Dabei wird die Lösung oder das unlösliche Material zwischen zwei Emulsionen gebracht. Man hat im Fall der Lösung die Möglichkeit einer absoluten Konzentrationsmessung unabhängig von Eichlösungen und ohne die Fehlerquelle durch Konzentrationsunterschiede beim Auftropfen. Im Fall unlöslichen Materials kann man ebenfalls die Gesamtaktivität messen, soweit keine störende Selbstabsorption im Probenmaterial auftritt.

Die einfache Auflagemethode mit direktem Kontakt zwischen Präparat und einer Emulsionsoberfläche ist eine verfeinerte Autoradiographie unter Verwendung einer Kernspur-Emulsion. Man spricht daher auch von Spuren-Autoradiographie. Es ist allerdings erforderlich, zwischen Präparat und Emulsion eine dünne Folie zu bringen, um unerwünschte, z.B. chemische oder Luminescenz-Einwirkungen des Präparats auf die Emulsion auszuschließen, wenn dies nicht durch die Art des Präparats bereits gewährleistet ist. Die Kernspur-Emulsion wird quantitativ ausgewertet. Im Gegensatz dazu ist die Autoradiographie im engeren Sinne oder Kontrast-Autoradiographie, die mit empfindlichen normalen oder Röntgen-Filmen durchgeführt wird, fast stets nur ein qualitatives Verfahren. Sie dient dazu, die Verteilung der Aktivität in einer Probe zu untersuchen. Solche Autoradiographien haben einen wesentlichen Anstoß zu den Untersuchungen über „heiße" Teilchen mit hoher Aktivitätskonzentration nach gewissen Atombombenexplosionen gegeben. In diesem Zusammenhang sind auch quantitative Messungen durchgeführt worden [Kern, Schumann (7)], deren Genauigkeit allerdings beschränkt ist und wegen der über mehrere Größenordnungen reichenden Aktivitätswerte erhebliche Schwierigkeiten bietet.

III. Radiochemische Methoden

1. Verhalten unwägbarer Substanzmengen

Bei Untersuchungen über radioaktive Stoffe in der Geophysik kommt man oft nicht ohne chemische Abtrennung bestimmter Nuklide aus. Dabei können sich manchmal die winzigen Mengen radioaktiven Materials anders verhalten als größere Mengen entsprechender stabiler Isotope, trägerfreie Nuklide anders als solche mit Träger. Oft gibt man daher inaktive Isotope des zu untersuchenden Elements dem Präparat zu. Quantitative Bestimmung dieser Beimengung am Ende der chemischen Operationen ermöglicht es, deren Ausbeute zu ermitteln. Dabei wird angenommen, daß der Gehalt der Probe an dieser inaktiven Beimengung gegenüber der zugegebenen Menge vernachlässigt werden kann, sowie gleiches Verhalten der aktiven und inaktiven Isotope.

In speziellen Fällen verwendet man nicht-isotope Träger, etwa wenn man für die Reaktionen große Trägermengen braucht, die beim Präparat wegen Selbstabsorption stören. Zum Beispiel kann man Sr zunächst mit Ca-Träger anreichern und dann mit kleinen Mengen Sr-Träger auskommen. Auch ist Ba ein oft benutzter Träger für Ra, bei dem keine inaktiven Isotope zur Verfügung stehen.

Bei Elementen, die in verschiedenen Wertigkeitsstufen auftreten, z.B. J, Ru, muß man die aktiven und inaktiven Isotope in dieselbe Stufe bringen. Abweichungen vom normalen chemischen Verhalten können auch bei Elementen auftreten, die leicht hydrolisieren, z.B. Zr [Herrmann (1)].

Unter Umständen kann man sich das abweichende Verhalten zunutze machen [Hahn]. Bei niedrigen Konzentrationen ist es möglich, daß bei Zugabe der üblicherweise zur Fällung verwendeten Reagentien, selbst wenn sie im Überschuß erfolgt, das Löslichkeitsprodukt nicht überschritten wird. Trotzdem läßt sich oft durch eine Fällung, die makrochemisch das betreffende Element gar nicht erfassen würde, die geringe Menge des radioaktiven Stoffes mit ausscheiden. Dabei kann echte oder anomale Mischkristallbildung auftreten, oder die Mitfällung erfolgt durch Adsorption. Dafür ist besonders $Fe(OH)_3$ geeignet. Will man Mitfällung vermeiden, wendet man Rückhalteträger an, inaktive Ionen, welche die kleinen Mengen des radioaktiven Stoffes bei der Adsorption verdrängen.

Ein Sonderfall sind die Radiokolloide. Vor Erreichung des Löslichkeitsprodukts können Niederschläge auftreten, die unsichtbar sind, aber durch Filtrieren oder Zentrifugieren aus der Lösung entfernt werden können. Für kolloidale Struktur sprechen Autoradiographien, anomales Verhalten bei Elektrolyse und Adsorption. Man findet Radiokolloide nur in schwach sauren oder alkalischen Lösungen.

Wegen der Möglichkeit, daß einzelne Substanzen unerwartetes Verhalten aufweisen, ist der chemische Trennungsgang kein Beweis dafür, daß die am Schluß erhaltene Aktivität wirklich die gesuchte ist. Vielmehr muß das Nuklid stets mittels der physikalischen Eigenschaften der emittierten Strahlung identifiziert werden.

Bei der Adsorption aktiver Stoffe an Glas (SCHÖNFELD et al.) handelt es sich sehr wahrscheinlich um Kationenaustausch. Sie ist nur teilweise reversibel. Man hat versucht, Glasgefäße mit Silikonen auszukleiden, doch sind solche Überzüge wenig haltbar (MÜNZEL). Am ehesten sind Kunststoff-Gefäße brauchbar, jedoch wegen ihrer meist geringen Temperaturbeständigkeit nur beschränkt. Vorbehandlung von Gefäßen durch Spülen mit Säuren oder Lösungen ändert die Adsorptionsbedingungen, gibt aber nicht immer Gewähr gegen Adsorptionsverluste.

Auch mit radioaktiver Verunreinigung der Reagentien muß man rechnen. Zum Beispiel wurden an Säuren (höchster kommerzieller Reinheitsgrad, pro analysi) folgende α-Aktivitäten gefunden: 1,4 cpm/l in HCl, 2,0 cpm/l in $HClO_4$, 1,25 cpm/l in HNO_3 (STOEPPLER).

Allgemein sollten die chemischen Prozeduren möglichst einfach sein und sich auf eine möglichst kleine Zahl von Schritten beschränken. Jede Reaktion birgt Fehlermöglichkeiten und, was bei kleinen Aktivitäten besonders ins Gewicht fällt, Gefahr von Aktivitätsverlusten. Man wird freilich oft einen Kompromiß eingehen müssen, denn vielfach ist hoher Reinheitsgrad nur auf Kosten der Ausbeute oder hohe Ausbeute nur auf Kosten der Reinheit zu erzielen.

2. Arbeitsmethoden

Neben den klassischen Methoden haben Ionenaustauscher erhebliche Bedeutung für die Abtrennung von radioaktiven Stoffen aus Gemischen. Sie erlauben auch Trennung chemisch sehr ähnlicher Stoffe wie etwa seltener Erden (TOMPKINS et al.), die bei den Spaltprodukten eine Rolle spielen. Durch Komplexbildner (z.B. Citrate, EDTA) kann man oft die Trennung erleichtern. Sie sollen die Absorption der für die Weiterarbeit nicht benötigten Ionen im Austauscher verhindern oder herabsetzen bzw. die Elution der gesuchten Ionen fördern. Die Trennung wird durch Verlängerung der Austauschersäule, Benutzung kleiner Korngrößen und Verringerung der Durchflußgeschwindigkeit verbessert. Auch Anionenaustauscher haben Bedeutung für Trennung radioaktiver Nuklide, z.B. Bi und Pb (KRAUS et al.), J oder S.

Die Papierchromatographie wurde für geophysikalische Untersuchungen bisher wenig benutzt. Immerhin sind Trennungen von Pb, Bi, Po sowie U und Th durchgeführt worden.

Trennung durch Verflüchtigung kommt bei den geophysikalisch interessanten Stoffen vor allem in Betracht für J, ferner S, P und auch Po, das bei etwa 200° C anfängt, flüchtig zu werden. Andererseits muß man bei der Probenverarbeitung berücksichtigen, daß diese Elemente bei Anwendung zu hoher Temperaturen teilweise verloren gehen können.

Extraktion mit organischen Lösungsmitteln wird häufig auch bei für die Geophysik wichtigen Nukliden angewendet. Wegen der Löslichkeit des Uranylnitrats in Äther kann man die Äther-Extraktion zur Trennung des U von Verunreinigungen oder Folgeprodukten benutzen (HAHOFER et al.). Ra, Ac, Th können durch Extraktion getrennt werden, da Radiumnitrat in Alkohol und Pyridin unlöslich, Actiniumnitrat in beiden löslich, Thoriumnitrat nur in Alkohol löslich ist (HAISSINSKY). Dreiwertiges Fe, das wie erwähnt häufig zu Mitfällungen verwendet wird, kann mittels Ätherextraktion aus stark salzsauren Lösungen von den meisten gefällten Nukliden getrennt werden (BRODA).

Organische Komplexbildner wurden schon bei den Ionenaustauschern erwähnt. Darüber hinaus kann man z.B. mit Dithizonextraktion durch Änderung des p_H-Wertes zahlreiche Trennungen durchführen (Sandell), so Pb, Bi, Po. Die Komplexbildung mit TTA (α-Thenoyltrifluoroaceton) ist für verschiedene Elemente zum Teil erheblich verschieden. Auf diese Weise läßt sich Be[7] aus Gemischen abtrennen [Goel (1)], Th aus U-Lösungen (Hagemann et al.).

Elektroabscheidung ist auch für kleinste Mengen anwendbar, z.B. für Po aus RaD, E, F in saurer Lösung auf Ag, danach für Bi bei höheren Temperaturen an Ni, Pb bleibt in Lösung. Wichtig ist dabei Vermeidung der Adsorption von Ionen, die in Lösung bleiben sollen, besonders da das Gleichgewicht sich bei Adsorption schnell einstellt, bei Elektroabscheidung langsam.

Das Verfahren der Isotopenverdünnung ist für Geochemie und Geophysik wichtig. Dabei wird im allgemeinsten Fall einem Gemisch das gesuchte Element in spezieller Isotopenzusammensetzung zugegeben und nach der Abtrennung der Anteil einzelner Isotope bestimmt. Daraus ergibt sich die Ausbeute des chemischen Prozesses. Spezialfälle sind Zugabe von inaktivem Träger oder aktivem Tracer. Man kann auch der Messung schlecht zugängliche radioaktive Nuklide, z.B. solche mit sehr langer Halbwertszeit oder besonders energiearmer Strahlung, durch für die Messung besser geeignete markieren, z.B. Th durch Th[234], das eine Halbwertszeit von 24 Tagen besitzt. Isotopenverdünnung ist prinzipiell für stabile wie für aktive Isotope möglich, erfordert natürlich im ersten Fall ein Massenspektrometer für die Analyse.

Die Aktivierungsanalyse, vor allem mit Neutronenaktivierung, hat in der Geochemie beträchtliche Bedeutung. Voraussetzung ist genaue Kenntnis der kernphysikalischen Reaktionen, ihrer Wirkungsquerschnitte und der entstehenden Aktivitäten. Man bedient sich des Verfahrens, um Nuklide zu erzeugen, die sich leichter physikalisch messen lassen als die ursprünglich vorhandenen. Fehlerquellen (Plumb et al.) sind Inhomogenität in Neutronenfluß (vor allem bei Bestrahlung im Reaktor) und Schwächung des Flusses in der Bestrahlungs-Anordnung. Monitor-Vergleichsmessungen sind betreffs zeitlicher und räumlicher Verschiedenheiten nötig und sollten bei möglichst gleichen Schwächungsbedingungen durchgeführt werden wie für die Probe. Beurteilung des Wirkungsquerschnitts der gewünschten Reaktion bzw. der Konkurrenz-Reaktionen erfordert möglichst gute Kenntnis des Neutronenspektrums. Bei der Aktivierung können nebenher so hohe Aktivitäten entstehen, daß für die Verarbeitung besondere Strahlenschutzmaßnahmen nötig werden.

3. Spezielle Nuklide

Die Anwendung der Methoden auf spezielle geophysikalisch wichtige Nuklide kann hier nur an Beispielen skizziert werden.

Verhältnismäßig einheitlich ist der Bereich der in den Mineralen vorkommenden natürlich radioaktiven Substanzen. Bei der chemischen Verarbeitung von U-Proben bedient man sich häufig der Äther-Extraktion. Zur Reinigung kann man von Ionenaustauschern Gebrauch machen. UO_2^- bildet bei p_H 4,25 bis 5,25 im Gegensatz zu anderen in Frage kommenden Elementen (z.B. Th) einen negativ geladenen Acetatkomplex, der am Anionenaustauscher absorbiert wird (Hecht et al., Tomic et al.). Mittels Kationenaustauschers mit Ascorbinatkomplexen kann man U und Th voneinander und von den die Endbestimmung störenden und Ascorbinatkomplexe bildenden Elementen Zr, W, Mo trennen. Auch Papierchromatographie ist für die U-Abtrennung verwendet worden (Hahofer et al.).

Bei Th-Bestimmung mit Kationenaustauscher (SACKETT et al.) wird mit Oxalsäure eluiert. α-Zählung über längere Zeit gibt Anhaltspunkte für die Beiträge der einzelnen Isotope.

Mittels Neutronenaktivierung lassen sich die Nachweisgrenzen zu erheblich kleineren Werten verschieben. Bei Untersuchungen an Meteoriten [BATE (2)] konnte ein Th-Gehalt von $6 \cdot 10^{-12}$ g in der Probe bestimmt werden. Jeweils ein Stück von etwa 1 g wurde in einem Quarzgefäß 3 Wochen lang in einem Neutronenfluß von $3 \cdot 10^{14}$/cm² sec aktiviert zusammen mit einem Monitorgefäß, in das eine Standard-Th-Lösung eingedampft war. Die radiochemische Analyse erfolgte über Pa^{233} (27 Tage Halbwertszeit), das Folgeprodukt des bei der Akti-

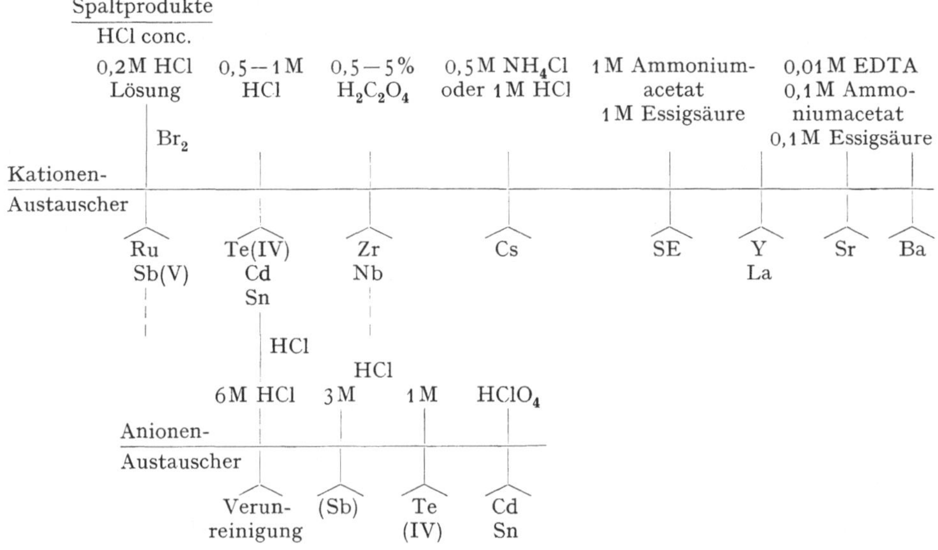

Fig. 5. Trennung von Spaltprodukten mit Ionenaustauschern (MINAMI et al.)

vierung entstehenden Th^{233} (23 min). Bei der U-Bestimmung mittels Neutronenaktivierung kann man die U-Spaltungen mit der Ionisationskammer zählen [FACCHINI (1)]. Die Bestimmung ist auch in Anwesenheit von Th möglich, weil Th^{232} nicht durch thermische Neutronen gespalten wird (I. JOLIOT-CURIE). Besser ist die Aktivierung an einzelnen Spaltprodukten zu messen, z.B. Ba^{139} (BRODA), Ba^{140} (SMALES), Te^{132} (FISHER et al.). REED u. a. benutzten Np^{239} als Nachweis für U^{238} und Ba^{140} für U^{235}. Die Nachweisgrenze dürfte prinzipiell in derselben Größenordnung liegen wie bei Th. Doch leidet die Genauigkeit unterhalb 10^{-10} g/g. Als Monitoren wurden benutzt Lösungen der hauptsächlichsten Metalle der Probe zusammen mit bekanntem U-Zusatz in Quarzgefäß oder ein Stück Gestein mit einem Bohrloch, in das die U-Lösung gefüllt wurde. Eine Auftrennung aller Produkte der U- und Th-Reihe gibt ROSHOLT. Die Trennung von Pb, Bi, Po, die üblicherweise zusammen in den Proben enthalten sind, kann auf verschiedene Weise erfolgen, z.B. durch Anionenaustausch (ISHIMORI, KIRSCH), Papierchromatographie (FRIERSON et al.), Dithizon-Extraktion (BOUSSIÈRES et al.), Elektro-Abscheidung, von Po auch durch Verflüchtigung.

Noch wesentlich komplizierter liegen die Dinge bei den von der kosmischen Strahlung erzeugten Kernarten und vor allem bei den Spaltprodukten. Für die erstgenannten, die wegen ihrer geringen Konzentrationen immer chemisch abgetrennt werden müssen, werden jeweils mehr oder weniger geläufige Verfahren

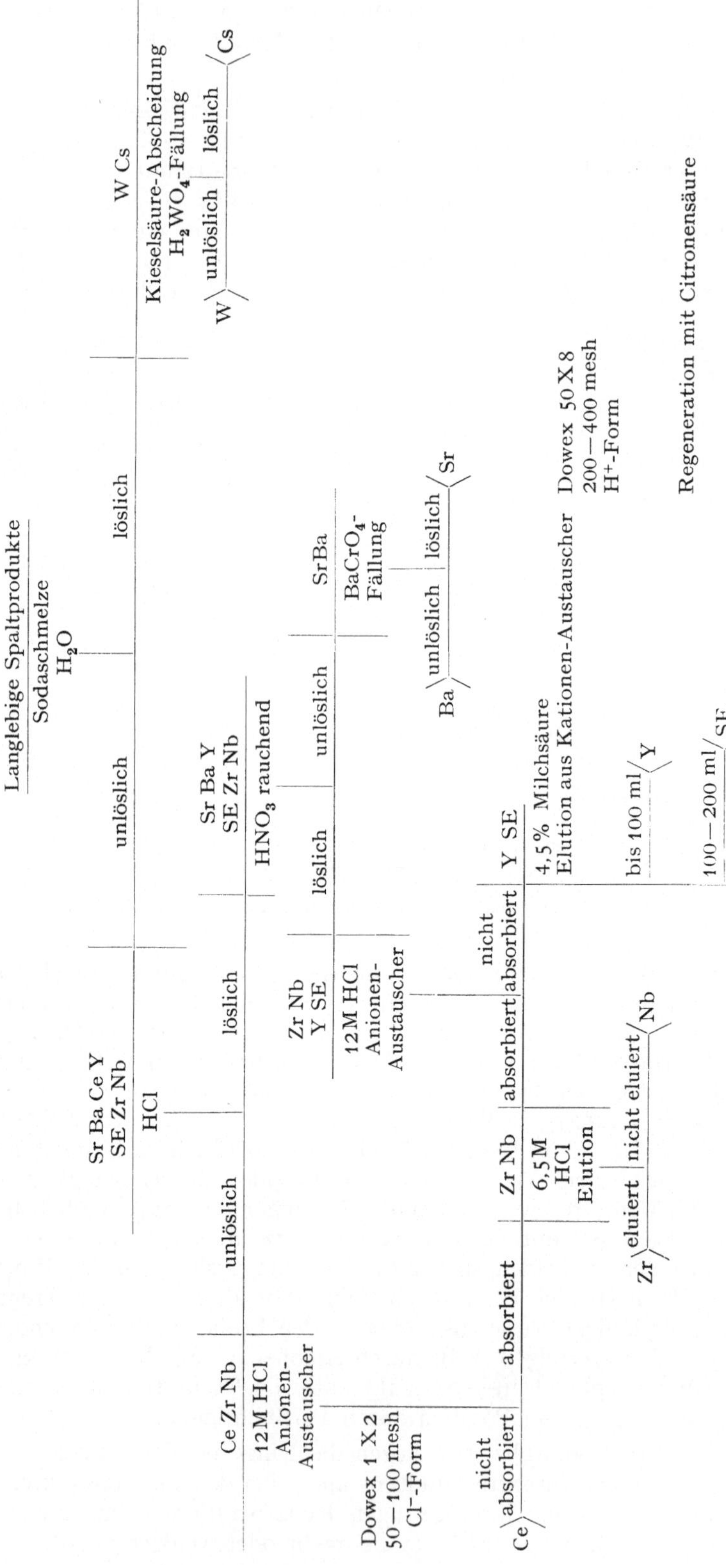

Fig. 6. Trennung langlebiger Spaltprodukte und des W^{185} [WELFORD (1)]

der analytischen Chemie für den Einzelfall benutzt, z.B. Reinigung des Be durch TTA- [GOEL (1)] oder Chloroform- (MCMILLAN) Extraktion, Bestimmung von S^{35} als elementarer Schwefel [GOEL (2)], von Cl-Isotopen als AgCl (WINSBERG), der P-Isotope als Magnesium-Ammonium-Phosphat (MARQUEZ et al.).

Die Chemie der Spaltprodukte mußte zum guten Teil völlig neu entwickelt werden (CORYELL u. SUGARMAN). Da in der Geophysik die langlebigen das Hauptinteresse haben, werden hier nur einige Hinweise auf solche gegeben. Sr^{90} u. Sr^{89} haben vor allem für Untersuchungen atmosphärischer Vorgänge Bedeutung. Die Abtrennung des Sr erfolgt zunächst mittels Fällung durch rauchende HNO_3 (HAHN u. STRASSMANN). Dabei werden außer Sr nur Ba, Ra, Pb, Tl gefällt. Ferner bleibt etwas Ca im Niederschlag, das besonders bei calciumreichen Proben stören kann. In der Probe können auch Ba^{140} mit dem Folgeprodukt La^{140} und verschiedene Ra-Isotope enthalten sein ebenso wie Pb^{210}. Abtrennung des Sr von den übrigen Elementen erfolgt z.B. durch Chromatfällung und Bestimmung als Strontium-Oxalat (GLENDENIN) oder -Carbonat (HOAGLAND) oder mit Kationenaustauscher (EULITZ). Für Untersuchung von Bodenproben stehen mehrere Aufschlußverfahren zur Verfügung: Extraktion mit HCl, Sinteraufschluß mit Soda (besonders bei sulfathaltigen Böden), Ammoniumacetat-Extraktion (F.J. BRYANT et al., MARTELL, HAMADA et al.). Besonders schwierig ist die Abtrennung von Sr aus Böden mit hohem Ca-Gehalt. Die Messung des Sr^{90} erfolgt zweckmäßig mittels des nach Einstellung radioaktiven Gleichgewichts abgetrennten Y^{90}, das bequem zu identifizieren ist. Zur Sr^{89}-Bestimmung wird das bei der chemischen Trennung erhaltene Sr-Präparat über mehrere Wochen gemessen und der Sr^{89}-Gehalt durch Differenzbildung aus der Gesamtaktivität und der über Y^{90} bestimmten Sr^{90}-Konzentration ermittelt. Ionenaustauscher sind auch für die Auftrennung eines ganzen Spaltprodukt-Gemisches verwendet worden (MINAMI et al.), zum Teil in Verbindung mit der normalen analytischen Gruppentrennung (ISHIBASHI et al.). Ein für die Routine-Messung langlebiger Spaltprodukte sowie des bei einigen amerikanischen Bomben-Explosionen als Tracer zugesetzten W^{185} angegebenes Trennschema gibt Fig. 6 [WELFORD (1)]. Weitere Untersuchungsmethoden für langlebige Spaltprodukte finden sich z.B. bei OSMOND et al. (Sr, Cs, Ce), für langlebige Spaltprodukte und langlebige natürliche Nuklide (Pb^{210}, Be^7) bei BAUS et al. und bei LOCKHART et al.

IV. Probenahme

1. Allgemeine Gesichtspunkte

Für geophysikalische Untersuchungen müssen die Proben möglichst repräsentativ sein. Alle geophysikalisch unwesentlichen Einflüsse sollen bei der Probennahme tunlichst ausgeschaltet werden. Hierzu gehören ausgesprochen lokale Bedingungen, seien sie geologischer oder orographischer, natürlicher oder anthropogener Art.

Bei Entnahme von Proben der atmosphärischen Luft in Bodennähe besteht die Gefahr, daß man sich in der Nähe von natürlichen oder künstlichen Quellen eines radioaktiven Nuklids befindet. Auf tiefreichenden Gesteinsspalten können anomale Emanationsmengen an die Oberfläche treten. Die Industrie kann radioaktives Material verbreiten, ohne daß sie bestehende gesetzliche Vorschriften verletzt, weil es hier auf winzige Mengen ankommt. Zum Beispiel wird man in der Umgebung einer Pb-Zn-Hütte immer eine Kontamination von RaD, E, F finden. Es gibt auch den umgekehrten Fall, daß die normale Konzentration an radioaktivem Material durch lokale Quellen verdünnt wird. Natürliche Quellen von sehr altem CO_2, in dem C^{14} bereits abgeklungen ist, oder auch anthropogene

Quellen wie Abgase von Kohle und Öl, deren Alter ebenfalls groß ist gegen die Halbwertszeit des C^{14}, können den natürlichen Gehalt der Atmosphäre an C^{14} verringern. Das gilt auch für die Biosphäre in diesem Bezirk, da diese in ständigem Austausch mit der Atmosphäre steht.

Die Zeit der Probenahme muß ebenfalls hinsichtlich Allgemeingültigkeit geprüft werden. So spielen in der Atmosphäre Tages- und Jahresgänge sowie meteorologische Bedingungen eine Rolle. Zum Beispiel ist die Rn-Konzentration bei ungestörten Witterungsverhältnissen in den frühen Morgenstunden am höchsten, die stratosphärische Fallout-Komponente im Frühjahr. Für die durch die kosmische Strahlung erzeugten Nuklide sind die periodischen und nichtperiodischen Schwankungen der erzeugenden Strahlung wichtig.

Bei inhomogener Verteilung der Aktivität in der Probe ist diese unter Umständen nicht mehr repräsentativ für die Bestimmung einer mittleren Aktivitätskonzentration. Bei Gesteinsproben kann fast die ganze Aktivität in wenigen Körnern U-Mineral enthalten sein. Wasser- und Bodenproben können kleinste Lebewesen enthalten, die bestimmte radioaktive Stoffe anreichern. Aerosolproben aus der Atmosphäre enthielten nach gewissen Atombomben-Explosionen Einzelteilchen extrem hoher spezifischer Aktivität.

Bei einer aus meßtechnischen Gründen manchmal erwünschten Unterteilung von Proben ist daher Vorsicht geboten. Selbst die Verteilung in chemischen Niederschlägen kann unregelmäßig sein. Höchstens bei Lösungen ist man sicher, wenn nicht Adsorption oder Ionenaustausch an den Gefäßwänden stattfindet oder Radiokolloide auftreten.

Für globale Abschätzungen ist die räumliche Verteilung der Proben schwierig. In der Atmosphäre sind weitaus die meisten Proben auf die bodennahen Schichten beschränkt. Für größere Höhen reicht bisher weder die Zahl der entnommenen Proben, noch konnten die Entnahmestellen und -zeiten hinreichend systematisch ausgewählt werden. Auch für die Hydrosphäre ist die Verteilung der Entnahme-Stellen und -Zeiten unregelmäßig und im ganzen gegenüber Atmosphäre und Lithosphäre zu gering, vor allem was die Weltmeere angeht. Bei der Lithosphäre ist man weitgehend beschränkt auf die Erdoberfläche. Entnahme aus größeren Tiefen können nur mit praktischen Arbeiten, z.B. Ölbohrungen, Bergwerken, Tunnelbauten, kombiniert werden und sind daher auf wenige Stichproben beschränkt.

Bei der Probenahme muß die Aktivitätsmeßmethode und ihre Empfindlichkeitsgrenze berücksichtigt werden. Da die Konzentrationen sehr gering sind, muß man oft an die äußerste Grenze der Probemenge gehen, die man in der Verarbeitung beherrscht. Zum Beispiel sind Bodenproben bis zu 1 kg für manche Spaltprodukte erforderlich, Niederschlagsproben von mehreren m^3 bei der Suche nach seltenen von der kosmischen Strahlung erzeugten Nukliden verwendet worden.

2. Gasproben

Bei der Messung des Rn- und Tn-Gehaltes der Atmosphäre wird eine entsprechende Luftmenge unter den nötigen Vorsichtsmaßnahmen hinsichtlich Reinigung und Trocknung der Meßkammer zugeführt. Man kann gegebenenfalls das Rn unter Anreicherung isolieren und mit dem Füll- bzw. Zählgas in die Meßkammer überführen.

Zur Untersuchung des Emanationsgehaltes der Bodenluft kann man die Kollektor-Elektrode der Ionisationskammer direkt in ein Bohrloch einführen. Sicherer ist das Abpumpen der Luft aus dem Bohrloch in eine Kammer üblicher Bauart. Bei durchfeuchtetem Boden können Fehler auftreten infolge Sperrung der Bodencapillaren durch das Wasser und dessen lösende Eigenschaften für die Emanation. Die Exhalation des Bodens läßt sich dadurch messen, daß ein Luftstrom von einigen cm^3 pro sec zwischen der Bodenoberfläche und einer in etwa 1 cm Abstand darüber angebrachten Platte hindurch zur Reinigungsanlage und dem Meßgerät geleitet wird, wobei er das exhalierte Rn mitnimmt.

Die in Wasser enthaltene Emanation wird mit Luft oder besser N_2 oder mit dem Zählgas ausgeperlt und nach entsprechender Reinigung in die Meßkammer überführt.

Zur direkten Entnahme des C^{14} aus der Luft wird Barytlauge verwendet (RAFTER et al.). Mit einer absorbierenden Oberfläche von 900 cm^2 einer 0,5 m NaOH-Lösung und 2 mMol $BaCl_2$/l erhält man in drei Tagen 300 mMol $\pm$ 10% CO_2 [MÜNNICH (2)]. Man ist allerdings bei C^{14} nicht auf die direkte Probennahme angewiesen. Wegen des ständigen Austausches zwischen Atmosphäre und Biosphäre bildet der C^{14}-Gehalt der Pflanzen ein unmittelbares Maß für den der Atmosphäre. Die Entnahme von C^{14}-Proben aus Ozeanwasser wird von BROECKER et al. beschrieben.

Zur Entnahme des H^3 aus dem atmosphärischen Wasserstoff muß dieser verlustfrei gewonnen werden. Bei der Luftverflüssigung bleibt neben Ne und He auch der Wasserstoff unverflüssigt. Wenn man die Ne-He-Fraktion zusammen mit dem Wasserstoff über Kupferoxyd leitet, wird dieser zu Wasser oxydiert. In den meisten anderen geophysikalisch interessanten Fällen liegt das H^3 schon als Wasser (HTO) vor, z.B. in Niederschlägen, Oberflächen- und Grundwasser, und kann in der üblichen Weise verarbeitet werden.

3. Aerosolproben

Für Abscheidung von Aerosolen aus der Atmosphäre zur Untersuchung auf Radioaktivität kommen hauptsächlich Schwebstoffilter und elektrische Präzipitatoren in Betracht. Nur für Spezialfälle ist gelegentlich der Impactor benutzt worden (JECH, DREVINSKY et al.).

Als Schwebstoffilter [HAXEL (2), SCHUMANN (3)] eignen sich Faser- und Membranfilter. Gute Faserfilter enthalten heute Fasern $< 1\ \mu$ Durchmesser, oft sogar 0,1 μ und darunter, je nachdem ob es sich um Asbestfasern handelt, die in Cellulose eingelagert werden, oder um Glas- oder Kunststoffasern. Das Filter kann fest oder beweglich gegenüber der Saugöffnung exponiert werden (diskontinuierliche oder kontinuierliche Probenahme). Membranfilter sind Trockengele aus Celluloseestern. Die Porendurchmesser sind bei den einzelnen Typen ziemlich einheitlich (etwa auf einen Faktor 2 um den Mittelwert) und in der Herstellung über einen weiten Bereich zu variieren (etwa 0,1 bis 1 μ). Die Filter sind in einigen organischen Lösungsmitteln (Aceton, Methylacetat, Ketone) löslich. Geringe Mengen Lösungsmittel, aber auch z.B. Kollodium-Lösung oder Immersionsöl machen die Filter transparent. Für chemische Verarbeitung sind sie in Anbetracht des verschwindenden Aschegehaltes besser geeignet als Asbest- und Glasfaserfilter (GOETZ).

Der Abscheidemechanismus in Schwebstoffiltern ist sehr komplex. Gravitationswirkung scheidet bei dem für geophysikalische Aktivitätsuntersuchungen nötigen hohen Luftdurchfluß praktisch aus. Einfache Siebwirkung, weil die Teilchen größer sind als die Poren, spielt nur bei Membranfiltern eine wesentliche Rolle. Beim Sperreffekt (direct interception) folgt das Teilchen den hydrodynamischen Stromlinien, doch ist sein Radius r größer als der Abstand der Stromlinie von der abscheidenden Oberfläche. Beim Trägheitseffekt erfolgt Abscheidung, weil das Teilchen wegen hoher Masse oder hoher Geschwindigkeit den Stromlinien in den stark verwinkelten Filterporen bzw. um die wirr durcheinander liegenden Filterfasern nicht folgt. Bei sehr kleinen Teilchen und nicht zu hohen Geschwindigkeiten erfolgt Abscheidung durch Diffusion. Über etwaige Beteiligung elektrostatischer Effekte ist weder bei Faser- noch bei Membranfiltern etwas bekannt.

Der Abscheidegrad η_f ist eine Funktion von r und der Anström-Geschwindigkeit v. Alle bisher in genügend weiten Bereichen untersuchten Filter haben ein Minimum von η_f für ein bestimmtes r bei konstantem v und für ein bestimmtes v bei konstantem r. Zwischen dem Bereich des Trägheitseffekts und dem der Diffusion liegt ein für die Abscheidung weniger günstiger Bereich.

Bei geophysikalischen Untersuchungen ist das Aerosolspektrum sehr breit (Junge). Es ist anzunehmen, daß sich die aktiven Atome an Teilchen aller Größen anlagern. Selektive Teilchengrößen-Effekte (Wilkening) bei der Anlagerung konnten nicht bestätigt werden [Jacobi (1), Lassen (2)]. Man müßte η_f mindestens von 1 µ abwärts bis in die Größenordnung 0,01 µ kennen. Überprüfungen in Abhängigkeit von r erfordern monodisperse Aerosole. Bei Prüfung mit normalem atmosphärischem Aerosol kann man gegebenenfalls zusätzliche Beladung mit Emanationsfolgeprodukten vornehmen [Lassen (1)]. Freilich ist das Größenspektrum des atmosphärischen Aerosols eine Funktion von Ort und Zeit. Trotzdem scheint die Methode, wenn die Prüfung am Meßort unter durchschnittlichen Bedingungen vorgenommen wird, akzeptabel.

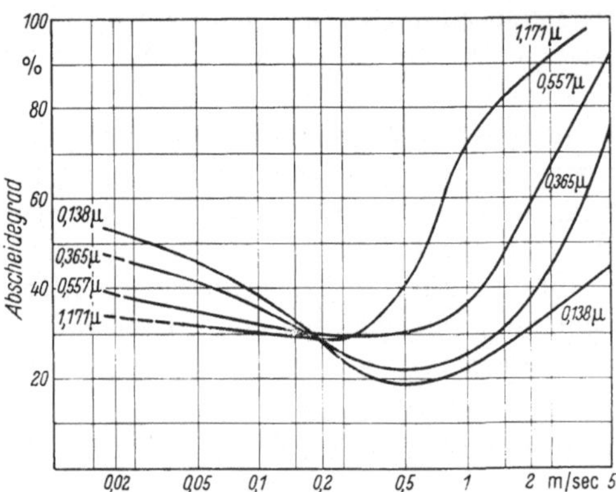

Fig. 7. Abscheidegrad eines Schwebstoffilters für verschiedene Teilchengrößen (Polystyrol-Kügelchen) und Geschwindigkeiten (IPC-Filter) (Stern et al.)

Die Möglichkeit selektiver Effekte bei der Abscheidung hinsichtlich chemischer Natur, Oberfläche usw. der Teilchen kann nicht ausgeschlossen werden. Darüber ist sehr wenig bekannt, da genaue Untersuchungen nur mit speziellen Testaerosolen durchgeführt worden sind, z.B. Dioctylphthalat, Latex, bestimmten Bakterien (La Mer, Stern).

Hohes η_f sichert auch bei starken Durchlässigkeits-Schwankungen (Faktor 2 Änderung von 99,5 auf 99,0%) etwa infolge Schwankungen im Teilchengrößenspektrum, der chemischen Zusammensetzung des Aerosols, der Luftfeuchtigkeit usw. vergleichbare Ergebnisse. Die Konzentration als Funktion der Eindringtiefe kann als Exponentialfunktion vorausgesetzt werden. Die Halbwerts-Eindringtiefe wurde an Faserfiltern für repräsentative Bedingungen zu 1 bis 2 mg/cm² ermittelt [Schumann (1)]. Sie ist für die Selbstabsorption der zu messenden Aktivität wichtig. Bei Membranfiltern ist sie noch geringer. Hinzu kommt der Einfluß der Grobstaub-Konzentration und der Saugzeit bzw. ihr Produkt. Die Staubansammlung erhöht nicht nur die Dicke der Schicht, in der sich die Aktivität befindet. Sie verursacht auch Herabsetzung des Luftdurchflusses sowie Verengung der Filterporen und damit Änderung der Strömungsgeschwindigkeit, also der Abscheide-Charakteristik. In Fällen, wo solche Änderungen bedenklich werden, muß man den Luftdurchfluß durch regeltechnische Maßnahmen konstant halten.

Eine unangenehme Begleiterscheinung hoher η_f ist großer Luftwiderstand. Der Druckabfall kann bei Durchflußgeschwindigkeiten von 1 m/s bereits mehrere

Zehntel Atmosphären betragen. Besonders hoch sind die Werte für Membranfilter, sonst streuen sie je nach Filtertyp über einen großen Bereich.

Bei der elektrostatischen Abscheidung [ALIVERTI (1)] ist der einfachste Fall die Zylindersymmetrie, man kann aber als Abscheide-Elektrode ebenso eine ebene Fläche benutzen, was z.B. die Verwendung einer Kernphotoplatte an dieser Stelle ermöglicht [ALIVERTI (2)]. Geräte mit kontinuierlich ablaufendem Metallband als Abscheide-Elektrode sind ebenfalls gebaut worden (BERGSTEDT).

Aufladung der Aerosolteilchen erfolgt mittels Korona-Entladung. Die Ladung q, die in der Zeit t von einem kugelförmigen Teilchen mit Radius r und Dielektrizitätskonstante ε im elektrischen Feld E aufgenommen wird, ist

$$q = n\,e = E\left(1 + 2\,\frac{\varepsilon - 1}{\varepsilon + 2}\right)r^2\,\frac{t}{t + \tau}\,,$$

wo τ die Halbwertszeit für die Erreichung der Maximalladung und n die Zahl der Elementarladungen e ist (PAUTHENIER et al.). Bei Teilchen unter $1\,\mu$ ist ferner mit Aufladung durch Ionendiffusion zu rechnen, wobei die Ladung in erster Näherung proportional r wird. Die Diffusionstheorie setzt allerdings längeren Aufenthalt im Ionisationsraum voraus, was bei den hier interessierenden Abscheidern nicht erfüllt ist, wo er nur in der Größenordnung 10^{-2} sec liegt. Deshalb reicht der Bereich überwiegend zu r^2 proportionaler Aufladung wohl bis zu kleineren Teilchengrößen, als man theoretisch erwarten würde. Jedenfalls wurde das Vorherrschen der r^2-Beziehung noch bis herunter zu $r = 0{,}26\,\mu$ experimentell verifiziert (FUCKS et al., GOYER et al.). Die Abscheidelänge z gemessen von der Korona ab ist für Teilchengrößen, bei denen vorherrschend Beladung durch Ionenbewegung im elektrischen Feld erfolgt, proportional $1/r$. Mit zunehmender Aufladung durch Diffusion weicht sie von dieser Proportionalität nach kleineren Werten ab.

Der Abscheidegrad eines elektrischen Präzipitators ist (W. DEUTSCH)

$$\eta_e = \frac{n_0 - n}{n_0} = 1 - e^{-S\,v/Q}$$

n_0, n Zahl der Teilchen pro Volumeneinheit durchgesaugte Luft am Eingang bzw. Ausgang, S Abscheidefläche, v Teilchengeschwindigkeit, Q Luftdurchfluß. Theoretisch läßt sich nahe 100% erreichen, und für gewisse ausgeführte Konstruktionen werden solche η_e genannt. Sie beziehen sich jedoch fast durchweg auf Geräte zur Luftreinigung, wo η_e aus dem Gewicht des abgeschlossenen Materials bestimmt wird, oder auf sehr kleinen Luftdurchfluß. Für kleine Teilchen, wie sie im normalen atmosphärischen Aerosol vorherrschen, läßt sich bei vernünftigen Feldstärken und Abmessungen nur ein wesentlich niedrigeres η_e realisieren, das oft bei 50% und darunter liegt [BERGSTEDT, LANDT, JACOBI (2)].

Ungünstig für die Aktivitätsmessung ist es zudem, daß es bei schwachen Konzentrationen schwierig ist, auf für den Detektor hinreichend kleinen Flächen die Aerosole einer hinreichend großen Luftmenge niederzuschlagen. Es können auch schwer erfaßbare Schwankungen von η_e auftreten dadurch, daß kleine Teilchen überhaupt nicht auf die Abscheide-Elektrode gelangen oder große wieder weggerissen werden. Geringe Schwankungen von Q können empfindlich stören. Ein großer Vorteil des elektrischen Abscheiders ist dagegen der praktisch vernachlässigbare Druckabfall, der es im Gegensatz zum Schwebstofffilter ermöglicht, mit geringem Energieaufwand sehr große Luftmengen zu verarbeiten. Die Möglichkeit von selektiven Effekten bei der Abscheidung kann hier ebenso wie beim Filter nicht ausgeschlossen werden. Es ist schon vorgekommen, daß Meßergebnisse aus solchen Gründen unzulässig verallgemeinert wurden (WILKENING).

Verschiedentlich wird als Vorteil des elektrostatischen Abscheiders hervorgehoben (Hultqvist), daß die abgeschiedene Aerosolschicht auf der glatten Fläche der Abscheide-Elektrode besser definiert ist als auf einem Schwebstofffilter, in das die Aerosole teilweise eindringen, und daß die Selbstabsorption geringer ist. Das gilt, solange die Aerosolkonzentration niedrig ist. Wenn die abgeschiedene Schichtdicke ein gewisses Maß überschreitet, geht dieser Vorteil weitgehend verloren, weil die Dickenzunahme, die in erster Linie durch das Grobaerosol bedingt ist, bei Filter und Elektroabscheider praktisch gleichmäßig erfolgt.

4. Niederschlag und Wasser

Das Sammelgefäß soll möglichst chemisch neutral sein und keine Angriffspunkte für Reaktionen, Adsorptionserscheinungen usw. bieten. Als Material geeignet ist besonders Polyäthylen, daneben andere Kunststoffe oder Metalle mit Kunststoff- oder Ceresin-Überzügen, ferner rostfreier Stahl.

Als Auffanggefäße für Niederschläge finden Becken und Trichter Verwendung. Beim Trichter sind die Spritzverluste geringer. Erfahrungsgemäß gibt jede Gefäßform etwas andere Meßwerte. Es gibt keine Idealform, sondern man kann nur eine Konvention treffen, um vergleichbare Werte zu erhalten, genau wie man sich auf eine bestimmte Form des meteorologischen Regenmessers geeinigt hat. Der genormte Regenmesser hat nur mit 200 cm² eine zu kleine Auffangfläche für Aktivitätsuntersuchungen. Die Regenmenge für die Bestimmung der Niederschlagsaktivität pro m² Erdoberfläche ist in jedem Fall der meteorologischen Messung zu entnehmen, da die Niederschlagsmengenwerte der anderen Auffanggefäße sehr streuen. Um Verdunstungsverluste zu vermeiden, sorgt man wie beim üblichen Regenmesser für ständigen Abfluß aus dem Auffangbecken in eine Vorratsflasche.

Vergleichsversuche mit verschiedenen Gefäßformen über 1 Jahr [Welford (2)] ergaben nur Abweichungen von etwa ±5%. Im allgemeinen wird man mit etwas höheren Differenzen rechnen müssen. Bei dem Vergleich wurde die vermutlich bessere Anreicherung durch Ionenaustauscher benutzt. Bei Anreicherung durch Eindampfen können die Meßwerte im allgemeinen nur eine Genauigkeit von 20% beanspruchen (Sittkus).

Da Niederschläge stets unlösliches Material enthalten, das leicht an den Gefäßwänden haftet und Aktivität enthalten kann, muß man die Gefäße bequem mechanisch und chemisch reinigen können. Die relative Verteilung der zu untersuchenden Aktivität auf gelöste und unlösliche oder an unlöslichem Material adsorbierte Anteile schwankt an vielen Orten stark [Schumann (5)]. Daher ist es im allgemeinen abgesehen von der Untersuchung spezieller Nuklide nicht möglich, sich auf den gelösten oder ungelösten Anteil zu beschränken. Von einzelnen Nukliden ist bekannt, daß sie praktisch nur in löslicher Form vorkommen, so von Sr⁹⁰ und Sr⁸⁹ (Martell).

Es gibt noch eine sehr primitive Methode, langlebige Niederschlagsaktivität zu sammeln, indem man geeignet präparierte Haftfolien (gummed paper, z.B. Unterlage Celluloseacetat mit Kleber auf Gummibasis) auslegt, auf denen sich der Staub aus der Luft absetzt und die Aktivität aus Niederschlägen, sofern deren Intensität nicht zu groß ist [etwa 60% der in den Niederschlägen gemessenen Aktivität in USA (Eisenbud et al.)]. Über längere Zeiten läßt die Haftfähigkeit nach. Das Verfahren ist sehr ungenau, hat sich aber für Übersichts-Messungen in Beobachtungsnetzen mit vielen Stationen und zentraler Auswertung bewährt. Für das Auffangen des trocken aus der Luft herabfallenden Staubes ist eine Wasserfläche sehr geeignet. Die Trennung zwischen Niederschlag und Staub kann durch eine Automatik bewirkt werden, die bei Niederschlagsbeginn das Staubfangbecken schließt und das Niederschlagsbecken öffnet und bei Niederschlagsende umgekehrt [KEGELMANN (1)].

5. Mineral- und Bodenproben

Die Verteilung der aktiven Substanzen in einem Gestein kann ziemlich gleichmäßig sein oder sich auf die Oberfläche von Einzelkörnern oder auf isolierte Körner beschränken. Prüfungsmöglichkeit bietet die Autoradiographie. Solche Inhomogenitäten können nicht nur die Ermittlung eines repräsentativen Mittelwertes der Aktivität beeinträchtigen, sondern auch die Messung selbst je nach der Detektor-Geometrie.

Aus diesen Gründen sind pulverförmige Proben günstiger als Schliffe. Sie lassen sich auch, was für α-Spektroskopie wichtig ist, in dünnen Schichten ausbreiten. Beim Zermahlen von Gesteinsproben entweicht ein Teil der Emanation. Verwendet man eine Meßmethode, bei der die Emanation oder ihre Folgeprodukte einen Beitrag liefern, so muß man die pulverisierte Probe in einen gasdichten Behälter schließen und warten, bis sich das radioaktive Gleichgewicht hinreichend eingestellt hat.

Die Frage des Gleichgewichts in den Zerfallsreihen muß in jedem Einzelfall geprüft werden.

Besteht Gleichgewicht zwischen U und Ra, oder will man speziell den Ra-Gehalt bestimmen, so wird man meist die Emanation messen. Für die Gewinnung aus Gesteinen gibt es verschiedene Verfahren. Die Emanation wird frei beim Erhitzen auf etwa 2000° C oder beim Schmelzen mit Alkalikarbonaten. Durch Erhitzen mit Ammoniumbifluorid in Au-Gefäßen läßt sich eine Gesteinsprobe soweit zersetzen, daß die Emanation frei wird. Schließlich kann man das Gestein quantitativ lösen und die Emanation aus der Lösung in der üblichen Weise (Durchperlen z.B. mit N_2 oder Zählgas) gewinnen.

Bei Untersuchungen von lockeren Böden kann man eine repräsentative Probe dadurch erzielen, daß man viele kleine über eine gewisse Fläche verteilte Proben zusammenfaßt. Bei der Entnahmetiefe muß berücksichtigt werden, daß landwirtschaftliche Nutzung eine jährliche Durchmischung mindestens der obersten 20 cm mit sich bringt. Außerdem spielt die Vegetation eine Rolle, da einmal gewisse Pflanzen dem Boden bevorzugt bestimmte Stoffe entziehen und andererseits manche Pflanzen aktives Material des Fallout direkt durch ihre oberirdischen Teile aufnehmen.

V. Messungen an Gasen

1. Edelgase

Die älteste Meßmethode für Emanation ist die Ionisationskammer-Messung. Die kurzlebigen Folgeprodukte des Rn haben eine effektive Halbwertszeit von etwa 30 min, stellen also keine bleibende Kontamination dar. Von den langlebigen Folgeprodukten hat man erst nach längerem Gebrauch der Kammer nachweisbare Effekte zu erwarten, weil die Halbwertszeit des Pb[210] etwa 20 Jahre beträgt. Trocknung des Rn bzw. der Rn-haltigen Luft vor Einführung in die Kammer ist erforderlich.

Wenn die Emanation nach Abfilterung der stets an Aerosole angelagerten Folgeprodukte in die Kammer gelangt, besteht kein radioaktives Gleichgewicht. Seine mehr oder weniger erfolgte Einstellung ist bei den Messungen zu berücksichtigen.

Absolutmessung ist schwierig, weil alle Abweichungen der tatsächlichen Ionisation von der bei voller Energieabgabe aller α-Teilchen theoretisch zu erwartenden genau untersucht werden müssen. Hierher gehören vor allem Wandeffekte, aber auch Säulenrekombination und dergleichen. Man eicht die Kammer deshalb zweckmäßig mit einem Standardpräparat, das meist einer Ra-Eichlösung entnommen wird. Auch dabei ist Vorsicht geboten, weil nach einiger Zeit die Emanierfähigkeit solcher Lösungen durch spurenweisen Ausfall von Radiumsalzen beeinträchtigt werden kann.

Die einfache Ionisationskammer mit Integralstrom-Messung erreicht die für geophysikalische Untersuchungen nötige Empfindlichkeit nur in Ausnahmefällen besonders hoher Rn-Konzentrationen, wie sie vor allem in Bodenluft (Bohrlöchern oder Bergwerken) auftreten. Mit 1 Liter Kammervolumen kommt man auf eine Nachweisgrenze von 10^{-9} C/m³ [Delibrias (1)]. Bei Durchströmungsbetrieb kann man Rn und Tn trennen. Bei Abstellen des Gasstromes sinkt der Meßwert mit der Halbwertszeit des Tn (54 sec) auf den vom Rn herrührenden Ionisationsstrom [H. Israël (2)]. Selbstverständlich muß die Zeitkonstante der Apparatur klein genug sein. Für Tn-Messung ist in jedem Fall Durchflußbetrieb nötig.

Die Doppelkammer mit Kompensationsschaltung ist in der Lage, noch $2 \cdot 10^{-11}$ C/m³ Rn zu messen. Evans benötigte dazu 13 Std Meßzeit, wobei der Nulleffekt der Einzelkammer sechs Ionenpaare je sec und cm³ betrug, von denen je ein Drittel auf die kosmische Strahlung, die γ-Strahlung der Umgebung und die α-Emission der Wände zurückzuführen waren. Das Emanometer von H. Israël (1) besteht aus zwei Kammern von je 21 l Volumen und gestattet ohne weiteres Absolutmessungen. Rajewski et al. erreichten die gleiche Nachweisgrenze mit Kammern von 10 l aus Edelstahl (s. a. Muth et al.: reinstes Elektrolytkupfer). Die letztgenannte Apparatur ist mit Registrierung durch Schwingkondensator-Elektrometer-Verstärker ausgerüstet.

Mit Impuls-Ionisationskammern kann man, wenn für extrem niedrigen Nulleffekt gesorgt wird, die Nachweisgrenze noch weiter nach unten verschieben. Mit einer Kammer aus korrosionsfreiem Stahl und einer Eigen-α-Emission von 0,3/cm² Tag ließ sich bei 20stündiger Meßzeit noch eine Aktivität von einigen 10^{-12} C/m³ bestimmen [Bate (1)]. Hier besteht die Möglichkeit, Tn in Anwesenheit von Rn dadurch zu bestimmen, daß man die Doppel-α-Impulse Tn-ThA registriert [Hurley (1)].

Auch der Proportionalzähler ist für Rn-Messungen geeignet. Eine sorgfältige Analyse gibt Hultqvist, der einen mit 1 at $A + CO_2$ (9:1) gefüllten Zähler zur Absolutmessung und Eichung von Ionisationskammern verwendet. Tn muß wegen seiner kurzen Halbwertszeit auch hier im Durchflußverfahren gemessen werden. Als Zählgas eignet sich in diesem Fall das für Durchflußzähler auch sonst häufig verwendete CH_4. Dasselbe Gas kann auch im abgeschlossenen Zähler zur Rn-Messung dienen (Begemann).

Der Szintillationszähler hat gegenüber Ionisationskammer oder Proportionalzähler den Vorteil der Unempfindlichkeit gegen Verunreinigungen sowohl gasförmiger Art als auch in Form von Aerosolen, sofern es sich um inaktive Stoffe handelt, die keine nennenswerte Absorption, Streuung oder Verkleinerung der Reichweite der α-Teilchen durch Erhöhung des Gasdrucks verursachen. Man kann also bei diesem Verfahren auch Rn in die Kammer bringen, das mit seinen Folgeprodukten im Gleichgewicht ist. Eine einfache zylindrische Kammer aus Glas, deren Innenseite mit ZnS, deren Außenseite mit einem Reflektor belegt ist, hat eine Nachweisgrenze von einigen 10^{-10} C/m³ bei 1 Std Meßzeit [Damon (1)]. Kunststoffe wie Plexiglas haben einen niedrigeren Nulleffekt, doch können durch

Eindiffundieren in solche Stoffe Rn-Verluste eintreten, die bei einer 120 cm³-Kammer in einigen Stunden etwa 1% ausmachten (van Dilla et al.).

Versuche zur Erhöhung der Ansprechwahrscheinlichkeit mit einer zylindrischen Messingkammer, in der koaxiale Celluloidrohre eingesetzt und alle inneren Oberflächen mit ZnS belegt waren, brachten keine Verbesserung gegenüber einfacheren Ausführungen (Malvicini). Dagegen wurde mit einer kugelförmigen Kammer aus Plexiglas von 15,6 cm Durchmesser (2 l Inhalt), die mit einem Fenster von 5 cm Durchmesser versehen und abgesehen von diesem Fenster vollständig mit ZnS belegt war, eine Nachweisgrenze von $7 \cdot 10^{-11}$ C/m³ $\pm 20\%$ bei 4 Std Meßzeit (Kaul) erreicht.

Ein tragbares Gerät für Schnellmessungen an Bodenluft oder sonstiger Luft mit Rn-Anreicherung, bei dem die Probe nach 40 sec automatisch wieder aus der Kammer ausgeschleust wird, hat eine Nachweisgrenze von $5 \cdot 10^{-8}$ C/m³ (Jehanno et al.).

Man kann bei solchen Szintillationszählern die positive Ladung der RaA-Atome ausnutzen und diese elektrostatisch abscheiden (J. Bryant et al.), bei kleinen Aktivitäten eventuell unter Zuhilfenahme einer Korona-Entladung (Drigo). Als Abscheide-Elektrode kann eine Al-Folie über der ZnS-Schicht dienen, oder man kann diese mit Al bedampfen. Als Nachweisgrenze wurde erreicht $1 \cdot 10^{-10}$ C/m³ bei 4 Std Meßzeit mit einer Genauigkeit von 20% (Kaul).

Zur Erhöhung der Nachweisempfindlichkeit bei allen Meßmethoden kann man natürlich das Rn in der Probe anreichern, z.B. durch Adsorption an Kohle oder Silicagel (Becker et al.) oder durch Ausfrieren.

Abgesehen von den Emanationen müssen hier noch die künstlichen Edelgasisotope erwähnt werden, die in der Atmosphäre auftreten können. Dazu gehören verschiedene Xe-Isotope, die allerdings alle recht kurzlebig sind, ferner von den Kr-Isotopen, die bei der Kernspaltung entstehen, vor allem das Kr⁸⁵ [de Vries (3)] wegen seiner Halbwertszeit von 10 Jahren sowie einige A-Isotope. Praktische Erfahrungen liegen abgesehen von Messungen für Reaktor-Sicherheits-Maßnahmen nur sehr wenige vor. Kr⁸⁵ wurde als Zählgas im Proportionalzähler gemessen sowie in kondensiertem Zustand die Maximalenergie des β-Spektrums mit Szintillationszähler ermittelt [Delibrias (2)]. A³⁹ wurde als Füllung mit Zusatz von Buten im Geiger-Zähler gemessen (Fireman) bzw. in einer Cuvette mit 7 mg/cm² Abdeckfolie mittels Fensterzähler (0,9 mg/cm²) (Wänke et al.) jeweils in Antikoinzidenzkranz. Im zweiten Fall erfolgte Identifizierung durch Absorptionsmessungen.

2. C¹⁴ und H³

Der Gaszähler ist heute im geophysikalischen Bereich das hauptsächlichste Meßgerät für C¹⁴ und H³. C¹⁴ wird dabei als CO_2, C_2H_2 oder CH_4 in ein Proportionalzählrohr gebracht und die Aktivität in einem Impulshöhenbereich gemessen, der mit Rücksicht auf das Spektrum des C¹⁴ und das des Nulleffekts eine möglichst günstige Ausbeute liefert. In CO_2 ist die Wanderungsgeschwindigkeit der Elektronen um eine Größenordnung niedriger als in CH_4 oder in Argon-Molekülgas-Gemischen und daher die Gefahr der Anlagerung an etwa vorhandene elektronegative Gase höher [Sharpe, Münnich (1)]. In C_2H_2 liegt die Sammelzeit zwischen den Werten für CO_2 und CH_4 [Scholz (1)]. CO_2 erfordert deshalb sorgfältige Reinigung, die Handhabung ist aber sicherer als bei C_2H_2 und einfacher als die Umwandlung zu CH_4. CH_4 hat außerdem den Nachteil, daß es sich nicht völlig kondensieren und daher nicht quantitativ in das Zählrohr überführen läßt. Man kann mit höheren Gasdrücken arbeiten, um die Ansprechwahrscheinlichkeit zu steigern. Bei CO_2 werden vielfach 3 at verwendet, auch über Versuche mit 10 at ist berichtet worden. Bei CH_4 bietet die Messung bei hohem Druck

Tabelle 8. C^{14}-Gaszähler

Volumen 1	Probe g C	Gas	Druck at	Rezentwert cpm	Nulleffekt cpm	Literatur
0,4	0,6	CO_2	3	5,6	3,9	DE VRIES (1)
0,7		CO_2	3	14,6	0,6	DE VRIES (4)
7,7	11,2	CO_2	3	103	14	FERGUSSON
0,7	6	CO_2	10	45,2	13,5	BRANNON
4	2,4	CO_2	1	24,5	16,5	MÜNNICH (1)
4,9		C_2H_2	1	65	12	KULP (2)
1		C_2H_2	1	15,8	2,3	SUESS
3	2,8	C_2H_2	1	64,5	30	CRATHORN
1,5		C_2H_2	1	15,2	0,8	HOUTERMANS (1)
0,9	0,85	CH_4	2	10	4,6	BURKE et al.
1	2,3	CH_4	5	30,6	10	DIETHORN

keinerlei Schwierigkeiten. In Tabelle 8 ist eine Übersicht über C^{14}-Gaszähler zusammengestellt, in Fig. 8 eine solche über Präparationsmethoden.

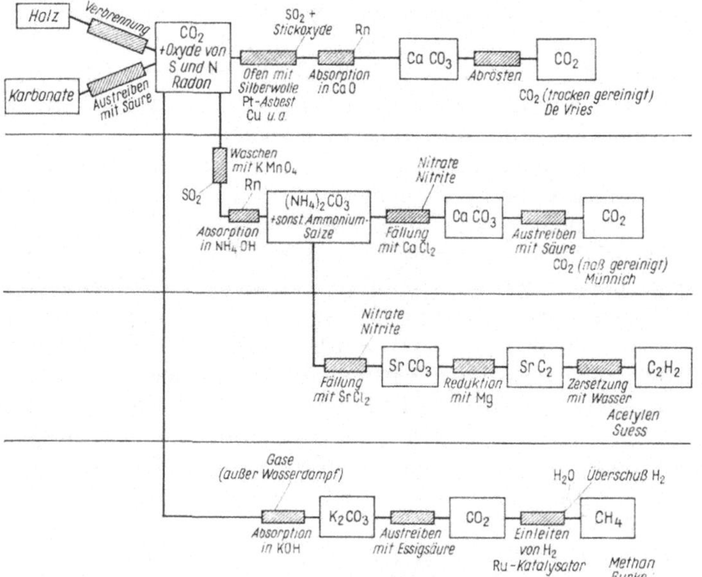

Fig. 8. Chemische Aufbereitung von C^{14}-Proben [erweitert nach HAXEL (4)]

Der ursprünglich für die C^{14}-Zählung von LIBBY u. Mitarb. benutzte screen wall counter (ANDERSON et al.), ein Geiger-Zähler, bei dem das Präparat aus festem Kohlenstoff bestand, ist durch die Entwicklung des Gaszählers ganz in den Hintergrund getreten sowohl wegen der umständlichen Probenpräparation und der besseren Ausbeute der Gaszähler als auch wegen der grundsätzlichen Vorteile des Proportionalzählers. Dagegen ist neuerdings die Messung mit flüssigen Szintillatoren, die ursprünglich bei Untersuchungen von organischen Substanzen mit künstlicher C^{14}-Markierung entwickelt worden war, in ihrer Empfindlichkeit und Betriebssicherheit so verbessert worden, daß sie auch für den geophysikalischen Bereich Interesse bekommen hat. In Tabelle 9 ist eine Übersicht zum Vergleich mit den Gaszählern gegeben. Die verwendeten Szintillatoren sind die auch sonst üblichen. Eine Komplikation gegenüber den Gaszählern besteht darin, daß das C^{14} in organische Stoffe überführt werden muß, deren Beimischung zum

Tabelle 9. *C^{14}-Zählung mit flüssigen Szintillatoren*

Einbau des C^{14} in	Ausbeute	Volumen	Probe	Ansprech-wahrschein-lichkeit	Rezent-wert	Null-effekt	Literatur
	%	cm³	g C	%	cpm	cpm	
33% Äthanol	50	100	14,2	25	54	26	ARNOLD
Hexan + Oktan in Toluol	30	100	4,7	25	182	26	ARNOLD
CH_3 an Toluol	50	20	2,3	50	15,5	3,3	FUNT (1)
Methanol in Toluol . .	50 bis 90	20	1,9	30	14,2	6,2	PRINGLE et al.
C_2H_2 in Toluol	65	80	7	45	45	40	AUDRIC et al.
p-Cymol		90	67		470	60	HAYES (2)
Paraldehyd in Toluol .		20	8		47,8	9,1	KOECHLIN et al.
Benzol	10	20	15	40		13	TAMERS

Szintillator möglich ist. Beim p-Cymol handelt es sich um einen leicht aus natürlichen Terpenen gewinnbaren Stoff, wie sie in Pflanzen weit verbreitet sind.

Durch Anreicherung des Isotops C^{14} in Thermodiffusionskolonnen konnte die
Empfindlichkeit des C^{14}-Nachweises um einen Faktor 10 erhöht werden (HARING
et al.). Der Aufwand lohnt sich allerdings nur in Ausnahmefällen.

Wegen der Isotopentrenneffekte, die bei den Kohlenstoffisotopen noch merklich sein können, sollte die C^{14}-Messung, sofern die Möglichkeit besteht, mittels
massenspektrometrischer C^{13}-Messung korrigiert werden. Die Korrekturen
erreichen jedoch im Höchstfall wenige Prozent und bleiben daher oft innerhalb
der Meßfehler.

Sehr viel wesentlicher können solche Korrekturen bei H^3 sein. Deshalb kommt
der H^2-Messung an den untersuchten Proben erhöhte Bedeutung zu. Bei Proben
geringer Aktivität ist eine Anreicherung des H^3 nicht zu umgehen. Sie kann erfolgen durch Elektrolyse (KAUFMANN et al.) oder Thermodiffusion (GONSIOR), wobei
andere Verfahren prinzipiell nicht ausgeschlossen, aber bisher kaum angewendet
worden sind. Die gewöhnlich in Form von HTO vorliegenden Proben werden an
Zn, Mg oder W bei hohen Temperaturen zu H_2 reduziert.

Für die H^3-Messung finden ebenfalls Gaszähler Verwendung, die teils als
Geiger-Zähler, teils als Proportionalzähler betrieben werden. H_2 selbst ist als
Zählgas nur verwendbar in Verbindung mit einem Dampf-Zusatz, wobei Toluol
gelegentlich bevorzugt Einsatz fand [FALTINGS (2)]. Man kann auch den Wasserstoff direkt in die Zählgas-Moleküle einbauen, etwa Äthan [FALTINGS (1)]. Mehr
Verwendung gefunden hat die Beimischung des H_2 zum Zählgas. Bei LIBBY
u. Mitarb. wurde Argon + Äthylen benutzt (WOLFGANG et al.) und z. B. mit 60 Torr A
und 18 Torr C_2H_4 bei einer Zumischung von 20 Torr H_2 ein Plateau von 100 V
Länge und 6%/100 V Steigung erhalten. Beim Proportionalzähler kann man
zusätzlich einen Bereich aussondern, in dem das Spektrum des H^3 gegenüber dem
des Nulleffekts am günstigsten gezählt werden kann. Zwischen 2 und 12 keV
(Maximalenergie des H^3 18 keV) betrug der Nulleffekt eines mit aktivitätsfreiem
CO_2 gefüllten, sorgfältig abgeschirmten Proportionalzählrohres nur 3 cpm.
Direkte Messung erlaubt dieses Zählrohr bis zu einem T:H-Verhältnis von 10^{-15}
herab, sonst muß Anreicherung stattfinden. Die Ansprechwahrscheinlichkeit
beträgt 70% (GONSIOR). Eine Mischung von 85% A + 15% CH_4 ist weniger
empfindlich gegen Elektronenanlagerung als CO_2. Damit kann man bei Aus-
blendung des Energiebereichs 0,2 bis 14 keV auf eine Ansprechwahrscheinlichkeit von 93% kommen (G. ISRAËL).

Mit der Entwicklung der Technik des Messens wäßriger Lösungen mit
flüssigen Szintillatoren haben diese auch für die H^3-Messung zunehmende Bedeutung gewonnen. Die Ansprechwahrscheinlichkeit bleibt noch merklich hinter der

des Gaszählers zurück. Eine Übersicht über die auf diesem Gebiet untersuchten Möglichkeiten gibt Tabelle 10.

Tabelle 10. *H^3-Zählung mit flüssigen Szintillatoren*

Lösungs- bzw. Verdünnungs-Mittel	Naph-thalin-Zusatz	H_2O-Anteil %	Ansprech-wahrschein-lichkeit %	Literatur
Dioxan	—	3	3,8	Farmer et al.
	+	20	8,6	Werbin et al.
Dioxan + Xylol.	+		12 bis 14	Jacobson et al.
Äthanol + Toluol	—	1	4,6	Hayes (1)
5 T. Xylol + 5 T. Dioxan + 3 T. Äthanol.	—	7,7	1,9	Kinard
	+	7,7	4,2	Kinard
20 T. Xylol + 9 T. Äthanol	±	3,5	2,8	Hayes (1)
6 T. Dioxan + 1 T. Anisol + 1 T. Dimethyloxyäthan	—	20	4	Davidson et al. Branson et al.

VI. Messungen an Aerosolen und Wasserproben

1. Arbeiten mit Aerosolproben

Eine Filterprobe kann zur Aktivitätsmessung entweder direkt an den Detektor gebracht oder vorher verascht oder chemisch aufbereitet werden. Im zweiten Fall kann man dem Präparat nahezu jede gewünschte Form geben und sich den Detektor beliebig wählen. Im ersten Fall wird man durch Abstimmung der Filterfläche auf den Detektor eine möglichst günstige Geometrie schaffen [Schumann (1)]. Für kontinuierlich bewegte Abscheideflächen kommt nur diese erste Methode in Betracht. Als Detektoren werden meist Geiger- oder Szintillations-Zähler verwendet. Doch kann man je nach dem speziellen Zweck der Messung Detektor und Verfahren wählen, z.B. auch spektroskopische Methoden. Weitgehend automatisierte Geräte für α-, β- oder γ-Messung sind gebaut worden, auch solche mit mehrfacher Messung in festen und variablen Zeitabständen, hauptsächlich für Spaltprodukte.

Die Autoradiographie hat Bedeutung für Untersuchung einzelner hochaktiver Aerosolteilchen. Quantitative Auswertung ist schwierig, weil sich die Aktivitätsbeträge pro Teilchen von 10^{-9} C abwärts über viele Größenordnungen erstrecken und eine Eichung nur bei den höchsten Werten leicht möglich ist [Kern, Schumann (7)]. Bei der Probenpräparation kann man in gewissen Größenbereichen die Zuordnung zwischen Schwärzung und Teilchen dadurch sicherstellen, daß man mit Membranfilter arbeitet, dieses nach Explosionsende unverrückbar mit dem Film verbindet und es nach der Filmentwicklung durchsichtig macht. Dann hat man im Mikroskop Teilchen und zugehörige Schwärzung direkt übereinander (Quan). Ähnlich kann man mit elektrostatisch auf Glas niedergeschlagenen Präparaten verfahren (La Rivière et al.).

Für Aktivitätsmessung an Filtern ist auch vorgeschlagen worden, das Filter mit flüssiger Szintillatorlösung zu befeuchten. Dabei wird als Lösungsmittel für die Szintillatorsubstanz statt des üblichen Tuluol das Monoisopropylbiphenyl empfohlen, weil es einen um einen Faktor 10^3 kleineren Dampfdruck hat [Funt (3)].

Die Aktivitätsmessung der elektrostatisch abgeschiedenen Proben erfolgt meist direkt, indem man die Abscheide-Elektrode ganz oder einen herausnehmbaren Teil an den Detektor bringt. Die Elektrode kann selbst als Detektor ausgebildet sein, z.B. eine Kernphotoplatte enthalten. Bei Konstruktionen mit kontinuierlich bewegtem Abscheideband wird dieses an einem oder mehreren

Detektoren vorbeigeführt wie bei Filtern. Eine Sonderkonstruktion benutzt analog zur Doppelionisationskammer zwei Präcipitatoren mit Bandelektroden zur Abscheidung aus normaler und eventuell kontaminierter Luft. Die Messung erfolgt mit α-Szintillatoren, wobei elektronisch die Differenz der Zählraten beider Kanäle gebildet wird. Das Prinzip der Anordnung (KNOWLES) ist in entsprechender Abwandlung auch für die Suche nach geophysikalisch interessanten α-Strahlern unter Ausschaltung des Untergrundes der kurzlebigen Rn-Folgeprodukte anwendbar.

Die Schwierigkeit der Messung an sehr großen Abscheideflächen läßt sich dadurch überwinden, daß man die Flächen mit einer Folie überzieht, die nach Exposition mit dem abgeschiedenen Material abgenommen und verascht wird (RIEZLER et al.).

Für Untersuchungen mit Photoverfahren sind Aerosolproben in Einzelfällen auch auf andere Weise hergestellt worden. SIKSNA benutzte die alte Aktivierungsmethode von ELSTER et al. und GEITEL et al., JECH den Impactor, ERDIK einfache Abscheidung aus einem vorbeistreichenden Luftstrom.

Die Übertragung der für Emanations-Folgeprodukte mittels Aerosol-Abscheidung gewonnenen Konzentrationswerte auf Rn oder Tn setzt radioaktives Gleichgewicht voraus. Zwar liefern die Filtermessungen die richtige Größenordnung für die Rn-Konzentration (in Heidelberg im Mittel über mehrere Jahre $2 \cdot 10^{-10}$ C/m³). Doch treten mindestens an gewissen Orten und zu bestimmten Zeiten Abweichungen vom Gleichgewicht auf [JACOBI (1)]. Nach amerikanischen Messungen wird die Abweichung bei Mittelwerten über Zeiten von der Größenordnung 4 Std und mehr auf maximal einen Faktor 2 geschätzt (HARLEY). Bei Tn ist wegen der Halbwertszeit-Verhältnisse kein Gleichgewicht mit ThB zu erwarten.

2. Messungen an Flüssigkeiten

Direkte Aktivitätsmessung an Flüssigkeiten bei kleinen Konzentrationen begegnet großen Schwierigkeiten, ganz besonders kontinuierliche Messung strömender Flüssigkeiten. α-Messung ist wegen der starken Absorption in der Flüssigkeit nur sehr beschränkt möglich. Für β- und γ-Messungen und geophysikalisch interessante Konzentrationen ist die Ansprechwahrscheinlichkeit der in der Kerntechnik üblichen Meßanordnungen (Tauchzähler, Becherzähler usw.) meist zu gering. Auch direkte Messung an Elutionslösungen beim Arbeiten mit Ionenaustauschern ist nur bei ausgesprochener Anreicherung möglich.

Für β-Teilchen ist die Absorption in Flüssigkeiten beträchtlich, so daß die Flüssigkeitsschicht dünn sein muß. Andererseits muß man bei niedrigen Aktivitätskonzentrationen ein großes Volumen Flüssigkeit an den Detektor zu bringen. Man ist deshalb zu einem Kompromiß genötigt. Mit etwa 4 mm dicker Wasserschicht bei 30 cm³ Flüssigkeitsvolumen zwischen zwei Methan-Durchfluß-Zählern mit Antikoinzidenzkranz und 10 cm Pb-Abschirmung läßt sich eine β-Nachweisempfindlichkeit erzielen, die über die konventioneller Anordnungen hinausgeht ($1{,}8 \cdot 10^{-13}$ C/cm³ für K⁴⁰, $3{,}6 \cdot 10^{-12}$ C/cm³ für S³⁵, $1{,}5 \cdot 10^{-13}$ für Sr/Y⁹⁰) [KEGELMANN (1)].

Für γ-Strahlung ist beim Zählrohr die Ansprechwahrscheinlichkeit klein. Man braucht große Flüssigkeitsmengen und kommt mit dem Szintillationszähler besser zum Ziel, besonders mit Bohrlochkristall oder Ringgefäß. In dieser Ausführung erreicht man zwar mit der Nachweisgrenze nicht ganz den Wert 10^{-13} C cm³, kommt ihm aber näher als mit anderen einfachen Apparaturen (NEUERT et al., JORDAN).

β-Zählung mit großflächigen Plastik-Szintillatoren beiderseits einer dünnen Wasserschicht oder zahlreicher laminar strömender Wasserfäden (JORDAN) bleibt um eine Größenordnung gegenüber dem vorgenannten Verfahren zurück.

Dünnen Schlauch aus Kunststoff-Szintillator-Material kann man zu einer Spirale entsprechend der Photokathodenfläche wickeln (Kimbel et al.) und auch zwischen zwei Multiplier bringen. Doch sind die ausnutzbaren Flüssigkeitsvolumina klein. Man kommt bei weichen β-Strahlern nicht unter 10^{-9} C/cm^3 und auch sonst nicht weiter herunter als mit anderen günstigen Anordnungen. Ferner ist es praktisch unmöglich, Kontaminationen des Schlauches zu beseitigen.

Flüssige Szintillatoren mit beigemischter Probeflüssigkeit sind nur verwendbar, wenn diese bereits chemisch vorbehandelt wurde.

Im Normalfall ist man also auf Aktivitäts-Anreicherung angewiesen. Prinzipiell kann man die Probe nach erfolgter Anreicherung auch in flüssiger Form messen, indem man jeweils auf ein bestimmtes Volumen einengt, das dem Detektor angepaßt ist [Houtermans (2)]. Trotzdem wird meist die Einengung bis zur Trockne fortgesetzt.

3. Anreicherung von Flüssigkeitsproben

Die einfachste Anreicherungsmethode ist das Eindampfen. Die Eindampftemperatur muß so niedrig sein, daß keine Dampfblasen durch die Flüssigkeitsoberfläche entweichen. Denn jede solche Blase reißt eine Anzahl kleiner Flüssigkeitströpfchen mit und möglicherweise auch Aktivität. Verluste durch direkte Verdampfung aktiver Substanzen könnten nur bei J (in frischen Spaltprodukten) eintreten. Bei mehrfachem Wechsel der Eindampfgefäße während der Verarbeitung einer Probe erhöht sich die Gefahr von Adsorptionsverlusten, auch wenn die Gefäße vorbehandelt sind.

Am günstigsten ist es, wenn das Eindampfen direkt in das für die Aktivitätsmessung verwendbare Schälchen erfolgt. Im allgemeinen ist das allerdings nur näherungsweise möglich, weil man das Absetzen unlöslichen Materials an den Wänden des Flüssigkeitsbehälters nicht ganz vermeiden kann, was mechanische Reinigung und Nachspülen erfordert.

Eine andere Möglichkeit der Anreicherung bieten die Ionenaustauscher. Dabei müssen die unlöslichen Beimengungen möglichst vor der Austauschersäule abfiltriert und getrennt gemessen werden. Die Säule selbst sollte, obwohl es in erster Linie um Kationen geht, auch einen Anionenaustauscher enthalten. Die Durchflußgeschwindigkeit kann gegenüber Austauscherarbeiten für chemische Trennungen erhöht werden, hat aber mit Rücksicht auf die tragbaren Dimensionen der Säule ihre Grenzen. Man kann den Austauscher selbst mit der absorbierten Aktivität an den Detektor bringen. Entweder diskontinuierlich (A. Hinzpeter), indem man kleine Austauschersäulen oder -tabletten nach Durchfluß der zu untersuchenden Flüssigkeit mißt. Schütteln des Austauschers mit der Flüssigkeit genügt nur in Sonderfällen für grobe Messungen. Oder kontinuierlich mit einer großen Säule und einem Detektor, der laufend deren Aktivität mißt. Emmons u. a. führten ein Zählrohr koaxial in die Säule ein, Neuert et al. seitlich einen NaJ-Kristall.

Aktivitäts-Anreicherung aus Wasserproben ist auch durch Mitfällung zu erzielen. Die besten Ergebnisse wurden mit kombinierter $Ca_3(PO_4)_2$—$Fe(OH)_3$-Fällung erhalten (Dannecker et al.). Relativ zur Eindampfmethode wurden 90 bis 100% der Uranspaltprodukte; 96 bis 99% Sr/Y^{90}, Ce^{141}, P^{32}; 80% Cs^{137}; 50% J^{131} mitgefällt. Das Verfahren ist auch auf die natürlich aktiven Pb-, Bi-, Po-Isotope anwendbar.

Zur kontinuierlichen Messung ist eine Methode vorgeschlagen worden (Buchner), bei der die Flüssigkeit mit Hilfe gefilterter inaktiver Heißluft versprüht und das beim Verdampfen der winzigen Tröpfchen entstehende Aerosol auf einem Filter angereichert und dann gemessen wird.

4. Analyse des Abfalls

Bei der Analyse des aus der Atmosphäre aufgesammelten Gemisches radioaktiver Aerosole hat man für eine aktive Substanz, die während der Exposition des Abscheiders nicht mehr erzeugt wird und unmittelbar in eine stabile Kernart übergeht [Schumann (1)],

$$N\lambda = f\,C\,(1 - e^{-\lambda T}),$$

N Zahl der Atome dieser Substanz auf dem Abscheider, C Zahl solcher Atome pro Volumeneinheit der verarbeiteten Luft, λ Zerfallskonstante, T Expositionsdauer, f Saugleistung.

$$\text{Für } T \gg 1/\lambda \quad \text{wird} \quad N\lambda = f\,C,$$
$$\text{für } T \ll 1/\lambda \quad \text{wird} \quad N\lambda = f\,C\,\lambda\,T.$$

Für Glieder einer Zerfallsreihe erhält man

$$dN_i/dt = f\,C_i + N_{i-1}\,\lambda_{i-1} - N_i\,\lambda_i.$$

Wenn auf dem Abscheider durch Zerfall ebenso viele Atome verloren gehen wie durch Ablagerung hinzukommen, wird

$$N_i\,\lambda_i = f\sum_{r=1}^{i} C_r,$$

und wenn noch radioaktives Gleichgewicht in der verarbeiteten Luft herrscht $(C_i\lambda_i = C_k\lambda_k = A)$

$$N_i\,\lambda_i = f\,A \sum_{r=1}^{i} \lambda_r^{-1}.$$

An kontinuierlich bewegten Abscheideflächen erhält man eine zeitliche Auflösung (Stebler) für Konzentrationsänderungen, die von Expositionsfläche und Transportgeschwindigkeit abhängt. Einen Sonderfall bilden kurzzeitige sehr hohe Aktivitätsspitzen, wie sie beim Auffangen einzelner Fallout-Teilchen großer spezifischer Aktivität, sog. heißer Teilchen, beobachtet wurden (Bosch et al.).

Bei den Tn-Folgeprodukten überwiegt die Halbwertszeit des ThB größenordnungsmäßig alle übrigen, so daß der Abfall praktisch exakt mit dieser Halbwertszeit erfolgt. RaB und C haben sehr viel kleinere, alle übrigen in der Atmosphäre vorkommenden Aktivitäten sehr viel größere Halbwertszeiten. Die Abklingkurve des am Abscheider erhaltenen Gemisches weist daher von etwa 3 Std nach Expositionsende bis etwa 20 Std einen vom ThB herrührenden exponentiellen Abfall auf. Daß dieser nicht genau der Halbwertszeit 10,6 Std des ThB entspricht, hat seinen Grund im Vorhandensein der langlebigen Aktivitäten — vor allem Spaltprodukte —, deren Abfall in den betrachteten 20 Std vernachlässigt werden kann.

Aus der Abweichung der scheinbaren Halbwertszeit von den 10,6 Std ermittelt man den Beitrag a der langlebigen Aktivitäten zu [Schumann (2)]

$$a = (a_2 - a_1\,e^{-\lambda_B t_{12}})/(1 - e^{-\lambda_B t_{12}}).$$

(a_1, a_2 gemessene Gesamtaktivitäten zu zwei möglichst weit auseinander liegenden Zeiten auf diesem Teil der Abfallkurve, t_{12} Zeitdifferenz, λ_B Zerfallskonstante des ThB), aus der auf das Ende der Expositionszeit rückextrapolierten reinen ThB-Kurve durch Differenzbildung die Aktivität der Rn-Folgeprodukte RaB, C, deren Halbwertszeiten 26 und 19 min nur wenig voneinander abweichen und daher keine einfache Abklingkurve geben. Sofern die langlebigen Aktivitäten gegenüber den Tn-Folgeprodukten stark überwiegen oder umgekehrt, werden die Fehler dieser Analyse schnell größer, weil man dann den einen Betrag als kleine Differenz zweier großer Zahlen erhält.

Für die natürlichen Aktivitäten ist es gleichgültig, ob man α-, β- oder γ-Aktivitäten mißt, da diese in bekanntem Verhältnis zueinander stehen. Die Spaltprodukte sind β-Strahler, und der relative Anteil der γ-Aktivität ist verschieden je nach Zusammensetzung des Gemisches.

Das bekannte $\alpha:\beta$-Verhältnis der Emanations-Folgeprodukte kann man benutzen, um sie von den übrigen Aktivitäten zu trennen, ohne den zeitlichen Abfall abzuwarten. Spaa hat während der Exposition auf der Ansaugseite des Filters einen α-Zähler, auf der Unterseite einen β-Zähler angeordnet, deren Messungen von zwei Ratemetern registriert werden. Durch elektronische Maßnahmen werden die Rn- und Tn-Folgeprodukte von anderen α- und β-Aktivitäten getrennt und diese direkt bestimmt.

Bei den Spaltproduktgemischen von Bombenexplosionen handelt es sich um sehr zahlreiche Nuklide mit Halbwertszeiten bis zu Jahrzehnten. Ihre Zusammensetzung ist für verschiedene spaltbare Substanzen und verschiedene Energien der spaltungsauslösenden Neutronen als Funktion der Zeit berechnet worden (Hunter et al., Löw et al., Malý et al.). Abweichungen durch Fraktionierung treten auf (Freiling).

Wegen der zahlreichen Halbwertszeiten, die sich bei einem normalen Spaltproduktgemisch überlagern, ist der Aktivitätsabfall eine Überlagerung sehr vieler Exponentialfunktionen und erfolgt proportional $t^{-(1+x)}$, wo x von der Größenordnung 0,1 und zeitlich nicht ganz konstant ist. Theoretisch erhält man bei statistischer Verteilung der Halbwertszeiten $t^{-1,2}$, während empirisch auf Grund der erwähnten Rechnungen für die ersten vier Tage $x=0,25$, für die Zeit von 4 bis 100 Tagen $x=0,03$ und für 100 Tage bis 3 Jahre $x=0,6$ angegeben wird. Der letzte Wert ist an den atmosphärischen Produkten kaum zu überprüfen, weil nach 100 Tagen die Spaltprodukte des betreffenden Ereignisses nur noch mit denen älterer Explosionen gemischt vorkommen. In den ersten drei Monaten nach einer Explosion ist die Abweichung des Exponenten von 1 so gering, daß man in guter Näherung Proportionalität zu $1/t$ annehmen kann. Die Darstellung $1/a=f(t)$ [Haxel (3)] bietet daher die beste Möglichkeit, den Abfall der Aktivität eines Spaltproduktgemisches zu analysieren, sofern man keine Abtrennung von Einzelnukliden vornimmt.

Die Spaltprodukt-Aktivität kann kurze Zeit nach einer Explosion überwiegend von diesem Einzelereignis herrühren. Oft aber folgten die Explosionen einander so schnell, daß sich Produkte von mehreren nur kurze Zeit zurückliegenden Ereignissen mischten. Franke hat die Überlagerung des Aktivitätsabfalls von Gemischen untersucht, die von einzelnen verschiedenen Explosionen stammen. Geophysikalisch besonders interessant wäre die Trennung der troposphärischen und der stratosphärischen Komponente. Eine solche ist aber nur möglich, wenn man das Datum der Explosion kennt, von dem die troposphärische Komponente stammt. Ist das nicht der Fall, so läßt die Kurve $1/a=f(t)$ zwar qualitativ einen jungen und einen alten Beitrag erkennen, doch ist es unmöglich, ihr Verhältnis quantitativ zu ermitteln.

Zur Analyse mittels Zerfallszeiten gehört auch die Ausnutzung sehr kurzer Halbwertszeiten in Zerfallsreihen zu Koinzidenz- oder Antikoinzidenz-Registrierung (Peirson, Jehanno et al.). Bei den Emanations-Folgeprodukten hat ThC' eine Halbwertszeit von $3\cdot10^{-7}$ s und RaC' eine solche von $1,6\cdot10^{-4}$ s. Man kann daher Rn-, Tn-Folgeprodukte und sonstige Aktivitäten voneinander trennen, indem man α-β-Koinzidenzen mit einer Auflösungszeit von etwa 1 μs und um einige μs verzögerte Koinzidenzen mit einer Auflösungszeit von einigen 10^{-4} s registriert.

5. Energiediskriminierung

Analyse eines Gemisches von α-Strahlern aus dünnen Präparaten ist einfach. Leider hat man es meist mit dicken Präparaten zu tun. Das behindert zwar die α-Zählung nur in einem gewissen Ausmaß. Die α-Spektroskopie dagegen wird beträchtlich erschwert. Eine Verbesserung ist durch teilweise Kollimation mittels durchlöcherter Folien zu erreichen. Doch ist dem eine Grenze gesetzt dadurch, daß man bei schwachen Präparaten die Ausbeute nicht zu stark herabsetzen darf. Man kann also nur durch geschickte Diskriminierung einzelne, durch Energie oder Intensität besonders ausgezeichnete α-Linien aus dem Spektrum ermitteln.

Für β-Spektren ist die Lage schlecht. Bei einem Gemisch von β-Strahlern in dicken Präparaten wie den normalen Spaltproduktgemischen der atmosphärischen Aerosole und Niederschläge ist die Identifizierung eines einzelnen reinen β-Strahlers durch Energiediskriminierung unmöglich.

Für γ-Strahler spielt die Dicke der Präparate keine Rolle, und die niedrigen Aktivitäten erfordern nur eine Verlängerung der Meßzeiten. Aber die gegenüber den Teilchenstrahlen ziemlich komplizierte Wechselwirkung erschwert die Ermittlung einzelner Nuklide in einem Gemisch zahlreicher Aktivitäten. Photoeffekt, Compton-Effekt, bei hohen Energien auch Paarbildung, bei niedrigen escape peaks komplizieren das Bild, das die Impulshöhenanalyse der Detektoreffekte liefert. Deshalb kann man nicht sehr viele verschiedene Aktivitäten eines Gemisches

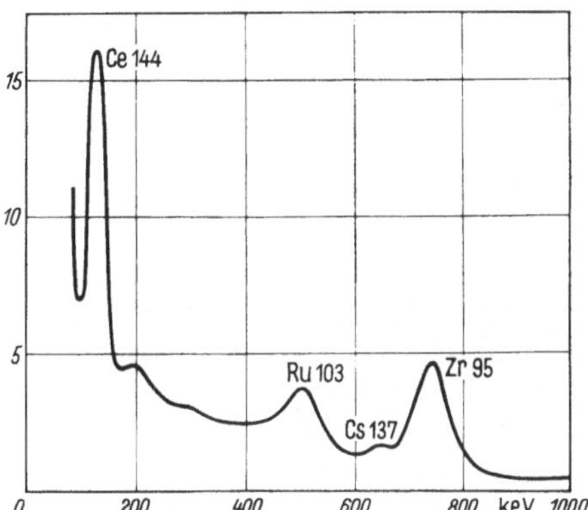

Fig. 9. γ-Spektrum der Spaltprodukt-Aktivität einer Luftfilterprobe [15. 4. 59, KEGELMANN (2)]

quantitativ bestimmen. Oft ist es unmöglich, ein Nuklid quantitativ zu bestimmen, dessen Photolinie in der Umgebung des Maximums der Compton-Verteilung eines anderen liegt und dessen Intensität beträchtlich kleiner ist als diese energiereichere Aktivität. Cs^{137} ist z. B. erst etwa neun Monate nach Entstehung eines Spaltproduktgemisches erkennbar, wenn der Anteil an der Gesamtaktivität einige Prozent erreicht hat und kurzlebige γ-Strahler (Ru/Rh^{103}, Zr/Nb^{95}) stärker abgefallen sind. Trotz dieser Einschränkungen ist die γ-Spektroskopie von atmosphärischen Aerosolen eine brauchbare Meßmethode, die manchmal die radiochemische Analyse erspart. Da die Aktivitäten oft so schwach sind, daß man mit einem Vielkanal-Spektrometer Meßzeiten von 24 Std benötigt, sind Einkanal-Geräte nur beschränkt verwendbar, z. B. wenn man viele Proben auf ein einziges Nuklid überprüfen will.

VII. Messungen an Mineral- und Bodenproben

1. Messungen von Totalaktivitäten

Bei den radioaktiven Bestandteilen von Mineralen spielen U und Th mit ihren Folgeprodukten die Hauptrolle. Im allgemeinen muß von Fall zu Fall entschieden werden, ob innerhalb einer Zerfallsreihe Gleichgewicht herrscht. Insbesondere bei den Emanationen kann das Gleichgewicht gestört sein, wenn die Möglichkeit des Austritts der Emanation aus dem Material bestanden hat. Aber auch die anderen Glieder der Zerfallsreihen können Verluste aufweisen infolge Verwitterung und Auslaugung (s. z. B. LANDSBERG). Im Zusammenhang mit den Meßmethoden interessieren diese Probleme nur insoweit, als an Hand der Gleichgewichtsüberlegungen zu entscheiden ist, ob und wie die Ergebnisse von Folgeprodukten auf Vorprodukte übertragen werden können.

Die übrigen in der festen Erdkruste vorkommenden natürlich radioaktiven Substanzen sind relativ seltene Isotope von stabilen Elementen, so daß die spezifische Aktivität gering ist. Das einzige häufige Element darunter ist K, dessen aktives Isotop K^{40} in unverarbeiteten Proben manchmal nachweisbar ist. Sonst kommt man ohne chemische Abtrennung der betreffenden Stoffe nicht aus. Man benutzt dazu normale chemische Verfahren, zur Untersuchung sehr seltener Stoffe neuerdings öfters die Neutronenaktivierung.

Zur Untersuchung auf U und Th ist die Messung der α-Strahlung direkt an Schliffen, Pulverproben und im Fall vorhergegangener chemischer Aufbereitung an chemischen Niederschlägen möglich. Oft ist es zweckmäßig, durch Überdeckung der großflächigen Proben mit Metallfolie oder -gitter dafür zu sorgen, daß keine unerwünschten Verzerrungen des elektrischen Feldes eintreten. Mit einem Durchfluß-Zähler von 20 cm² Nutzfläche und 4 cph Nulleffekt wurden bei aktivitätsärmsten Kalksteinen (1 bis $2 \cdot 10^{-6}$ g/g U bzw. Th) 16 cph und mehr gezählt [Adams (2)]. Eine Möglichkeit der Ermittlung des Verhältnisses Th:U bietet die Kombination einer Messung der Gesamt-α-Aktivität mit fluorimetrischer oder chemischer U-Bestimmung. Man benötigt für diese Methode nur 5 g Probematerial, die Nachweisgrenze liegt bei 10^{-7} g/g U und Th.

Will man Gleichgewichtsfragen vermeiden, kann man die einzelnen α-Strahler chemisch voneinander trennen und jede Aktivität für sich messen. Eine einfache α-Zählung ist mit dem Szintillationszähler durchzuführen, wobei die chemischen Niederschläge mit ZnS vermischt und aufgeschlämmt werden (Rosholt).

Bei α-Messungen kann die Verteilung der Aktivität in der Probe eine Rolle spielen. Trotzdem wurde an einer größeren Zahl von Proben sehr verschiedener Aktivität gute Übereinstimmung zwischen γ-Spektroskopie, für die das unwesentlich ist, und α-Zählung an dicken Präparaten kombiniert mit Fluorimetrie gefunden [Adams (2)]. Das spricht dafür, daß die α-Methode durchaus anwendbar ist trotz Schwierigkeiten der Absorptionskorrektur infolge möglicherweise inhomogener Verteilung der Strahler in der Probe und trotz der wegen der Emanationserzeugung im Innern dicker Präparate vorgebrachten Bedenken [Picciotto (3)]. Mit beiden Methoden erhält man eine Genauigkeit für das Th:U-Verhältnis von besser als $\pm 20\%$ bei Th:U $\geqq 3$. Bei Th:U $\leqq 1$ ist die Genauigkeit geringer.

Messungen mit dem Geiger-Zähler gestatten nicht ohne weiteres eine einfache Deutung. Es gibt aber ein Verfahren kombinierter β- und γ-Messungen für die U-Reihe, das von der Frage des Gleichgewichts weitgehend unabhängig ist und es sogar ermöglicht, Gleichgewichtsabweichungen zu ermitteln [Thommeret, Eichholz (2)]. Dabei geht man davon aus, daß praktisch γ-Quanten über 100 keV lediglich von Bi^{214} emittiert werden, β-Strahlen oberhalb 200 keV nur von Pa^{234}, Bi^{214} und Bi^{210}. Gleichgewicht zwischen Th^{234} und Pa^{234} ist wegen der kleinen Halbwertszeit des Pa^{234} von 1,2 min immer vorhanden. So erhält man die Aktivität von Th^{234} als Differenz der Gesamt-β-Aktivität und der β-Aktivität der Rn-Folgeprodukte Bi^{214} und Bi^{210}. Da die letztere in einem konstanten Verhältnis zur γ-Aktivität des Bi^{214} steht, läßt sich die Bestimmung der Aktivität des Th^{234} zurückführen auf eine Messung der Gesamt-β-Aktivität und eine solche der Gesamt-γ-Aktivität. Dabei ist nur darauf zu achten, daß die energiearmen Elektronen und Quanten unterhalb der genannten Grenzen die Detektoren nicht erreichen. Der Beitrag der γ-Strahlung zur Zählrate des β-Zählers kann vernachlässigt werden. Die Abweichung des Verhältnisses der Gesamt-β-Aktivität zur Gesamt-γ-Aktivität von seinem Gleichgewichtswert gibt Informationen über etwa vorliegende Gleichgewichtsstörungen. Die Methode kann unmittelbar als Bestimmung von U selbst angesehen werden, wenn U^{238} und Th^{234} im Gleichgewicht sind, d.h. gegenwärtig (Halbwertzeit des Th^{234} 24 Tage) keine das Gleichgewicht beeinträchtigenden Vorgänge stattfinden. Dies Verfahren zur U-Bestimmung ist auch anwendbar in Anwesenheit von Th, wie EICHHOLZ u.a.(2) unter Benutzung empirischer Werte für die Ansprechwahrscheinlichkeit dargelegt haben. Genaue Th-Bestimmung ist auf diesem Wege jedoch nicht möglich in Proben, die neben Th auch die Stoffe der U-Reihe enthalten, wenn diese nicht im Gleichgewicht sind. Denn Th führt zu einer Erhöhung der γ-Zählrate relativ zur β-Zählrate, Rn-Verlust dagegen zu einer Verminderung.

Bei Untersuchung von Bodenproben auf Spaltprodukte muß in den meisten Fällen chemische Abtrennung des betreffenden Nuklids erfolgen. Die Messung kann, wenn mit ausreichender Isolierung zu rechnen ist (etwa bei Sr^{90}), durch Bestimmung der Gesamt-β-Aktivität erfolgen.

2. α- und γ-Spektroskopie

Für α-Spektroskopie zur Bestimmung des U- und Th-Gehaltes von Gesteinen braucht man hohe Auflösung, weil die Zahl der in den Proben enthaltenen α-Strahler beträchtlich ist. Mit der Gitterionisationskammer konnten an dünnen Präparaten die α-Linien von Th, UI, ThX, Tn, ThA, RaC', ThC' isoliert werden, während die von Io und UII und Ra, von RdTh und Po und Rn, von RaC und ThC gruppenweise unaufgelöst blieben. Die Gesamtmenge der mit Wasser und Haftmittel aufgeschlämmten Mineralproben betrug in diesem Falle maximal 20 mg auf einer Fläche von 14 cm Durchmesser. Als Nachweisgrenze wird 10^{-4} g/g U und Th genannt bei einigen Stunden Meßzeit [FACCHINI (4)].

Man kann α-Spektroskopie und Isotopenverdünnung kombinieren. Die Auflösung der Gitterionisationskammer genügt zur Trennung von Th^{232} (3,98 MeV), Th^{230} (4,68 MeV) und Th^{228} (5,3 MeV). Zur U-Bestimmung in Abwesenheit von Th wird Th^{232} zugegeben und dann Thoriumnitrat extrahiert, in Äthylalkohol gelöst und mit Zaponlack auf die Präparatfläche gebracht. Aus dem α-Spektrum ergibt sich das Verhältnis $Th^{230}:Th^{232}$, bei Voraussetzung des Gleichgewichts zwischen Th^{230} und U^{238} läßt sich U bestimmen. Zur Bestimmung von Th in Abwesenheit von U wird U zugegeben z.B. in Form von Pechblende, wo U und Th^{230} mit größter Wahrscheinlichkeit im Gleichgewicht sind und der Th^{232}-Anteil vernachlässigt werden kann. Aus dem Verhältnis $Th^{230}:Th^{232}$ wird Th^{232} bestimmt. Zur gleichzeitigen Bestimmung von U und Th wird aus dem α-Spektrum ohne Zugabe eines Verdünnungsisotops das Th:U-Verhältnis ermittelt und aus dem Vergleich dieses Spektrums mit dem bei Zugabe des Verdünnungsisotops der Th-Gehalt der ursprünglichen Probe. Die Methode ist weit empfindlicher als einfache α-Spektroskopie, weil man mit reinem Thoriumsalz arbeitet. Unterhalb von Konzentrationen der Größenordnung 10^{-4} g/g U_3O_8 oder ThO_2 wird eventuell das chemische Arbeiten schwierig (HOWARD).

Beschränkt man die Untersuchung auf die Emanationsfolgeprodukte, wobei die Gleichgewichtsfrage gesondert zu klären ist, so bietet das Verhältnis der α-Linien von RaC' und ThC' eine gute Bestimmungsmöglichkeit für das Verhältnis U:Th. Die α-β-Koinzidenzmethode für RaC/C' und ThC/C' (vgl. VI.4) ist auch hier anwendbar. Auch die Emanationsmessung kann mit anderen Methoden kombiniert werden. Mit Rn-Bestimmung und fluorimetrischer U-Messung lassen sich 10^{-9} g/g U, mit Tn-Bestimmung und ThC'-α-Messung 10^{-8} g/g Th nachweisen (DALTON).

Für die Bestimmung von Ra ist schon immer die Messung der Emanation benutzt worden. Bei Gleichgewicht ist daraus auch der Schluß auf den U-Gehalt möglich. Die Ra-Bestimmung kann natürlich auch radiochemisch ausgeführt werden [Herrmann (2)].

Zur Bestimmung von U und Th mittels γ-Spektroskopie kann man sich verschiedener γ-Strahler in beiden Zerfallsreihen bedienen, wobei je nach ihrer Stellung in den Reihen wieder die Gleichgewichtsfragen zu stellen sind. Whitham benutzte die γ-Diskriminierung zwischen Bi^{214} und Bi^{212} zur Ermittlung des U:Th-Verhältnisses. Hurley (2) verwendete bei der Messung an 200 g-Proben in etwa $\frac{1}{2}''$ dicker Schicht im Ringgefäß um den Kristall die 188 keV-Linie des Ra^{226} und die 238 keV-Linie des Pb^{212}. Die Probenschicht darf nicht zu dick sein, damit die Compton-Effekte innerhalb der Probe nicht zu häufig werden. Die γ-Absorption in der Probe muß man gegebenenfalls berücksichtigen. Bei Strahlungsenergien unter 1,5 MeV kann unter Umständen der K-Gehalt der Probe stören. Deshalb wählten Adams et al. (2) für ihre Messungen die 1,76 MeV-Linie des Bi^{214} und die 2,62 MeV-Linie des Tl^{208}. Sie verglichen auch die so gewonnenen Ergebnisse mit solchen unter Verwendung der 609 keV-Linie des Bi^{214} und fanden Übereinstimmung auf $\pm 15\%$. Benutzt wurden 600 g Probenmaterial, die in einem $\frac{1}{2}''$ dicken Ringgefäß um einen $3'' \times 3''$ NaJ-Kristall untergebracht waren, das noch $1''$ über diesen hinausragte. Zur Eichung dienen Standard-Präparate z.B. von Pechblende oder Carnotit.

Bei Untersuchung von Bodenproben auf Spaltprodukte kann die γ-Spektroskopie eventuell herangezogen werden, falls die Aktivitäten ausreichen. Sie kann auch zur Identifizierung einzelner abgetrennter γ-Strahler (z.B. Cs^{137}) dienen.

3. Messungen mit Photoemulsionen

Bei der Untersuchung von Mineralproben mittels Kernspur-Emulsionen spielen die α-Sterne mit bis zu fünf Spuren eine wichtige Rolle. Fünffach-Sterne kommen in der U-Reihe praktisch nicht vor, wie sich auch theoretisch aus den Zerfallskonstanten zeigen läßt. Dagegen finden sich in der Th-Reihe solche Sterne, bei denen die Spuren von Th^{228}, Ra^{224}, Rn^{220}, Po^{216}, Bi^{212} $(\frac{1}{3})$ oder Po^{212} $(\frac{2}{3})$ herrühren, und in der Ac-Reihe, wo die Spuren zu Th^{227}, Ra^{223}, Rn^{219}, Po^{215}, Bi^{211} gehören. Die Produkte der Ac-Reihe, die bei allen anderen Meßverfahren wegen ihrer Seltenheit bzw. der kurzen Halbwertszeiten keine Rolle spielen, sind mit der photographischen Methode nachweisbar. Das Verhältnis der Zahl der Fünffach-Sterne des Th^{228} zu der der entsprechenden Sterne des Th^{227} ist leicht zu bestimmen, weil $\frac{2}{3}$ der Th^{228}-Sterne durch die extrem langen α-Spuren des Po^{212} kenntlich sind [Picciotto (2)].

Eine Schwierigkeit bringt die Diffusion der Rn-Atome in der Emulsion mit sich. Rn^{222} kann wegen seiner langen Halbwertszeit von 3,8 Tagen beträchtliche Strecken zurücklegen und sogar aus der Emulsion verloren gehen (Koczy et al.). Rn^{220} mit 54 sec Halbwertszeit kann immerhin noch einige μ wandern. Daher findet man oft Th^{228}-Sterne, bei denen die Spuren des Th^{228} und des Ra^{224} von einem Punkt ausgehen, während die des Rn^{220}, des Po^{216} und des Bi^{212} oder Po^{212} in einem ein klein wenig dagegen verschobenen Punkt ihren Ursprung haben [Eichholz (1)]. Die Halbwertszeit des Rn^{219} mit 4 sec ist zu kurz, als daß die Diffusionseffekte merklich würden. Th^{232} und Th^{230}, U^{238} und U^{235} sowie Po^{210} liefern praktisch immer Einzelspuren.

Die Häufigkeit der verschiedenen Mehrfach-Sterne als Funktion der Expositionszeit bei gegebener U- und Th-Konzentration läßt sich berechnen (Bremner; Schneider et al.). Auf dieser Grundlage ermittelt man aus der beobachteten Zahl z.B. der Fünffach-Sterne bei gegebener Expositionszeit die Th-Konzentration in der Probe.

Man kann auch Spurenlängen-Messungen dazu benutzen, den U- und den Th-Gehalt oder ihr Verhältnis zu bestimmen. Dies Verfahren ist jedoch recht mühsam. Es begegnet auch beträchtlichen meßtechnischen Schwierigkeiten, besonders da es sich in den meisten Fällen um dicke Präparate handelt. Relativ günstig ist noch die Aussonderung bestimmter Spurenlängen-Gruppen, z.B. derjenigen über 40 μ, die nur von Po^{212} stammen können, und der zwischen 33 und 40 μ, die außer von Po^{212}-α-Teilchen, die einen Teil ihrer Energie außerhalb der Emulsion abgegeben haben, noch von Po^{214} und Po^{215} herrühren können. Auf Grund des Zahlenverhältnisses dieser beiden Gruppen kann man das Th:U-Verhältnis abschätzen [I. JOLIOT-CURIE (1), COPPENS]. Jedoch ist die Genauigkeit nicht allzu gut.

Günstiger sind die Bedingungen bei der Untersuchung chemisch abgetrennter Fraktionen, z.B. von Th (KOCZY et al.). Dort muß man nur die Einzelspuren von Th^{232} und Th^{230} durch Längenmessung voneinander trennen. Th^{228} und Th^{227} werden aus der Zahl der Fünffach-Sterne bestimmt. Freilich muß man von den chemischen Prozessen her mit kleinen Beträgen von Verunreinigungen rechnen, die sich aber durch Spurenmessungen eliminieren und unter Umständen identifizieren lassen.

4. Fluorimetrie

Ein spezielles Verfahren zur U-Bestimmung ist die fluorimetrische Methode (HERNEGGER et al.). Wenn man U-Salze mit NaF schmilzt, fluoreszieren sie bei UV-Bestrahlung (etwa 3650 Å) intensiv grüngelb. Die Intensität dieser Fluorescenz ist in weiten Bereichen proportional zur U-Konzentration. GRIMALDI et al. gibt als Nachweisgrenze 10^{-10} g U an gegenüber 10^{-5} g für die polarographische und die kolorimetrische Methode. Deshalb ist das Verfahren gerade für niedrige Konzentrationen besonders geeignet. Die Messung der Fluorescenz kann mikrophotometrisch erfolgen oder einfacher mit Photozelle und Galvanometer. Als Flußmittel ist auch Borax verwendbar. Besonders günstig ist ein Gemisch von 9% NaF und je 45,5% Na_2CO_3 und K_2CO_3, dessen Schmelzpunkt bei 650° C liegt gegenüber 992° C für NaF. Fluorescenzgifte sind hauptsächlich Fe, Cr, Mn, Co, Ni, Cu, Ce, V. 0,3 mg Fe_2O_3/g NaF erniedrigen die Fluorescenz um 15%, 2 mg CeO_2/g NaF um 30%. Bei den übrigen Elementen sind 10 mg/g NaF noch tragbar (ERLENMEYER et al.).

Die Genauigkeit des fluorimetrischen Verfahrens ist $\pm 15\%$ oder etwas besser (LARSEN et al.). Im Vergleich zu U-Bestimmungen mittels γ-Spektroskopie scheinen die Werte im Mittel ein wenig tiefer zu liegen, was auf Extraktionsverluste und Lichtverluste bei der Fluorescenz zurückzuführen sein könnte. Andererseits liegen die fluorimetrischen Werte eher etwas höher als die aus Rn-Bestimmungen ermittelten [ADAMS (2)].

Literatur

ADAMS, J.A.S. (1) in H. FAUL (ed.): Nuclear geology, p. 89. New York 1954; (2) —, J.E. RICHARDSON and C.C. TEMPLETON: Geochim. et Cosmochim. Acta 13, 270 (1958). — AGRANOFF, B.W.: Nucleonics 15, No. 10, 106 (1957). — ALBOUY, G., et H. FARAGGI: J. Phys. Radium 10, 105 (1949). — ALIVERTI, G.: (1) Nuovo Cim. 15, 66 (1938); (2) Nuovo Cim. 12, 270 (1954). — ANDERSON, A.C., J.R. ARNOLD and W.F. LIBBY: Rev. Sci. Instrum. 22, 225 (1951). — ARNOLD, J.R.: Science 119, 155 (1954). — AUDRIC, B.N., and J.V.P. LONG: Nature, Lond. 173, 992 (1954). — BAKER, R.G., and L. KATZ: Nucleonics 11, No. 2, 14 (1953). — BARENDSEN, G.W.: 2nd Conf. Atomic Energy, Geneva, Proc. 21, 150 (1958). — BATE, G.L. (1) —, H.L. VOLCHOK and J.L. KULP: Rev. Sci. Instrum. 25, 153 (1954); (2) —, H.A. POTRATZ and J.R. HUIZENGA: Geochim. et Cosmochim. Acta 14, 118 (1958). — BAUS, R.A., P.R. GUSTAFSON, R.L. PATTERSON and A.W. SAUNDERS: US Naval Res. Lab. Memorandum Rep. 758 (1957). — BEARDEN, J.A.: Rev. Sci. Instrum. 4, 271 (1933). — BECKER, A., u.

K. H. Stehberger: Ann. Phys., Lpz. **1**, 529 (1929). — Begemann, F.: Helv. phys. Acta **27**, 451 (1954). — Bergstedt, B. A.: J. Sci. Instrum. **33**, 142 (1956). — Bernstein, W., and J. Ballentine: Rev. Sci. Instrum. **21**, 158 (1950). — Blifford, I. H., H. Friedman, L. B. Lockhart and R. A. Baus: US Naval Res. Lab. Rep. 4760 (1956). — Borkowski, C. J.: Oak Ridge Nat. Lab. Progress Rep. 1160 (1951). — Bosch, J., u. G. Ulsenheimer: Nukleonik **1**, 157 (1958). — Boulenger, R., et E. Gourski: 2nd Conf. Atomic Energy, Geneva, Proc. **23**, 362 (1958). — Boussières, G., et C. Ferradini: Analyt. chim. Acta **4**, 610 (1950). — Brannon, H. R.: Rev. Sci. Instrum. **26**, 269 (1955). — Branson, B. M., G. R. Hagee, A. S. Goldin, G. J. Karches and C. P. Straub: J. Amer. Water Works Ass. **51**, 438 (1959). — Bremner, J. W.: Proc. Phys. Soc. Lond. A **64**, 25 (1951). — Broda, E.: (1) J. Sci. Instrum. **24**, 136 (1947); (2) —, N. Feather and D. H. Wilkinson: Phys. Soc. Conf. Rep. Camb. 1947, p. 114. — Broecker, W. S., C. S. Tucek and E. A. Olson: Internat. J. appl. Radiation and Isotopes **7**, 1 (1959). — Bryant, J., and M. Michaelis: Brit. Min. Supply Radiochem. Centre RCC/R 26 (1952). — Bryant, F. J., A. C. Chamberlain, A. Morgan and G. S. Spicer: Atomic Energy Res. Establ. Harwell HP/R 2056 (1956). — Buchner, W.: Atompraxis **3**, 405 (1957). — Burke, H., and M. Meinschein: Rev. Sci. Instrum. **26**, 1137 (1955). — Coppens, R.: J. Phys. Radium **11**, 21 (1950). — Coryell, C. D., and N. Sugarman: Nat. Nucl. Energy Series Div. IV, Vol. 9 (1951). — Crathorn, A. R.: Nature, Lond. **172**, 632 (1953). — Curtiss, L. F.: J. Res. Nat. Bur. Stand. **30**, 157 (1943). — Dalton, J. C.: Geochim. et Cosmochim. Acta **3**, 272 (1953). — Damon, P. E. (1) — and H. J. Hyde: Rev. Sci. Instrum. **23**, 766 (1952); (2) — and R. F. Overman: Nucleonics **10**, No. 1, 64 (1952). — Dannecker, A., H. Kiefer u. R. Maushart: Nukleonik **1**, 319 (1959). — Davidson, J. D., and P. Feigelson: Int. J. Appl. Rad. and Isotopes **2**, No. 1, 1 (1957). — Delibrias, G.: (1) J. Phys. Radium **15**, 78 A (1954); (2) — et C. Jehanno: Bull. Inform. Sci. Tech. No. 30, 14 (1959). — Deutsch, S. (1) — et E. C. Dodd: Nuovo Cim. **10**, 858 (1953); (2) —, Nuovo Cim. **3**, 1166 (1956); (3) —, F. G. Houtermans et E. Picciotto: Geochim. et Cosmochim. Acta **10**, 166 (1956); (4) — et E. Picciotto: Nuovo Cim. **6**, 953 (1957). — Deutsch, W.: Ann. Phys., Lpz. **68**, 335 (1922). — Diethorn, W.: US Atomic Energy Commission Rep. NYO 6628 (1956). — Dilla, M. A. van, and D. H. Taysum: Nucleonics **13**, No. 2, 68 (1955). — Dratz, A. F.: Nucleonics **15**, No. 8, 83 (1957). — Drevinsky, P. J., C. E. Junge, I. H. Blifford, M. I. Kalkstein and E. Martell: US Air Force Camb. Res. Center TN-58-652 (1958). — Drigo, A.: Nuovo Cim. **7**, 501 (1950). — Duuren, K. van, W. K. Hofker and J. Hermsen: 2nd Conf. Atomic Energy Geneva Proc. **14**, 339 (1958). — Eichholz, G. G. (1) — and F. C. Flack: J. Chem. Phys. **19**, 363 (1951); (2) —, J. W. Hilborn and C. McMahon: Canad. J. Phys. **31**, 613 (1953). — Eisenbud, M., and J. Harley: Science **124**, 251 (1956). — Emmons, A. H., and R. A. Lauderdale: Nucleonics **10**, No. 6, 22 (1952). — Erdik, E.: J. Phys. Radium **15**, 445 (1954). — Erlenmeyer, H. u. a.: Helv. chim. Acta **88**, 25 (1950). — Eulitz, G.: Nukleonik **2**, 85 (1960). — Evans, R. D.: Rev. Sci. Instrum. **6**, 99 (1935). — Facchini, U. (1) — and L. Orsoni: Nuovo Cim. **6**, 241 (1949); (2) — and A. Malvicini: Nucleonics **13**, No. 4, 36 (1955); (3) — —: Nucleonics **14**, No. 5, 96 (1956); (4) —, M. Forte, A. Malvicini and T. Rossini: Nucleonics **14**, No. 9, 126 (1956). — Faltings, V. (1) — u. P. Harteck: Z. Naturforsch. **5a**, 438 (1950); (2) —, Naturwissenschaften **40**, 409 (1953). — Farmer, E. C., and I. A. Bernstein: Science **117**, 279 (1953). — Fergusson, G. J.: Nucleonics **13**, No. 1, 18 (1955). — Fireman, E. L.: Nature, Lond. **181**, 1613 (1958). — Fisher, C., et J. Beydon: Bull. Soc. chim. France **1953**, C 102. — Franke, T.: Atomkernenergie **4**, 308 (1959). — Freiling, F. C.: US Naval Radiol. Defense Lab. TR-385 (1959). — Frierson, W. J., and J. W. Jones: Analyt. Chem. **23**, 1447 (1951). — Fucks, N., I. Petrjanoff and B. Rotzeig: Trans. Faraday Soc. **32**, 1131 (1936). — Funt, B. L. (1) —, S. Sobering, R. W. Pringle and W. Turchinetz: Nature, Lond. **175**, 1042 (1955); (2) —, Nucleonics **14**, No. 8, 83 (1956); (3) — and A. Hetherington: Science **131**, 1608 (1960). — Geiger, H., u. O. Haxel: Deutsches Reichspatent 682657, 1937. — Geiss, J., C. Gfeller, F. G. Houtermans and H. Oeschger: 2nd Conf. Atomic Energy, Geneva, Proc. **21**, 147 (1958). — Glendenin, L. E.: Nat. Nucl. Energy Series, Div. IV, Vol. 9, Pap. 236. — Goel, P. S.: (1) Nature, Lond. **178**, 1458 (1956); (2) —, S. Jha, D. Lal, P. Ramakrishna and T. Rama: Nuclear Phys. **1**, 196 (1956). — Goetz, A.: Amer. J. Publ. Health **43**, 150 (1953). — Gonsior, B.: Z. angew. Phys. (im Druck). — Goyer, G. G., R. Gruen and V. K. LaMer: J. Phys. Chem. **58**, 137 (1954). — Greisen, K.: Phys. Rev. **61**, 212 (1942). — Grimaldi, F. S. u. a.: US Geol. Survey Bull. 1006 (1954). — Hagemann, F., L. I. Katzin, M. H. Studier, G. T. Seaborg and A. Ghiorso: Phys. Rev. **79**, 435 (1950). — Hahn, O.: Applied radiochemistry. Ithaca, N.Y.: Cornell Univ. Press 1936. — Hahn, O., u. F. Strassmann: Naturwissenschaften **27**, 89 (1939). — Hahofer, E., u. F. Hecht: Mikrochim. Acta **1954**, 417. — Haissinsky, M.: C. R. Acad. Sci., Paris **196**, 1788 (1933). — Hamada, G. H., E. P. Hardy and L. T. Alexander: US Atomic Energy Commission HASL 33 rev. (1960). — Haring, A., A. E. de Vries and H. de Vries: Science **128**, 472 (1958). — Harley, J.: Nucleonics **11**, No. 7, 12 (1953). — Haxel, O. (1) — u. F. G. Houtermans: Z. Physik **124**, 705 (1948); (2) —, Z. angew. Phys.

5, 241 (1953); (3) — u. G. Schumann: Naturwissenschaften **40**, 458 (1953); (4) —, Naturwissenschaften **44**, 163 (1957). — Hayes, F.N. (1) — and R.G. Gould: Science **117**, 480 (1953); (2) —, E.C. Anderson and J.R. Arnold: Conf. Atomic Energy, Geneva, Proc. **14**, 188 (1955); (3) —, B.S. Rogers and W.H. Langham: Nucleonics **14**, No. 3, 48 (1956). — Hecht, F., J. Korkisch, R. Patzak u. F. Thiard: Mikrochim. Acta **1956**, 1283. — Heintze, J.: (1) Diss. Heidelberg 1955; (2) — u. H. Fischbeck: Z. Physik **147**, 277 (1957). — Hermsen, J., A.M. J. Jaspers, P.Kraayeveld and K. van Duuren: Conf. Atomic Energy, Geneva, Proc. **14**, 275 (1955). — Hernegger, F., u. B. Karlik: Wien. Ber. II a **144**, 217 (1935).— Herrmann, G.: (1) Strahlenschutz **10**, 23 (1959); (2) —, H. Hauser u. H. Riedel: Nukleonik **1**, 305 (1959). — Hinzpeter, A.: Naturwissenschaften **44**, 611 (1957). — Hirschfelder, J. O., J. L. Mager and M.H. Hull: Phys. Rev. **73**, 852 (1948). — Hoagland, E. J.: Nat. Nucl. Energy Series Div. IV, Vol. 9, Pap. 237 (1951). — Houtermans, F.G. (1) — u. H. Oeschger: Helv. phys. Acta **31**, 117 (1958); (2) — u. C. Mühlemann: 2nd Conf. Atomic Energy, Geneva, Proc. **18**, 559 (1958). — Howard, L.E.: Nucleonics **16**, No. 2, 112 (1958). — Hultquist, B.: Kungl. Svenska Vetensk. Akad. Handl., Ser. IV **6**, No. 3 (1956). — Hunter, H.F., and N.E. Ballou: Nucleonics **9**, No. 5, C2 (1951). — Hurley, P.M. (1) — and R.R. Shorey: Trans. Amer. Geophys. Un. **33**, 722 (1952); (2) —, Bull. Geol. Soc. Amer. **67**, 395 (1956). — Ishibashi, M., T. Shigemitsu, T. Ishida, S. Okada, T. Nishi, H. Takahashi, C. Matsumoto, S. Shimizu, T. Hyodo, F. Hirayama and S. Okamoto: Bull. Inst. Chem. Res. Kyoto Univ., Suppl., 39 (1954). — Ishimori, T.: Bull. Chem. Soc. Jap. **28**, 432 (1955). — Israel, G.: Dipl.-Arb. Heidelberg 1961. — Israël, H.: (1) Balneologe **1**, 318 (1934); (2) Wiss. Abh. Reichsamt Wetterdienst **2**, No. 10 (1937). — Jacobi, W. (1) —, A. Schraub, K. Aurand u. H. Muth: Beitr. Phys. Atmosph. **31**, 244 (1959); (2) — u. H. Stephan: Hahn-Meitner-Inst. Berlin HMI-B4 (1959). — Jacobson, H. J., G.N. Gupta, C. Fernandez, S. Hennix and E.V. Jensen: Arch. Biochem. Biophys. **86**, 89 (1960). — Jaffey, A.H.: Nat. Nucl. Energy Series Div. IV, Vol. 14 A, p. 596 (1949). — Jech, C.: Phys. Rev. **76**, 1731 (1949). — Jehanno, C., A. Blanc, C. Lallemant et G. Roux: 2nd Conf. Atomic Energy, Geneva, Proc. **23**, 372 (1958). — Johnson, R.G., O.E. Johnson and L.M. Langer: Phys. Rev. **102**, 1142 (1956). — Joliot-Curie, I.: (1) J. Phys. Radium **7**, 313 (1946); (2) — et H. Faraggi: C. R. Acad. Sci., Paris **232**, 959 (1951). — Jordan, K.: Atompraxis **3**, 398 (1957). — Junge, C.E.: Ann. Meteor. 1952, Beiheft. — Kaufmann, S., and W.F. Libby: Phys. Rev. **93**, 1337 (1954). — Kaul, A.: Dipl. Arb. Max-Planck-Inst. Biophys. Frankfurt (Main) 1960. — Kegelmann, G.: (1) unveröffentlicht; (2) — u. G. Schumann: 1959, unveröffentlicht. — Kern, W.: (1) Bull. Schweiz. Akad. Med. Wiss. **14**, 401 (1958); (2) Nukleonik **2**, 203 (1960). — Ketelle, B.H.: Phys. Rev. **80**, 758 (1950). — Kiefer, H., u. R. Maushart: Atomwirtschaft **4**, 247 (1959). — Kimbel, K.H., u. J. Willenbrink: Naturwissenschaften **45**, 567 (1958). — Kinard, F.E.: Rev. Sci. Instrum. **28**, 293 (1957). — Kirsch, E.: Dipl.-Arb. Heidelberg 1958. — Knowles, D. J.: Nucleonics **13**, No. 6, 98 (1955). — Koczy, F.F., E. Picciotto, G. Poulaert et S. Wilgain: Geochim. et Cosmochim. Acta **11**, 103 (1956). — Koechlin, Y., L. Koch, A. Lansart and G. Pietri: 2nd Conf. Atomic Energy, Geneva, Proc. **21**, 174 (1958). — Korkisch, J., A. Thiard u. F. Hecht: Mikrochim. Acta **1956**, 1422. — Kraus, K.A. and F. Nelson: J. Amer. Chem. Soc. **76**, 5916 (1954). — Kulp, J.L. (1) — and L.E. Tryon: Rev. Sci. Instrum. **23**, 296 (1952); (2) —, Nucleonics **12**, No. 12, 19 (1954). — LaMer, V.K.: US Atomic Energy Commission Rep. NYO 512 (1951). — Landsberg, H. (1) — and A.I. Ingham: Oil Weekly **90** (4), 26 (1938); (2) — and M.R. Klepper: Amer. Inst. Mining an Metallurg. Eng. Tech. Publ. No. 1103 (1939). — Landt, E.: Schwebstofftechn. Arbeitstag. Univ. Mainz 1956, S. 10. — Larsen, E.S., and G. Phair in H. Faul (ed.): Nuclear geology. New York: Wiley 1954. — Lassen, L.: (1) Diss. Heidelberg 1958; (2) —, u. G. Rau: Z. Physik **160**, 504 (1960). — Lattes, C.M.G., P.H. Fowler and P. Cüer: Proc. Phys. Soc. Lond. **59**, 883 (1947). — Leutz, H.: Z. Physik **159**, 462 (1960). — Lockhart, L.B., R.A. Baus, R.L. Patterson and I.H. Blifford: US Naval Res. Lab. Rep. 5208 (1957). — Loevinger, R., and M. Berman: Nucleonics **9**, No. 1, 26 (1951). — Löw, K., and R. Björnerstedt: Ark. Fysik **13**, 85 (1957). — Malvicini, A.: Nuovo Cim. **12**, 821 (1954). — Malý, J., J. Kutzendörfer, V. Machaček, V. Kóuřim u. V. Jeřábek: Czech. J. Phys. **7**, 45 (1957) (Orig. russ.). — Marquez, L., e N.L. Costa: Nuovo Cim. **2**, 1038 (1955). — Martell, E.: US Atomic Energy Commission Rep. AECU 3262 (1956). — McDaniel, E.W., H. J. Schaeefr and J.K. Colehouse: Rev. Sci. Instrum. **27**, 864 (1956). — McMillan, E.M.: Phys. Rev. **72**, 591 (1947). — Miller, C.E. (1) —, I.D. Marinelli, R.E. Rowland and J.E. Rose: Nucleonics **14**, No. 4, 40 (1956); (2) —, H.A. May and L.D. Marinelli: US Atomic Energy Commission Rep. TID 7577 (1959). — Minami, E., M. Honda and Y. Sasaki: Bull. Chem. Soc. Japan **31**, 372 (1958). — Münnich, K.O. (1) Diss. Heidelberg 1957; (2) — u. J.C. Vogel: C¹⁴-Symposium Groningen 1959. — Münzel, H.: Strahlenschutz **6**, 147 (1957). — Muth, H., A. Schraub u. K.Aurand: Strahlentherapie, Sonderbd. **35**, 227 (1955). — Nervik, W.E., and P.C. Stevenson: Nucleonics **10**, No. 3, 18 (1952). — Neuert, H., u. H. Henschen:

Industrie-Elektronik **5**, 1 (1957). — Nilsson, G., and G. Aniansson: Nucleonics **13**, No. 2, 38 (1955). — *Nuclear Enterprises Ltd.*, Bull. No. 21 (1960). — Nydal, R., and R. S. Sigmund: Appl. Sci. Res. B **6**, 393 (1957). — Osmond, R. G., A. G. Prachett and J. B. Warricker: Atomic Energy Res. Establ. Harwell C/R 2165 (1957). — Pauthenier, M., et M. Morreau-Hanot: J. Phys. Radium **3**, 590 (1932). — Peirson, D. H.: Proc. Phys. Soc. Lond. B **64**, 876 (1951). — Picciotto, E.: (1) C. R. Acad. Sci., Paris **228**, 2020 (1949); (2) — and S. Wilgain: Nature, Lond. **173**, 632 (1954); (3) — e F. Salvetti: Nuovo Cim. **3**, 815 (1956). — Plesset, M. S., and S. T. Cohen: J. Appl. Phys. **22**, 350 (1951). — Plumb, R. C., and J. E. Lewis: Nucleonics **13**, No. 8, 42 (1955). — Pringle, R. W., W. Turchinetz and B. L. Funt: Rev. Sci. Instrum. **26**, 859 (1955). — Quan, J. T.: US Naval Radiol. Defense Lab. TR-229 (1958). — Raeth, C. H., B. J. Sevold and C. N. Pederson: Rev. Sci. Instrum. **22**, 461 (1951). — Rafter, T. A., and G. J. Fergusson: 2nd Conf. Atomic Energy, Geneva, Proc. **18**, 526 (1958). — Rajewski, B., H. Muth, H. J. Hantke u. K. Aurand: Strahlentherapie **104**, 157 (1957). — Rapkin, E.: Packard Instr. Co. Inc., Tech. Bull. July 1960. — Reed, G. W., H. Hamaguchi and A. Turkevich: Geochim. et Cosmochim. Acta **13**, 248 (1957). — Riezler, W., u. W. Kern: Nukleonik **1**, 191 (1959). — La Rivière, D., and S. K. Ichiki: Nucleonics **10**, No. 9, 22 (1952). — Rosholt, J. N.: 2nd Conf. Atomic Energy, Geneva, Proc. **2**, 230 (1958). — Sackett, W. M., H. A. Potratz and E. D. Goldberg: Science **128**, 204 (1958). — Sandell, E. B.: Colorimetric determination of traces of metals. New York: Interscience 1944. — Schiff, L. J., and R. D. Evans: Rev. Sci. Instrum. **7**, 456 (1935). — Schneider, W., u. T. Matitsch: Wien. Ber. II a **161**, 131 (1952). — Schönfeld, T., u. E. Broda: Mikrochem. **36/37**, 485 (1951). — Scholz, T. (1): 1957, unveröffentlicht; (2) Dipl.-Arb. Heidelberg 1961. Auszug im Druck. — Schraub, A.: Strahlenschutz **6**, 149 (1957). — Schumann, G.: (1) Arch. Meteor. Geophys. u. Bioklim., Ser. A **9**, 204 (1956); (2) Z. angew. Phys. **8**, 361 (1956); (3) Ann. Int. Geophys. Year **5**, 355 (1957); (4) Strahlenschutz **6**, 31 (1957); (5) Strahlenschutz **6**, 90 (1957); (6) Beitr. Phys. Atmosph. **30**, 189 (1958); (7) Strahlenschutz **12**, 106 (1959). — Shapira, J., and W. H. Perkins: Science **131**, 414 (1960). — Sharpe, J.: Nuclear radiation detectors. London: Methuen 1955. — Siksna, R.: Ark. Geophys. **1**, 123 (1950). — Sittkus, A.: Strahlenschutz **6**, 71 (1957). — Smales, A. A.: Analyst **77**, 778 (1952). — Souch, A. E., and D. R. Sweetman: Rev. Sci. Instrum. **29**, 794 (1958). — Spaa, J. H.: 2nd Conf. Atomic Energy, Geneva, Proc. **21**, 169 (1958). — Stebler, A.: Strahlenschutz **6**, 46 (1957). — Steinberg, D.: Nature, Lond. **182**, 740 (1958). — Stern, S. C., H. W. Zeller and A. J. Shekman: J. Colloid. Science **15**, 546 (1960) — Stoeppler, M.: Dipl.-Arb. Univ. Mainz 1958. — Suess, H. E.: Science **122**, 415 (1955). — Sugihara, T. L., R. I. Wolfgang and W. F. Libby: Rev. Sci. Instrum. **24**, 511 (1953). — Suttle, A. D., and W. F. Libby: Analyt. Chem. **27**, 921 (1955). — Tamers, M. A.: Science **132**, 668 (1960). — Thomas, A.: Nucleonics **6**, No. 2, 50 (1950). — Thommeret, J.: J. Phys. Radium **10**, 249 (1949). — Tomic, E., I. M. Ladenbauer u. M. Pollak: Z. analyt. Chem. **161**, 28 (1958). — Tompkins, E. R., J. X. Khym and W. E. Cohn: J. Amer. Chem. Soc. **69**, 2769 (1947). — Volchok, H. L., and J. L. Kulp: Nucleonics **13**, No. 8, 49 (1955). — Vries, Hl. de (1) — and G. W. Barendsen: Physica, Haag **19**, 987 (1953); (2) —, Nuclear Phys. **1**, 477 (1956); (3) Appl. Sci. Res. B **5**, 387 (1956); (4) Nuclear Phys. **3**, 65 (1957); (5) —, M. Stuiver and I. Olsson: Nuclear Instr. **5**, 111 (1959). — H. Wänke u. E. Vilcsek: Z. Naturforsch. **14**a, 929 (1959). — Weiner, E. V., and R. E. Peterson: Nucleonics **13**, No. 7, 54 (1955). — Welford, G. A. (1) —, W. R. Collins, D. C. Sutton and R. S. Morse: US Atomic Energy Commission Rep. HASL 57 (1958); (2) — and W. R. Collins: Science **131**, 1791 (1960). — Werbin, H., I. L. Chaikoff and M. R. Imada: Proc. Soc. Exp. Biol., N.Y. **102**, 8 (1959). — White, G., and S. Helf: Nucleonics **14**, No. 10, 46 (1956). — Whitham, K.: Ph.D. Thesis Toronto Univ. 1951. — Wiedenbeck, M. L.: Phys. Rev. **72**, 974 (1947). — Wilkening, M. H.: Nucleonics **10**, No. 6, 36 (1952). — Winsberg, L.: Geochim. et Cosmochim. Acta **9**, 183 (1956). — Wolfgang, R. L., and W. F. Libby: Phys. Rev. **85**, 437 (1951).

Zusammenfassende Darstellungen kernphysikalischer Nachweismethoden: Baranow, W. J.: Radiometrija, Izd. Akad. Nauk SSSR Moskwa 1956: deutsche Übersetzung. Leipzig, Teubner 1959. — Fünfer, Ewald, u. H. Neuert: Zählrohre und Szintillationszähler, 2. Aufl. Karlsruhe: Braun 1959. — Hecht, Friedrich, u. M. K. Zacherl (ed.): Handbuch der mikrochemischen Methoden, Bd. II: Broda, E., u. T. Schönfeld: Radiochemische Methoden der Mikrochemie; Bernert, T., B. Karlik u. T. Lintner: Messung radioaktiver Strahlen; Lauda, H.: Photographische Methoden in der Radiochemie. Wien: Springer 1955. — Joos, Georg, u. Erwin Schopper: Grundriß der Photographie und ihrer Anwendungen besonders in der Atomphysik. Frankfurt: Akademische Verlags-Gesellschaft 1958. — Price, William J.: Nuclear radiation detection. New York: McGraw-Hill 1958. — Sharpe, J.: Nuclear radiation detectors. London: Methuen 1955. — Wilkinson, D. H.: Ionization chambers and counters. Cambridge: Cambridge Univ. Press 1950. — Yagoda, Herman: Radiation measurements with nuclear emulsions. New York: Wiley 1949.

Die Kosmische Strahlung in der Geophysik[1]

von

A. EHMERT

Mit 23 Figuren

A. Einleitung

Die Kosmische Strahlung (KS) stellt eine Kernstrahlung in anderer Bedeutung des Wortes dar, als es im Titel des Buches gemeint ist. Die Primärstrahlung, wie sie aus dem Weltraum einfällt, besteht aus völlig ionisierten Atomkernen hoher Geschwindigkeit und damit hoher Energie. Die höchsten vorkommenden Energien, welche allerdings recht selten sind, werden heute nach ausgedehnten Teilchenschauern in der Atmosphäre auf etwa 10^{19} eV geschätzt. Aber 90% der Primärteilchen haben Energien unter 10^{10} eV und sind damit in dem Bereich, der heute von den Großgeräten zur künstlichen Teilchenbeschleunigung beherrscht wird.

Da die Energie der Teilchen zum Eindringen in Atomkerne ausreicht, und auch noch große Überschußenergie zur Verfügung steht, führte die Untersuchung der KS zur Entdeckung einer Reihe von Elementarteilchen wie Positron, μ-Meson oder Muon, π-Meson oder Pion, K-Meson, sowie Hyperon, welche schwerer als Protonen sind und als angeregte Protonen aufgefaßt werden können. Auch die Antiteilchen der Mesonen gehören dazu. Das Antiproton wurde allerdings bereits mit künstlich erzeugter Strahlung überzeugend nachgewiesen. Bezüglich der für die Theorie der Teilchen besonders wichtigen hoch energetischen Prozesse mit Erzeugung vieler Teilchen ist man heute noch auf die kosmische Strahlung angewiesen, und dies wird sich nicht mehr wesentlich ändern können. Dieser Aspekt der KS für die Teilchenphysik soll aber hier nur erwähnt werden.

Für die Geophysik erlangte die Erzeugung radioaktiver Isotope durch die kosmische Strahlung Bedeutung, da auf ihr die physikalischen Altersbestimmungen beruhen. Dieser Erscheinung ist ein besonderer Beitrag dieses Buches gewidmet.

Hier sollen nur die Erscheinungen der KS besprochen werden, welche mit dem Geomagnetismus zusammenhängen. Einmal bietet die Bewegung der Primärteilchen im Magnetfeld der Erde die Möglichkeit zu einer Vermessung des magnetischen Außenfeldes, welche weniger von lokalen magnetischen Störungen beeinflußt wird als die Ausmessung des Erdfeldes am Erdboden. Tatsächlich bestehen Differenzen, deren Deutung jedoch schwierig ist. Beide, die magnetische Feldstärke an der Erdoberfläche und die Intensität der kosmischen Strahlung zeigen Variationen mit der Zeit, welche zum Teil in enger Beziehung stehen, da beide Variationen von der Sonnentätigkeit gesteuert werden. Wir wissen heute, daß diese Steuerung der KS im interplanetaren Raum vor sich geht. Auch ein Teil der magnetischen Störungen scheint von dort — wenn auch aus erdnäheren Bezirken — einzuwirken, während andere aus dem direkten Einfall solarer Wellen- und Corpuscularstrahlung in die Erdatmosphäre resultieren. Hier läßt sich Geophysik und Astrophysik nicht mehr getrennt behandeln.

---

[1] Die englische Zusammenfassung dieses Beitrages findet sich auf S. 382.

Genau so ist es mit der solaren KS[1], welche in gleichem Maß als Sonde für das äußere Erdfeld als auch für interplanetare Felder herangezogen werden kann.

Der hier zur Verfügung stehende Raum reicht eben für einen knappen Überblick über die beim derzeitigen Stand bestehenden Probleme und Ergebnisse, und die Anführung der Originalliteratur kann nur sehr beschränkt durchgeführt werden.

B. Die primäre galaktische Kosmische Strahlung

Unter galaktischer KS versteht man die nicht im Sonnensystem erzeugte KS. Ihre räumliche Energiedichte ist zu hoch, als daß man eine ebenso gleichmäßige Erfüllung des außergalaktischen Raumes annehmen könnte. Vielmehr wird eine magnetische Speicherung in unserer Galaxis angenommen.

In Erdnähe weist die Primärstrahlung bereits die erwähnte Modulation durch die Sonnentätigkeit auf, welche eine Veränderung der Energieverteilung und auch der Anteile von Protonen, α-Teilchen und schwereren Kernen bewirkt. Diese Teilchen können bereits in Höhen um 30 km gezählt und nach ihrer Kernladungszahl unterschieden werden. Ihre ionisierende Wirkung in Materie ist eine Funktion ihrer Geschwindigkeit und bei relativistischen Geschwindigkeiten dem Quadrat ihrer Ladung (d.h. aber ihrer Kernladungszahl, solange sie voll ionisiert sind) proportional. Die spezifische Ionisierung kann mit Proportionalzählern festgestellt werden, welche beim Durchgang eines Teilchens einen Stromimpuls liefern, welcher der von dem passierenden Teilchen im Gasraum des Zählers gebildeten Ionenzahl proportional ist. Ebenso wird in dünnen Szintillatoren ein Lichtblitz erzeugt, welcher der Anzahl der Ionisierungsvorgänge durch ein passierendes Teilchen proportional ist; in Photoemulsionen entsteht entlang der Teilchenbahn durch die Ionisierung eine Spur von entwickelbaren Körnern. Die Zahl dieser Körner auf eine passende Länge bezogen, ist ebenfalls ein Maß für die spezifische Ionisierung der Teilchen. Die Geschwindigkeit der Teilchen bestimmt zusammen mit ihrer Ruhemasse ihre Energie. Die Reichweite ist eine Funktion von Energie und Ladung. Die Geschwindigkeit kann auch aus magnetischer Ablenkung, aus Streuerscheinungen und auch aus der Čerenkow-Strahlung bestimmt werden, welche beim Passieren einer durchsichtigen Substanz mit größerem Brechungsindex von den geladenen Teilchen erzeugt wird.

So kann z.B. in einer geeichten Photoemulsion aus der hinterlassenen Körnerspur auf Masse, Ladung, Energie und magnetische Steifigkeit des Teilchens geschlossen werden, wenn man eine ausreichend dicke Schicht darbietet. Diese wird in der Praxis aus einem ganzen Paket von Emulsionsfilmen zusammengesetzt, das mittels hochfliegender Ballone der Primärstrahlung exponiert werden kann und nachher zur Ausmessung der Spuren geborgen werden muß.

Sehr erfolgreich für die Zählung von Protonen und α-Teilchen bei direkter Bestimmung ihrer Energie im Bereich kleiner Energien waren auch Ballongeräte, in welchen für jedes passierende Teilchen die Čerenkow-Strahlung und die Ionisierung gemessen und beide Ergebnisse zum Boden gefunkt wurden. Damit war der Erfolg nicht an die Bergung des Gerätes gebunden, welche in arktischen Gebieten schwierig ist. Bei Teilchen hoher Energie ist die Energiemessung auf diese Weise nicht möglich, jedoch die Bestimmung ihrer Ladung. Solche Messungen in verschiedenen Breiten ergeben mit Hilfe der Störmerschen Theorie auch eine integrale Energieverteilung der Primärstrahlung.

Solche Energieverteilungen für verschiedene Zeiten des letzten Sonnenaktivitätscyclus zeigt Fig. 1 nach Messungen von MacDonald und Webber [1]

---

[1] Die Bezeichnung „Solare Kosmische Strahlung" ist ein Widerspruch, der dadurch zustande kam, daß im englischen Sprachbereich „cosmic rays" auf den ganzen Bereich hoch energetischer Teilchen ausgedehnt wurde.

in einer Zusammenstellung von EHMERT [2]. Es ist als Abszisse der Logarithmus der magnetischen Steifigkeit gewählt. Als Ordinate der Logarithmus des Teilchenflusses aller Teilchen der jeweiligen Art mit einer magnetischen Steifigkeit größer als derjenigen, welcher für dieselbe Teilchensorte die zugehörige Abszisse bestimmt.

Dieses Diagramm gilt in seiner Gesamtheit für hohe geomagnetische Breiten bei senkrechtem Einfall. In diesem Fall hat das Magnetfeld der Erde keinen Einfluß. Dasselbe ist außerhalb des Einflusses der Erde auf der Bahn der Erde zu erwarten.

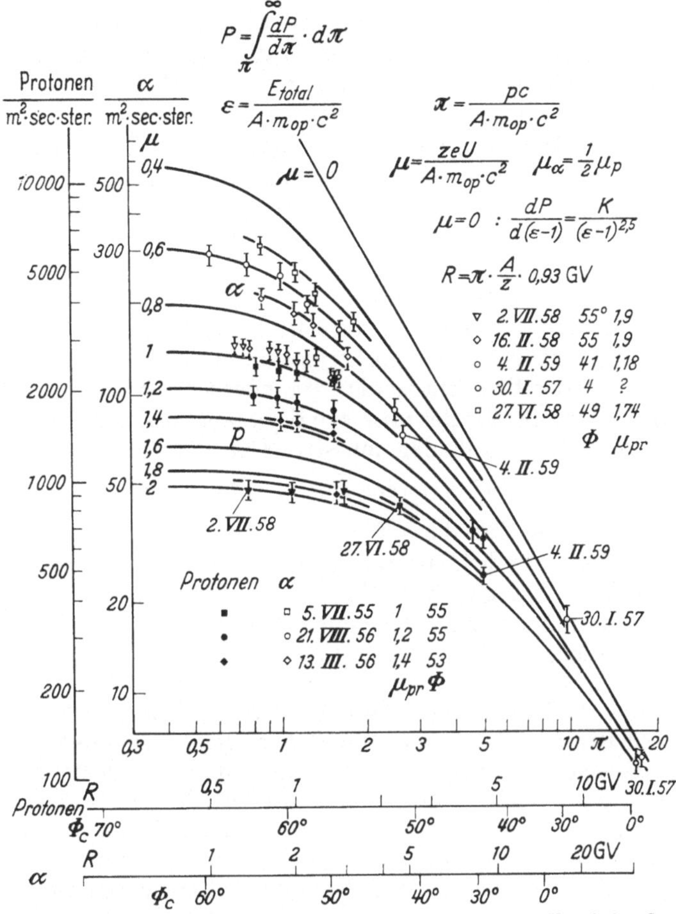

Fig. 1. Die integrale Impulsverteilung der Protonen und der α-Teilchen der primären Kosmischen Strahlung nach Messungen von MCDONALD und WEBBER [1] zu verschiedenen Zeiten. Die Meßpunkte schließen sich gut an die eingezeichneten Kurven an, welche EHMERT für das Modell der elektrischen Modulation für verschiedene Modulationsspannungen berechnete. Der Parameter μ hat eine sehr gute Korrelation mit den Intensitäten, welche am Erdboden zur selben Zeit gemessen wurden. Für die α-Teilchen ist μ immer halb so groß wie für die Protonen

Die Skala der Teilchensteifigkeit ist für α-Teilchen um log 2 nach links gerückt, so daß die Meßpunkte über der Größe $\pi=pc/Mc^2$ erscheinen. Setzt man $M=A\cdot M_{proton}$ und die Ladungszahl der Teilchen gleich Z, so ist die magnetische Steifigkeit $R=\pi\cdot\frac{A}{Z}\cdot 0,93$ GV. Setzt man die Maßzahl $\varepsilon=$ Gesamtenergie eines Teilchens/Ruheenergie, so ist die kinetische Energie in Einheiten der Ruheenergie gewählt $\varepsilon_K=\varepsilon-1$ und es bestehen die bekannten Beziehungen

$$\varepsilon-1=\varepsilon_{Kin},\qquad \varepsilon^2=\pi^2+1,\qquad \pi=\varepsilon_{Kin}\sqrt{1-\frac{2}{\varepsilon_{Kin}}}$$

$$\beta=\frac{v}{c}=\frac{\pi}{\varepsilon},\qquad \pi=\frac{\beta}{\sqrt{1-\beta^2}},\qquad \varepsilon=\frac{1}{\sqrt{1-\beta^2}}\cdot$$

Die in der Fig. 1 eingezeichneten Kurven mit dem Parameter μ sind von Ehmert [2] unter Zugrundelegung der Hypothese berechnet, daß in großem Abstand von der Sonne für beide Teilchenarten ein differentielles Energiespektrum $dP/d\varepsilon_K = K/(\varepsilon-1)^{2,5}$ bis herab zu mindestens $\varepsilon_K = 0,41$ (entsprechend $\pi = 1$) vorliege. Dem entspricht in Fig. 1 die π-Abhängigkeit unter dem Parameter $\mu = 0$. Nach Rossi u. Mitarb. [3] ist aus Messungen in Raketen von Photoemulsionen im Energiebereich bis zu 10^{13} eV nach Messungen in großen Tiefen unter der Erde bis zu 10^{16} eV und nach Messungen an großen Teilchenschauern in der Atmosphäre bis zu 10^{18} eV bekannt, daß das obige Spektrum in diesem ganzen weiten Energiebereich gilt. Bei größeren Energien unterscheidet sich ε und ε_{Kin} ganz unwesentlich, im Bereich der Fig. 1 jedoch erheblich.

Man sieht, daß für beide Teilchenarten das gemessene integrale Spektrum (über π) bei kleinen Impulsen nicht mehr ansteigt. Das differentielle Spektrum also ein Maximum hat und zu kleineren Teilchenimpulsen hin wieder steil abfällt und daß 1955 (bei geringer Sonnentätigkeit) bei beiden Teilchenarten die Gesamtzahl, d.h. alle Teilchen mit $\pi > 0,3$ um fast den Faktor 3 größer war als am 2. 7. 58 bei starker Sonnentätigkeit. Bei $\pi = 10$ ist der Unterschied nur noch 15% und bei hohen Energien verschwindet er ganz.

Die eingezeichneten Kurven sind von Ehmert unter der weiteren Hypothese berechnet, daß die Modulation des Spektrums durch die Sonnentätigkeit dadurch erfolgt, daß am Ort der Erde ein zeitlich veränderliches elektrisches Potential U gegen den Raum in großer Entfernung von der Sonne wirksam ist, gegen welches die galaktischen Teilchen anlaufen müssen. Der Parameter μ der Kurven ist $\mu = \dfrac{Z \cdot e \cdot U}{A \cdot M_\mu \cdot c^2}$, d.h. die dabei verlorene Energie in Einheiten der Ruheenergie der Teilchen. Wir kommen auf diese Hypothese im Kapitel E zurück. Sie wird durch den Zusammenhang der Variationen bei hohen Energien und den Verlauf der Spektren bei kleinen Energien nach Fig. 1 gefordert. Der Wert von μ bzw. U kann aus Registrierung am Erdboden laufend abgelesen werden. Der Parameter μ ist immer für α-Teilchen halb so groß wie für Protonen, da U für beide gleich sein muß.

Die unmodulierte Primärstrahlung enthält bei gleichem π 20mal soviel Protonen als α-Teilchen und damit auch im Bereich hoher Energien, wo $\pi \approx \varepsilon$ auch 20mal soviel, bei gleicher magnetischer Steifigkeit wegen des anderen Verhältnisses A/Z nur 6,3mal soviel Protonen und dieses Verhältnis ist auch im modulierten Bereich der Fig. 1 noch fast konstant. Die Unterschiede nach den Kurven der Fig. 1, welche ja mit elektrischer Modulation berechnet sind, sind geringer als die Meßgenauigkeit. MacDonald und Webber stellten an Hand ihrer Messungen fest, daß beide Teilchenarten immer dieselbe Verteilung über die magnetische Steifigkeit besitzen, also sich ihre Spektren bei der Modulation durch die Sonnentätigkeit konform ändern.

Im Sonnenfleckenminimum ist die Gegenspannung $U \approx 1$ GV. Auch das unmodulierte Spektrum der KS muß einmal von der Kurve für $\mu = 0$ zu kleineren Werten abbiegen. Wir wissen nicht wo. Aber mit wachsendem Abstand von der Sonne ist auf jeden Fall eine weitere Zunahme der KS zu erwarten, und in dieser Richtung deuten nach brieflicher Mitteilung die bei CERN mit der induzierten Radioaktivität in Meteoriten gemachten Erfahrungen.

Die primäre KS enthält weiterhin noch eine geringe Zahl von schwereren Kernen. Diese sind in Photoemulsionen ausgemessen worden. Ihre Massenverteilung ist ein Schlüssel zur Erforschung der Vorgeschichte der KS, welche wir hier nicht behandeln wollen. Wir verweisen auf vorliegende Zusammenfassungen [4]. Ein neuer Aspekt von Biermann und Davies [5] geht dahin, daß die heute bestehende KS schon etwa zur Zeit der Entstehung des Sonnensystems und des Milchstraßensystems auf ihre Energie beschleunigt wurde und seither im Milchstraßensystem und seinem Halo mit etwa 100 000 Lichtjahren Radius gespeichert ist.

Die nachstehende Tabelle gibt nach [4] (etwas modifiziert) eine Zusammenstellung der Elementhäufigkeit in der kosmischen Strahlung und derjenigen der kosmischen Materie nach UREY und SUESS [6]. Die Energieverteilung der schweren Kerne ist etwa die gleiche wie die der α-Teilchen, auch bei der Modulation. Z/A ist bei den stabilen Isotopen durchweg ebenfalls nahe bei $\frac{1}{2}$. Sofern man auf gleiche Energie pro Nukleon bezieht, verhält sich ein schwerer Kern

Tabelle. *Häufigkeitsverteilung der Kernarten in der kosmischen Strahlung und im Weltall, jeweils bezogen auf 10^5 Protonen und bei der kosmischen Strahlung auf gleiche Energie pro Nukleon*

Kernladungszahl	Kern	Gruppe	KS	Kosmische Materie
1	H		10^5	10^5
2	He		$5 \cdot 10^3$	$7,7 \cdot 10^3$
3	Li			
4	Be	L (leicht)	170	$3,6 \cdot 10^{-4}$
5	B			
6	C			
7	V	M	520	80
8	O	(mittelschwer)		
9	F			
11 bis 30	Na bis Zn	S (schwer)	170	30
(26)	(Fe)		(30)	(1,5)
30 bis 92	Ga bis U	Rest	1	10^{-3}

mit dem Atomgewicht A wie $A/4$ gekoppelte α-Teilchen, bei großen Kernen mit Neutronenüberschuß wäre die Modulation wieder etwas schwächer zu erwarten. Bezogen auf die gleiche kinetische Energie der ganzen Kerne stimmt die Größenordnung der beiden letzten Spalten in der Tabelle auch für die mittelschweren und die schweren Kerne wieder überein, wenn man ebenfalls eine integrale $E^{-1,5}$-Verteilung annimmt. Das Fehlen der leichten Gruppe (L) in der kosmischen Materie rührt vom hohen Wirkungsquerschnitt dieser Kerne für thermonucleare Prozesse in den Sternatmosphären her.

C. Der Einfluß des Magnetfeldes der Erde auf die galaktische Kosmische Strahlung

I. Die Bewegung der Teilchen

Die Teilchen der kosmischen Strahlung werden wegen ihrer Ladung in Magnetfeldern abgelenkt. Ist der Impuls des Teilchens $\mathfrak{p} = m_0 \cdot \mathfrak{v}/\sqrt{1 - \beta^2}$, wobei m_0 seine Ruhemasse und $\mathfrak{v}$ seine Geschwindigkeit sei $\left(\beta = \frac{|\mathfrak{v}|}{c}\right)$, so lautet die Differentialgleichung der Bahn im Magnetfeld $\mathfrak{H}$ bei seiner elektrischen Ladung $Z \cdot e$

$$\frac{d}{dt}\mathfrak{p} = \frac{Z \cdot e}{c}[\mathfrak{v}\,\mathfrak{H}]. \tag{1}$$

Der Absolutbetrag des Impulses $p = m_0\, c^2\, \beta/c\sqrt{1 - \beta^2} = A \cdot m_p \cdot c \cdot \pi$ (π vgl. Abschnitt B) bleibt dabei konstant. Im Falle einer Bewegung senkrecht vom

Magnetfeld entsteht eine Kreisbahn mit dem Krümmungsradius ϱ und es ist die magnetische Steifigkeit

$$R = H\varrho = \frac{pc}{Z \cdot e} = \frac{A}{Z} \cdot \frac{m_p \cdot c^2}{e} \cdot \pi = \frac{A}{Z} \cdot \frac{m_p c^2}{e} \cdot \varepsilon_{\text{Kin}} \cdot \sqrt{1 + \frac{2}{\varepsilon_{\text{Kin}}}}.$$

Wird ϱ in cm und H in Gauß gemessen, so ist

$$R = H\varrho = \frac{A}{Z} \cdot \pi \cdot 3{,}1 \cdot 10^6 \text{ Gauß cm}.$$

Die Bewegung im Feld der Erde ist jedoch kompliziert. Ihr sei hier ein etwas breiterer Bericht gewidmet, da sie die Grundlage für die Verknüpfung von kosmischer Strahlung und Geomagnetismus bildet.

Das Erdfeld wurde von Störmer [7] als das Feld eines Dipols genähert. Hat dieser ein magnetisches Moment M, so ist seine Feldstärke als Ortsfunktion

$$\mathfrak{H} = -\operatorname{grad}\left(\frac{M}{r^3}\right). \tag{2}$$

Da p und v sich entlang einer Teilchenbahn nicht ändern, führte Störmer als zweckmäßige Längeneinheit die Strecke

$$l_0 = \sqrt{\frac{ZeM}{pc}} = \sqrt{\frac{\text{magnetisches Moment des Dipols}}{\text{magnetische Steifigkeit des Teilchens}}} \tag{3}$$

ein und an Stelle der Zeit die Bogenlänge längs der Bahn ein: $ds = v \cdot dt/r$. Durch Übergang zu Zylinderkoordinaten: z als Höhe über der Ebene durch die Mitte des Dipols senkrecht zu dessen Achse, ϱ als Abstand von der Dipolachse und λ als Winkel der Ebene durch die Dipolachse und das Teilchen gegen eine feste Ebene durch die Dipolachse, kam er aus (1) und (2) zu

$$\frac{d}{ds}\left(\varrho^2 \frac{d\Omega}{d\varrho} + \frac{\varrho^2}{(\varrho^2 + z^2)^{\frac{3}{2}}}\right) = 0 \tag{4}$$

mit dem Integral

$$\varrho^2 \frac{d\lambda}{d\varrho} + \frac{\varrho^2}{(\varrho^2 + z^2)^{\frac{3}{2}}} = 2\gamma, \tag{4a}$$

wobei die Integrationskonstante 2γ eine anschauliche Bedeutung hat. Sie ist der doppelte Wert des Produktes aus der Impulskomponente des Teilchens senkrecht zum Dipol in unendlicher Entfernung des Teilchens von demselben und dem Abstand der Bahntangenten im Unendlichen von der Dipolachse, der in Einheiten von l_0 zu messen ist.

Die Separierung der Differentialgleichung gelang Störmer durch die Einführung einer neuen Funktion

$$P(\varrho, z, \gamma) = 1 - \left(\frac{\varrho}{(\varrho^2 + z^2)^{\frac{3}{2}}} - \frac{2\gamma}{\varrho}\right)^2. \tag{5}$$

Gl. (4a) geht damit über in $\varrho^2 (d\lambda/ds)^2 = 1 - P(\varrho, z, \gamma)$ und wegen der Eigenschaft der Zylinderkoordinaten

$$\left(\frac{d\varphi}{ds}\right)^2 + \left(\frac{dz}{ds}\right)^2 + \varrho^2\left(\frac{d\lambda}{ds}\right) = 1$$

wird

$$P(\varrho, z, \gamma) = \left(\frac{d\varrho}{ds}\right)^2 + \left(\frac{dz}{ds}\right)^2 \tag{6}$$

und damit kann (1) aufgespalten werden in

$$\frac{d^2\varrho}{ds^2} = \frac{1}{2}\frac{\delta P}{\delta\varrho} \quad \text{und} \quad \frac{d^2 z}{ds^2} = \frac{1}{2}\frac{\delta P}{\delta z}. \tag{7}$$

Die Bahn des Teilchens ist somit zerlegt in eine Bewegung in einer Ebene, welche dauernd das Teilchen und die Dipolachse enthält (wir nennen sie die Meridianebene des Teilchens) und die Drehung dieser Ebene. (7) weist aus, daß die für jeden festen Wert von γ berechenbare Funktion $P_{(\varrho,\,z)}$ den Charakter eines Potentials hat.

Stellt man P als Relief über der ϱ, z-Ebene dar und läßt darin eine Kugel mit der Geschwindigkeit $v = \dfrac{1}{4\gamma^2}$ an einem tief gelegenen Ausgangspunkt (z.B. in großer Entfernung oder im Ursprung) abrollen, so beschreibt die Projektion ihrer Bahn eine entsprechende Teilchenbahn in der Meridianebene.

STÖRMER hat das Niveaulinienbild von $P(\varrho, z, \gamma)$ für viele Werte von γ gegeben und durch numerische Integration zahlreiche Bahnen in der Meridianebene wie auch im Raum berechnet.

Außerdem hat STÖRMER aus (4) eine Übersicht über erlaubte Bereiche abgeleitet. Führt man den Winkel Θ zwischen der Bahntangente im Teilchenpunkt P (Fig. 2) und der Normalen zur Meridianebene des Teilchens ein, so ist aus geometrischen Gründen $\cos\Theta = -\varrho\, d\lambda/ds$ und damit geht (4a) über in

$$- \cos\Theta = \frac{2\gamma}{\varrho} - \frac{\varrho}{(\varrho^2 + z^2)^{\frac{3}{2}}}. \qquad (8)$$

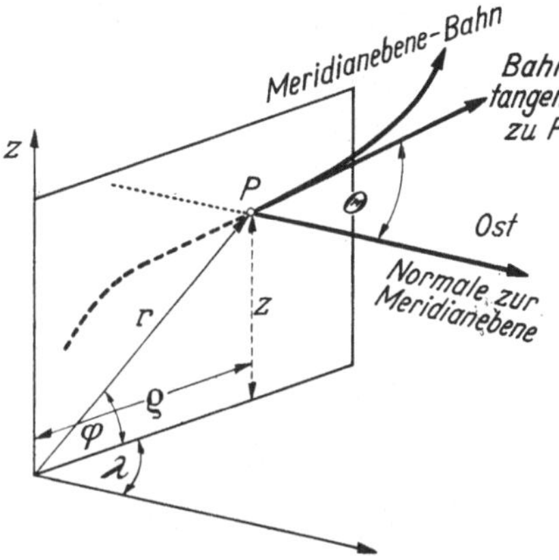

Fig. 2. Zur Definition des Winkels Θ in der Störmerschen Theorie

Da $-1 \leqq \cos\Theta \leqq +1$ sein muß, ergibt sich die Einschränkung der Bereiche, in welchen bei einem jeweils festen γ überhaupt Bahnen möglich sind zu

$$-1 \leqq \frac{2\gamma}{\varrho} - \frac{\varrho}{(\varrho^2 + z^2)^{\frac{3}{2}}} \leqq +1. \qquad (9)$$

Fig. 3 zeigt für verschiedene γ-Werte die erlaubten (hell) und die verbotenen Teile der Meridianebenen. Rotation um die Dipolachse gibt die entsprechenden Räume.

Bei $\gamma = 1,0$ schnürt sich ein erlaubtes Gebiet ab. Es ist für größere Drehimpulse γ der Teilchen vom erlaubten Außenraum getrennt. Es kann also keine kosmische Strahlung von außen eindringen und darin erzeugte Teilchen bleiben gefangen. Damit war die Theorie der Strahlungsgürtel schon lange gegeben. Man glaubte nur nicht, daß dort tatsächlich Teilchen geeigneter Energie entstehen oder eingeschleust werden.

Die Erde hat in den Bildern der Fig. 3 einen nach (3) von der magnetischen Steifigkeit der Teilchen und vom magnetischen Moment des Dipols abhängigen Radius. Mit $M = 8,1 \cdot 10^{25}$ Gauß cm³ und dem Erdradius $r_e = 6,38\,(1 + \Delta) \cdot 10^8$ cm wird der einzuzeichnende Radius

$$r_0 = r_e \cdot \sqrt{\frac{\mu\,c}{Z\,e \cdot \text{Masse}}} \cdot (1 + \Delta) \qquad (10)$$

oder umgekehrt der zu einem r_0 in Störmer-Einheiten gehörigen Impuls

$$p = 59{,}7\ \text{GeV} \cdot r_0 \cdot \frac{Z}{c}\,(1 + \Delta) \quad \text{oder} \quad \pi = 61{,}7 \cdot \frac{Z}{A} \cdot r_e^2\,(1 + \Delta)^2, \qquad (10a)$$

wobei π die im Abschnitt B benutzte Maßzahl für pc ist, welche in Fig. 1 auftritt. Drückt man die Koordinaten ϱ und z in r_0 und der geomagnetischen Breite φ aus, so geht (8) über in

$$ - \cos \Theta = \frac{2\gamma}{r_0 \cos \varphi} - \frac{\cos \varphi}{r_0^2} . \tag{11} $$

Für konstantes r_0 und γ ist der Teilcheneinfall verboten bis Θ gerade 180° wird. Damit können eben $+$-Teilchen genau aus Westen den Meßort erreichen. Dann nimmt Θ ab und $\Omega = 180° - \Theta$ nimmt zu. Der Kreiskegel mit Ω um die Westlinie ist mit Teilchen erfüllt, und wenn $\Omega = 180°$ geworden ist, sind alle Richtungen erlaubt. Man nennt diesen Kegel den Störmer-Konus.

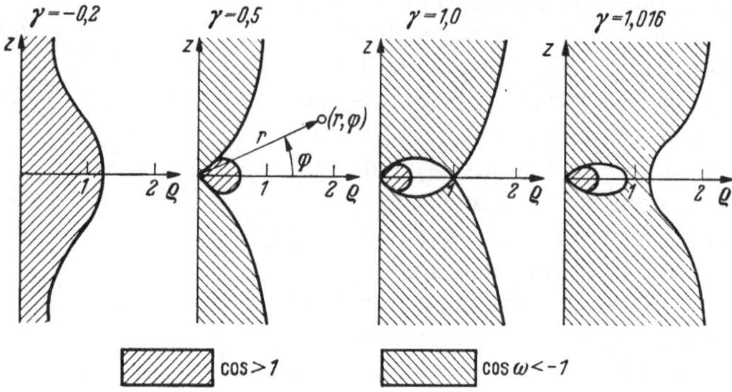

Fig. 3. Erlaubte und verbotene Zonen in der $\varrho\, z$-Meridianebene für verschiedene γ-Werte. $\cos \Theta > 1$ und $\cos \Theta < -1$ sind nicht möglich. (Nach R. Lüst in W. Heisenberg, Kosmische Strahlung, S. 9—20. Springer 1953)

Für senkrechten Einfall auf die Erdoberfläche ist $\cos \Theta = 0$ und damit aus (11) $r_0 = 2\gamma \cos^2 \varphi$ und nach (10a) können damit alle Teilchen mit

$$ \pi \geqq 61{,}7\, \frac{Z}{A}\, (1 + \varDelta) \cdot \frac{\cos^4 \varphi}{4\gamma}\bigg|_{\gamma=1} \to 15{,}4 \cdot \frac{Z}{A}\, (1 + \varDelta) \cdot \cos^4 \varphi \tag{12} $$

einfallen. In der Äquatorebene ist dabei $\gamma = 1$ die Grenze für γ.

Tatsächlich liegen die Dinge komplizierter. Ein Teil der vom Störmer-Kegel zugelassenen Bahnen wäre nur bei völlig transparenter Erde möglich. Diese mathematischen Bahnen tauchen in die Erde ein und wieder aus und werden so abgeschattet.

Besonders stark ist dies bei Bahnen, welche den Meßort aus der Richtung des näher gelegenen Pols unter größeren Zenitwinkeln erreichen. Man kann hier direkt einen Schattenkegel definieren. Dann folgt von der Grenze des Störmer-Kegels nach Westen ein breites Gebiet, in welchem ein Teil der Bahnen durch die Erde abgeschattet sind, während dazwischen andere Bahnen an der Erde vorbeiführen. Man nennt dieses Gebiet der Einfallsrichtung an einem bestimmten Ort Halbschattenzone oder nach Lemaître [8] Penumbra. Sie wird weiter nach Westen durch den Hauptkegel begrenzt, in welchem keine Abschattung mehr eintritt. Die Penumbra ist in niederen Breiten sehr wenig besetzt, in hohen Breiten fast ganz besetzt und dazwischen sehr kompliziert.

Der Teilchenfluß, ausgedrückt als Teilchen/cm² sec sterad für eine bestimmte Energie, ist entlang einer erlaubten Bahn immer konstant, wie Störmer an Hand des Liouvilleschen Theorems für rein magnetische Ablenkung (ohne Energieverlust) gezeigt hat und ist deshalb an jeder Stelle im Feld genau so groß wie er außerhalb des Feldes war, wo wir für die galaktische kosmische Strahlung weitgehende Richtungsisotropie annehmen müssen. Für einen endlichen Raumwinkel ist also nur der Prozentsatz der darin zugelassenen Richtungen für den resultierenden mittleren Teilchenfluß maßgebend.

Fig. 4 zeigt nach VALLARTA [9] die Breitenabhängigkeit der kritischen Energie für Protonen verschiedene Zenitwinkel ϑ sowohl für den Störmer-Kegel als untere Kurve und für den Hauptkegel. Der Grad der Schraffur ist als Maß dafür gedacht, wie stark die Abschattung dazwischen ist. Alle unter der kritischen Energie E_1 liegenden Energien sind verboten, alle über E_2 liegenden erlaubt. Die oben benutzten π-Werte werden aus den angegebenen Energien durch Division mit 0,93 GeV erhalten. Man erkennt, daß das Halbschattengebiet schmal genug ist, um in der Praxis mit einer mittleren kritischen magnetischen Steifigkeit zu rechnen.

Weiter erkennt man, daß in geringen Breiten bei gleichem Zenitwinkel aus Westen die Energiegrenze niedriger liegt und deshalb nach der Energieverteilung (Fig. 1) ein Strahlungsüberschuß aus Westen besteht. Er konnte noch in den Sekundärstrahlungen am Erdboden nachgewiesen werden und erlaubte schon in den dreißiger Jahren zusammen mit dem Breiteneffekt den Schluß, daß man es mit einer Primärstrahlung überwiegend positiver Ladung zu tun hat, denn für negative Ladung der Teilchen wäre ein umgekehrter „Ost-West-Effekt" zu erwarten gewesen.

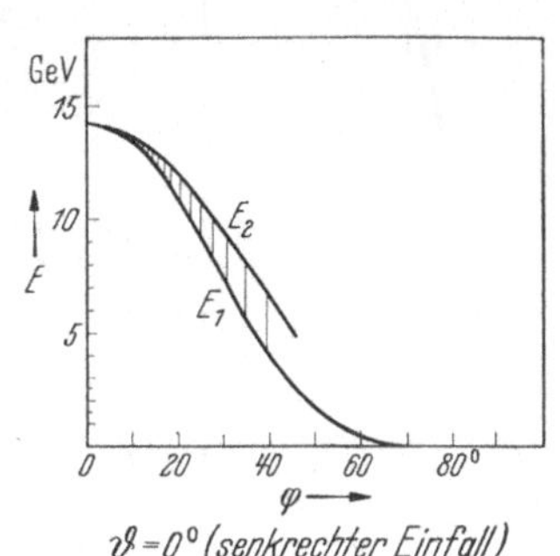

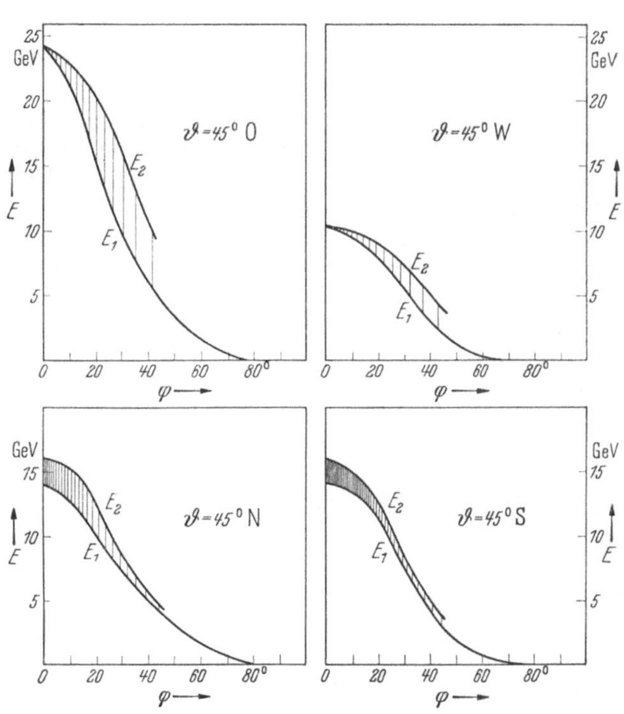

Fig. 4. Die kritische Energie E des Störmer-Kegels und E_2 des Hauptkegels in Funktion der geomagnetischen Breite bei verschiedenen Zenitwinkeln ϑ. (Nach R. LÜST in W. HEISENBERG, Kosmische Strahlung, S. 9—20. Springer 1953)

II. Der Breiteneffekt

Die Breitenabhängigkeit der von der kosmischen Strahlung am Boden bewirkten Ionisation hat CLAY [10] zuerst auf Schiffsreisen nach Afrika nachgewiesen. Sie war gering, etwa 20%, weil in diese Meßgröße hohe Primärenergien stark bevorzugt eingehen. Die in Fig. 1 gezeigten Messungen bezogen sich auf die annähernd senkrecht einfallende Primärstrahlung selbst. Die Meßpunkte bei hohen π-Werten stellen den gesamten Teilchenfluß der betreffenden Art über

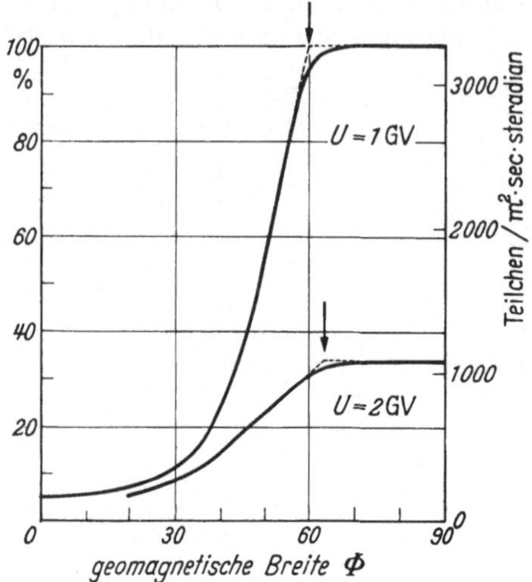

Fig. 5. Der Breiteneffekt der Primärstrahlung im Sonnenfleckenminimum und im Sonnenfleckenmaximum

der für den Meßort berechneten magnetischen Steifigkeit dar, wobei jedoch bereits eine verbesserte Darstellung des Erdfeldes zugrunde gelegt wurde, auf die wir noch zurückkommen. Zu den Skalen der magnetischen Steifigkeit haben wir die Skalen für die geomagnetische Breite hinzugefügt, in welcher jeweils alle Teilchen einfallen können, welche ein größeres R haben. Man erkennt, daß der Breiteneffekt von der Modulation durch die Sonne abhängt.

Wir haben in Fig. 5 die Breitenabhängigkeit der Gesamtteilchenzahl nach den Kurven der Fig. 1 bei $\mu = 2$ im Sonnenfleckenminimum in linearem Maßstab aufgetragen. Charakteristisch ist der steile Anstieg im Gebiet mittlerer Breiten und der als Knie bekannte Knick, der hier in Breiten von 50—70°

auftritt und sich mit zunehmender Modulation etwas zu größeren Breiten verschiebt. Er ist nach unserer Meinung nur eine Folge der Modulation und ihrer Entfernung energiearmer Teilchen aus dem Primärspektrum. Für die unmodulierte Strahlung mit $\mu = 0$ würde die Teilchenzahl bis zum Pol ansteigen.

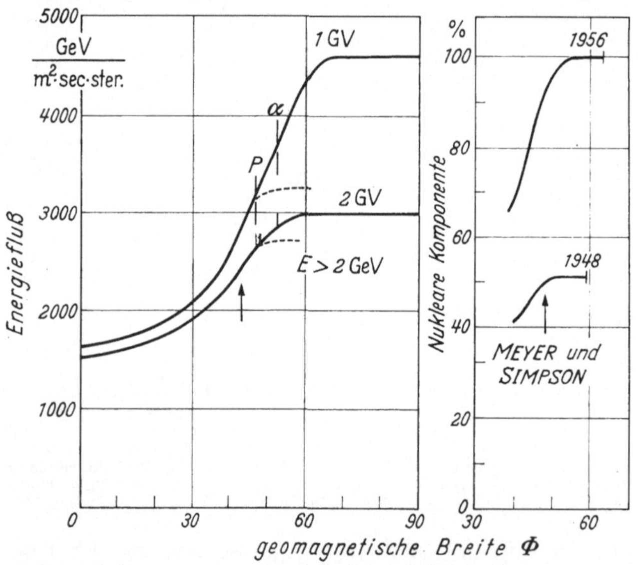

Fig. 6. Der Breiteneffekt des Energieflusses der Kosmischen Strahlung bei senkrechtem Einfall (links) und dazu rechts Messungen der nuclearen Komponente nach Meyer und Simpson bei 310 g/cm² Luftdruck

Fig. 6 zeigt die an Hand der in Fig. 5 dargestellten Kurven berechnete Breitenabhängigkeit des senkrecht einfallenden Energieflusses für das Sonnenfleckenminimum $(U = 1$ GeV$)$ und das Maximum $(\mu = 2)$. Die Kurven geben den Integralwert des Produkts aus Teilchenzahl und Energie. Die α-Teilchen tragen bei allen Breiten 9% zum Energiefluß bei.

Das Knie dieser Kurve für 2 GeV ist gegenüber Fig. 5 nach Süden

gerückt. Es verschiebt sich weiter zu kleineren Breiten, wenn man alle Primärteilchen unterhalb einer festen Energie E_1 wegläßt. Dies ist in Fig. 6 durch gestrichelte Linien für $E_1 = 2$ GeV dargestellt. Dies ist von Bedeutung, wenn Sekundärkomponenten am Boden oder in der Atmosphäre gemessen werden, zu deren Erzeugung eine Mindestenergie erforderlich ist. Allerdings steigt die Ausbeute der Erzeugung der Sekundären dann nicht sprunghaft sondern mit zunehmender Energie nur langsam an, so daß das Knie stark abgerundet wird und der Gesamteffekt kleiner bleibt. Wir haben in Fig. 6 rechts Messungen der kernaktiven Komponente (Abschnitt F) im Flugzeug bei einem Luftdruck von 310 g/cm² nach MEYER und SIMPSON [11] dazugezeichnet. Der Ordinatenmaßstab ist willkürlich gewählt, aber für beide Kurven gleich. Diese Meßgröße ist in dieser Höhe etwa der Energie der Primärstrahlung proportional, doch ist noch ein Energieverlust abzuziehen und außerdem entsprechen die Daten von U 1956 im Mittel der Messungen und derjenigen 1948 nicht gerade 1 und 2 GV. Man erkennt aber die Ähnlichkeit. Damals glaubte man auf eine Verschiebung der Abschneideenergie durch das Erdfeld schließen zu müssen. Die linke Seite von Fig. 6 ist jedoch mit konstanten Abschneideenergien gerechnet. Die Modulation des Primärspektrums ist die maßgebende Veränderung. Weitere Beispiele werden bei den Sekundärkomponenten behandelt.

III. Der Längeneffekt

Auch dieser wurde zuerst von CLAY gefunden. Er besagt, daß der Breiteneffekt auf verschiedenen Meridianen der Erde unterschiedlich ist. Man konnte ihn auf eine Exzentrizität der Magnetachse der Erde zurückführen. Sie geht 400 km am Erdmittelpunkt vorbei und ist außerdem gegen die Erdachse um einen Winkel $\alpha = 11{,}5°$ verdreht. Die Exzentrizität hat zur Folge, daß auf einer Seite der Erde der Abstand der Oberfläche vom Dipolmittelpunkt um 12,5% weiter entfernt ist als auf der anderen Seite. Wir haben deshalb in den Formeln (10), (10a) und (12) den Erdradius mit $r_e \cdot (1 + \Delta)$ bezeichnet. Es ergibt sich eine Verschiedenheit der magnetischen Grenzsteifigkeit, welche den Längeneffekt verursacht. Er muß in die Analyse einbezogen werden.

IV. Die Verteilung über die Erde, die Isokosmen

Es ist nun möglich für jeden Ort auf der Erde die Grenze der magnetischen Steifigkeit zu berechnen, mit welcher eben noch Teilchen senkrecht einfallen können. Man kann dazu das oben genannte Feld eines exzentrischen Ersatzdipols als Erdfeld einführen oder auch noch weitere Verbesserungen hinzufügen wie etwa Quadrupolglieder (VALLARTA). Die Ausführung der Integrationen mit elektronischen Rechenmaschinen erlaubt sehr eingehende Prüfungen. Fig. 7 zeigt nach WEBBER [12] eine Karte der Linien gleicher Grenzsteifigkeit für das exzentrische Dipolfeld bester Näherung an die magnetischen Messungen. Die eingetragenen Werte gelten für Teilchen mit $A = 2Z$, also α-Teilchen. Für Protonen sind die Werte zu verdoppeln. Entlang den Linien gleicher Grenzsteifigkeit ist konstante Intensität der kosmischen Strahlung zu erwarten. Die Linien sind also „Isokosmen". Solche Karten haben COMPTON u. Mitarb. u. a. schon vor 30 Jahren auf Grund der Ionisationsmessungen auf Schiffen und an einem Stationsnetz veröffentlicht. Die Genauigkeit war nicht sehr groß, doch hatten diese Isokosmen einen ähnlichen Verlauf wie die Linien in Fig. 7, welche sich in höheren Breiten den Linien gleicher Nordlichthäufigkeit gut anpassen.

Die nucleare Komponente, welche mit Neutronenzählern gemessen wird (man vgl. Abschnitt E), hat einen wesentlich größeren Breiteneffekt als die Ionisation durch die μ-Mesonen, wie sie mit Ionisationskammern gemessen wird. Insbesondere wenn man im Flugzeug bei konstantem Luftdruck mißt, kann man bei einem Flug entlang einem Meridian die Breite mit geringster Stärke der nuclearen Komponente recht genau bestimmen, wenn man die im Abschnitt E besprochenen zeitlichen Variationen eliminiert oder Zeiten geringer Variationen dieser Art auswählt. Fig. 8 zeigt den von verschiedenen Gruppen [13], [14] so 1956 bis 1958 gemessenen „Äquator für die kosmische Strahlung" (I) im Vergleich zu dem aus Fig. 7 zu entnehmenden für das exzentrische Dipolfeld erwarteten

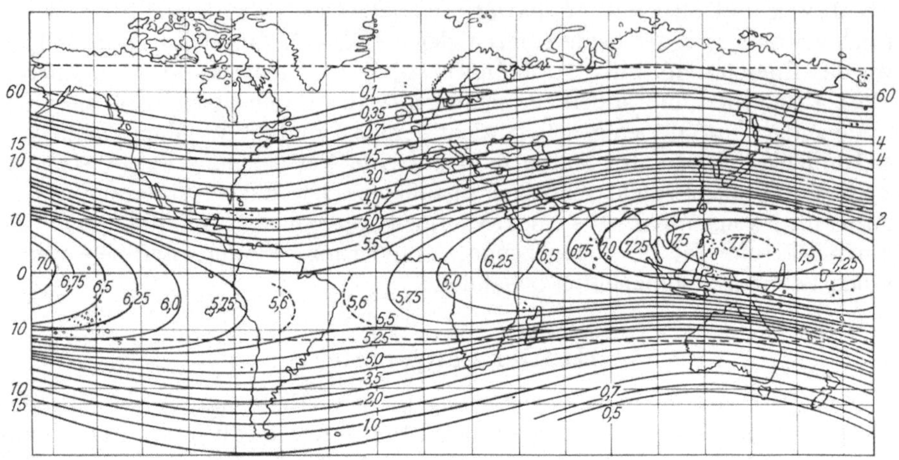

Fig. 7. Karte der Linien gleicher magnetischer Grenzsteifigkeit, welche gleichzeitig als Isokosmen zu erwarten sind, nach Webber [12] für das exzentrische Dipol-Modell für das magnetische Erdfeld im Jahre 1955. Die eingetragenen Werte gelten für α-Teilchen, für Protonen sind sie zu verdoppeln

„Äquator" (II) und die Linie (III) mit der Inklination 0 des erdmagnetischen Feldes an der Erdoberfläche nach den magnetischen Messungen im Jahre 1945. Man erkennt, daß im Bereich Afrika—mittlerer Atlantik eine kräftige auch in der Magnetik zum Ausdruck kommende Störkomponente des erdmagnetischen Feldes vorliegt, welche Neuberechnungen für die Bewegung der kosmischen Strahlung erfordert.

Eine ideale Messung dieser Art wäre die direkte Messung der Primärteilchenzahl (getrennt nach Protonen und α-Teilchen) in einem künstlichen Erdsatelliten, dessen Bahn möglichst hoch gegen den Äquator geneigt ist und durch die Elliptizität der Bahn einen gewissen Höhenbereich bestreicht.

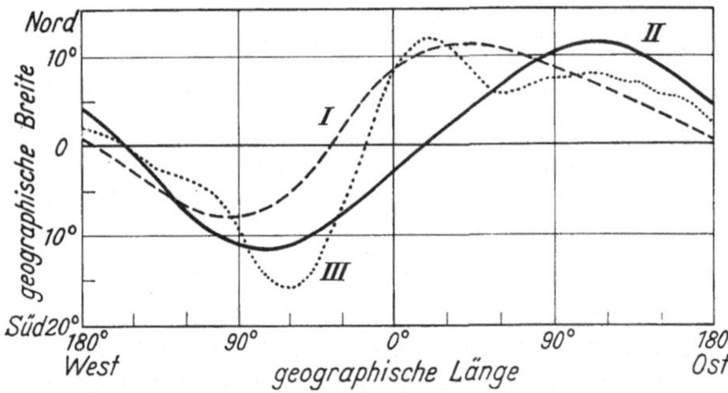

Fig. 8. I KS-Äquator (geringste Intensität in jedem Meridian). II Zu erwarten nach dem exzentrischen Dipol-Modell. III Linie mit horizontalen magnetischen Kraftlinien (nach Katz, Meyer und Simpson)

Die empfindliche Richtung des Meßelementes, etwa eines kombinierten Szintillatorteleskopes, mit Messung der Čerenkow-Strahlung der Teilchen müßte dauernd registrieren und annähernd zum Zenit ausgerichtet werden. Durch Feststellung etwa der Sonnenrichtung müßte die Ausrichtung kontrolliert werden. Natürlich müßte man die drei Komponenten des Erdfeldes ebenfalls mitmessen. Schließlich sollte dieses Experiment zur Zeit des Minimums der Sonnenaktivität ausgeführt werden und die Teilchen des Strahlungsgürtels müßten durch einen Bleipanzer von einigen Zentimeter Dicke eliminiert werden. Die elektronische Auswertung der Meßdaten in Abhängigkeit von den geographischen Daten und der Höhe würde eine sehr genaue Analyse zulassen. Von besonderem Interesse wären dann die Differenzen, welche sich bei höherer Sonnenaktivität einstellen. Solche Messungen rücken in den Bereich des Möglichen.

Bisher war die Ausrichtung eines Satelliten nicht möglich. Es wurde mit allseitig empfindlichen Detektoren, z. B. Geiger-Müller-Zählern, gearbeitet und dabei der Strahlungsgürtel der Erde entdeckt. Soweit die Satelliten-Bahnen unterhalb desselben verlaufen, ist auch eine Ausweitung der Messungen für das vorliegende Problem im Gange. Auf eine andere Art der Untersuchung des erdmagnetischen Feldes mit Hilfe der Bewegung solarer kosmischer Strahlung kommen wir im Abschnitt F zurück.

D. Die Sekundärstrahlung

Die Primärteilchen der kosmischen Strahlung stoßen in der Atmosphäre auf Kerne der Luft und erzeugen dabei in hochenergetischer Wechselwirkung neben schweren Kerntrümmern α-Teilchen und Protonen und Neutronen mit begrenzter Reichweite, auch schnelle Neutronen und vor allem Pionen (man gebraucht auch den Ausdruck π-Mesonen) hoher Energie. Diese sind instabil. Sie wandeln sich nach einer mittleren Lebensdauer von $2{,}5 \cdot 10^{-8}$ sec in ein Müon (μ-Meson) um, sofern sie elektrische Ladung haben. Da sie gegen unser erdfestes System große Geschwindigkeit haben, erscheint ihre Lebensdauer in unserem System um einen Faktor $(1 - (v/c)^2)^{-\frac{1}{2}}$ verlängert und es kommen Wegstrecken von Metern bis zu einigen Kilometern vor, letztere allerdings selten. Inzwischen können die Pionen selbst wieder mit etwa gleicher Wahrscheinlichkeit, welche von der Luftdichte abhängig ist, auf neue Kerne stoßen und dort wieder neue Kernwechselwirkungen mit Erzeugung neuer Pionen auslösen. Gerade bei hohen Energien überwiegt wegen der längeren Lebensdauer in unserem System die Wahrscheinlichkeit neuer Kerntreffer die Wahrscheinlichkeit des Zerfalles. So bilden sich ganze Kaskaden von π-Mesonen aus, welche auch von Neutronen, Protonen und Elektronen begleitet sind.

Die geladenen Teilchen verlieren auf ihrem Weg durch Ionisation der Luft laufend an Energie, die Neutronen ebenso durch Kernstöße. Auf diese Weise scheiden die langsamen Teilchen in der Luft rasch aus.

Die ungeladenen Pionen können ebenfalls weitere Kernzertrümmerungen auslösen. Sie zerfallen aber ebenfalls ohne Wechselwirkung in zwei γ-Quanten gleicher Energie und diese Quanten hoher Energie erzeugen ihrerseits in Wechselwirkung mit den Atomen der Luft Elektron-Positron-Paare. Diese Teilchen erzeugen wieder in der Luft Bremsstrahlung, also γ-Quanten, welche bei genügender Energie wiederum Elektronenpaare erzeugen usw., bis die Energie nicht mehr zur Paarerzeugung ausreicht. Außerdem verlieren die Quanten Energie durch Compton-Prozesse, d.h. durch Energieübertragungen an Elektronen und die Elektronen werden durch ihre Ionisation der Luft gebremst. Bei extrem hoher Primärenergie können durch diese kaskadenartigen Vermehrungen so große Teilchenschauer erzeugt werden, daß man die von den vielen Teilchen in der Luft erzeugte Ionisation mit Radar feststellen kann. Diese Schauer haben einen Kern von 10 bis 50 m Durchmesser am Erdboden, in welchem die Teilchen hoher Energie konzentriert sind. Teilchen und Quanten geringerer Energie sind über eine Fläche von der Größenordnung eines Quadratkilometers und mehr zerstreut.

Aus dem in Abschnitt B besprochenen Energiespektrum ersieht man aber, daß diese hohen Primärenergien recht selten sind. Die große Mehrzahl der Primärteilchen erzeugt nur kleine Kaskaden, deren Energie rasch verbraucht wird. Deshalb hat die Teilchenzahl, welche mit zunehmendem Luftdruck durch die

Sekundärteilchen ansteigt, bei 100 mb Luftdruck, also in einer Höhe von 18 km, ein Maximum und fällt dann mit wachsendem Luftdruck wieder stark ab. Dies gilt für die annähernd vertikal fliegenden Teilchen. Schräg unter dem Zenitwinkel α einfallende Strahlung hat entsprechend der durchlaufenen Luftschicht ihr Maximum schon beim Luftdruck von (100 mb) $\cdot$ cos α. Dadurch hat die Teilchenzahl aus allen Richtungen ein äußerst flaches Maximum und beim Vorliegen sehr energiearmer Primärstrahlung, wie z.B. in hohen Breiten während des Sonnenfleckenminimums, rückt das Maximum zu geringerem Luftdruck. Man vgl. Fig. 10.

Die durch die Umwandlung der geladenen Pionen entstehenden Müonen haben nur einen sehr geringen Wirkungsquerschnitt für Kernwechselwirkungen. Für sie entfällt die weitere Energieaufteilung durch Kaskadenbildung. Sie verlieren nur ständig Energie durch Ionisierung von Luft. Dies macht bei senkrechter Durchquerung der Luftschicht etwa 2 GeV aus. Sie bringen deshalb die meiste Energie der kosmischen Strahlung durch die Luftschicht und sind noch in Tiefen von 1000 m Wasser nachweisbar. Allerdings ist ihre Zahl dort sehr gering. Die Abnahme ihrer Zahl mit der Wassertiefe gibt über ihre Energieverteilung Aufschluß. Am Erdboden stellen sie mit ihren Sekundärteilchen den weit überwiegenden Anteil der Teilchen der kosmischen Strahlung, welche durch einen Schirm aus Blei von einigen Zentimeter Dicke dringen können.

Die Müonen sind ebenfalls instabil und zerfallen mit einer mittleren Lebensdauer von 2 μsec in ein Elektron (oder bei positiver Ladung in ein Positron) und ein Neutrino bzw. Antineutrino. Die Zerfallselektronen werden durch die Kaskadenbildung wieder wesentlich rascher absorbiert, können aber örtlich bei der Passage einer Bleischicht von einigen Millimeter Dicke zu einem sehr kleinen Teil jedoch ansehnliche Elektronenschauer auslösen. Durch die von ihrer Energie abhängige relativistische Zeitdehnung haben die Müonen in der Atmosphäre einen mittleren Laufweg von einigen Kilometern. Die meisten zerfallen in der Atmosphäre. Aber von denjenigen, welche dank ihrer Energie bis zu 50 m Wassertiefe durchdringen können, zerfällt bereits nur ein kleiner Bruchteil auf dem Weg durch die Atmosphäre.

Der Zerfall der Pionen und der Müonen hat zur Folge, daß auch bei unverändertem Primärspektrum und sogar bei unverändertem Luftdruck am Boden die Müonenzahl am Boden von der Temperaturverteilung in der Atmosphäre abhängt. Zwar nur in geringem Maße, doch immerhin soviel, daß die Messungen der Mesonenzahl zur Feststellung von Variationen des Primärspektrums einer Korrektur bedürfen. Wird die Luftschicht von 15 bis 25 km wärmer, so finden in der geringeren Luftdicke die Pionen auf ihrer mittleren Wegstrecke bis zu ihrem Zerfall weniger Gelegenheit zu neuen Kerntreffern. Dadurch steigt die Wahrscheinlichkeit der Müonenbildung etwas an. Man mißt am Boden etwas mehr Müonen. Allerdings sehr wenig, etwa 1 % bei den in fester geographischer Position vorkommenden Änderungen, aber merklich in verschiedenen Breiten.

Erwärmt sich eine tiefer liegende Luftschicht durch Advektion, etwa durch eine Warmfront, so hebt sich die Schicht der Müonenerzeugung und die Müonen haben einen längeren Weg. Damit kommen unten weniger an. Dies bedingt in unseren Breiten zwischen Sommer und Winter und bei Wetterfronten Effekte bis zu einigen Prozent. Bei Kenntnis der Atmosphäre aus Radiosondenmessungen kann man diesen Einfluß heute rechnerisch zuverlässig eliminieren. Es ist bemerkenswert, daß dieser Einfluß unabhängig vom Zenitwinkel ist, unter welchem die Strahlung einfällt. Die Verlängerung des Weges wird durch längere Lebensdauer der unten ankommenden Müonen wettgemacht, weil diese in den entsprechenden Luftschichten höhere Energien hatten.

Eine vollständige Beschreibung der kosmischen Strahlung in der Atmosphäre und auch am Erdboden oder im Wasser verlangt für jede geographische Position eine Angabe der Teilchenzahl und Energieverteilung der Primär- und Sekundärstrahlung in Funktion von der Höhe bzw. Tiefe und der Zeit. Für die geophysikalische Anwendung benutzt man einfachere Parameter, welche den Meßmethoden angepaßt sind und auch ausreichen.

E. Die Meßmethoden

Die klassische Meßmethode ist die Ionisationskammer. Sie liefert ein Maß für alle ionisierenden Teilchen, welche die Kammer passieren. Soweit es sich um relativistische Teilchen handelt, ist ihre spezifische Ionisation mit etwa 100 Ionenpaaren pro cm in Normalluft weitgehend einheitlich. Die Empfindlichkeit der Kammer ist in jeder Richtung gleich. Gegen die Strahlung radioaktiver Substanzen in ihrer Umgebung muß die Kammer peinlich abgeschirmt werden. Andererseits kann ihre Empfindlichkeit durch die genau bekannte Strahlung einer radioaktiven Substanz geeicht und über lange Zeiten genau kontrolliert werden.

Das Geiger-Müller-Zählrohr liefert bei jedem einzelnen Ionisierungsakt in seinem Innern einen zählbaren Stromimpuls, zu dessen Auslösung nur wenige, bei guten Typen sogar nur ein einzelnes im Innern erzeugtes Elektron genügt. Ein kohärenter Strahlenschauer mit sehr vielen ionisierenden Teilchen wird genauso wie ein einzelnes ionisierendes Teilchen gezählt.

Proportionalzählrohre erlauben dagegen in bekannter Weise die Messung der Zahl der Ionen, welche in ihnen bei einem Ereignis gebildet werden und ermöglichen dadurch eine Unterscheidung von Strahlenarten.

Dasselbe trifft für die bekannten Szintillationszähler zu, welche letztlich ebenfalls ein Maß für die Ionisierung in der Szintillationssubstanz beim einzelnen Ereignis liefern. Ihre zeitliche Auflösung ist wesentlich besser.

Als Koinzidenzteleskop bezeichnet man Geräte, welche den Durchgang eines KS-Teilchens durch mehrere Zählrohre oder Szintillatoren erfassen.

Diese von BOTHE und KOLHÖRSTER eingeführte Methode hat große Vorteile. Man kann leicht radioaktive Einflüsse und Teilchen geringer Reichweite ausschalten und hat insbesondere die Möglichkeit durch passende Anordnung der einzelnen Detektoren, einen Winkelbereich festzulegen, aus welchem die zur Messung kommenden Teilchen einfallen müssen. Daher die etwas anspruchsvolle aber übliche Bezeichnung Zählrohr-„Teleskop". Beispielsweise erwies sich oft die Beschränkung auf Strahlung um den Zenit als nützlich. Beim heutigen Stand der Elektronik und der Detektoren ist es möglich, sehr genaue Absolutmessungen des Teilchenflusses durchzuführen und durch Anwendung großer Flächen so viele Teilchen/sec einzufangen, daß genügende statistische Genauigkeit erreicht wird. Für die ständige Registrierung der Mesonenkomponente wurde im Internationalen Geophysikalischen Jahr 1957/58 an vielen Stationen die Standardapparatur benutzt, welche aus zwei horizontalen quadratischen, aus Zählrohren zusammengesetzten Flächen besteht, deren vertikaler Abstand gleich der Seitenlänge der Zählflächen besteht. Damit ist die Zenitwinkelverteilung der Empfindlichkeit einer solchen Anordnung definiert. Zwischen den Zählflächen wurde aus meßtechnischen Gründen eine weitere Zählfläche und außerdem eine 10 cm dicke Bleischicht angebracht, welche alle Elektronen und Mesonen geringer Energie ausschaltete. Bei einer Seitenlänge von 1 m beträgt der mittlere statistische Fehler des Mittelwertes der Teilchenzahl in Meereshöhe über 1 Std etwa $1,5^0/_{00}$.

Für Ballonmessungen zur Erfassung der Primärstrahlung genügen Flächen von der Größe eines Quadratdezimeters.

Neutronen können in Paraffinschichten von einigen Zentimeter Dicke auf thermische Geschwindigkeiten gebremst werden und dann mit Proportionalzählrohren erfaßt werden, welche meist mit Bortrifluorid mit hochangereichertem B^{10} gefüllt sind. Der B^{10}-Kern hat einen großen Einfangquerschnitt für thermische Neutronen und der entstehende Zwischenkern zerfällt in einen Lithiumkern und ein α-Teilchen, wobei eine Energie von 2,78 MeV frei wird. Diese stark ionisierenden Teilchen ergeben in dem als Proportionalzählrohr betriebenen Zählrohr Stromimpulse, die weit größer sind als die von relativistischen Teilchen oder Compton-Elektronen ausgelösten Stromimpulse. Sie können deshalb leicht abgetrennt werden, so daß man ein spezifisches Meßgerät mit sehr geringem Störeffekt (durch α-Teilchen aus den Wänden des Zählrohrs) zur Verfügung hat.

Man kann mit solchen Geräten die Dichte der Neutronen in der Atmosphäre messen. Diese Neutronen werden von der besprochenen kernaktiven Komponente der Sekundärstrahlung durch Kernzertrümmerungen erzeugt. Diese Komponente nimmt dabei beim Durchdringen von 100 g Luft/cm², also einer Luftschicht von etwa 1 km Dicke, auf die Hälfte ab. Dementsprechend sind die erzeugten Neutronen zerstreut. Dasselbe kann man in einer 1 cm dicken Bleischicht erreichen. Dann sind aber die entstehenden Neutronen auf engem Raum konzentriert. Sie verlassen das Blei mühelos und können in einer 10 cm dicken Paraffinschicht gebremst und mit Neutronenzählrohren der beschriebenen Art (oder mit Szintillatoren mit chemisch eingebauten B^{10}-Atomen) gemessen werden. Auf diesem Prinzip hat Simpson mit seiner Gruppe in Chicago ein neues Standardgerät zur Messung und auch zur laufenden Registrierung der kernaktiven Strahlungskomponente entwickelt, das heute an vielen Stationen verwendet wird. Auch die in Fig. 6 rechts gezeigte Kurven und die in Fig. 17 wiedergegebenen Variationen an verschiedenen Stationen wurden auf diese Weise gemessen. Die im Internationalen Geophysikalischen Jahr verwendeten Standardgeräte benutzten bei einer Fläche von etwa 2 m² einige Tonnen Blei und Paraffin. Jetzt ist man zur Erhöhung der statistischen Genauigkeit bestrebt, eine um ein Vielfaches größere Fläche mit entsprechend mehr Material zu benutzen; wobei die Zahl der Bortrifluoridzähler und der elektronischen Einrichtungen ebenfalls entsprechend größer wird.

Wegen der starken Absorption der kernaktiven Komponente mit einem Absorptionskoeffizienten von 1/(140 g/cm²) hat man im Hochgebirge eine etwa zehnmal so große Zahl von Ereignissen, welche Neutronen erzeugen und dementsprechend einen dreimal kleineren relativen statistischen Fehler.

F. Die Komponenten der galaktischen Strahlung und ihre Sekundärstrahlung in der Atmosphäre

Die kurze Beschreibung der Vorgänge im Abschnitt D zeigt, daß in der Atmosphäre die verschiedensten Sekundärstrahlungen nebeneinander vorliegen und die meisten Meßmethoden erfassen ein Gemisch. Man bezieht dabei oft den Ausdruck Komponente auch auf das, was mit einem bestimmten Instrument gemessen wird.

Mit einem zum Zenit gerichteten Zählrohrkoinzidenzteleskop mißt man in Funktion des Luftdrucks die „Pfotzer Kurve", Pfotzer hat als erster solche Messungen gemacht. Wir haben den Verlauf für drei geomagnetische Breiten und für die Zeit des Sonnenfleckenminimums ($U=1$ GV, vgl. Abschnitt B und G) und des Sonnenfleckenmaximums ($U=2$) in Fig. 9 angegeben, um den großen Bereich aufzuzeigen, der wirklich vorkommt. Man kann recht gut auf Grund der Registrierungen am Boden und der magnetischen Steifigkeit, bei welcher das magnetische Erdfeld das Spektrum der Primärstrahlung am Meßort

abschneidet, den Verlauf der Kurve voraussagen und damit an Hand von Bodenregistrierungen und Aufstiegen an verschiedenen Orten die dortige Grenzsteifigkeit genauer messen als dies mit Bodenmessungen allein möglich ist [16].

Die Fig. 9 und 10 beziehen sich auf einen Meridian durch Europa. In Amerika ist der Breiteneffekt noch stärker, über Japan schwächer wie man aus Fig. 7 ersieht. Der Anstieg bei kleinen Drucken ist so gut linear, daß man auf den Druck 0, d.h. aber auf die Primärstrahlung aus Ballonhöhen gut extrapolieren kann [17]. Allerdings mißt man auch etwas zurückgestreute Strahlung mit. Man kann leicht solche Albedostrahlung ausschalten, welche direkt von unten kommt. Schwierig ist es mit den Teilchen, welche an anderen Orten in der Atmosphäre zurückgestreut werden und nun nach

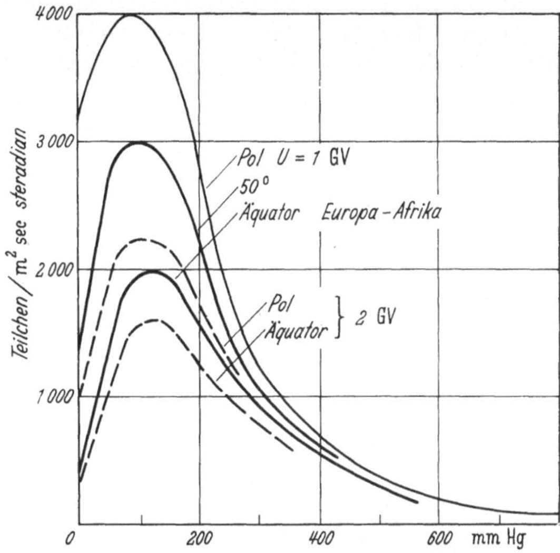

Fig. 9. Senkrechter Teilchenfluß in der Atmosphäre

magnetischer Ablenkung am Meßort von oben einfallen. Ihre Zahl ist jedoch gering, wenn man sich nicht für die kleinsten Energien interessiert.

Schaltet man durch Bleifilter zwischen den Zählrohren eines Koinzidenzteleskops die „weichen" Teilchen aus, so erhält man schon mit 2 cm Blei statt der in Fig. 10 gezeigten Kurve für die Gesamtstrahlung die mit „hart" bezeichnete Kurve als Absorptionskurve harter Strahlung. Bei dickeren Filtern (etwa bis 9 cm) erhält man eine kaum meßbar verschiedene Kurve. Diese Kurven haben ein nur ganz schwach ausgeprägtes, in hohen Breiten überhaupt kein Maximum. Dies liegt daran,

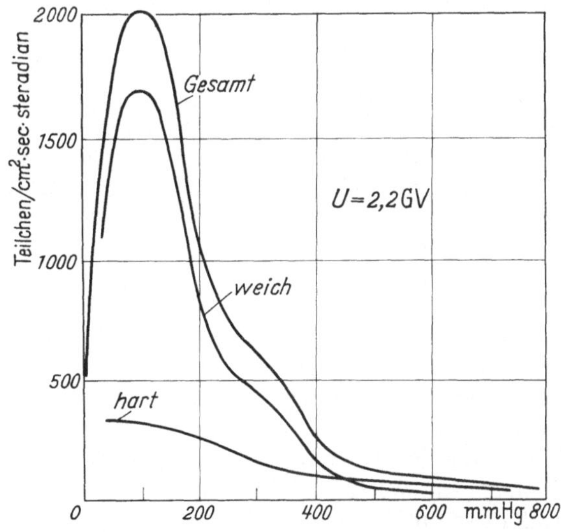

Fig. 10. Weiche und harte Teilchen in der Atmosphäre aus der galaktischen Strahlung im Sonnenfleckenmaximum nach EHMERT [Z. Physik 115, 326 (1940)]

daß in dem Blei durch eindringende Primärteilchen erst entstehende Sekundärmesonen bereits mitgemessen werden, weil das erste Zählrohr vom auslösenden Primärteilchen oder Pion zum Ansprechen gebracht wird.

Mit dünnwandigen Ionisationskammern liefert die galaktische Strahlung einen sehr weiten Bereich, wie man aus der Übersicht in Fig. 11 sieht. Dieselben Kurven ergeben sich mit einzelnen Zählrohren vom Auslösetyp. Die *mittlere* spezifische Ionisation der Teilchen ist nach Regener und Pfotzer [19] im Bereich der oberen Atmosphäre 103 Ionenpaare/cm unter Normalbedingungen und selbst in großen Wassertiefen ist sie immer noch nach Ehmert [20] 94 Ionenpaare/cm.

Aus den Kurven der Fig. 11 können durch Umrechnung auf die jeweilige Luftdichte die Kurven für die jeweilige Ionisierung in der freien Atmosphäre berechnet werden, wie sie etwa zur Berechnung der Leitfähigkeit in der Stratosphäre und Troposphäre benötigt werden.

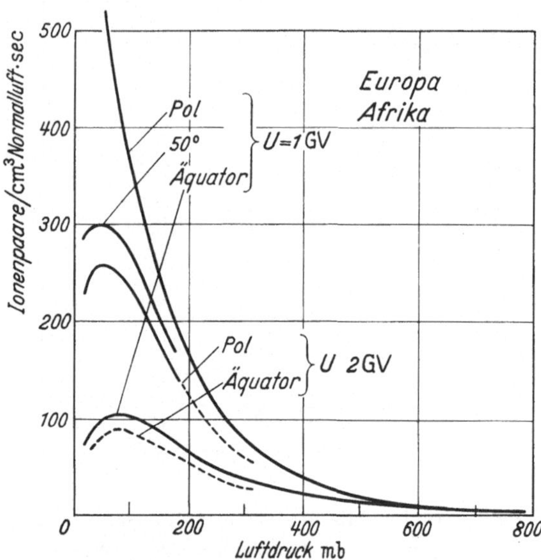

Fig. 11. Ionisierungsstärke in der Atmosphäre, d.h. Ionenbildung in einer Kammer mit Luft unter der Dichte am Erdboden (bei 0° C). Nach Messungen von Regener [19] und Neher [18]

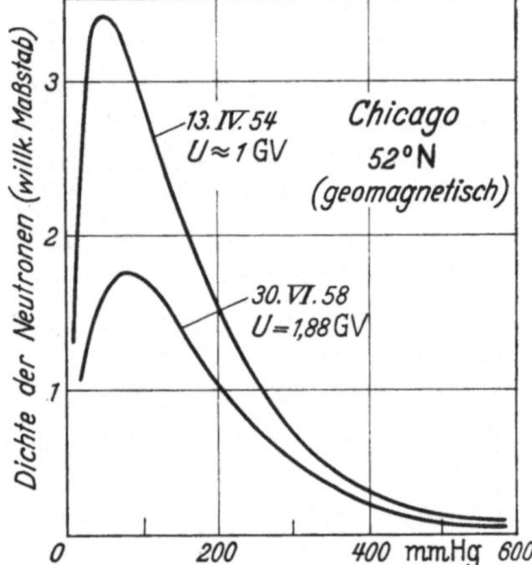

Fig. 12. Die kernaktive Komponente in der Atmosphäre nach Simpson unter 52° geomagnetischer Breite bei verschiedener Modulation nach J.A. Simpson [Astrophys. J. Suppl. **44** (IV), 378 (1960)]

Die kernaktive oder nucleare Komponente zeigt Fig. 12 nach Messungen von Simpson über Chicago für verschiedene Sonnenaktivität.

Weitere Messungen, wie etwa die der Strahlenschauer oder der Elektronen sollen im Zusammenhang mit der Geophysik übergangen werden.

G. Die Modulation der galaktischen Strahlung durch die Sonnentätigkeit

Zu den in Fig. 1 wiedergegebenen experimentell bestimmten integralen Energieverteilungen haben wir unsere mit der in Abschnitt B geschilderten Hypothese berechneten Verteilungen gezeichnet, welche offenbar die Veränderungen des Verlaufs der Spektren richtig wiedergeben. Man kann aus jeder solchen Messung einen Wert für den Parameter μ und damit für die hypothetische Bremsspannung U angeben, welche die Teilchen auf ihrem Weg zur Erde überwinden müssen. Daß diese für Protonen und α-Teilchen gleich herauskommt, spricht für die Hypothese. Man kann sie aber auch jeder Registrierung am Erdboden entnehmen, wenn diese einmal an solche Messungen der Primärstrahlung oder an eine andere bereits an die Primärstrahlung angeschlossene Registrierung ihrerseits angeschlossen wird. Dieser Anschluß ist denkbar einfach, da wir bis jetzt in den vorkommenden Bereichen der Registrierungen nur lineare Abhängigkeit beobachten konnten. Fig. 3 zeigt über den aus Fig. 1 (und anderen Messungen) entnommenen Werten von μ_α und μ_p und von U, teils (bis 1956) die Ionisation an der äquatornahen Hochgebirgsstation Huancayo in den Einheiten die Forbush [21] benutzt und welche auch in Fig. 14 auftreten, teils (ab 1956) die mit dem Neutronenmonitor unserer Station Weißenau jeweils zur Zeit der Messungen in der Stratosphäre registrierten Neutronenzahlen in den Einheiten, in welchen diese veröffentlicht wurden. Die Tagesmittel beider Größen haben eine sehr enge Korrelation, welche nur zeitweise durch nicht korrigierte Einflüsse der Atmosphärentemperatur über Huancayo etwas gestört ist. Der Regressionskoeffizient $0{,}24\%$ $I_{\text{Huancayo}} : 1\%$ $N_{\text{Weissenau}}$ und der absolute Anschluß stecken in den Skalen der Fig. 13. Die lineare Extrapolation ergibt $N_{\text{Weissenau}}$ $(U=0)=140\%$ und I_{Huancayo} $(U=0)=112\%$. Beides Werte, welche nie vorkommen. Die höchsten Werte aus 1936 mit 108% entsprechen $U=0{,}70$ GV, im letzten Sonnenfleckenminimum 1954 wurde nur 0,95 erreicht.

Dauerregistrierungen der kosmischen Strahlung mit dauernd konstanter Empfindlichkeit, und dauernd konstanten Basiswerten sind nicht leicht. Zunächst standen nur Ionisationskammern zur Verfügung. Hier sind die Variationen auf 12% beschränkt, weil energiearme Primäre sich hier gar nicht mehr auswirken. Dagegen sind der Einfluß des variierenden Luftdrucks und temperaturabhängige und andere Empfindlichkeitsschwankungen zu korrigieren und diese Korrekturen können erst aus langen Meßreihen mit einiger Sicherheit bestimmt werden. Dazu kam die Korrektur durch den μ-Mesonenzerfall, welche von dem Temperatur

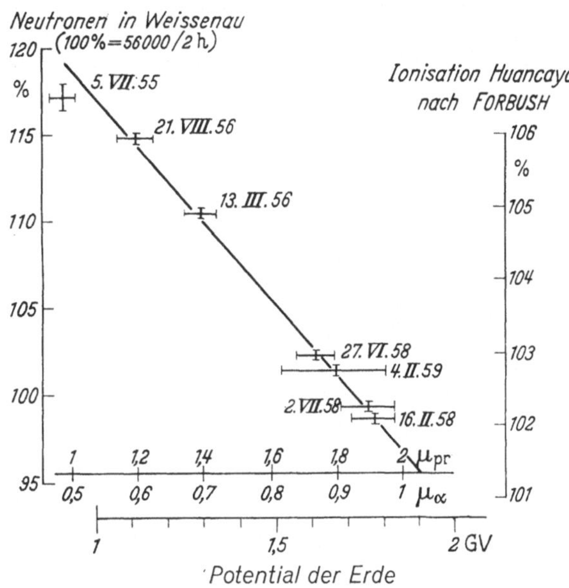

Fig. 13. Die μ-Werte von Fig. 1 sind eine lineare Funktion der gleichzeitig in Weißenau registrierten Neutronenzahl und der in Huancayo registrierten Ionisation. Unten ist das entsprechende Bremspotential der Erde angegeben

verlauf über die Höhe in der Atmosphäre abhängig ist. Dieses Problem wurde erst in den letzten Jahren befriedigend gelöst durch den Anschluß der Zählrohrkoinzidenzmessungen und Ionisationsmessungen an die Neutronenregistrierungen. So blieben die Unsicherheiten lange Zeit in der Größenordnung der echten Variationen. Nur auf Hochgebirgsstationen lagen die Verhältnisse günstiger. Das von Hess und Steincke eingerichtete europäische Stationsnetz wurde in den dreißiger Jahren wieder aufgegeben. Das von Millican, Bowen, Compton, Neher und Forbush eingerichtete amerikanische Stationsnetz ist, von Forbush vorbildlich betreut und ausgestattet, heute noch in Betrieb. Ihm verdanken wir die längsten Meßreihen. Kolhörster führte die Registrierung mit Zählrohrteleskopen ein, welche Verfasser und Duperier in England um dieselbe Zeit 1939 aufnahmen. Hier waren im Interesse konstanter Empfindlichkeit noch viele Schwierigkeiten zu überwinden, bis einwandfreie Zählrohre zur Verfügung standen. Es zeigte sich, daß mit diesen nur aus Zenitnähe kommende Mesonen registrierenden Geräten, die echten Variationen etwa die doppelten Prozentamplituden hatten [24].

Die im Abschnitt E bereits geschilderten Neutronenmonitoren von Simpson ergaben in geomagnetischen Breiten um $50°$ wiederum über doppelt so große Prozentamplituden als die Zählrohrteleskope. Dies rührt daher, daß die Erzeugung von μ-Mesonen erst bei höheren Mindestenergien der Primärteilchen möglich ist als die Erzeugung der nuclearen Komponente, die sich am Erdboden noch auswirkt. Dies zeigt sich überzeugend am Breiteneffekt der nuclearen Komponente, der mit 45 % etwa dreimal so groß wie der Breiteneffekt des vertikalen Mesonenflusses (bei Korrektur der Temperatureinflüsse) ist [25].

Aber die nucleare Komponente hat noch den weiteren großen Vorteil eines kaum nachweisbaren Einflusses des Temperaturverlaufs in der Atmosphäre. Man muß deshalb nur den allerdings sehr großen Luftdruckeinfluß von etwa $K_L \approx 0{,}1/1$ cm Hg nach der Formel

$$N_{\text{korrigiert}} = N_{\text{gemessen}} \cdot e^{K_L(H-H_0)}$$

korrigieren. $N =$ Neutronenzahl/Zeiteinheit, $H =$ Luftdruck, $H_0 =$ Luftdruck auf welchen korrigiert wird. K_L ist etwas breitenabhängig und im Sonnenflecken-

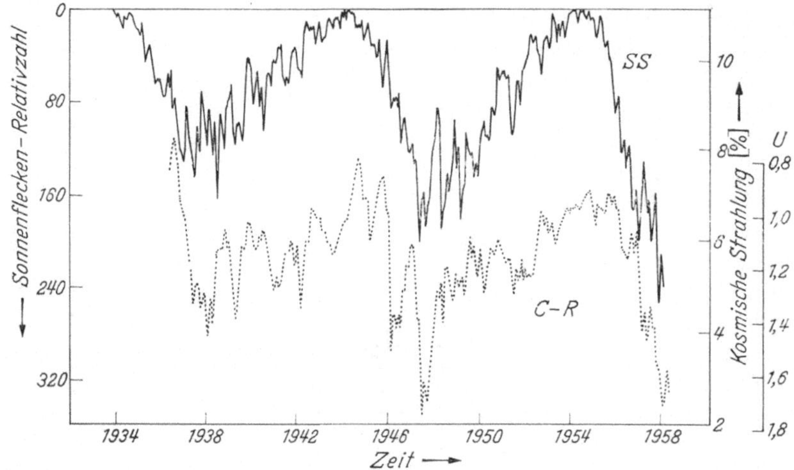

Fig. 14. Kosmische Strahlung ($C - R$ Monatsmittel in % über 100% entsprechend einem willkürlich gewählten Festwert) nach Ionisationskammermessungen von Forbush [27] an der äquatorischen Gebirgsstation Huancayo und Sonnenfleckenrelativzahl (SS Nos, gegenläufig aufgezeichnet) zeigen den hier nur einige Prozent ausmachenden Einfluß der Sonnentätigkeit auf die kosmische Strahlung

maximum etwas kleiner als im Sonnenfleckenminimum, d.h. eine schwach veränderliche Funktion $K(H_0, \Phi, U)$. Heute noch lassen sich die meisten Unstimmigkeiten zwischen den Ergebnissen verschiedener Stationen auf ungenügende Luftdruckkorrektur zurückführen.

Nachstehend befassen wir uns nur mit den auf die atmosphärischen Einflüsse korrigierten Werten, obgleich diese Korrektur nicht immer ganz zuverlässig ist.

Fig. 14 gibt nach Forbush Monatsmittel der Ionisation in Huancayo über 24 Jahre zusammen mit den Sonnenfleckenrelativzahlen. Eine Korrelation ist evident. Zunehmende Sonnentätigkeit führt zu einer Verminderung der kosmischen Strahlung. Den Charakter dieser Modulation zeigt Fig. 15 an Hand der Tagesmittel über 4 Jahre. Kleinste, die Meßgenauigkeit kaum übersteigende Variationen im Sonnenfleckenminimum treten große Intensitätseinbrüche im Maximum mit langsamer Erholung gegenüber. Forbush erkannte zuerst die Kopplung dieser ,,Forbush-Effekte" mit erdmagnetischen Stürmen [27]. Wir behandeln dieses Phänomen weiter in Abschnitt J. Fig. 16 zeigt die Variation der korrigierten Tagesmittel der Neutronenzahl in Weißenau für 1957 und 1958.

Die Auswirkung dieser Modulation an einer Reihe von Stationen zeigt Fig. 17 nach einer Zusammenstellung des japanischen Datenzentrums für das Geophysikalische Jahr. Wenn man von einigen Störungen wohl technischer Art absieht, erkennt man einen überall gleichen Gang mit zum Äquator hin abnehmender Amplitude. Sie ist dort nur $^1/_3$ der Amplituden in hohen Breiten und die Abnahme der Amplituden der Stationen in Meereshöhe erfolgt in der gleichen Weise wie die Abnahme der Neutronenzahl durch den Breiteneffekt. Diese Feststellung war der eigentliche Schlüssel zu der besprochenen Modulationshypothese mit dem elektrischen Gegenfeld [2]. Dazu kam unsere als Ergebnis zahlreicher Ballonaufstiege mit kleinen Zählrohrteleskopen über Weißenau getroffene Feststellung, daß *dort* einer Veränderung der Primärstrahlung um 1% eine Änderung der Neutronenzahl am Boden um 0,46% und der Mesonenzahl am Boden um 0,216% (und der Ionisation um 0,10%) entspricht und daß die Verhältnisse 1:0,46:0,216 recht genau dem Verhältnis der Zunahme dieser Komponente in Weißenau gegenüber dem Äquatorwert der Komponenten entsprechen [28].

Mit anderen Worten: Die absolute Änderung der Primär und Sekundärkomponente an verschiedenen Orten bei einem bestimmten Modulationsereignis ist in erster Näherung dem Quadrat der Komponentenstärke am jeweiligen Ort proportional. Dies läßt sich aus den Kurven in Fig. 1 an Hand der Ausbeutefunktionen tatsächlich verifizieren.

Auf die zahlreichen Theorien der Modulation wollen wir

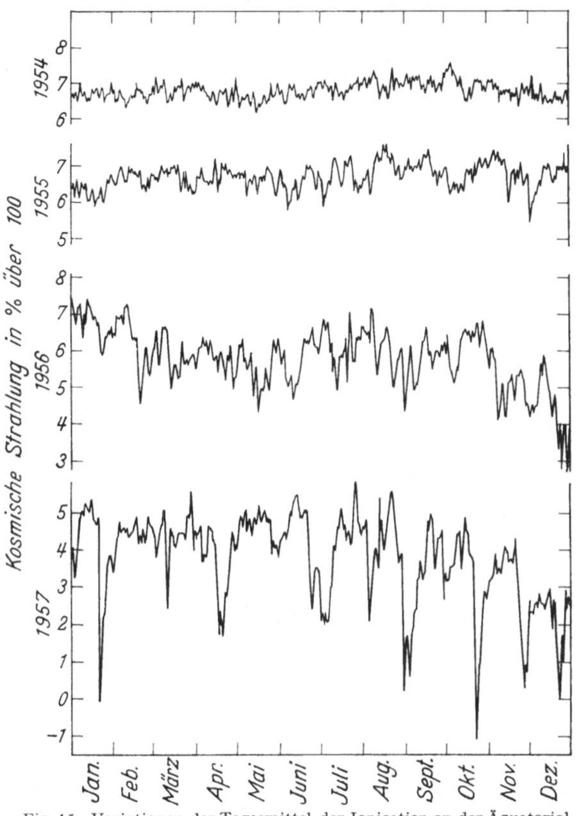

Fig. 15. Variationen der Tagesmittel der Ionisation an der Äquatorialgebirgsstation Huancayo nach Forbush und Lange [27] beim Übergang vom Sonnenfleckenminimum (1954) zu starker Sonnentätigkeit (1957)

hier nicht eingehen. Doch ist zu bemerken, daß die Spannung U keineswegs von einer Ladung der Erde herrühren kann. Die Feldstärken würden sich nach diesem Modell untragbar groß erweisen. Als zirkumsolares Feld aufgefaßt, ergibt sich auf der Erdbahn immer noch eine Feldstärke von 10^{-4} V/cm. Auch das ist weit mehr als die Leitfähigkeit des ionisierten interplanetaren Gases zuläßt. Es handelt sich also kaum um ein einfaches elektrostatisches Feld, sondern um eine Einwirkung allein auf die schnellen Teilchen der kosmischen Strahlung, die jedoch mit den zunächst naheliegenden rein magnetischen Einwirkungen nicht gedeutet werden kann. Wir wollen es hier mit der phänomenologischen Beschreibung durch U bewenden lassen [34].

Daß die Modulation nicht an die Erde gebunden ist, oder, wie man lange vermutete, durch das Zusammenspiel von erdmagnetischem Feld und solaren Plasmaströmen bedingt ist, sondern ein sehr weiträumiger Vorgang ist, ergab sich zunächst daraus, daß von Allen und Frank [35] außerhalb des Strahlungsgürtels der Erde einen Teilchenfluß an kosmischer Strahlung fanden, der unserem für

dieselbe Zeit über dem Pol berechneten gleich war [*36*]. FAN, MEYER und SIMP-
SON [*37*] registrierten zwischen 37000 und 48000 km Entfernung vom Erdmittelpunkt mit einem allseitig empfindlichen Zählrohrkoinzidenzgerät, das mit einem

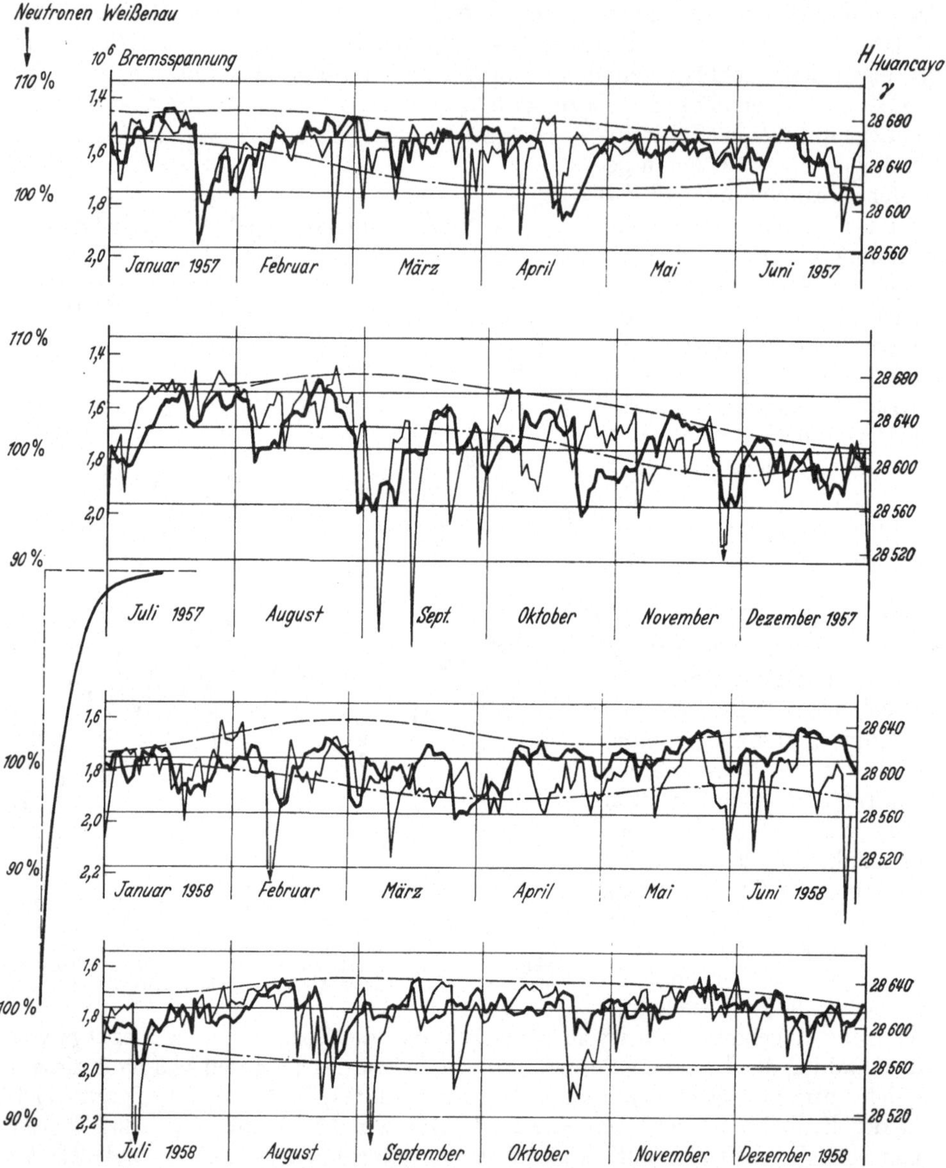

Fig. 16. Tagesmittel der Neutronenzahl in Weißenau (nach EHMERT und PFOTZER) und der magnetischen Horizontalintensität der Station Huancayo nach vorläufigen Mitteilungen von Herrn Dr. GIESEKE. Die Werte der Horizontalintensität
sind um die Differenz zwischen der oberen gestrichelten Linie und der strichpunktierten angehoben (s. Text) um den Einfluß
der einzelnen Stürme besser vergleichen zu können

Bleipanzer versehen war, welcher die Teilchen des Strahlungsgürtels fernhielt,
während eines magnetischen Sturmes genau den gleichen Forbush-Effekt wie
wir ihn aus den Bodenregistrierungen dieses Ereignisses für den Teilchenfluß
über dem Pol berechnen können.

Dieselben Forscher registrieren mit einem gleichartigen Gerät den Teilchenfluß der kosmischen Strahlung in der amerikanischen Venusrakete Explorer VI

bis zu 6 Millionen km Abstand von der Erde in Richtung zur Sonne [*38*], [*39*].
20 Tage nach dem Start trat ein kräftiger Forbush-Effekt ein, nach welchem auf
der Erde eine sehr langsame Erholung im Lauf von 30 Tagen eintrat. Wir haben
an anderer Stelle [*34*] gezeigt, daß man im Einklang mit allen anderen Ergebnissen die Registrierkurve in der Rakete als Superposition ansehen muß eines
Gradienten des allseitigen Teilchenflusses der kosmischen Strahlung und eines
zeitlichen Verlaufs der Modulation wie er an der Erde zur gleichen Zeit beobachtet

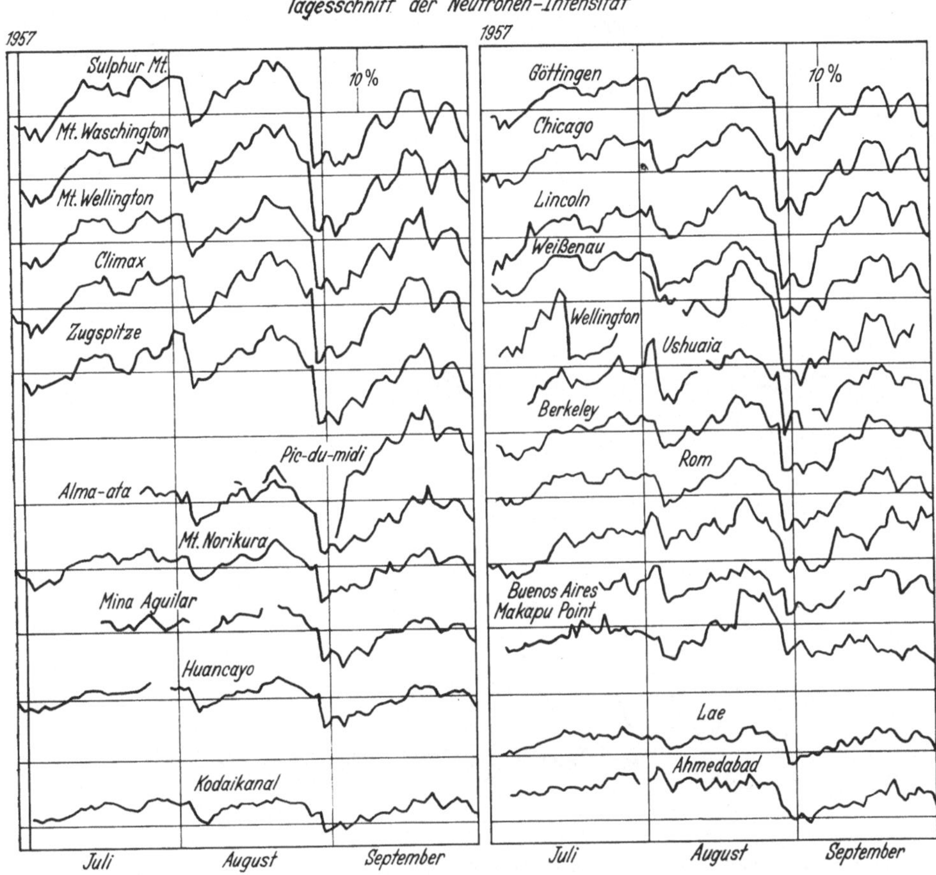

Fig. 17. Die Tagesmittel der korrigierten Neutronen an verschiedenen Stationen nach der Zusammenstellung des japanischen Datenzentrums für das Internationale Geophysikalische Jahr. Es ist nur ein Teil der Stationen als Beispiel
wiedergegeben

wurde. Aus dem Gradienten berechneten wir den Verlauf von U in Abhängigkeit
vom Sonnenabstand R für die Zeit der Messung in Erdbahnnähe zu $U = U_E \times$
$(R/R_E)^{-0,86}$, wobei U_E die Modulationsspannung auf der Erde und R_E die Entfernung Sonne—Erde ist [*34*]. Die genannten Autoren sind jedoch der Ansicht,
daß kein oder nur ein sehr kleiner Gradient bestehe und die Modulation im freien
Raum anders verläuft. Weitere Messungen dieser Art werden dies entscheiden.

H. Isotrope und anisotrope Modulation

Die Modulation der kosmischen Strahlung ist weltweit. Man erhält für die
Tagesmittel die besprochene Verteilung der Änderungen über die ganze Erde.
Es treten aber beim Vergleich der Modulation der nuclearen Komponente und

der Mesonenkomponente zeitweise kleine Verschiedenheiten auf in dem Sinn, daß für die letztere bei starken magnetischen Stürmen überraschend eine etwas größere Variation von U herauskommt als dem allgemeinen Zusammenhang entspricht. In diesen Fällen weist die Intensität kleine Tagesgänge auf, welche im Durchschnitt ein Maximum um die Mittagszeit haben, das aber an einzelnen Tagen um Stunden früher oder später auftreten kann. Fig. 18 zeigt einige

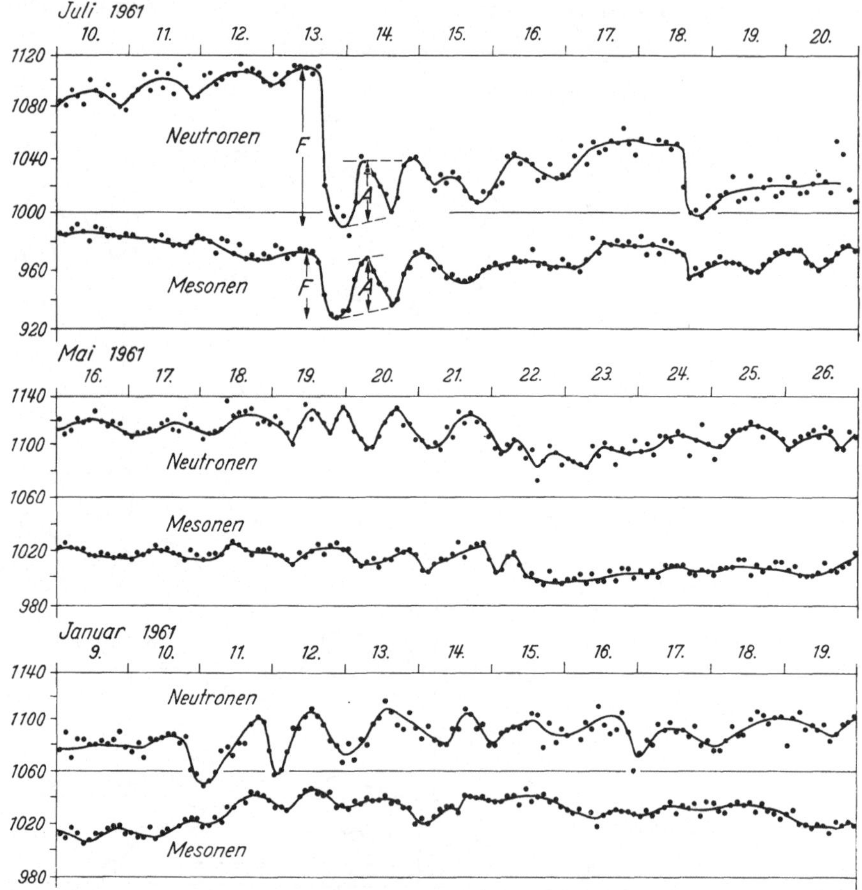

Fig. 1 . Korrigierte 2 Std-Werte der Neutronen und Mesonen in Lindau. Typische Tagesgänge durch anisotrope Modulation. $F_N:F_M=2{,}7:1$, $A_N:A_M=1{,}25$ am 13. 7. 61. Die zwischen die Meßpunkte gezeichneten Kurven veranschaulichen den wahrscheinlichen Verlauf

charakteristische Beispiele mit und ohne gleichzeitigen Forbush-Effekt. Bei der ersten sicheren Feststellung aus direkten Registrierungen ohne Mittelbildung über viele Tage [40] konnte gezeigt werden, daß die Erscheinung an Stationen verschiedener geographischer Länge das Maximum zur selben Ortszeit zeigt. Es liegt also eine Anisotropie der Strahlung vor. Die Amplitude ist meist in mittleren Breiten am größten. Am 10. bis 17. 7. 61 sind nach Fig. 18 die prozentualen Amplituden dieser Tagesgänge in Lindau für die Neutronen um den Faktor 1,25 größer als für die Mesonen, während die mit F bezeichnete prozentuale Forbush-Abnahme der Intensitäten im Verhältnis 2,7:1 steht wie es in Lindau sowohl für die langsame Variation im Sonnenfleckencyclus als auch für solche Forbush-Effekte gilt, welche nicht von solchen Anisotropen begleitet sind. Bei den letzteren tritt die Abnahme auch weltweit zum selben Zeitpunkt ein. Typisch ist ebenso die leichte Abnahme der Mesonen im Lauf von 2 Tagen vor dem Sturm am 13. 7. 61 während die Neutronen in dieser Zeit etwas zugenommen haben. Dies ist nicht auf atmosphärische Veränderungen zurückzuführen, da diese sehr konstant waren. Die großen Maxima lagen um 6.00 und um 24.00 Uhr am 14. 7. 61. Die Richtung der Anisotropie hat sich demnach rasch geändert. Manchmal tritt mit der Forbush-Abnahme gleichzeitig ein deutlicher Phasensprung des Tagesganges auf. Die relative Größe der Amplitude in der Mesonenzahl wie auch die Gleichzeitigkeit der Maxima und Minima beider Komponenten weisen darauf hin, daß die Anisotropie in diesem Fall für einen begrenzten Energiebereich ein Maximum hat, der wesentlich über 10^{10} eV liegt. Denn die Erscheinung kann schwer auf eine innerhalb des Erdfeldes wirkende Ursache zurückgeführt werden. Ist die Anisotropie aber schon außerhalb vorhanden, so werden Teilchen verschiedener Energie im Erdfeld verschieden abgelenkt. Tatsächlich treten die Maxima im allgemeinen bei stärkeren Effekten dieser Art nach Untersuchungen in Lindau am frühesten unter 40° geomagnetischer Breite, am Äquator 1 Std später und in hohen Breiten sogar einige Stunden später auf, was beides sehr gut aus der Ablenkung im Erdfeld zu verstehen ist. Ebenso daß bei großen Amplituden im Winter die Maxima um Stunden früher eintreten. Dies läßt sich auf die andere Lage der magnetischen Erdachse zur Sonnenrichtung zurückführen. Da die Analyse dieser komplexen Erscheinung an mehreren Orten im Gange ist, soll sie hier nicht weiter dargestellt werden. Wir glauben, daß diese Anisotropien mit starken Krümmungen der Flächen konstanten Potentials U der isotropen Modulation zusammenhängen, die notwendig entstehen, wenn solare Plasmawolken oder Plasmastrahlen die Flächen gleichen Potentials um die Sonne in der Erdnähe stark nach außen aufbeulen, so daß für die Erde ein höheres Bremspotential vorliegt, welches die Forbush-Abnahme der galaktischen Strahlung verursacht. Wie weit diese Beschreibung zutrifft, wird man durch ausgedehntere Messungen von Teilchenzahl und Isotropie bzw. Anisotropie in großem Abstand von der Erde mit Raumsonden klären können.

J. Modulation der Kosmischen Strahlung und erdmagnetischem Feld

Es war wiederum FORBUSH [27], der zuerst auf eine Korrelation zwischen Forbush-Abnahmen und erdmagnetischen Stürmen aufmerksam gemacht hat. Fig. 16 zeigt als dick ausgezogene Kurve die Tagesmittel der Neutronenzahl in unserer früheren Station Weißenau für die Jahre 1957 und 1958. Als Maßstab ist links neben den veröffentlichten Prozentzahlen gleich der daraus nach Fig. 13 berechnete Wert der Bremsspannung U angegeben, welche als ein Maß für die Modulationstiefe der kosmischen Strahlung benutzt werden kann. Die mit einer dünnen Linie verbundenen Meßpunkte stellen die magnetische Horizontalintensität der äquatorialen Station Huancayo im Mittel von 5 Std um die lokale Mitternacht dar. Sie ist in γ (1 $\gamma = 10^{-5}$ Gauß) angegeben. Die Differenz zwischen diesem Feld und dem Normalwert vor den Stürmen wird als Ringstromfeld bezeichnet [41]. Man kann solche Störungen als Feld eines ausgedehnten Ringstromes um die Erde beschreiben und dieser Ringstrom wird aufgebaut, wenn nach einer chromosphärischen Eruption auf der Sonne die Wolke ausgestoßenen Plasmas nach einer Laufzeit zwischen 10 und 30 Std die Erde erreicht. In den Tagesstunden ist die magnetische Feldstärke durch das Feld ionosphärischer Ströme erhöht. Zu Beginn eines magnetischen Sturmes, insbesondere wenn Nordlichter auftreten und in der Polarlichtzone ein starker ionosphärischer Ost-West-Strom entsteht, tritt auch zur Nachtzeit am Äquator eine zusätzliche Abnahme des Magnetfeldes durch ionosphärische Störströme auf. Diese Erscheinung äußert sich in Fig. 16 durch die spitzen tiefen Ausschläge der Kurve zu negativen Werten. Sie gehören also nicht zu dem besprochenen Ringstrom. Sie deuten in jedem Fall an, daß solares Plasma (ionisiertes Gas mit einer Dichte von der Größenordnung 100 Ladungspaare pro cm^3) die Erde umspült hat. Die Größe solcher Wolken wird nach der Dauer der magnetischen Einwirkung und der Geschwindigkeit auf die Größenordnung 10^{10} cm geschätzt. Zweifellos kommen große Unterschiede vor.

Ringströme um die Erde sind durch Messungen der magnetischen Feldstärke in amerikanischen Raumsonden von SONETT u. Mitarb. [42] nachgewiesen worden. Sie liegen in einer Entfernung von 6 bis 8 Erdradien vom Erdmittelpunkt. Es ist noch offen, wie weit sie einen dauernden Anteil besitzen.

Die Maßstäbe sind in Fig. 16 so gewählt, daß 1% so groß wie 10 γ wird. Wir hatten bei dem einzelnen großen magnetischen Sturm am 27. 7. 46 festgestellt [43], daß während der sechstägigen von keinen weiteren Stürmen gestörten Erholungszeit die Mitternachtswerte der Horizontalintensität in Huancayo und die Ionisation in Cheltenham genau den gleichen Verlauf mit einem Maßstabsverhältnis von 17,8 γ/1% haben. Mit dem bekannten Maßstabsverhältnis von Ionisation in Cheltenham und Neutronenzahl in Weißenau für isotrope Modulation berechnet sich daraus das oben angegebene und in Fig. 16 benutzte Verhältnis. Man erkennt z.B. am Verlauf beider Kurven Ende Januar und im Februar 1957, daß dieses Verhältnis sinnvoll ist, obgleich wir hier schon eine Reihe von aufeinanderfolgenden Stürmen haben.

Wir wollen untersuchen, wie weit diese oft zitierte und auch oft kritisierte Korrelation geht. Nun sind in Fig. 16 nicht die Originalwerte der magnetischen Feldstärke aufgezeichnet. Sie sind vielmehr um den Abstand zwischen der gestrichelten Linie und der strichpunktierten Linie nach oben gerückt worden, um einen übersichtlichen Vergleich der Stürme selbst zu gewinnen. Mit diesen beiden Linien hat es folgende Bewandtnis: Auf der linken Seite ist eine e-Funktion angezeichnet, nach der sich im Mittel beide Störungen erholen, wenn man die Tage der großen Spitzen der magnetischen Störung wegläßt. Man kann diese Funktionen an die Kurvenstücke zwischen zwei Stürmen anlegen und den O-Wert, auf welchen die Erholung jeweils ausgerichtet ist, ablesen. Man bekommt so eine Reihe von Punkten, welche durch eine glatte Kurve verbunden wurden. Dies ist für die kosmische Strahlung die gestrichelte Linie. Dieselbe Prozedur auf die originalen magnetischen Werte angewandt, liefert ziemlich eindeutig die strichpunktierte Linie. Die Variation der Differenz ist als eine langsame magnetische Variation aufzufassen, welche nicht mit der kosmischen Strahlung verknüpft ist. Fig. 19 zeigt das Mittel beider Linien über die vier Jahre 1957 bis 1960. Es ergibt sich jeweils eine klare Halbjahreswelle, welcher noch eine kontinuierliche Abnahme der Horizontalintensität um 40 γ/Jahr überlagert ist. Diese letztere ist als Säkularvariation der Station bekannt und hängt mit Vorgängen im Erdinnern zusammen. Sie tritt in allen vier Jahren gleichmäßig in der Differenz auf, während die kosmische Strahlung in diesem Zeitabschnitt gerade ein Minimum im Rahmen der Variation im Sonnenfleckencyclus durchläuft und im Mittel neben der Halbjahreswelle keinen Gang zeigt. Diese Kurven stellen also die Variationen nach Befreiung von den einzelnen Stürmen dar. Mit den Originalwerten und Mittelung über viele Jahre würde man sehr ähnliche Kurven mit durch die Stürme verstärkten Amplituden erhalten. Wir schließen daraus, daß die Variation der Horizontalintensität zusammengesetzt ist aus:

    1. Der Säkularvariation.

    2. Einer Halbjahreswelle mit dem Minimum (als Maximum des von außen wirkenden Störfeldes) etwa 10 Tage nach den Äquinoktien. Wir wollen diese

Variation A-Variation nennen. Ihre Amplitude A betrug zur Zeit des Sonnenfleckenmaximums 1957 bis 1960 33 γ und scheint im Sonnenfleckenminimum 1954 nicht merklich kleiner gewesen zu sein. Die kosmische Strahlung zeigt eine genau gegenläufige Halbjahreswelle.

3. Einen dem Erdfeld entgegen gerichteten Ringstromanteil, welcher sowohl im Sonnenfleckencyclus als auch bei großen erdmagnetischen Störungen mit der kosmischen Strahlung gegenläufig variiert. Wir wollen sie B-Variation nennen. Das magnetische Störfeld, welches H_{Huancayo} vermindert und die Modulationsspannung U, welche die kosmische Strahlung vermindert, variieren also in guter Korrelation gleichsinnig.

4. Die Einflüsse der Ionosphärenströme.

Die B-Variation hat wie die Modulation der kosmischen Strahlung ihre Ursache im „solaren Wind", wie man das Ausströmen der solaren Materie bezeichnet. Da die Erdachse gegen die Ekliptik geneigt ist, ist auch der Vektor des solaren Windes zur Zeit der Äquinoktien senkrecht zu ihr, zur Zeit der Solstitien dagegen geneigt. In Wirklichkeit kommt es auf den Winkel zur magnetischen Dipolachse der Erde an, die sich jedoch im Laufe des Tages ändert, wobei wegen der längeren Zeitkonstanten des Ringstromes ein Mittelwert gemessen wird. Man kann deshalb aus dem Unterschied von 33 γ bei einem Winkel von 23,45° schließen, daß die gesamte A-Komponente im letzten Sonnenfleckenmaximum etwa 430 γ ausmachte. Im Sonnenfleckenminimum verschwindet die

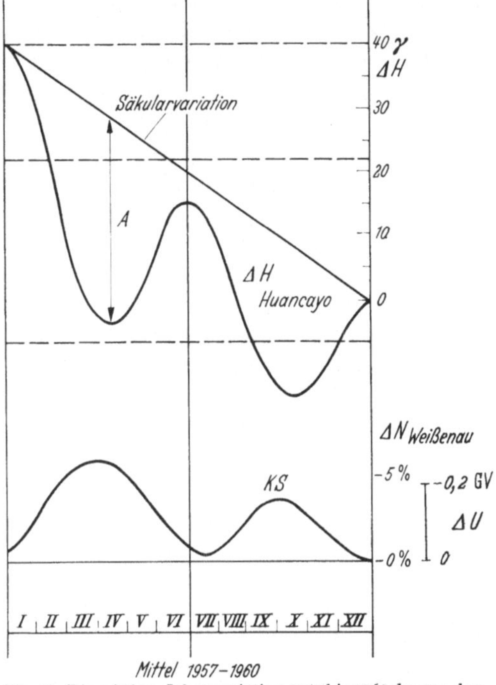

Fig. 19. Die mittlere Jahresvariation 1957 bis 1960 der von dem Einfluß der einzelnen magnetischen Stürmen nach Fig. 16 befreiten erdmagnetischen Horizontalintensität, deren Verlauf für 1957 und 1958 in Fig. 16 als strichpunktierte Linie dargestellt ist (oben) und der entsprechenden Variation der Neutronenzahl in Weißenau bzw. der zugehörigen Bremsspannung U

Amplitude der A-Variation bei gleicher Bestimmung nicht. Es müßte also nach dieser Deutung auch zu dieser Zeit noch ein erheblicher solarer Wind vorliegen. Darauf deutet auch unabhängig das Verhalten der kosmischen Strahlung.

Da die Modulation der kosmischen Strahlung ein großräumiger Vorgang ist, würde man hier zunächst keinen Einfluß der Stellung der Erdachse zum solaren Wind erwarten. Die vorliegende Halbjahreswelle von 5 % in der Neutronenzahl von Weißenau, ist in ihrer Phase auch genau entgegengesetzt zum normalen Zusammenhang, da das Maximum auf die Äquinoktion mit Minimum der Horizontalkomponente fällt, während sich sonst ja beide gleichsinnig ändern. Man kann sich das aus einer Verschiebung der Grenzsteifigkeit für die Meßstation Weißenau durch das wechselnde Ringstromfeld erklären. Nach Rechnungen von OBAYASHI und HAKURA [44] ist für ein dem Erdfeld superponiertes Ringstromfeld bei 50° geomagnetischer Breite eine solche Verschiebung zu erwarten. Demnach setzt sich die Variation der kosmischen Strahlung zusammen aus:

1. der außerterrestrischen Modulation durch den solaren Wind und

2. einer kleineren gegenläufigen Variation durch die Steuerung der Abschneideenergie des Spektrums durch den solaren Wind über das Ringstromfeld.

Die Halbjahreswelle liefert den Zusammenhang mit dem Ringstromfeld in Huancayo zu $-5,5\,\gamma/1\%$ der Neutronenzahl in Weißenau. Sicher ist dieser Einfluß auch den anderen unter 1. zusammengefaßten Variationen überlagert, so daß der zunächst auf der Erde für die letztere festgestellte Zusammenhang von $+10\,\gamma/1\%$ zu korrigieren ist, damit man den reinen Zusammenhang zwischen Ringstromfeld und U erhält. Es ergibt sich $+15,5\,\gamma/1\%$ oder $34\,\gamma/0,1$ GV. Lineare Extrapolation über den Erfahrungsbereich hinaus führt zu $340\,\gamma$ Ringstromfeld in Huancayo bei einer Bremsstrahlung von $1\,GV$. Damit ist der oben aus der Magnetik berechnete Wert von $430\,\gamma$ zu vergleichen, welcher sich jedoch auf das Sonnenfleckenmaximum bezieht mit einem mittleren $U = 2$ GV. Beide unabhängige Schätzungen führen etwa zum selben Wert für den vom solaren Wind verursachten Anteil in der magnetischen Horizontalintensität in Huancayo. Wir sehen darin auch eine weitere Stütze der Vermutung auf Grund von Fig. 1, daß die kosmische Strahlung auch im Sonnenfleckenminimum bereits einer Modulation durch ein Bremsfeld von etwa 1 GV unterliegt.

Die genaue Analyse all dieser Dinge erfordert noch viel Arbeit. Schon jetzt hat man jedoch ein genaueres Bild über den Charakter dieser Variationen, das bisher durch die Verschiedenheiten der beiden Kurven in Fig. 16 (und noch mehr bei Verwendung der magnetischen Originalwerte) ziemlich undeutlich war. Dies hat einen einfachen Grund. Die beiden Kurven in Fig. 16 zeigen neben manchen auffallenden Übereinstimmungen auch unübersehbare Verschiedenheiten trotzdem bereits die früher nie untersuchten nicht korrelierten langsamen veränderlichen Anteile des Feldes schon eliminiert sind. Es sind tiefe magnetische Einbrüche vorhanden, mit welchen nur eine kleine Abnahme der kosmischen Strahlung, ja sogar manchmal eine kleine Zunahme verbunden ist. Auch der umgekehrte Fall kommt vor. So Ende Oktober 1957 oder Ende März 1958.

Die erdmagnetische Variation tritt nur auf, wenn eine Plasmawolke oder ein Plasmastrahl die Erde trifft. Geht er in größerem Abstand an der Erde vorbei, so hat er keine erdmagnetische Wirkung. Der Ringstrom kann nicht verstärkt werden. Trotzdem kann die Ausbeulung der Flächen $U = $const um die Sonne so breit sein, daß sie die weiträumige Modulation der kosmischen Strahlung auch noch am Ort der Erde ändert. Ein schönes Beispiel ist in der zweiten Hälfte des April 1957 in Fig. 36 zu finden. In diesem seltenen Beispiel muß die Wirkung des Plasmastrahls einer Fleckengruppe in einem Kegel von 45° Öffnungswinkel um die sich mit der Sonne drehende Strahlachse auf die kosmische Strahlung wirksam gewesen sein, während die magnetische Wirkung naturgemäß auf die Strahlbreite beschränkt bleibt. Andererseits sind offensichtlich gar nicht alle Plasmastrahlen bezüglich der kosmischen Strahlung wirksam. Vermutlich hängt dies mit ihrer magnetischen und räumlichen Struktur und eventuell auch mit ihrer Geschwindigkeit zusammen. Dies ist noch zu verifizieren.

Auf jeden Fall hat man im Pioneer V nach COLEMANN, DAVIS und SONETT [45] außerhalb des Erdfeldes eine ständige Magnetfeldkomponente von $3\,\gamma$ quer zur Sonnenrichtung gemessen, der zahlreiche jeweils nur kurze Zeit andauernde Stöße der Feldstärke in der Größenordnung $10\,\gamma$ überlagert waren. Sie erreichten maximal $50\,\gamma$ und waren dann auch von einem kräftigen magnetischen Sturm auf der Erde begleitet, da sich das Meßgerät zwischen Sonne und Erde noch nahe der letzteren befand. Am folgenden Tag lag im Pioneer wieder die normale Feldstärke vor, während auf der Erde nur eine sehr langsame, sich fast über einen Monat hinziehende Erholung beobachtet wurde. Die Feldkomponente in der Richtung zur Sonne wurde bei diesen Messungen nicht erfaßt.

Es ist nach all dem sehr wahrscheinlich, daß die Modulation der Kosmischen Strahlung und die erdmagnetische Störung beide vom solaren Wind abhängen.

K. Die solare Kosmische Strahlung
I. Beobachtungen

Die Modulation der galaktischen Strahlung erholt sich nach den Forbush-Effekten langsam. Man hatte auf jeden Fall lange Zeit keine abrupten Anstiege beobachtet und sich an die weitgehende Konstanz der Strahlung mit Einflüssen von Luftdruck und Erdmagnetik gewöhnt. Nur ZIRKLER [46] beobachtete mit einer nur gegen den Erdboden aber nicht nach oben gepanzerten Ionisations-kammer zahlreiche plötzliche Ionisationsanstiege bis zu 100% und schrieb sie chromosphärischen Eruptionen der Sonne zu. Der Zusammenhang war allerdings statistischer Art und wir wissen heute, daß auch vor der Verseuchung der Luft mit künstlicher Radioaktivität bei heftigen Regenfällen oft die mit solchen Kammern gemessene Ionisation durch ausgewaschene natürliche Radioaktivität vorübergehend stark ansteigt.

REGENER beobachtete am 29. 3. 33 in der Stratosphäre eine Ionisation, welche bei 35 mm Hg um 15% höher als sonst war und vermutete, daß dies mit einer den Zentralmeridian passierenden Fleckengruppe zusammenhing, „welche eine zusätzliche weiche Strahlung geliefert hatte".

Wir selbst registrierten [48] mit einem Zählrohrteleskop im Jahr 1941 jeweils einige Stunden anhaltende Zunahmen bis zu 20%, meist zur Nachtzeit, aber auch einmal nach einer Eruption der Größenklasse 2$^+$ am 20. 6. 41 eine Zunahme bis zu 76% für einen Stundenwert und 65% in der folgenden Stunde. Ebenso am 11. 7. 41, ebenfalls nach einer Eruption der Klasse 2 einen mehrere Stunden dauernden Anstieg mit maximal 15% Zunahme. Alle diese Effekte hatten ein sehr abruptes Ende, teilweise mit anschließendem kleinem Forbush-Effekt. Durch die Kriegsverhältnisse hatten wir keine Vergleichsmöglichkeit und es gab so viele stärkere Eruptionen ohne jeden Begleiteffekt dieser Art. Wir wissen heute, daß diese Effekte in Amerika nicht registriert wurden. Auch am 7. 3. 42 beobachteten wir zwischen 5 und 6 Uhr Weltzeit einen um 28% zu hohen Strahlungswert, dem in der nächsten Stunde noch ein um 10% zu hoher Wert folgte und konnten auch hier keine gleichzeitige Eruption und keinen magnetischen Sturm aus-machen. Erst 1946 erfuhren wir, daß LANGE und FORBUSH [49] in Cheltenham, Godhaven und Christchurch mit einer Ionisationskammer zur gleichen Zeit einen Effekt von maximal 8% registriert hatten, wie auch BERRY und HESS [50] und ebenso SÖHRN am 28. 2. 42 nach einer sehr kräftigen Eruption eine ganz ähnlich verlaufene Zunahme der Ionisation registriert hatten. Hier war der Zusammen-hang ganz offensichtlich. Aber in unserer Station in Friedrichshafen war an diesem Tage nicht der geringste Effekt aufgetreten. Wir haben daraus geschlossen [48], daß es sich um Teilchen mit positiver Ladung gehandelt haben muß. Solche wur-den, wenn sie von der Sonne kamen, zwar nicht in Amerika, wohl aber in Europa durch das Magnetfeld der Erde am Einfall verhindert. Für negative Teilchen wäre es umgekehrt zu erwarten gewesen. In Huancayo war der Effekt ebenfalls nicht aufgetreten. Die Energie der Teilchen war zu schwach, um dort das Erd-feld zu durchdringen. Auch in London wurde nur der Effekt vom 7. 3. 42 von DUPERIER [51] registriert, der vom 28. 2. 41 war ebenfalls ausgefallen. Dagegen waren auch in Japan die Bedingungen am 28. 2. günstig und ISHÜ, SEKIDO und MIYAZAKI [52] registrierten 8% Zunahme der Ionisation. Der nächste Fall trat erst wieder mit zunehmender Sonnentätigkeit am 25. 7. 46 auf und wurde an mehreren Stellen registriert [53] bis [55], jedoch ebenfalls nicht in Huancayo am Äquator [53].

Noch mehr Registrierstationen waren im Gang als am 19. 11. 49 ein neuer Ausbruch wiederum im Zusammenhang mit einer Eruption der Intensität 3 erfolgte, deren Maximum 10^{34} GT beobachtet wurde. Wiederum war wie am 28. 2. 42 die Intensität in Nordamerika wesentlich größer. Die Ionisationskammer-Stationen der Carnegie Institution hatten Zunahmen von 40% in Cheltenham, 30% in Godhaven auf Grönland und auf der Gebirgsstation Climax (3500 m) sogar von 180% der normalen Intensität beobachtet und daraus einen Absorptionskoeffizienten abgeleitet, der genau denjenigen der nuclearen Komponente entspricht. In Huancayo war wieder der Effekt ausgefallen. Die Energien dieser Strahlung waren zu gering. In Kanada wurden mit Teleskopen Zunahmen von 170% gemessen. In Japan [56] und in Europa [58] bis [61] wurden dagegen mit Teleskopen etwa 15%, mit Ionisationskammern 8% gefunden. Dagegen fanden Adams und Braddick [61] mit einer Anordnung zur Registrierung der von der kosmischen Strahlung erzeugten Neutronen in Manchester eine Zunahme um etwa 250% und einen Absorptionskoeffizienten der Zusatzstrahlung in Blei von $^1/_{147}$ g/cm². Sie hatten Bortrifluorid-Zählrohre mit verschiedenen Bremssubstanzen, wie Paraffin und Graphit umgeben und darüber Blei geschichtet, dessen Dicke sie zur Erlangung eines Optimums variierten. Sie hatten so den Barometereffekt der nuclearen Komponente gemessen. Es war somit offensichtlich, daß diese Komponente wesentlich empfindlicher ist. Außerdem dauert es in dieser Registrierung über 12 Std bis der Effekt abgeklungen war, gegen 3 Std in den anderen Registrierungen in Europa.

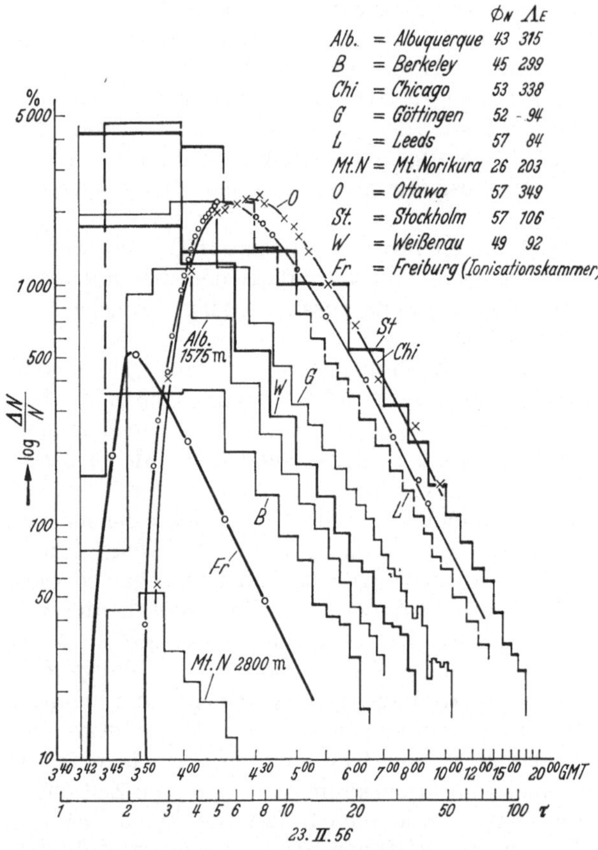

Fig. 20. Der Verlauf der von solarer kosmischer Strahlung an verschiedenen Stationen am 23. 2. 56 erzeugten nuclaren Komponente. Außerdem ist der Verlauf der Ionisation in Freiburg [65] mit aufgenommen, um den früheren Einsatz in der Trefferzone (Freiburg) gegenüber Chicago und Ottawa zu verdeutlichen. Der logarithmische Maßstab sowohl für die Intensität als auch für die Zeit (vom Zeitpunkt der Eruption auf der Sonne ab gezählt) läßt das Abklinggesetz $\sim t^{-2}$ besser erkennen. τ ist die Laufzeit der Teilchen in Einheiten der Laufzeit des Lichtes [64]

findlicher ist. Außerdem dauert es in dieser Registrierung über 12 Std bis der Effekt abgeklungen war, gegen 3 Std in den anderen Registrierungen in Europa.

Bis zum nächsten großen Einfall solarer Strahlung am 23. 2. 56 waren während des Sonnenfleckenminimums bereits im Hinblick auf das Internationale Geophysikalische Jahr zahlreiche Stationen mit der von Simpson vorgeschlagenen Neutronenanlage ausgerüstet. Eine Übersicht über den Verlauf der durch das solare Ereignis hervorgerufenen Neutronenzahlen zeigt Fig. 20 nach einer Zusammenstellung von Ehmert und Pfotzer [62], [63] in einer Darstellung, welche sowohl für die Intensität als auch für die Zeit logarithmischen Maßstab benutzt. Als 0-Zeit ist 3^{32} GMT angenommen; es ist die Zeit es Aufleuchtens der Eruption in H_α. 3^{40} traf auf der Erde UV- und Röntgenstrahlung ein und führte zu

Ionosphärenstörungen. Diese Strahlung wurde 3^{32} auf der Sonne erzeugt [64]. Die Zeit soll von der Entstehung der solaren Strahlung ab gezählt werden. Die Treppenstufen sind Mittel über die Ableseintervalle. Aus Gründen der übersichtlichen Darstellung wurden die Kurven für Chicago und Ottawa anders gezeichnet. Ebenso die Kurve für die Ionisationsregistrierung in Freiburg. Die Zahl τ gibt an, wievielmal länger die Laufzeit der eintreffenden Strahlung war als die des Lichtes [64]. Man erkennt wichtige Tatsachen:

1. Alle Kurven fallen schließlich über zum Teil mehr als 10 Std lang nach dem von EHMERT und PFOTZER angegebenen Abklinggesetz $I(t) \sim t^{-2}$ ab, das EHMERT [64] auch für andere ältere Ausbrüche verifiziert hat.

2. Der Breiteneffekt und damit auch die Energieverteilung der auf die Erde einfallenden solaren Strahlung bleibt während der Abklingphase konstant.

3. Da sich die Erde während der Abklingphase erheblich weiterdreht, ist die Einhaltung des Abklinggesetzes nur bei weitgehender Isotropie der Strahlung verständlich.

4. Die europäischen Stationen hatten einen deutlich früheren Einsatz und erreichten schneller das Abklinggesetz. Besonders deutlich wird dies aus dem Vergleich der Kurven von Ottawa und Chicago mit der deshalb in diese Figur aufgenommenen Ionisationskammerregistrierung in Freiburg, aus deren kontinuierlicher Aufzeichnung SITTKUS, KÜHN und ANDRICH [65] den ersten Einsatz und Anstieg vergleichbar genau ausmessen konnten wie MEYER, PARKER und SIMPSON [63] und ROSE und KATZMANN [63] aus der Ablesung je Minute der Neutronensumme in Chicago und Ottawa.

Man versteht dieses Verhalten, wenn man annimmt, daß in den ersten Minuten nach 3^{42} nur Strahlung ankam, welche sich geradlinig von der Sonne zur Erde bewegte. Sie konnte nur bestimmte Zonen, die Einfallszonen, in Europa treffen. Mit wachsender Zeit traf Strahlung ein, welche zwar zur selben Zeit erzeugt wurde, jedoch durch einen Umweg Zeit verloren hatte, denn ihre Geschwindigkeit unterschied sich nicht meßbar von der zuerst ankommenden Strahlung.

Gleichzeitig verbreiterte sich der Raumwinkel, aus welchem die Teilchen in das Erdfeld einfallen. Man konnte direkt nachweisen, wie die Grenze des Einfalls über die Erde huschte [64]. Schließlich erhielt nach 8 min jede noch so abgelegene Station Strahlung, aber offensichtlich noch nicht aus dem vollen Raumwinkel bzw. mit allen erlaubten Energien. Erst wenn dieser erfüllt war, trat das Abklinggesetz in Aktion. Merkwürdigerweise war dies in Ottawa merklich später als in Chicago der Fall und an der Station Godhavn im hohen Norden war es erst gegen 8.00 Uhr GMT; also nach mehr als 4 Std soweit. In Jakutsk bei gleicher Anfangskurve für ein Zählrohrteleskop, also Zenitstrahlung um 4^{15}, für eine Ionisationskammer erst nach 5.00 Uhr [64]. Vor allem wird es darauf angekommen sein, daß zu jeder Station speziell die Primärteilchen geringer Energie Zugang hatten, die sich in der betreffenden Registrierung maximal auswirken.

Die weitere Frage ist, wie die Teilchen so lange gespeichert wurden. SIMPSON [66] nahm an, daß ein vorausgegangener magnetischer Sturm einen Raum um die Sonne von etwa dem $1^{1}/_{2}$fachen Erdbahndurchmesser von magnetischen Feldern leergefegt hat und an der Grenze dieses Raumes die solaren Partikel reflektiert und so in dem Raum zusammengehalten wurden, so daß mehr und mehr Isotropie entstand und gleichzeitig durch das Entkommen von Teilchen eine Abnahme ihres Flusses mit der Zeit t nach einem $t^{-\frac{3}{2}}$-Diffusionsgesetz zustande kam.

Wir entwickelten die Vorstellung [64], daß ein schwaches Magnetfeld im interplanetaren Raum wirksam war, in welches die Teilchen unter verschiedenen Winkeln α gegen die Feldrichtung eingeschossen wurden. Sie führen dann je nach diesem Winkel Spiralen großer bis zu sehr kleiner Steigung aus. Im Fall eines von der Sonne zur Erde gerichteten Feldes werden die zuerst ankommenden Teilchen gar nicht abgelenkt. Die spät eintreffenden kommen auf einem Kreiskonus um die Magnetfeldlinie durch die Erde mit demselben Winkel α an, unter dem sie in das Feld eintraten. Die verlangten Speicherzeiten, das Abklinggesetz, die Konstanz des Spektrums und die Verbreiterung des Einfallwinkels ergibt sich damit automatisch. Auch der allgemeinere Fall eines schräg zur Sonnenrichtung verlaufenden Feldes wurde behandelt [64]. In diesem Fall ist bei homogenem Feld eine Mindestenergie der Teilchen erforderlich, während bei direkt zur Erde gerichtetem Magnetfeld eine starke Inhomogenität desselben im Sinne einer Abnahme der Feldstärke die Teilchen parallel zur Feldrichtung ausrichten und damit eine Speicherung verhindern würde. Eine Superposition verschiedener Felder wird wohl der Wirklichkeit am besten gerecht werden. In diesem Zusammenhang ist zu bemerken, daß die auslösenden Eruptionen am 7. 3. 42 und 25. 7. 46 am westlichen Sonnenrand und die am 19. 11. 49 74° westlich vom Zentralmeridian lagen. Die Eruption am 28. 2. 42 war östlich vom Zentralmeridian und die vom 12. 11. 60. Neuerdings wurde von Gold [67] der Gesichtspunkt einer Speicherung nach dem Prinzip der magnetischen Flasche durch passende Feldinhomogenitäten am Anfang und Ende solarer Plasmastrahlen hinzugefügt, der uns nicht ohne weiteres zum Abklinggesetz zu passen scheint, das sich in einer Reihe von neueren Fällen prompt einstellte und auch für die früheren Fälle von uns nachgewiesen wurde [64].

Diese Fragen des interplanetaren Raumes sollen hier nicht weiter behandelt werden.

Unter der Annahme wirklicher Isotropie der Strahlung während der Abklingphase kann mit Hilfe des Breiteneffekts sowohl einerseits das Spektrum dieser Strahlung (das ja lange Zeit konstant blieb) und auch andererseits die Lage des wirksamen Dipols der Erde empfindlicher kontrolliert werden. Pfotzer [68] fand, daß mit einem differentiellen Energiespektrum $f(E)\,dE \sim E^{-6}$ (für Protonen) und dem exzentrischen Dipol (des konventionellen geomagnetischen Systems) ohne weitere Verdrehung der Dipolrichtung gute Übereinstimmung mit den Messungen erzielt wird. Dieses Spektrum ist wesentlich steiler als das der galaktischen Strahlung mit dem Exponenten 2,5. Daraus erklärt sich die besondere Eignung der nuclearen Komponente zum Nachweis solarer Strahlung am Grunde der Atmosphäre.

Inzwischen wurden mehrere Fälle solarer kosmischer Strahlung geringeren Ausmaßes und zwei weitere größere Ausbrüche kosmischer Strahlung von der Sonne am 12. und 15. 11. 60 beobachtet. In diesen Fällen reichte die Energie dieser Strahlung nicht zur Erzeugung von Mesonen in merklichem Umfang aus. Daher konnte der Untergrund der galaktischen Strahlung recht genau abgezogen werden. Dies war um so wichtiger, als die während des Effektes am 12. 11. 60 durch einen Forbush-Effekt erheblichen Schwankungen unterlag [69].

Fig. 21 zeigt den verbleibenden Anteil der solaren Strahlung an den Neutronen in der Registrierung in Lindau, geomagnetische Breite 51, 55°, Grenzsteifigkeit der Teilchen 2,45 GV. Der Beginn der auslösenden Eruption der Größenklasse 3$^+$ ist mit einem Pfeil markiert. Darüber sind eine unempfindliche erdmagnetische Sturmregistrierung des Göttinger Geophysikalischen Instituts [70] und eine Registrierung der magnetischen Horizontalintensität in Colombo, welche wir einer Mitteilung von Herrn Prof. Marsden in Leeds entnommen haben, aufgezeichnet. Etwa 20 min nach der Beobachtung der Eruption auf der Erde setzt die Strahlung ein und erreicht ein erstes Maximum um 16.00 Uhr, nach welchem sie zunächst schwach abnimmt. Von 18^{05} bis 18^{30} erneuter langsamer

Anstieg, dem ein sprunghafter Anstieg um 18³⁰ folgt. Die untere und stärkere gestrichelte Kurve ist in der Abklingphase ein t^{-2}-Gesetz. Ebenso die obere gestrichelte Linie in der Abklingphase. Sie ist durch die Höchstwerte der Kurve gezogen. Für die Minima wurde das Abklinggesetz gegen 20 Uhr erreicht, für die obere nach 22 Uhr.

Der zweite Anstieg um 18 Uhr ist von keiner neuen Eruption eingeleitet. Vielmehr beginnt er mit dem Einsetzen der Hauptphase des magnetischen Sturmes.

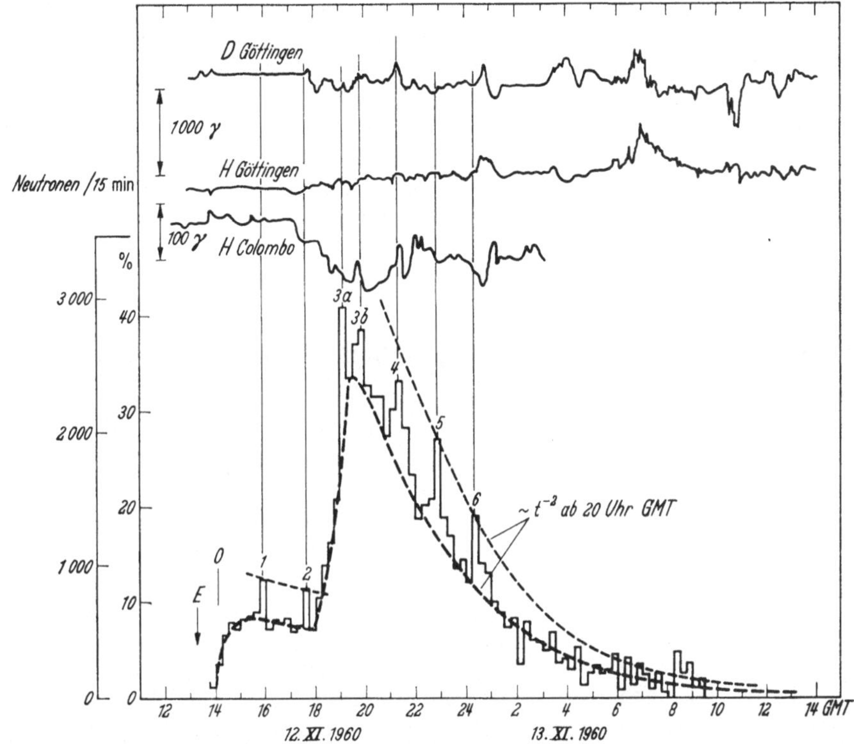

Fig. 21. Erdmagnetische Variationen in Göttingen und Colombo und der durch solare Strahlung bedingte Anteil der Neutronenzahl in Lindau (Harz) am 12. und 13. 11. 60. *E* Zeitpunkt der auslösenden Eruption. Die erdmagnetischen Variationen gehen auf eine frühere Eruption zurück

Die der unteren Begrenzung des Verlaufs aufgesetzten Maxima, welche wir mit 1, 2, 3a und 3b, 4, 5, 6 numeriert haben, sind erstaunlich scharf. Die letzteren sind äquidistant mit einem Abstand von 90 min, der Abstand der ersten drei Maxima beträgt je 100 min und ebenso der Abstand vom Beginn des ersten Anstiegs bis zum ersten Maximum.

In den Registrierungen von Prag und London sind die Maxima 3a, 3b, 4, 5 und 6 ebenfalls sehr ausgeprägt, in Leeds sind sie sehr verwaschen und in Uppsala nicht mehr zu erkennen. Dagegen ist dort das Maximum 1 stark verbreitert vorhanden.

Man hat den Eindruck, daß die Maxima auch in der Magnetik eine gewisse Parallele haben. Wegen des steilen Energiespektrums der Strahlung könnte eine Verschiebung der Grenzsteifigkeiten an den Stationen durch ausgedehnte magnetische Störfelder solche Zunahmen bewirken. Es wäre dann verständlich, daß der Effekt in hohen Breiten verschwindet, weil dort die geringsten erdmagnetisch zugelassenen Mindestenergien sowieso nicht mehr ausreichen, um zur Neutronenauslösung unter der Atmosphäre etwas beizutragen. Die auffallende Periodizität könnte aber auch durch die Führung der Teilchen in einem Magnetfeld bedingt sein, wo Teilchen, die gerade eine ganzzahlige Anzahl von Umläufen um die Feldrichtung gemacht haben, unter Umständen eine Art Fokussierung erfahren. Dies kann hier nicht weiter behandelt werden.

Das Maximum erreichte in Uppsala und Leeds fast 100% des Wertes der galaktischen Strahlung. Jungfraujoch war die Station mit geringster Breite, welche eben noch einen Anstieg, nämlich 3%, zu verzeichnen hatte.

Eine andere sehr schwache solare Strahlung haben Carmichael und Steljes [71] nach einer Eruption am 17. 7. 59 nachgewiesen. Sie zeichneten die Variationen zahlreicher Stationen übereinander, jeweils in solchem Maßstab auf, daß die Variationen überall gleich groß wurden. (Man kann auch sagen, man ermittelt aus jeder Registrierung den Modulationswert U und zeichnet diesen auf.) Dann ergibt sich ein schwarzes Band von der Breite der statistischen und sonstigen Ungenauigkeit der Meßwerte. Am 17. 7. 59, am Tage vor einem neuen magnetischen Sturm erhoben sich nun alle Kurven für Stationen mit kleinerer Abschneidesteifigkeit als 1 GV um 5% und bei Gebirgsstationen bis 10% über das Band der anderen Kurven. Dieselbe Strahlung wurde über Lindau in großen Höhen gemessen (vgl. K III).

II. Bewegung der solaren Kosmischen Strahlung im Erdfeld

Im Gegensatz zur galaktischen kosmischen Strahlung, welche aus allen Richtungen einfällt, kommt die solare Strahlung im Idealfall nur aus einer Richtung, der Richtung der Sonne in das Erdfeld. Dabei sind alle γ-Werte (Abschnitt C) zugelassen. Es ist die Frage, welche Zonen der Erde sie erreichen. Mit Zählrohrteleskopen und noch mehr bei Registrierung der nuclearen Komponente werden nur die Teilchen aus einem ziemlich engen (etwa 30°) Bereich um den Zenit erfaßt. Bei Ballonmessungen und erst recht bei Messungen in Satelliten wird dagegen Strahlung aus einem großen Winkelbereich und in entfernten Raumsonden aus allen Richtungen registriert. Hier benötigt man eine große Anzahl von Bahnen im Erdfeld. Dabei ist die richtige Approximation des letzteren besonders wichtig. Ebenso die genaue Einordnung des Meßpunktes. Die von A. Schmidt [72] berechnete zweite Näherung der Feldverhältnisse unter Berücksichtigung von fünf weiteren Gauß-Koeffizienten, von Bartels [73] eingehend diskutiert, ergibt den im Abschnitt C bereits erwähnten exzentrischen Dipol, dessen Achse parallel zur Achse des zentrischen Dipols ist. Yory [74] hat nach den Daten von Vestine und Lange [75] die Koordinaten des Zentrums neu berechnet zu

$$\lambda_2 = 154° \, \mathrm{E}, \quad \varphi_2 = 14° \, \mathrm{N}, \quad R_2 = 6{,}29 \cdot 10^{-2} R_e,$$

wobei R_2 der Abstand vom Erdmittelpunkt und R_e der mittlere Erdradius ist.

Die aus Näherung mit drei Gauß-Koeffizienten berechnete Achse des zentrischen Dipols ergibt eine Drehung um 11,5° gegen die Rotationsachse der Erde auf dem Meridian 190° Ost. Auf diesen zentrierten Dipol wird das übliche geomagnetische Koordinatensystem bezogen. Sind λ und φ die geographische Länge und Breite der Stationen und λ_0 und φ_0 die Koordinaten des borealen Durchstoßpunktes der Achse des zentrischen magnetischen Dipols ($\lambda_0 = 290{,}9$ E; $\varphi_0 = 78{,}5$ N), so erhält man die geomagnetischen Koordinaten Λ und Φ nach den Formeln

$$\sin \Phi = \cos \varphi \cdot \cos \varphi_0 \cos (\lambda - \lambda_0) + \sin \varphi \sin \varphi_0$$

$$\cos \Lambda = \frac{\cos \varphi \cdot \sin \varphi_0 \cdot \cos (\lambda - \lambda_0) - \sin \varphi \cdot \cos \varphi_0}{\cos \Phi}.$$

Wir haben im Abschnitt C II gesehen, daß diese Näherung auch bei Berücksichtigung der Exzentrizität noch nicht den „Äquator" der kosmischen Strahlung richtig wiedergibt. Analytische Korrekturen für denselben liegen noch nicht vor.

STÖRMER [76], DWIGHT [77] und SCHLÜTER [78] haben nun eine große Anzahl von Bahnen im Dipolfeld berechnet und MALMFORS [79] und BRUNBERG [80] haben mit Modellversuchen (leider nur für höhere Energien) die Asymptoten vieler Bahnen für vorgegebene Einfallsrichtungen auf der Erde unter verschiedenen Breiten experimentell im Modellversuch bestimmt.

Man kann sich für eine vorgegebene Station für senkrechten (oder anderen) Einfall die Herkunftsrichtung als Funktion $\lambda_s(\varphi_s)$ auftragen und entlang der

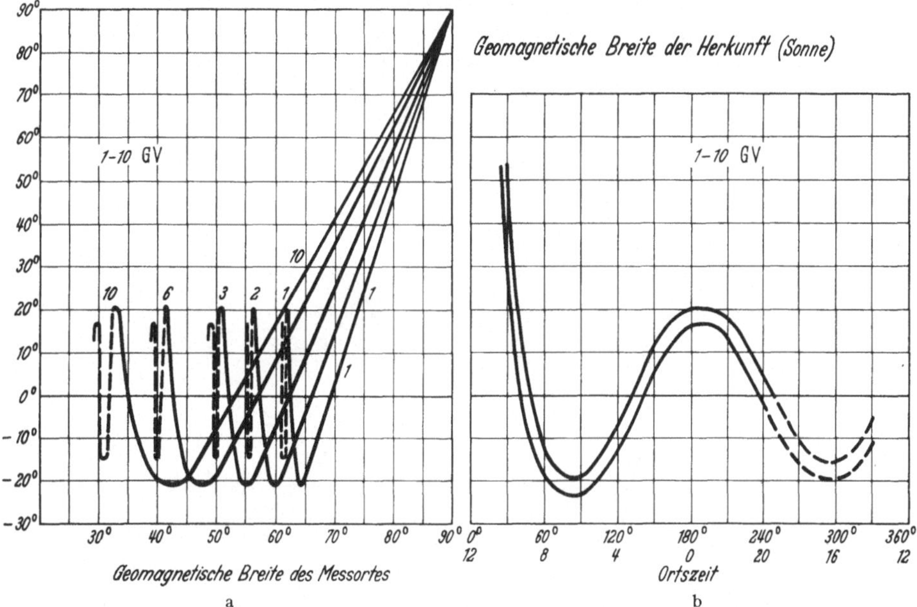

Fig. 22a u. b. Zur Herkunft der gerichteten Strahlung des Meßortes kann links für jede Energie die geomagnetische Breite entnommen werden, aus welcher die Strahlung in das Erdfeld einfallen muß. Mit dieser Breite geht man in die rechte Figur und kann die entsprechende geomagnetische Länge der Herkunftsrichtung zwischen den beiden Kurven ablesen. Zwischen +20° und −20° gibt es mehrere Möglichkeiten

Kurve eine Energieskala anbringen, wie wir dies zuerst zur Beurteilung der Einfallszonen an Hand einiger Bahnen getan haben [48], welche STÖRMER [81] für unsere Station Friedrichshafen berechnet hatte. Man kann dann den Stand der Sonne eintragen und ersieht, welche Energie ohne Ablenkung im interplanetaren Raum die Station durch das Erdfeld erreichen kann. Ist eine solche Energie aus einer Messung bekannt, so gewinnt man die Ablenkung dieser Teilchen auf dem Weg von der Sonne zur Erde. FIROR [82] hat dies für alle Lagen auf der Erde in zwei Diagrammen zusammengestellt, welche wir in Fig. 22 nebeneinander gezeichnet haben. Man kann dann mit dem Sonnenstand aus dem linken Teil die Energie, aus dem rechten Teil die Ortszeit der Trefferzone entnehmen. Die geomagnetischen Längen sind von den geographischen wenig verschieden, weshalb die Ortszeit eingetragen werden kann. Die geomagnetischen Breiten der Meßorte unterscheiden sich zum Teil von den geographischen beträchtlich und die geomagnetische Breite λ_s der Sonne schwankt im Lauf des Tages um 23°. Die Punkte für Bahnen verschiedener Teilchenenergien winden sich im rechten Teil der Fig. 22 alle in dem Bereich zwischen den beiden eingezeichneten Kurven. Dieser Bereich heißt im Fachjargon die „Firorschlange". Sie windet sich noch mehrmals um den Erdball und ergibt so ein ganzes Band von Einfallsmöglichkeiten, wobei die Intensität zu früheren Ortszeiten abnimmt. Man ersieht auch aus dem linken Teil der Fig. 22 sofort, daß für jede Energie unabgelenkte Teilchen von der Sonne in geomagnetischen Breiten über 62° für 10 GV und über 75° für 1 GV nicht einfallen können.

Fig. 23 zeigt als Beispiel für den 23. 2. 56 die Trefferzone zu verschiedenen Zeiten für senkrechten Einfall und die Lage der Stationen [62]. Die Verzögerung

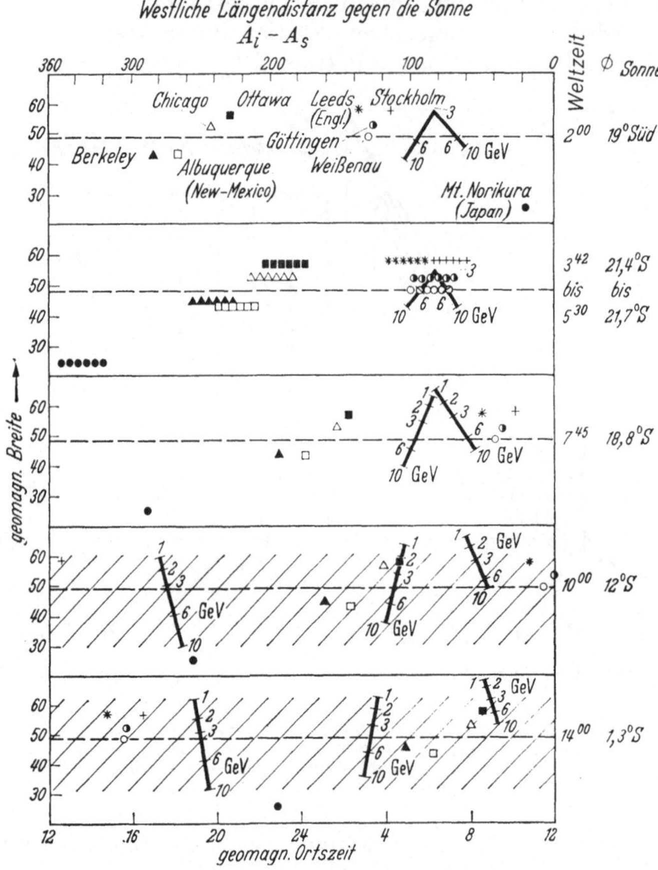

Fig. 23. Entwicklung der Trefferzonen während des solaren Strahlungsausbruches am 23. 2. 56 und ihre relative Lage zu verschiedenen Registrierstationen. Sonnenstand $\varphi_{geogr} = 10,3°$ S. Nach Ehmert und Pfotzer [62]. Mann kann daraus die zusätzliche Ablenkung der Strahlung bis zu ihrem Einfall in das Erdfeld entnehmen

des Einfalls in Amerika kann also, wie oben angegeben, durch eine rasche Ausweitung des Einfallswinkels beschrieben werden, welche nur einer Ablenkung der später ankommenden Teilchen außerhalb des Erdfeldes zugeschrieben werden kann.

III. Solare Kosmische Strahlung geringer Energie

Um diese Strahlung zu messen, muß man in hohe Breiten gehen und durch Ballonmessungen die Absorption der Atmosphäre ausschalten. Die in Abschnitt K I erwähnte solare Strahlung am 15. 7. 59, welche sich, nur in höheren Breiten in den Neutronenregistrierungen auswirkte, war bei einem Ballonaufstieg über Lindau mit einem Zählrohr in Höhen über 20 km gemessen worden [82]. Die normale Druckabhängigkeit für die galaktische Strahlung ist auf Grund der

Neutronenzahl am Boden an Hand unseres Erfahrungsmaterials angebbar. Sie hat ein Maximum bei 80 mm Hg und fällt dann mit abnehmendem Druck ab. An diesem Tage nahm jedoch die gemessene Strahlung von 50 mm Hg ab mit abnehmendem Druck zu und die Differenz zu der berechneten galaktischen Strahlung ergibt eine Absorptionskurve der solaren Strahlung von der Form $J = J_0 \cdot (p/p_0)^{-1,45}$, woraus auf ein differentielles Energiespektrum $\sim E^{-3,6}$ geschlossen wurde [82]. Das Spektrum reichte meßbar bis zu 400 MeV. Solche Protonen wären in einem ungestörten Magnetfeld der Erde in dieser Breite verboten gewesen. In der galaktischen Strahlung sind sie nur mit sehr geringer Zahl vorhanden. Diese Strahlung war gleichzeitig von BROWN und D'ARCY [83] in ähnlicher Weise und von NEY [84] mit Emulsionen beobachtet worden. Ebenfalls mit Emulsionen war von NEY, WINCKLER und FREYER [85] im Mai 1954 bei einem ähnlichen Effekt festgestellt worden, daß es sich vorwiegend um Protonen handelt.

ANDERSON und ENEMARK [86] führten am 18. und 19., 21./22. und 24./25. und 26. 27. Juli in Resolute Bay in hoher Breite ähnliche Ballonmessungen durch und fanden, daß diese energiearme solare Strahlung, welche auf eine sehr helle chromosphärische Eruption am Nachmittag des 16. Juli zurückzuführen war, trotz ihrer raschen Abnahme sehr lange nachzuweisen war, weil sie zu Anfang einen 200mal größeren Teilchenfluß als die galaktische Strahlung lieferte. Ihr Abklinggesetz wird von den Autoren $\sim t^{-3}$ angegeben, wobei jedoch die Zeit von einem Zeitpunkt 16 Std nach der Eruption gezählt wird. Zählt man vom Zeitpunkt der Eruption ab, so findet man, daß der von ihnen angegebene Verlauf mehr als 7 Tage lang sehr genau ein Abklinggesetz $\sim t^{-3,5}$ erfüllt. Dies ist um so erstaunlicher, als am Morgen des 18. Juli ein erdmagnetischer Sturm eingetreten war und die Neutronenzahl in Lindau um 11% abgesunken war, jedoch das Abklinggesetz der energiearmen solaren Strahlung nicht beeinflußt wurde. Die Fleckengruppe, in welcher die Eruption am 16. 7. 59 aufgetreten war, verschwand am 21. Juli hinter dem westlichen Sonnenrand. Auf den Verlauf des Teilchenflusses hatte auch dies keinen Einfluß.

Solare Korpuskelstrahlung dieses Energiebereiches erzeugt in der Atmosphäre eine Ionisation, welche im Höhenbereich über 50 km eine Verstärkung der unteren ionosphärischen D-Schicht am Einfallsort verursacht. Sie äußert sich in einer erhöhten Absorption des kosmischen Rauschens im Meterwellenbereich, weil diese Ionisation auch im Bereich hoher Stoßzahlen (zwischen den Elektronen und den Luftmolekülen), erzeugt wird, so daß man aus Intensitätsabnahmen des kosmischen Rauschens am Boden auf eine derartige Ionisierung schließen kann [88], [89]. Allerdings wird eine ähnliche, jedoch auf kürzere Zeiten beschränkte Ionisation auch durch solare Röntgenstrahlung (bei Eruptionen) und in hohen Breiten durch Röntgenstrahlung im Energiebereich bis zu 200 keV erzeugt. Diese Röntgenstrahlung hängt mit Elektronen zusammen, welche aus dem Strahlungsgürtel stammen und nicht zur kosmischen Strahlung gehören. Diese Ionisation, welche auch oft direkt gemessen wurde, ist sehr unruhig und tritt auch in großen Stößen von etwa 10 min Dauer auf [86]. Sie ist die häufigere Erscheinung. Bei Ballonmessungen ist es zweckmäßig, durch die Verwendung von Ionisationskammern, Einzelzählrohren und Koinzidenzteleskopen oder Szintillationszählern die Art der jeweiligen Strahlung festzustellen. Man kann so auch aus den Absorptionsmessungen mit dem Riometer, wie die zur Messung benutzte Empfangs- und Registriereinrichtung [90] genannt wird, vom Boden aus eine dauernde Überwachung durchführen und mit einem Stationsnetz auch die Einfallszonen und ihre zeitlichen Veränderungen feststellen.

Da die meisten Effekte nur an den Polkappen auftreten ist die Bezeichnung PCA (polar cap absorption) für diese Erscheinung eingeführt worden. Sie hat sich als wichtiges Glied bei der Erfassung der solaren kosmischen Strahlung geringer Energie erwiesen, welche im Sonnenfleckenmaximum immerhin rund zehnmal pro Jahr auftrat.

Die Absorption wirkt sich auch auf die längeren Wellen aus, welche die Ionosphärenlotungsstationen benutzen. Sie führt dort zu einem Versagen der Meßmethodik beim „polar black out", wenn wegen der Absorption keine reflektierten Wellen mehr zurückkommen, weil sie zweimal die absorbierende Schicht durchqueren müssen. Langwellen werden dagegen in der D-Schicht selbst reflektiert. Tatsächlich wurde bei der starken solaren Strahlung am 23. 2. 56 auf verschiedenen Übertragungsstrecken eine Dämpfung der Langwellen beobachtet und auch eine Änderung der Übertragung von den Längstwellen, welche von Blitzen ausgeht [91].

Schließlich ist noch eine direkte Messung solarer kosmischer Strahlung zu erwähnen, welche Fan, Meyer und Simpson [37] in großem Abstand von der Erde mit dem im Abschnitt G erwähnten Gerät in der Venusrakete Pionier V in etwa 500000 km Entfernung von der Erde am 1. und 2. 4. 60 registrierten. Das Gerät erfaßte bei allseitigen Empfindlichkeiten Protonen bis herab zur Energie von 75 MeV. Der Gesamtteilchenfluß stieg dort um 170% an und fiel gleichmäßig im Laufe eines Tages auf den Ausgangswert zurück. Auf der Erde wurde nichts beobachtet. Das Ereignis trat während eines Forbush-Effektes auf, der für die dort registrierte Strahlung 19% ausmachte.

IV. Die Ausbreitung solarer Kosmischer Strahlung

Solare kosmische Strahlung tritt offensichtlich nach heftigen chromosphärischen Eruptionen auf. Auch Eruptionen sehr nahe dem westlichen Sonnenrand waren mehrfach wirksam, während von ihnen keine erdmagnetisch wirksame Partikelstrahlung die Erde erreicht.

Die Radiostrahlung der Sonne im Meterwellenlängenbereich steigt bei großen Eruptionen um mehrere Größenordnungen an und ist nach ihrer zirkularen Polarisation als Synchrotronstrahlung beschleunigter Elektronen in Magnetfeldern zu deuten [92]. Man unterscheidet nach Frequenzspektrum und zeitlichem Verlauf verschiedene Typen. Ein Typ, mit Typ IV bezeichnet, zeigt ein langsameres Abklingen und reicht zu längeren Wellenlängen. Die beobachtete solare kosmische Strahlung geht auf solche Ausbrüche zurück. Aber es wird natürlich nicht bei jeder solchen Eruption auch auf der Erde solare kosmische Strahlung beobachtet.

Die Statistik ergibt insbesondere für die energiearme solare kosmische Strahlung eine Bevorzugung von Eruptionen auf der westlichen Sonnenseite. Dies hängt sicher mit der Bewegung von solarem Plasma zusammen.

Besonders auffallend ist, daß nach Fig. 22 am 12. 11. 60 solare kosmische Strahlung auf der Erde zunächst nur einen geringeren Wert erreichte und ihr Teilchenfluß mit dem Eintreffen einer solaren Wolke von einer früheren Eruption gegen 18 Uhr GMT an allen Stationen rasch auf einen mehrfach höheren Wert anstieg und dann zum Abklinggesetz einbog. Dies zeigt, daß eine Führung in solchen Plasmawolken erfolgt und in diesem Fall der Teilchenfluß vor der Wolke geringer war als in derselben. Nach Alfvén u. a. erwartet man, daß in der Wolke ein stärkeres Magnetfeld herrscht als vor derselben. Durch hohe Leitfähigkeit in einer solchen Wolke wird dies erhalten, wenn sie bei ihrer Entstehung schon ein Magnetfeld hatte. Und daß dies der Fall ist, kann man auf Grund der Radiostrahlung erwarten. Ist die Hauptrichtung des Feldes in derselben etwa nach den Annahmen von Alfvén [93] radial zur Sonne, so ist für die in einem solchen Feld auf Spiralen geringer Steigung fliegenden Teilchen eine Feldabhängigkeit des Teilchenflusses derart zu erwarten, daß er bei größerem Feld ebenfalls größer ist, weil dieselben Teilchen ein kleineres Volumen erfüllen und somit weniger verdünnt sind.

Die Ausdehnung des Einfalls energiearmer Strahlung über mehrere Tage, wenn auch mit so starker Abnahme, daß bei einer entsprechenden Abnahme der Teilchen hoher Energie längst die Nachweisgrenze unterschritten ist, wirft für das Problem der Speicherung erhebliche Schwierigkeiten auf. In dieser Zeit erholt sich auch die Modulation der galaktischen Strahlung weitgehend und man könnte die Frage aufwerfen, ob man es nicht mit einer länger andauernden Erzeugung solarer Strahlung zu tun hat und die Erzeugung ihrerseits das Abklinggesetz bedingt. Es erscheint nicht ausgeschlossen, daß das Bremsfeld der elektrischen Modulation der galaktischen Strahlung für die solare Strahlung eine beschleunigende Rolle spielt und die besprochene anisotrope Modulation mit zeitweise beschleunigendem Charakter ist vielleicht nur ein ähnlicher Vorgang an den Teilchen aus der galaktischen Strahlung ohne Nachlieferung genügend rascher solarer Teilchen. Die größeren Intensitäten solarer kosmischer Strahlung traten im Laufe eines Forbush-Effektes auf, also bei örtlich erhöhtem Bremsfeld und Einschuß vieler solarer Teilchen bei großen Eruptionen.

L. Schlußbetrachtung

Beim Entwurf dieses Beitrages war noch ein Kapitel über den Strahlungsgürtel der Erde vorgesehen. Es muß aus äußeren Gründen entfallen und wir verweisen den Leser auf die Darstellung von G. PFOTZER [94] und deuten hier nur die Zusammenhänge an.

Der Strahlungsgürtel wurde von VAN ALLEN u. Mitarb. [95] entdeckt, als sie im Satelliten Explorer I mit einem Geiger-Zählrohr den Breiteneffekt der primären kosmischen Strahlung untersuchen wollten. Sie fanden in Äquatornähe in Höhen über 1000 km eine Zone mit 1000fachem Teilchenfluß. Es zeigte sich später bei Messungen mit der Fernrakete Pionier III und dem Satelliten Explorer IV, daß zwei Gebiete mit besonders hohen Intensitäten vorliegen. Das Innere mit dem Maximum 10000 km vom Erdmittelpunkt entfernt ist nach FAN, MEYER und SIMPSON [37], ARNOLDY, HOFFMANN und WINCKLER [96] und ROSEN, FARLEY und SONETT [97] mit Protonen und Elektronen erfüllt. Das äußere etwa doppelt so weit vom Erdmittelpunkt entfernte Gebiet maximalen Teilchenflusses enthält Elektronen. Die Teilchen bewegen sich nahezu senkrecht zu den magnetischen Kraftlinien und gelangen bei Annäherung an die Erde entlang den Kraftlinien in Gebiete mit größerer Feldstärke, wobei die Spiralen ihrer Bahn immer mehr senkrecht zu den Kraftlinien verlaufen. Verschwindet schließlich die Längskomponente der Bewegung des Teilchens, so führt die Bahn wieder in Spiralen wachsender Steigung in das Gebiet kleinerer Feldstärken zurück und darüber hinaus zum konjugierten Reflexionspunkt auf der anderen Seite des magnetischen Äquators. Sie sind in einem magnetischen Käfig gefangen, den wir auf S. 349 bereits im Zusammenhang mit der Störmerschen Theorie erwähnten. Die Frage, wie die Teilchen hineinkommen wurde von mehreren Autoren gleichzeitig beantwortet [98]. Sie fliegen zumindest zum Teil als Neutronen hinein. Diese stammen aus der Erdatmosphäre, wo sie von der kosmischen Strahlung durch Kernzertrümmerungen erzeugt werden. Zerfällt ein Neutron im Gebiet des Strahlungsgürtels, so können sowohl das Proton als auch das Elektron dort gefangen bleiben. DESSLER und KARPLUS [99] führen den ganzen Strahlungsgürtel auf diesen Vorgang zurück, andere vermuten, daß durch Feldstörungen bei magnetischen Stürmen solare Teilchen eingeschleust und anschließend durch Magnetfeldänderungen beschleunigt werden und so zum äußeren Teil des Strahlungsgürtels beitragen.

Die Intensität des äußeren Teiles erwies sich als sehr wechselnd. Bei erdmagnetischen Stürmen nimmt sie stark ab, weil viele Teilchen durch die Störung des Feldes in die Atmosphäre eindringen und abgebremst werden, wobei auch die im Abschnitt K III erwähnte Röntgenstrahlung und Nordlichter entstehen [96]. Nachher steigt sie wieder sehr stark an.

Die Teilchen müssen wegen der Inhomogenität des Feldes neben ihrer Bewegung entlang den magnetischen Kraftlinien auch eine Drift quer zu den Kraftlinien ausführen und zwar Protonen und Elektronen in verschiedener Richtung. Das Ergebnis ist ein Ringstrom, dessen Stärke von der Besetzung des Strahlungsgürtels abhängt. Und diese ist wieder von erdmagnetischen Störungen einerseits und von der kosmischen Strahlung andererseits abhängig, und die enge Korrelation zwischen derselben und dem Ringstromfeld, welche wir im Abschnitt J besprochen haben, könnte so doch entgegen den ursprünglichen Annahmen z. T. auf der Modulation der kosmischen Strahlung als Ursache und dem Ringstromfeld als sekundärer Erscheinung beruhen. Dies setzt ein ständiges Gleichgewicht zwischen dem Teilchenverlust des Strahlungsgürtels und der Nacherzeugung durch die kosmische Strahlung voraus, was Dessler und Karplus [99] ja auch annehmen. Auch die Gesamtzahl der im Erdfeld abgelenkten Primärteilchen der kosmischen Strahlung stellt bereits einen Ringstrom dar.

Wir ersehen, daß die kosmische Strahlung und die erdmagnetischen Variationen ein weiteres enges Bindeglied zwischen solaren und geophysikalischen Vorgängen darstellen und daß die Geophysik hier aus der direkten und aus der indirekten Weltraumforschung entscheidende Impulse erhält.

Summary

A brief survey is given about the nature, the energy distribution and the variations of primary galactic cosmic rays. The influences of the earth's steady magnetic field on them and the reverse possibility of investigating this field by cosmic ray observation. The secondary components in the atmosphere and their dependence on the air pressure are reviewed. The modulation of the primary galactic radiation and of its energy distribution is discussed more in detail, especially the description of the isotropic modulation by a single parameter resulting from the hypothesis of an electric field modulation. This model proofs to give a thorough description of the variations at any point on the earth as well as of those in interplanetary space as they are measured in a few examples until now. A special comparison of the modulation of cosmic rays by the suns activity with earth magnetic disturbances of ring current field type shows as well a high correlation between both quantities demonstrating the similarity of both variations governed by the solar wind, as also characteristic differences that result from a wider influence of solar beams on galactic cosmic rays. This correlation allows an estimation of the total magnetic field component originating outside the earth and this estimation is confirmed by the amplitude of the semi-annual wave of this magnetic field component if it is ascribed to the varying exposure of the earth's magnetic axis to the direction of the solar wind. Anisotropic modulation of cosmic ray intensity by such beams is briefly discussed and finally the characteristics of solar cosmic rays are treated, especially the guidance of solar cosmic rays by solar plasma beams, the phenomena of storage, the intensity-time law, the influence of the geomagnetic field on solar cosmic rays and the main feature, that these don't show the modulation observed for galactic cosmic rays.

Literatur

[1] MacDonald, B., and W. R. Webber: Phys. Rev. **115**, 194 (1959); — Space Res. COSPAR Meeting Nizza 1960, p. 968.

[2] Ehmert, A.: Space Research, COSPAR-Meeting Nizza 1960, p. 1000. IUPAP, Moskau Meeting 1959, vol. IV, p. 142, 1960.

[3] Rossi, B.: Rev. Mod. Phys. **21**, 104 (1949).

[4] Ginzburg, V.L.: Progress in Elementary Particles and Cosmic Ray Physics, Bd. IV, 1958. IUPAP Moskau-Meeting 1959, vol. III, p. 196, 1960.

[5] Biermann, L., u. L. Davies jr.: Proceedings IUPAP, Moskaw Meeting 1959, vol. III, p. 282, 1960.

[6] Suess, H.E., and H.C. Urey: Rev. Mod. Phys. **28**, 53 (1956).

[7] Störmer, C.: Z. Astrophysik **1**, 237 (1930).

[8] Lemaître, G.: Nature, Lond. **140**, 23 (1937).

[9] Vellarta, M. S.: Phys. Rev. **47**, 647 (1935); **74**, 1837 (1948).

[10] Clay, J.: Proc. Amsterdam **30**, 1165 (1927).

[11] Meyer, P., e J.A. Simpson: Nuovo Cim. Suppl. Nr. 2, 233 (1958); Nr. 8, 277 (1958).

[12] Webber, W.R.: Nuovo Cim. Suppl. **10**, Nr. 2, 532 (1958).

[13] Rothwell, P., u. J. Quenby: Nuovo Cim. Suppl. **8**, 249 (1958).

[14] Pomerantz, M.A., A.E. Sandström e D.C. Rose: Nuovo Cim. Suppl. **8**, 257 (1958).

[15] Pfotzer, G.: Z. Physik **102**, 23, 41 (1936).

[16] Ehmert, A., u. H. Erbe: IUPAP Moskaw-Meeting 1959, vol. IV, p. 80, 1960.

[17] Erbe, H.: Mitteilungen aus dem Max-Planck-Institut für Aeronomie Nr. 2. Springer 1959.

[18] Neher, H.V.: Man vergleiche die Zusammenfassung in Progress in Cosmic Rays, p. 243, 1952.

[19] Regener, E., u. G. Pfotzer: Phys. Z. **35**, 779 (1934).

[20] Ehmert, A.: Z. Physik **115**, 326 (1940).

[21] Forbush, S.E.: J. Geophys. Res. **63**, 651 (1958).

[22] Über die Datazentren des AGI.

[23] Ehmert, A.: IUPAP Moskaw-Meeting 1959, vol. IV, p. 25, 1960. Dort finden sich noch zahlreiche russische Arbeiten zu diesem Thema.

[24] — Bagnères Konferenz 1953, Recueil des Travaux du Lab. d Pic du Midi, p. 15, 1953.

[25] Simpson, J.A., F. Yori and M. Pyka: J. Geophys. Res. **61**, 11 (1956).

[26] Forbush, S.E.: J. Geophys. Res. **63**, 651 (1958).

[27] — Terr. Mag. **43**, 207 (1938) magnetische Stürme.

[28] Erbe, H.: Mitteilungen aus dem Max-Planck-Institut für Aeronomie Nr. 2, 1959. Dissertation. Stuttgart: Springer.

[29] Morrison, P.: Phys. Rev. **101**, 1397 (1956).

[30] Parter, E.N.: Phys. Rev. **109**, 1734 (1958); **110**, 1445 (1958).

[31] Meyer, P., E.N. Parker and J.A. Simpson: Phys. Rev. **104**, 708 (1956).

[32] Kitamura, M.: Proc. IUPAP Moscow Konferenz 1959, vol. IV, p. 128, 1960. — Kitamura, M., u. M. Kodama: IUPAP Moscow Konferenz 1959, vol. IV, p. 202, 1960.

[33] Elliot, H.: Nature, Lond. **186**, 299 (1960); — Phil. Mag. **5**, 60 (1960).

[33a] Dorman, L.I.: Proc. IUPAP Moscow Konferenz, vol. IV, p. 320, 1960.

[33b] McCracken, K.G., and R.A. Palmeira: J. Geophys. Res. **65**, 2673 (1960).

[33c] Singer, S.F.: Nuovo Cim. Suppl. **2**, 334 (1958).

[34] Ehmert, A.: Physiker Tagg Wiesbaden 1960, S. 13. Mosbach: Physik Verlag 1961.

[35] Allen, I.A. van, and L.A. Frank: Nature, Lond. **183**, 430 (1959).

[36] Ehmert, A.: Proc. Space Research, Proc. COSPAR Meeting Nizza 1960, p. 1000.

[37] Fan, C.Y., P. Meyer and J.A. Simpson: Space Research, COSPAR-Meeting, Nizza 1960, p. 950, 1960.

[38] — — — Phys. Rev. Letters **5**, 289 (1960).

[39] — — — Phys. Rev. Letters **5**, 272 (1960).

[40] Ehmert, A., u. A. Sittkus: Z. Naturforsch. **6a**, 618 (1951).

[41] Chapman, G., and J. Bartels: Geomagnetism. Oxford: Clarendon Press 1960.

[42] Sonett, C.P., E.J. Smith, D.L. Judge and P.J. Coleman: Phys. Rev. Letters **4**, 161 (1960).

[43] Ehmert, A.: Naturwissenschaften **41**, 312 (1954).

[44] Obayashi, T., and Y. Hakura: Space Research, COSPAR Meeting Nizza 1960, p. 665.

[45] Coleman jr., P.J., L. Davis and C.P. Sonett: J. Geophys. Res. **65**, 1856 (1960).

[46] Zirkler, J.: Z. Geophys. **18**, 126 (1943).

[47] Regener, E.: Phys. Z. **34**, 820, 880 (1933).

[48] Ehmert, A.: Z. Naturforsch. **3a**, 264 (1948).

[49] Lange, I., and S.E. Forbush: Terr. Magn. **47**, 185, 331 (1942).

[50] Berry, E. B., and V. F. Hess: Terr. Magn. 47, 251 (1942).
[51] Duperier, H.: Proc. Phys. Soc. 57, 468 (1945).
[52] Ishii, Y., Y. Sekido and Y. Miyazaki: J. Phys. Soc. Inst. 4, 353 (1949).
[53] Forbush, S.E.: Phys. Rev. 70, 771 (1946).
[54] Neher, H.V., and W.C. Roesch: Rev. Mod. Phys. 20, 350 (1948).
[55] Forbush, S.E., T.B. Stinchcomb and M. Schein: Bull. Amer. Phys. Soc. 25, 15 (1950); — Phys. Rev. 79, 501 (1950).
[56] Miyazaki, Y., N. Wada u. I. Kondo: Rep. Ion. Ses. Japan 4, 176 (1950).
[57] Rose, C.D.: Phys. Rev. 78, 181 (1950).
[58] Clay, J., and M.F. Jongen: Phys. Rev. 79, 909 (1950).
[59] Elliot, H., and D.W.N. Dolbeer: Proc. Roy. Soc. 63, 137 (1950).
[60] Ehmert, A., A. Sittkus, H. Salow, O. Augustin and W. Menzel: J. Atm. Terr. Phys. 1, 39, 40, 41, 47 (1950).
[61] Adams, N.: Phil. Mag. 41, 503 (1950). — Adams, N., and H. J. I. Braddick: Phil. Mag. 41, 505 (1960); — Z. Naturforsch. 6a, 592 (1951).
[62] Ehmert, A., u. G. Pfotzer: Mitteilungen aus dem Max-Planck-Institut für Physik der Stratosphäre, Weißenau Nr. 6, 1956. Z. Naturforsch. 11a, 322 (1956).
[63] Die damals im Austausch freundlichst mitgeteilten Ergebnisse sind heute publiziert:
Albuquerque: Brown, R.A.: J. Atmos. Terr. Phys. 8, 278 (1956).
Berkeley: Brode, R.B., and A. Goodwin jr.: Phys. Rev. 103, 377 (1956).
Chicago: Meyer, P., E.N. Parker and J.A. Simpson: Phys. Rev. 104, 768 (1956).
Göttingen: Meyer, B.: Private Mitteilung.
Leeds: Marsden, P. L., J. W. Berry, P. Fieldhouse u. J. G. Wilson: J. Atmosph. Terr. Phys. 8, 278 (1956).
Mt. Norikura: Sekido, Y., Y. Ishii u. Y. Miyazaki: Private Mitteilung.
Stockholm: Sandström, A. E.: Private Mitteilung.
Ottawa: Rose, C.D., u. J. Katzmann: National Research Council Ottawa (private Mitteilung).
[64] Ehmert, A.: Electromagnetic Phenomena in Cosmical Physics, Symposium No. 6 der Int. Astronomischen Union, Paper 42, p. 404, 1958, Cambridge-University Press.
[65] Sittkus, A., W. Kühn u. E. Andrich: Z. Naturforsch. 11a, 325 (1956).
[66] Simpson, J.A.: Electromagnetic Phenomena in Cosmical Physics, Symposium No. 6 der Int. Astronomischen Union, Paper No. 38, p. 355, 1958.
[67] Gold, T.: COSPAR Meeting in Florenz 1961. Space Research II, p. 828. 1961. Amsterdam: North-Holland Publishing Company 1961.
[68] Pfotzer, G.: Mitteilungen aus dem Max-Planck-Institut für Physik der Stratosphäre, Weißenau bei Ravensburg, Nr. 7 u. 9, 1956.
[69] Ehmert, A., H. Erbe u. G. Pfotzer: COSPAR Meeting Florenz 1961. Space Research II, p. 778. 1961.
[70] Bartels, J.: Mitteilungen aus dem Max-Planck-Institut für Aeronomie (im Druck).
[71] Carmichael, H., and J.F. Steljes: Symposium on the July 1959 events Helsinki 1960, p. 10.
[72] Schmidt, A.: Beitr. angew. Geophys. 41, 346 (1939).
[73] Bartels, J.: Terr. Magnetism and Atm. Electricity 41, 225 (1936).
[74] Yori, F. S.: Phys. Rev. 102, 1167 (1956).
[75] Vestine, E.H., and L. Lange: Carnegie Inst. Washington, Publ. Nr. 578 u. 580, 1949.
[76] Störmer, C.: Z. Astrophysica Norwegica 1, 1 (1934).
[77] Dwight, K.: Phys. Rev. 78, 40 (1950).
[78] Schlüter, A.: Z. Naturforsch. 6a, 613 (1951).
[79] Malmfors, K.G.: Arkiv Mat. Astron. Fysik A 32, Nr. 8 (1945).
[80] Brunberg, E.: J. Geophys. Res. 58, 272 (1953).
[81] Störmer, C.: Astrophysica Norwegica 2, 193 (1937).
[82] Ehmert, A., H. Erbe, G. Pfotzer, C.D. Anger and R.R. Brown: J. Geophys. Res. 65, 2683 (1960).
[83] Brown, R.R., and R.G. D'Arcy: Phys. Rev. Letters 3, 390 (1959).
[84] Ney, E.P.: Midwest Cosmic Ray Conference 1937.
[85] — J.R. Winckler and P.S. Freyer: Phys. Rev. Letters 3, 183 (1959).
[86] Anderson, K.A., and D.C. Enemark: Space Research, COSPAR-meeting Nizza 1960, p. 702, 1960.
[87] Winckler, J.R.: J. Geophys. Res. 65, 1331 (1960).
[88] Reid, G.C., and H. Leinbach: J. Geophys. Res. 64, 1801 (1959); — Symposium July 59 events, Helsinki 1960, UGGI-Monographie No. 7, 1960.
[89] Beiley, D.K.: Proc. Inst. Radio Eng. 47, 255 (1959).
[90] Little, G.C., and H. Leinbach: Proc. Inst. Radio Eng. 47, 315 (1959).
[91] Ehmert, A., u. K. Revellio: Z. Geophysik 23, 113 (1957).

[92] Kiepenheuer, K.O.: Nature, Lond. **158**, 340 (1946).
[93] Alfvén, H.: Tellus **8**, 8 (1956).
[94] Pfotzer, G.: Physikertag Wiesbaden, p. 42. Mosbach: Physik Verlag 1961.
[95] Allen, I.A. van, H.G. Ludwig, E.C. Ray and C.L. McIlwain: Jet Propulsion **28**, 588 (1958).
— Corpuscular Radiation in Space. Radiation Res. **14**, 540 (1961).
[96] Arnoldy, R., E. Hoffmann and I.R. Winckler: J. Geophys. Res. **65**, 1361 (1960).
[97] Rosen, A., T.A. Farley and C.F. Sonett: Space Research, COSPAR-Meeting, Nizza 1960, p. 938, 1960.
[98] Vernow, S.N.: CSAGI-Meeting, Moskau 1958, p. 797, Amsterdam 1960.
Allen, I.A. van, C.E. McIlwain and G.E. Ludwig: J. Geophys. Res. **64**, 271 (1959).
Kellogg, P.J.: Nuovo Cim. **10**, 48 (1959).
Singer, S.F.: Phys. Rev. Letters **1**, 181, 1958, Space Research, COSPAR-Meeting Nizza 1960, p. 797, 1960.
[99] Dessler, A.I., and R. Karplus: Phys. Rev. Letters **4**, 271 (1960).

Einschlägige Literatur

Heisenberg, W.: Vorträge über kosmische Strahlung. Berlin-Göttingen-Heidelberg: Springer 1953.
Jánossy, L.: Cosmic Rays. Oxford: Clarendon Press 1948.
Dauvillier, A.: Les Rayons Cosmiques. Paris: Dunod 1954.
Handbuch der Physik, Bd. 46/1. Berlin-Göttingen-Heidelberg: Springer 1961. — Bd. 46/2 u. 49 (im Druck).
Chapman, S., and J. Bartels: Geomagnetism. Oxford: Clarendon Press 1940. 2 Volumes. Neudruck 1951.
Wilson, J.G.: Progress in Cosmic Ray Physics. Amsterdam: North Holland Publishing Company 1952.
Wilson, J.G., and S.A. Wouthuysen: Progress in Elementary Particles and Cosmic Ray Physics II—IV, 1958. — Vol. V in Vorbereitung.

Anhang

(Tabellen)

Tabelle 1. *Übersicht über die Uran-Radium-, die Uran-Actinium- und die Thorium-Reihe*
(Entnommen aus GRIMSEHL, Lehrbuch der Physik, 12. Aufl., Bd. IV. Leipzig 1959)

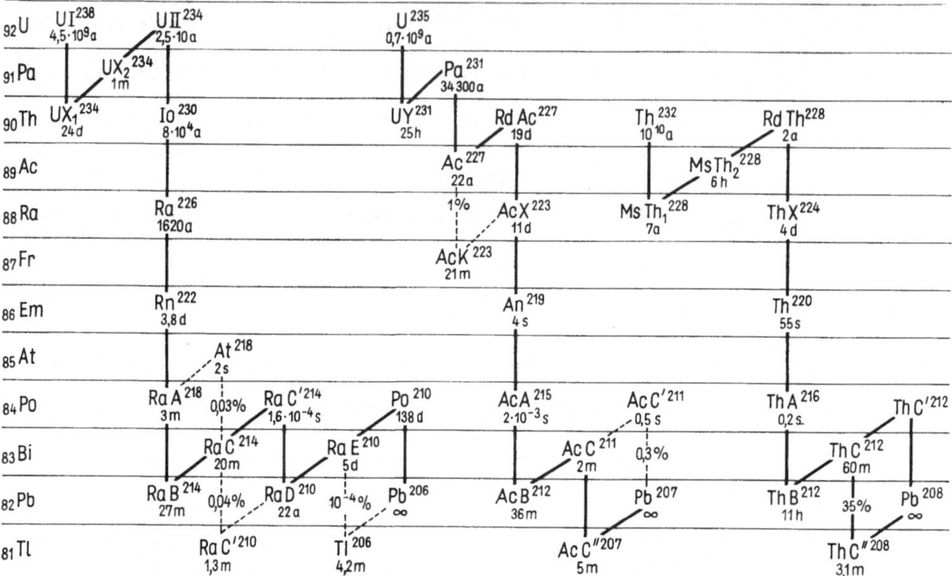

Da die Ordinate des Diagramms die Ordnungszahl Z darstellt, führt ein α-Zerfall zwei Zeilen abwärts, ein β-Zerfall eine Zeile aufwärts. Die Massenzahl bleibt beim β-Zerfall konstant und nimmt beim α-Zerfall um vier Einheiten ab. Bei dualen Zerfällen, z.B. RaC, AcC, ThC, ist der Hauptweg stark ausgezogen, der Nebenweg punktiert; bei letzterem ist auch die Prozentzahl für die Häufigkeit des Nebenweges angegeben: Zum Beispiel zerfallen von den ThC-Kernen 65% in ThC′, 35% in ThC‴.

Tabelle 2. *Die drei natürlichen radioaktiven Zerfallsreihen und ihre Konstanten**

Element		Halbwertszeit	Zerfallskonstante in sec⁻¹	Angaben zur Strahlung
herkömmliche Bezeichnung	Isotopen-Bezeichnung			
UI	$_{92}U^{238}$	$4,51 \cdot 10^9$ a	$4,88 \cdot 10^{-18}$	α (4,27 MeV) bei etwa 23% der ZA (Zerfallsakte) schwache γ-Strahlung ($< 0,05$ MeV)
UX₁	$_{90}Th^{234}$	24,10 d	$3,33 \cdot 10^{-7}$	β (0,19 MeV) bei etwa 26% der ZA schwache γ-Strahlung ($< 0,1$ MeV)

* Die in der Tabelle aufgeführten Werte entstammen größtenteils dem Artikel von H. DANIEL: Die vier radioaktiven Zerfallsreihen (1959) und sind z.T. ergänzt nach J.M. HOLLANDER, I. PERLMAN und G.T. SEABORG „Table of Isotopes" (1955) und nach D. STROMINGER, J.M. HOLLANDER und G.T. SEABORG „Table of Isotopes" (1958). Die in der letzten Kolonne aufgeführten „Angaben zur Strahlung" enthalten die Werte der beim α- und β-Zerfall freiwerdenden Energien Q_α und Q_β (disintegration energies) (auf zwei Dezimalen abgerundet) sowie Zahlen über begleitende γ-Strahlung (obere Energie-Grenze) — soweit solche bei mehr als 1% der Zerfallsakte auftritt. Zusätzliche Strahlungssymbole in Klammern — (α), (β), (γ) — geben an, daß diese Strahlung zusätzlich in 0,1 bis 1% der Zerfallsakte auftritt. Bei noch geringeren Prozentsätzen sind Angaben nur bei dualem Zerfall gemacht. Völliges Fehlen anderer Strahlungsanteile, wie es nur in einigen wenigen Fällen eintritt, ist besonders vermerkt. Zerfallsschemata siehe in der angegebenen Literatur.

Tabelle 2. (Fortsetzung)

Element		Halbwertszeit	Zerfalls-konstante in sec^{-1}	Angaben zur Strahlung
herkömmliche Bezeichnung	Isotopen-Bezeichnung			
UX$_2$	$_{91}$Pa^{234}	1,18 m	$9,8 \cdot 10^{-3}$	β (2,31 MeV) bei etwa 4% der ZA γ-Strahlung (bis 1,83 MeV)
UZ*	$_{91}$Pa^{234}	6,66 h	$2,94 \cdot 10^{-5}$	β (bis 1,2 MeV) γ (bis 0,9 MeV)
U II	$_{92}$U^{234}	$2,50 \cdot 10^5$ a	$8,8 \cdot 10^{-14}$	α (4,85 MeV) bei etwa 28% der ZA schwache γ-Strahlung ($<$ 0,1 MeV)
Io	$_{90}$Th^{230}	$8,2 \cdot 10^4$ a	$2,68 \cdot 10^{-13}$	α (4,76 MeV) bei etwa 23% der ZA schwache γ-Strahlung ($<$ 0,07 MeV)
Ra	$_{88}$Ra^{226}	1620 a	$1,36 \cdot 10^{-11}$	α (4,87 MeV) bei etwa 6% der ZA γ-Strahlung ($<$ 0,2 MeV)
RaEm(Rn)	$_{86}$Em^{222}	3,825 d	$2,10 \cdot 10^{-6}$	α (5,59 MeV)
RaA	$_{84}$Po^{218}	3,05 m	$3,78 \cdot 10^{-3}$	α (6,11 MeV) ($\approx$ 100%) β (0,35 MeV) (0,03%)
RaB	$_{82}$Pb^{214}	26,8 m	$4,31 \cdot 10^{-4}$	β (0,99 MeV) bei 94% der ZA γ-Strahlung (bis 0,5 MeV)
**	$_{85}$At^{218}	2 sec	0,347	α (6,75 MeV) (9,99%) β (2,75 MeV) (0,1%)
RaC	$_{83}$Bi^{214}	19,7 m	$5,86 \cdot 10^{-4}$	α (5,61 MeV) (0,04%) davon bei 55% der ZA schwache γ-Strahlung (bis 0,06 MeV) β (3,26 MeV) ($\approx$ 100%) davon bei 81% γ-Strahlung (bis 3,0 MeV)
+	$_{86}$Em^{218}	0,019 s	36,5	α (7,26 MeV) (γ)
RaC′	$_{84}$Po^{214}	$1,63 \cdot 10^{-4}$ s	4250	α (7,83 MeV) (keine weitere Strahlung bekannt)
RaC″	$_{81}$Tl^{210}	1,32 m	$8,7 \cdot 10^{-3}$	β (5,40 MeV) γ (bis 2,6 MeV)
RaD	$_{82}$Pb^{210}	22 a	$1,00 \cdot 10^{-9}$	β (0,06 MeV) bei 84% der ZA schwache γ-Strahlung ($<$ 0,05 MeV)
RaE	$_{83}$Bi^{210}	5,01 d	$1,61 \cdot 10^{-6}$	α (5,05 MeV) (10^{-4}%) β (1,17 MeV) ($\approx$ 100%)
RaF (Po)	$_{84}$Po^{210}	138,4 d	$5,80 \cdot 10^{-8}$	α (5,41 MeV)
***	$_{81}$Tl^{206}	4,23 m	$2,73 \cdot 10^{-3}$	β (1,51 MeV) (keine weitere Strahlung bekannt)
RaG (Uranblei)	$_{82}$Pb^{206}	stabil		

Uran-Aktinium-Reihe

Element		Halbwertszeit	Zerfalls-konstante in sec^{-1}	Angaben zur Strahlung
AcU	$_{92}$U^{235}	$7,1 \cdot 10^8$ a	$3,10 \cdot 10^{-17}$	α (4,64 MeV) bei 93% der ZA γ-Strahlung (bis 0,38 MeV)
UY	$_{90}$Th^{231}	25,6 h	$7,52 \cdot 10^{-6}$	β (0,39 MeV) bei 58% der ZA γ-Strahlung (bis 0,3 MeV)

* 0,4% der Umwandlungen des UX$_1$.
** 0,03% aller RaA-Zerfälle
*** 10^{-4}% der RaE-Zerfälle
+ 0,1% der $_{85}$At^{218}-Zerfälle

Tabelle 2. (Fortsetzung)

Element		Halbwertszeit	Zerfalls-konstante in sec^{-1}	Angaben zur Strahlung
herkömmliche Bezeichnung	Isotopen-Bezeichnung			
Pa	$_{91}$Pa^{231}	$3,4 \cdot 10^4$ a	$6,48 \cdot 10^{-13}$	α (5,14 MeV) bei 90% der ZA γ-Strahlung (bis 0,38 MeV)
Ac	$_{89}$Ac^{227}	21,8 a	$1,01 \cdot 10^{-9}$	α (5,03 MeV) (1,2%) β (0,05 MeV) (98,8%)
RdAc	$_{90}$Th^{227}	18,6 d	$4,32 \cdot 10^{-7}$	α (6,14 MeV) bei 81% der ZA γ-Strahlung (bis 0,64 MeV)
AcK	$_{87}$Fr^{223}	22 m	$5,25 \cdot 10^{-4}$	α (5,44 MeV) $(5 \cdot 10^{-3}\%)$*
	1,2% aller Ac-Umwandlungen			β (1,15 MeV) ($\approx$ 100%)
AcX	$_{88}$Ra^{223}	11,6 d	$6,92 \cdot 10^{-7}$	α (5,97 MeV) bei 99% der ZA γ-Strahlung (bis 0,58 MeV)
An	$_{86}$Em^{219}	3,92 s	0,177	α (6,93 MeV) bei 31% der ZA γ-Strahlung (bis 0,59 MeV)
AcA	$_{84}$Po^{215}	$1,83 \cdot 10^{-3}$ s	379	α (7,52 MeV) ($\approx$ 100%) β (0,75 MeV) $(4 \cdot 10^{-4}\%)$*
AcB	$_{82}$Pb^{211}	36,1 m	$3,20 \cdot 10^{-4}$	β (1,40 MeV) bei 20% der ZA γ-Strahlung (bis 0,83 MeV)
AcC	$_{83}$Bi^{211}	2,16 m	$5,35 \cdot 10^{-3}$	α (6,75 MeV) (99,7%) β (0,60 MeV) (0,3%)
AcC'	$_{84}$Po^{211}	0,52 s	1,32	α (7,57 MeV)
	0,3% der AcC-Umwandlungen			(γ)
AcC''	$_{81}$Tl^{207}	4,77 m	$2,42 \cdot 10^{-3}$	β (1,44 MeV) (γ)
DcA	$_{82}$Pb^{207}	stabil		

Thorium-Reihe

Element		Halbwertszeit	Zerfalls-konstante in sec^{-1}	Angaben zur Strahlung
Th	$_{90}$Th^{232}	$1,41 \cdot 10^{10}$ a	$1,56 \cdot 10^{-18}$	α (4,08 MeV) bei 23% der ZA γ-Strahlung (0,04 MeV)
MsTh I	$_{88}$Ra^{228}	6,7 a	$3,28 \cdot 10^{-9}$	β (0,05 MeV) (keine weitere Strahlung bekannt)
MsTh II	$_{89}$Ac^{228}	6,13 h	$3,14 \cdot 10^{-5}$	β (2,16 MeV) bei 88% der ZA γ-Strahlung (bis 1,7 MeV)
RdTh	$_{90}$Th^{228}	1,91 a	$1,15 \cdot 10^{-8}$	α (5,52 MeV) bei 29% der ZA γ-Strahlung (bis 0,21 MeV)
ThX	$_{88}$Ra^{224}	3,64 d	$2,20 \cdot 10^{-6}$	α (5,79 MeV) bei 5% der ZA γ-Strahlung (bis 0,24 MeV)
Tn	$_{86}$Em^{220}	52 s	$1,33 \cdot 10^{-2}$	α (6,40 MeV) (γ)
ThA	$_{84}$Po^{216}	0,16 s	4,33	α (6,90 MeV) (wahrscheinlich keine weitere Strahlung)
ThB	$_{82}$Pb^{212}	10,64 h	$1,81 \cdot 10^{-5}$	β (0,59 MeV) bei 88% der ZA γ-Strahlung (bis 0,41 MeV)

* Diese nehmen einen anderen, hier nicht aufgeführten Zerfallsweg.

Tabelle 2. (Fortsetzung)

| Element | | Halbwertszeit | Zerfalls-konstante in sec^{-1} | Angaben zur Strahlung |
herkömmliche Bezeichnung	Isotopen-Bezeichnung			
ThC	$_{83}Bi^{212}$	60,5 m	$1,91 \cdot 10^{-4}$	α (6,20 MeV) (35%) davon in 73% der ZA γ-Strahlung (bis 0,62 MeV) β (2,26 MeV) (65%) davon in 35% der ZA γ-Strahlung (bis 2,2 MeV)
ThC'	$_{84}Po^{212}$	$3 \cdot 10^{-7}$ s	$2,31 \cdot 10^6$	α (8,95 MeV) (keine weitere Strahlung bekannt)
ThC''	$_{81}Tl^{208}$	3,1 m	$3,73 \cdot 10^{-3}$	β (5,00 MeV) γ (2,61 MeV)
ThD	$_{82}Pb^{208}$	stabil		

Anhang-Literatur

DANIEL, H.: Die vier radioaktiven Zerfallsreihen. Ergebn. exakt. Naturw. **32**, 118—179 (1959).

HOLLANDER, J.M., I. PERLMAN and G.T. SEABORG: Table of isotopes. Reactor handbook physics (US Atomic Energy Commission), S. 158—350. New York-Toronto-London: McGraw Hill Book Comp. 1955.

STROMINGER, D., J.M. HOLLANDER and G.T. SEABORG: Table of isotops. Rev. Mod. Phys. **30**, 585—904 (1958).

Namenverzeichnis

Sachverzeichnis
(Deutsch-Englisch)

Subject Index

(English-German)

Atmosphere, Distribution of fission-products with height in the —, *Höhenverteilung von Kernspaltungsprodukten in der Atmosphäre* 90

—, Distribution of natural radioactive elements with height in the —, *Höhenverteilung der natürlich-radioaktiven Elemente in der Atmosphäre* 85

—, Global circulation of radioactivity in the —, *Globale Zirkulation der Radioaktivität in der Atmosphäre* 151

—, Global distribution of radioactivity in the —, *Globale Verteilung der Radioaktivität in der Atmosphäre* 136

—, Global transfer of radioactivity in the —, *Globaltransport von Radioaktivität in der Atmosphäre* 151

—, Intermediate scale transfer and circulation of radioactivity in the —, *Regional-Transport und -Zirkulation der Radioaktivität in der Atmosphäre* 139

—, Large scale transfer and circulation of radioactivity in the —, *Global-Transport und -Zirkulation von Radioaktivität in der Atmosphäre* 139

—, Natural Radioactivity in the —, *Natürliche Radioaktivität der Atmosphäre* 76

—, Radioactivity in the — formed by cosmic radiation, *Erzeugung atmosphärischer Radioaktivität durch kosmische Strahlung* 156

—, Radioactivity in the surface layers of the —, *Radioaktivität in den unteren Atmosphärenschichten* 144

—, Radon content in the —, *Radon-Gehalt in der Atmosphäre* 83, 84

—, Residence time of a radioactive cloud in the —, *Aufenthaltsdauer einer radioaktiven Wolke in der Atmosphäre* 220

—, — — of CO_2 in the —, *Aufenthaltsdauer von CO_2 in der Atmosphäre* 159, 160, 164

—, Small-scale turbulence in the free —, *Kleinturbulenz in der freien Atmosphäre* 149

—, Thoron content in the —, *Thoron-Gehalt der Atmosphäre* 84

—, Transfer and circulation of radioactivity in the —, *Transport und Zirkulation radioaktiver Stoffe in der Atmosphäre* 136

Atmospheric circulation, Use of tritium for — — studies, *Tritium und atmosphärische Zirkulation* 160

Atmospheric transfer, Small-scale — — of radioactivity, *Kleinräumiger Transport von Radioaktivität in der Atmosphäre* 139

Attachment, Influence of charge on the —, *Ladungseinfluß auf den Anlagerungsprozeß* 180, 190

—, Theory of —, *Theorie der Anlagerung* 182

—, Velocity of —, *Anlagerungsgeschwindigkeit* 181

Autoradiography, *Autoradiographie* 310, 313, 325

—, Track —, *Spuren-Autoradiographie* 313

"austausch", Coeffizient of —, *Austauschkoeffizient* 136, 137

Background, Nuclear emulsion —, *Nulleffekt bei Kernphotoemulsionen* 312

— of a counter, *Nulleffekt eines Zählers* 296, 305

—, Radiation —, *Nulleffekt (bei Strahlungsmessungen)* 298

—, Scintillation counter —, *Nulleffekt von Szintillationszählern* 308, 309

Backscattering, *Rückstreuung* 300

Beta-radiation, Range of —, *Reichweite der Beta-Strahlung* 301, 302

Beta-spectroscopy, *Beta-Spektroskopie* 311, 335

Beryllium in ocean water, *Beryllium im Meerwasser* 28

— in red clay cores, *Beryllium im roten Tiefseeton* 28

Beryllium 7 production, *Beryllium-7 Produktion* 111, 112

— — in rain water, *Beryllium-7-Gehalt des Regenwassers* 112, 114

Beryllium 10-age method, *Beryllium-10-Altersbestimmung* 39

— — in deep sea sediments, *Beryllium-10-Gehalt von Tiefseesedimenten* 114, 115

Biological hazards from heavy primaries, *Biologische Gefahren von schweren Primärteilchen (der kosmischen Strahlung)* 255

Biological significance of high energy radiations, *Biologische Bedeutung energiereicher Strahlung* 243

Biological transport of radioactive elements (sea), *Biologischer Transport radioaktiver Elemente (im Meer)* 35

Bio-radioactivity, *Bio-Radioaktivität* 247, 251

Biotite, *Biotit* 10

Bone, Alpha activity of human — ashes, *Alpha-Aktivität der Asche von menschlichen Knochen* 250

—, Sr^{90} analysis of human — for Sr^{90}, *Untersuchung des Sr^{90}-Gehaltes in den menschlichen Knochen* 267, 268, 269

"Box models", „*Schachtel-Modelle*" 33

Cancer, *Krebs* 253

Capillary condensation, *Kapillarkondensation* 237

Carbon 14, Age determination by — —, *C^{14}-Altersbestimmung* 106

— — analyses on ocean water, *C^{14}-Datierung von Meerwasser* 28

— — chronometer, *C^{14}-Chronometer* 62

— —, content of — — in the atmosphere, *Atmosphärischer C^{14}-Gehalt* 105

— — dating of bones, *C^{14}-Datierung von Knochen* 63

— — dating of deep sea sediments, *C^{14}-Datierung von Tiefseesedimenten* 63

— —, Dilution of — — by fuel combustion, „*Industrie-Effekt*" *im C^{14}-Gehalt* 109

— —, Dose to the human body due to — —, *Ganzkörper-Strahlendosis durch C^{14}* 272

— — exchange, *C^{14}-Austausch* 107

— —, Exchange times of — —, *C^{14}-Austauschzeiten* 109

Cosmic rays, Frequency of nuclei in the
— —, *Häufigkeitsverteilung der Kern-
arten in der kosmischen Strahlung* 347
— —, Galactic — —, *Galaktische kosmische
Strahlung* 344
— —, Global distribution of — —, *Vertei-
lung der kosmischen Strahlung über die
Erde* 353
— —, "History" of — —, *Vorgeschichte der
kosmischen Strahlung* 346
— —, Ionizing power of — —, *Spezifische
Ionisierung der kosmischen Strahlung* 344
— —, Isocosmes of — —, *Isokosmen der kos-
mischen Strahlung* 353
— —, Isotropic modulation of — —, *Iso-
trope Modulation der kosmischen Strah-
lung* 365
— —, Latitude effect of — —, *Breiteneffekt
der kosmischen Strahlung* 351, 352
— —, Longitude effect of — —, *Längeneffekt
der kosmischen Strahlung* 353
— —, Methods of measurement for — —,
Meßmethoden für kosmische Strahlung 357
— —, Modulation of the — — by the earth'
magnetic field, *Modulation der kosmischen
Strahlung durch das erdmagnetische Feld*
367
— —, Movement of particles of — —, *Bewe-
gung der Teilchen der kosmischen Strah-
lung* 347
— —, Particle flux of — —, *Teilchenfluß in
der kosmischen Strahlung* 345, 359, 364
— —, Primary — —, *Primäre kosmische
Strahlung* 343, 344
— —, Secondary — —, *Sekundärstrahlung
der kosmischen Strahlung* 355
— —, Spectrum of particles in — —, *Teil-
chenspektrum der kosmischen Strahlung*
345
— —, Unisotropic modulation, *Anisotrope
Modulation der kosmischen Strahlung* 365
Counter, Flow —, *Durchflußzähler* 297, 306,
326, 336
—, Geiger —, *Geiger-Zähler* 328, 329, 336
—, Halogen —, *Halogen-Zähler* 306
—, Proportional —, *Proportionalzähler* 307,
326, 327, 329
—, Scintillation —, *Szintillationszähler* 308,
326, 336
—, Screen wall —, „*Screen-wall-counter*"
328
Counting efficiency, *Ansprechwahrscheinlich-
keit* 296, 297
Counting statistics, *Zählstatistik* 299
Cs^{137} in ground level air, Cs^{137} *in bodennaher
Luft* 262
— in human beings, Cs^{137} *im menschlichen
Körper* 271
— in the human diet, Cs^{137} *in der mensch-
lichen Nahrung* 270
Cs^{137}/potassium ratios "in vivo", Cs^{137}/K-
Verhältnisse im lebenden Organismus 270,
271
Cunningham's correction, *Cunninghamsche
Korrektur des Fallgesetzes* 208

Dating, Potassium-40-argon-40 —, *Ka-
lium-40-Argon-40-Datierung* 65
—, Radiocarbon —, *C 14-Datierung* 26, 62
Daughter products, Short-lived — in the
oceans, *Kurzlebige Zerfallsprodukte im
Meer* 23
Decay of fission products, *Zerfall der Spalt-
produkte* 334
Decay series, Natural — —, *Natürliche Zer-
fallsreihen* 386
Deep-sea clays, Thorium in — —, *Thorium
in Tiefseetonen* 32
— —, Uranium in — —, *Uran in Tiefsee-
tonen* 32
Deep-sea sediments, Beryllium 10 in — —,
Beryllium-10 in Tiefsee-Sedimenten 114,
115
— —, C^{14} dating of — —, C^{14}-*Datierung
von Tiefsee-Sedimenten* 63
— —, Ionium in — —, *Ionium in Tiefsee-
Sedimenten* 36
Deposition, Dry weather —, *Trocken-Fall-
Out* 204
—, Mechanismus of —, *Ablagerungsmecha-
nismen* 288
—, Wet —, *Naß-Fall-Out* 204
Diet, Contamination of human —, *Verseu-
chung der menschlichen Nahrung* 266
—, Cs^{137} in the human —, Cs^{137} *in der mensch-
lichen Nahrung* 270
—, Sr^{90}/Ca ratio in the —, Sr^{90}/Ca-*Verhältnis
in der Nahrung* 266
Diffusion coefficient of emanation in soil air,
*Diffusionskoeffizient der Emanation in
Bodenluft* 81, 82
Diffusion diagram, *Diffusionsdiagramm* 141
Diffusion processes in ocean water, *Diffusions-
vorgänge im Ozeanwasser* 19
Diffusion from a point source, *Diffusion von
einer Punktquelle* 144
— from a widespread source, *Diffusion von
einer Flächenquelle* 146
— Theory of —, *Diffusionstheorie* 139
—, Thermal —, *Thermal-Diffusion* 215
Discolorations, *Verfärbungen* 11
Dose, Radiation — due to I^{131}, *Strahlendosis
durch* I^{131} 272
— to the human body due to C^{14}, *Ganzkörper-
Strahlendosis durch* C^{14} 272
—, Maximum permissible —, *Maximal zu-
lässige Dosis* 278
Dose rate due to fallout, *Strahlendosis durch
Fallout* 274
Doses, Maximum permissible — in an
emergency, *Maximal zulässige Dosen bei
Unglücksfällen* 285
—, Permissible genetic —, *Zulässige geneti-
sche Dosen* 259, 275, 276
— to tissue, *Strahlendosen im Gewebe*
275
"Doubling dose", *Verdopplungs-(Mutations)-
Dosis* 256
Droplet, Collection efficiency of a condens-
ing —, „*Sammelwirkung*" *eines konden-
sierenden Tröpfchens* 227

REPRINT FROM

NUCLEAR RADIATION IN GEOPHYSICS
KERNSTRAHLUNG IN DER GEOPHYSIK

EDITED BY

H. ISRAËL
AACHEN

A. KREBS
LOUISVILLE

SPRINGER-VERLAG / BERLIN · GÖTTINGEN · HEIDELBERG 1962
(PRINTED IN GERMANY)

RADIOACTIVITY OF THE LITHOSPHERE

BY

JOHN A. S. ADAMS

WITH 1 FIGURE

REPRINT FROM

NUCLEAR RADIATION IN GEOPHYSICS

KERNSTRAHLUNG IN DER GEOPHYSIK

EDITED BY

H. ISRAËL
AACHEN

A. KREBS
LOUISVILLE

SPRINGER-VERLAG / BERLIN · GÖTTINGEN · HEIDELBERG 1962
(PRINTED IN GERMANY)

RADIOACTIVITY IN OCEANOGRAPHY

BY

F. F. KOCZY AND J. N. ROSHOLT

WITH 1 FIGURE

REPRINT FROM

NUCLEAR RADIATION IN GEOPHYSICS
KERNSTRAHLUNG IN DER GEOPHYSIK

EDITED BY

H. ISRAËL
AACHEN

A. KREBS
LOUISVILLE

SPRINGER-VERLAG / BERLIN · GÖTTINGEN · HEIDELBERG 1962
(PRINTED IN GERMANY)

RADIOACTIVITY IN HYDROLOGY

BY

ERIK ERIKSSON

WITH 2 FIGURES

REPRINT FROM

NUCLEAR RADIATION IN GEOPHYSICS
KERNSTRAHLUNG IN DER GEOPHYSIK

EDITED BY

H. ISRAËL
AACHEN

A. KREBS
LOUISVILLE

SPRINGER-VERLAG / BERLIN · GÖTTINGEN · HEIDELBERG 1962
(PRINTED IN GERMANY)

RADIOACTIVE METHODS OF AGE DETERMINATION

BY

WALTER R. ECKELMANN

WITH 1 FIGURE

SONDERDRUCK AUS

KERNSTRAHLUNG IN DER GEOPHYSIK
NUCLEAR RADIATION IN GEOPHYSICS

HERAUSGEGEBEN VON

H. ISRAËL
AACHEN

A. KREBS
LOUISVILLE

SPRINGER-VERLAG / BERLIN · GÖTTINGEN · HEIDELBERG 1962
(PRINTED IN GERMANY)

DIE NATÜRLICHE UND KÜNSTLICHE RADIOAKTIVITÄT DER ATMOSPHÄRE

VON

H. ISRAËL

MIT 7 FIGUREN

SONDERDRUCK AUS

KERNSTRAHLUNG IN DER GEOPHYSIK
NUCLEAR RADIATION IN GEOPHYSICS

HERAUSGEGEBEN VON

H. ISRAËL
AACHEN

A. KREBS
LOUISVILLE

SPRINGER-VERLAG / BERLIN · GÖTTINGEN · HEIDELBERG 1962
(PRINTED IN GERMANY)

ERZEUGUNG RADIOAKTIVER KERNARTEN DURCH DIE KOSMISCHE STRAHLUNG

VON

O. HAXEL UND **G. SCHUMANN**

MIT 10 FIGUREN

REPRINT FROM

NUCLEAR RADIATION IN GEOPHYSICS
KERNSTRAHLUNG IN DER GEOPHYSIK

EDITED BY

H. ISRAËL
AACHEN

A. KREBS
LOUISVILLE

SPRINGER-VERLAG / BERLIN · GÖTTINGEN · HEIDELBERG 1962
(PRINTED IN GERMANY)

TRANSFER AND CIRCULATION OF RADIOACTIVITY IN THE ATMOSPHERE

BY

BERT BOLIN

WITH 16 FIGURES

SONDERDRUCK AUS

KERNSTRAHLUNG IN DER GEOPHYSIK
NUCLEAR RADIATION IN GEOPHYSICS

HERAUSGEGEBEN VON

H. ISRAËL
AACHEN

A. KREBS
LOUISVILLE

SPRINGER-VERLAG / BERLIN · GÖTTINGEN · HEIDELBERG 1962
(PRINTED IN GERMANY)

RADIOAKTIVE AEROSOLE

VON

CHR. E. JUNGE

MIT 5 FIGUREN

REPRINT FROM

NUCLEAR RADIATION IN GEOPHYSICS
KERNSTRAHLUNG IN DER GEOPHYSIK

EDITED BY

H. ISRAËL
AACHEN

A. KREBS
LOUISVILLE

SPRINGER-VERLAG / BERLIN · GÖTTINGEN · HEIDELBERG 1962
(PRINTED IN GERMANY)

RADIOACTIVE PRECIPITATIONS AND FALLOUT

BY

L. FACY

WITH 23 FIGURES

REPRINT FROM

NUCLEAR RADIATION IN GEOPHYSICS
KERNSTRAHLUNG IN DER GEOPHYSIK

EDITED BY

H. ISRAËL
AACHEN

A. KREBS
LOUISVILLE

SPRINGER-VERLAG / BERLIN · GÖTTINGEN · HEIDELBERG 1962
(PRINTED IN GERMANY)

BIOLOGICAL ASPECTS

BY

A. KREBS AND **N. G. STEWART**

WITH 14 FIGURES

SONDERDRUCK AUS

KERNSTRAHLUNG IN DER GEOPHYSIK
NUCLEAR RADIATION IN GEOPHYSICS

HERAUSGEGEBEN VON

H. ISRAËL
AACHEN

A. KREBS
LOUISVILLE

SPRINGER-VERLAG / BERLIN · GÖTTINGEN · HEIDELBERG 1962
(PRINTED IN GERMANY)

MESSMETHODEN

VON

G. SCHUMANN

MIT 9 FIGUREN

SONDERDRUCK AUS

KERNSTRAHLUNG IN DER GEOPHYSIK
NUCLEAR RADIATION IN GEOPHYSICS

HERAUSGEGEBEN VON

H. ISRAËL
AACHEN

A. KREBS
LOUISVILLE

SPRINGER-VERLAG / BERLIN · GÖTTINGEN · HEIDELBERG 1962
(PRINTED IN GERMANY)

DIE KOSMISCHE STRAHLUNG IN DER GEOPHYSIK

VON

A. EHMERT

MIT 23 FIGUREN

REPRINT FROM

NUCLEAR RADIATION IN GEOPHYSICS
KERNSTRAHLUNG IN DER GEOPHYSIK

EDITED BY

H. ISRAËL
AACHEN

A. KREBS
LOUISVILLE

SPRINGER-VERLAG / BERLIN · GÖTTINGEN · HEIDELBERG 1962
(PRINTED IN GERMANY)

INTRODUCTION

BY

ROBLY D. EVANS

Handbuch der Physik · Encyclopedia of Physics

Herausgegeben von / edited by S. FLÜGGE, Freiburg / Br.

54 Bände mit Beiträgen in deutscher, englischer und französischer Sprache. Jeder Band ist einzeln käuflich. Das Gesamtwerk kann zum Subskriptionspreis bezogen werden.
54 volumes with contributions in English, French and German. Each volume is available separately. The complete Encyclopedia may be obtained at a subscription price.

46. Band, 1. Teil
Kosmische Strahlung I - Cosmic Rays I

In englischer Sprache. Mit 150 Figuren. VI, 333 Seiten Gr.-8°. 1961
Ganzleinen DM 98,—; Subskriptionspreis DM 78,40

The origin of cosmic rays. By PH. MORRISON. — Theory of the geomagnetic effects of cosmic radiation. By M. S. VALLARTA. — Experimental results of flights in the stratosphere. By E. C. RAY. — Penetrating showers. By K. SITTE. — Extensive air showers. By G. COCCONI. — The hard component of μ-mesons in the atmosphere. By G. N. FOWLER and A. W. WOLFENDALE.

47. Band
Geophysik I - Geophysics I

Redaktion J. BARTELS. Mit 289 Figuren. VIII, 659 Seiten (davon 383 Seiten in englischer und 129 Seiten in französischer Sprache) Gr.-8°. 1956
Ganzleinen DM 118,—; Subskriptionspreis DM 94,40

The rotation of the earth. By Sir H. SPENCER-JONES. — Séismométrie. Par J. COULOMB. — Seismic wave transmission. By K. E. BULLEN. — Surface waves and guided waves. By W. M. EWING and F. PRESS. — L'agitation microséismique. Par J. COULOMB. — Seismic prospecting. By W. M. EWING and F. PRESS. — Messung elastischer Eigenschaften von Gesteinen. Von H. BAULE und E. MÜLLER. — Gravity and isostasy. By G. D. GARLAND. — Structure of the earth's crust. By W. M. EWING and F. PRESS. — Forces in the earth's crust. By A. E. SCHEIDEGGER. — Radioactivity and age of minerals. By J. T. WILSON, R. D. RUSSELL, and R. MacCUNN FARQUHAR. — The earth's interior. By J. A. JACOBS. — Electricité tellurique. Par L. CAGNIARD. — Magnetization of rocks. By S. K. RUNCORN. — The magnetism of the earth's body. By S. K. RUNCORN. — Figur der Erde. Von K. JUNG.

48. Band
Geophysik II - Geophysics II

Redaktion J. BARTELS. Mit 497 Figuren. VIII, 1046 Seiten (davon 581 Seiten in englischer und 43 Seiten in französischer Sprache) Gr.-8°. 1957
Ganzleinen DM 198,—; Subskriptionspreis DM 158,40

Dynamic meteorology. By A. ELIASSEN and E. KLEINSCHMIDT. — Strahlung in der unteren Atmosphäre. Von F. MÖLLER. — Vision through the atmosphere. By W. E. K. MIDDLETON. — Polarization of skylight. By Z. SEKERA. — Diffusion des radiations par les gouttes d'eau en suspension dans l'atmosphère. Par J. BRICARD. — Ozon in der Erdatmosphäre. Von H.-K. PAETZOLD und E. REGENER †. Geophysical aspects of meteors. By A. C. B. LOVELL. — Sound propagation in air. By E. F. COX. — The physics of clouds. By F. H. LUDLAM and B. J. MASON. — Atmosphärische Elektrizität. Von R. MÜHLEISEN. — Oceanography. By H. U. SVERDRUP. — Oberflächen-Wellen des Meeres. Von H. U. ROLL. — Gezeitenkräfte. Von J. BARTELS. — Tides of the solid earth. By R. TOMASCHEK. — Flutwellen und Gezeiten des Wassers. Von A. DEFANT. — Atmosphärische Gezeiten. Von W. KERTZ. — Physical volcanology. By S. SAKUMA and T. NAGATA.

Gesamtübersicht auf Anfrage — Leaflet on all volumes upon request

SPRINGER-VERLAG · BERLIN · GÖTTINGEN · HEIDELBERG